6th International High Performance Buildings Conference at Purdue 2021

AA005193

West Lafayette, Indiana, USA
24-28 May 2021

ISBN: 978-1-7138-3071-9

Printed from e-media with permission by:

Curran Associates, Inc.
57 Morehouse Lane
Red Hook, NY 12571

Some format issues inherent in the e-media version may also appear in this print version.

Printed with permission by Curran Associates, Inc. (2021)

For permission requests, please contact Ray W. Herrick Laboratories
at the address below.

Ray W. Herrick Laboratories
School of Mechanical Engineering
Purdue University
177 S. Russell Street
West Lafayette, IN 47907-2099, USA.

Phone: +1 (765) 494-2132
Fax: +1 (765) 494-0787

rhlab@ecn.purdue.edu

Additional copies of this publication are available from:

Curran Associates, Inc.
57 Morehouse Lane
Red Hook, NY 12571 USA
Phone: 845-758-0400
Fax: 845-758-2633
Email: curran@proceedings.com
Web: www.proceedings.com

TABLE OF CONTENTS

BUILIDING SIMULATION AND OPTIMIZATION

HVAC, ENERGY MANAGEMENT AND HEAT PUMP TECHNOLOGIES

INTELLIGENT BUILDING CONTROLS AND MPC

THERMAL STORAGE SYSTEMS, DEVICES AND MATERIALS

Fault "Auto-correction" for HVAC Systems: A Preliminary Study

Guanjing Lin1[1], Marco Pritoni[1], Yimin Chen[1], Christopher Weyandt[1], Raphael Vitti[1], Jessica Granderson[1]*

[1]Lawrence Berkeley National Laboratory,
Berkeley, CA, USA
Contact Information (gjlin@lbl.gov, mpritoni@lbl.gov, yiminchen@lbl.gov, cweyandt@lbl.gov, cweyandt@lbl.gov, JGranderson@ lbl.gov)

* Corresponding Author

ABSTRACT

A Fault Detection and Diagnostics (FDD) tool is a type of energy management and information system that is designed to continuously identify the presence of faults and efficiency improvement opportunities through a 1-way interface to the building automation system and application of automated analytics. It is estimated that 5-30% energy saving can be achieved by employing FDD tools and implementing efficiency measures based on FDD findings. Although the potential of this technology is high, actual savings are only realized when an operator takes an action to fix the problem. There is a subset of faults that can be potentially addressed automatically by the system, without operator intervention. Automating this fault "correction" can significantly increase the savings generated by FDD tools and reduce the reliance on human intervention.

This paper presents preliminary efforts towards delivering automated fault correction. It describes nine fault auto-correction algorithms for heating ventilation and air conditioning (HVAC) systems that were developed to automatically correct faults or improve controls operation. It also presents preliminary testing results of one auto-correction algorithm (improve air handling unit static pressure setpoint reset) in a commercial building, located in Berkeley, California, US. The auto-correction algorithms and implementation frameworks of this initial study provide a foundation for future auto-correction algorithm development and novel schemes for improving building operation performance and reliability.

1. INTRODUCTION

Fault detection and diagnostics (FDD) uses building operational data to identify the presence of faults or efficiency opportunities and isolate their root causes. Over the last three decades the development of FDD methods for building heating ventilation and air conditioning (HVAC) systems has been an area of active research. Well-cited review publications in the HVAC FDD area include two International Energy Agency Annex Reports (Hyvarinen *et al.*, 1996, Dexter *et al.*, 2001) and literature reviews by Katipamula *et al.* (2005a, 2005b) and Kim *et al.* (2018).

Commercially available FDD tools provide a means of monitoring-based commissioning, through which instances of operational inefficiency can be continuously identified, isolated, and surfaced for resolution by operations and maintenance staff. FDD tools represent one of the fastest growing markets in technologies for building analytics. There are over 30 FDD products available in the US and new software products continue to enter the market (Kramer *et al.*, 2020). Today's FDD technology has been documented to enable whole building savings of 9 percent on average, across users (Kramer *et al.*, 2020). Although FDD tools are being used to enable cost-effective energy savings, there is a capability gap in current products. Today's FDD technologies operate in an open loop manner. Faults are identified by the FDD tools, which may also provide a report of the duration and frequency of faults, cost and/or energy impacts, and relative priority levels. However, the identified faults must be corrected through manual human intervention. Building Automation System (BAS) and automated HVAC control optimization technologies offer closed loop supervisory control, but do not provide full-fledged, robust, continuous FDD. In practice, the need for human intervention to fix faults once they are identified often results in delay or inaction, causing additional operations and maintenance costs or deteriorating comfort conditions. This capability gap is not only technical, but also represents market-relevant desired functionality on behalf of FDD users and vendors (Granderson *et al.*, 2017). Therefore, this work seeks to develop automated fault correction approaches and integrate them with commercial FDD technology offerings, thereby closing the loop between passive diagnostics and active control. Automating the correction of these types of faults can increase the savings realized through the use of FDD tools, and reduce the extent to which savings are dependent upon human intervention.

Minimal research and development has been performed on self-correcting or fault-tolerant controls for buildings and HVAC systems. Fernandez *et al.* (Fernandez *et al.*, 2009a; Fernandez, *et al.*, 2009b) and Brambley *et al.* (Brambley *et al.*, 2011) developed passive and proactive fault auto-correction algorithms for an air-handler unit (AHU) and a variable-air-volume (VAV) box. The methods are proposed to correct faults occurring in temperature and humidity sensors, dampers, control hunting, and manual overrides. A subset of these algorithms (sensor bias and minimum outdoor air damper position) were implemented and tested in a laboratory experiment. Related to the concept of fault correction is a body of work in the building control literature that focuses on fault tolerant control. The purpose of a fault tolerant controller is to maintain proper operation of a system despite the presence of faults (Zhang and Jiang, 2008). Wang *et al.* (2002) developed a supervisory control scheme that adapts to the presence of a measurement error in outdoor air flow rate. Hao *et al.* (2005) employed principal component analysis to develop fault tolerant control and data recovery in the HVAC monitoring system. Bengea *et al.* (2015) developed a fault-tolerant optimal control strategy for a HVAC system integrating FDD and model predictive control. While the literature focuses on the development of these advanced controllers, it is not yet readily implemented in today's buildings control systems and also does not explore their integration with existing FDD technologies.

This paper presents a preliminary study of automated fault correction. It describes the development of fault auto-correction algorithms that are designed to be integrated with commercial FDD tools. It also presents early testing results of one auto-correction algorithm (improve air handling unit static pressure setpoint reset). In the testing, the algorithm was implemented in a commercial FDD product and successfully changed the static pressure setpoint in two AHUs of a commercial building in Berkeley, California, US.

2. FAULT AUTO-CORRECTION ALGORITHMS

2.1 Fault Auto-correction Algorithms

The primary objective of this study is to develop automated fault correction algorithms that can be integrated with existing commercial FDD and BAS products. Therefore, the developed auto-correction algorithms are decoupled from the FDD algorithms embedded in the existing FDD tools. This permits applicability and feasibility of the developed correction algorithms across a variety of FDD technologies that employ different FDD rules and algorithms. Further, it is assumed that the FDD tools are able to detect the faults of focus, as they represent some of the more commonly encountered faults in commercial buildings.

Nine innovative fault auto-correction algorithms for HVAC system were developed to automatically correct the faults or improve the operation, including: (1) HVAC schedules are incorrectly programmed, (2) revert manual overrides, (3) AHU outdoor air or supply air temperature sensor bias, (4) control hunting, (5) rogue zone, (6) improve economizer high lockout temperature setpoint, (7) improve zone temperature setpoint setback, (8) improve AHU static pressure setpoint reset , and (9) improve AHU supply air temperature setpoint reset (Lin *et al.,* 2020). Figure 1 shows a flow chart of the general auto-correction process. In this process, after the FDD algorithm generates a fault flag of a specific fault, the fault auto-correction algorithm is initiated to correct this fault with approval from the building operator. A correctable variable (referred to as Control_variable_being_overwritten), is the key element in the auto-correction process. The algorithm overwrites this variable (Control_variable_being_overwritten_current) via BACnet or other protocol to a new value (Control_variable_being_overwritten_new). The control_variable_being_overwritten_current is the one identified in the FDD algorithm to be associated with the problematic value (fault) or potential to improve (opportunity). The control_variable_being_overwritten_new is the same variable that has the correct value (fault) or optimized value (opportunity). All of the auto-correction algorithms developed in this work follow this structure, with different control variables overwritten in BAS, and different ways to determine the correct or improved value of the variable.

For example, the correctable variable in algorithm "revert manual overrides" is the variable that indicates the equipment manual control status or equivalent flag: 1 – equipment is in manual control, 0 – equipment is in automatic control.

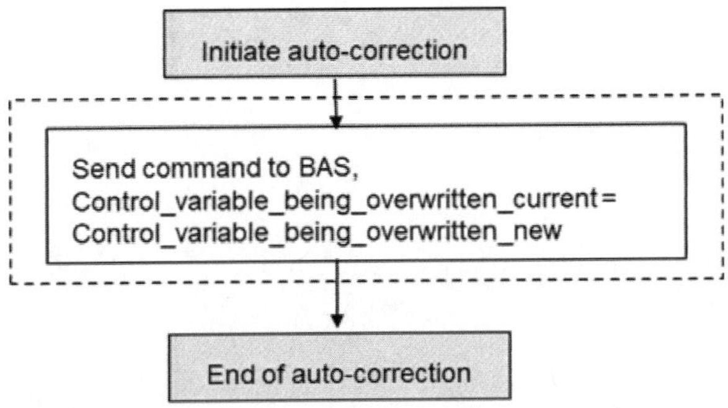

Figure 1: Flow chart of the general auto-correction process

2.2 Auto-correction of AHU Static Pressure Setpoint

The auto-correction algorithm "Improve AHU static pressure setpoint reset" is presented here as an example. The other eight algorithms are described in detail in Lin *et al.* (2020). AHU static pressure reset is one of the top ten efficiency measures implemented by organizations through use of FDD technology (Kramer *et al.*, 2020). The auto-correction algorithm for this opportunity is closely related to ASHRAE High-Performance Sequences of Operation Guideline 36 (ASHRAE, 2018), but deployed via the FDD tool. The algorithm corrects the fault "continuously" as it adjusts control variables to optimize equipment operation (e.g., setpoints). It is relevant for AHUs with downstream VAV boxes without sophisticated reset strategies, such as no reset or simple resets based on the average or maximum of downstream VAV damper positions.

The AHU static pressure setpoint is the correctable variable. The auto-correction algorithm uses the ASHRAE High-Performance Sequences of Operation Guideline 36 (ASHRAE, 2018) "Trim and Respond" logic for the new static pressure setpoint. To optimize the operation of the AHU and minimize discomfort, the static pressure setpoint is continually reset using Trim and Respond logic between a minimum and maximum setpoint. When the supply air fan turns on, after a period of inactivity, the setpoint starts from an initial value. The reset logic is active while the supply air fan is proven on. When active, for every time step t, when the sum of pressure requests (R) from the downstream zones is less than or equal to a defined number of ignored requests (I), the setpoint is trimmed by a fixed trim amount. If R is more than I, the setpoint changes by a respond amount based on R, but no more than the maximum response per time interval.

3. PRELIMINARY TESTING

This section illustrates the test results of one auto-correction algorithm: "improve AHU static pressure setpoint reset" (Section 2.2). The algorithm was deployed in a commercial FDD product and tested on two AHUs between March 4th and August 4th, 2020. The goal of this preliminary test was to determine whether the enhanced FDD solution is able to correct the fault without adverse operational effects.

3.1 Description of the Testing Site and Equipment

The test building is a commercial building operated by Lawrence Berkeley National Laboratory (LBNL) and utilized for offices and laboratories. Under normal conditions, the building is operated all week and its lab zones have 24/7 ventilation. Offices are ventilated and conditioned from 4 am to 9 pm. The HVAC system is primarily composed of a water-cooled chilled water plant, a gas-powered heating hot water plant, multiple AHUs and Variable Air Volume (VAV) boxes. AHU01 and AHU02, that serve 90% of the building floor area, and their connected zones (n=83 and n=80) are used for the test of this algorithm. Starting from March 17th the occupancy of the buildings decreased significantly due to COVID-19, but the Labs kept running 24/7. Figure 2 shows the BAS graphics (i.e. native dashboard) for one of the two AHUs.

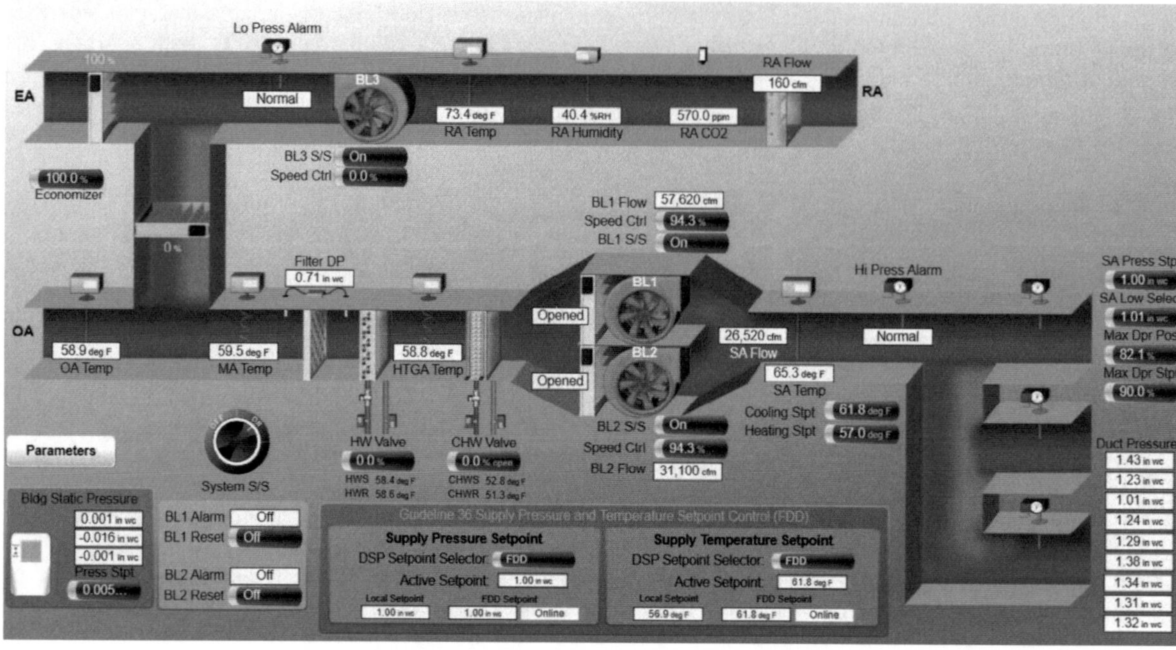

Figure 2: BAS graphics for AHU02 at the LBNL test site

Both AHU01 and 02 are controlled by a "sequence of operation" implemented in the native control language and hosted on local controllers. Each AHU is controlled independently. The baseline Static Pressure Setpoint (SSP_SP) Reset used by AHU01 and AHU02 is coded in the AHU controllers and it is summarized below in English. The supply air fan speed of each AHU is regulated by a direct-acting proportional–integral–derivative loop to keep static pressure at its setpoint, determined by the following reset logic.

Baseline Static Pressure Setpoint Reset Strategy

• If the AHU is enabled (based on schedules, normally 24/7):
  ○ Calculate the max damper position (MAX_DPR) for all the zones served by an AHU.
  ○ Use a direct-acting proportional–integral (PI) control loop to reset the Supply Static Pressure Setpoint (SSP_SP) between a minimum and maximum setpoint 249 pa and 373 pa respectively) to keep the max damper position at 90%. The setpoint is reset continuously.

3.2 Implementation of Auto-correction into the FDD Software

The limitation of the baseline strategy is that using the maximum damper position to trigger the reset can easily be negatively impacted by rogue zones (i.e., a stuck damper would cause the SSP_SP to be at its max value the whole time). This opportunity can be addressed by implementation of the correction algorithm "Improve AHU static pressure setpoint reset" into the FDD software. The implementation process includes (1) Change settings in the FDD tool to enable the write capability to the BAS setpoints (2) Program the auto-correction routine in the analytics engine of FDD tool (3) Commission the algorithm, review the auto-correction algorithm outputs, and verify the point value to be auto-corrected.

The FDD tool connected to the BAS is a commercial product managed by the sustainability department of the site. The tool allows for custom programming and bi-directional communication to the BAS via the BACnet network. In contrast to the BAS, the FDD tool coding language is a modern scripting language with the ability to use high-level functions that allow portability of the code between buildings and equipment. The auto-correction algorithm was coded using this platform and tested on the two AHUs. The architecture of the FDD tool and the BAS to support fault auto-correction is presented in Figure 3. The solid line shows the original infrastructure and the red dash line shows the upgrade. Two auto-corrective commands are sent to the BAS BACnet network and then to the writable properties used to control the static pressure. One is the new static pressure setpoint SSP_SP calculated from the correction algorithm (Section 3.3), the other is a "heartbeat" message indicating the FDD tool's status (online or offline). If the heartbeat is not received by the BAS every 15 minutes (adjustable), the BAS reverts back to the baseline pressure reset control program. On the BAS side, some programming changes are also required. A setpoint variable (highlighted in red and marked with 2 in Figure 4) was added to receive the new SSP_SP coming from the FDD tool and then overwrite the static pressure setpoint calculated by the baseline control logic. A watchdog input (1 in Figure 4) was also added to pair with the heartbeat command from the FDD tool. Finally, an enable input (3 in Figure 4) was added to provide the BAS operator with the ability to switch between the baseline setpoint and the FDD autocorrected setpoint.

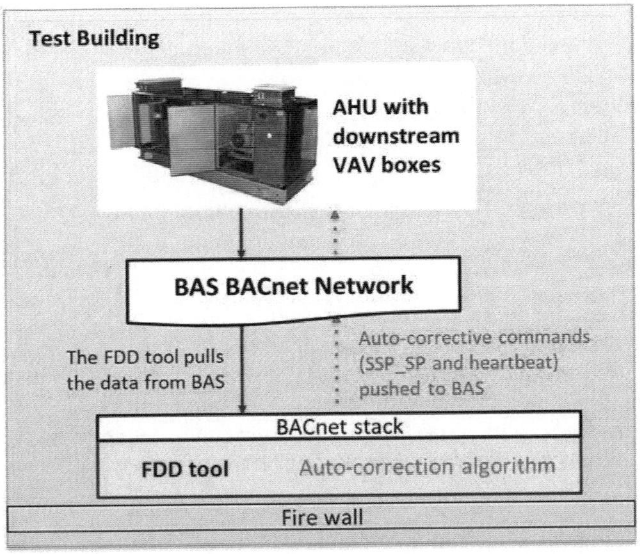

Figure 3: The FDD-BAS architecture to support auto-correction

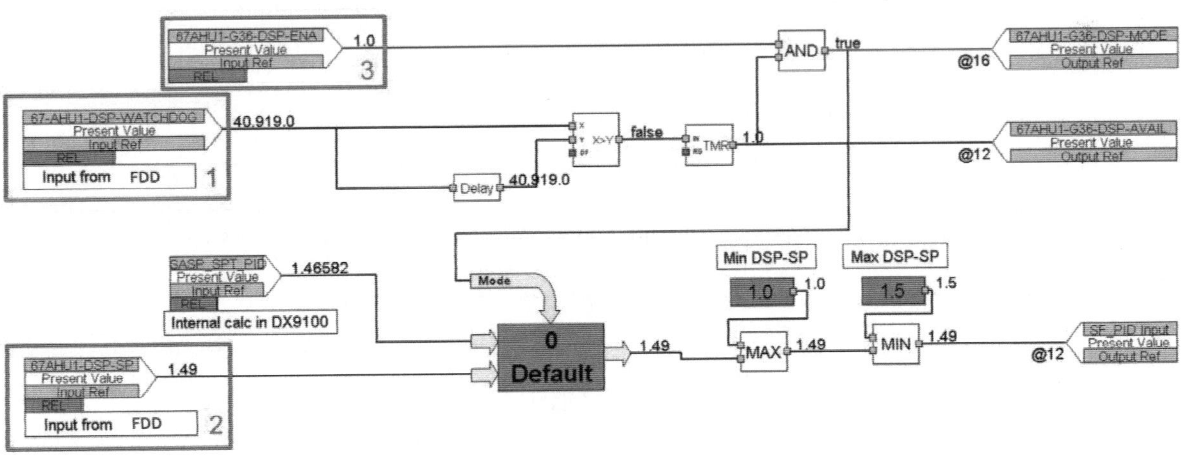

Figure 4: BAS logic that determines what algorithm is in control of the static setpoint. If the enable (3) and the watchdog (1) flag (that checks whether the FDD toolis online) are both TRUE the FDD tool setpoint is used, otherwise the BAS setpoint is used.

3.3 Auto-Correction Code Developed for FDD Tool

The code adopts the algorithm in Section 2.2. The static pressure setpoint (SSP_SP) is continually reset using "Trim and Respond" logic between a minimum and maximum setpoint (SSP_SP_MIN = 249 pa and SSP_SP_MAX =373 pa). When the supply air fan is turned on, the initial setpoint is set to SSP_SP0 = 249 pa and the reset logic is active immediately. When active, every time step t = 5 min, the following algorithm controls the SSP_SP. The advantages of this new algorithm are its simplicity, and the ability to operate efficiently regardless of rogue zones (i.e., these can be handled by increasing the number of ignored zones I)

New SPP Reset Strategy

Variables definition and instantiation:
• Supply Static Pressure Actual: SSP
• Supply Static Pressure Setpoint: SSP_SP
• Min Supply Static Pressure: SSP_MIN = 249 pa
• Max Supply Static Pressure: SSP_MAX = 373 pa
• Initial Supply Static Pressure Setpoint: SSP_SP0 = 249 pa
• VAV Damper Position: DMP
• Requests: R
• Ignores: I =1
• Net Requests: R*
 	
Logic:
• If the AHU is enabled (based on schedules, normally 24/7):
  o For each zone, every 5 min calculate the number of requests based on the VAV damper position (using hysteresis conditions):
    ▪ (trigger condition) if DMP_i > 95% generate a request
    ▪ (release condition) if DMP_i < 85% remove the request
  o Sum the requests R for all the zones
  o Subtract the fixed number of ignored requests I=1
  o Calculate the net number of requests R* = R - I
  o If R* <=0
    ▪ SSP_SP is trimmed by a fixed amount (SSP_trim = 2.5 pa) until the SSP_SP reaches SSP_MIN
  o If R* >0
    ▪ SSP_SP is increased (respond) by an amount proportional to R* (SSP_res = 5 x R* pa with a maximum response of 25 pa) until the SSP_SP reaches SSP_MAX

3.4 Testing Procedure

The algorithm was tested on AHU 01 and AHU 02 between March 4th and August 4th 2020. The test procedure is the following:

- Verify the ability to override all setpoints/parameters to be tested.
- Document pre-implementation behavior.
- Execute the auto-correction routine.
- Observe and document the effect of the automated fault correction

3.5 Testing Results

3.5.1 Baseline reset strategy

An example of the baseline behavior of the SSP_SP reset is displayed in the lower panel (b) of Figure 5. The setpoints profile is smooth because it is continuously reset by a PI loop. The power use of the two fans is depicted in the upper panel (a) of Figure 2. Given the high load due to occupancy and laboratory process load, the pressure setpoint reaches its max between 10-11 am and between 5-6 pm. The fan speed increases over the entire 4 AM - 9 PM period to maintain the static pressure at setpoint with an increased demand from VAV dampers that are only open during the day. Fan speed further increases during the day to meet the increased static pressure setpoint when the demand is even higher, e.g. for meeting the cooling load in some spaces.

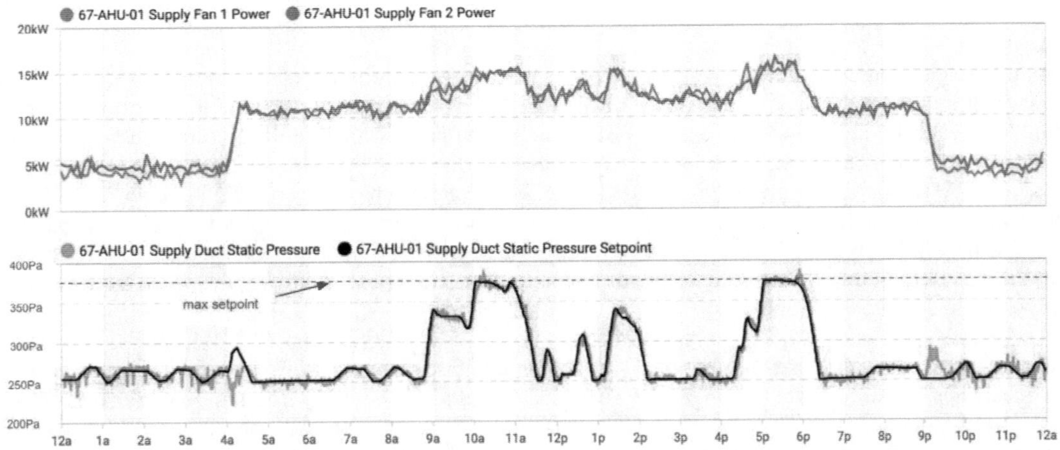

Figure 5. SSP setpoint of AHU01 with the baseline control strategy algorithm (Feb 13, 2020). a) Upper panel: power draw of the two fans controlled using the static pressure; b) Lower panel: SSP_SP using the baseline algorithm (black), actual SSP (light blue).

3.5.2 Test of the Auto-Correction Algorithm

The auto-correction algorithm "Improve AHU static pressure setpoint reset" successfully changed the setpoint of AHU01 and AHU02. Figure 6 shows the successful implementation of the algorithm for one day in the testing period, and the SSP_SP changes followed the routine described in Section 3.3. The total number of requests for all the zones served by AHU01 is displayed in the upper panel of Figure 6. When the net number of requests R* goes above zero the "respond" logic is triggered, increasing the SSP_SP as illustrated in the lower panel of the figure. The first case of this behavior happens around 4am. The actual pressure, SSP follows the setpoints after some oscillations. In this case, the SSP_SP never reaches its max value, but it stops increasing when R* becomes zero again. At that point SSP_SP is constantly "trimmed" until it reaches its minimum SSP_SP_MIN more than an hour later. The same behavior repeats other three times during the same day, although the slope of the "respond" phase is lower, because R* is lower compared to the first case.

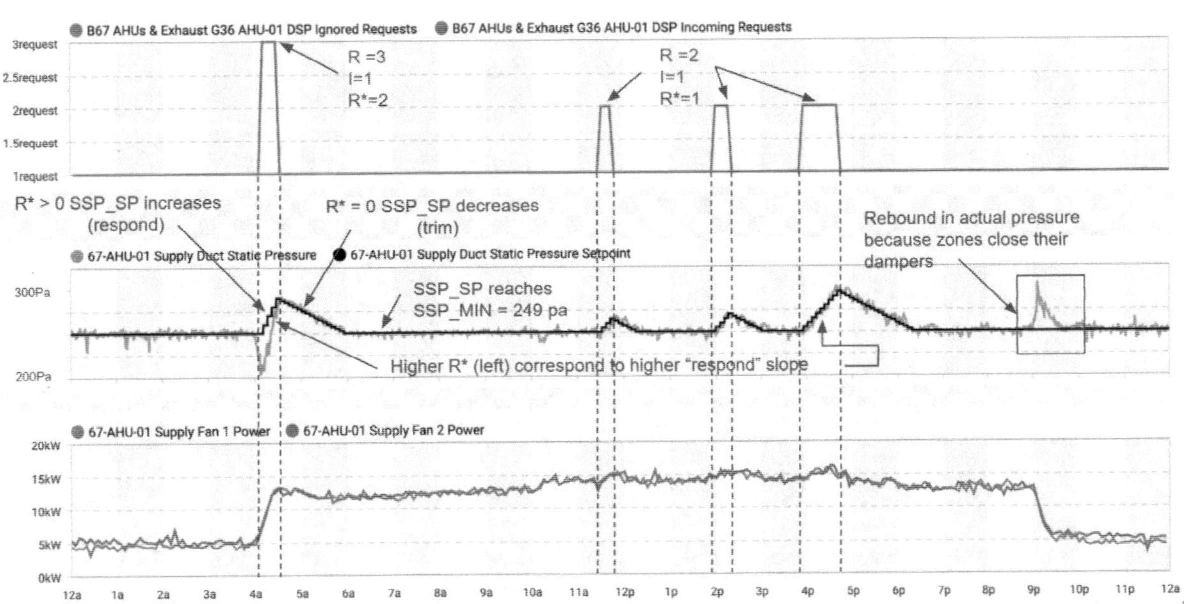

Figure 6: SSP setpoint of AHU01 after the execution of the auto-correction algorithm (Jul 14, 2020). a) Upper panel: number of requests generated by the zones (red). b) Lower panel: SSP_SP using the new algorithm (black) and SSP (light blue).

3.6 Discussion

Since both the baseline logic and the new logic use feedback loops, a quantitative comparison between the two is not possible without modeling the dynamic behavior of the system or collecting enough data to perform a system-level evaluation. However, Figure 7 shows an example of the improved performance of the automated correction routine, in case of a rogue zone. The rogue zone essentially breaks the feedback loop by keeping its damper 100% open regardless of the AHU static pressure. During June 28th 2020, a zone damper was consistently open at 100% (Figure 7, bottom panel). This single zone drove the SSP_SP of the baseline strategy (in red in Figure 7, top panel) to its maximum for the whole day, while the new strategy performed consistently better. The baseline strategy was not in control at the time, but in this situation we know that the baseline strategy setpoint would indeed have stayed at the maximum value of 373 pa.

The auto-correction algorithm has been running for five months without issues and no occupant complaints, although occupancy in the building was significantly reduced due to COVID-19. The building managers have been encouraged by the outcomes of this algorithm. Therefore, they decided to remain the algorithm active in regular operation, because it works better than the original sequence.

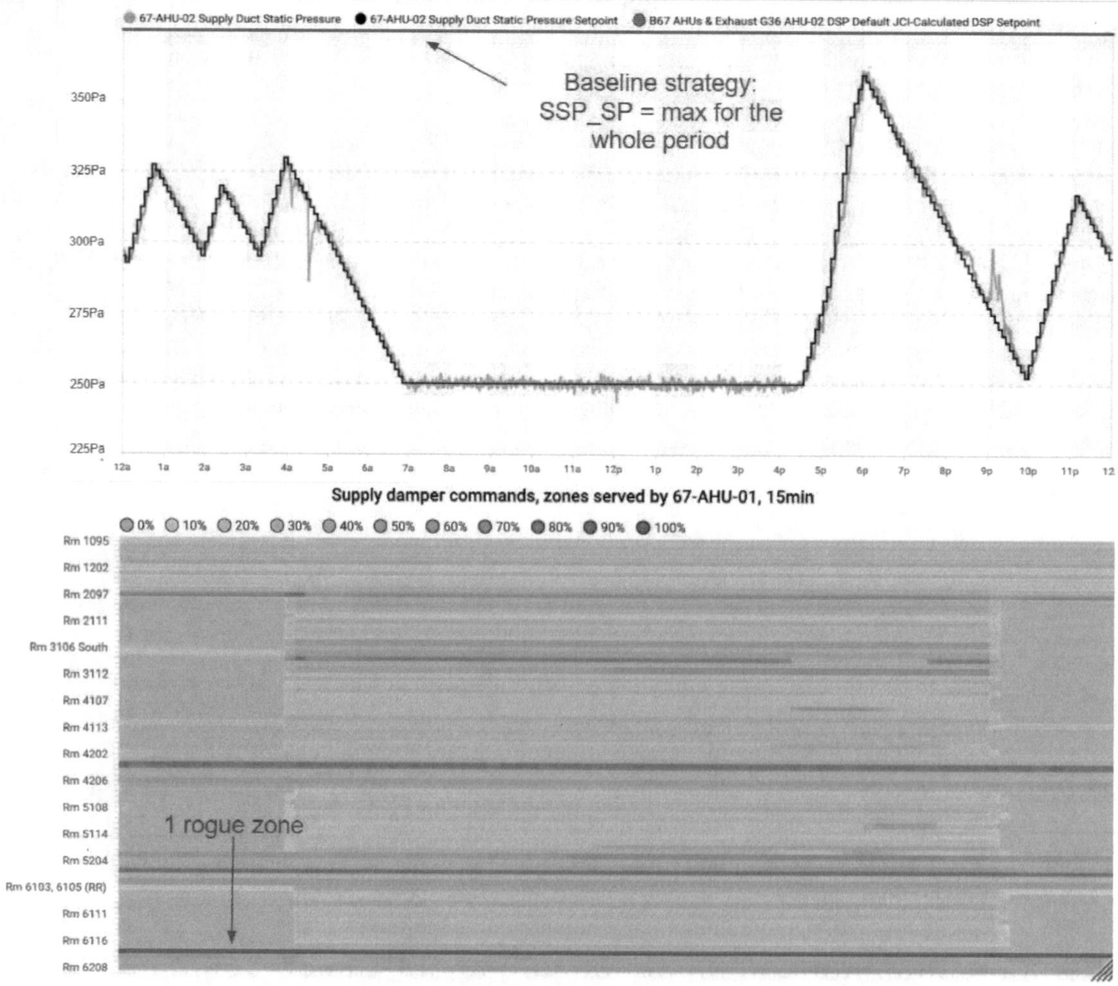

Figure 7: Example of 1 rogue zone (lower panel) driving the original SSP_SP setpoint to remain at 373 pa for the entire day (upper panel, in red), while the new strategy generates lower SSP_SP.

The development and deployment of this algorithm stimulated an interesting discussion among the partners and advisors of the project about the role of the FDD and the BAS. Typically, commercial FDD tools are developed as a software layer on top of the existing BAS. There exists a natural separation of roles in this arrangement, in which the BAS actively controls the building and the FDD tool observes its operation and provides insights and recommendations to the building manager. However, some consider FDD tools as a good supplement of BAS that can take over some of its functionalities when it is necessary. In this preliminary testing, implementing the correction algorithm to address the static pressure setpoint opportunity directly in the BAS would be a difficult task. Zone controllers at this testing site cannot directly calculate "requests" necessary for the implementation of the algorithm because both the lab controllers and the office controllers can't easily be modified to run customized logic. Alternatively, calculating requests and rogue zones for each zone in the AHU controller would be overly complex, since each AHU serves more than 80 zones. The existing AHU controllers would likely not have enough computing power to run such calculations. For these reasons we used the FDD tool to implement the algorithm and achieved success.

4. CONCLUSION AND FUTURE WORK

This paper presents preliminary efforts towards delivering automated fault correction. It describes nine fault auto-correction algorithms for heating ventilation and air conditioning (HVAC) systems that were developed to automatically correct faults or improve controls operation. It also presents preliminary testing results of one auto-correction algorithm (improve air handling unit static pressure setpoint reset). In the testing, the algorithm was implemented in a commercial FDD product and successfully changed the static pressure setpoint in two AHUs. The preliminary test confirmed the efficacy of the algorithm, as tested in a real commercial building. It provided a foundation of future auto-correction algorithm development and novel schemes for improving building operation performance and reliability.

In the future, more field testing of the FDD integrated correction algorithms with different FDD tools will be performed in a cohort of commercial buildings. This will include evaluation of the technical efficacy and performance of each correction routine, evaluation of the operations and maintenance benefits for each site in cohort, and characterization of challenges and best practices. A second area of future work will entail design and execution of a techno-economic analysis to quantify the broader market opportunity to inform ongoing commercialization efforts.

REFERENCES

American Society of Heating Refrigerating and Air Conditioning Engineers(ASHRAE) Guideline 36. (2018). High Performance Sequences of Operation for HVAC Systems.

Bengea, S. C., Li, P., Sarkar, S., Vichik, S., Adetola, V., Kang, K., Lovett, T., Leonardi, F., Kelman, A. D. (2015). Fault-tolerant optimal control of a building HVAC system. Science and Technology for the Built Environment, 21(6), 734-751.

Brambley, M. R., Fernandez, N., Wang, W., Cort, K. A., Cho, H., Ngo, H., & Goddard, J. K. (2011). *Final project report: Self-correcting controls for VAV system faults Filter/Fan/Coil and VAV box sections* (No. PNNL-20452). Pacific Northwest National Lab.(PNNL), Richland, WA (United States).

Dexter, A., & Pakanen, J. (Eds.). (2001). *Demonstrating automated fault detection and diagnosis methods in real buildings*. Technical Research Centre of Finland (VTT).

Fernandez, N., Brambley, M. R., & Katipamula, S. (2009). *Self-correcting hvac controls: Algorithms for sensors and dampers in air-handling units* (No. PNNL-19104). Pacific Northwest National Lab.(PNNL), Richland, WA (United States).

Fernandez, N., Brambley, M. R., Katipamula, S., Cho, H., Goddard, J. K., & Dinh, L. H. (2010). *Self-Correcting HVAC Controls Project Final Report* (No. PNNL-19074). Pacific Northwest National Lab.(PNNL), Richland, WA (United States).

Granderson, J., Singla, R., Mayhorn, E., Ehrlich, P., Vrabie, D., & Frank, S. (2017). Characterization and survey of automated fault detection and diagnostics tools. *Report Number LBNL-2001075 (Lawrence Berkeley National Laboratory, 2017).*

Hao, X., Zhang, G., & Chen, Y. (2005). Fault-tolerant control and data recovery in HVAC monitoring system. *Energy and buildings*, *37*(2), 175-180.

Hyvarinen, J., & Karki, S. (1996). Building optimization and fault diagnosis source book. *IEA Annex*, *25*(1996.8).

Katipamula, S., & Brambley, M. R. (2005). Methods for fault detection, diagnostics, and prognostics for building systems—a review, part I. *HVAC&R Research*, *11*(1), 3-25.

Katipamula, S., & Brambley, M. R. (2005). Methods for fault detection, diagnostics, and prognostics for building systems—A review, part II. *HVAC&R Research*, *11*(2), 169-187.

Granderson, J., Kramer, H., Lin, G., Curtin, C., & Crowe, E. (2020). Proving the business case for building analytics. Lawrence Berkeley National Laboratory, October 2020. https://doi.org/10.20357/B7G022

Kim, W., & Katipamula, S. (2018). A review of fault detection and diagnostics methods for building systems. *Science and Technology for the Built Environment*, *24*(1), 3-21.

Lin, G., Pritoni, M., Chen, Y., & Granderson, J. (2020). Development and Implementation of Fault-Correction Algorithms in Fault Detection and Diagnostics Tools. *Energies*, *13*(10), 2598.

Wang, S., & Chen, Y. (2002). Fault-tolerant control for outdoor ventilation air flow rate in buildings based on neural network. *Building and Environment*, *37*(7), 691-704.

Zhang, Y., & Jiang, J. (2008). Bibliographical review on reconfigurable fault-tolerant control systems. *Annual reviews in control*, *32*(2), 229-252.

ACKNOWLEDGEMENT

This work was supported by the Assistant Secretary for Energy Efficiency and Renewable Energy, Building Technologies Office, of the U.S. Department of Energy under Contract No. DE-AC02-05CH11231. The authors wish to acknowledge Harry Bergmann for his guidance and support of the research. We also thank the fault detection and diagnostics technology and service providers who participated in this study.

Superheat Prediction & Fault Diagnostics of HVAC from Simple Temperature Measurements Using Big Data Approach

Mehari K. Tesfay[1*], Mostafa Rafaie[1], , Kevin Sinkar[2], Parthiban Arunasalam[2], Dhaman Besarla[2,] Fadi Alsaleem[1]

[1]Durham school of architectural and construction Engineering, University of Nebraska-Lincoln

[2]DunAn Microstaq, Inc., Austin, TX, 78723, United States

[*]Corresponding Author

ABSTRACT

New advancements in data & algorithms have pushed new techniques and methods to the forefront in optimizing energy efficiency as well as keeping the thermal comfort of residents in intelligent buildings research. HVAC elements, being ubiquitous and fundamental elements in buildings today, their diagnostics maintenance, operational functionalities, and control are essential aspects in this regard. The tremendous amount of data generated from buildings every day and recent developments with data science tools have changed the control and monitoring of these units from exhausting physical modeling and operation to data-driven techniques that are more reliable and efficient. The massive streaming data generated by smart building sensors have inspired new ways of controlling and diagnosing faults in comfort systems using machine learning and big data analytics. In this work, we present a big-data driven approach to model the dynamic of two similar HVAC (but, healthy and faulty) systems from simple temperature measurements collected over an extended period. The model showed good accuracy in predicting the system superheat for both systems. This demonstrates the potential of big data approach to substitute the need for having the expensive pressure sensor to measure superheat.

1. INTRODUCTION

It has been reported that HVACs currently consume more than 40% of total electricity use in the U.S [1]–[6]. Due to their significant impact on system efficiency, energy consumption and occupant comfort, faults in building HVAC systems need to be detected appropriately and hence diagnostic measures are taken to restore it to ideal working states. Being responsible for consuming a significant portion of global energy, HVACs require a reliable Fault Detection and Diagnosis (FDD) method to ensure an efficient equipment operation in order to achieve optimal energy usage and deliver a good performance. The figure below shows the classification of FDD methods[7]

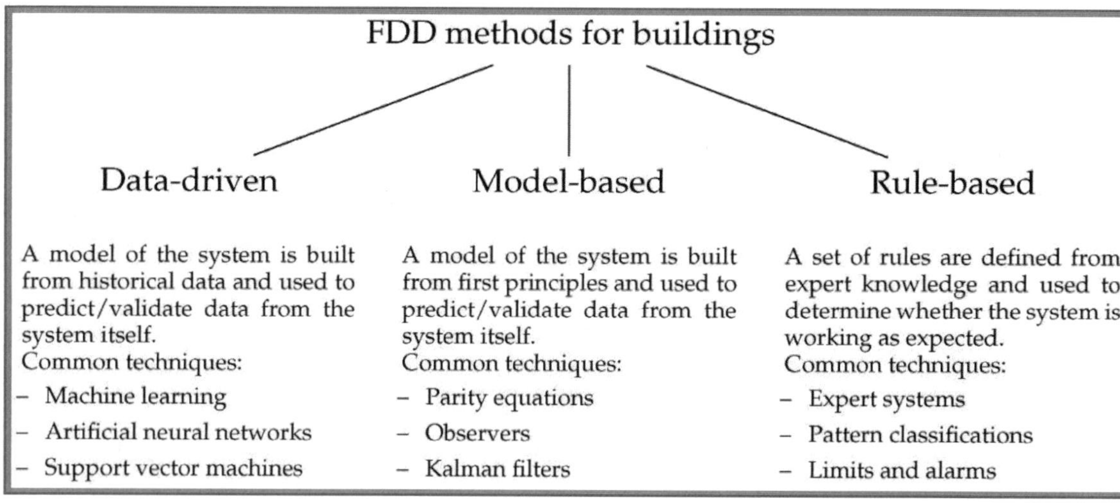

Figure 1: Different classifications of Automated Fault Detection and Diagnostics methods

This study categorizes AFDD methods in HVAC areas into three main categories— namely process history-based, qualitative model-based, and quantitative model-based methods.

Rule-based & qualitative physics-based modeling techniques use a priori knowledge to set up thresholds and rules to identify fault symptoms and make a final decision about the state of a system. On the other hand, quantitative model-based methods rely on established & explicit mathematical models to represent each component of the plant and simulate the steady and transient behaviors of the system to reach an analytical outcome to detect and diagnose the cause of faults[8]. This analytical result is followed by experimental data for validation. Process history-based methods, such as black-box models, rely on the relationship between inputs and outputs of a process without any consideration for physical significance. Those models include Bayesian networks [7]–[12] and artificial neural networks (ANNs) [13]–[19]

Another work in [20] relies on experimental data-driven approaches and algorithms for modeling, optimizing, and controlling HVAC systems. In this work, supply air duct static pressure and supply air temperature are used as control variables to improve energy efficiency and maintain thermal comfort. Data-driven approaches are used in [21] to estimate HVAC specific energy consumption in buildings using an improved Fourier series decomposition. Having no prior knowledge about abnormal phenomena that occurred in the system, researchers in [22] designed a semi-supervised data-driven approach, namely Principal Component Analysis (PCA). The PCA was employed for fault detection and isolation by distinguishing anomalies from normal operation variability as well as isolating variables related to faults.

2. METHODOLOGY

In this field experiment, we have a healthy and faulty unit. For the healthy unit, a baseline approach was used to monitor the deviation of the system from normal working conditions. Here, a reliable model has been designed to learn and predict superheat values of an HVAC unit from temperature measurements only, namely suction, outdoor and liquid line temperatures, excluding pressure variables. In this approach, a prediction with high accuracy has been obtained. Having reliably designed baseline model & predicted the superheat of a healthy unit, it was applied to new data collected from another unit with fault for successive months to establish & study a machine learning-based diagnostics nature of the HVAC units. Accordingly, the fault in the system has been identified. Based on this approach, control algorithms could be applied to bring superheat values to acceptable ranges or a decision could be made to replace the elements depending on the degree of severity and malfunction, to maintain the comfort of occupants.

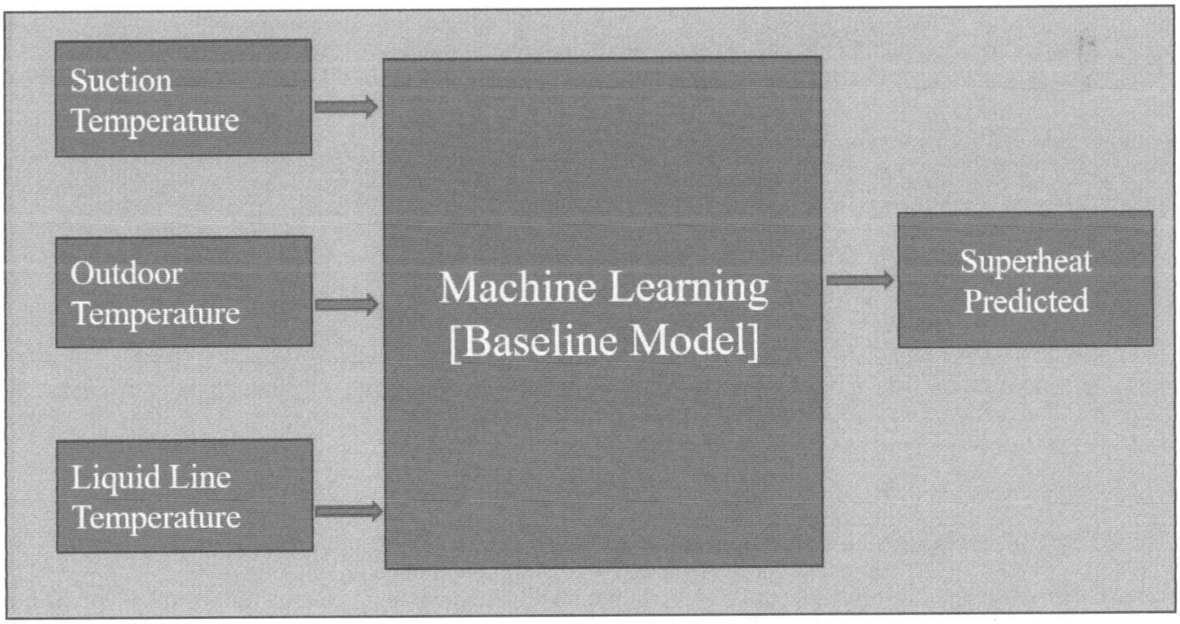

Figure 2: Baseline model design and predictors

3. THE PROPOSED DESIGN APPROACH

After massive data has been collected from our remote site, a local database has been set up to clean and process it. Since separating the ON and OFF data is necessary to analyze the performance of the HVAC, cleaning the data was the primary task to do before starting an in-depth study on the acquired sensory data. The dynamics of superheat throughout the operation time of the HVAC units has been chosen as a good variable to investigate the behavior and health of the element. To this end, leveraging the power of the huge data collected and many machine learning techniques, a model has been trained to learn the superheat from all temperature inputs. Superheat values have been predicted from these models and compared with actual measurements to verify their validity.

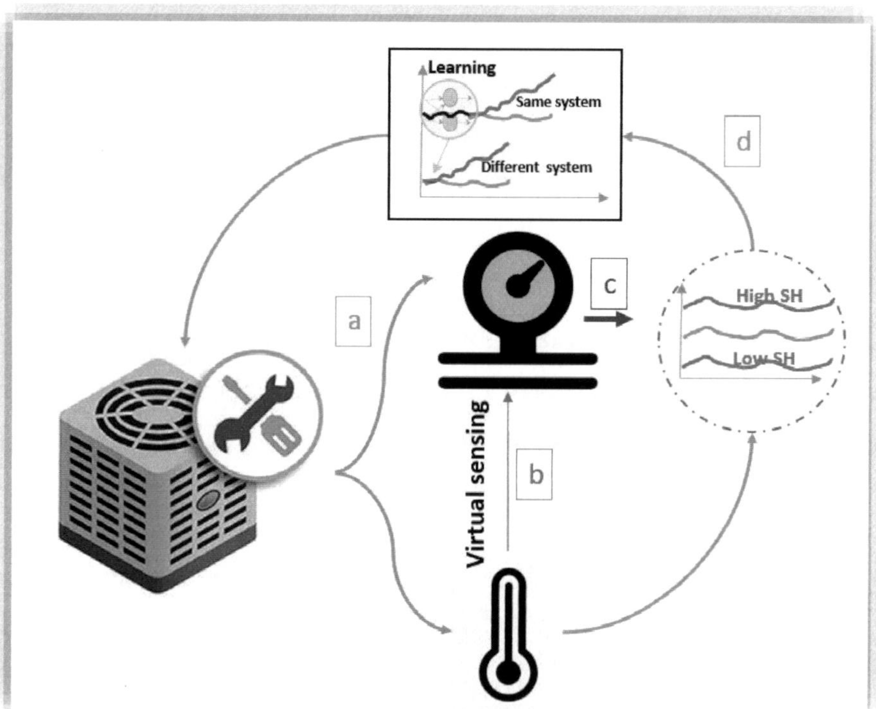

Figure 3: A comprehensive proposed design approach

Once a reliable model has been designed & the healthy pattern has been identified, it has been established as a baseline model against which past and future measurements of the same unit or other units are compared. As a continuation of this work, once the optimal superheat values of a sufficiently healthy system are identified, this range of values can be fed as set point to a control system that functions to maintain the comfort level of dwellers. The feedback from the resident, in the form of continuous data captured via wearable devices or intermittent reports on their comfort states, completes the cycle.

The existing approach in HVAC performance characteristics is mainly virtual sensing where pressures values at the inlet and outlets of compressors are estimated from temperatures. In another case, expensive equipment and sensors are deployed in the field to acquire real-time data to analyze HVAC behavior and performance. This is economically demanding and complicates the work of personnel. Though it is neither easy to calculate nor cheap to measure, superheat of a system, and sometimes Subcooling, is known as a very important indicator whether a system is healthy or faulty.

In this work, leveraging the availably of big data and many machine learning methods, we have come up with a novel way of estimating the superheat from simple temperature measurements. Utilizing this superheat and other predictor variables, a comprehensive baseline model has been designed for one of our healthy units, against which future and previous measurements on the state of the system are compared & hence detect whether faulty or not. Moreover, this same model can be used to identify & diagnose faults in other units where their data is unreliable to build models, *cross-unit detection*.

4. MACHINE LEARNING ALGORITHMS USED

Decision tree

The decision tree algorithm, specifically ***rpart*** function, is a non-parametric supervised learning method and is one of the simplest and yet most successful forms of machine learning for classification and regression. It has a tree-like graph representation that can be trained as a classifier to decide from multiple possible choices. The depth of the tree is one of the main parameters that can be tuned to enhance learning performance.

Gradient Boosting Regressor

The Gradient Boosting Regressor, XGBoost, is another ensemble learning technique used for classification and regression. This algorithm is known as a robust method to avoid overfitting[23].

Random Forest Regressor

Random Forest is based on utilizing the aggregation of decision trees built from various sub-samples of the datasets and their averages to improve the predictive accuracy Similar to the previous classifiers, it was employed while varying the estimator count to achieve better accuracy

5. RESULTS AND DISCUSSION

On a complete departure from previous trends, in this approach we devised a mechanism to depend on a system that we know is healthy from other previous studies and build a model based on which current and future data gathered from that system are measured against to detect fault.

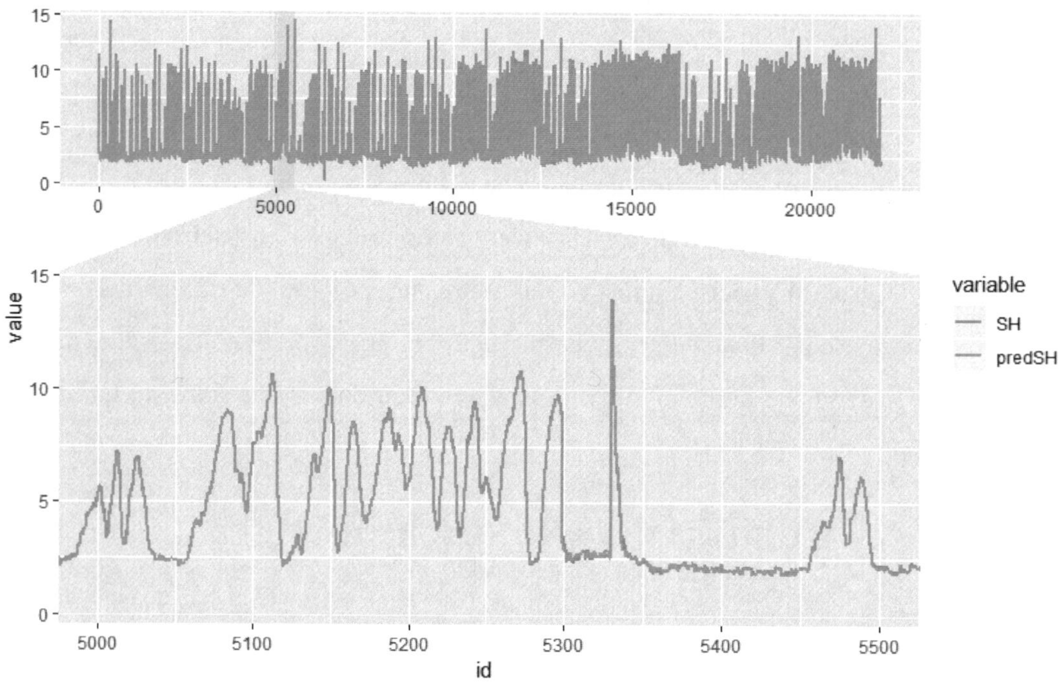

Figure 4: Predicting superheat of the healthy unit using the baseline model[°C]

In this model, the accepted ranges of variables of a properly functioning HVAC are learnt and in a later use the model takes field measured values of important predictors and predicts the superheat values. These values are then compared to the actual measured values of superheat to determine the drift in performance of the system and hence its health. In this study, the model exhibited excellent prediction accuracy and the mean error values are very small. Fig4 below shows, the actual prediction residual levels.

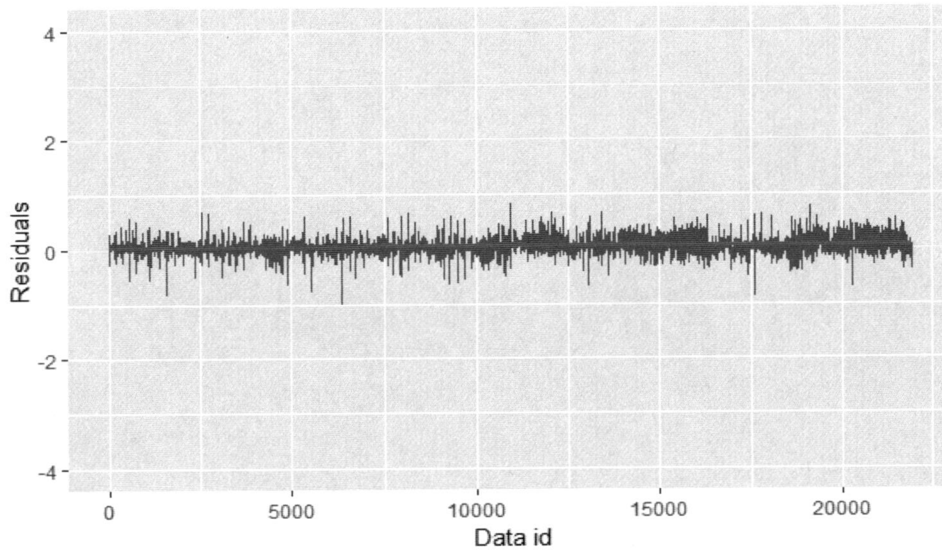

Figure 6: The superheat prediction errors are very small for healthy unit [°C]

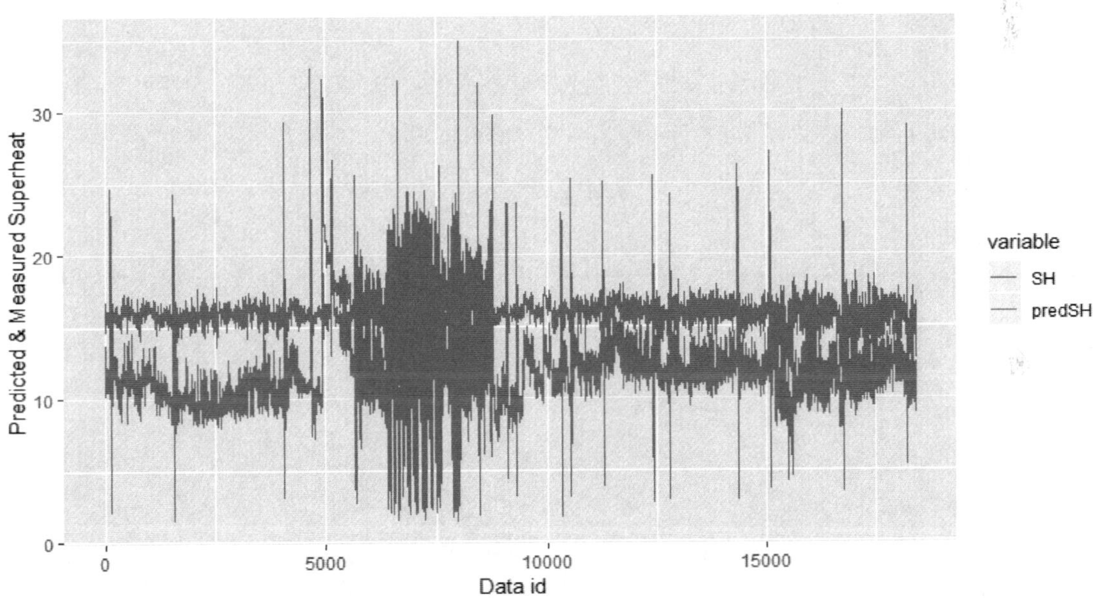

Figure 5: Predict SH of Unit-2 from Baseline Model trained by Unit-1 Data

However, the biggest contribution of this work is in its comprehensive capability in analyzing, diagnosing and monitoring malfunctions on other HVAC units, hence cross-unit detection. In instances where there is no enough data to learn the behavior of systems or when the data acquired from these units is deemed unreliable for various reasons, the general model built from a healthy & reliable unit can be employed to make comparative studies.

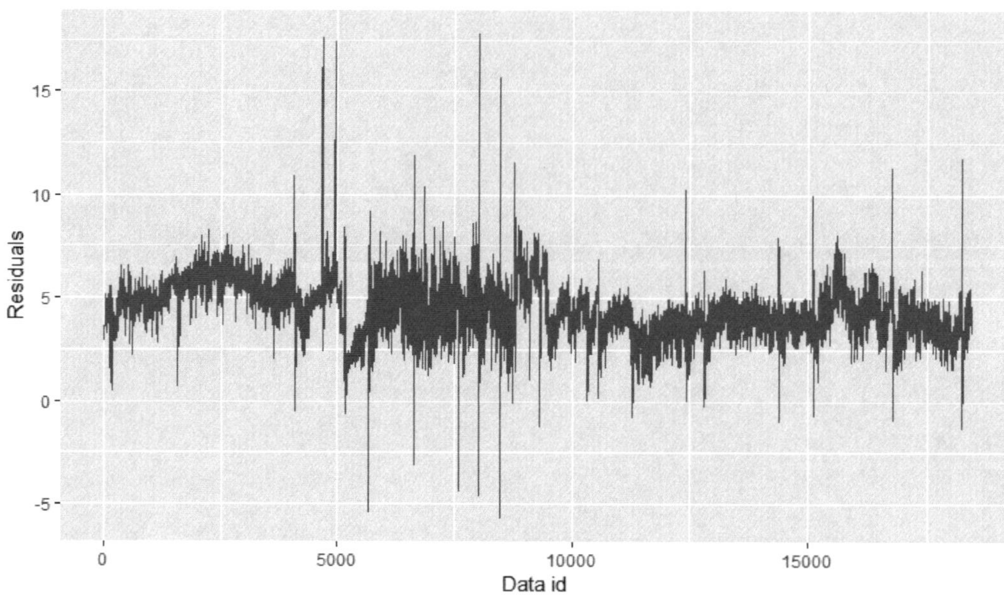

Figure 7: Actual Computed mean Residuals[°C]

This is exactly what we have implemented in this work. Figures 6 and 7 demonstrates the superheat values predicted for a faulty system using the baseline model. The predicted values are 5-10 degrees lower than those predicted, indicating that the HVAC unit had operational malfunction.

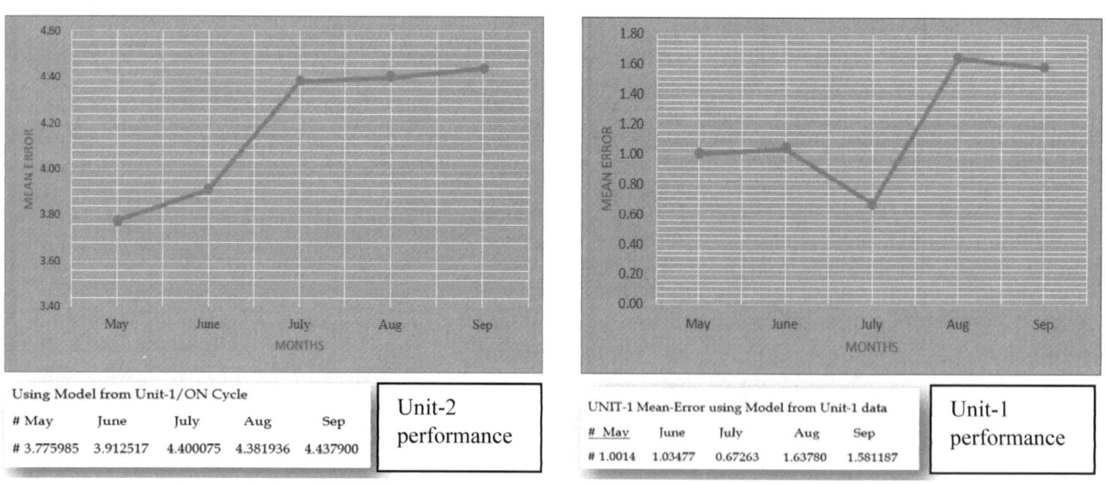

Figure 8: Mean prediction errors of unhealthy and healthy units

The two figures above show the comparative performance of the models for both the healthy and faulty systems. The mean superheat prediction error for the unit based on which the model was designed is approximately in the range of [1 to 1.6]. On the other hand, the error for the defective unit upon which the base-model was applied is within the ranges of [3.5 to 4.5]

6. CONCLUSION

In this work, by using a simple temperature sensor, we were able to build reliable machine learning models that successfully detect and diagnose faults in HVAC systems. Moreover, the accuracy of predictions of these models was investigated for both the defective and healthy units. These results confirmed the preliminary assessment that was conducted based on few superheat and subcooling values. Among the machine learning algorithms applied, Gradient Boosting Regressor showed better performance.

REFERENCES

[1] T. Wei, Y. Wang, and Q. Zhu, "Deep Reinforcement Learning for Building HVAC Control," in *Proceedings - Design Automation Conference*, 2017, vol. Part 12828, pp. 1–6, doi: 10.1145/3061639.3062224.

[2] H. Hao, A. Kowli, Y. Lin, P. Barooah, and S. Meyn, "Ancillary service for the grid via control of commercial building HVAC systems," in *2013 American control conference*, 2013, pp. 467–472.

[3] T. Wei, Q. Zhu, and M. Maasoumy, "Co-scheduling of HVAC control, EV charging and battery usage for building energy efficiency," in *2014 IEEE/ACM International Conference on Computer-Aided Design (ICCAD)*, 2014, pp. 191–196.

[4] B. Balaji, J. Xu, A. Nwokafor, R. Gupta, and Y. Agarwal, "Sentinel: Occupancy based HVAC actuation using existing wifi infrastructure within commercial buildings," in *SenSys 2013 - Proceedings of the 11th ACM Conference on Embedded Networked Sensor Systems*, 2013, pp. 1–14, doi: 10.1145/2517351.2517370.

[5] M. Avci, M. Erkoc, and S. S. Asfour, "Residential HVAC load control strategy in real-time electricity pricing environment," in *2012 IEEE Energytech, Energytech 2012*, 2012, pp. 1–6, doi: 10.1109/EnergyTech.2012.6304636.

[6] L. Pérez-Lombard, J. Ortiz, and C. Pout, "A review on buildings energy consumption information," *Energy Build.*, vol. 40, no. 3, pp. 394–398, 2008, doi: 10.1016/j.enbuild.2007.03.007.

[7] C. G. Mattera, J. Quevedo, T. Escobet, H. R. Shaker, and M. Jradi, "A method for fault detection and diagnostics in ventilation units using virtual sensors," *Sensors (Switzerland)*, vol. 18, no. 11, 2018, doi: 10.3390/s18113931.

[8] W. Kim and S. Katipamula, "A review of fault detection and diagnostics methods for building systems," *Sci. Technol. Built Environ.*, vol. 24, no. 1, pp. 3–21, Jan. 2018, doi: 10.1080/23744731.2017.1318008.

[9] A. L. Dexter and D. Ngo, "Fault diagnosis in air-conditioning systems: A multi-step fuzzy model-based approach," *HVAC R Res.*, vol. 7, no. 1, pp. 83–102, 2001, doi: 10.1080/10789669.2001.10391431.

[10] C. G. Mattera, J. Quevedo, T. Escobet, H. R. Shaker, and M. Jradi, "A Method for Fault Detection and Diagnostics in Ventilation Units Using Virtual Sensors †," 2018, doi: 10.3390/s18113931.

[11] F. Xiao, Y. Zhao, J. Wen, and S. Wang, "Bayesian network based FDD strategy for variable air volume

terminals," *Autom. Constr.*, vol. 41, pp. 106–118, 2014, doi: 10.1016/j.autcon.2013.10.019.

[12] Z. Wang, L. Wang, Y. Tan, and J. Yuan, "Fault detection based on Bayesian network and missing data imputation for building energy systems," *Appl. Therm. Eng.*, vol. 182, Jan. 2021, doi: 10.1016/j.applthermaleng.2020.116051.

[13] S. Wang and Y. Chen, "Fault-tolerant control for outdoor ventilation air flow rate in buildings based on neural network," *Build. Environ.*, vol. 37, no. 7, pp. 691–704, Jul. 2002, doi: 10.1016/S0360-1323(01)00076-2.

[14] Z. Du, B. Fan, X. Jin, and J. Chi, "Fault detection and diagnosis for buildings and HVAC systems using combined neural networks and subtractive clustering analysis," *Build. Environ.*, vol. 73, pp. 1–11, Mar. 2014, doi: 10.1016/j.buildenv.2013.11.021.

[15] Z. Du, B. Fan, J. Chi, and X. Jin, "Sensor fault detection and its efficiency analysis in air handling unit using the combined neural networks," *Energy Build.*, vol. 72, pp. 157–166, Apr. 2014, doi: 10.1016/j.enbuild.2013.12.038.

[16] Y. Zhu, X. Jin, and Z. Du, "Fault diagnosis for sensors in air handling unit based on neural network pre-processed by wavelet and fractal," *Energy Build.*, vol. 44, no. 1, pp. 7–16, Jan. 2012, doi: 10.1016/j.enbuild.2011.09.043.

[17] B. Fan, Z. Du, X. Jin, X. Yang, and Y. Guo, "A hybrid FDD strategy for local system of AHU based on artificial neural network and wavelet analysis," *Build. Environ.*, vol. 45, no. 12, pp. 2698–2708, Dec. 2010, doi: 10.1016/j.buildenv.2010.05.031.

[18] Z. Wang, R. S. Srinivasan, and J. Shi, "Artificial intelligent models for improved prediction of residential space heating," *J. Energy Eng.*, vol. 142, no. 4, Dec. 2016, doi: 10.1061/(ASCE)EY.1943-7897.0000342.

[19] Z. Du, X. Jin, and Y. Yang, "Fault diagnosis for temperature, flow rate and pressure sensors in VAV systems using wavelet neural network," *Appl. Energy*, vol. 86, no. 9, pp. 1624–1631, 2009, doi: 10.1016/j.apenergy.2009.01.015.

[20] G. Xu, "HVAC system study: a data-driven approach." University of Iowa, pp. xii, 87 pages.

[21] F. Niu, Z. O'Neill, and C. O'Neill, "Data-driven based estimation of HVAC energy consumption using an improved Fourier series decomposition in buildings," in *Building Simulation*, 2018, vol. 11, no. 4, pp. 633–645, doi: 10.1007/s12273-018-0431-2.

[22] A. Beghi, R. Brignoli, L. Cecchinato, G. Menegazzo, M. Rampazzo, and F. Simmini, "Data-driven Fault Detection and Diagnosis for HVAC water chillers," *Control Eng. Pract.*, vol. 53, pp. 79–91, 2016, doi: 10.1016/j.conengprac.2016.04.018.

[23] A. Almaghrebi, F. Aljuheshi, M. Rafaie, K. James, and M. Alahmad, "Data-Driven Charging Demand Prediction at Public Charging Stations Using Supervised Machine Learning Regression Methods," *Energies*, vol. 13, no. 16, p. 4231, 2020, doi: 10.3390/en13164231.

Fault Detection and Diagnosis for Residential HVAC Systems Using Transient Cloud-based Thermostat Data

Fangzhou GUO, Bryan P. RASMUSSEN

Texas A&M University, Mechanical Engineering,
College Station, Texas, U.S.
Email: {fzguo, brasmussen}@tamu.edu

ABSTRACT

Fault detection and diagnosis (FDD) using aggregated smart thermostat data is a relatively new research field, but one with immediate practical application to residential indoor climate control. This paper analyzes a cloud-based dataset which contains thermostat history records of nearly 370,000 distinct residential HVAC systems in the U.S. The large, diverse, and growing dataset enables novel methods for detecting and diagnosing faults on systems with limited sensor data. This paper proposes a statistics-based FDD method for non-variable speed heat pump and air conditioning units, and demonstrates the effectiveness with several case studies. The proposed method identifies systems within a similar climate region, and then segments and classifies the time series data based on operational mode and behavior. Various data features are then extracted from the time series segments to identify systems that exhibit poor transient behavior. Additional features are used to refine and classify the problem severity. Statistical methods are then used to compare system performance to the entire population and identify outlier behavior due to operational faults that affect system efficiency and occupancy comfort. The resulting algorithm demonstrates the potential of big data fault detection for air conditioning systems using limited cloud-based sensor information.

1. INTRODUCTION

Space heating and air conditioning accounts for a large amount of residential households' energy expenditures. From the Residential Energy Consumption Survey (EIA, 2015), the average annual energy expenditure for a household is $1,856, among which $543 (29.3%) spent on space heating and $265 (14.3%) spent on air conditioning. In order to reduce residential households' energy consumption on space heating and air conditioning, researchers have been working on improving system efficiency, employing better control strategies, and improving preventive maintenance using fault detection and diagnosis (FDD). By detecting operational faults caused by system aging or improper commissioning early, failures can be prevented and system inefficiencies avoided.

HVAC system faults have a considerable impact on system efficiency and occupant comfort. Studies report that the average performance of residential HVAC systems is at least 17% below their design performance (Proctor and Downey, 1999). Correcting only refrigerant charge and air flow problems has an estimated 17% demand savings potential and 7% peak demand savings potential (Neme *et al.*, 1999). Incorporating FDD into the preventive maintenance routines has the potential for increasing occupant comfort and reducing energy expenditures. Additionally, manufacturers could improve brand loyalty by ensuring high-performance and reliable systems; service companies would be able to provide better maintenance to their customers. With regard to FDD performed on building systems, Kim and Katipamula (2018) pointed out that only 16% studies are associated with small commercial and residential buildings, such as rooftop packaged units and split systems, while the majority of research focus is on variable air volume air handling units (VAV-AHUs), chillers, cooling towers, and overall building diagnosis. Thus traditional FDD methods are usually designed for one large system and require installing additional sensors (Rogers *et al.*, 2019). But residential split systems are produced in large quantities, with limited sensors, and with a variety of system types and custom installation. For many systems, the addition of the sensors required for traditional FDD techniques is prohibitively expensive.

Cloud-based data from smart thermostats provides an opportunity for alternative FDD methods that do not require additional sensors, but utilize historical records of basic system operation. The popularity of smart thermostats that upload data to a remote server is increasing, allowing for computationally intensive fault detection on distributed platforms. Many manufacturers are establishing a cloud database, remotely monitoring thousands of installed residential systems. Typically, this cloud-based thermostat data includes indoor temperature and relative humidity measured by built-in sensors, system mode status, cooling and heating temperature setpoints, and outdoor temperature acquired from a third party. The resulting big data creates an opportunity for novel FDD approaches.

FDD methods for residential HVAC systems with smart thermostat data is still a relatively new research field, and few published studies are currently available. Ham *et al.* (2016) proved the capability of thermostat data to evaluate the thermal characteristics of a house, such as insulation levels, thermal capacity, and thermal time constants. With these indicators, houses showing inferior thermal response characteristics, known as construction-level faults, could be detected. Turner *et al.* (2017) developed a recursive least-squares model to detect faults on cooling, heating and ventilation equipment. However, the method was only validated by simulated indoor and outdoor temperature data, and focused mainly on total equipment failures. Jain *et al.* (2019) trained a linear model to learn the normal behavior, and then applied the model to monitor capacity degradation faults such as refrigerant leakage. Although the model was originally developed for cold rooms in retail outlets, the author mentioned that its application could be extended to residential buildings. Besides, combining smart thermostat data with building assessor's data and energy data, Do and Cetin (2019) showed the potential to predict the electricity demand of residential HVAC systems, and then perform fault detection by comparing the predicted demand with the actual demand. In short, these studies all focused on a small group of systems, training models and performing fault detection for individual systems. In contrast, this paper seeks to leverage large-scale cloud-based thermostat data of thousands of systems to determine hard and soft faults. Service companies are generally more interested in finding soft faults, such as capacity degradation and control problems, rather than hard faults or total failures, since soft faults are generally difficult to detect and may not be recognized by occupants for an extended period of time (Rogers *et al.*, 2019).

We propose a statistics-based FDD method that finds anomalous operation behavior and preliminarily categorizes fault types by comparing fault indicators (known as *features*) among multiple systems. In this way, systems having soft faults would be identified as outliers exhibiting anomalous operational behavior. The general analytical process is described as follows. First, the method identifies systems within a similar climate region by relating the system location information to the U.S. climate zones, according to the International Energy Conversion Code climate regions and moisture regimes shown in Figure 1 (International Code Council *et al.*, 2012). Second, from a particular climate region, thermostat time-series data of a group of systems is queried from the database and preprocessed. From the resulting data, raw system cycling mode data (i.e., cooling on, heating on, system off) are transformed to more specific categories of operating modes, namely regulating, tracking and free response modes. The time-series data is segmented and classified with these new mode labels (Rogers *et al.*, 2020). Based on the type of abnormal behavior to be identified, multiple features will be selected and extracted from the segmented data. Generally, the features characterize system performance, indicate faults, and problem severity, or describe the overall condition of a house and its ambient environment. Third, statistics methods are applied to compare the features among all systems in the selected group to identify outliers. Finally, labeled systems with operational faults affecting system efficiency and occupancy comfort are outputted.

Using this general approach, the performance of a system can be deeply inspected, including both the pseudo steady-state and transient behavior in cooling and heating modes. This paper focuses on the time periods that exhibit poor *transient* behavior, known as poor tracking periods. Both cooling mode and heating mode can exhibit this type of behavior, but this study only focuses on the cooling mode. A poor tracking period is defined as a single cooling on-cycle that usually occurs during high cooling load, and during which the indoor temperature can increase a few degrees above the cooling setpoint, unable to be controlled although the system is operating continuously. Note that the term 'high cooling load' is regarded as relative to the system capacity. If the capacity of a system is far below its rated capacity, even normal loads may well be incapable for it to deal with and thus will be considered as high loads. Therefore, this situation is unlikely to happen for a properly sized fault-free system, and is often related to the capacity degradation fault.

In this paper, Section 2.1 and 2.2 discuss two necessary and innovative data preprocessing steps: data cleansing and mode labeling, which are applied prior to every statistics-based FDD algorithm we proposed. After this, the algorithm will analyze the poor tracking periods specifically. Section 2.3 will introduce a list of features which help to find this poor transient behavior and are able to indicate its severity. Section 2.4 will discuss an unsupervised machine learning method, hierarchical clustering, to identify the poor tracking period using some of the extracted features. Then, in

Section 3, statistics tests on the features indicating problem severity will be performed among systems. In order to provide further implications of the features and recommend thresholds to label faulty systems, the statistical distribution of the features are estimated by the kernel density estimation. Finally, Section 4 will show statistics of the fault detection results as well as a couple of case studies, illustrating the effect of operational faults on system efficiency and occupancy comfort.

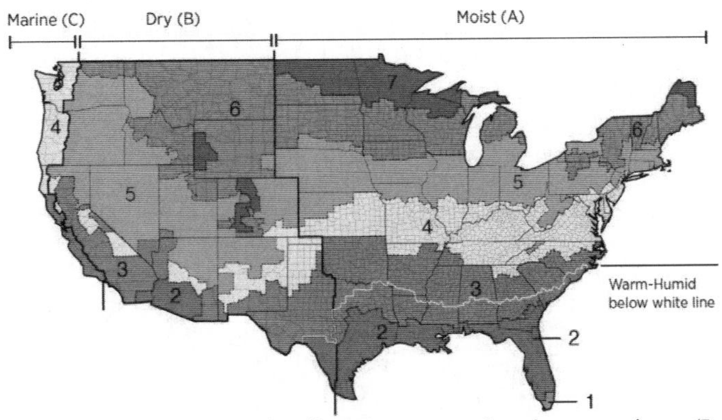

Figure 1: International Energy Conservation Code climate zones and moisture regimes (International Code Council *et al.*, 2012). Climate zones: 1~7 in different colors. Moisture regimes: marine, dry, and moist from west to east.

2. INNOVATIVE DATA PREPROCESSING PROCEDURES

Preprocessing procedures are indispensable in order to perform correct and accurate statistical tests for fault detection. Once raw thermostat data is queried from the database, four preprocessing procedures will be conducted sequentially, namely data cleansing, labeling modes of operation, feature extraction, and identification of poor tracking periods. In brief, data cleansing removes data points with sensor faults and time periods where the thermostat is offline. Then the mode labeling transforms raw data into a more useful dataset by defining three system operation modes. Afterwards multiple useful data features are extracted from each operational mode. Finally, poor tracking periods are identified from the cooling tracking mode, one of the three operation modes, by using an unsupervised machine learning method.

2.1 Integrity Verification: Data Cleansing

The raw thermostat data is event-based, which means that the database is updated only when a new event occurs. For example, when the indoor/outdoor temperature changes more than a quantized amount, the cooling/heating setpoint is changed, or the system starts/stops. The event-based dataset records the event and corresponding time, and takes up less digital storage space than the uniformly sampled time series data. Sometimes, however, sensor integrity issues and network communication issues can occur (e.g. the sensor is unplugged, the thermostat goes offline or otherwise fails to communicate data). These issues could strongly affect the analytical process. For example, if a setpoint change event is not recorded, then the thermostat may seem to control the indoor temperature to value different from the supposed setpoint; or if a cooling off event is missed, then the system may appear to continue running for a long time. Therefore, after the raw data is queried, the first preprocessing step is to cleanse the data and remove these issues. This algorithm uses a combination of filters and logical tests to identify and remove these data.

2.2 Data Partitioning and Classification: Labeling Modes of Operation

After the raw data is cleansed, the mode labeling algorithm is applied to transform the raw operational data (cooling, heating, off) into a more useful dataset based on mode: regulating mode, tracking mode, and free response mode (Rogers *et al.*, 2020). Note that both heating and cooling have respective regulating and tracking modes.

- In the regulating mode, a system cycles on and off to maintains a relatively constant indoor temperature around the setpoint. The space being cooled or heated is assumed to be in a pseudo steady-state condition.
- In the tracking mode, a system remains powered on to reach the setpoint and exhibits transient behavior. Typically, the cooling tracking mode is activated after a drop in cooling setpoint or during high cooling load; the heating tracking mode is activated after a rise in heating setpoint or during high heating load.

- In the free response mode neither setpoint is active. A free response mode may occur after a rise in cooling setpoint or a drop in heating setpoint, or when both cooling and heating loads are low such that the indoor temperature remains between setpoints for an extended period.

Figure 2 provides an example of labeling modes of operation. In Figure 2a, the cyan color shows the cooling regulating mode when the indoor temperature is fluctuating and maintained relatively constant, and the red color shows the cooling tracking mode when the indoor temperature drops significantly. In Figure 2b, the green color shows the free response mode when indoor temperature rises above the heating setpoint and the orange color shows the heating regulating mode when the indoor temperature drops to the heating setpoint.

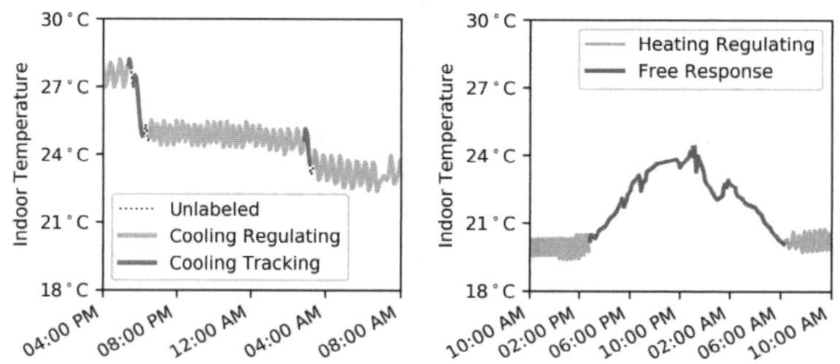

(a) An example of cooling regulating and cooling tracking modes.

(b) An example of heating regulating and free response modes.

Figure 2: An example of labeled modes of operation. Subplot (a) shows the cooling regulating and cooling tracking modes, and subplot (b) shows the heating regulating and free response modes.

2.3 Feature Generation: Extracting Features from the Cooling Tracking Mode

Within the operating modes introduced above, this paper studies the poor transient behavior, corresponding to the cooling tracking mode activated during high cooling load. Note that even when the cooling tracking mode is activated by a drop in cooling setpoint, a system can sometimes exhibit poor transient behavior if the cooling load is too large for the system. The features listed in Table 1 are extracted from the cooling tracking periods. Among all of the features, the first three are calculated within each cooling tracking period, and will be applied to identify the poor transient behavior; the fourth and fifth features are comprehensive indicators averaged over all poor tracking periods for each system, and will be used to characterize the problem severity.

Table 1: Supporting features for the cooling tracking mode

Symbol	Feature name	Definition
$T_{i,inc}$	Indoor temperature increase [°C]	The maximum indoor temperature increase*/** in a cooling tracking period.
Δt_{inc}	Increase time [hr]	The time required to reach the maximum indoor temperature increase.
$\dot{T}_{i,inc}$	Indoor temperature increase rate [°C/hr]	The rate at which indoor temperature increases; defined as $\dot{T}_{i,inc} = T_{i,inc}/\Delta t_{inc}$.
$\overline{DH}$	Average degree hour [°C·hr]	The integral of the difference between the indoor temperature and the cooling setpoint, during a time segment exhibiting poor transient behavior, averaged over the whole cooling season; defined as: $\overline{DH} = \frac{1}{N}\sum_1^N \int_0^{24} I \cdot \max[0, (T_{id} - T_{sp,cool})]\, dt$; where N is the number of days in a cooling season; I equals 1 in poor tracking periods, and 0 in other cases.
$\bar{T}_{i,max}$	Average maximum indoor temperature [°C]	The average of a fixed number of highest $T_{i,max}$ in all poor tracking periods** after removing a few extreme cases.

Notes: * A searching algorithm is applied to find the maximum increase in a time series.
** If a poor cooling tracking period is longer than 24 hours, then it will be segmented with one period a day. Generally, there is at most one poor tracking period each day, and the maximum indoor temperature occurs in the afternoon when cooling load is the highest. Please see Figure 7 and 8 in case studies for reference.

During normal cooling tracking periods activated by a decrease in cooling setpoint, the indoor temperature should decrease (or increase a very small amount) after cooling starts, such that both the value of the indoor temperature increase and the temperature increase time should be very small. For the poor tracking periods, however, both of these two features could show very high values.

2.4 Behavior Recognition: Identification of Poor Tracking Periods

In this step, the features extracted above will be analyzed to identify all poor transient behavior of a system. In total four types of transient behavior are common in the cooling tracking mode. Recall that a cooling tracking period is just one single cooling on-cycle. Figure 3a shows normal transient behavior (the most common), where the cooling system starts after a drop in setpoint and remains on until reaching the new setpoint. Figure 3b shows poor transient behavior, where the indoor temperature increases over 5°C before returning, despite the cooling system operating continuously for nearly 24 hours. Both Figure 3c and 3d show interesting transient behavior where the indoor temperature first increases precipitously after the cooling system turns on. The behavior in Figure 3c may be caused by lack of insulation on the return air duct in the attic, such that warm air is initially blown into the house when the fan starts after a long downtime; the behavior in Figure 3d may be caused by the internal heat load of the thermostat itself on the sensor, such that indoor temperature seems to rise as soon as the fan starts (after cooling starts) and the air circulation direction of the room changes.

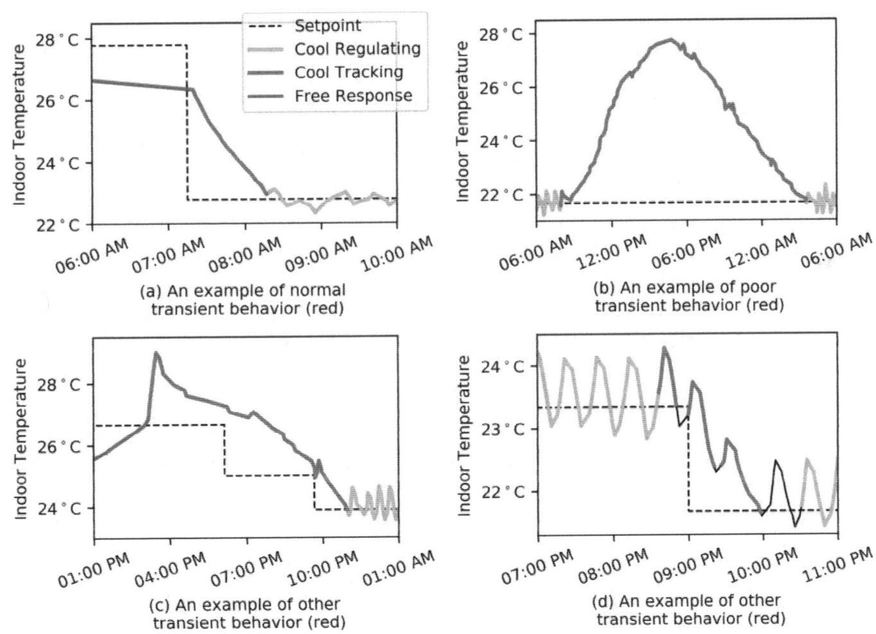

Figure 3: Four main types of transient behavior exhibited in the cooling tracking mode (red color).

Three features, namely indoor temperature increase ($T_{i,inc}$), increase time (Δt_{inc}), and increase rate ($\dot{T}_{i,inc}$), are used together to separate the different causes of poor transient behavior. As a first step, normal transient behavior should exhibit $T_{i,inc} \approx 0$, and hence these normal periods can be eliminated by a simple range filter, e.g., $T_{i,inc} > 0.56$°C (1°F). Distinguishing between the types of undesirable transient behavior, however, is more difficult. Poor transient behavior could exhibit long Δt_{inc} and low $\dot{T}_{i,inc}$, or vice versa. Figure 4a shows the kernel density estimation of all tracking periods after removing those identified to be normal, with darker regions having more data. The data forms two groups, shown by Figure 4b. The cluster on the left could only correspond to the poor tracking periods due to its long increase time, and the other on the bottom could only correspond to other types of behavior due to its fast increase rate. Only data between two groups in the lower left region has some ambiguity. However, based on an assumption that if a system has some confirmed poor tracking periods, then the ambiguous ones are more likely to be poor tracking periods as well, such that the hierarchical clustering method can be applied to group the two behavior for each system.

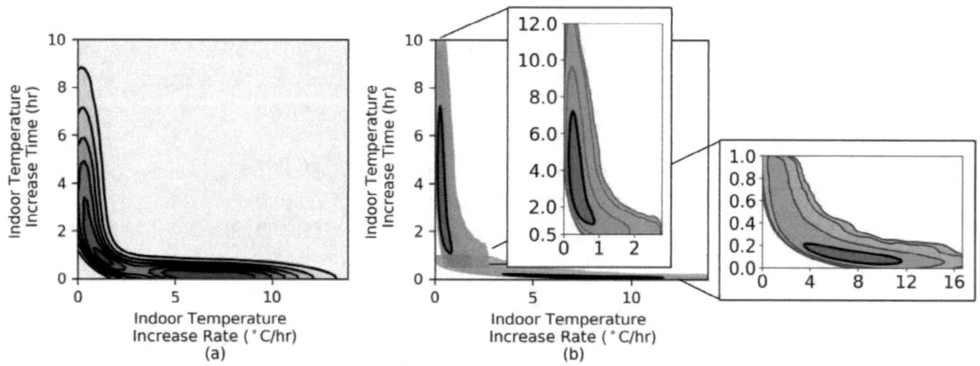

Figure 4: (a) Kernel density estimation of the remaining tracking periods after the normal ones are removed. (b) The data forms two groups. The group in red corresponds to poor transient behavior, and the group in blue corresponds to other transient behavior. The black circles show the core region (50% of mass) of each group.

In this study, Ward's agglomerative hierarchical clustering method (Ward and Joe, 1963) is applied to build a hierarchy of clusters from the transient data of each system. This method starts with each data point in its own cluster. Then as the algorithm moves up the hierarchy, it seeks to merge pairs of clusters which can minimize the variance of the clusters. In order to exactly find the two clusters of behavior, some generated data from the two black circles in Figure 4b is added into the real data. Note that the black circles correspond to the core regions of each cluster, taking up 50% of the mass, with known true labels. Also, features on both axes are normalized such that the Euclidean distance is applicable. By this means, the algorithm iterates over each system to label the tracking periods mixed with generated data. Since the two clusters are defined by expertise and experience, this approach can be regarded as setting soft thresholds between the two behaviors. After checking the results, the method is found to be stable and robust, and can correctly label the data of either type of behavior. Therefore, all cooling tracking periods labeled with poor transient behavior can be separated from the other behavior.

3. STATISTICS-BASED FAULT DETECTION METHOD

This section presents the proposed statistics-based fault detection method. In order to compare system performance to the entire population and identify outlier behavior, the statistical distribution of two features, namely average degree-hour above setpoint and average maximum indoor temperature, will be investigated among all systems by means of kernel density estimation. These two features could strongly indicate the existence and the severity of operational faults, and thresholds for identifying faulty systems will be recommended

3.1 Statistical Distributions of Features

Once all poor tracking periods are identified, a statistics-based fault detection algorithm can be applied using two additional features, namely average degree hour above setpoint and average maximum indoor temperature. The average degree hour above setpoint ($\overline{DH}$, unit: [°C·hr]) averages the area between indoor temperature and cooling setpoint for only all poor tracking periods (see Figure 3b as an example) each day in the whole summer cooling season, which represents the system effectiveness and occupant comfort. For instance, a system with $\overline{DH} = 12$°C·hr may on average have indoor temperature 3°C above setpoint lasting for 4 hours every day in the summer, even though cooling runs in full power. The average maximum indoor temperature ($\overline{T}_{i,max}$) is calculated by averaging ten highest indoor temperatures during poor tracking periods, after first removing the five most extreme instances in the cooling season. Thus, this feature represents the most severe instances of occupant discomfort. For instance, a system with $\overline{T}_{i,max} =$ 30°C will have indoor temperature rising over 30°C on average for at least 15 days in the summer, even though cooling runs in full power. These two metrics characterize the severity of the detected problem.

To exemplify the distributions of the two features, approximately 10,000 residential single-stage cooling systems in 2A IECC climate zone (see Figure 1) in Florida are queried with data from June 1st to September 30th in 2019. Among them, 904 systems have been identified with more than 15 (a hard threshold) poor tracking periods. The kernel density estimation of the average degree hour above setpoint and average maximum indoor temperature are shown in Figure 5. From Figure 5a and 5b, most systems have average degree hour above setpoint less than 15°C·hr and average maximum indoor temperature less than 30°C. From the 2-D distribution in Figure 5c, one can find that generally $\overline{DH} > 8$°C·hr along with $\overline{T}_{i,max} > 27$°C is abnormal.

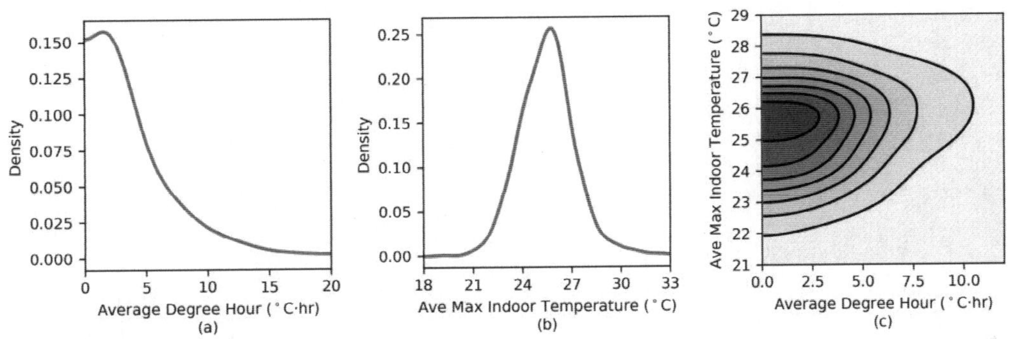

Figure 5: Kernel density estimation of (a) the average degree hour above setpoint, (b) average maximum indoor temperature, and (c) both features together, for all 904 systems identified with more than 15 poor tracking periods.

3.2 Thresholds Recommendation for Fault Detection

Although high $\overline{DH}$ and high $\overline{T}_{i,max}$ are not equivalent to faults, they can strongly indicate the existence or severity of faults. We hope to use the distributions of these two metrics to set thresholds for fault and estimate fault severities. Practically the real numbers of faulty systems are unknown and the exact boundary between faulty and fault-free systems is indeterminate. Furthermore, all systems exhibiting poor transient behavior could have some level of operational faults. Nevertheless, to be conservative, in this study the thresholds are adjusted such that about 5% systems with the poor transient behavior are classified as faulty. Note that approximately 9% systems are identified with poor transient behavior, so this threshold will label about 0.5% among all systems to be faulty. The threshold is flexible and subject to change based on the service companies' preferences. The study finally takes 80% quantile of the $\overline{DH}$ and $\overline{T}_{i,max}$ in the dataset as thresholds, respectively 6.9°C·hr and 26.7°C, calculated from the estimated probability density function in Figure 5a and 5b. Any systems with both features above thresholds are flagged. Two case study examples are shown in the next section.

4. RESULTS AND DISCUSSION

Using the approach outlined above, systems that exhibit faulty behavior are flagged. These faults could be related to either capacity degradation faults or undersizing issues. In this section, general fault detection results are summarized, and two typical cases are presented, with each representing one type of common behavior. The results demonstrate the potential of large data fault detection methods for residential HVAC systems using cloud-based thermostat data.

4.1 Summarization of Fault Detection Results

Figure 6 briefly summarized the fault detection results. Figure 6a shows how faulty systems are flagged layer by layer. The Venn diagram in Figure 6b shows a total of 181 systems are flagged by each feature using the proposed thresholds, and the 68 systems at the intersection of the two groups are labeled as faulty. The next three scatterplots illustrate the behavior of these 68 labeled faulty systems in the 2019 summer cooling season. Figure 6c shows that in most of these homes the indoor temperature often reaches to around 29°C although the cooling setpoint is usually set to around 23°C, and there are some extreme systems having very high indoor temperature and degree hour above setpoint. Figure 6d compares the previously defined degrees hours with the degrees hours when the indoor temperature is 1.1°C (2°F) above setpoint. The latter feature can be calculated by the equation in Table 1 assuming the setpoint increased by 1.1°C (2°F). The result shows strong linear relationship, indicating higher degree hours means both higher temperature and longer time. Occupants living in a high-degree-hour conditioned space have to endure high indoor temperature well above the setpoint value they selected for several hours every day. Finally, Figure 6e shows that most faulty systems spend between 400 and 2000 hours with poor tracking during the cooling season. No particular relationship can be found between maximum indoor temperature and total time on poor tracking due to the difference in setpoint values for each system.

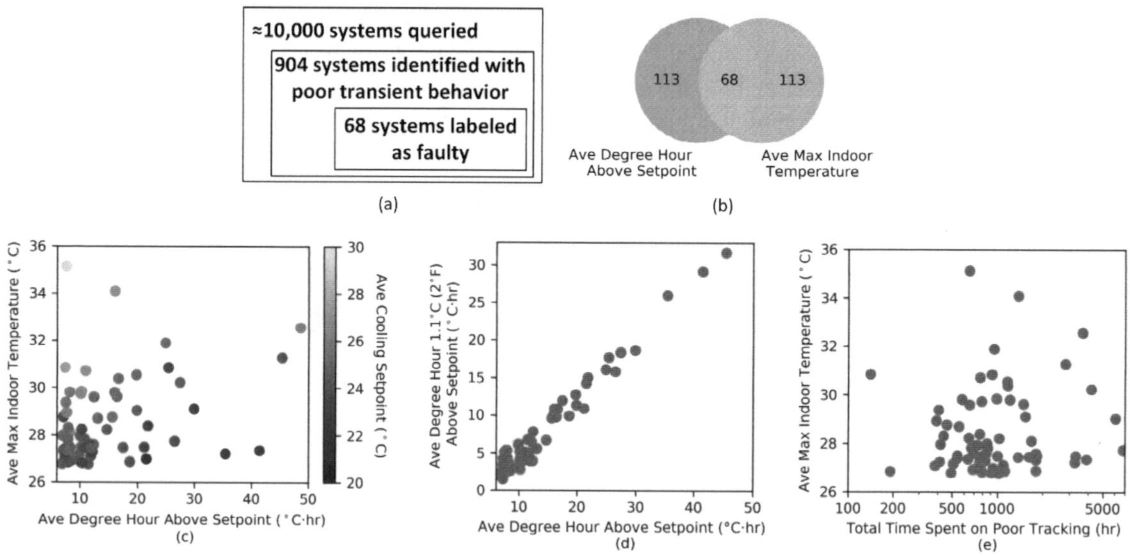

Figure 6: Illustration of the statistics-based fault detection results. (a): General statistics. (b): A Venn diagram showing the numbers of flagged systems. (c) to (e): Statistics of the 68 faulty systems in the summer cooling season.

4.2 Case Study I: Significant Time above Setpoint, High Maximum Indoor Temperature

Figure 7 displays a part of daily operation of a selected faulty system, recorded from June 10th to June 14th. Note that this is the most common behavior among all labeled faulty systems. The first plot records the indoor temperature variation and cooling setpoint changes with modes labeled in red, cyan and green color. The second plot records the cooling load factor (i.e. on/off status) of this single-stage cooling unit. The third plot records the outdoor temperature variation. From this figure, the system frequently shows long periods of poor transient behavior, and the indoor temperature can spike from 24°C to above 30°C when the outdoor temperature approaches 32°C in the afternoon. Typically, every day the system operates all the time from the sunrise to sunset, until the outdoor temperature drops below 26°C. It works very little during the night, because the cooling setpoint is above the outdoor temperature and the house is cooled passively by the ambient environment.

Statistics shows the average degree hour above setpoint is 25.4°C·hr and average maximum indoor temperature is 30.8°C. These numbers could be understood in the following way: the occupants in the house on average experience 3.2°C above setpoint every day for 8 hours, and the maximum indoor temperature in a day approaches 30.8°C on average for at least 15 days in summer. Clearly the comfort level is very low, and the system capacity does not meet the cooling load.

4.3 Case Study II: Significant Time above Setpoint, Low Maximum Indoor Temperature

Another example shown in Figure 8 also has many long periods of poor tracking and even have one poor tracking period lasting for several days (e.g., the period from September 8th to September 14th). The average degree hour above setpoint is 46.4°C·hr, much higher than the former example. This case, however, is very different from the former one in that the occupant comfort is not totally lost. The average maximum indoor temperature is 25.3°C, and the cooling setpoint is adjusted 19.5°C on average when the system exhibits poor transient behavior. Because of the low indoor temperature, this example is *not* labeled as faulty. Nonetheless, this behavior is very common, characterized by an occupant driven issue in addition to possible capacity degradation issues.

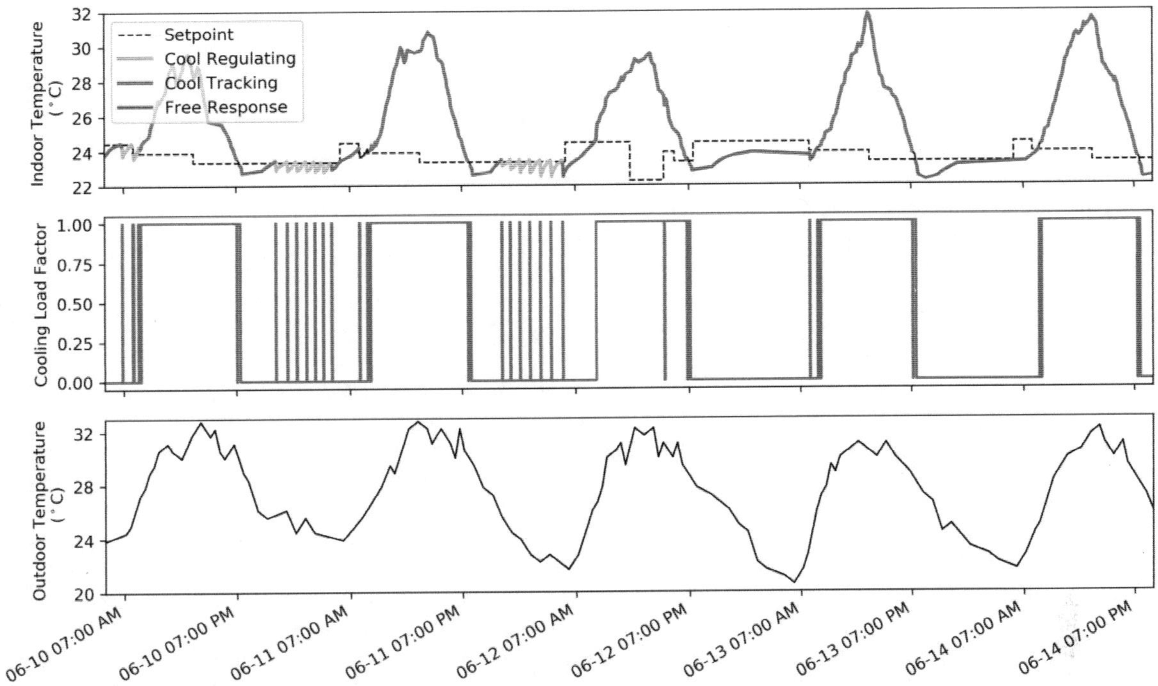

Figure 7: Daily operating condition of system I with labeling modes of operation, from June 10th to June 14th. Red: poor cooling tracking periods; cyan: cooling regulating periods; green: free response periods; dotted line: cooling setpoint.

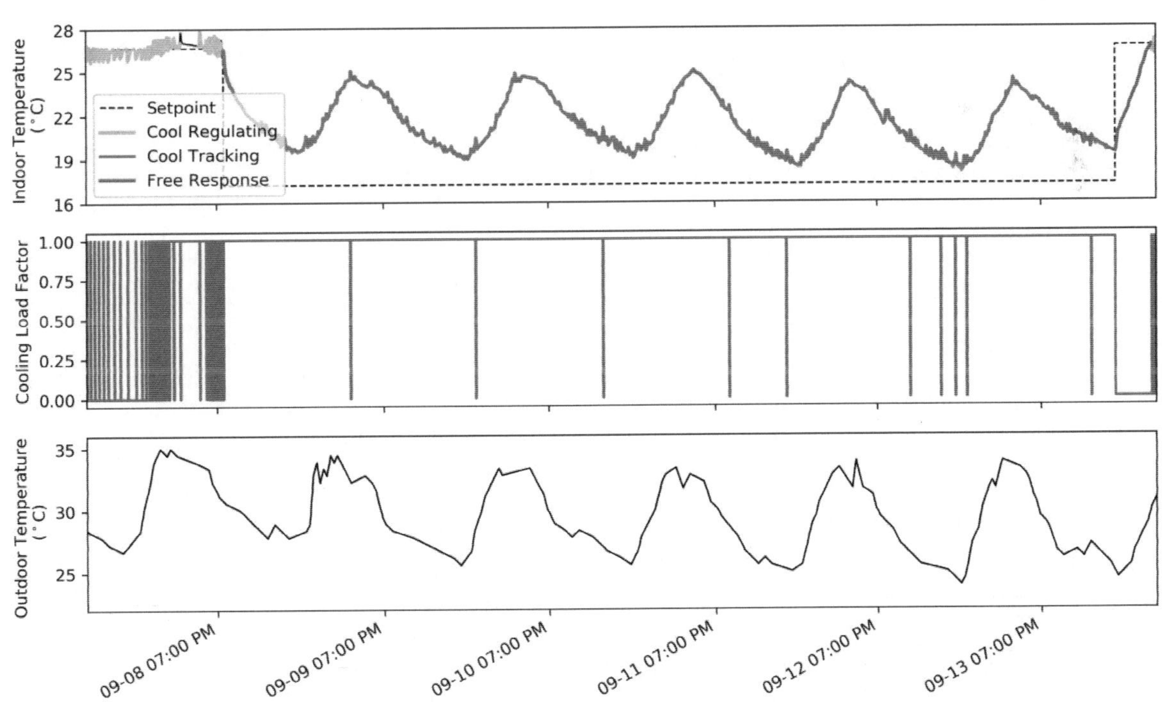

Figure 8: Daily operating condition of system II with labeling modes of operation, from September 8th to September 14th. Red: poor cooling tracking periods; cyan: cooling regulating periods; dotted line: cooling setpoint.

Figure 8 captures a segment of the occupant driven issue. On the evening of September 8th, the occupant deliberately lowered the cooling setpoint to 17.2°C (63°F) in order to receive more cooling, even though he or she understood that the indoor temperature might never reach the setpoint. From the figure, the system precools the home at night to approximately 19°C, and during the daytime the indoor temperature can be well-controlled below 25°C. In other words, the occupants were forced to decrease the setpoint because he or she had realized that the system capacity was not large enough, and if the system did not operate during the nighttime, the indoor temperature would probably rise above 27°C in every afternoon. Additionally, we speculate that nobody was at home before and after the displayed poor tracking period, so the occupant could adjust the setpoint at 26.7°C (80°F) to reduce cooling and the associated electric bill. However, this precooling method is not always effective and is symptomatic of faulty equipment. A better solution for the home owner will be to contact service companies and request maintenance.

5. CONCLUSIONS

By means of cloud-based smart thermostat data of thousands of residential HVAC systems, the proposed fault detection method can successfully identify faulty systems from analysis of their transient behavior, and infer the problem severity using specific data features. The flagged faulty systems exhibit low cooling capacity compared to the load, which is the result of system degradation, improper sizing, or improper commissioning, and the occupants either experience discomfort or are forced to compensate by artificially adjusting the cooling setpoint.

This paper only introduces two features indicating severity of capacity degradation fault. Other features extracted from the cooling tracking mode can potentially indicate additional faults observed only from the transient behavior. For instance, indoor temperature increase ($T_{i,inc}$) and increase time (Δt_{inc}) can be applied to diagnose problems such as lack of insulation in the house attic, thermostat misplacement issues, and smart thermostat internal heat load issues. The potential of large data fault detection methods requires both domain expertise and sound statistical analysis.

REFERENCES

Do, H., & Cetin, K. S. (2019). Data-Driven Evaluation of Residential HVAC System Efficiency Using Energy and Environmental Data. *Energies, 12*, 188.

EIA. (2015). *Residential Energy Consumption Survey (RECS).* U.S. Energy Information Administration.

Ham, W., Klein, M., Tabatabaei, S. A., Thilakarathne, D. J., & Treur, J. (2016). Methods for a Smart Thermostat to Estimate the Characteristics of a House Based on Sensor Data. *Energy Procedia, 95*, 467-474. doi:https://doi.org/10.1016/j.egypro.2016.09.067

International Code Council, Building Officials, Code Administrators International, International Conference of Building Officials, & Southern Building Code Congress International. (2012). *International energy conservation code.* International Code Council.

Jain, M., Gupta, M., Singh, A., & Chandan, V. (2019, 3). Beyond Control: Enabling Smart Thermostats for Leakage Detection. *Proc. ACM Interact. Mob. Wearable Ubiquitous Technol., 3*, 14:1--14:21. doi:10.1145/3314401

Kim, W., & Katipamula, S. (2018). A review of fault detection and diagnostics methods for building systems. *Science and Technology for the Built Environment, 24*, 3-21. doi:10.1080/23744731.2017.1318008

Neme, C., Nadel, S., & Proctor, J. (1999). Energy savings potential from addressing residential air conditioner and heat pump installation problems.

Proctor, J., & Downey, T. (1999). Transforming routine air conditioner maintenance practices to improve equipment efficiency and performance. *Proceedings of the 1999 international energy program evaluation conference*, (pp. 1-12).

Rogers, A. P., Guo, F., & Rasmussen, B. P. (2019). A review of fault detection and diagnosis methods for residential air conditioning systems. *Building and Environment, 161*, 106236. doi:https://doi.org/10.1016/j.buildenv.2019.106236

Rogers, A. P., Guo, F., Martinez, J., & Rasmussen, B. P. (2020). Labeling Modes of Operation and Extracting Features for Fault Detection with Cloud-Based Thermostat Data. *2020 ASHRAE Annual Conference.*

Turner, W. J., Staino, A., & Basu, B. (2017). Residential HVAC fault detection using a system identification approach. *Energy and Buildings, 151*, 1-17. doi:10.1016/j.enbuild.2017.06.008

Ward Jr, J. H. (1963). Hierarchical grouping to optimize an objective function. *Journal of the American statistical association, 58*, 236-244.

A Modeling Approach to Characterize the Demand Flexibility of the Air Distribution System in Commercial Buildings

Elvin VINDEL*, Mario BERGES, Burcu AKINCI

Carnegie Mellon University, Department of Civil and Environmental Engineering
Pittsburgh, Pennsylvania, USA
Email: {evindel, marioberges, bakinci}@cmu.edu

* Corresponding Author

ABSTRACT

Commercial VAV systems have been a studied as potential flexible power consumption resources to provide these services. A significant limitation is the development of control oriented models for these systems that accurately control power consumption across timescales. To this end, we develop a modeling approach that characterizes the demand flexibility of the air distribution system in commercial buildings. The model maps all feasible terminal zone damper configurations to power consumption by identifying a system resistance model connected to a pressure-flow model of the supply fan. Under this model structure, the operational range of an installed system can be quantitatively bounded by the extreme resistance curves the air distribution system encounters. We also propose a simplified control algorithm that leverages this model structure to emulate a demand flexibility event. We validate this approach by calibrating a model in an experimental testbed and additionally testing the results in a simulation environment. A brief discussion of important properties and observations that can be derived from this approach as well as avenues for improvements is presented.

1. INTRODUCTION

The sustainable integration of large shares of renewable energy in the power grid requires more sources of flexibility to balance the grid's operation (Ulbig & Andersson, 2015). Buildings have a valuable untapped potential to provide some of this flexibility through reliability-based grid services (Neukomm, Nubbe, & Fares, 2019). In particular, commercial VAV systems have received attention due to their significant energy consumption and pre-existing automation capabilities (Goddard, Klose, & Backhaus, 2014). Recent research has demonstrated, through a variety of control algorithms, that the thermal capacity of zones can allow for flexible power consumption. The majority of this research has focused on simulation models that characterize demand flexibility uniquely as a property of the building's envelope and thermal mass (Kathirgamanathan, De Rosa, Mangina, & Finn, 2021). However, to accurately model the effect of advanced control strategies on power consumption, a more detailed model of the building mechanical system with high predictive accuracy is necessary (Yin, Kiliccote, & Piette, 2016).Within the scope of commercial VAV systems, air distribution system modeling has been labeled a deficient core modeling capability in modern building energy modeling (Roth & Reyna, 2020). Moreover, recent studies highlight the limitations of existing modeling approaches in supporting experimental observations, particularly for complex demand flexibility events (MacDonald, Vrettos, & Callaway, 2020). This is a challenge because the deployment of this technology for advanced grid services requires performance guarantees in predicting the duration, ramp time, and magnitude of power deviations.

Due to the relatively large impact of fans on energy consumption in buildings served by large central multizone HVAC systems (20 to 80% according to (ASHRAE, 2016)), our goal is to develop a model for the supply fan in a VAV system so as to accurately control and characterize its power consumption for more stringent demand flexibility applications (i.e. contingency reserves, regulation, ramping, etc.). Based on the common structure of most VAV systems, and shared control architecture, several control inputs are candidates to modify the power consumption of the supply fan. These control inputs, and their respective control loops can be visualized in Figure 1. At the AHU level, two control inputs can be actuated through the fan controller: the rotational speed of the fan, (Macdonald, Kiliccote, Boch, Chen, & Nawy, 2014), and the static pressure setpoint, $\hat{p}$, (Adetola et al., 2018). Although high accuracy models have been developed with these inputs, it is possible to cause undesirable effects on downstream components (Goddard et al., 2014). At the zone level, three control inputs can be modified having an indirect effect on fan power consumption. Figure 1 shows the controls of one of such zones; multizone VAV systems have N such zone subsystems. Temperature setpoint reset, $\hat{\mathbf{T}}$, is commonly used in practice and in literature due to its simplicity (Goddard et al., 2014). However, given that several control loops separate this input and the AHU, it has been shown to have a complex and sometimes negligible effect on power (Keskar, Anderson, Johnson, Hiskens, & Mathieu, 2019). Zone airflow setpoint, $\hat{\mathbf{q}}$, is a popular choice connecting to both the zone thermal dynamics and power through the fan laws (Wang, Huang, Wu, & Lu, 2020). This modeling approach is implemented at the expense of simplifying the pressure dynamics of the fan-duct system. To accurately control power consumption, a model should include both the pressure and flow dimensions (known as the operating point) (Stein & Hydeman, 2004). Zone damper positions, $\boldsymbol{\theta}$, actively influence the pressure dynamics of the duct network. In most VAV systems, the supply fan is a pressure-controlled device, meaning that zone dampers are the unfiltered connection between the control inputs at the zone level and the supply fan. This property makes damper position control stand out as a control input choice to modify the fan power consumption.

Models that focus on the relation between damper positions and the pressure dynamics are commonly used for static pressure reset analysis. A statistical model was developed to predict the energy savings from pressure reset strategies using historical building automation system (BAS) data (Tukur & Hallinan, 2017). More recently, a mathematical model coupled with newer machine learning techniques was developed to find the optimal reset strategy for energy savings (Jing et al., 2019). Although these approaches capture the pressure-flow dynamics, they have not been applied to the context of control for demand flexibility. Nonetheless, because of the modeling need described above, recent works have revisited models that include damper position control, as it is a significant factor in the effectiveness of demand flexibility strategies (Blum & Norford, 2014). A detailed model was developed for supply fan power consumption that include the pressure-flow dynamics (Hydeman, Taylor, Stein, Kolderup, & Hong, 2003). Although this model characterizes the operational range of the system, a connection to zone damper positions is not explored. A generic polynomial multi-regression model was developed by taking as inputs pressure, flow, and damper positions to predict fan energy consumption (Dong et al., 2019). The model's predictive capabilities were evaluated, but the authors do not elaborate on how this model can be used for fan control. Pertinent to our work, a single-zone mathematical model from damper positions, and their pressure drop characteristics, to fan power was developed (Raisoni, Raman, Barooah, Munk, & Im, 2018). Due to the model complexity, this approach required the identification of 8 parameters per zone, scaling linearly with the number of zones. Furthermore, the identification process required targeted experiments beyond nominal operation making it unclear whether this approach is generalizable or scalable.

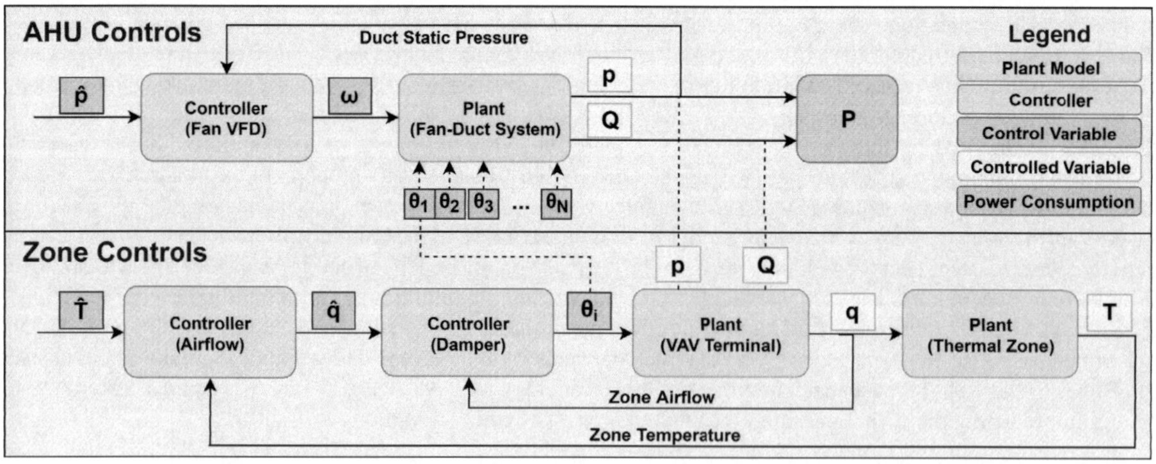

Figure 1: Diagram of common control architecture in multizone VAV systems

From prior work we observe three areas for improvement. First, the proposed model structures do not characterize the operational range that can be actuated for an installed system to modify fan power. Defining these bounds provides a quantitative measure of the technical capability of the system to modify its power consumption (i.e. demand flexibility). Our approach proposes using the extreme system resistance curves as a method to characterize the range of power consumption. Second, current work is primarily validated for predictive capabilities, and does not elaborate on the effect of dynamic control inputs on power consumption. We propose the use of two low complexity models (system resistance and pressure-flow) to map the relation between terminal damper positions and fan power consumption. In addition, we present an algorithm that uses this structure to control the system resistance using damper position inputs and predicting the future operating point, hence the power consumption. Under this framework, the dynamics of the model is primarily driven by the performance of the fan controller. Third, the referenced approaches required targeted experiments for identification. The framework we propose can be identified primarily using historical undisturbed BAS data. However, in some systems, due to varying design capacity, targeted experiments might be necessary.

2. METHODOLOGY

2.1 Preliminaries

The supply fan in commercial VAV systems is designed to generate both pressure and airflow to deliver conditioned air to multiple zones. Terminal units regulate airflow by adjusting zone damper positions in response to individual cooling and heating requirements. The interaction between the supply fan and terminal units is the primary consideration for fan design which accounts for aspects such as: size, type, controller, noise, vibration, surge avoidance, and system effects (ASHRAE, 2016). A tool used in design is the *system curve*, which is a graphical representation of the total head loss of a particular system at its design airflow. For turbulent air systems (such as VAV systems), the operation follows the law in Equation 1 closely. By rearranging Equation 1 we obtain Equation 2 describing a quadratic function parametrized by the resistance coefficient. These curves are drawn as functions intersecting the origin in Figure 2a. The resistance coefficient, R, is a continuous variable that lumps all of the pressure losses downstream of the fan for a fixed configuration. This term incorporates pressure losses from both static sources (e.g. duct geometry, leakage) and dynamic sources (e.g. terminal dampers). Fan controllers are designed to overcome these pressure losses and maintain constant duct static pressure across all operational system curves. Variable frequency drives (VFD) are commonly used to modulate fan speed, effectively changing the fan performance curves shown in Figure 2a. The intersection between the fan performance curve and the system curve is referred to as the operating point. For a particular fan, determining the operating point is sufficient information to accurately predict power consumption (Stein & Hydeman, 2004).

$$(p_1/p_2) = (Q_1/Q_2)^2 \tag{1}$$

$$p = RQ^2 \tag{2}$$

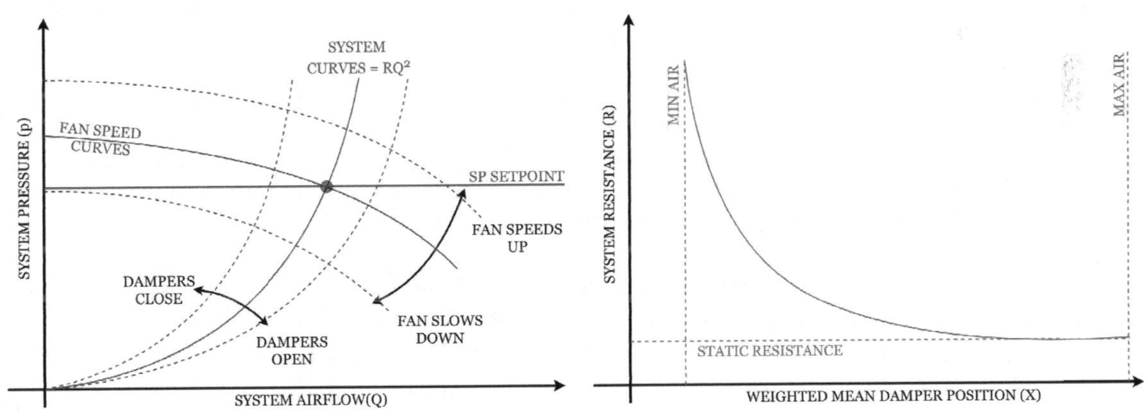

Figure 2: (a) Pressure-flow model: Fan Performance and System Curves (b) System resistance model: resistance coefficient vs weighted mean damper position

2.2 Model Description

The model approach we seek to develop is based on characterizing and bounding the range of system curves that supply fan is subjected to. Given that dampers are the source of variable resistance, we seek to find a model that maps all feasible configurations of damper positions to a particular system resistance. To do so, we adopt the relation between damper positions ($\boldsymbol{\theta}$) and system resistance (R) suggested by (Tukur & Hallinan, 2017). To account for varying VAV box sizes, the damper positions are weighted by their nominal airflow, $\mathbf{w}$. In their work , they implement a dimensionality reduction technique to map the vector of all zone damper positions to a single weighted average value (X). We refer to this relation as the system resistance model shown in Figure 2b. The referenced work leverages this model as a statistical tool to estimate the energy savings of a static pressure reset strategy. We extend the use of this approach to actively control the power consumption of the supply fan. The output of this model can be used in Equation 2 to find the operating point in a pressure-flow balance model of the supply fan which ultimately outputs power. Figure 3 shows the complete modeling framework from damper positions to power. The underlying principle is that for each particular system curve the supply fan will tend towards a known operating point, hence power consumption. This operating point will lie at the intersection between the current system curve and the static pressure setpoint. The dynamics of how the operating point moves are tied to both the changing damper configuration and the performance of the fan controller. One important property is that the system resistance values are bounded by the configurations of minimum and maximum resistance. The lower bound (i.e. minimum resistance) is reached at the point when all dampers are a their most open configuration. This system configuration, otherwise known as the design airflow, is hardly ever reached by most VAV systems by design (Hydeman et al., 2003). The upper bound (i.e. maximum resistance) is reached when all zones are operating at their minimum air setting simultaneously.

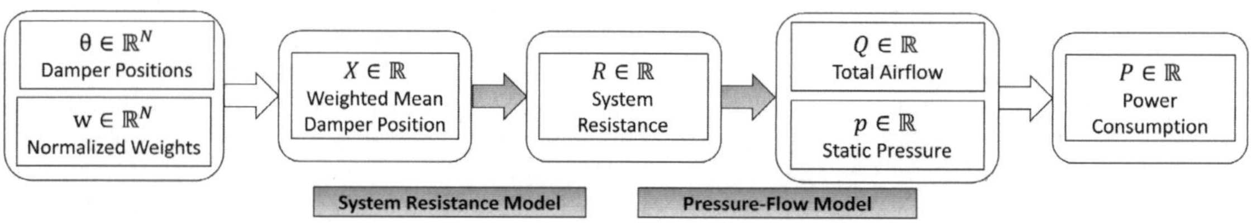

Figure 3: Proposed model structure with damper positions as inputs and power consumption as output

Once the two models are calibrated, fan power consumption can be determined for a particular damper configuration. To show how this can be achieved we developed an algorithm to follow a specified power consumption (**P**) for a time horizon (H). The desired power consumption is converted to a sequence of system resistances using the pressure-flow model, assuming static pressure setpoint is maintained. The weighted mean damper position (X) is obtained by using the system resistance model. Finally, we solve a sequence of optimization problems to determine the damper positions that will output the desired sequence of system resistances. Note that in this simplified case, the optimization problems are in quadratic form having an equality constraint for the weighted mean damper position and a set of inequality constraints to respect each damper's operational limits, $\mathcal{X}$. This simplified algorithm has two important assumptions. The first assumption is that the fan controller ideally maintains the static pressure setpoint. In real systems, this controller is commonly a proportional-integral controller reading static pressure and outputting a speed command to the fan motor. Hence, fluctuations around the static pressure setpoint are expected, but the process variable is expected to reach the setpoint asymptotically. The second assumption is that the zone thermal constraints are not modeled in the action selection process. To provide a complete solution, this approach, which focuses on the mechanical system, needs to be coupled to other constraints that are imposed by a thermal model of the zones.

Algorithm 1: Supply Fan Power Control through System Resistance (Damper Positions)

Input: Initial weighted damper positions $\mathbf{x}_0$, and desired power consumption **P**.

Output: A sequence $\mathbf{x}_\mathbf{k}^* \ \forall \ k = 1, \ldots, H$ of commanded damper positions to achieve the desired power deviation **P**.

```
Solve for (Q*_k, p̂) using P_k, ∀ k = 1,...,H          // solve for operating point at static pressure set point
Solve for R*_k using (Q_k*, p̂) ∀ k = 1,...,H           // solve for sequence of system resistances
Solve for X*_k using R̂_k ∀ k = 1,...,H                 // solve for sequence of weighted damper positions
for k ← 1 to H do
    Find x*_k that minimizes ||x_k − x_0||^2            // follow predicted damper positions
    such that:
    X_k = X̂*_k ∀ k = 1,...,H                            // match system resistance
    x_k ∈ 𝒳 ∀ k = 1,...,H                               // damper operational limits
end
return x*_k
```

3. IMPLEMENTATION

3.1 Experimental Calibration

To test the proposed model, we applied an identification process in a three-room testbed located in Carnegie Mellon University's Porter Hall (a 3D render of the testbed is shown in Figure 4). The three zones are served by an exclusive VAV system providing an isolated testbed. The dedicated AHU system is connected to the building automation platform from which remote commands were sent. To validate the model, we covered all combinations of feasible damper positions discretely, and maintained each setting for 5 minutes to allow the system to reach steady state. Note that historical BAS data is also an immediate option as described in Section 4.3. Table 1 shows a summary of the 320 system configurations explored. The nominal airflows used to weigh damper positions for the zones A,B, and C are 0.35, 0.35, and 0.21 m^3/s, respectively. During this experiment the fan speed was maintained constantly at 100% to make a clear distinction between system curves. The rooms were completely isolated from users due to possible significant fluctuations in zone temperatures. For this particular system, a weighted mean damper position beyond $X = 0.75$ (i.e. low system resistance) is the maximum value for which the supply fan is capable of maintaining the duct static pressure setpoint ($\hat{p}$ = 250Pa). This can be graphically verified by the intersection of the max fan speed curve and the static pressure setpoint in Figure 6. For completeness, we also recorded data points beyond this setting to validate the asymptotic portion of the model where the static losses dominate the change in system resistance. Figure 5 shows the results from the set of experiments previously described. Note that the system resistance parameter was normalized by its maximum value for clarity. The key takeaway from Figure 5 is that it validates the dimensionality reduction technique from $\boldsymbol{\theta} \in \mathbb{R}^N$ to $X \in \mathbb{R}$. Additionally, we fitted a third order polynomial to the feasible range of the model after uniformly sampling the data . The model choice was made to simply select a low-order continuous model that shows a good fit (i.e. $R^2 = 98.5\%$).

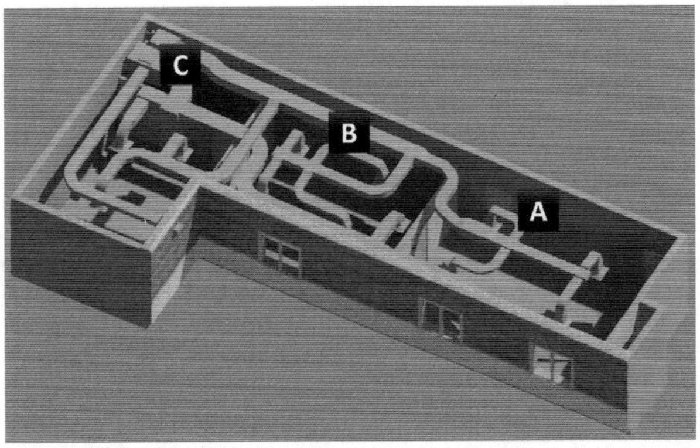

Figure 4: Porter Hall: 3-zone VAV Experimental Testbed

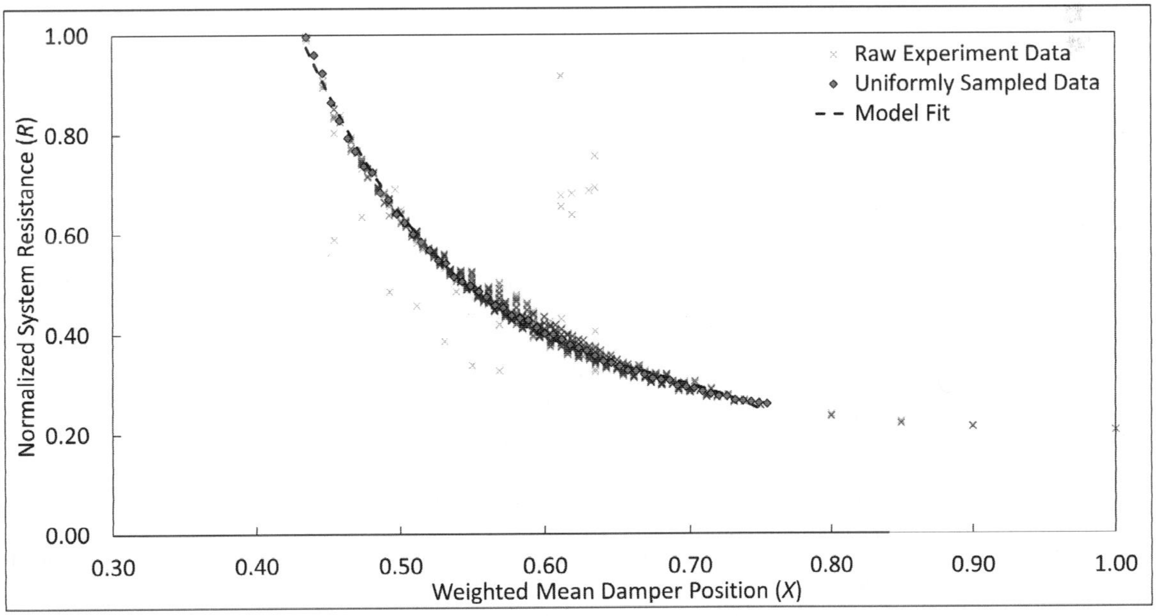

Figure 5: Experimental System Resistance Model from Porter Hall Testbed

Table 1: Set of tested damper positions for identification experiments

ZONE	1	2	3	4	5	6	...	315	316	317	318	319	320
A	0.40	0.40	0.40	0.40	0.40	0.40	...	0.75	0.75	0.75	0.75	0.75	0.75
B	0.40	0.40	0.40	0.40	0.40	0.45	...	0.70	0.75	0.75	0.75	0.75	0.75
C	0.55	0.60	0.65	0.70	0.75	0.40	...	0.75	0.55	0.60	0.65	0.70	0.75

A second set of experiments were performed to identify the fan's pressure-flow characteristics. Although the manufacturer provided performance curves can be used, these curves are generated using standardized laboratory conditions (ASHRAE, 2016). Hence, they do not include the performance of the system under the installed conditions, which is in most cases different than the design conditions. In this case, 8 characteristic system curves were tested for varying fan speeds to create the pressure-flow model. Similar to the system resistance model, this set of experiments are performed to support the validity of the model. For each system configuration the fan speed (ω) was varied in steps of 10%. This set of curves included both the minimum and maximum system resistance curves encountered by this unit. The system curves were fitted with second-order polynomials closely following the expected theoretical results from Equation 2 (i.e. $R^2 > 0.99$). A plot of these results is shown in Figure 6. Interestingly, the 8 system curves from the second experiment can also be used to build the system resistance model, yielding virtually identical model fit to the one shown in Figure 5. For this unit, the combination of the model in Figure 5 and Figure 6 bound and characterize the available demand flexibility of the air distribution system.

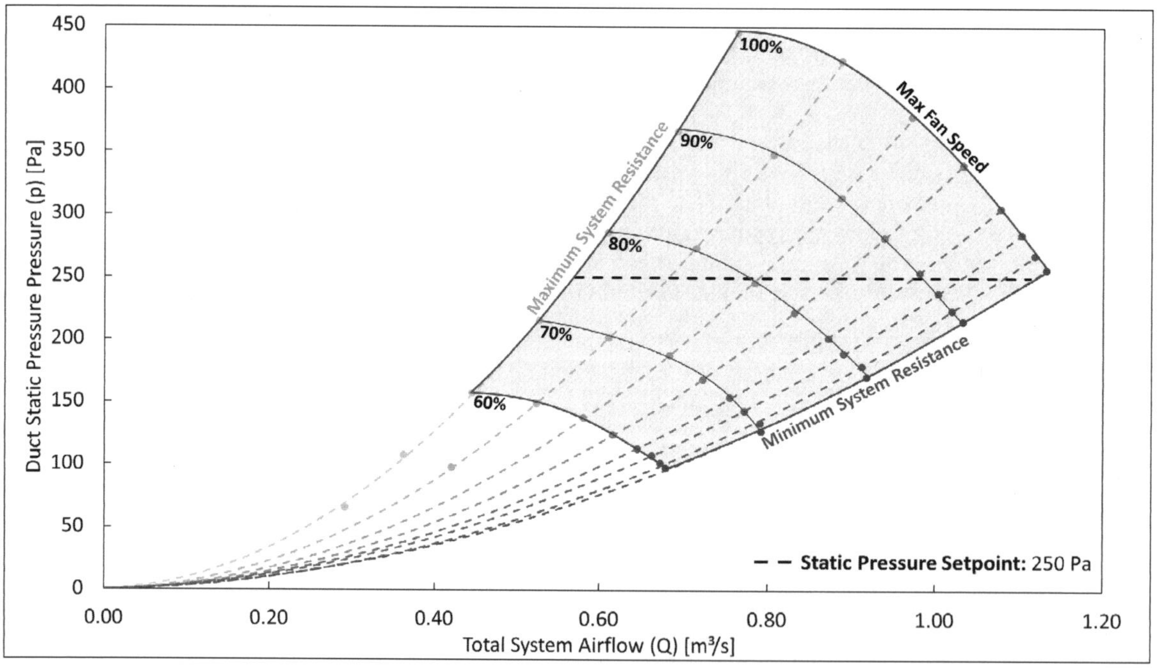

Figure 6: Experimental Fan Performance and System Curves to form the Pressure-Flow Model for Porter Hall Testbed

3.2 Simulation Results

To test the control algorithm on a dynamic setting, we created a model of the experimental setting using the Modelica Buildings Library (Wetter, Zuo, Nouidui, & Pang, 2014). The simulation model is based of a VAV system example the library includes using the control sequence where the supply fan is regulated based on the duct static pressure. Using the experimental data, the model of the HVAC system and the three thermal zones was created. We first build an offline system resistance model using simulated data gathered from a week of undisturbed operation. Similarly, the fan pressure-flow model is calibrated with the experimental results. Then we selected a power deviation trajectory for the supply fan to follow. In this setting we used PJM's RegA test signal to use a low frequency signal to follow for an hour (Macdonald et al., 2014). The choice of signal is based on selecting a standard power deviation signal rather than suggesting the exclusive use for this model for frequency regulation. We also selected a arbitrarily predicted trajectory for the three dampers over the one hour horizon, which simulated dynamic cooling demand. Figure 7 shows the results of this simulation. The left pane of the figure shows the predicted and actual damper positions for the three zones as well as the damper operational limits. This figure shows that the control algorithm follows closely the predicted damper positions. All zone dampers respect the operational limits for the entire event. The right pane shows the results of the system resistance model for this simulated unit. Although not included in the optimization algorithm, the temperature deviations in the three zones are less than 0.50°C. This continues to reinforce the property that zones have the thermal capacity to not experience significant temperature deviations for fast changes in airflow. The results comparing the desired power deviation and measured power deviation are not shown as they are almost identical meaning that the algorithm accurately tracked the demand flexibility event. However, the simulation environment reflects an idealized HVAC system and further studies are needed to validate the accuracy of this approach in an experimental setting.

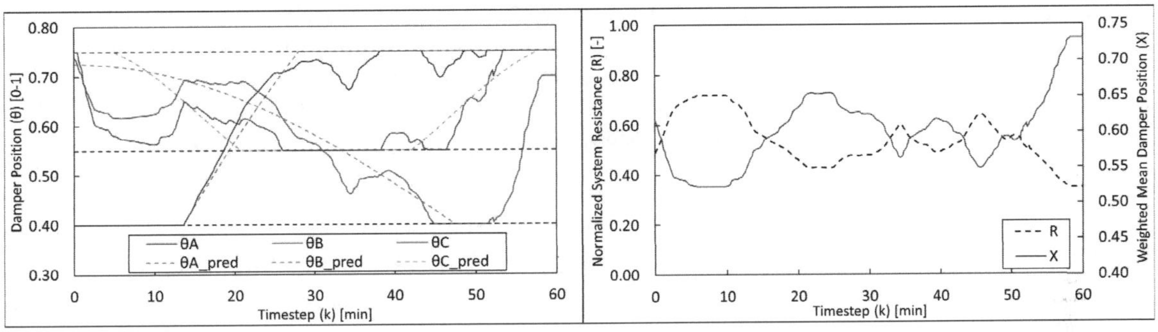

Figure 7: Simulation results (a) damper positions for the three zones denoted as A,B,C in the figure (b) system resistance model results for demand flexibility event

4. RESULTS AND DISCUSSIONS

4.1 System Bounds

An important property of this modeling approach is to quantitatively define the bounds on which an installed system would be subjected to, independent of zone-level control input. This, in turn, defines the available range power can be actuated for a particular system. As VAV systems develop from the original design intent to operation, intermediate choices lead to unique operational range across buildings. These bounds can be observed in the limits of the system resistance model. For example, for the Porter unit, the extreme system resistance values occur within the range $0.44 \leq X \leq 0.75$. For comparison, we show a different 5-zone VAV system, referred to as Gate (see Figure 8), whose system bounds occur within the range $0.10 \leq X \leq 0.30$. This range is unique to the control sequence and capacity each installed system and can be used to qualify certain buildings for different types of grid services. For example, the normal operation of the Porter unit shows that it runs at minimum air over 90% of the time, indicating an over-designed capacity. This observation shows that, although the zones inherently have thermal capacity, this system is currently not suitable for participating in load shedding reliability-based services such as contingency reserves. On the other hand, the Gate unit nominally operates across its available range, as depicted historical data in Figure 8. For this unit, the range derived from this modeling structure, is significantly lower than the range that would be derived form using airflow setpoint as a control variable. Hence, airflow setpoint control would misrepresent the demand flexibility of the Gate system.

4.2 Control Performance

In a complementary study of 35+ AHUs, we analyzed the qualification of airflow as a proxy for dynamic fan power control in VAV systems. Due to the imperfect operation of control loops and controllers, a simplified polynomial relation between airflow setpoints and power does not show high accuracy in predicting power consumption. This is likely because changes in airflow setpoint are dependent on the zone airflow controllers and are subject to latency in reaching the fan controller at the AHU level. Note that for certain applications, such as energy efficiency and unmeasured load shedding, the airflow-to-power assumption might be sufficient. However, more demanding grid services require sequential control inputs, sometimes in a timescale of minutes, that accurately shape power consumption. An important property of our modeling approach, and others that control damper position, is that there are no intermediate control loops that distort or delay the effect of control inputs. This means the effect of dynamic control inputs on fan power are primarily explained by the performance of the fan controller. For example, the Porter unit has a poorly tuned fan controller making it respond slowly to changes in static pressure. The performance of this individual controller restricts the use of this unit for grid services that require fast changes in power consumption. More importantly, the limitation can be measured and pinpointed to the parameters of the proportional integral (PI) fan controller. Under the assumptions of existing modeling approaches, this limitation would not be highlighted causing to misrepresent the demand flexibility of the Porter unit.

4.3 Generalizability

Another property we wanted to hold with the presented modeling approach is the ability to identify the model with undisturbed historical BAS data. In the complementary study referenced in Section 4.2, we find that in most systems the recorded data covers the minimum and maximum resistance damper configuration. The Gate unit, in Figure 8, shows how historical BAS data can be used to fully characterize the system resistance model. One important aspect to note is that although the Gate system resistance model was built using historical data, there was significant preprocessing before raw data can be used. For reference, Figure 8 only shows 20% of the raw data collected for a 3 month period at a 1 min resolution. This is because a large portion of the raw data was excluded because it was generated from modes outside the expected operation of the unit. For example, portions of the data showed the static pressure drop significantly below the setpoint, because the fan capacity was not able to meet the airflow requests from the VAV terminal dampers. This is a evidence of a mechanical system limitation that would not be captured by using airflow control at the zone level. Further work needs to be done to design a method that processes raw BAS data to build models, such as the one presented, using domain informed decisions based on the expected behavior of the system.

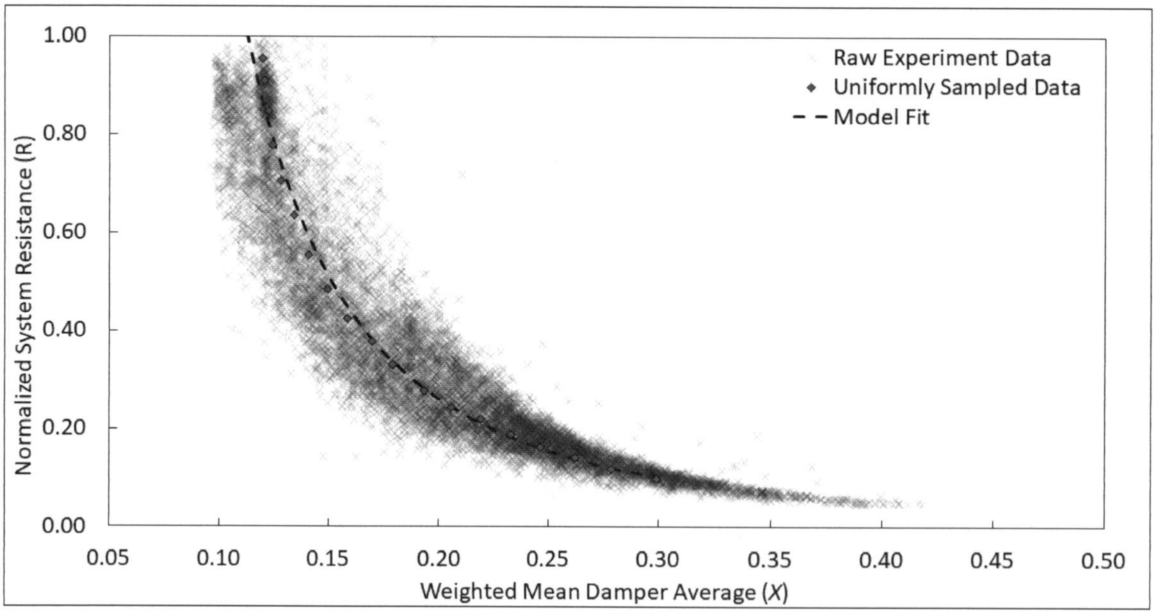

Figure 8: System Resistance Model for Gate Unit (Generated from undisturbed historical BAS data)

5. CONCLUSIONS

We presented a modeling approach to characterize the demand flexibility of the supply fan in VAV systems. The model maps all terminal zone damper configurations to a particular system resistance connected to a pressure-flow model of the supply fan to control its power consumption. Using this framework, the operational range of an existing unit can be quantitatively bounded by the minimum and maximum resistances the fan operation is subjected to. A control algorithm was proposed that leverages this model structure taking in a desired power deviation and outputting a sequence of damper configurations to achieve it. There are various avenues for continued work. We plan to extend and improve this modeling approach through experimental validation in different units. Another extension is to include data from a dedicated fan power meter to complete the pressure-flow model. This would include the effect of varying efficiencies in fan performance curves. We also plan to improve and extend the control algorithm we presented to include the zone thermal dynamics as constraints and incorporate predictive capabilities. Another avenue of investigation is the use of this approach, to extend compatibility with more advanced control sequences with trim & respond logic such as the ASHRAE Guideline 36 (ASHRAE, 2018). Given that this approach consists of relatively simple relations, we expect it can serve as a complementary model to other methods that focus on thermal comfort. Finally, another line we are exploring is the use of the theoretical foundations of this model for fault detection and diagnosis applications.

NOMENCLATURE

$\boldsymbol{\theta} = \{\theta_1, \theta_2, \ldots, \theta_N\}$	zone damper positions	(-)
$\mathbf{w} = \{w_1, w_2, \ldots, w_N\}$	VAV box nominal airflow	(m^3/s)
$\mathbf{x} = \{x_1, x_2, \ldots, x_N\}$	weighted damper positions	(-)
$\mathbf{q} = \{q_1, q_2, \ldots, q_N\}$	zone airflow	(m^3/s)
$\mathbf{P} = \{P_1, P_2, \ldots, P_H\}$	power consumption sequence	(kW)
$\mathcal{X} := \{\ \mathbf{x}\ \lvert \underline{\mathbf{x}} \leq \mathbf{x} \leq \overline{\mathbf{x}}\}$	feasible damper positions	(-)
X	weighted mean damper position	(-)
R	system resistance coefficient	$(Pa/(m^3/s)^2)$
Q	system total airflow	(m^3/s)
p	system static pressure	(Pa)

* hat (ˆ) notation used to denote a setpoint

Subscript

N	number of zones
H	time horizon
k	timesteps

REFERENCES

Adetola, V., Lin, F., Adetola, V., Lin, F., Yuan, S., & Reeve, H. (2018). Building Flexibility Estimation and Control for Grid Ancillary Services. In *International high performance buildings conference.*

ASHRAE. (2016). *2016 ASHRAE Handbook - Fundamentals (SI Edition).* American Society of Heating, Refrigerating and Air-Conditioning Engineers.

ASHRAE. (2018). *Guideline 36-2018 – High-Performance Sequences of Operation for HVAC Systems* (No. May). American Society of Heating, Refrigerating and Air-Conditioning Engineers.

Blum, D. H., & Norford, L. K. (2014, 11). Dynamic simulation and analysis of ancillary service demand response strategies for variable air volume HVAC systems. *HVAC and R Research*, *20*(8), 908–921. doi: 10.1080/10789669.2014.958975

Dong, J., Im, P., Huang, S., Chen, Y., Münk, J., & Kuruganti, T. (2019). Development and calibration of an online energy model for AHU fan. *ASHRAE Transactions*, *125*, 341–349.

Goddard, G., Klose, J., & Backhaus, S. (2014, 7). Model Development and Identification for Fast Demand Response in Commercial HVAC Systems. *IEEE Transactions on Smart Grid*, *5*(4), 2084–2092. Retrieved from `http://ieeexplore.ieee.org/lpdocs/epic03/wrapper.htm?arnumber=6839121` doi: 10.1109/TSG.2014.2312430

Hydeman, M. M., Taylor, S. T., Stein, J., Kolderup, E., & Hong, T. (2003). *Advanced Variable Air Volume System Design Guide.*

Jing, G., Cai, W., Zhang, X., Cui, C., Yin, X., & Xian, H. (2019). Modeling, air balancing and optimal pressure setpoint selection for the ventilation system with minimized energy consumption. *Applied Energy*, *236*(September 2018), 574–589. Retrieved from `https://doi.org/10.1016/j.apenergy.2018.12.026` doi: 10.1016/j.apenergy.2018.12.026

Kathirgamanathan, A., De Rosa, M., Mangina, E., & Finn, D. P. (2021). Data-driven predictive control for unlocking building energy flexibility: A review. *Renewable and Sustainable Energy Reviews*, *135*(January 2020), 110120. Retrieved from `https://doi.org/10.1016/j.rser.2020.110120` doi: 10.1016/j.rser.2020.110120

Keskar, A., Anderson, D., Johnson, J. X., Hiskens, I. A., & Mathieu, J. L. (2019). Do commercial buildings become less efficient when they provide grid ancillary services? *Energy Efficiency*. doi: 10.1007/s12053-019-09787-x

Macdonald, J., Kiliccote, S., Boch, J., Chen, J., & Nawy, R. (2014). Commercial Building Loads Providing Ancillary Services in PJM. *2014 ACEEE Summer Study on Energy Efficiency in Buildings*, 192–206.

MacDonald, J. S., Vrettos, E., & Callaway, D. S. (2020). A Critical Exploration of the Efficiency Impacts of Demand

Response From HVAC in Commercial Buildings. *Proceedings of the IEEE*, 1–17. Retrieved from https://ieeexplore.ieee.org/document/9143166/ doi: 10.1109/JPROC.2020.3006804

Neukomm, M., Nubbe, V., & Fares, R. (2019, 4). *Grid-Interactive Efficient Buildings* (Tech. Rep. No. April). EERE Publication and Product Library. Retrieved from http://www.osti.gov/servlets/purl/1508212/ doi: 10.2172/1508212

Raisoni, R., Raman, N. S., Barooah, P., Munk, J. D., & Im, P. (2018). A Control-Oriented Dynamic Model of Air Flow in a Single Duct HVAC System. In *5th international high performance buildings conference at purdue* (pp. 1–10).

Roth, A., & Reyna, J. (2020, 11). *Innovations in Building Energy Modeling: Research and Development Opportunities for Emerging Technologies* (Tech. Rep. No. November). Golden, CO (United States): National Renewable Energy Laboratory (NREL). Retrieved from https://www.osti.gov/servlets/purl/1710155/ doi: 10.2172/1710155

Stein, J., & Hydeman, M. M. (2004). Development and testing of the characteristic curve fan model. *2004 Winter Meeting - Technical and Symposium Papers, American Society of Heating, Refrigerating and Air-Conditioning Engineers*, 303–312.

Tukur, A., & Hallinan, K. P. (2017). Statistically informed static pressure control in multiple-zone VAV systems. *Energy and Buildings*, *135*, 244–252. Retrieved from http://dx.doi.org/10.1016/j.enbuild.2016.11.032 doi: 10.1016/j.enbuild.2016.11.032

Ulbig, A., & Andersson, G. (2015). Analyzing operational flexibility of electric power systems. *International Journal of Electrical Power and Energy Systems*, *72*, 155–164. Retrieved from http://dx.doi.org/10.1016/j.ijepes.2015.02.028 doi: 10.1016/j.ijepes.2015.02.028

Wang, J., Huang, S., Wu, D., & Lu, N. (2020). Operating a Commercial Building HVAC Load as a Virtual Battery through Airflow Control. *IEEE Transactions on Sustainable Energy*, *3029*(c), 1–1. doi: 10.1109/tste.2020.2988513

Wetter, M., Zuo, W., Nouidui, T. S., & Pang, X. (2014, 7). Modelica Buildings library. *Journal of Building Performance Simulation*, *7*(4), 253–270. Retrieved from http://www.tandfonline.com/doi/abs/10.1080/19401493.2013.765506 doi: 10.1080/19401493.2013.765506

Yin, R., Kiliccote, S., & Piette, M. A. (2016, 7). Linking measurements and models in commercial buildings: A case study for model calibration and demand response strategy evaluation. *Energy and Buildings*, *124*, 222–235. Retrieved from http://dx.doi.org/10.1016/j.enbuild.2015.10.042 doi: 10.1016/j.enbuild.2015.10.042

ACKNOWLEDGMENT

The authors would like to gratefully acknowledge the support provided by the Pennsylvania Infrastructure Technology Alliance (PITA). The authors are also thankful for the assistance in data and building automation control access to the Facilities Management and Campus Services team at Carnegie Mellon University.

Assessing the Value of Uncertainty-Aware Transactive Control Framework for Commercial and Residential Buildings

Min Gyung YU[1,2], Gregory S. PAVLAK[1,3*]

[1] The Pennsylvania State University,
University Park, PA, U.S.A

[2] Contact Information (mzy57@psu.edu)
[3] Contact Information (gxp93@psu.edu)
* Corresponding Author

ABSTRACT

With the increasing adoption of renewable energy and electric vehicles in the power grid, dealing with uncertainty in both supply and demand is critical to ensuring reliable and efficient operations. In this study, we discuss the value of a two-stage stochastic control framework for an aggregator to address such problems by promoting improved decision-making and performance despite inherent uncertainty. An uncertainty-aware transactive control framework was developed to account for uncertainties in future conditions due to occupancy patterns, weather conditions, on-site power generation, and real-time pricing schemes. In the day-ahead period, the aggregator decides the electricity procurement plan considering the possible real-time control strategies for operation of the commercial building thermal energy storage (TES) assets and residential building electric water heaters. During real-time operations, the aggregator modulates controllable loads based on transactive market mechanisms with model predictive control (MPC). In order to evaluate the performance, this study quantified the expected value of perfect information (EVPI) and the value of the stochastic solution (VSS) to analyze the cost of uncertain information and potential benefits of solving the stochastic optimal control problem. This paper demonstrates how the stochastic solution of the developed framework can provide useful information for customers and grid operators in the management of uncertain situations to support grid reliability and sustainability.

1. INTRODUCTION

As new policies are introduced to address climate change, increased investments in renewable power sources are anticipated in support of bringing U.S. greenhouse gas emissions down to net-zero (Galvin and Healy, 2020). With a high penetration of renewable energy sources, the continued addition of clean energy production can become incrementally less useful unless it can be paired with energy storage or complimentary shifts in demand. Thus, potential solutions lie in technical and institutional changes on both demand and supply-side sources of flexibility (Cochran et al., 2015). As a form of flexibility, energy storage can compensate for the fluctuating output of renewable energy sources. An analysis by the National Renewable Energy Laboratory showed that adding controllable behind-the-meter energy storage can manage building peak demand and reduce electricity costs (Neubauer and Simpson, 2015).

In the meantime, the supply and demand of electricity must match in real-time. Not all systems have the flexibility to respond instantaneously, which means some level of centralized coordination is necessary. The grid operators need to know ahead of time to schedule the system operation for the following day. Thus, two-settlement market approaches with day-ahead and real-time settlement periods have been applied for managing electric grid operations. The day-ahead market allows for a credible plan to be developed based on anticipated future conditions, while the real-time market allows for the balancing of any differences between the day-ahead plan and actual conditions.

Although demand side flexibility can serve as an important resource for the grid, the availability and response magnitude can vary across time, building types, and system types, among other factors (Walnum et al., 2019). Correspondingly, it can be challenging to predict a building's potential flexibility given the uncertainty in occupant behavior, weather conditions, and on-site renewable energy generation. In our previous work, we developed an uncertainty-aware transactive control framework for commercial buildings with thermal energy storage (TES) assets participating in both the day-ahead and real-time market that can account for uncertainties in future conditions (Yu and Pavlak, 2020; Yu and Pavlak, 2021). In this study, we extend the framework to consider portfolios of residential buildings with controllable water heaters, and also incorporate behind-the-meter variable renewable generation as additional sources of uncertainty. In this work, the role of the aggregator is to decide the day-ahead electricity procurement plan and real-time operation strategy for residential and commercial building controllable thermal energy storage assets, accounting for uncertainty in occupant behavior, weather, variable generation, and grid prices. The performance of the framework is evaluated in terms of the value of the stochastic solution.

2. RELATED WORK

As a distributed control approach, transactive control has been introduced to manage electric grid operations (Liu et al., 2017). With increasing variability on both side of the meter, stochastic planning based on transactive control has been developed to contend with uncertainties in future conditions (Liu et al., 2018). A significant amount of past work has demonstrated the potential benefits of coordinating multiple energy resources with two-settlement market approaches over uncertain scenarios (Liu et al., 2018).

For the building sector, Hao et al. (2016) and Corbin et al. (2016) considered the application of transactive controls to commercial building Heating, Ventilation, and Air-Conditioning (HVAC) systems. Starke et al. (2019) and Adhikari et al. (2016) studied a transaction-based strategy to analyze the effectiveness in controlling residential building HVAC systems and water heater. Correa-Flrez et al. (2019) proposed an optimization model based on a transactive environment for a residential aggregator with PV, electric water heaters and batteries by controlling the settings of flexible devices. Robust optimization was used to consider uncertainty and a comparative study was performed. The results showed that using robust optimization allowed strategic bidding to capture uncertainties while complying with obligations in the wholesale and local markets. To integrate the flexibility potential of both residential and commercial buildings, Golmohamadi et al. (2019) proposed a central demand response provider to coordinate the plans of industrial and residential demand response aggregators. The aggregator used the thermal and electrical storage systems linked with on-site PV systems. The integrated flexibility was traded in the electricity market to maximize the profit of the market participants in a competitive environment. The results showed that the integrated flexibility could safeguard the future of power systems against intermittent power.

Based on the reviewed literature, there is an identified need to evaluate the value of the uncertainty-aware framework as well as the integrated flexibility potential of on-site PV generation in the portfolio. The value of the stochastic solution, which depends on the variation of the uncertain load scenarios, can be analyzed to maximize the potential benefits of the controllable demand side flexibility and support grid resiliency. This study evaluated the transactive control framework for an aggregator of both commercial and residential buildings with building emulator in real-time operation. The performance of the stochastic solution that accounts for uncertainties from the occupancy behavior, weather, PV generation, and market prices was analyzed to characterize the demand-side flexibility.

3. UNCERTAINTY-AWARE TRANSACTIVE CONTROL FRAMEWORK

3.1 Architecture of the control framework

In this paper, an uncertainty-aware transactive (UA-Tx) control platform with two separate market systems is proposed for an aggregator to dispatch energy to one group of commercial buildings and another group of residential buildings with thermal energy storage resources. Figure 1 highlights the architecture of the platform.

The control platform has two stages, one for day-ahead planning and one for real-time operation. In the day-ahead planning, the aggregator purchases electricity for the following day based on a two-stage stochastic optimization. The optimal procurement is determined to minimize the total cost by considering the expected operation of buildings with multiple uncertain scenarios. A set of scenarios is generated considering weather conditions, occupancy presence and absence patterns, domestic hot water use, appliance and electric equipment use, PV power generation, and real-time power prices as the sources of uncertainty. Scenarios were generated using a Monte-Carlo method. The detailed two-stage stochastic formulation is described in Section 3.2. In the real-time operation, the aggregator determines the operation of the TES charging/discharging for commercial buildings and water heater setpoint temperature changes for residential buildings based on the transactive market between the aggregator and buildings. The procured electricity, real-time power prices, and the actual building states are key factors for the aggregator to decide the operation and maximize its profit. Each building broadcasts its response curve representing constraints of the control and preferences as in Figure 1. As the aggregator clears the market price based on the information, the operation of building thermal energy storage resources is determined following the response curve of each building. A model predictive controller

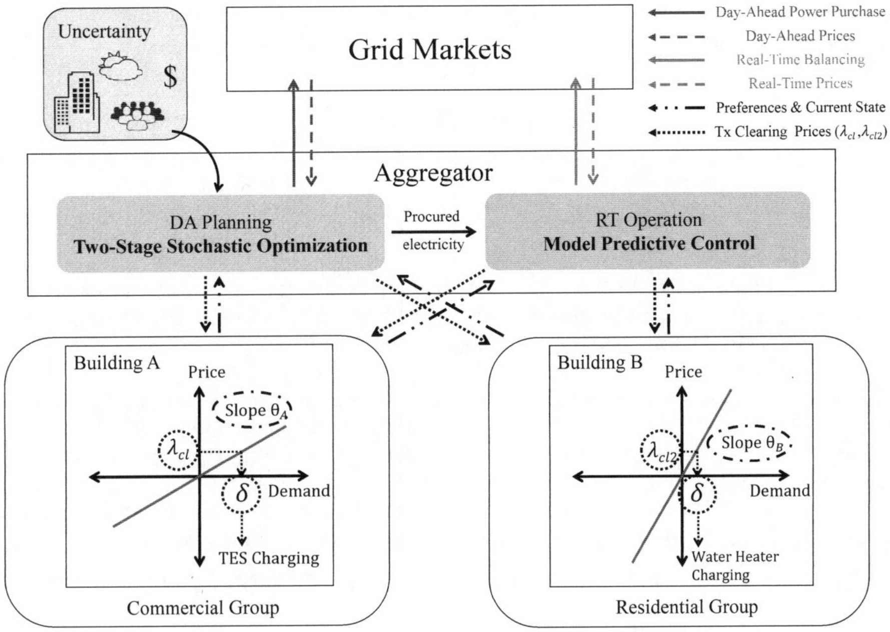

Figure 1: Architecture of the uncertainty-aware transactive building energy control platform.

(MPC) is used to modulate the amount of TES charging/discharging for the buildings in the commercial groups and the temperature setpoint of the water heaters for residential domestic hot water. In every time interval, the optimal control strategy is utilized in the energy management system (EMS) of the building emulator to simulate the response. In this work, detailed EnergyPlus models were used as the building emulators to verify the operation strategy from the control-oriented model. EnergyPlus EMS code was developed to implement the control strategies in the model. After executing the control strategy, the actual state is updated with the aggregator to solve the next time interval. The operation scheme of each commercial building with TES, described by the response curve, can be formulated as Eqs. (1)–(3) below,

$$\delta_{b,t}^{c} = \theta_{b}^{c} \lambda_{t}^{cl,c} \tag{1}$$

$$q_{b,t}^{c*} = q_{b,t}^{c} + \delta_{b,t}^{c} COP \tag{2}$$

$$p_{b,t}^{c*} = p_{b,t}^{c} + p_{b,t}^{oth,c} - p_{b,t}^{pv,c} + \delta_{b,t}^{c} = q_{b,t}^{c*}/COP + p_{b,t}^{oth,c} - p_{b,t}^{pv,c} \tag{3}$$

where, $\delta_{b,t}^{c}$ is the power compensation of the plant system by TES operation; θ_{b}^{c} is the coefficient for customer preference, representing the response curve's slope; $\lambda_{t}^{cl,c}$ is the cleared price between the aggregator and commercial building group; $p_{b,t}^{c*}$ is the actual power of commercial building in time interval t; $p_{b,t}^{c}$ is the commercial building electric demand before TES operation; and $p_{b,t}^{pv,c}$ is the power generation by the PV on the commercial building.

In every time interval, $\delta_{b,t}^{c}$ is determined by the response curve slope θ_{b}^{c} and the cleared price $\lambda_{t}^{cl,c}$ as in Equation (1). $q_{b,t}^{c*}$ is the modified heat transfer rate of plant system, which is defined as the sum of the original heat transfer rate $q_{b,t}^{c}$ and the power compensation multiplied by the coefficient of the system (COP) as in Equation (2). In TES charging mode, $q_{b,t}^{c*}$ is higher than the original rate since it considers the additional heat transfer rate of the chiller to storage. When it comes to TES discharging mode, $q_{b,t}^{c*}$ is lower than the original rate as the base chiller is operating at lower level. $p_{b,t}^{c*}$ is defined as the sum of the original electric demand $p_{b,t}^{c}$, other electrical loads from light and electric equipment $p_{b,t}^{oth}$ and the power compensation $\delta_{b,t}^{c}$ minus power supplied by PV generation on the site $p_{b,t}^{pv,c}$ as in Equation (3). In this work, the constant COP is used in the control-oriented model because of potential computational challenges of a more detailed model that accounts for dynamic changes in efficiencies. The value was set to reflect the efficiency of all components of the cooling plant (Henze et al., 2004). Although a constant COP is used in formulating a control-oriented model, the detailed building emulator captures the non-linear characteristics. The operation scheme of residential buildings with the water heater by the response curve can be formulated as Eqs. (4)–(6) below,

$$\delta^r_{b,t} = \theta^r_b \lambda^{cl,r}_t \tag{4}$$

$$q^{r*}_{b,t} = q^r_{b,t} + \delta^r_{b,t}/\mu \tag{5}$$

$$p^{r*}_{b,t} = p^r_{b,t}\mu + p^{oth,r}_{b,t} - p^{pv,r}_{b,t} + \delta^r_{b,t} = q^{r*}_{b,t}\mu + p^{oth,r}_{b,t} - p^{pv,r}_{b,t} \tag{6}$$

where, $\delta^r_{b,t}$ is the power compensation of the water heater; θ^r_b is the coefficient for customer preference; $\lambda^{cl,r}_t$ is the cleared price between the aggregator and residential building group; $p^{r*}_{b,t}$ is the actual power of residential building; and $p^r_{b,t}$ is the original residential building electric demand.

In every time interval, $\delta^r_{b,t}$ is determined by the response curve slope θ^r_b and the cleared price $\lambda^{cl,r}_t$ as in Equation (4). $q^{r*}_{b,t}$ is the modified heating demand of the water heater, which is defined as the sum of the original heating demand $q^r_{b,t}$ and the power compensation divided by the efficiency μ as in Equation (5). When the water heater is in charging mode, $q^{r*}_{b,t}$ is higher than the original rate since the heater is heated to set to the increased set temperature. In discharging mode, $q^{r*}_{b,t}$ is lower than the original rate as the water heater is not heated as much as it used to for the decreased set temperature. $p^{r*}_{b,t}$ is defined as the sum of the original electric demand $p^r_{b,t}$, other electrical loads from light and electric equipment $p^{oth}_{b,t}$ and the power compensation $\delta^r_{b,t}$ minus power supplied by PV generation $p^{pv,r}_{b,t}$ as in Equation (6).

3.2 Two-stage stochastic optimization framework

3.2.1 Scenario generation

To perform a two-stage stochastic optimization in day-ahead planning, a set of scenarios for the expected operation of the following day is needed. In this section, the generation of building load scenarios, weather scenarios, PV generation scenarios, and real-time power price scenarios is described.

The building load scenarios are generated with the detailed building energy simulation program EnergyPlus by changing the uncertain variables such as occupancy schedule, equipment usages, and weather conditions based on the building models. For the uncertain occupancy schedules of the commercial buildings, a Monte-Carlo approach is utilized to select a random occupancy schedule for the planning day from a year of occupancy schedule. The base full year of occupancy schedule data is created by using the stochastic occupancy model developed by Lawrence Berkeley National Laboratory (2017). When sampling a random schedule, holidays, and Sundays are filtered out. For the commercial buildings, the occupancy schedules are also used to schedule the lighting and other electric equipment usages in the building. The uncertain occupancy schedules of the residential buildings are generated based on Richardson's model, which is available for free download. The model is discussed in Richardson et al. (2008, 2010).

The occupant transition probabilities in the model are updated with the American occupant behaviors, which are extracted from the American Time Use Survey (ATUS) dataset from the Bureau of Labor Statistics. A Python script was developed to process the stochastic occupancy profiles into a format that can be updated in the building energy model of EnergyPlus. The domestic hot water (DHW) use and electricity use in the residential buildings are based on separate schedules of DHW usage scenarios and active occupancy by assuming their inactive (sleeping) time period from 11 PM to 5 AM. For the DHW event schedule, a Python script was developed to generate each DHW event schedule for dish washer, shower, sink, bath, and clothes washer, based on Hendron's probability distribution functions (PDF) (Hendron and Engebrecht, 2010). The number of DHW events during the planning day, start time, and duration are randomly selected based on the PDF and the occupants' active schedule. The generated schedule is written in a format of an hourly factor to utilize as schedules in EnergyPlus. For variation in the weather, the actual weather data from 2018 in Chicago, IL, is used. A 30-day period, centered on the planning day, is taken as the window from which to sample random weather scenarios. A weather file with the new random weather data for the planning day is read in EnergyPlus as input data. Based on the updated uncertain variables, one building load scenario is generated by running EnergyPlus. The weather variation is also used to create PV generation scenarios. The Model Chain function of the Python PV library was used to calculate the system output at a specific location (Holmgren et al., 2018).

For the real-time power prices, this work used a bootstrap approach based on the hourly distributions of deviations between day-ahead and real-time prices of the historical power prices to consider a reliable correlation between day-ahead and real-time power prices. The hourly price deviations are added to the day-ahead price of the planning day to create an uncertain real-time power price.

3.2.2 Two-stage stochastic formulation for day-ahead planning

In the two-stage stochastic optimal planning, the first-stage is related to the day-ahead procurement decision and the second-stage is representing building operating decisions based on the generated set of scenarios. In the optimal day-ahead planning, the procurement is decided to minimize the electricity procurement cost and expected real-time operation cost as in Equation (7),

$$\min_{x} \quad (\lambda^{DA})^T x + \mathbb{E}[\mathcal{Q}(x,\omega)] \tag{7}$$

where x is the vector of procured electricity in the day-ahead market; λ^{DA} is the vector of the power prices in the day-ahead market; $\mathbb{E}$ is the expected value of the real-time operation to calculate the probability-weighted sum over all the scenarios. The first term in Equation (7) is the day-ahead procurement cost and the second term is the expected cost in real-time operation. The second-stage objective can be expressed as Equation (8),

$$\begin{aligned}\mathcal{Q}(x,\omega) = \min \quad & (\lambda_{\omega}^{cl,c})^T \sum_{b\in B_c} \delta_{b,\omega}^{c} + (\lambda_{\omega}^{cl,r})^T \sum_{b\in B_r} \delta_{b,\omega}^{r} + \lambda^{DR} 1^T (\sum_{b\in B_c} |\delta_{b,\omega}^{c}| + \sum_{b\in B_r} |\delta_{b,\omega}^{r}|) \\ & -\lambda^{ba} 1^T (\sum_{b\in B_c} p_{b,\omega}^{c,buy} + \sum_{b\in B_r} p_{b,\omega}^{r,buy} - \sum_{b\in B_c} p_{b,\omega}^{c,sell} - \sum_{b\in B_r} p_{b,\omega}^{r,sell}) + (\lambda_{\omega}^{rt'})^T (\sum_{b\in B_c} p_{b,\omega}^{c*} + \sum_{b\in B_r} p_{b,\omega}^{r*} - x) \\ & +\eta_1 [\ \max(0, 1^T(x - 1.25(\sum_{b\in B_c} p_{b,\omega}^{c*} + \sum_{b\in B_r} p_{b,\omega}^{r*})))^2 + \max(0, 1^T(0.75(\sum_{b\in B_c} p_{b,\omega}^{c*} + \sum_{b\in B_r} p_{b,\omega}^{r*}) - x))^2\ ]\end{aligned} \tag{8}$$

where λ^{DR} is the demand response (DR) price in the transactive market; λ^{ba} is the base electricity price charged to the buildings; $\lambda_{\omega}^{rt'}$ is the uncertain real-time prices in scenario ω; $p_{b,\omega}^{c,buy}$ and $p_{b,\omega}^{r,buy}$ are the actual amount of power that the commercial and residential buildings buy from the aggregator; and $p_{b,\omega}^{c,sell}$ and $p_{b,\omega}^{r,sell}$ are the power that commercial and residential buildings sell to the aggregator. The first three terms in Equation (8) represent the cost that the aggregator pays to customers to reduce the building demand via the transactive market. Here, different clearing prices $\lambda_{\omega}^{cl,c}$, $\lambda_{\omega}^{cl,r}$ are provided to commercial and residential group, respectively, using the separate transactive markets. A fixed demand response price λ^{DR}, in addition to the clearing prices, provides additional compensation to customers for use of their TES resources. The fourth term is that the aggregator makes profits by selling the power to the buildings for their actual usage at a fixed price λ^{ba}, as well as the aggregator, purchases their extra electricity generated from the PV at the same price λ^{ba}. The fifth term is the balancing cost of aggregator in the real-time market. Depending on the procurement decision and actual power usage by customers, the aggregator may purchase more power or sell the extra power in the real-time market. In the day-ahead planning, the uncertain real-time price $\lambda_{\omega}^{rt'}$ is used. The sixth term is a penalty term to make sure the total day-ahead procured energy is within 25% of the sum of energy use in the expected real-time operation.

The first stage problem is subject to the procured electricity being positive and lower than the daily aggregated peak demand of the buildings as in Equation (9). p_b^h represents the peak demand of building in all scenario.

$$0 \le x \le \sum_{b\in B} p_b^h \tag{9}$$

The constraints of the second stage are described in Eqs. (10)–(15) for commercial buildings, and Eqs. (16)–(21) for residential buildings,

$$SOC_b^{min} \le SOC_{b,t,\omega} \le SOC_b^{max} \tag{10}$$

$$SOC_{b,t,\omega} = (q_{b,t,\omega}^{c*} - q_{b,t,\omega}^{c})/CAP_b^{tes} + SOC_{b,t-1,\omega} \tag{11}$$

$$p_{b,\omega}^{c*} = p_{b,\omega}^{c,buy} - p_{b,\omega}^{c,sell} \tag{12}$$

$$q_{b,\omega}^{c*}, p_{b,\omega}^{c,buy}, p_{b,\omega}^{c,sell} \ge 0 \tag{13}$$

$$q_{b,\omega}^{c*} - q_{b,\omega}^{c} \le \min(CAP_b^{ch}, (SOC_b^{max} - SOC_{b,t,\omega})CAP_b^{tes}) \tag{14}$$

$$\max(-q_{b,t,\omega}^{c}, (SOC_b^{min} - SOC_{b,t,\omega})CAP_b^{tes}) \le q_{b,t,\omega}^{c*} - q_{b,\omega}^{c} \tag{15}$$

$$T_b^L \leq T_{b,t,\omega}^{dhw} \leq T_b^H \tag{16}$$

$$T_{b,t,\omega}^{dhw} = \frac{(q_{b,t,\omega}^{r*} - q_{b,t,\omega}^{r})}{c_p \dot{m}} + T_{b,t-1,\omega}^{dhw} \tag{17}$$

$$q_{b,t,\omega}^{r*} - q_{b,t,\omega}^{r} \leq \min(q_{max}^{dhw}, c_p \dot{m}(T_{b,t}^H - T_{b,t,\omega}^{dhw})) \tag{18}$$

$$c_p \dot{m}(T_b^L - T_{b,t,\omega}^{dhw}) \leq q_{b,t,\omega}^{r*} - q_{b,t,\omega}^{r} \tag{19}$$

$$p_{b,t,\omega}^{r*} = p_{b,t,\omega}^{r,buy} - p_{b,t,\omega}^{r,sell} \tag{20}$$

$$q_{b,t,\omega}^{r*}, p_{b,t,\omega}^{r,buy}, p_{b,t,\omega}^{r,sell} \geq 0 \tag{21}$$

where $p_{b,\omega}^h$ is the hourly peak demand of building over all the scenario; $SOC_{b,t,\omega}$ is the state-of-charge (SOC) levels of TES in scenario ω; CAP_b^{tes} is the thermal energy storage capacity of the building b; CAP_b^{ch} is the capacity of the dedicated thermal energy storage chiller of the building b; $T_{b,t,\omega}^{dhw}$ is the set temperature of the water heater in scenario w; T_b^H is the maximum temperature of the water heater, T_b^L is the minimum temperature of the water heater, and q_{max}^{dhw} is the water heater capacity of the residential building.

The SOC levels of the TES are constrained within the specified lower and upper limit as in Equation (10). Also, the SOC level in each time interval is calculated with the TES heat transfer rate and the SOC level in the previous time interval as in Equation (11). The maximum TES charging rate cannot be over the minimum of the dedicated TES chiller capacity and the extra heat capacity of TES as in Equation (12). The maximum TES discharging rate is constrained by the maximum of the building cooling load and the rest thermal energy in TES as in Equation (13). The actual power of building $p_{b,\omega}^{c*}$ is defined as the power that the customers purchase from the aggregator and the power that they can sell to the aggregator. If the PV generates more than the building electric demand, the actual power $p_{b,\omega}^{c*}$ can be a negative value, which means they are selling their extra power to the aggregator as in Equation (14). Equation (16) constrains the temperature of the water heater within the specified range with a lower and upper limit. Equation (17) defines the water heater temperature in each time interval as calculated by the modified heating demand from the original demand. Equation (18) keeps the maximum charging energy limit by taking the minimum of the heater capacity and the extra heat capacity of the water heater considering the maximum temperature. Equation (19) restricts the maximum discharging rate by the rest of the thermal energy that the water heater can discharge considering the minimum temperature limit.

In the real-time operation, the aggregator modulates the operation of thermal energy storage resources of multiple buildings to minimize its total cost based on procurement in the day-ahead market and real-time power prices. The formulation of the MPC is nearly equivalent to the second stage of the two-stage stochastic optimization model with the exception that the penalty term is removed. The constraints for MPC model are the same as Eqs. (10)–(21).

4. CASE STUDY

4.1 Simulation conditions in case study

Simulation case studies were performed to evaluate the performance of the proposed control framework. The building models used in this study are three commercial and three residential buildings located in Chicago, IL. The commercial building model is the 5-zone example office model in EnergyPlus (U.S. Department of Energy, 2019), and a residential prototype building energy model with electric water heater and heat pump from the Department of Energy is used for the heating system. Then, the parameters of the building models were varied to generate three different commercial and residential building models. The main plant system configuration of the commercial building model consists of a base chiller (87.8 kW), TES dedicated chiller (23.4 kW), and TES (266 kWh). This work set the initial, minimum, and maximum SOC of TES as 100%, 0%, and 100% of TES capacity, respectively. The water heater tank of the residential building has a volume of 0.197 m^3 and the heater has a maximum capacity of 5.5 kW.

The case studies were performed for June 2018 and actual weather data from 2018 were used. The electricity prices in the day-ahead market were obtained from the historical data in 2018. In this work, the base electricity price λ^{ba} was set to be \$0.03/kWh. For the case studies presented in this work the demand response price λ^{DR} is was set to \$0.01/kWh. In this work, the performances of the framework for the aggregator with PV generation and the aggregator without considering PV generation were compared. The stochastic optimization and MPC controller model were developed in Python, and the solver MOSEK was used to solve the problem. This study utilized the expected value of perfect information (EVPI) and the Value of Stochastic Solution (VSS) analysis to evaluate the stochastic framework. EVPI is defined as the difference between the wait-and-see (WS) solution and the recourse problem (RP). WS represents the average profit that could be achieved if decisions could be made with knowledge of the uncertain variables, thus, EVPI can be thought of as a reasonable price to pay for a perfect forecast. VSS is defined as the difference between

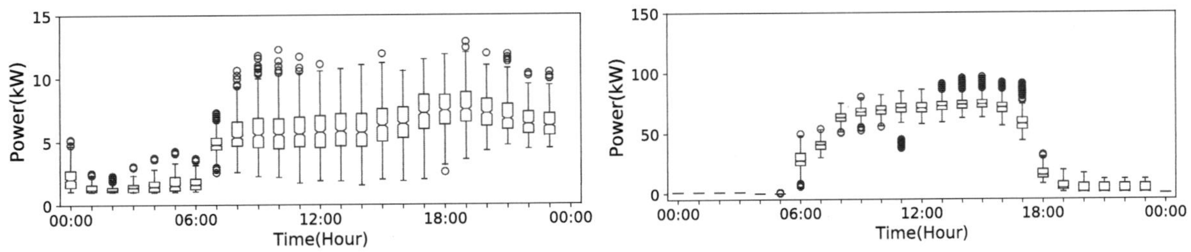

Figure 2: Stochastic load profile of residential (left) and commercial building (right) without PV (6/7).

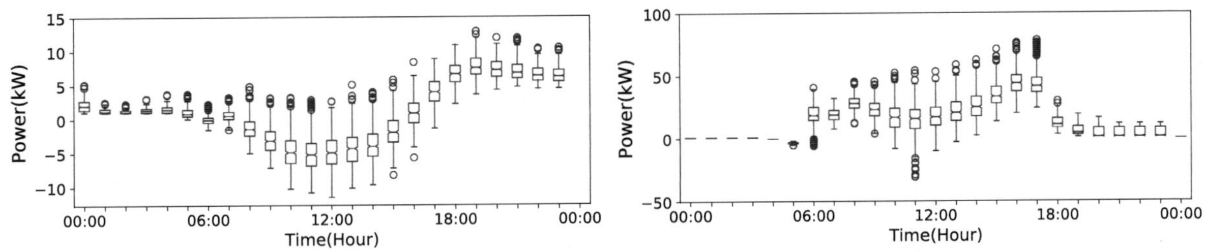

Figure 3: Stochastic load profile of residential (left) and commercial building (right) with PV (6/7).

the expected result of using the expected value (EEV) solution and RP. VSS represents the possible gain by solving the stochastic model, rather than simply adopting the mean value solution.

Figs. 2 and 3 provide samples of the stochastic load profile of 3 residential and 3 commercial buildings for one day. The left subfigure of Figure 2 represents the stochastic load profile of residential buildings without PV generation has a wide range of variation from 8:00 until midnight. It represents the uncertain occupant behavior for going to work and other activities. The stochastic load profile of commercial buildings without PV generation in the right subfigure has a relatively small variation in every hour during the operation. When the residential buildings install PV generation and widely use it for its electricity usage, the residential building has a variation in negative values during the daytime, since the PV output is greater than the required building load as in left subfigure of Figure 3. For the commercial building with PV generation in the right subfigure, it shows a relatively wider variation than the load profile without PV generation. Especially, when there is a peak PV generation at 11:00, it is expected that the commercial buildings can also sell the extra power to the aggregator.

4.2 Results

Table 1 highlights the overall results of case studies, in terms of WS, EEV, RP, EVPI, and VSS. For WS, EEV, and RP in Table 1, negative values indicate profit for the aggregator, and positive values indicate a loss. EVPI is quite large for each case, thus, the aggregator could benefit greatly from reducing uncertainty through gathering more information or improving forecasts. EVPI of the case with PVs can reach around 377% of RP as more uncertainties from the PV generation lead to much larger potential for profit and loss. Comparing both cases shows that the VSS/RP increased from 19% to 21% with increased variation.

Table 1: Summary of case study results with EVPI and VSS values.

	RP ($)	WS ($)	EEV ($)	EVPI ($)	EVPI/RP	VSS ($)	VSS/RP
No PV	-57.28	-199.2	-46.13	141.9	248 %	11.15	19 %
PV	-31.48	-137.7	-24.98	106.2	337 %	6.5	21 %

Figure 4 shows the detailed results of a single day (June 7th) for both cases. On that day, the real-time price spiked significantly from 15:00 to 18:00 as in Figure 6. The left and right subfigures represent the no PV and PV case, respec-

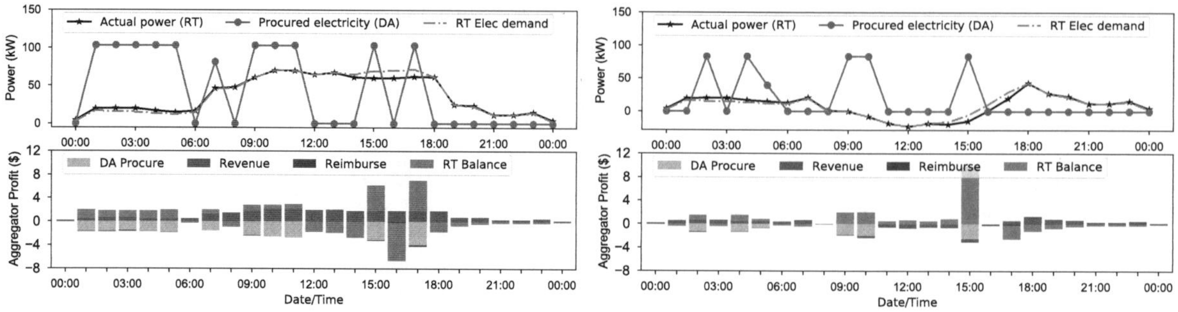

Figure 4: Detailed results of no PV case (left) and PV case (right) for June 7th.

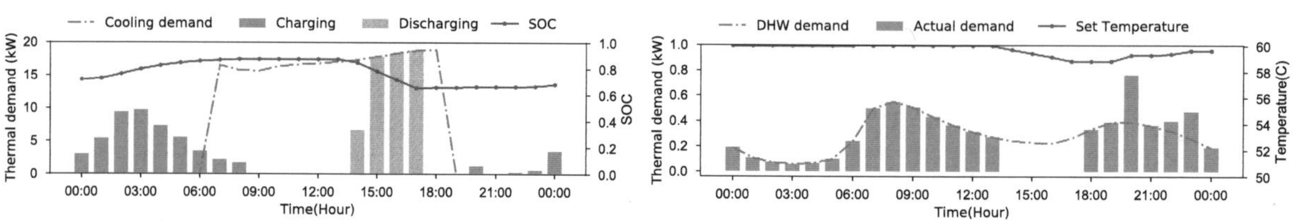

Figure 5: Stochastic load profile of residential (left) and commercial building (right) with PV (6/7).

tively. The top subfigures show the comparison of procured electricity, total electric demand (i.e., before operation of storage), and actual power (i.e., after operation of storage). Without consideration of PV generation on-site in the left subfigure, the aggregator purchased a large amount of power during morning hours and around 20% of electricity for the afternoon hours when the price spikes. When PV is included in the right subfigure, the majority of the power was procured during the early morning and afternoon. In this case, the actual power and the total electric demand are negative values from 10:00 to 15:00, which means the PV generated more power than the buildings needed. With the gap between them, it was confirmed that TES was utilized during the high priced hours. The second subfigure from the top shows the hourly objective function components of the aggregator for each hour of the day. The aggregator without PV made revenue by selling the power to customers during operating hours. Even though the aggregator had to purchase the power for the building demand at 16:00 when the price was high, the aggregator was able to make profits by selling the excess procured power in the real-time market. The interesting finding from the aggregator with PV is that the revenue values from 10:00 until 15:00 are negative as the aggregator purchased the power from the customers. It was able to make high profits by selling the power from the customers as well as the extra procured power in the real-time balancing market.

Figure 5 shows the TES operation of one commercial building in the left subfigure and the water heater operation of one residential building in the right subfigure. As expected, the TES is discharged when the real-time price is highest. The water heater also dropped the set temperature for discharging and increased it once the real-time price dropped below the day-ahead price.

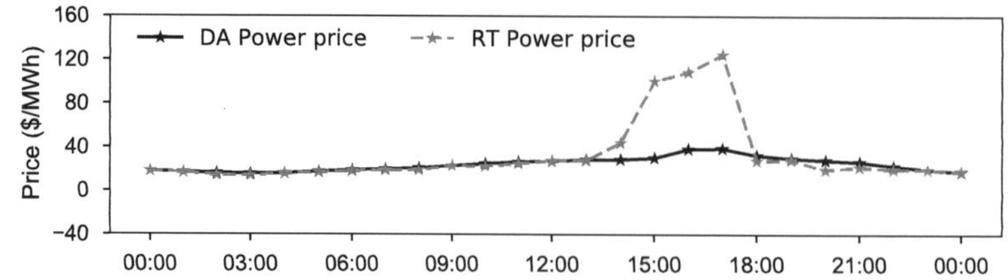

Figure 6: Day-ahead and real-time power price (6/7).

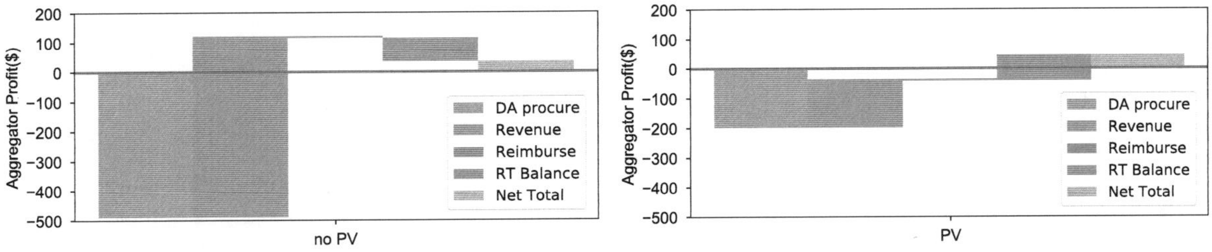

Figure 7: Stochastic load profile of residential (left) and commercial building (right) with PV (6/7).

Figure 7 shows the total surplus of the aggregator. Both cases made profits with different strategies. The aggregator without PV system in the left subfigure purchased a large amount in the day-ahead market and made large revenue by serving the power to the customers in real-time operation. However, it also needed to make purchases in the real-time balancing market to provide enough power to customers. For the case with PV in the right subfigure, the aggregator procured less power in the day-ahead market, made lower revenue from customers, and increased its profits from the real-time balancing market.

5. CONCLUSIONS

This work addresses the electricity procurement problem of the aggregator of both commercial and residential buildings with thermal energy storage assets. A two-stage stochastic programming approach has been proposed to account for uncertainties in future conditions with occupancy patterns, weather conditions, on-site power generation, and real-time pricing schemes in the decision process. To evaluate the capability of the proposed controller, the study conducted simulation experiments for portfolios with and without PV. It was observed that the proposed framework could offer the optimal procurement decisions for both cases and could offer the optimal TES operation scheme based on the customer's preference, and inherent uncertainty in the information available when making day-ahead decisions. From the EVPI and VSS analysis, it was observed that substantial benefits could be created by solving the stochastic program, especially when more uncertainties are included in the portfolio. The portfolio with PVs had higher EVPI/RP and VSS/RP than the portfolio without the PV generation on site. It was likely that the increased variability and uncertainty would potentially influence the results. This work highlights the increasing importance of accounting for uncertainty in operational decision making as load variability increase for future electric grids and microgrids with numerous distributed energy resources.

Future work can quantify the additional building-level benefits compared to individually optimized buildings. When PV is considered in the portfolio, the customers can sell their extra generated power to the aggregator at a constant rate regardless of the real-time power prices. The aggregator can shield customers from any price risk in the market. On the other hand, the individuals would deal with the risks of the fluctuating power prices in the market. Future work can also consider risk management to be beneficial for both the aggregator and customers against any financial risk.

REFERENCES

Adhikari, R., Pipattanasomporn, M., Kuzlu, M., and Rahman, S. (2016). Simulation study of transactive control strategies for residential hvac systems. In *2016 IEEE PES Innovative Smart Grid Technologies Conference Europe (ISGT-Europe)*, pages 1–5.

Cochran, J., Denholm, P., Speer, B., and Miller, M. (2015). Grid integration and the carrying capacity of the u.s. grid to incorporate variable renewable energy. Technical report, National Renewable Energy Laboratory.

Corbin, C. D., Makhmalbaf, A., Huang, S., Mendon, V. V., Zhao, M., Somasundaram, S., Liu, G., Ngo, H., and Katipamula, S. (2016). Transactive control of commercial building hvac systems. Technical report, Pacific Northwest National Laboratory.

Correa-Flórez, C., Michiorri, A., and Kariniotakis, G. (2019). Comparative analysis of adjustable robust optimization alternatives for the participation of aggregated residential prosumers in electricity markets. *Energies*, 12:1–27.

Galvin, R. and Healy, N. (2020). The green new deal in the united states: What it is and how to pay for it. *Energy Research Social Science*, 67:101529.

Golmohamadi, H., Keypour, R., Bak-Jensen, B., and Pillai, J. (2019). A multi-agent based optimization of residential and industrial demand response aggregators. *International Journal of Electrical Power Energy Systems*, 107:472–485.

Hao, H., Corbin, C., Kalsi, K., and Pratt, R. (2016). Transactive control of commercial buildings for demand response. *IEEE Transactions on Power Systems*, 32:1–1.

Hendron, R. and Engebrecht, C. (2010). Building america research benchmark definition: Updated december 2009. Technical report, National Renewable Energy Laboratory.

Henze, G., Felsmann, C., and Knabe, G. (2004). Evaluation of optimal control for active and passive building thermal storage. *International Journal of Thermal Sciences*, 43:173–183.

Holmgren, W. F., Hansen, C. W., and Mikofski, M. A. (2018). pvlib python: a python package for modeling solar energy systems. *Journal of Open Source Software*, 3:884.

Lawrence Berkeley National Laboratory (2017). Occupancy simulator.

Liu, Z., Wu, Q., Huang, S., and Zhao, H. (2017). Transactive energy: A review of state of the art and implementation. In *2017 IEEE Manchester PowerTech*, pages 1–6.

Liu, Z., Wu, Q., Ma, K., Shahidehpour, M., Xue, Y., and Huang, S. (2018). Two-stage optimal scheduling of electric vehicle charging based on transactive control. *IEEE Transactions on Smart Grid*, PP:1–1.

Neubauer, J. and Simpson, M. (2015). Deployment of behind-the-meter energy storage for demand charge reduction. Technical report, National Renewable Energy Laboratory.

Richardson, I., Thomson, M., and Infield, D. (2008). A high-resolution domestic building occupancy model for energy demand simulations. *Energy and Buildings*, 40:1560–1566.

Richardson, I., Thomson, M., Infield, D., and Clifford, C. (2010). Domestic electricity use: A high-resolution energy demand model. *Energy and Buildings*, 42:1878–1887.

Starke, M., Munk, J., Zandi, H., Kuruganti, T., Buckberry, H., Hall, J., and Leverette, J. (2019). Agent-based system for transactive control of smart residential neighborhoods. In *2019 IEEE Power Energy Society General Meeting (PESGM)*, pages 1–5.

U.S. Department of Energy (2019). Energyplus engineering reference. *The Reference to EnergyPlus Calculations*.

Walnum, H., Justo Alonso, M., Clauß, J., and Lindberg, K. (2019). Characterization of heat load profiles in buildings and their impact on demand side flexibility. *IOP Conference Series: Materials Science and Engineering*, 609:052035.

Yu, M. G. and Pavlak, G. S. (2020). Two-stage stochastic planning for control of building thermal energy storage portfolios with transactive controls. In *2020 American Control Conference (ACC)*, pages 2339–2344.

Yu, M. G. and Pavlak, G. S. (2021). Assessing the performance of uncertainty-aware transactive controls for building thermal energy storage systems. *Applied Energy*, 282:116103.

ACKNOWLEDGEMENT

We gratefully acknowledge Penn State Institutes of Energy and the Environment for supporting this work. In addition, computations for this research were performed on the Pennsylvania State University's Institute for Computational and Data Sciences' Roar supercomputer.

Analysis of Automated Fault Detection and Diagnostics Records as an Indicator of HVAC Fault Prevalence: Methodology and Preliminary Results

Amir EBRAHIMIFAKHAR[1]*, David YUILL[1], Amanda D. SMITH[2], Jessica GRANDERSON[3], Eliot CROWE[3], Yimin CHEN[3], Hayden REEVE[2]

[1]University of Nebraska – Lincoln, Architectural Engineering,
Omaha, NE, USA
amir.ebrahimifakhar@huskers.unl.edu, dyuill@unl.edu

[2]Pacific Northwest National Laboratory, Electricity Infrastructure & Buildings Division,
Richland, WA, USA
amanda.d.smith@pnnl.gov, hayden.reeve@pnnl.gov

[3]Lawrence Berkeley National Laboratory, Building Technology and Urban Systems Division,
Berkeley, CA, USA
JGranderson@lbl.gov, ECrowe@lbl.gov, YiminChen@lbl.gov

* Corresponding Author

ABSTRACT

Faults in commercial buildings can cause energy waste and other performance problems such as reduced occupant comfort, reduced equipment longevity, and increased noise. However, it is currently unknown how commonly faults occur in different equipment types. This paper describes a method to estimate the prevalence of faults in air handling units, air terminal units, and rooftop units and the use of three metrics for summarizing results. This method was developed by the authors as part of a study which includes data from several automated fault detection and diagnostics (AFDD) data providers, providing a large sample with a wide range of building types, geographical locations, and equipment types. This dataset includes fault diagnoses from thousands of buildings throughout the United States, as well as anonymized metadata describing the building and equipment characteristics. The number of fault records is on the order of 10^6. We describe here how the data from different data providers can be processed and unified using a common taxonomy, and illustrate three metrics that can provide insights using this type of data. The methods developed for this study are illustrated here with preliminary data. This work supports a multi-year, multi-institutional project that will provide insight into the drivers of fault prevalence; for example, whether prevalence is correlated with characteristics like building type, building size, and geographical location (including related factors like local climate and utility rates). We discuss some of the challenges of harmonizing disparate outputs from multiple AFDD providers, the usefulness of applying a unifying fault taxonomy, and provide preliminary figures that illustrate three fault prevalence metrics.

1. INTRODUCTION

Commercial buildings consume approximately 18% of total energy and 37% of electrical energy in the United States (EIA, 2018). Heating, ventilation, and air-conditioning (HVAC) systems are one primary end use in these buildings. Unfortunately, these systems often operate far from their optimal efficiencies because of design, installation, and operational problems. HVAC faults, or deviation from the expected operating conditions of an HVAC system or component, can increase a building's energy consumption and operational costs; may prevent the building from receiving needed services for HVAC; may negatively affect other interconnected energy systems; and could increase equipment maintenance or replacement costs (Ebrahimifakhar *et al.*, 2020).

Automated fault detection and diagnosis (AFDD) tools use building automation system data to detect the presence of HVAC faults and support diagnosis of their root causes. Applying AFDD tools in commercial buildings and correction of the identified faults can save 9% of energy consumption (Kramer *et al.*, 2020). Faults in U.S. commercial buildings waste approximately 0.9–2.7 quads of energy annually (Frank *et al.*, 2019). However, this energy waste estimate is based on uncertain estimates of actual fault prevalence in the field. There is a lack of reliable data about which HVAC faults appear how frequently by building and system type. The purpose of this study is to fill the gap in the current state of knowledge about HVAC fault prevalence.

Researchers and AFDD providers have largely focused on evaluating AFDD performance building by building, and quantifying costs or other impacts. They often propose approaches that purport to improve the accuracy of fault detection, but by necessity will limit their investigations to simulated data (Li and O'Neill, 2019), a single building, or a small collection of buildings. A study exploring the use of automated methods for identifying "non-routine events" (possible faults) found success in streamlining measurement and verification processes, but recommended further work analyzing a larger set of buildings, including data from multiple real-world buildings and projects (Touzani *et al.*, 2019). However, no unified dataset has been published on the observed prevalence of faults that could inform future studies. An exploratory study (limited to 12 buildings) that informed the current project was the first of its kind to attempt to harmonize AFDD data from multiple buildings and identify the necessary steps and the barriers to doing so (Newman *et al.*, 2020). One key challenge was the lack of a common taxonomy across individual studies. This was addressed by Chen *et al.* (2020) presenting a standardized taxonomy for HVAC faults related to air handling unit (AHU), air terminal unit (ATU), and rooftop unit (RTU) systems, which is described in Section 2.2.

Several studies have been conducted for finding the frequency of faults in refrigeration and air conditioning systems. Stouppe and Lau (1989) examined 15,760 failure records occurring between 1980 and 1987 on different air conditioning systems by analyzing insurance claims. They found that in hermetic air conditioning systems 76.6% of faults were electrical, 18.9% of faults were mechanical, and 4.5% of faults were attributed to a malfunction in the refrigerant circuit. Breuker and Braun (1998) estimated the frequencies of occurrence and the service costs of different RTU faults by analysis of service records of a company from 1989 to 1995. They found that 60% of failures were electrical or control problems, while 40% of faults were mechanical. They also found that although compressor failures do not happen as frequently as other faults, they have the highest service costs in RTUs. Comstock *et al.* (2002) conducted a fault survey among four major American chiller manufacturers to identify the most common faults in chillers. They reported that most common faults happened in control box and starter sections. Refrigerant leakage was the second most commonly cited fault in chillers.

Felts and Bailey (2000) monitored and analyzed over 250 RTUs in northern California in various climate zones. This study showed that 40% of the RTUs were more than 25% oversized, and 10% of the RTUs were more than 50% oversized. It was also shown that economizers generally did not operate correctly. Downey and Proctor (2002) collected and analyzed performance data on over 13,000 air conditioners in residential and commercial buildings in California. Their analysis concluded that 57% of the units had improper refrigerant charge, and 21% of the units had low airflow rate through the indoor coil. Cowan (2004) investigated data from 503 RTUs at 181 commercial buildings sites in 5 states, gathered in four field studies. It was found that 46% of the units had improper refrigerant charge, 64% of the units had economizer problems, 42% of the units had airflow problems, 58% of the units had thermostat problems, and 20% of the units had sensor problems. Madani (2014) analyzed the fault reports provided to heat pump manufacturers and insurance companies in Sweden. This study showed that control and electronics faults are the most common and costliest faults in heat pump systems.

Yoshida *et al.* (1996) conducted a survey among HVAC experts in Japan to identify the ten most important faults in variable air volume (VAV) air handling systems based on their experience. The faults were ranked not only on frequency of occurrence, but also other factors such as environmental impacts, energy impacts, difficulty of detection, causing physical damage, and repair costs. The survey suggested that faults that occur in outdoor air damper and VAV box sections are fairly common. Qin and Wang (2005) conducted a site survey in a large commercial building in Hong Kong with 1,251 pressure independent VAV terminal units. Their investigation showed that zone temperature sensor error and local direct digital control error are the most common faults in VAV terminals. Gunay *et al.* (2019) developed a text-mining algorithm to extract information about fault frequency of HVAC systems from computerized maintenance management systems databases in Canada. Analyzing a central heating and cooling plant dataset showed that the average annual warning/failure rate was 4.5 for a chiller, while it was 6.5 for a boiler. From a building cluster dataset, they found that approximately 50% of the warning/failure events were related to room/zone/floor level systems.

Recently, Shoukas *et al.* (2020) analyzed the fault data collected from AFDD tools provided by four companies, representing over 28,000 RTUs, to determine the frequency of the reported faults. Since different companies use different formats, fault definitions, diagnostics, and reporting, they were not able to compare between AFDD tools, and results were presented separately for each data provider. They concluded that the frequency of the faults depends on the fault definitions and the diagnostics methods. They found that RTU faults occurred most commonly on economizer dampers, sensors, communications, and cooling systems.

This paper presents a method that is being developed to estimate the prevalence of faults in AHU, ATU, and RTU systems. It incorporates AFDD data from providers who monitor existing buildings and provide information on detected faults. The preliminary fault prevalence estimates presented in this paper are limited to commercial buildings in the U.S. whose AHU, ATU, and/or RTU systems are monitored by one of three AFDD providers. The faults are categorized using a standard taxonomy, and the results of a pilot study that provides preliminary illustrative values from analysis of a subset of data for estimated fault prevalence are presented here. Future work will analyze how the prevalence of specific faults is related to factors expected to affect the likelihood of observing faults such as building type, building size, climate, or utility costs, and will include a comparison with observed data garnered from manual inspection of individual buildings monitored by an AFDD provider. A key motivation in presenting the method in this forum is to dialogue with experts, in order to gain insight and new ideas for improving the study while it is in progress.

2. METHODOLOGY

The primary source of data for this study comes from commercial AFDD software tools. This is because commercial AFDD software outputs can be obtained for a large number of buildings and HVAC systems at a relatively low cost. Since AFDD software outputs are subject to error, i.e., they might have some level of false negative, false positive, and misdiagnosis rates, some of the AFDD software outputs will be verified in our future work using manual inspection of buildings.

2.1 Data Overview

The preliminary fault data received for this study is sourced from three commercial AFDD software tools. Data from at least four additional data providers will be added in the future. The study dataset includes at least twelve months of data for each building. The study includes three classes of system: AHU, ATU and RTU, and includes analysis of components of these systems, such as a supply air temperature sensor for an AHU. Table 1 shows the number of buildings, HVAC systems, and daily fault records for each of the data providers. A "daily fault record" constitutes the presence of a specific fault on a unique piece of equipment on a single day. A fault flagged multiple times in a single day constitutes one daily fault. For example, an RTU flagged with a stuck economizer damper fault every hour in 2019 would generate 365 daily fault records in the study dataset. During that same time period the same RTU could generate other daily fault records relating to other fault types. Tables 2 and 3 show that the sample space of data obtained from these three providers represents multiple building types and climate zones.

Table 1: AFDD data sources

Data Source	# of Buildings	# of AHUs	# of ATUs	# of RTUs	# of Daily Fault Records
Provider A	131	964	18,896	0	3,246,379
Provider B	131	0	0	2,174	2,944,853
Provider C	1103	0	0	5,843	348,911

Table 2: Number of buildings by building type

Building Type	Mercantile	Other	Office	Health Care	Food Service	Food Sales	Public Assembly	Service	Religious Worship	Education	Lodging	Warehouse and Storage
# of Buildings	840	159	82	82	66	57	36	12	11	8	7	5

Table 3: Number of buildings by Building America climate zone

Climate Zone	Cold	Marine	Mixed-Humid	Hot-Dry	Hot-Humid	Mixed-Dry	Very Cold
# of Buildings	381	328	254	238	159	3	2

Curating and analyzing data from a number of different sources is complicated by the diversity of data formats, fault naming conventions, and metadata and file structures that the AFDD software tools employ. The first, and most intensive, step is to prepare the data by cleaning and normalizing it by mapping it to a common fault taxonomy. Data preparation includes the following steps:

- Cleaning data to identify and resolve missing, mislabeled, empty fields, erroneous data, etc.
- Anonymizing data to ensure that any sensitive information that may identify buildings or partners is removed
- Normalizing data to a standard format using a common fault taxonomy

Fault data from each partner are converted to a standard format, which is called binary daily fault (BDF) data. Table 4 shows a sample of BDF data. HVAC fault prevalence metrics are calculated from the BDF data.

Table 4: Standard binary daily fault data

Fault record	Building ID	Equipment ID	Equipment type	Date	Fault name mapped
1	A0001	A-AHU00001	AHU	20190101	AHU-Heating-Coil_valve-Leakage
2	A0002	A-AHU00002	AHU	20190101	AHU-Cooling-Coil_valve-Stuck
3	B0001	B-ATU00001	ATU	20190101	ATU-Discharge_air-Damper-Stuck
4	B0002	B-ATU00002	ATU	20190103	ATU-Discharge_air-Airflow-Abnormal
5	C0001	C-RTU00001	RTU	20190102	RTU-Outside_air-Airflow-Abnormal
6	C0002	C-RTU00002	RTU	20190104	RTU-Mixed_air-Temperature_sensor-Frozen

2.2 Standardized Taxonomy for HVAC Faults

Each AFDD tool uses different fault names to refer to the same fault in an HVAC system. For example, in one commercial AFDD tool, an "economizer damper hunting" fault is reported to show a malfunctioning damper control, but in another tool, this fault may be reported as an "economizer damper short cycling" fault or an "unstable economizer damper" fault. Therefore, a unifying taxonomy for HVAC faults in AHUs, ATUs, and RTUs in commercial buildings was developed (Chen *et al.*, 2020). The developed fault taxonomy contains 134, 39, and 115 unique fault names for AHUs, ATUs, and RTUs, respectively. Mapping functions were created for each AFDD tool to convert their fault reports to this unifying taxonomy. Table 5 shows a selection of some of the HVAC faults in the taxonomy.

There are three different fault categories based on how the faults are presented: condition-based, behavior-based, and outcome-based (Frank *et al.*, 2019). Condition-based faults are improper or undesired physical conditions in HVAC systems such as stuck dampers, leaky valves, and biased sensors. Behavior-based faults present improper or undesired behavior during the operation of HVAC systems. Examples of behavior-based faults are economizer damper hunting, and simultaneous heating and cooling. Outcome-based faults are states in which an outcome or performance of the HVAC systems deviates from expected values, such as excessive energy consumption or insufficient ventilation rate. The HVAC fault taxonomy applied in the current project only includes condition-based and behavior-based faults, since they are most commonly used in AFDD software tools.

An important feature of the taxonomy is that it supports flexible analysis based upon multiple levels of equipment class. For example, prevalence can be calculated for specific faults related to RTU supply air temperature sensors, supply air temperature sensors in general, temperature sensors in general, or sensors in general. Similarly, prevalence could be calculated for all heating faults, all damper faults, all stuck damper faults, and so on.

Table 5: Example list of the HVAC faults in the developed taxonomy

Equipment	Component	Fault Name	Fault ID	Fault Type*
AHU	Air economizer	Economizer damper hunting	AHU-Economizer-Damper_control-Hunting	BB
	Cooling coil valve	Cooling coil valve stuck	AHU-Cooling-Coil_valve-Stuck	CB
	Outside air temperature sensor	Outside air temperature sensor bias	AHU-Outside_air-Temperature_sensor-Bias	CB
ATU	Reheat coil valve	Reheat coil valve leakage	ATU-Reheat-Coil_valve-Leakage	CB
	Discharge air damper	Discharge air damper hunting	ATU-Discharge_air-Damper_control-Hunting	BB
	Discharge air temperature sensor	Discharge air temperature sensor drift	ATU-Discharge_air-Temperature_sensor-Drift	CB
RTU	Air economizer	Economizer damper stuck	RTU-Economizer-Damper-Stuck	CB
	Supply air temperature sensor	Supply air temperature sensor frozen	RTU-Supply_air-Temperature_sensor-Frozen	CB
	Compressor	Compressor short cycling	RTU-Compressor-Unassigned-Short_cycling	BB

*BB=Behavior-based, CC=Condition-based

2.3 Metric Definitions

There are many different ways to express fault prevalence. To determine the priority HVAC fault prevalence metrics to be calculated in this study, we identified several questions that we expect to be of most interest to the study's target audience of AFDD providers, users, regulators, and researchers. These questions include:

1. What percentage of units are observed to be faulted at any given point in time?
2. Which faults are most often observed to be present?
3. How many faults are observed to be present each month for a given building?

To quantitatively characterize the HVAC fault prevalence, the following metrics are defined.

2.3.1 Metric 1 (Monthly Fault Presence)

This metric gives the percentage of equipment that experiences the presence of fault type 'x' on one or more days, for each month of the year, and is expressed as a percentage of all equipment. For a given piece of equipment, if fault 'x' is present for at least one day in a given month, that month is denoted as a "1" binary value, and considered one "fault_month". If the fault is observed to be present in multiple years for a given piece of equipment (e.g., present in February 2018 and in February 2019), each case will be considered a distinct value for this metric (e.g., February 2018 = 1, and February 2019 =1, a total of two "fault_months" for February).

This metric is calculated by:

$$monthly\_fault\_presence = \frac{fault\_months}{equipment\_months} \quad (1)$$

where $fault\_months$ is the accumulated number of monthly fault occurrences for one type of fault in a calendar month across different years, and $equipment\_months$ is the number of monitored pieces of a specific type of equipment in one calendar month, or in a calendar month over a range of years. For example, if 100 dampers are monitored for two full years, the damper $equipment\_months$ count for June would be 200.

The $fault\_months$ is calculated by:

$$fault\_months = \sum_{i=1}^{num\_calender\_year} \sum_{j=1}^{equipment\_months} Fault\_OC_{month} \quad (2)$$

where $Fault\_OC_{month}$ is the monthly fault occurrence. If there is at least one fault record in the AFDD report within the month, then $Fault\_OC_{month} = 1$. The $num\_calender\_year$ is the number of all years that may cover the time range of interest (e.g., the month of January appears in our dataset for a piece of equipment across two years, hence $num\_calender\_year$ would equal 2.

Figure 1 illustrates the calculation of $Fault\_OC_{month}$ and $fault\_months$ under a selected time period. There are three AHUs, each monitored for two years. In January three out of six pieces of equipment had a fault flagged at least once during the month (so that $Fault\_OC_{month} = 1$ for these three), hence there is a total of three $fault\_months$ for January of six possible. This represents a monthly fault presence of 50 percent for January.

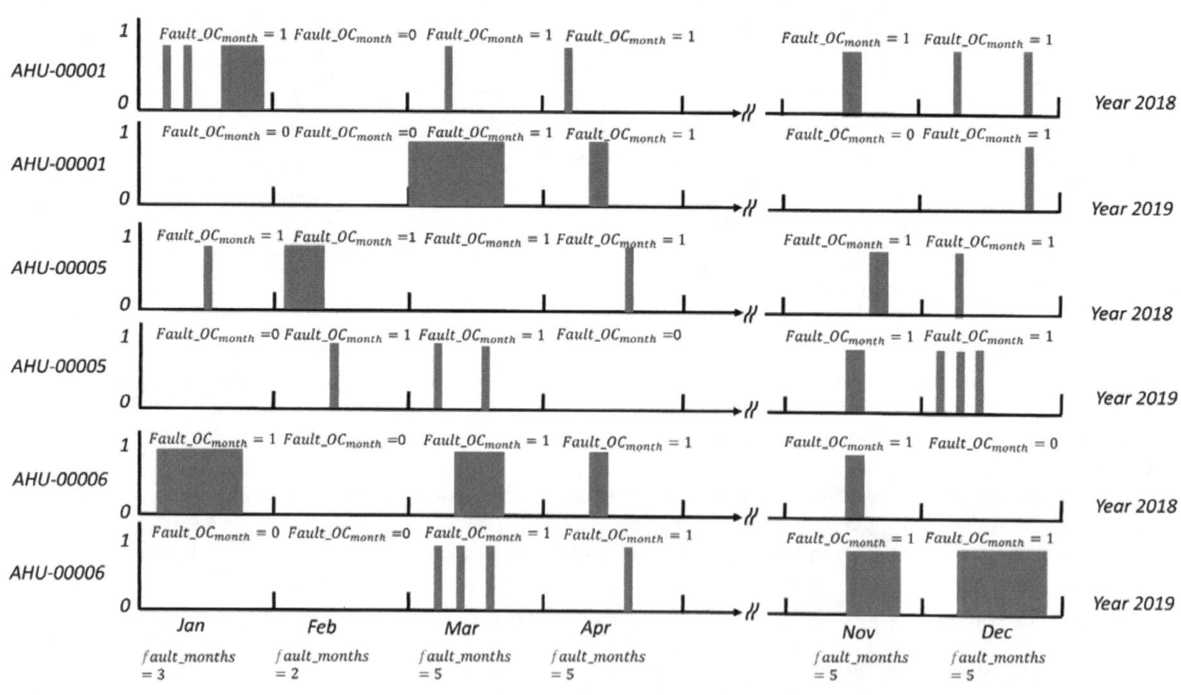

Figure 1: Graphical depiction of Metric 1 (Monthly Fault Presence)

2.3.2 Metric 2 (Average Monthly Fault Presence)
Metric 2 is closely related to Metric 1, and shows the percentage of equipment that experiences the presence of a given fault type on one or more days in a month, averaged across all months (whereas Metric 1 presents a different fault presence value for each month). This metric shows which fault types are most often present in the data.

2.3.3 Metric 3 (Mean Number of Faults per Building per Month)
This metric shows how many faults are observed to be present (at the building level) each month, among the set of faults considered in this study. The calculation steps of this metric are:

1. Establish total unique faults for each month, for one building
2. Calculate mean value across all months for that building
3. Repeat for all buildings
4. Calculate mean of all building-specific mean values

3. PRELIMINARY RESULTS AND DISCUSSION

A total of 6,540,143 daily fault records of AHUs, ATUs, and RTUs were analyzed from the three AFDD providers. Values for metrics 1 to 3 have been generated. There are several questions that will be explored by further analysis of the results in future work. For example: does fault prevalence vary with climate zone (perhaps correlated to energy costs, for example)? Does fault prevalence vary with building type, season, building size, building type, or other factors? Do different AFDD providers detection rates vary significantly? It is possible that the sample size will be too small, in many cases, to provide statistically significant answers to these types of questions.

Figures 2a and 2b show the monthly fault presence (Metric 1) for two AHU fault types: "heating coil valve leakage" (condition-based) and "outside air damper position abnormal" (behavior-based). The analysis will explore, for example, whether the apparent seasonal trend in Figure 2a does represent a genuine trend. There could be competing factors driving this trend: reduced usage of heating systems in summer; and difficulty diagnosing a leak in winter (when there *should* be flow much of the time). The fault in Figure 2b has higher overall rates, and also is likely to be correlated to season, but there is less apparent seasonal trend. Interestingly, the shoulder months for both fault types have the highest prevalence values. These preliminary results illustrate the potential use of this metric and further analysis to be done.

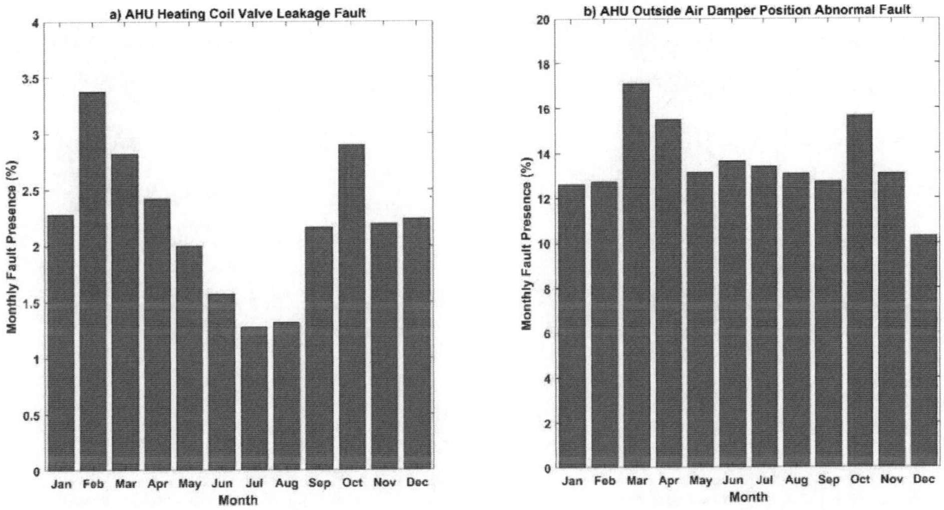

Figure 2: Monthly fault presence (Metric 1) for two AHU faults (preliminary illustrative result)

Figures 3a and 3b show the monthly fault presence (Metric 1) for two different ATU faults. "Discharge air damper stuck" is a condition-based fault and "reheat coil valve hunting" is behavior-based. Figure 3a shows that somewhat fewer ATU dampers were diagnosed as being stuck in summer months, but with a range of 0.5% to 2.5%, these differences from month to month may not be significant. 8% to 9% of reheat coil valves were diagnosed to be hunting each month, but with no obvious seasonal trend.

Figures 4a and 4b show the 10 most common AHU and ATU faults. These faults are selected out of 34 AHU faults and 13 ATU faults that were successfully mapped to the fault taxonomy. Average monthly fault presence is a useful way to sort the relative prevalence of all individual fault types, and can also help in understanding the most problematic system components (e.g., dampers, sensors) or functional elements (e.g., cooling, heating). Five AHU and six ATU faults in the 10 most common faults lists are behavior-based.

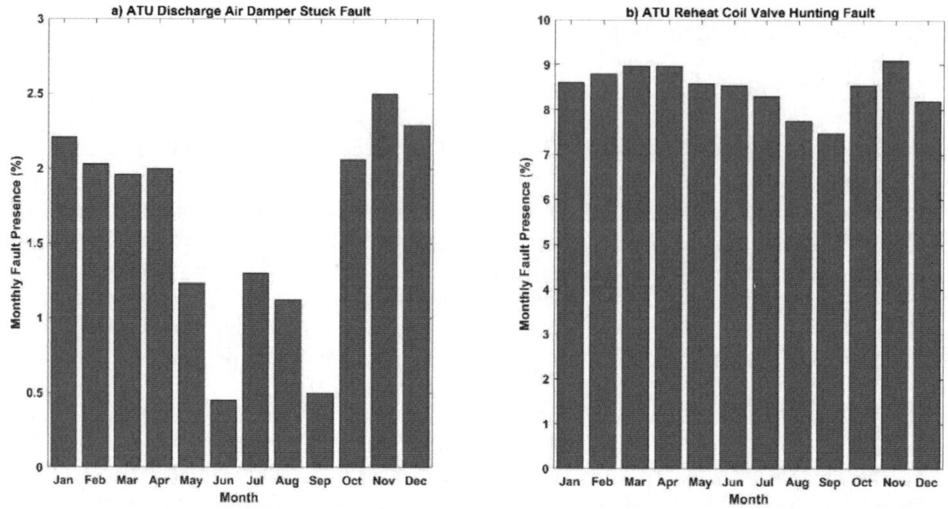

Figure 3: Monthly fault presence (Metric 1) for two ATU faults (preliminary illustrative result)

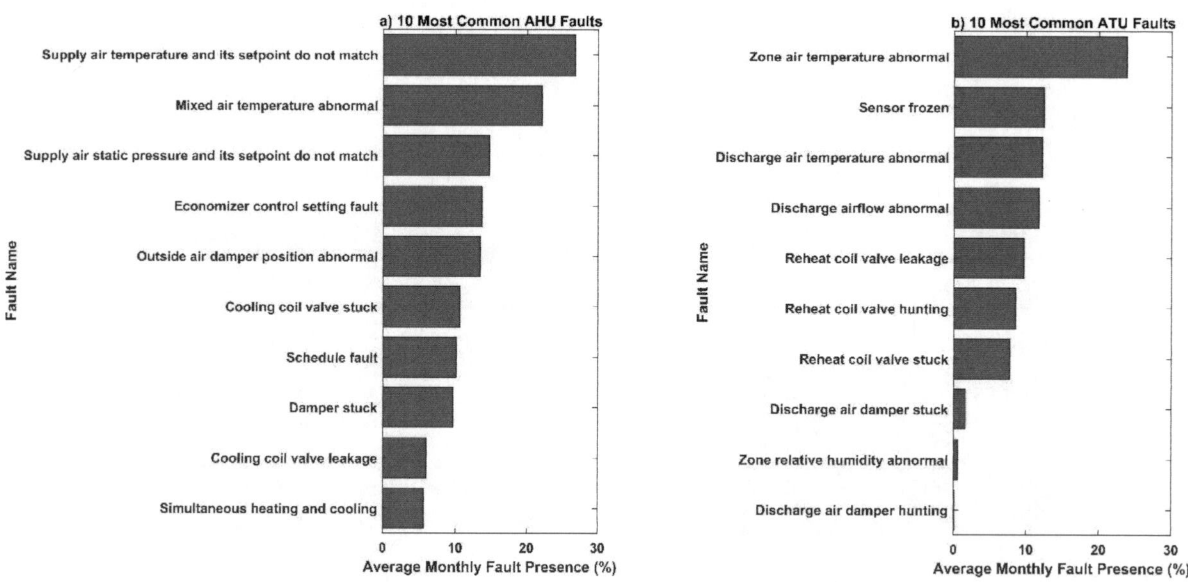

Figure 4: Average monthly fault presence (Metric 2) for 10 most common AHU (left) and ATU (right) faults

Figure 5 shows the distribution of the mean number of faults per building per month. As can be seen, 77.4% of the buildings were in the range of 0-50 faults per month, 10.8% were in the range of 50-100, 5.6% were in the range of 100-150, 2.0% were in the range of 150-200, and 4.2% had higher than 200 faults per month. It should be noted that the number of faults in each building includes all the AHU, ATU, and RTU faults. As we expected, buildings with higher quantities of equipment had higher quantities of faults. One health care (inpatient) building in a hot-dry climate zone with 38 AHUs and 834 ATUs had 1071 faults per month which was the highest number among all the buildings. This is an example of a metric where it could make more sense to normalize, and the study will consider normalizing factors such as the number of pieces of equipment in the building and the number of fault detection rules programmed into the AFDD tool, or other related factors affecting the number of reported faults.

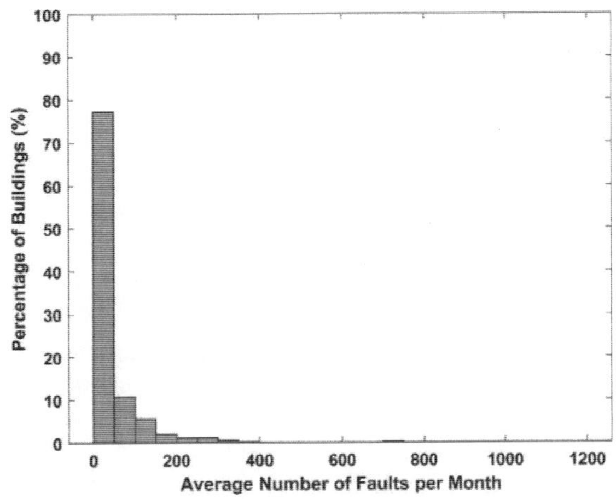

Figure 5: Mean number of faults per building per month distribution

3.1 Ongoing Challenges and Questions

A key element of the work in this project is development of fault prevalence concepts and systematic methods for quantifying and communicating prevalence. To illustrate: one challenge in fault prevalence is that for a given condition-based fault, a number of behavioral symptoms could arise. Conversely, a behavior-based fault may arise from multiple condition-based faults. For example, behavior-based faults, such as "supply air temperature and its setpoint do not match" and "simultaneous heating and cooling" may each be caused by the "cooling coil valve stuck" condition-based fault. This study acquires data generated by several AFDD software tools, that contain a mixture of condition-based and behavior-based faults. Therefore, it is important to address any potential overlap or duplication with a well-designed taxonomy and careful mapping to this taxonomy from the AFDD.

Another question concerns the relationship between fault presence and fault detection. In a temporal sense, a fault that is not addressed could be flagged by an AFDD tool intermittently over time. A stuck economizer damper fault might be undiagnosable with some diagnostic approaches when the system is not calling for economizing; a valve leak may only be detected when a threshold of flow leakage is exceeded; etc. AFDD approaches also may deliberately or inadvertently miss faults that have a small severity. The planned field verification portion of the project is one approach that will help to address this challenge, but other data-driven approaches also will be needed.

Preliminary data review and analysis is proving insightful as we gain a more granular understanding of HVAC fault prevalence. The study team is also working to address many data-related challenges, with one in particular being the interpretation of fault absence. For instance, the absence of a fault record may indicate fault-free operation but could potentially be due to the lack of a specific component type (e.g., an AHU without a heating coil cannot have a heating coil valve leakage fault), or an AFDD software tool was not programmed to identify that fault. Each AFDD dataset is being validated separately, in collaboration with the data provider, to address these types of issues.

4. CONCLUSIONS

The results presented here are preliminary and illustrate how fault prevalence can be assessed in various dimensions. These results are not intended to indicate the final representative fault prevalence for any specific fault type. The team continues to add new data, quality check the data, and validate the data in partnership with the AFDD providers. The process for unifying data from disparate AFDD tools is labor-intensive. The authors have developed a method for mapping this data to a common taxonomy of HVAC faults that facilitates unification and comparison of the data and has the ability to allow related faults to be aggregated together.

Preliminary metrics have also been developed to provide information related to specific questions that we believe that the study's audience will find useful. As the study proceeds, we will analyze data by fault type, by building type, and by HVAC system type. We will gather additional data, including data from new data providers; perform statistical analyses to assess national representativeness, precision and confidence, and drivers of prevalence; and validate a subset of results using field study data from manual site inspections. We are also implementing additional metrics that can provide new insights about the data.

This study will conclude in 2022 and is on track to generate the largest empirical study to date on HVAC fault occurrence rates in existing commercial buildings. Future work could potentially apply the same methods to other equipment types (e.g., chillers); evaluate the persistence of fault resolution; analyze longer term fault trends; and assess the relationship between fault rates, false positives and negatives, and the use of different fault detection algorithms.

REFERENCES

Breuker, M. S., & Braun, J. E. (1998). Common faults and their impacts for rooftop air conditioners. *HVAC and R Research, 4*(3), 303–318. https://doi.org/10.1080/10789669.1998.10391406

Chen, Y., Crowe, E., Lin, G., & Granderson, J. (2020). What's in a name? Developing a standardized taxonomy for HVAC system faults. *Lawrence Berkeley National Laboratory.*

Comstock, M. C., Braun, J. E., & Groll, E. A. (2002). A survey of common faults for chillers. *ASHRAE Transactions, 108*, 819-825.

Cowan, A. (2004). Review of recent commercial roof top unit field studies in the Pacific Northwest and California. *Northwest Power and Conservation Council and Regional Technical Forum.*

Downey, T., & Proctor, J. (2002). What can 13,000 air conditioners tell us? *Proceedings of the 2002 ACEEE Summer Study on Energy Efficiency in Buildings 1*, 53–67.

Ebrahimifakhar, A., Kabirikopaei, A., & Yuill, D. (2020). Data-driven fault detection and diagnosis for packaged rooftop units using statistical machine learning classification methods. *Energy and Buildings, 225*, 110318.

EIA (U.S. Energy Information Administration) (2018). Commercial Buildings Energy Consumption Survey. https://www.eia.gov/consumption/commercial/.

Felts, D., & Bailey, P. (2000). The state of affairs - packaged cooling equipment in California. *Proceedings of the 2000 ACEEE Summer Study on Energy Efficiency in Buildings, 3*, 137-147.

Frank, S., Lin, G., Jin, X., Singla, R., Farthing, A., Zhang, L., & Granderson, J. (2019). Metrics and methods to assess building fault detection and diagnosis tools. *National Renewable Energy Laboratory.*

Gunay, H. B., Shen, W., & Yang, C. (2019). Text-mining building maintenance work orders for component fault frequency. *Building Research & Information, 47*(5), 518–533.

Kramer, H., Lin, G., Curtin, C., Crowe, E., & Granderson, J. (2020). Proving the business case for building analytics. *Lawrence Berkeley National Laboratory.* http://dx.doi.org/10.20357/B7G022

Li, Y., & O'Neill, Z. (2019). An innovative fault impact analysis framework for enhancing building operations. *Energy and Buildings, 199*, 311–331. https://doi.org/10.1016/j.enbuild.2019.07.011

Madani, H. (2014). The common and costly faults in heat pump systems. *Energy Procedia, 61*, 1803–1806.

Newman, S., Lerond, J., Reeve, H., Vrabie, D., Belew, S., & Tucker, J. (2020). Pilot study for determining HVAC fault prevalence from fault monitoring data. *ACEEE 2020 Summer Study on Energy Efficiency in Buildings.*

Qin, J., & Wang, S. (2005). A fault detection and diagnosis strategy of VAV air-conditioning systems for improved energy and control performances. *Energy and Buildings, 37*(10), 1035–1048.

Shoukas, G., Bianchi, M., & Deru, M. (2020). Analysis of fault data collected from automated fault detection and diagnostic products for packaged rooftop units. United States. doi:10.2172/1660228

Stouppe, D. E., & Lau, Y. S. (1989). Air conditioning and refrigeration equipment failures. *National Engineer, 93*(9), 14–17.

Touzani, S., Ravache, B., Crowe, E., & Granderson, J. (2019). Statistical change detection of building energy consumption: Applications to savings estimation. *Energy and Buildings, 185*, 123–136. https://doi.org/10.1016/j.enbuild.2018.12.020

Yoshida, H., Iwami, T., Yuzawa, H., & Suzuki. M. (1996). Typical faults of air-conditioning systems, and fault detection by ARX model and extended Kalman filter. *ASHRAE Transactions 102*(1), 557-564.

ACKNOWLEDGEMENT

This work was supported by the Assistant Secretary for Energy Efficiency and Renewable Energy, Building Technologies Office, of the U.S. Department of Energy under Contract No. DE-AC02-05CH11231. The authors would like to thank Erika Gupta from the U.S. Department of Energy for her support, as well as the AFDD providers who contributed data to this study.

Simulation-based Energy and Daylighting Impact Assessment of Integrated Shading Devices and Lighting Controls in Commercial Buildings

Niraj Kunwar, Mahabir Bhandari*

Oak Ridge National Laboratory, Buildings and Transportation Science Division,
Oak Ridge, TN 37831, USA

* Correspondence: bhandarims@ornl.gov; Tel.: +1-865-574-0989; Fax: +1-865-574-3851

ABSTRACT

Fenestration systems are estimated to be responsible for 1,790 PJ of primary energy consumption annually in commercial buildings in the US and have significant impact on occupants' visual and thermal comfort. Window shading devices such as shades and blinds are cost-effective means to reduce energy losses/gains from fenestration systems and provide a comfortable environment in the buildings. Besides being cost effective, these products are less intrusive to install when retrofitting the buildings. To maximize the benefits, however, these products require integrated control strategies. Due to complex modeling process involved, existing control strategies used by shade automation industries, generally are not evaluated for energy and daylighting benefits. This paper presents the impacts of shading devices on various aspects of built environment using a comprehensive simulation study. The simulation is performed using a framework enabling automation of daylight and energy simulation and their interaction as well as result post-processing. The results from the simulation show that the control strategies enabled annual cooling and lighting energy savings of up to 40% and 25% respectively. The impact of different shading types, controls and climate conditions on the energy savings results is analyzed. The results of the impact of this technology on energy consumption, daylighting, and occupant comfort from this study will help building owners, designers, engineers, and utilities make informed decisions.

1 INTRODUCTION

Fenestrations in the United States attributes to 1,790 PJ of primary energy consumption in commercial buildings from heating and cooling loads (Sawyer, 2014). Generally, fenestrations have lower thermal performance compared to the opaque envelope. In addition, unlike the opaque envelope, windows also transmit daylight/solar radiation into the building and thus have significant impact on occupant's visual and thermal comfort (Kunwar et al., 2018). Solar gain from the windows is one of the most significant cooling loads in the buildings. While, high efficiency windows can control some of these parameters such as thermal transmittance and solar heat gain, the window retrofits might be disruptive and not be economically viable option (Ariosto & Memari, 2013). Also, even highly efficient windows such as electrochromic glazing are not able to control glare without use of shading devices (Fernandes et al., 2013). Window attachments/shading devices can achieve multiple objectives of energy savings, daylight provision and occupant comfort. However, the effectiveness of these systems depends on their properties and the strategies used to control. Most of the shading devices used currently are manually controlled and hence are not optimal in terms of achieving energy savings and daylight provisions in the space. Therefore, significant research efforts have been directed towards different types of shading devices and their automation in the past two decades.

The studies on shading devices has been carried out both using simulation (Atzeri et al., 2018) and experimental studies (Kunwar et al., 2019). These studies have evaluated the energy savings potential and impact on daylighting and glare that occurs from shading control. Experimental testing was performed using a low cost camera to control the shading device to meet visual comfort requirements (Goovaerts et al., 2017). However, the qualification of the impact of general shading control strategies that are utilized by shade automation industries at present is missing. The quantification of impact of automated shading devices on various aspects of built environment is essential to enable adoption of such technologies in US building stock.

For addressing this gap of studies on general shading control strategies in office buildings, a comprehensive study of different shading devices and shading controls was performed in (Kunwar & Bhandari, 2020) by the authors. The paper modeled shading control strategies used by three shade automation industries and evaluated their impact in medium office prototype buildings (The U.S. Department of Energy (DOE), n.d.). In this paper, the analysis of the shading controls will be extended to 4 different types of buildings including medium office, secondary school, midrise apartment, and large hotel for new buildings vintage of 2016 (The U.S. Department of Energy (DOE), n.d.). Here, the medium office buildings used has a different window to wall ratio (WWR) than that used in (Kunwar & Bhandari, 2020). The control strategy for shading devices were chosen considering building type, occupancy, and the primary purpose of building. Combined occupancy and daylighting-based controls were used in medium office and secondary school buildings. While in midrise apartment and large hotel, the shading control was used only with occupancy-based lighting controls. In the next section, different control strategies used in different building types and their result in six different location in the US is presented. The conclusions based on the results and their analysis is presented in the subsequent section.

2 METHODOLOGY AND RESULTS

This section will first provide an overview of buildings and shading devices that were used for the simulations, Then, later for each building types, the different shading control strategies used and the results from their applications are provided in the sub-sections 2.1 to 2.4. The details of the 4 buildings that were used for the simulation is provided in Table 1. Among these buildings, both energy and daylight simulations were performed for the medium office building while for other three types of buildings only energy simulations were performed. Custom scripts were used to implement the different control strategies based on exterior illuminance by using the results from the daylight simulations, while some schedule-based control strategies were directly implemented in energy simulation. The shading and lighting state after implementation of control strategies were saved as schedules and used as text-based input into energy simulation. This framework of simulation has also been discussed in (Kunwar & Bhandari, 2020). Control strategies suitable to the different building types were used in the respective buildings as all shading control strategies are not suitable for all building types due to varying nature of occupancy and activities. The simulations were performed in six different climate zones/locations i.e., Houston (2A), Los Angeles (3B-CA), Washington DC(4A), Seattle (4C), Chicago (5A) and Minneapolis (6A).

Table 1: US DOE prototype building characteristics

Building type	Building area (m^2)	No. of floors	WWR	Heating, ventilation, and air-conditioning system	Building image
Medium office	4,982	3	48	Heating: Gas furnace and electric reheat Cooling: Packaged air-conditioning units	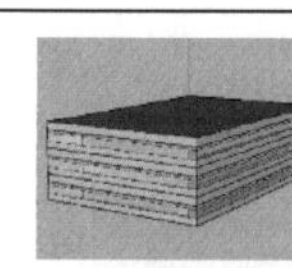
Midrise apartment	3,135	6	20	Heating: Gas furnace Cooling: Split system direct expansion	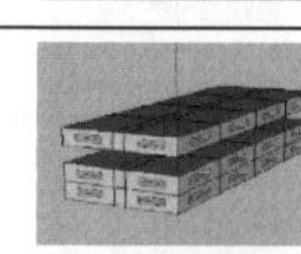
Large hotel	11,345	4	30	Heating: Gas-fired boiler Cooling: Air-cooled chiller	
Secondary School	19,592	2	35	Heating: Gas-fired boiler Cooling: Air cooled chiller	

Different shading devices used in these buildings are provided in Table 2. The properties of shading devices and secondary glazing are provided in the table along with the building type in which each of the shading devices are used for the simulation.

Table 2: Properties of shading devices and secondary glazing

Shade	Other Properties	Location	k (W/m-K)	VT (F/B)	T_{sol} (F/B)	R_{sol} (F/B)	Building types
Venetian blind	24 mm wide, 1.5 mm rise	interior	160	0.00	0.00	0.68	medium office, midrise apartment, secondary school
Roller shades A	light color, 1% OF	interior or exterior	0.3	0.12	0.18	0.74	medium office, midrise apartment, secondary school
Roller shades B	dark colored, 1 % OF	interior or exterior	0.12	0.00	0.00	0.04	large hotel
Secondary glazing (SG)	low e	interior	1.00	0.88	0.74	0.10	all

2.1 Medium office:

The medium office building used here is with increased WWR, i.e., from 33% of the US DOE medium office prototype building to 48% to cover the modern glass buildings with higher WWR. The control strategies in this case are the same control strategies which were used in (Kunwar & Bhandari, 2020) as provided below.

Baseline (B): No shading, no lighting control
Lighting control (LC): No shading, lighting dimming
Manual control (MC): Shading control based on different user type
Automated Control 1 (AC1): Roller shades closed to different heights based on exterior illuminance

- Exterior vertical illuminance > 53000 lux: Fully closed
- Exterior vertical illuminance <4300 lux: Fully opened
- Else shade deployed to at least 50% or greater height to prevent solar penetration depth of ~5 ft

Automated Control 2 (AC2): Roller shades open or closed based on cooling status or glare

- Daylight glare index >22 or zone in cooling mode: Fully closed
- Else: Fully opened

Automated Control 3 (AC3): Roller shades closed or opened based on exterior illuminance

- Exterior vertical illuminance > 20000 lux: Fully closed
- Exterior vertical illuminance <15000 lux: Fully opened

Automated Control 4 (AC4): Roller shades closed to different heights based on exterior illuminance

- Exterior vertical illuminance < 10000 lux: Fully opened
- Exterior vertical illuminance >50000 lux: 75% closed
- Exterior vertical illuminance > 30000 lux: 50% closed
- Exterior vertical illuminance >15000 lux: 25% closed

Secondary glazing (SG): Fixed secondary glazing
Venetian blinds (VC): Venetian blind controlled to block beam solar

- Blinds rotated to prevent direct sunlight from entering the space when external vertical irradiation on the façade is greater than 150 W/m^2

All the automated controls (AC1-AC4) and manual control (MC) used Roller shades A as shading device. Also, except the baseline case all the other controls strategies utilized lighting dimming controls to maintain a setpoint illuminance of 500 lux. The lights could be dimmed up to 30% of full power of the lighting device. The results from the application of the control strategies listed was then evaluated for energy savings and daylighting. The results for energy consumption for heating, cooling and lighting and the total energy savings from control strategies compared to B is shown in Figure 1. These results are provided for all the control strategies listed above for six different locations of

US. Here, the total energy savings refers to the sum of energy savings for heating, cooling and lighting. The total energy savings in the medium office buildings was in the range of 0.8 to 17.7 kWh/m^2-yr. From the figure it can be seen that higher energy savings is realized from use of roller shades (AC1-AC4, MC) and venetian blinds (VC) in cooling dominated climate of Houston and Los Angeles compared to heating dominated climate. The use of SG was able to achieve significant energy savings in all the climate zones. Among the automated strategies AC3 that uses exterior shades was able to achieve higher energy savings compared to interior shades. In all the climate zones, significant energy savings i.e., average of nearly 5 kWh/m^2-yr was from lighting control and reduction in lighting energy savings. The variation in energy savings was less for lighting energy consumption compared to cooling or heating energy across different climate zones as well as control strategies. The average lighting energy savings was approximately 25% which was lower than 50-75% savings from an experimental study (Shen & Tzempelikos, 2017). This difference comes from our assumption that the lowest dimming level for lighting was 30% of total lighting power, while in (Shen & Tzempelikos, 2017) the dimming capability was up to 1% of total power.

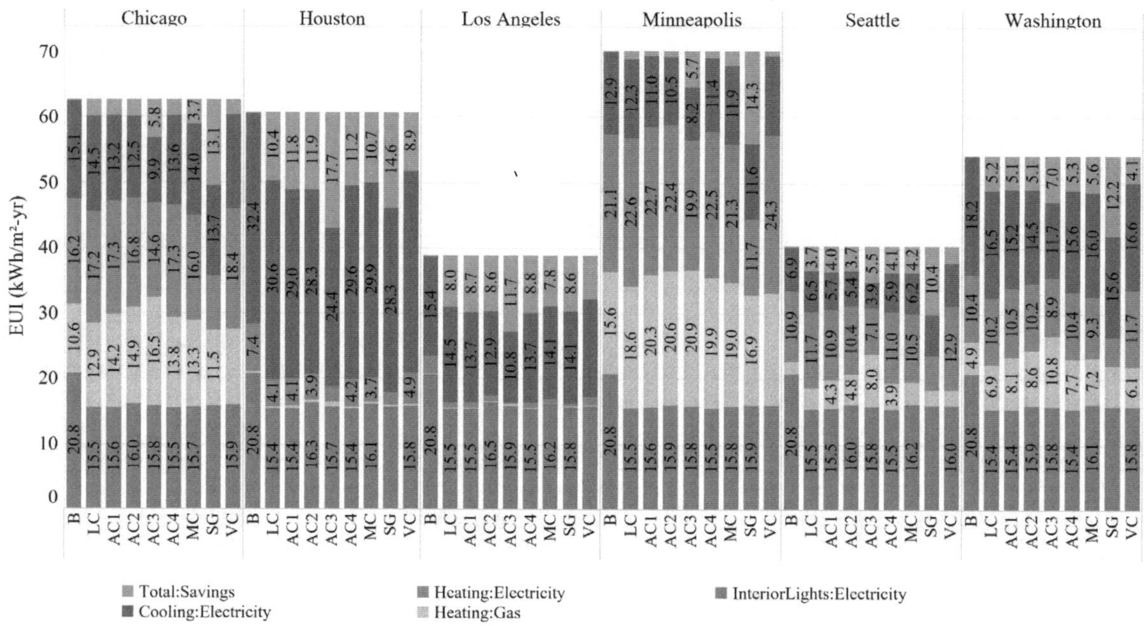

Figure 1: Energy consumption and savings in medium office building

The results for daylighting are provided for the four automated control strategies (AC1-AC4) and baseline (B) case. The results are provided in terms of spatial daylight autonomy (sDA) and annual sunlight exposure (ASE) (IES Daylight Metrics Committee, 2012) calculated based on grid setup of work plane illuminance sensors shown in Figure 2. The results for $sDA_{300,50}$ i.e. percentage of floor area with illuminance greater than 300 more than 50% of the occupied hours and ASE i.e. percentage of floor area which receives direct illuminance greater than 1000x for more than 250 occupied hours are provided in Table 3. The results in the table show that generally from the use of shading device there is decrease in excessive illuminance that could cause visual discomfort reflected by reduction in ASE. There is also reduction in useful amount of daylight that could be utilized shown by reduction in sDA. AC2 which includes cooling status in the control strategy shows more variation in daylight across the location compared to other control strategies. While, using AC2 higher sDA and ASE is there in cold climate of Chicago and Minneapolis compared to other locations.

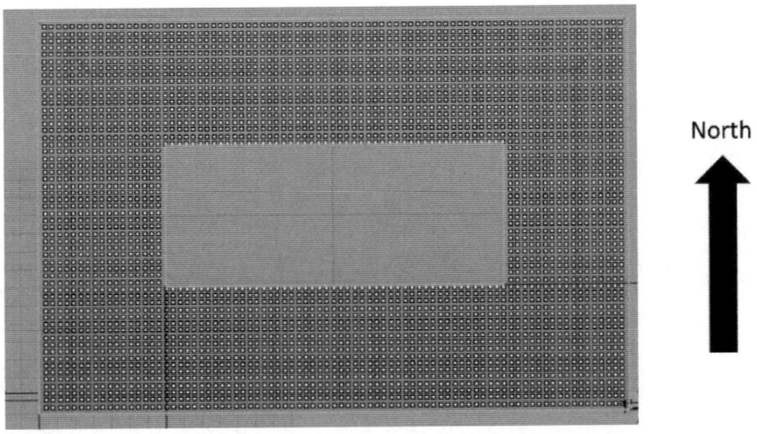

Figure 2: Grid setup for daylighting simulation in medium office building

Table 3: Daylighting results for medium office building using sDA and ASE

	ASE						$sDA_{300,50}$				
	B	AC1	AC2	AC3	AC4		B	AC1	AC2	AC3	AC4
Chicago	41	13	26	13	14		89	43	42	65	67
Houston	34	9	3	14	10		92	44	24	66	69
Los Angeles	38	13	7	10	15		93	46	21	67	69
Minneapolis	46	14	31	11	17		89	44	47	66	67
Seattle	40	9	7	11	14		83	40	30	62	65
Washington	35	11	19	12	14		91	45	39	66	68

2.2 Secondary School:

Secondary school has similar nature of occupancy as office building. Also, school building can utilize daylight and needs to ensure visual comfort for occupants' well-being and productivity like in the case of medium office. Considering this, the same shading devices and control strategies that were used for medium office were also used for secondary school. The results for the energy savings in secondary school from the use of those control strategies is provided in Figure 2. In case of the secondary school, the energy savings is in the range of 1.4 to 12.2 kWh/m^2-yr. There is less variation in energy savings across different locations compared to medium office buildings. However, the energy savings is higher from use of SG in secondary school like medium office buildings.

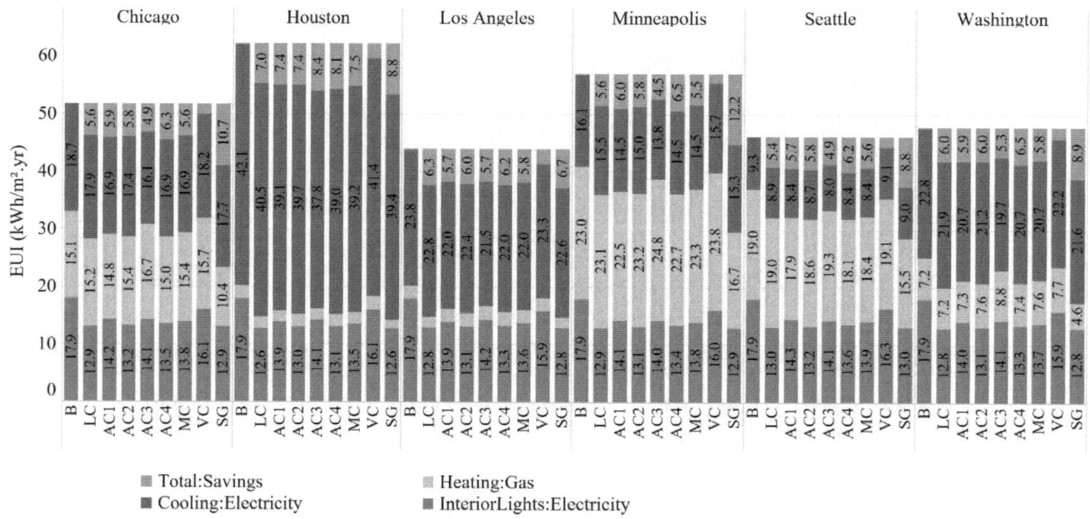

Figure 3: Energy consumption and savings in secondary school

2.3 Midrise apartment:

In residential buildings, the shading devices are mostly closed during the night-time for privacy reasons. Based on this control strategy "AC3" used for medium office was used in this building but the roller shades were closed completely after 6:00 pm in the evening until 8:00 am in the morning. Another strategy using same control strategy but using roller shades in the interior of the window "AC3-I" was tested in midrise apartment. Apart from this one case with secondary glazing "SG" and other two control strategy based on average schedule of shading state on residential building considering different time of day, day of week and season was applied for roller shades and venetian blinds namely "AER" and "AEV" respectively. This schedule was developed by (Peng & Curcija, 2017) based on a behavioral study of window and window attachments (DRI, 2013).Thus, not including the baseline case "B", five other scenarios of shading devices was tested. Since residential building are less likely to have lighting control and utilize daylighting, this building type was only evaluated for energy performance of shading device without using any lighting control. The results for energy consumption for midrise apartment are shown in Figure 4.

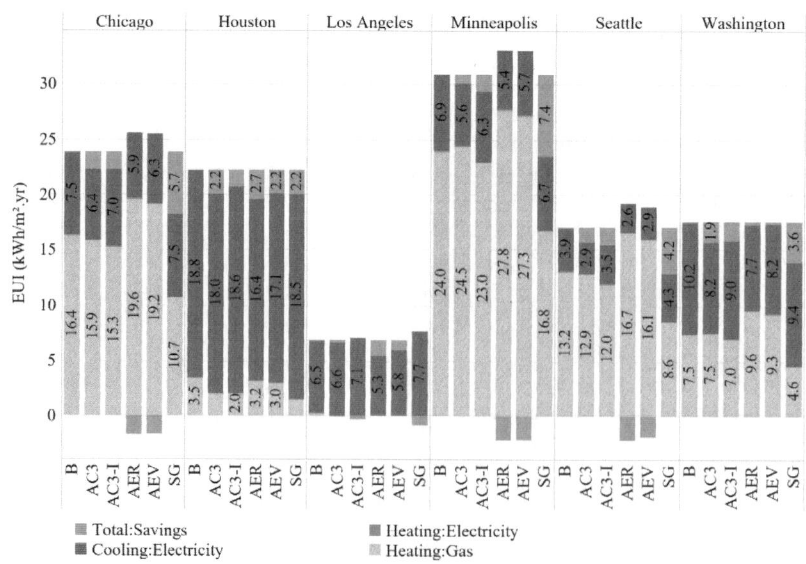

Figure 4: Energy consumption and savings in midrise apartment

As shown in the figure, for majority of the cases the shading control has contributed to energy savings while some of the cases also faced energy penalty while using shading device in mid-rise apartment without any lighting controls. The range of energy savings/penalty was -1.7 to 7.4 kWh/m^2-yr. Mostly the penalty is seen for control algorithms AER and AEV which provides the average representation of current operation of blinds by users in an area. Note that these two control algorithms cannot be directly applied in a household but can be used as average manual control and comparison with the automated shading control. In general, the energy performance of both AC3 and AC3-I was better than AER and AEV which represented weighted energy performance of manually operated roller shades and venetian blinds respectively. This also implies that the use of automated roller shades has a potential to reduce energy consumption in mid-rise apartment even without considering the influence on daylight, visual comfort, etc. compared to manually controlled shading devices.

2.4 Large hotel:

This is another building type where automated shading solution was utilized, and their performance was evaluated for energy consumption. Similar to residential building, large hotel was considered to have their shading devices closed during the nighttime after 6:00 pm until 8:00 am in the morning for all the control strategies. For the purpose of simplification one control strategy had shading devices always ON and another had shading devices ON during night (ON meaning closed) and open during the day. For each of these control, one case with interior shading device and other with exterior shading device was simulated in the large hotel building thus making up four different control strategies:

AOE- Always ON with exterior roller shades
AOI- Always ON with interior roller shades
NOE- ON during night with exterior roller shades
NOI- ON during night with interior roller shades

For all the four control strategies Roller shades B was used as the shading device. Other than these four control strategies one baseline case "B" and other case with secondary glazing "SG" was also simulated for large hotel. Although the closing of shading device during night is generally agreed upon there is potential of further research on operation of shading device during daytime and also based on the occupancy and other factors such as temperature, humidity etc. which would be area of future study. As for lighting, occupancy schedule-based weighting of lighting consumption was used for all the cases including baseline, but no daylighting dimming based control was used. The results for large hotel on energy consumption at six different locations are provided in Figure 5.

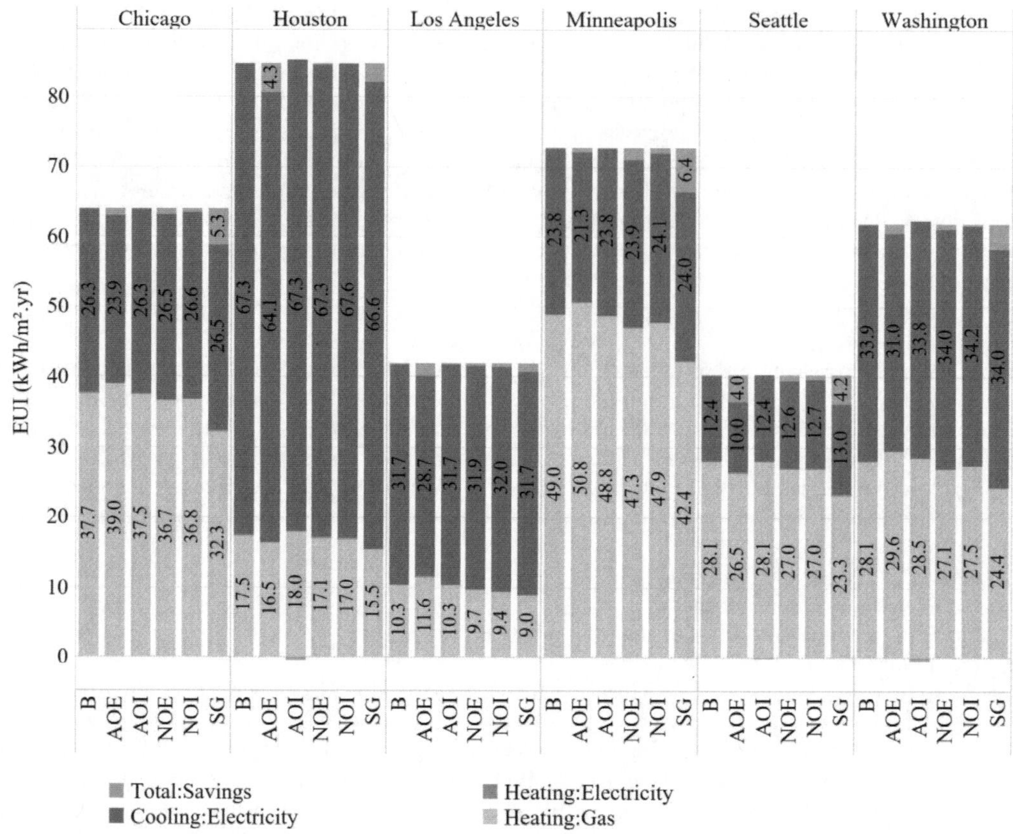

Figure 5: Energy consumption and savings in large hotel

The energy savings from heating and cooling ranges from approximately -0.5 to 6.4 kWh/m^2 in the large hotel. The results show that if left always ON, energy performance of exterior roller shades is better compared to the interior roller shades (i.e. AOE has higher savings compared to AOI). This is also seen for the shading devices closed only during night; the performance seems to be slightly better while using the exterior roller shades compared to the interior roller shades. Other than that, in large hotel, similar to medium office, SG provides the best energy savings in colder climate (Minneapolis, Chicago) and equivalent performance to exterior shades that is always closed (AOE) in hotter climate (Houston, Los Angeles). These results show that only using simple control strategy like closing the shading devices all the time some energy savings can be achieved. Further, if these are used to harvest solar heat gain during winter, considering the occupancy of each room then the energy savings could be higher.

3 CONCLUSIONS

Energy simulations were performed on 4 different types of commercial buildings at six different locations to evaluate the energy impact of different shading devices and control strategies. Daylight simulations were performed for the medium office buildings. Different control strategies were used in different building types considering the nature of their occupancy and activities in the buildings Energy savings up to 17.7 kWh/m^2-yr was seen in medium office buildings. Daylighting result showed that while use of automated shading devices enhanced visual comfort in the space by reducing excessive illuminance (shown by reduction in ASE), it also reduced some useful amount of daylight compared to the case without any shading device. The energy savings were higher in the buildings using both shading and lighting control compared to buildings using only shading control. This was seen by higher energy savings in medium office and secondary school buildings compared to midrise apartment and large hotel. The energy savings in secondary school was in the range of 1.4-12.2 kWh/m^2-yr while in case of midrise apartment and large hotel the energy savings was below 3 kWh/m^2-yr except for the case of SG. Without integration of lighting controls, it was seen that there could be some energy penalty if appropriate strategy is not used for shading control from results in midrise apartment and large hotel.

There are several limitations associated with this study. First, we use a single type of shading device for each of the buildings, thus the impact of shading devices on building from shading devices with different properties might vary than one presented in this study. Another limitation is that for midrise apartment and large hotel very simple shading strategy is used which does not consider the variation in climatic condition, occupancy, etc. More advanced shading control strategies and the integration of shading control with HVAC controls is an area that need more exploration in the future to tap the potential of integrated controls. Integrated controls that also includes properties of shading devices and fits the needs of the occupant in different building types, while providing energy benefits and grid service potential should be studied in the future.

NOMENCLATURE

F/B	Front and Back	(–)
k	Thermal conductivity	W/m-k
kWh	Kilowatt hour	(–)
m	meter	(–)
OF	Openness Factor	(–)
R_{sol}	Solar Reflectance	(–)
T_{sol}	Solar Transmittance	(–)
VT	Visible Transmittance	(–)
W	Watt	(–)
yr	Year	(–)

REFERENCES

Ariosto, T., & Memari, A. M. (2013). Evaluation of Residential Window Retrofit Solutions for Energy Efficiency. In *The Pennsylvania Housing Research Center* (Issue 111).

Atzeri, A. M., Gasparella, A., Cappelletti, F., & Tzempelikos, A. (2018). Comfort and energy performance analysis of different glazing systems coupled with three shading control strategies. *Science and Technology for the Built Environment*, *24*(5), 545–558. https://doi.org/10.1080/23744731.2018.1449517

DRI. (2013). *Residential Windows and Window Coverings : A Detailed View of the Installed Base and User Behavior*.

Fernandes, L. L., Lee, E. S., & Ward, G. (2013). Lighting energy savings potential of split-pane electrochromic windows controlled for daylighting with visual comfort. *Energy and Buildings*, *61*(February), 8–20. https://doi.org/10.1016/j.enbuild.2012.10.057

Goovaerts, C., Descamps, F., & Jacobs, V. A. (2017). Shading control strategy to avoid visual discomfort by using a low-cost camera: A field study of two cases. *Building and Environment*, *125*, 26–38. https://doi.org/10.1016/j.buildenv.2017.08.030

IES Daylight Metrics Committee. (2012). IES Spatial Daylight Autonomy (sDA) and Annual Sunlight Exposure (ASE), Daylight Metrics Committee. Approved Method IES LM-83-12. *Illuminating Engineering Society of North America*.

Kunwar, N., & Bhandari, M. (2020). *A Comprehensive Analysis of Energy and Daylighting Impact of Window Shading Systems and Control Strategies on Commercial Buildings in the United States*. https://doi.org/10.3390/en13092401

Kunwar, N., Cetin, K. S., & Passe, U. (2018). Dynamic Shading in Buildings: a Review of Testing Methods and Recent Research Findings. *Current Sustainable/Renewable Energy Reports*, *5*(1), 93–100. https://doi.org/10.1007/s40518-018-0103-y

Kunwar, N., Cetin, K. S., Passe, U., Zhou, X., & Li, Y. (2019). Full-scale experimental testing of integrated dynamically-operated roller shades and lighting in perimeter office spaces. *Solar Energy*, *186*(February), 17–28. https://doi.org/10.1016/j.solener.2019.04.069

Peng, J., & Curcija, D. C. (2017). *Energy Performance Indices EPC and EPH Calculation Methodology and Implementation in Software tool*.

Sawyer, K. (2014). *Window and Building Envelope Reserach and Development: Roadmap for Emerging*

Technologies. http://library1.nida.ac.th/termpaper6/sd/2554/19755.pdf

Shen, H., & Tzempelikos, A. (2017). Daylight-linked synchronized shading operation using simplified model-based control. *Energy and Buildings*, *145*, 200–212. https://doi.org/10.1016/j.enbuild.2017.04.021

The U.S. Department of Energy (DOE). (n.d.). *Commercial Prototype Building Models*. Retrieved April 15, 2019, from https://www.energycodes.gov/development/commercial/prototype_models

ACKNOWLEDGEMENT

We would like to thank Amy Jiron and Cedar Blazek (Commercial Building Integration Program, US Department of Energy) for the project support and review. We would also like to thank Lucy Albin (Attachments Energy Ratings Council), D. Charlie Curcija (Lawrence Berkeley National Laboratory), Brent Protzman and Craig Casey (Lutron), Alen Mahic and Kevin Van Den Wymelenberg (University of Oregon), and John Crowley (formerly Rollease), for their input and feedback at different stages of the project.

A Comparison of Interpolation Methods in Fast Fluid Dynamics

Eric LEONARD, Hongtao QIAO*, Saleh NABI

Mitsubishi Electric Research Laboratories,
Cambridge, MA, USA
rick.leonard8@gmail.com, qiao@merl.com, nabi@merl.com

* Corresponding Author

ABSTRACT

This paper examines the use of OpenFOAM's cellPoint and cellPointWallModified interpolation routines in fast fluid dynamics (FFD). In addition, an arrangement of OpenFOAM's data structures is proposed which speeds up FFD and an adjustment to FFD which incorporates the Boussinesq approximation is discussed. The findings of this paper show that the use of the cellPointWallModified interpolation routine in FFD produces results closer to reference data in higher Reynolds number flows than the use of the cellPoint interpolation routine, a result which highlights the important role that interpolation plays in the accuracy of FFD.

1. INTRODUCTION

The efficient use of air conditioners in buildings is an important, ongoing area of research. In order to develop accurate models of the performance of an air conditioner, it is necessary to incorporate information about the room that the air conditioner cools. The accuracy of a coupled simulation which models an air conditioner as well as the dynamics of the air flow in a room induced by the air conditioner can be enhanced through the use of computational fluid dynamics (CFD). By using a CFD simulation to generate detailed information about the dynamical and thermal patterns of air flow in a room generated by an air conditioner, researchers can, for instance, determine locations for the air conditioner which will lead to more efficient performance. Although there are great benefits to using CFD in a coupled simulation which models the performance of an air conditioner and the room that it cools, CFD simulations can be slow as described in Qiao et al. (2019).

Concerns around the speed of CFD have facilitated an interest in an alternative algorithm to traditional computational fluid dynamics called fast fluid dynamics (FFD) in which the solutions to the Navier-Stokes equations are approximated through the use of a time-splitting method which generates separate equations for the dynamics related to advection, diffusion, and pressure gradients as described in Zuo (2010). The authors of this paper have implemented a version of FFD using the open-source software toolbox OpenFOAM which is in line with the work of Liu et al. (2016). One particularly interesting aspect of FFD is that the advection equation that it generates necessitates the use of an interpolation routine to determine intermediate velocities at points not located at the cell centers of meshes designed to cover the domain of interest. Following this line of inquiry, the current paper compares and contrasts the use of two of OpenFOAM's interpolation routines - cellPoint and cellPointWallModified - in the FFD algorithm. In addition, this paper presents a novel use of OpenFOAM's data structures which speeds up FFD and discusses the incorporation of the Boussinesq approximation into FFD.

2. METHODOLOGY

The following section introduces FFD as well as describes a way in which OpenFOAM's data structures can be arranged to speed-up FFD. Incorporation of the Boussinesq approximation into FFD is also discussed.

2.1 Basic Fast Fluid Dynamics Algorithm

When considering only pressure and velocity in incompressible flow and with an index repeated in a term implying summation of that index over the number of relevant dimensions, the Navier-Stokes equations are given by Equation (1).

$$\frac{\partial U_i}{\partial x_i} = 0 \tag{1a}$$

$$\frac{\partial U_i}{\partial t} + U_j \frac{\partial U_i}{\partial x_j} = -\frac{1}{\rho}\frac{\partial p}{\partial x_i} + \nu \frac{\partial^2 U_i}{\partial x_j \partial x_j} + \frac{1}{\rho} B_i \tag{1b}$$

The fast fluid dynamics algorithm breaks up the momentum equation, Equation (1b), into three separate equations when in discretized form, one equation each for the effects of advection, diffusion, and pressure gradients. The time-splitting approach of FFD can be quantitatively described by Equation (2).

$$\frac{U_i^* - U_i^n}{\Delta t} = -U_j^n \frac{\partial U_i^*}{\partial x_j} \tag{2a}$$

$$\frac{U_i^{**} - U_i^*}{\Delta t} = \nu \frac{\partial^2 U_i^{**}}{\partial x_j \partial x_j} + \frac{1}{\rho} B_i \tag{2b}$$

$$\frac{U_i^{n+1} - U_i^{**}}{\Delta t} = -\frac{1}{\rho}\frac{\partial p^{n+1}}{\partial x_i} \tag{2c}$$

To ensure that the velocity field is divergence-free at time step $n + 1$, one can take the divergence of Equation (2c) and utilize Equation (1a) to derive a Poisson equation for the pressure as given in Equation (3).

$$\frac{\partial^2 p^{n+1}}{\partial x_i \partial x_i} = \frac{\rho}{\Delta t}\frac{\partial U_i^{**}}{\partial x_i} \tag{3}$$

FFD then solves Equation (2a) to obtain U_i^*, Equation (2b) to obtain U_i^{**}, Equation (3) to obtain p^{n+1}, and Equation (2c) to obtain U_i^{n+1}. While FFD solves Equation (2b) and Equation (3) by standard means, FFD solves Equation (2a) via a semi-Lagrangian scheme: Equation (2a) can be taken to represent the idea that a fluid particle's velocity will remain unchanged as it advects from a departure point $\mathbf{x}_d$ where it has velocity $\mathbf{U}^n$ to an arrival point $\mathbf{x}_a$ where it has velocity $\mathbf{U}^*$. The departure point $\mathbf{x}_d$ can be approximated using known variables as in Equation (4).

$$\mathbf{x}_d = \mathbf{x}_a - \Delta t \cdot \mathbf{U}^n(\mathbf{x}_a) \tag{4}$$

$\mathbf{U}^*$ at the arrival point $\mathbf{x}_a$, $\mathbf{U}^*(\mathbf{x}_a)$, can be related to $\mathbf{U}^n$ at the departure point $\mathbf{x}_d$, $\mathbf{U}^n(\mathbf{x}_d)$, as in Equation (5).

$$\mathbf{U}^*(\mathbf{x}_a) = \mathbf{U}^n(\mathbf{x}_d) \tag{5}$$

Using Equation (4), Equation (5) can be rewritten as in Equation (6).

$$\mathbf{U}^*(\mathbf{x}_a) = \mathbf{U}^n\left(\mathbf{x}_a - \Delta t \cdot \mathbf{U}^n(\mathbf{x}_a)\right) \tag{6}$$

By taking arrival points to be the cell centers of the mesh covering the domain of interest, Equation (6) can be used to determine $\mathbf{U}^*$ in the FFD algorithm. Because the departure point $\mathbf{x}_d$ will, in most instances, not coincide with a cell center as in Figure 1, $\mathbf{U}^n(\mathbf{x}_d)$ has to be determined through interpolation as in Liu et al. (2016).

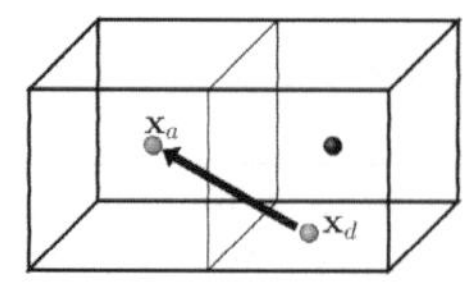

Figure 1: Illustration of a departure point (●) advecting to an arrival point (●) which coincides with a cell center; ● denotes a generic cell center

The authors of this paper implemented FFD using OpenFOAM following the work of Liu et al. (2016). OpenFOAM has a number of different interpolation routines which can be used to determine $\mathbf{U}^n(\mathbf{x}_d)$, two of which are the cellPoint and cellPointWallModified routines. Once the cell to which an interpolation point belongs is determined, the cellPoint method, as discussed in an OpenFOAMWiki (2010), works by covering each face of the cell with triangles, forming tetrahedrons composed of these triangles and the cell center, determining which tetrahdron encloses the interpolation point, and performing linear interpolation using inverse distance weights. The cellPointWallModified method, as documented in the OpenFOAM-7 Source Code (2019), is defined only when the interpolated variable is a vector (velocity, for instance) and works similarly to the cellPoint method except that the values of the interpolated variable at points on the boundaries of the domain are extrapolated from their values at the cell centers and then modified in such a way that they do not point out of the domain. Figure 2 compares and contrasts these two interpolation routines.

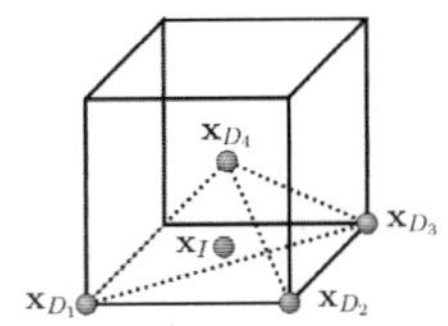

Figure 2: Data points $\mathbf{x}_{D_1}$ - $\mathbf{x}_{D_4}$ (•) used to evaluate a fluid property at an interpolation point $\mathbf{x}_I$ (•); if a data point lies on a boundary face, the cellPoint interpolation routine considers the value of the fluid property at that point to be as prescribed by the relevant boundary condition while the cellPointWallModified interpolation routine bases the value of the fluid property at the boundary data point on the value at the cell center

The cellPointWallModified interpolation routine significantly improves the accuracy of FFD in the following way: suppose, along the lines of Figure 2, $\mathbf{x}_{D_i}$, $i = 1, \ldots, 4$ are the positions of points used to perform interpolation in which $\mathbf{x}_{D_1}$, $\mathbf{x}_{D_2}$, and $\mathbf{x}_{D_3}$ are cell vertices of a face which lie on the boundary of the mesh, $\mathbf{x}_{D_4}$ is a cell center, and $\mathbf{x}_I$ is the interpolation point. Suppose also that f is a component of velocity which is to be interpolated and $d(\mathbf{x}_{D_i}, \mathbf{x}_I)$ denotes the distance between $\mathbf{x}_{D_i}$ and $\mathbf{x}_I$. With $\Delta \equiv \sum_{i=1}^{4} \frac{1}{d(\mathbf{x}_{D_i}, \mathbf{x}_I)}$, then, without loss of generality for the following argument, the value of the velocity component f at $\mathbf{x}_I$, f_I, can be taken to be based on the values of f at the four data points, f_{D_1} - f_{D_4}, as in Equation (7).

$$f_I = \frac{1}{\Delta}\left(\frac{f_{D_1}}{d(\mathbf{x}_{D_1}, \mathbf{x}_I)} + \frac{f_{D_2}}{d(\mathbf{x}_{D_2}, \mathbf{x}_I)} + \frac{f_{D_3}}{d(\mathbf{x}_{D_3}, \mathbf{x}_I)} + \frac{f_{D_4}}{d(\mathbf{x}_{D_4}, \mathbf{x}_I)}\right) \tag{7}$$

If, as previously specified, $\mathbf{x}_{D1}$, $\mathbf{x}_{D2}$, and $\mathbf{x}_{D3}$ lie on a boundary face and f_I lies close to this face and away from the cell center then, according to Equation (7), $\frac{1}{d(\mathbf{x}_{D_1}, \mathbf{x}_I)}, \frac{1}{d(\mathbf{x}_{D_2}, \mathbf{x}_I)}, \frac{1}{d(\mathbf{x}_{D_3}, \mathbf{x}_I)} >> \frac{1}{d(\mathbf{x}_{D_4}, \mathbf{x}_I)}$ and $f_I \approx \frac{1}{\Delta}\left(\frac{f_{D_1}}{d(\mathbf{x}_{D_1}, \mathbf{x}_I)} + \frac{f_{D_2}}{d(\mathbf{x}_{D_2}, \mathbf{x}_I)} + \frac{f_{D_3}}{d(\mathbf{x}_{D_3}, \mathbf{x}_I)}\right)$. If, in addition, the values of f_{D_1}, f_{D_2}, and f_{D_3} are taken to be those prescibed by the relevant boundary conditions, as in the cellPoint interpolation routine, then the value of f_I will not be accurately approximated in high Reynolds number flows because of a failure to take into account the steep velocity gradients that exist near boundaries as noted by Kundu et al. (2012). The cellPointWallModified interpolation routine rectifies this problem near the boundary by basing f_{D_1}, f_{D_2}, and f_{D_3} on the value of f in the cell center, f_{D_4}, when f_{D_1}, f_{D_2}, and f_{D_3} lie on a boundary face. Because f_{D_4} represents the value of f at an interior point in the flow, the cellPointWallModified routine better resolves the steep velocity gradients present near boundaries in high Reynolds number flows. In order to observe the differences in performance when used in FFD, the cellPoint and cellPointWallModified interpolation routines were tested on three different cases and the outcomes of these tests are presented in the Results section of this paper.

Because the present authors implemented FFD in OpenFOAM, an arrangement of OpenFOAM's data structures which sped-up the authors' implementation of FFD is now described. OpenFOAM enables its users to represent the velocity at the cell centers of a mesh through the use of data structures called *Fields*, as documented in the OpenFOAM Programmer's Guide (2015). In addition, the authors' implementation of FFD solved the matrix equations generated by Equation (2b) through the use of an iterative solver which used as its initial guess for $\mathbf{U}^{**}$ at time step $n + 1$ the value of $\mathbf{U}^{**}$ at time step n, as documented in the OpenFOAM User Guide (2019). By using separate *Fields* for $\mathbf{U}$, $\mathbf{U}^{*}$, and $\mathbf{U}^{**}$, the solution to Equation (2b) was sped-up because at each time step the initial guess for $\mathbf{U}^{**}$ was its value at the previous time step and as a simulation approached steady-state fewer iterations were needed to solve Equation (2b) as the change in $\mathbf{U}^{**}$ diminished.

2.2 Incorporation of the Boussinesq Approximation into FFD

When performing simulations of air flow in rooms related to the performance of air conditioners, it is clearly important to account for thermal effects. Use of the Boussinesq approximation allows for the incorporation of thermal effects into CFD without having to solve the fully compressible Navier-Stokes equations. In particular, the Boussinesq approximation assumes that the fluid's density is constant except when generated by buoyant forces and it is noteworthy that Gray and Giorgini (1976) demonstrated the validity of the Boussinesq approximation for air at room temperature. The Navier-Stokes equations with the Boussinesq approximation are given by Equation (8).

$$\frac{\partial U_i}{\partial x_i} = 0 \tag{8a}$$

$$\frac{\partial U_i}{\partial t} + U_j \frac{\partial U_i}{\partial x_j} = -\frac{1}{\rho}\frac{\partial p}{\partial x_i} + \nu \frac{\partial^2 U_i}{\partial x_j \partial x_j} - g_i \beta \tau \tag{8b}$$

$$\frac{\partial \tau}{\partial t} + U_j \frac{\partial \tau}{\partial x_j} = \alpha \frac{\partial^2 \tau}{\partial x_j \partial x_j} \tag{8c}$$

The relative temperature, τ, is defined as $\tau \equiv T - T_0$. In order to incorporate the change to the body force term generated by the Boussinesq approximation into FFD, the authors of this paper utilized a technique wherein at each time step after the semi-Lagrangian procedure was employed to determine $\mathbf{U}^*$ an iterative procedure was used which first solved Equation (2b), Equation (3), Equation (2c), and Equation (8c) for the intermediate variables $\widetilde{\mathbf{U}}^{**}$, $\tilde{p}^{n+1}$, $\widetilde{\mathbf{U}}^{n+1}$, $\tilde{\tau}^{n+1}$ and then solved Equation (2b), Equation (3), Equation (2c), and Equation (8c) again with the intermediate variables to determine $\mathbf{U}^{**}$, p^{n+1}, $\mathbf{U}^{n+1}$, τ^{n+1}. The first iteration of this procedure can be quantitatively described by Equation (9a) - Equation (9d).

$$\frac{\widetilde{U}_i^{**} - U_i^*}{\Delta t} = \nu \frac{\partial^2 \widetilde{U}_i^{**}}{\partial x_j \partial x_j} - g_i \beta \tau^n \tag{9a}$$

$$\frac{\partial^2 \tilde{p}^{n+1}}{\partial x_i \partial x_i} = \frac{\rho}{\Delta t}\frac{\partial \widetilde{U}_i^{**}}{\partial x_i} \tag{9b}$$

$$\frac{\widetilde{U}_i^{n+1} - \widetilde{U}_i^{**}}{\Delta t} = -\frac{1}{\rho}\frac{\partial \tilde{p}^{n+1}}{\partial x_i} \tag{9c}$$

$$\frac{\partial \tilde{\tau}^{n+1}}{\partial t} + \widetilde{U}_j^{n+1} \frac{\partial \tilde{\tau}^{n+1}}{\partial x_j} = \alpha \frac{\partial^2 \tilde{\tau}^{n+1}}{\partial x_j \partial x_j} \tag{9d}$$

The second iteration of this procedure can be quantitatively described by Equation (10a) - Equation (10d).

$$\frac{U_i^{**} - U_i^*}{\Delta t} = \nu \frac{\partial^2 U_i^{**}}{\partial x_j \partial x_j} - g_i \beta \tilde{\tau}^{n+1} \tag{10a}$$

$$\frac{\partial^2 p^{n+1}}{\partial x_i \partial x_i} = \frac{\rho}{\Delta t}\frac{\partial U_i^{**}}{\partial x_i} \tag{10b}$$

$$\frac{U_i^{n+1} - U_i^{**}}{\Delta t} = -\frac{1}{\rho}\frac{\partial p^{n+1}}{\partial x_i} \tag{10c}$$

$$\frac{\partial \tau^{n+1}}{\partial t} + U_j^{n+1} \frac{\partial \tau^{n+1}}{\partial x_j} = \alpha \frac{\partial^2 \tau^{n+1}}{\partial x_j \partial x_j} \tag{10d}$$

As a way of providing a visual summary, Figure 3 displays a flowchart of FFD with the Boussinesq approximation.

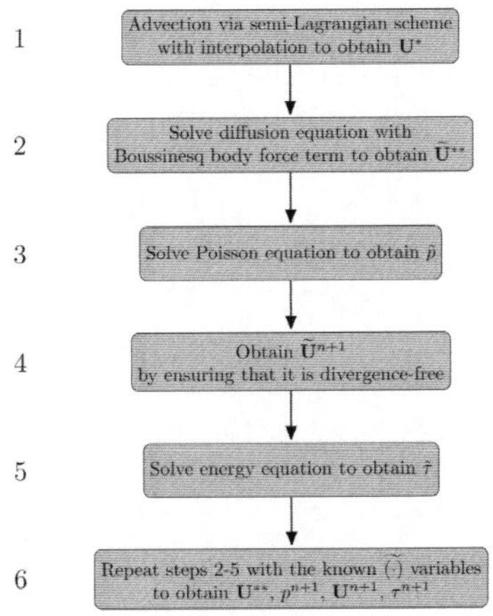

Figure 3: Flowchart of FFD with the Boussinesq approximation

3. RESULTS

The following section presents and discusses results from the authors' implementation of FFD in OpenFOAM for three different cases: flow in a 2D cavity, flow in a 2D forced convection cavity, and flow in a 3D mixed convection cavity. Table 1 displays the specifications for these cases while Table 2 displays the runtimes of FFD when using cellPoint interpolation as compared to cellPointWallModified interpolation. All subsequent velocity and temperature plots in this section are taken from the final time step of the simulations to which these plots belong. While the results and discussions in this section compare the accuracy of using cellPoint interpolation in FFD with the accuracy of using cellPointWallModified interpolation in FFD, examination of Table 2 demonstrates that the runtimes of these two different versions of FFD are roughly equivalent.

Table 1: Case Specifications

Case	Simulation Time (sec)	Δt (sec)	Re	Mesh
2D Cavity	10	0.01	100	64x64
2D Forced Conv.	400	1	~5000	44x27
3D Mixed Conv.	100	0.05	~2000	44x44x44

Table 2: Case Runtimes (in seconds) when using a 2 GHz processor

Interpolation Routine	2D Cav.	2D Forc. Conv.	3D Mix. Conv.
cellPoint	20.7	13.2	1629.36
cellPointWallModified	20.79	14.17	1692.01

3.1 2D Cavity

The 2D cavity case, pictured in Figure 4a, consists of flow in a two-dimensional square cavity driven by a top plate which moves with a constant velocity while the other boundaries remain stationary. Figure 4b displays the mesh used to run the simulation.

To check the current authors' implementation of FFD, Figure 5 compares velocity profiles generated by FFD using both the cellPoint and cellPointWallModified routines to reference data given in Ghia et al. (1982). Examination of Figure 5 demonstrates that, in this low Reynolds number flow, using both interpolation routines in FFD yield accurate results for the 2D cavity.

Figure 6 plots the number of iterations needed to solve the matrix equations for U_x^{**} and U_y^{**} as the simulation proceeded forward in time. Examination of Figure 6 demonstrates that the number of iterations does indeed decrease with time as the simulation approaches steady-state and the initial guesses to solve for U_x^{**} and U_y^{**} at each time step become closer to their true values.

(a) **(b)**

Figure 4: Illustration of the 2D cavity case: (a) Schematic (based on Figure 4-1 of Zuo (2010)) (b) Mesh

(a) **(b)**

Figure 5: Velocity profiles from the 2D cavity case: —, cellPoint; •, cellPointWallModified; ▪, Reference data from Ghia et al. (1982)

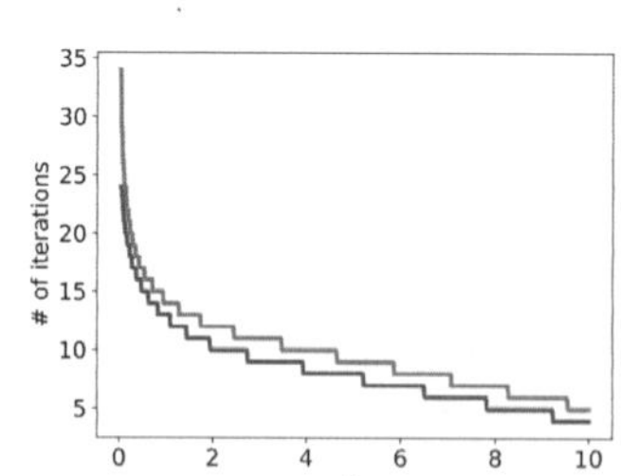

Figure 6: Number of iterations necessary to solve the matrix equations for U_x^{**} and U_y^{**} vs. simulation time (sec): —, U_x^{**}; —, U_y^{**}

3.2 2D Forced Convection

The 2D forced convection case, pictured in Figure 7a, consists of a 2D, rectangular cavity in which an inlet stream flows in from the upper left corner of the domain and an aperture in the bottom right corner of the domain allows for mass flow out of the cavity. Table 3 provides numerical values for the parameters specified in Figure 7a, while Figure 7b displays the mesh used to run the simulations. The values in Table 3 are the same as in Liu et al. (2016) while the mesh in Figure 7b is based on the mesh in Figure 2b of the same journal article.

Figure 8a displays instantaneous streamlines from the FFD simulation of the forced convection cavity with the cellPoint interpolation routine while Figure 8b displays instantaneous streamlines from the FFD simulation with the cellPoint-WallModified interpolation routine. Figure 8c displays a plot of how the streamwise velocity varies along the height of the forced convection cavity from two-thirds of the way down the length of the cavity while Figure 8d charts how the streamwise velocity varies along the length of the cavity starting in the middle of the inlet slot for the FFD simulations with the cellPoint and cellPointWallModified interpolation routines as compared against reference data from Nielsen et al. (1978). Examination of Figure 8c and Figure 8d demonstrates that the FFD simulation with the cellPoint-WallModified interpolation routine does indeed do a better job of capturing the steep velocity gradient present at the top of the domain in this high Reynolds number flow. In addition, comparison of Figure 8a and Figure 8b demonstrates that the FFD simulation with the cellPointWallModified interpolation routine generates a smoother, more well-formed recirculation bubble in the center of the cavity. Because the cellPointWallModified routine adjusts the treatment of the velocity at the wall, Figure 8 demonstrates that the way in which the conditions at the wall are incorporated into a solver affect not only the behavior of the fluid near the wall, but also the behavior within the interior of the domain as can be seen by the stark change in streamline pattern between Figure 8a and Figure 8b.

Figure 7: Illustration of the 2D forced convection case: (a) Schematic (based on Figure 4-9 of Zuo (2010)) (b) Mesh

Table 3: Numerical values of the parameters displayed in the schematic of the 2D forced convection case, Figure 7a

Parameter	L	H	h_{in}	h_{out}	U_{in}
Value	9 m	3 m	0.168 m	0.48 m	0.455 m/s

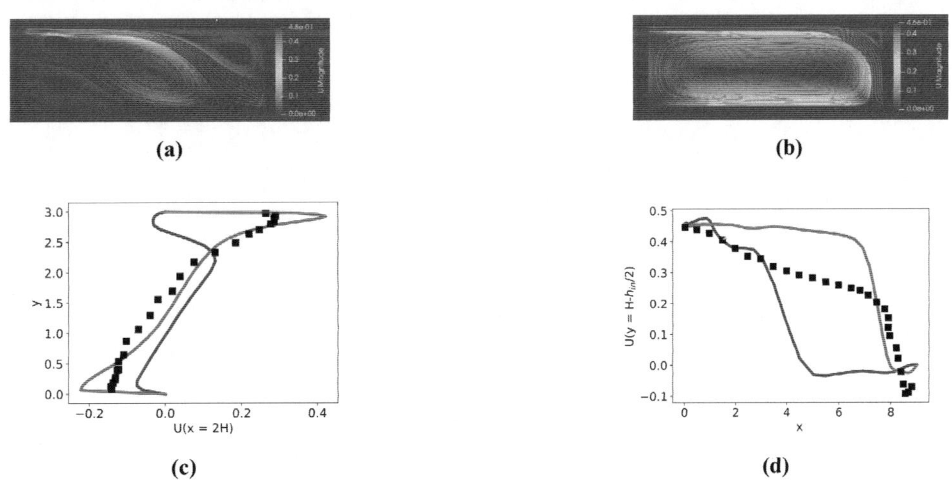

Figure 8: Results of the FFD simulation of the 2D forced convection case: (a) Streamlines from the simulation with the cellPoint interpolation routine color-coded by the magnitue of velocity (b) Same plot as (a) except cellPoint-WallModified was used (c) Plot of $U(x = 2H)$ vs. y —, cellPoint; —, cellPointWallModified; ■, Reference data from Nielsen et al. (1978) (d) Plot of x vs. $U(y = H - h_{in}/2)$ with the same color-coding as in (c)

3.3 3D Mixed Convection

The 3D mixed convection case, pictured in Figure 9a, consists of a cubic cavity in which there exists a solid cubic box at the bottom of the cavity. An inlet stream sends air into the cavity and there exists an outlet slot through which mass can leave the cavity. The walls of the cavity and the box are maintained at different temperatures so that it is necessary to account for thermal effects. Table 4 provides numerical values, equivalent to those in Liu et al. (2016), for the parameters relevant to this case. Because the temperature distribution is non-uniform in the 3D mixed convection cavity, FFD with the Boussinesq approximation as in Section 2.2 was used to simulate this case. Figure 9b displays the mesh used to conduct the simulations. The mesh in Figure 9b is based on Figure 9b of Liu et al. (2016). To analyze the results of the simulation Figure 10 plots the variation of a scaled magnitude of velocity along the y axis at four different locations while Figure 11 plots the variation of a scaled temperature at the same locations as in Figure 10. The points in the x-z plane from which the data from Figure 10 and Figure 11 are taken are visualized in Figure 9c.

Examination of Figure 10 demonstrates that the FFD simulation with the cellPointWallModified routine not only generates an inlet stream with more momentum than its counterpart simulation with the cellPoint interpolation routine, behavior which is in line with the other higher Reynolds number flow tested in this paper (2D forced convection), but also possesses greater momentum in the interior of the domain, results which are closer to the reference data. Examination of Figure 11 demonstrates that the FFD simulation with the cellPointWallModified routine does a better job of capturing the temperature in the interior of the domain, but also is more effective at capturing the trend of the temperature distribution near the inlet (Figure 11a).

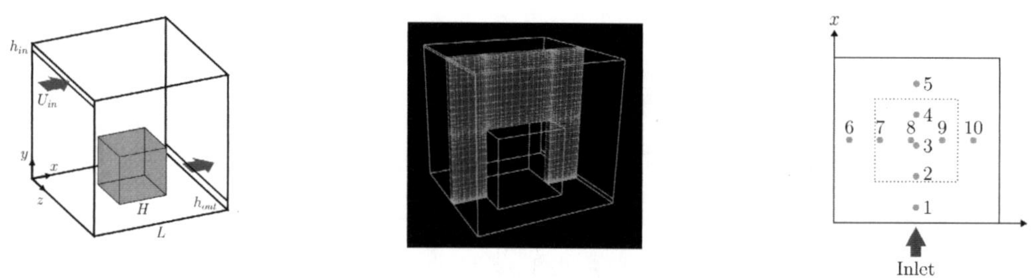

(a) (b) (c)

Figure 9: Illustration of the 3D mixed convection case based on Figure 9 of Liu et al. (2016): (a) Schematic (b) Mesh (c) View from above of the x-z locations from which the data in Figure 10 and Figure 11 are taken

Table 4: Numerical values of the parameters displayed in the schematic of the 3D mixed convection case, Figure 9a

Parameter	L	H	h_{in}	h_{out}	U_{in}
Value	2.44 m	1.22 m	0.03 m	0.08 m	0.455 m/s
Parameter	T_{in}	T_{box}	$T_{\text{cavity ceiling}}$	$T_{\text{cavity walls}}$	$T_{\text{cavity floor}}$
Value	22.2 °C	36.7 °C	25.8 °C	27.4 °C	26.9 °C

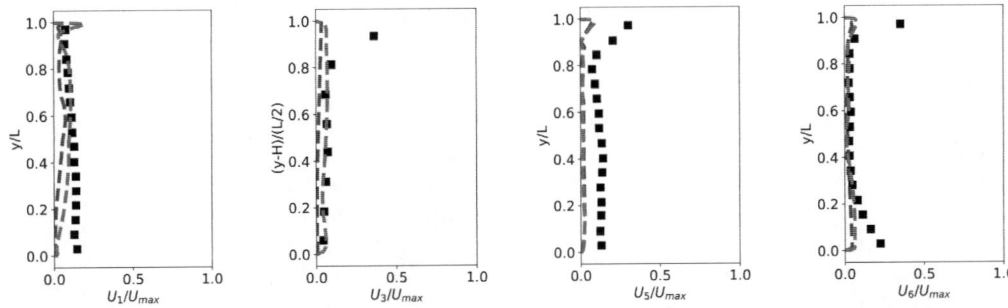

(a) (b) (c) (d)

Figure 10: Plots of $|\mathbf{U}|/U_{\text{max}}$ vs. y/L at locations (specified by Figure 9c): (a) 1 (b) 3 (c) 5 (d) 6; $U_{\text{max}} = 1.5$ m/s; - - -, cellPoint; - - -, cellPointWallModified; ■, Reference data from Wang and Chen (2009)

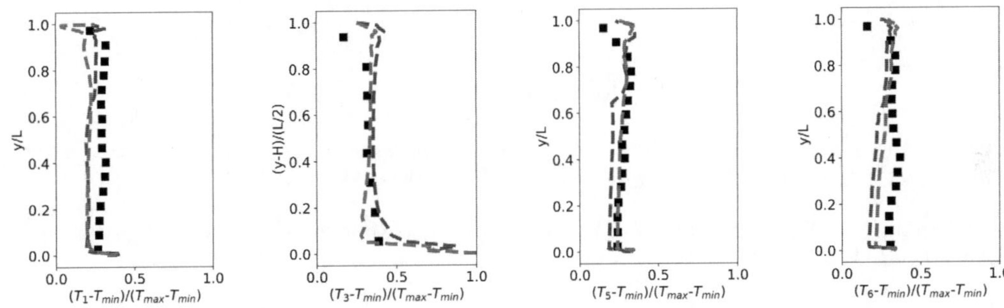

(a) (b) (c) (d)

Figure 11: Plots of $(T - T_{\text{min}})/(T_{\text{max}} - T_{\text{min}})$ vs. y/L at locations (specified by Figure 9c): (a) 1 (b) 3 (c) 5 (d) 6; $T_{\text{min}} = 22.2$ °C, $T_{\text{max}} = 36.7$ °C; - - -, cellPoint; - - -, cellPointWallModified; ■, Reference data from Wang and Chen (2009)

4. CONCLUSIONS

In this paper, FFD was implemented in OpenFOAM with two different interpolation schemes, cellPoint and cellPointWallModified, and applied to three different test cases. While the timings of FFD with cellPoint interpolation and cellPointWallModified interpolation were roughly equivalent, FFD with cellPointWallModified interpolation produced more accurate simulations in higher Reynolds number flows by adjusting the nature of the interpolation scheme near the boundaries of the domain. Given this information, it appears that the version of FFD with the cellPointWallModified interpolation scheme is superior to the version of FFD with the cellPoint interpolation scheme.

NOMENCLATURE

α thermal diffusivity ($m^2 s^{-1}$)
β coefficient of thermal expansion (K^{-1})
B body forces (N)
$d(\cdot,\cdot)$ distance (m)
Δ sum of reciprocal distances (m^{-1})
f generic velocity component ($m s^{-1}$)
f_D velocity component at data point ($m s^{-1}$)
f_I velocity component at interpolation point ($m s^{-1}$)
g gravitational acceleration ($m s^{-2}$)
ρ density ($kg m^{-3}$)
p pressure (Pa)
τ relative temperature (K)
T temperature (K)
T_0 reference temperature (K)
t time (s)
Δt time step (s)
U velocity ($m s^{-1}$)
ν kinematic viscosity ($m^2 s^{-1}$)
x spatial coordinates (m)
$\mathbf{x}_a$ arrival point (m)
$\mathbf{x}_d$ departure point (m)
$\mathbf{x}_D$ data point (m)
$\mathbf{x}_I$ interpolation point (m)

Superscript

n current time step
$n + 1$ succeeding time step
$*$, $**$ intermediate time variables
~ intermediate iteration variables

REFERENCES

Ghia, U., Ghia, K., and Shin, C. (1982). High-re solutions for incompressible flow using the navier-stokes equations and a multigrid method. *Journal of computational physics*, 48(3):387–411.

Gray, D. and Giorgini, A. (1976). The validity of the boussinesq approximation for liquids and gases. *International Journal of Heat and Mass Transfer*, 19(5):545–551.

Kundu, P., Cohen, I., and Dowling, D. (2012). Chapter 12-turbulence. *Fluid Mechanics (Fifth Edition), Academic Press, Boston*, pages 541–620.

Liu, W., Jin, M., Chen, C., You, R., and Chen, Q. (2016). Implementation of a fast fluid dynamics model in openfoam for simulating indoor airflow. *Numerical Heat Transfer, Part A: Applications*, 69(7):748–762.

Nielsen, P., Restivo, A., and Whitelaw, J. (1978). The velocity characteristics of ventilated rooms. *Journal of Fluids Engineering*, 100:291–298.

OpenFOAM-7 Source Code (2019). cellPointWallModified Header File. https://github.com/OpenFOAM/OpenFOAM-7/blob/master/src/finiteVolume/interpolation/interpolation/interpolationCellPointWallModified/interpolationCellPointWallModified.H.

OpenFOAM Programmer's Guide (2015). Section 2.3.2 Defining a geometricField in OpenFOAM. http://foam.sourceforge.net/docs/Guides-a4/ProgrammersGuide.pdf.

OpenFOAM User Guide (2019). Section 4.6.1.1 Solution tolerances. http://foam.sourceforge.net/docs/Guides-a4/OpenFOAMUserGuide-A4.pdf.

OpenFOAMWiki (2010). OpenFOAM guide/Interpolation (by cell). https://openfoamwiki.net/index.php/OpenFOAM_guide/Interpolation_(by_cell)#interpolationCellPoint.

Qiao, H., Nabi, S., and Laughman, C. (2019). Performance evaluation of hvac systems via coupled simulation between modelica and openfoam.

Wang, M. and Chen, Q. (2009). Assessment of various turbulence models for transitional flows in an enclosed environment (rp-1271). *Hvac&r Research*, 15(6):1099–1119.

Zuo, W. (2010). *Advanced simulations of air distributions in buildings*. Purdue University.

Development and Evaluation of Low-Cost CO_2 Sensors for Buildings

Zachary A. Siefker[1,2], Abhi Boyina[1,2], Xikang Zhao[3], John N. Hodul[2,4], James E. Braun[1,2], George T.-C. Chiu[1,2], Bryan W. Boudouris[3,4], Jeffrey F. Rhoads[1,2*]

[1] School of Mechanical Engineering
[2] Ray W. Herrick Laboratories
[3] Charles D. Davidson School of Chemical Engineering
[4] Department of Chemistry
Purdue University, West Lafayette, IN 47907

* Corresponding Author

ABSTRACT

There is a significant opportunity to improve building energy efficiency and indoor environmental quality by accurately monitoring CO_2 levels. However, current CO_2 sensors tend to be expensive or require regular recalibration. This work presents research related to the initial development and evaluation of two novel CO_2 sensors based on chemiresistive and resonant mass sensing techniques. Prototype sensors were assessed in a bench-top test chamber at temperatures, humidity levels, and CO_2 concentrations, typical of indoor environments. Under these conditions, prototype sensors required only 60 mW of power, or less. Further, each sensor was developed to have a footprint of less than 25 mm^2 and a cost of less than \$50. Given the relative low cost, small size, and potential for low power consumption, these sensors may serve as an attractive alternative to the commercial CO_2 sensors that are currently available.

1. INTRODUCTION

In the United States, buildings account for nearly 40% of total energy consumption (EIA, 2019). Although the average American spends nearly 90% of their time indoors (EPA, 1989), individual spaces in buildings typically have low or no occupancy much of the time. Thus, there is significant opportunity for reducing energy use by decreasing ventilation rates and adjusting thermostat temperatures according to measured occupancy. One common surrogate measurement for occupancy is a room's carbon dioxide (CO_2) level.

CO_2-based demand controlled ventilation (DCV) offers the possibility of saving energy by reducing building ventilation rates during periods of low occupancy, while also ensuring adequate levels of outdoor air ventilation when necessary. To ensure healthy indoor air, ASHRAE Standard 62.1-2010 suggests maintaining a 700 ppm difference between indoor and outdoor CO_2 concentrations, based on a ventilation rate of 7.5 Lps/person and assuming an occupant activity level of 1.2 MET (ASHRAE, 2010b). In the absence of occupancy data, many systems are designed to provide constant ventilation based on maximum expected occupancy levels. However, studies have shown that CO_2-based DCV can offer up to 44% energy savings as compared to constant-rate ventilation systems (O'Neill et al., 2020). Thus, DCV has the potential to substantially improve building energy efficiency and indoor air quality.

As the Internet of Things (IoT) begins to translate into the buildings sector, new opportunities are arising with the advent of distributed sensing networks to improve building operation (Gunay & Shen, 2017). For example, some studies have shown the possibility of estimating building air leakage by monitoring CO_2 concentration decay patterns (You et al., 2012; Ng & Wen, 2011). Others have suggested using CO_2 sensing for occupancy detection to provide localized comfort control (Candanedo & Feldheim, 2016; Ekwevugbe et al., 2013; Jin et al., 2018). However, the commercial CO_2 sensors that are currently available tend to be costly and difficult to install, especially if implemented as retrofits. Thus, there is a need for low-cost and easily deployable sensors for monitoring indoor CO_2 levels in buildings.

This work presents research related to the initial development and evaluation of low-cost, small, and low-power sensors, suitable for monitoring indoor CO_2 levels. Two different types of sensors are presented that have the potential to perform as well or better than, current commercially available CO_2 sensing technologies. The first is a chemiresistive sensor that uses a carbon nanotube (CNT) thin film in conjunction with a polymer blend of branched polyethylenimine

(PEI) and polyethylene glycol (PEG). The second is a resonant mass sensor functionalized with a similar polymer blend. Prototype sensors were assessed in a bench-top environmental test chamber at boundary points of the 0.5 clo and 1.0 clo comfort zones of ASHRAE standards (ASHRAE, 2010a). Prototype sensors were then benchmarked against commercially available CO_2 sensing technologies.

2. DEVICE FABRICATION AND THEORY OF OPERATION

2.1 Chemiresistive Sensor Fabrication

To prepare the functional polymer blend, PEI (M_n = 10,000 g mol^{-1}) and PEG (M_n = 300 g mol^{-1}) were dissolved in methanol to generate a solution of 1% PEI and 3% PEG. The CNT thin films were prepared by adding 3 mg of CNTs to 3 mL of chlorosulfonic acid and stirring the mixture for 3 days. Then, 20 μL of this solution were sandwiched between two glass slides, and the thin films were generated by manually pressing and sliding across the glass slides, in a manner similar to what has been reported previously (Li et al., 2013). To transfer the thin film, the glass slides were gently placed into water to peel the thin film, and the free-floating films (Figure 1a) were deposited onto pre-patterned electrodes of a printed circuit board (PCB) substrate (Figure 1b). Fabrication of the sensors was completed by drop-casting the functional polymer solution onto the CNT thin films, followed by the removal of methanol by drying for 5 hr in a fume hood.

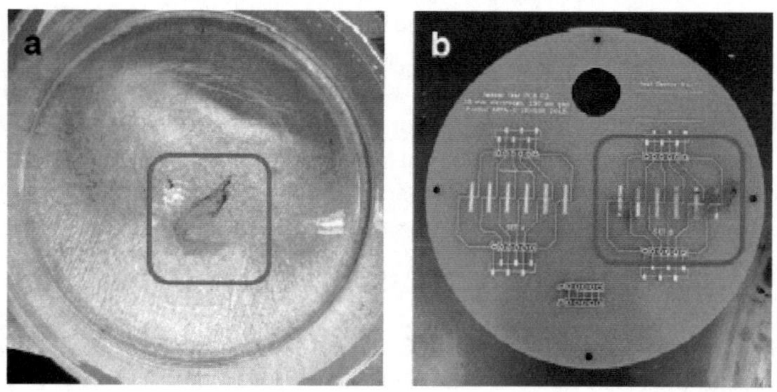

Figure 1: (a) The free-floating CNT film generated after casting a dispersion of the CNTs from a concentrated solution. The CNT film floating on water is highlighted by the red box. (b) The final CNT thin films after they have been laminated atop the pre-fabricated PCB substrates. The devices with the CNT thin films laminated on them are highlighted by the blue box. A separate set of bare electrodes are present, but are intentionally not covered with the CNTs for visual comparison.

2.2 Chemiresistive Sensing Mechanism

A functional mixture of PEI, PEG, and CNTs was utilized for the detection of CO_2. PEI, is a highly branched polymer with several secondary and tertiary amino groups along the polymer backbone. In the mixture, PEI n-dopes the CNTs with its electron negative lone pairs. In the presence of CO_2, the amino groups readily react to form electron deficient ammonium cations, which weakens the n-doping effect. This leads to a change in the carrier concentration within the material. Since this reaction is facilitated by water, PEG is added to serve as a vapor absorbing material.

2.3 Resonant Mass Sensor Fabrication

The resonant mass sensors consisted of a quartz crystal resonator embedded in a Pierce oscillator circuit that utilized a crystal oscillator driver, an inverter, two load capacitors (C_1 = 22 pF and C_2 = 22 pF), a feedback resistor (R_1 = 2 MΩ), and an isolation resistor (R_2 = 510 Ω), as shown in Figure 2a. These circuits were developed with a separate resonator board that connected to the rest of the oscillator circuit via spring pin connectors (Figure 2b). This two-part configuration allowed for easier functionalization of the resonators with the CO_2 sensitive polymer material.

To prepare the functional polymer blend, PEI (M_n = 25 kg mol^{-1}) and PEG (M_n = 100 kg mol^{-1}) were dissolved in methanol to generate a solution of 0.1% PEI and 0.3% PEG. To functionalize each resonator, 1μL of functional material was deposited onto each resonant element using a micro-pipette. The resonator board was then placed in a vacuum oven at 40 °C for 8 hours to remove methanol from the solution.

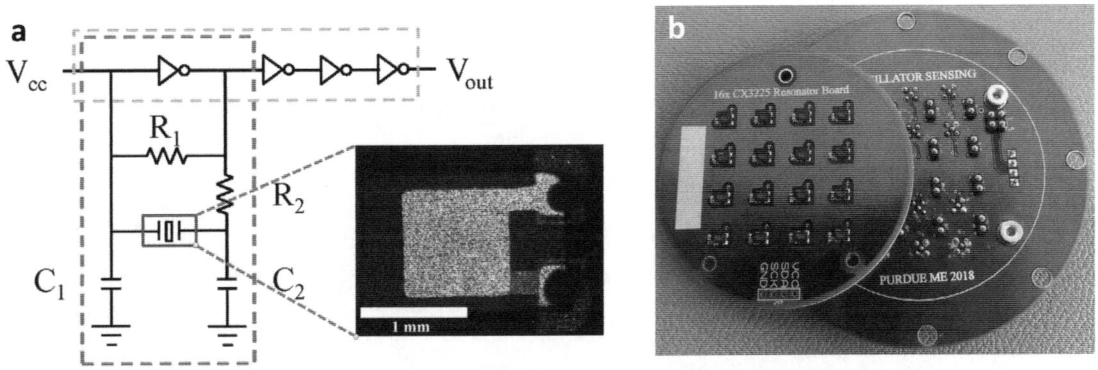

Figure 2: (a) The Pierce oscillator topology used for each resonant mass sensor, with the resonant element shown after the package cap has been removed. (b) A resonant mass sensing system with 16 Pierce oscillators. A resonator board (left) containing 16 resonant elements is shown offset from the instrumentation board (right) which completes the Pierce oscillator circuit.

2.4 Resonant Mass Sensing Mechanism

For bulk acoustic resonators, the added mass due to analyte/sensor interactions induces a shift in the device's resonant frequency. This can be modeled as a simple harmonic oscillator, with the frequency, f, defined as

$$f = \frac{1}{2\pi}\sqrt{\frac{k}{m}} \quad (1)$$

where k is the stiffness and m is the mass of the resonator. Thus, a change in the mass of the resonator, changes its resonant frequency. Similar to the chemiresistive device, a mixture of PEI and PEG was utilized as the CO_2 absorbing layer on the resonator. The sorption of CO_2 in this material results in an increase in the mass of the resonator. Thus, by measuring a given frequency shift, one can measure the presence of CO_2.

3. TESTING PROCEDURE

The bench-top testing system for both the chemiresistive and resonant mass sensors were organized in a similar fashion as seen in Figure 3. The test chamber used for the chemiresistive sensor was a 114 mm diameter and 20 mm tall cylindrical chamber with walls made of aluminum. A 95 mm diameter and 23 mm tall cylindrical aluminum test chamber was used to conduct the resonant mass sensor tests. Thermoelectric coolers (TEC) were attached to the exterior of each test chamber for temperature control of the test environment.

The chamber was equipped with an inline flow distribution system using two gas sources: pure nitrogen and a blend of 1% carbon dioxide balanced with nitrogen. The gas sources were connected to three mass flow controllers (MFC) in parallel. The pure nitrogen gas was connected to two 500 ccm rated MFCs, one of which was connected to a bubbler containing deionized water to introduce humidity to the test chamber. The gas mixture of carbon dioxide and nitrogen was connected to a 40 ccm rated MFC. The output of these three lines converged in a manifold which was connected directly to the testing chamber inlet.

Prior to starting each test, the PCB substrate (chemiresistive sensor) or functionalized resonator board (resonant mass sensor) was secured in the testing chamber. At the start of each test, baseline test conditions were developed by flushing the chamber with nitrogen at a flow rate of 500 ccm, and bringing the chamber to the test-specified temperature and humidity. The test-specified concentrations of CO_2 were then introduced, ensuring the volumetric flow rate was maintained at 500 ccm. For the chemiresistive sensor tests, the resistances were measured using bench-top digital multimeters and a custom measurement program, sampling at a rate of 0.5 Hz. The resonant mass sensor tests utilized frequency counters, implemented in a field programmable gate array (FPGA) within a commercial embedded device. A custom measurement program was designed to count the occurrence of the rising edge of the oscillator output voltage and provided a frequency readout accurate to ±1 Hz.

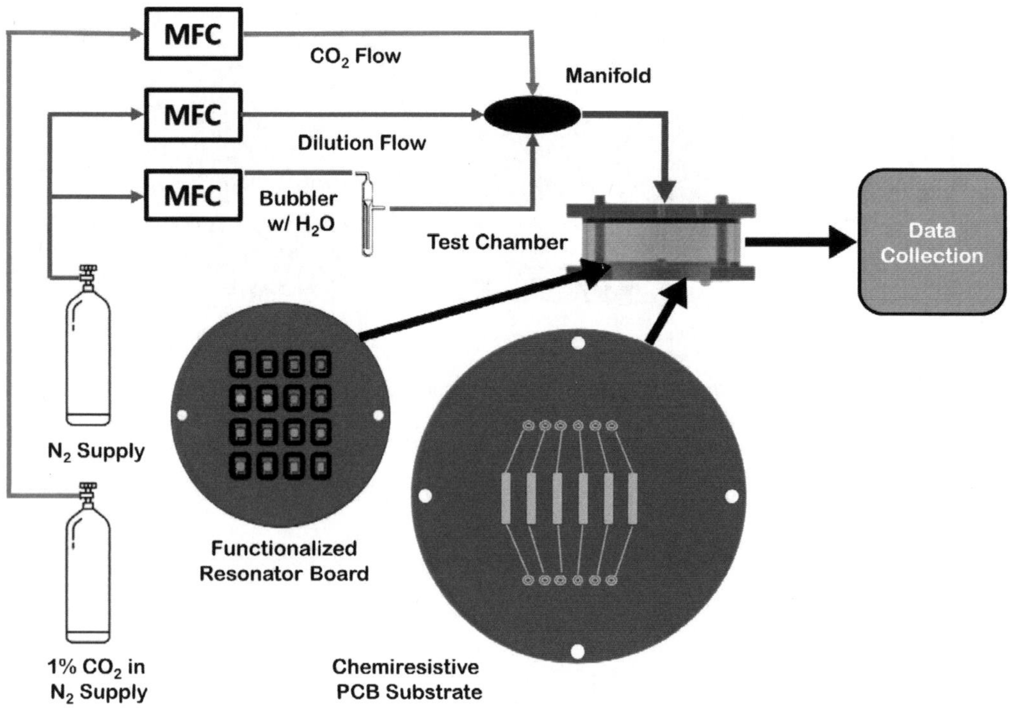

Figure 3: Schematic of gas the distribution system and the bench-top testing setup used to assess the performance of the prototype CO_2 sensors.

Sensor performance was measured at boundary points of the 0.5 clo and 1.0 clo comfort zones stipulated by ANSI/ ASHRAE Standard 55-2010 (ASHRAE, 2010a). These comfort zones approximately span 22–26 °C and from dry air to approximately 80% relative humidity (RH). This region of interest is depicted on the psychrometric chart in Figure 4. Testing was performed near the extremes of each of these regions, as well as near the center, to map sensor performance for typical indoor conditions. Additionally, testing was performed on a background of 400 ppm CO_2, which is approximately the concentration of outdoor air. Higher CO_2 concentrations were introduced, up to 2,000 ppm, to cover the relevant range of expected indoor CO_2 levels.

4. RESULTS AND DISCUSSION

4.1 Chemiresistive Sensor Performance

Figure 5 represents a snapshot of temporal test results for the chemiresistive sensor at each tested environmental condition. As the CO_2 concentration increases, the resistance of the device decreases.The resistance change (blue line) at each CO_2 concentration (red shaded regions) was calculated as

$$\Delta R_0 = \frac{R - R_0}{R_0} \tag{2}$$

where R is the measured electrical resistance, in ohms, and R_0 is the electrical resistance at the start of the temporal data set. The time series data provide a clear indication of sensor responsiveness at each of the environmental test conditions considered. At high humidity conditions, a larger and faster sensor response is demonstrated. At low humidity conditions, a weaker response to CO_2 is evident. More concretely, Figure 6 shows the mean response at each indicated CO_2 concentration as the change in resistance of the device relative to baseline conditions. This relative

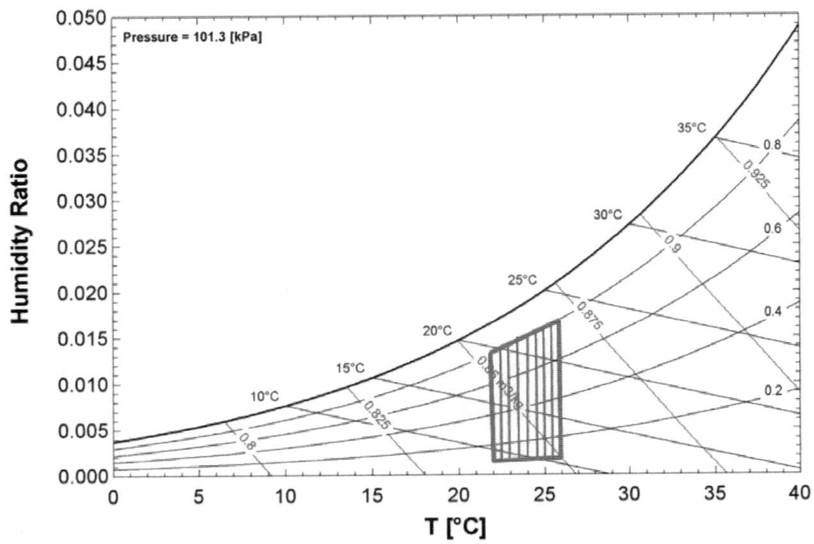

Figure 4: Psychrometric chart highlighting the range of environmental conditions for CO_2 sensor testing. This region spans across the 0.5 clo and 1.0 clo comfort zones stipulated by ANSI/ASHRAE Standard 55-2010 (ASHRAE, 2010a).

resistance change was calculated as

$$\Delta R_b = \frac{R - R_b}{R_b} \tag{3}$$

where R is the measured electrical resistance, in ohms, and R_b is the baseline electrical resistance at the start of each CO_2 pulse. A near-linear trend is present in these data and can be used as a calibration curve for each of the given temperature and RH conditions.

Observing these preliminary data, temperature had a minor influence on the sensor. However, tests across the humidity range show a large difference in the magnitude of the sensor's response to CO_2. In fact, the presence of humidity promotes the sensor's response. Notably, there is a slight signal drift that can be observed in the time series plots. It is hypothesized that this drift is a slight capacitive charging between the electrodes due to the continuous voltage applied during measurement. Thus, real-world applications would require either non-continuous measurement or calibration to the constant drift.

4.2 Resonant Mass Sensor Performance

Figure 7 represents a snapshot of time-series data at each environmental condition for the resonant mass sensor. As the CO_2 concentration increases, the frequency of the device decreases. The frequency change (blue line) at each CO_2 concentration (red shaded regions) was calculated as

$$\Delta f_0 = f - f_0 \tag{4}$$

where f is the measured frequency (Hz) and f_0 is the frequency at the start of the plot. The time series data provide an indication of the sensor's responsiveness, as well as the noise present at each test condition. As evidenced by these plots, a smooth signal is demonstrated at low humidity levels, whereas, higher humidity levels result in a choppier signal as device interaction with water occurs. Further, Figure 8 shows the mean response of the sensor at each indicated CO_2 concentration as the change in frequency relative to baseline conditions. The relative frequency shift was calculated as

$$\Delta f_b = f - f_b \tag{5}$$

where f is the measured frequency and f_b is the baseline frequency at the start of each CO_2 pulse. A near-linear trend is demonstrated in these data, which can serve as a calibration curve for each of the given temperature and RH conditions.

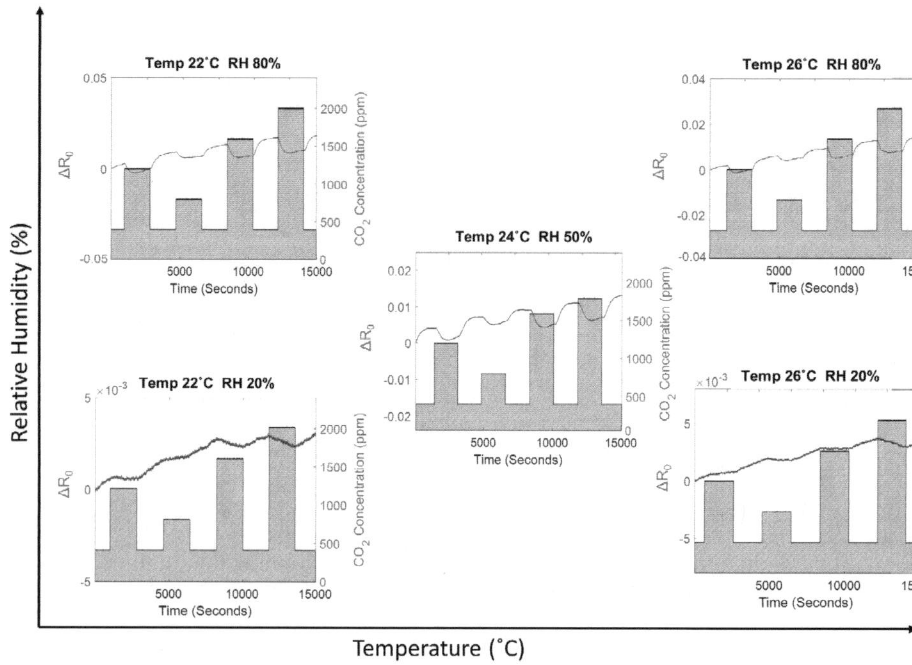

Figure 5: Time-series samples showing the change in resistance of the chemiresistive sensor at the "four corners and center" of the indoor comfort zone. This test zone spanned from approximately 22-26 °C and from dry air to 80% RH. The sensor response (blue line) is shown relative to changes in CO_2 concentration (red shaded regions).

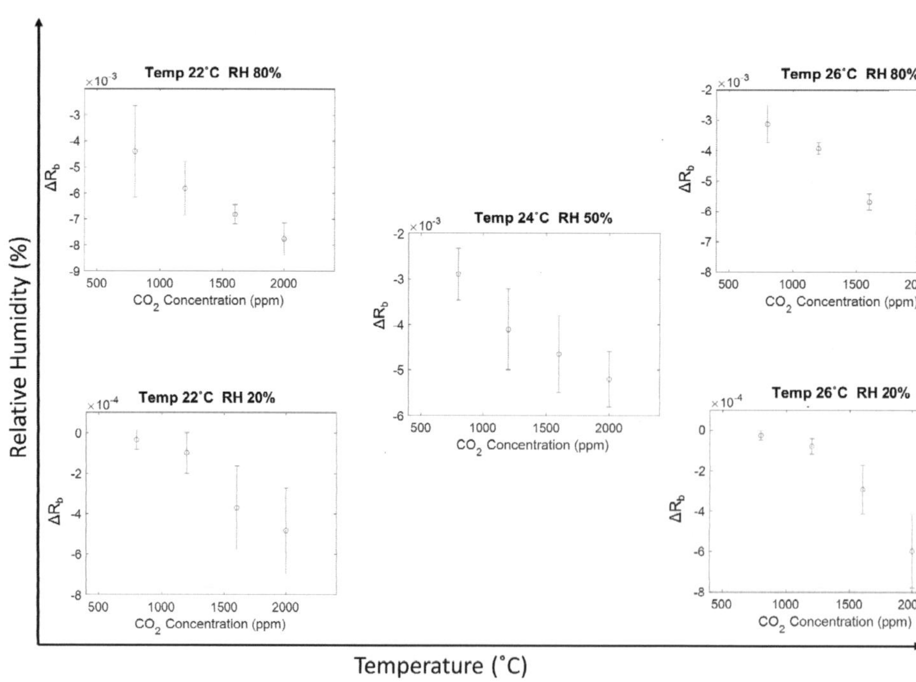

Figure 6: Chemiresistive sensor calibration curves at the "four corners and center" of the indoor comfort zone. The mean response at each CO_2 concentration is shown with error bars indicating one standard deviation in the data set.

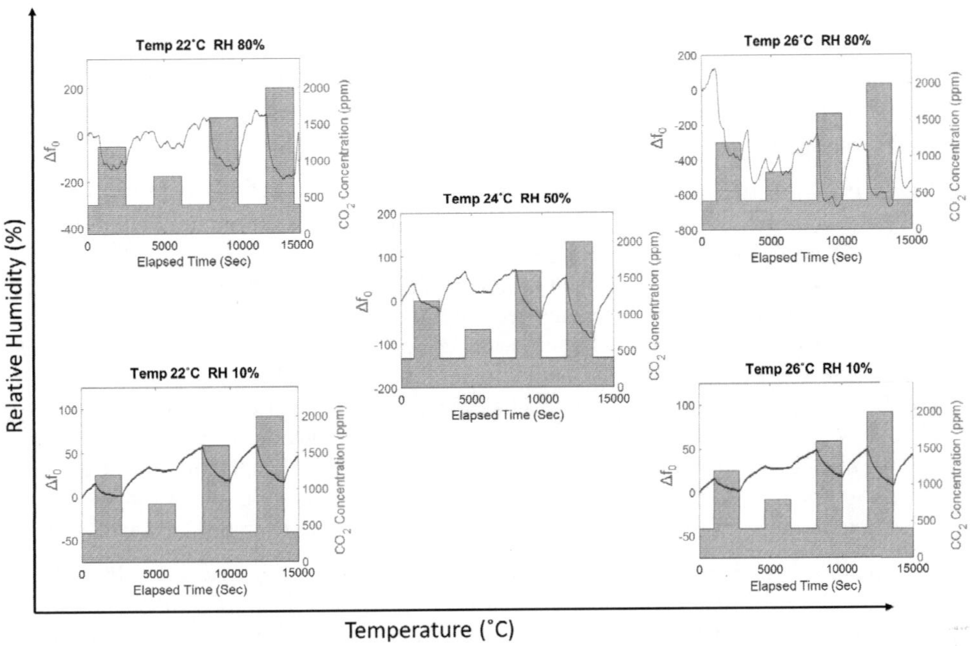

Figure 7: Time-series samples showing the frequency shift of the resonant mass sensor at the "four corners and center" of the indoor comfort zone. This test zone spanned from approximately 22-26 °C and from dry air to 80% RH. The sensor response (blue line) is shown relative to changes in CO_2 concentration (red shaded regions).

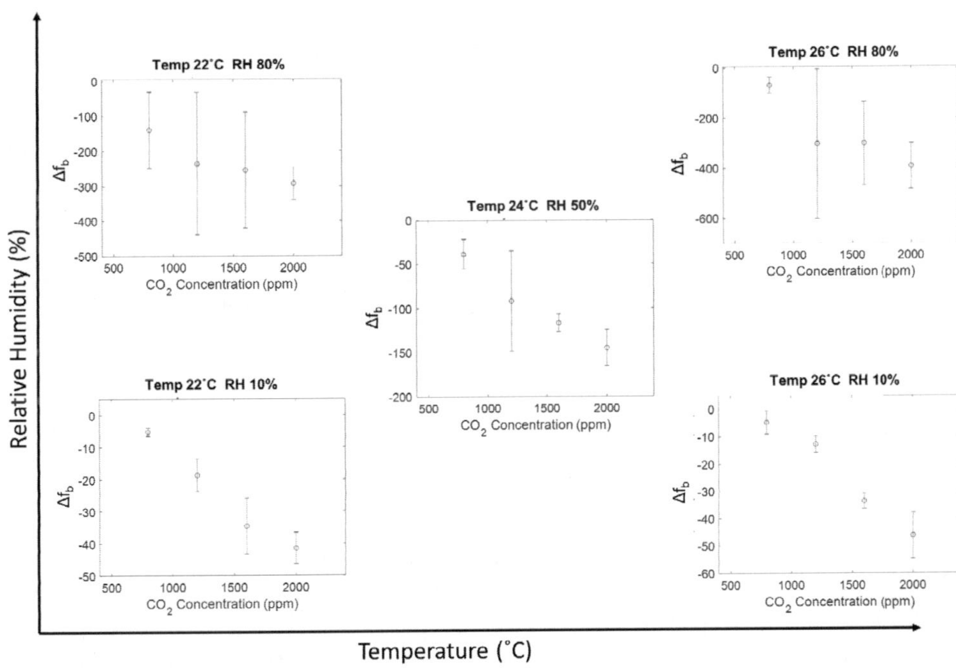

Figure 8: Resonant mass sensor calibration curves at the "four corners and center" of the indoor comfort zone. The mean response at each CO_2 concentration is shown with error bars indicating one standard deviation in the data set.

Similar to the chemiresistive sensor, the temperature range considered shows minor influence on the sensor. However, humidity plays a larger role. Though the magnitude of the sensor response is greater, the time-series signal clarity is reduced at high humidity levels. However, compared to the chemiresitistve device, the sensitivity of the resonant mass sensor is much higher at all of the considered environmental conditions, which is particularly important when a higher measurement resolution is necessary.

4.3 Comparison with Commercial Sensor Technologies

Commercially available low-cost CO_2 sensors include metal-oxide sensors (MOS) and most commonly nondispersive infrared (NDIR) sensors. Metal-oxide sensors can be utilized by measuring a change in resistance of a metal oxide film due to the adsorption of the target analyte. This change may result in an increase or decrease in resistance, depending on the target analyte and the sensing material being used. Metal-oxide sensors are a relatively low-cost option for CO_2 monitoring, however, accuracy below 2,000 ppm tends to be poor (Fine et al., 2010). NDIR sensors are most common for low-cost CO_2 monitoring below 2,000 ppm, such as in HVAC applications. NDIR sensors operate based on principles of molecular spectroscopy and their use in building applications, such as indoor air quality monitoring and demand-controlled ventilation have been demonstrated extensively (Labeodan et al., 2015; Emmerich & Persily, 2001; Yi et al., 2005; Schibuola et al., 2018; Ng et al., 2011). However, though still considered low-cost, these sensors typically market for \$100-\$200. As well, the infrared light inherent to NDIR sensors requires higher power consumption and a large sensor footprint, making them difficult to implement in embedded applications.

Comparatively, the power consumption of the chemiresistive and resonant mass sensors developed herein was relatively low in bench-top testing conditions. The chemiresistive devices were nominally 10 kΩ. For this range, the digital multimeter used for benchtop testing provides a load of about 100 μA; therefore, the chemiresistive devices in their current state require approximately 100 μW of power. During bench-top testing of the resonant mass sensors, each oscillator used about 18 mA when provided 3.3 V from a DC power supply. Therefore, the resonant mass sensors in their current state require approximately 60 mW of power. It is likely that the power consumption may be reduced if a commercial sensor package is realized and lab-grade measurements are no longer necessary. Even so, the chemiresistive and resonant mass CO_2 sensors consume far less power than typical commercial NDIR or MOS CO_2 sensors.

Table 1 provides a comparison of the prototype chemiresistive and resonant mass sensors, with typical commercial MOS and NDIR CO_2 sensors. For a comparable measurement range, the chemiresistive and resonant mass sensors are small, low-cost, and low-power. This is particularly important when considering wireless sensor deployment and high resolution indoor air monitoring. As such, these sensors are promising candidates for room-level CO_2 monitoring in both new and retrofit building applications.

Table 1: A comparision of prototype sensors with commercial sensing technologies

Sensor Type	Measurement Range	Footprint	Cost	Power
Chemiresistive	0-10,000 ppm	< 25 mm^2	<\$50	100 μW
Resonant Mass	0-10,000 ppm	< 25 mm^2	<\$50	60 mW
Metal-Oxide	2,000-10,000 ppm	< 25 mm^2	\$50-\$75	200 mW
Nondispersive infrared	0-10,000 ppm	3 cm^2	\$100-\$200	300 mW

5. CONCLUSIONS

Two novel CO_2 sensors utilizing a polymer blend of PEI and PEG were presented. The performance of prototype sensors in a bench-top environmental test system was demonstrated at CO_2 concentrations, temperatures, and humidity levels, typical of indoor conditions. A chemiresistive sensor utilizing CNTs and the polymer blend was shown to detect CO_2 with very low power consumption. Similarly, a resonant mass sensor, utilizing the polymer blend, was found to be more sensitive than its chemiresistive counterpart. However, this was achieved at the cost of slightly higher power consumption and reduced signal clarity at high humidity levels. Nevertheless, the power consumption and footprint of both of these sensors is much smaller than most of the commercial sensors currently available. Such metrics make wireless and ubiquitous CO_2 sensing in buildings relevant, which presents new opportunities in smart building controls.

With further development, these sensors may serve not only as an alternative to currently available commercial CO_2 sensors, but provide new data nodes throughout buildings that were previously not economically viable.

REFERENCES

ASHRAE. (2010a). *ASHRAE/ANSI Standard 55-2010 Thermal Environmental Conditions for Human Occupancy* (Tech. Rep.). Atlanta, GA: American Society of Heating, Refrigerating, and Air-Conditioning Engineers.

ASHRAE. (2010b). *ASHRAE/ANSI Standard 62.1-2010 Ventilation for Acceptable Indoor Air Quality* (Tech. Rep.). Atlanta, GA: American Society of Heating, Refrigerating, and Air-Conditioning Engineers.

Candanedo, L. M., & Feldheim, V. (2016). Accurate occupancy detection of an office room from light, temperature, humidity and CO_2 measurements using statistical learning models. *Energy and Buildings*, *112*, 28-39.

EIA. (2019). *Frequently Asked Questions on Buildings in the United States* (Report No. P-41). United States Energy Information Administration.

Ekwevugbe, T., Brown, N., Pakka, V., & Fan, D. (2013). Real-time building occupancy sensing using neural-network based sensor network. In *2013 7th IEEE International Conference on Digital Ecosystems and Technologies (DEST)* (p. 114-119).

Emmerich, S. J., & Persily, A. K. (2001). *State-of-the-Art Review of CO_2 Demand Controlled Ventilation Technology and Application* (Report No. 6729). NIST Interagency/Internal Report (NISTIR).

EPA. (1989). *Report to Congress on Indoor Air Quality* (Vol. 2; Report No. EPA/400/1-89/001C). United States Environmental Protection Agency.

Fine, G. F., Cavanagh, L. M., Afonja, A., & Binions, R. (2010). Metal oxide semi-conductor gas sensors in environmental monitoring. *Sensors*, *10*, 5469-5502.

Gunay, B., & Shen, W. (2017). Connected and distributed sensing in buildings: Improving operation and maintenance. *IEEE Systems, Man, and Cybernetics Magazine*, *3*, 27-34.

Jin, M., Bekiaris-Liberis, N., Weekly, K., Spanos, C. J., & Bayen, A. M. (2018). Occupancy detection via environmental sensing. *IEEE Transactions on Automation Science and Engineering*, *15*, 443-455.

Labeodan, T., Zeiler, W., Boxem, G., & Zhao, Y. (2015). Occupancy measurement in commercial office buildings for demand-driven control applications—a survey and detection system evaluation. *Energy and Buildings*, *93*, 303-314.

Li, X., Jung, Y., Sakimoto, K., Goh, T.-H., Reed, M. A., & Taylor, A. D. (2013). Improved efficiency of smooth and aligned single walled carbon nanotube/silicon hybrid solar cells. *Energy & Environmental Science*, *6*, 879-887.

Ng, L. C., & Wen, J. (2011). Estimating building airflow using CO_2 measurements from a distributed sensor network. *HVAC&R Research*, *17*, 344-365.

Ng, M. O., Qu, M., Zheng, P., Li, Z., & Hang, Y. (2011). CO_2-based demand controlled ventilation under new ASHRAE standard 62.1-2010: a case study for a gymnasium of an elementary school at West Lafayette, Indiana. *Energy and Buildings*, *43*, 3216-3225.

O'Neill, Z. D., Li, Y., Cheng, H. C., Zhou, X., & Taylor, S. T. (2020). Energy savings and ventilation performance from CO_2-based demand controlled ventilation: Simulation results from ASHRAE RP-1747 (ASHRAE RP-1747). *Science and Technology for the Built Environment*, *26*, 257-281.

Schibuola, L., Scarpa, M., & Tambani, C. (2018). CO_2 based ventilation control in energy retrofit: An experimental assessment. *Energy*, *143*, 606-614.

Yi, S., Park, Y., Han, S., Min, N., Kim, E., & Ahn, T. (2005). Novel NDIR CO_2 sensor for indoor air quality monitoring. In *The 13th International Conference on Solid-State Sensors, Actuators and Microsystems, 2005. Digest of Technical Papers. Transducers '05.* (Vol. 2, p. 1211-1214).

You, Y., Niu, C., Zhou, J., Liu, Y., Bai, Z., Zhang, J., … Zhang, N. (2012). Measurement of air exchange rates in different indoor environments using continuous CO_2 sensors. *Journal of Environmental Sciences*, *24*, 657-664.

ACKNOWLEDGMENT

The information, data, or work presented herein was funded in part by the Advanced Research Projects Agency-Energy (ARPA-E), U.S. Department of Energy, under Award Number DE-AR0000943. The views and opinions of authors expressed herein do not necessarily state or reflect those of the United States Government or any agency thereof.

ENVIRONMENTAL INDEX FOR REAL-TIME FEEDBACK ON BIOWALL INDOOR AIR QUALITY

William HUTZEL[1]*, Henry LIU[2], Jun Ho KIM[3]

[1]Purdue University, Department of Mechanical Engineering Technology,
West Lafayette, US
hutzelw@purdue.edu

[2]Purdue University, Department of Mechanical Engineering Technology,
West Lafayette, IN, USA
liu1372@purdue.edu

[3]Purdue University, Lyles School of Civil Engineering,
West Lafayette, IN, USA
kim1969@purdue.edu

*Corresponding Author

ABSTRACT

This paper describes an ongoing field evaluation of a simplified Environmental Index (EI) that provides a homeowner with an intuitive metric for assessing indoor air quality based on regular measurements of temperature, humidity, and Volatile Organic Compounds (VOC's). These discrete measurements are used to compute an overall weighted score that correlates with the comfort and health of the living space. Although this environmental index is not a rigorous scientific computation, it is an initial point to improve residents' understanding of Indoor Air Quality (IAQ) by providing active feedback. The context of this environmental index work is the Biowall, a living wall of plants that is integrated into the central air conditioning system of a home to improve IAQ. The environmental index is an opportunity to begin providing a homeowner with real time feedback on comfort, health, and the performance of the Biowall. A simplified graphic user interface was designed using User Experience (UX) principles to provide a graphical presentation of an environmental index score that is easy to interpret and act upon. As expected, the environmental index fluctuates based on occupancy and activities in the space, such as cooking or cleaning. It was found that VOC levels fluctuated the most and thus had the most significant impact on the Environmental Index.

1. INTRODUCTION

Sustainable building environments are designed to enhance the comfort and health of the occupants. Indoor Environmental Quality (IEQ) is a comprehensive term that represents the overall quality of living in the built environment. IEQ depends on air temperature and humidity, but it also includes many other factors like acoustics, lighting, odors, and the overall design of the space. Indoor Air Quality (IAQ) is an important subset of IEQ that specifically targets healthy air. ASHRAE standard 55 "Thermal Environmental Conditions for Human Occupancy" and ASHRAE standard 62.1 "Ventilation for Acceptable Indoor Air Quality" provide guidance on maintaining Indoor Air Quality to protect the health and comfort of building occupants.

IAQ monitoring systems are imperative as Americans spend more than 90% of their time indoors. A variety of indoor airborne contaminants has caused diverse health problems such as asthma, chronic pulmonary disease, and allergies (Alraddadi, 2016). The United States Environmental Protection Agency (EPA) has revealed that the concentration of some pollutants are 2 to 5 times higher indoors than outdoors. Part of that problem occurs due to an increase in the number of energy-efficient buildings that limit mechanical ventilation (Soreanu, 2013). Improving IAQ influences not only human health but also the quality of life and productivity of residents (Newkirk, 2014).

Since it is typically not possible to quantify all the factors that influence IAQ, many IAQ assessments are based on measurements of a few of the most important parameters. Temperature, humidity, CO_2 levels, CO levels, and Total Volatile Organic Compound (TVOC) levels, are some of the common measurements that indicate the quality of an indoor environment (Parkinson, 2019). The challenge becomes combining these different indicators into one comprehensive metric that represents comfort and health in the built environment.

Real-time IAQ measurements are often included with the sophisticated Building Automation Systems (BAS) that are the "brains" for controlling modern high performance buildings. The IAQ sensors within the BAS are permanently installed within the HVAC infrastructure of the building. As such they can deliver real-time monitoring of IAQ levels, indoor temperature, and humidity (Allocca, 2019). This information is available to facility managers but it is not always provided as active feedback to the occupants.

A new generation of devices attempt to provide continuous IEQ monitoring using a suite of sensors packaged within one low-cost system. In addition to air measurements, these instruments sometimes include measurements of noise levels. IEQ monitoring systems are easy to temporarily install in a room and typically include a wireless communication network so that the data from one or more sensors can be remotely monitored from a user-friendly app on a cell phone or other mobile device. Big data analysis is also contributing to this paradigm shift where low-cost smart automation can help maintain sustainable indoor building environments (Parkinson, 2019).

Software for the built environment has begun applying User Experience (UX) principles to improve how complex building performance data is presented and interpreted. As one example, UX principles have been applied to develop easy to understand energy dashboards to track the performance of net zero energy buildings (Switzer, 2018). Similar Graphic User Interfaces (GUI) have also been developed to present the numerical and graphical information regarding real-time TVOC, temperature and humidity levels to determine the occupant's living quality and satisfaction. One particular GUI closes the feedback loop by communicating with the users to operate, monitor and adjust the building's indoor environmental conditions. (Parkinson, 2019).

2. ENVIRONMENTAL INDEX

When considering the occupants in a home, a single numeric score that is related to IAQ is a solid starting point for beginning to provide information that is easy to understand and act upon. Many homeowners would not know how to interpret a large number of discrete measurements for temperature, humidity, etc. An Environmental Index (EI) attempts to accomplish this by computing one weighted score that combines temperature, relative humidity and VOC levels (Tom, 2008). A mathematical model to compute an Environmental Index is presented in equation (1) below. Equation (1) computes how far the actual measured value for VOC, temperature, and humidity are from their ideal value. This computation is expressed as a simple linear score with modifiable weights.

$$EI = \alpha \frac{\Delta VOC_{act}}{\Delta VOC_{max}} + \beta \frac{\Delta T_{act}}{\Delta T_{max}} + \gamma \frac{\Delta RH_{act}}{\Delta RH_{max}} \tag{1}$$

The key terms are equation (1) are:

ΔVOC_{actual} is the absolute value of the difference between measured and ideal VOC level. ΔVOC_{max} is the absolute value of difference between maximum allowable and ideal VOC level. ASHRAE standards and handbooks provide recommendations for VOC_{max} and VOC_{ideal}.

ΔT_{actual} is the absolute value of the difference between measured and ideal temperature. ΔT_{max} is the absolute value of the difference between maximum allowable and ideal temperature. ASHRAE standards and handbooks provide recommendations for T_{max} and T_{ideal}.

ΔRH_{actual} is the absolute value of the difference between measured and ideal relative humidity. ΔRH_{max} is the absolute value of the difference between maximum allowable and ideal relative humidity. ASHRAE standards and handbooks provide recommendations for RH_{max} and RH_{ideal}.

The Environmental Index is computed in a range from 0 to 100%. To accomplish this, three coefficients (α, β and γ) from equation (1) are defined as "priority factors" in equation (2). These three values create a weighted score that determines which parameter (temperature, relative humidity, or VOC level) is most important in the EI computation. In other words, α, β and γ are set by a user depending on one's priority. For example, if a user puts emphasis on VOC levels, α can be the largest multiplier such as 0.5 and β and γ are set to 0.25 each. These three coefficients can also be used to dampen out an exaggerated impact of one term, like VOC's, that can momentarily escalate to very high levels.

$$\alpha + \beta + \gamma = 1 \tag{2}$$

The EI is computed on a continuous basis to give a simple real time estimate of IAQ. This EI term is also integrated over an entire day to give a daily average. Both the instantaneous and average EI values are compared to arbitrary benchmarks to give home occupants a real time indication of air quality. For example, when the EI is above 75% the environmental quality level is categorized as "satisfied".

The EI value provided to the homeowner must be interpreted with caution. A direct linkage between EI and overall IAQ has not been established, at least partially because the EI computation does not include all the parameters that can influence IAQ. The EI value presented to the homeowner is a step in the right direction, but does not measure IAQ in a comprehensive way. In that sense the EI score sacrifices some level of scientific accuracy in order to focus on simplicity and overall ease of use.

3. RENEWW HOUSE AND BIOWALL

The ReNEWW House shown in Figure 1 is a partnership between Purdue University and Whirlpool Corporation for research into highly efficient and sustainable residential buildings. The term "ReNEWW" stands for Retrofitted, Net-zero Energy, Water and Waste. The house was built in 1928 near Purdue University and provides approximately 223 m^2 (2,400 ft^2) of living space. The home was used by university students for many years, but was remodeled starting in 2015 to serve as a testbed for new residential technologies. The home now features a variety of technologies like extensive air sealing and insulation, triple-paned windows, a geothermal heat pump system for heating and cooling, and solar photovoltaic panels on its roof.

Figure 1. Net-Zero High Performance Building, ReNEWW house, in Purdue Campus

One unique feature of the ReNEWW house is a Biowall, which is sustainable solution for improving IAQ using a living plant wall. As shown in Figure 2, the Biowall consists of four trays of plants that are integrated with the Central HVAC system of the house. Return air is pulled through Biowall on its way to being heated or cooled. As the HVAC system runs over time, all the air in the house is eventually pulled through the Biowall. On a typical day, the Biowall achieves at least 1 air change, roughly a quantity of air equal to the indoor volume of the entire house.

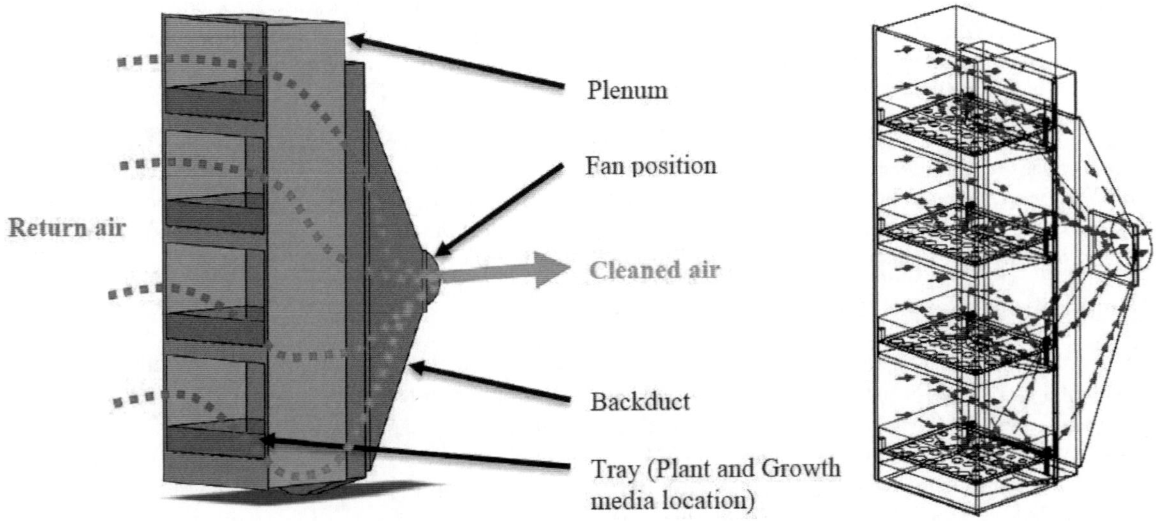

Figure 2. Biowall mechanism with air flow dynamics

When the air conditioning system operates, air gets pulled through loose soil in the plant trays of the Biowall. The microbe community in the plant's root zone slowly metabolizes the Volatile Organic Compounds (VOCs) in the air. Common and potentially harmful VOC's include benzene, formaldehyde and toluene that can be introduced through indoor human activities or chemical sources. Laboratory tests of the Biowall have shown its potential to reduce concentrations of toluene (Alraddadi, 2016).

The image to the left of Figure 3 is the overall Biowall device. It has a dimension of 0.96 m x 0.64 m (77" x 25 ½"). Four 559 mm x 445 mm (22" x 17 ½") plant trays, each with 1.0 m^2 (10.7 ft^2) surface area, are stacked inside. The Biowall frame was made of brushed stainless steel for appearance and to inhibit rust.

Figure 3. Biowall installed in ReNEWW house (left) and a single plant tray (right)

The image to the right of Figure 3 is a close-up of a single plant tray. Plant species are selected based on widely known air purifying plants such as Snake Plant (*Dracaena trifasciata*), Spider Plant (*Chlorophytum comosum*), Golden Pathos (*Epipremnum aureum*), or Philodendron (*hederaceum*). Additional plant varieties can be added to the trays to improve color and overall aesthetics.

A comprehensive Building Automation System monitors and controls Biowall operation in this prototype installation. A number of sensors monitor air temperature, humidity, pressure, and VOC levels. Actuators control LED lights, watering for the plants, and the amount of air passing through the Biowall. A computer algorithm determines how the Biowall operation based on inputs from the sensors. Work is underway to develop a smaller and less expensive microcontroller that has been specifically designed for managing the Biowall (Allocca, 2019).

3.1 Graphic User Interface for Environmental Index

Figure 4 is a prototype graphic user interface that was designed to convey real time information about the Environmental Index in a format useful to a typical homeowner, who may not have the time or the interest to interpret numeric engineering data. The goal is to present complex information in a compelling way that can be quickly viewed and understood from a smart phone or a touch screen controller. This customer-centered approach builds on UX (User Experience) design principles that are commonly used in other applications, but are sometimes overlooked during the implementation of building automation systems (Switzer, 2018).

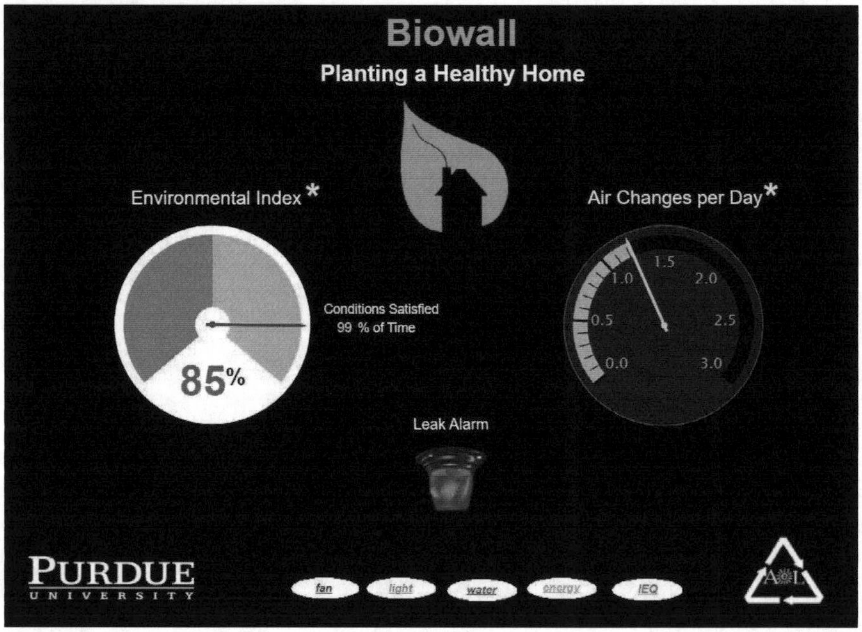

Figure 4. Graphical User Interface for Environmental Index

The dial gage in the upper left of Figure 4 is the Environmental Index, whose score ranges from 0 to 100%. The red area on the dial means the Environmental Index score is at an unacceptably low level (i.e. below 50%) while the green area represents a good Environmental Index (i.e. above 50%). The dial gives a homeowner a quick visual indication of current comfort levels in a format that is easy to understand and act upon, if necessary.

The middle of Figure 4 shows the percent of time that environmental conditions were satisfied for one day. This additional data informs the user about how comfortable conditions were maintained throughout the day. This additional information is helpful because it puts the current Environmental Index in the context of an entire day's worth of data. The percent of time conditions were satisfied could help avoid an overreaction in a situation where the Environmental Index is momentarily low, but would not cause alarm in the context of an entire day's worth of data.

The right of Figure 4 is another gage showing the air changes per day on a scale of 0 to 3. The air change rate is the volume of air passing through the Biowall as compared to the overall volume of the house. In other words, an air change of 1.0 means that the volume of air filtered by the Biowall is equal to the overall volume inside the house. Airflow through the Biowall typically occurs in conjunction with the operation of the HVAC system, so the air change rate tends to be higher when the windows are closed during hot weather or cold weather. A higher air change rate correlates with more air cleaning is taking place via the Biowall.

The lower middle part of Figure 4 has a leak alarm for the Biowall, which tracks the performance of the automatic plant watering system. Although a water leak is unlikely, it would it could be extremely detrimental to the Biowall and the home itself. When tripped, this alarm produces a flashing red signal that is impossible to overlook.

The bottom of Figure 4 has five interactive buttons labelled for the fan, light, water, energy and IEQ. These buttons allow the user to navigate further into the control system to get additional information on specific features of the Biowall. For example, clicking on the "fan" tab takes the user to a graph showing the fan operation over the past day. Serious users can use these features to further optimize Biowall operation by adjusting the fan, light and watering cycles.

The graphic interface in Figure 4 is still a prototype. It is deployed on a website hosted by a commercial building automation system that monitors and controls the Biowall. The graphic is optimized for viewing from a personal computer or tablet, although it can be viewed from a smartphone. Additional work is needed to transition to an interface designed primarily for mobile devices.

4. RESULTS

Table 1 is a small sample of data collected on September 11th, 2019 from the 5:00 AM to 7:00 AM. The data is automatically logged at regular intervals and stored in the controller, where it can be accessed and placed into a spreadsheet. The three middle columns of data represent the three input parameters for the EI computation. The VOC level has units of parts per million (ppm). The Zone Temperature (ZT) was measured in °F, but was converted to °C for consistency in the SI formatting of this paper. The Zone Relative Humidity (ZH) were measured in %. The last column contains the computed Environmental Index, which is a unitless numerical value ranging from 0 to 100%.

Table 1. Conditions of the ReNEWW house

Date	Time	VOC (ppm)	ZT (°C)	ZH (%)	EI (%)
Sep 11th	5:00 AM	955	24.4	64	90
	6:00 AM	1061	24.4	64	87
	7:00 AM	1370	24.4	64	42

Table 1 illustrates how quickly the EI value can change. The air quality in the house was "good" during hours 5 & 6 because the real time measurements of VOC, temperature, and humidity were within acceptable range. The resulting value computed for the EI was close to 90%, well above the 50% threshold for acceptable air quality. During the 7th hour, conditions changed, most likely due to cooking in the house at meal time. The human activities emitted particles to the space that contributed to a significant increase in the VOC level. The EI quickly dropped to 42, which is in the red "needs improvement" zone in terms of air quality.

Figure 5 shows how quickly the Environmental Index (EI) changes with respect to real time changes in VOC, temperature (Temp), or relative humidity (RH) levels. The x-axis shows one day of measurements. The y-axis on the left shows the Environmental Index, zone temperature, and zone humidity on the same scale because their values are the same order of magnitude. It is not standard practice to show parameters that have different units (e.g. temperature and %) on the same axis, but it was done in this instance to highlight the interactions between these terms. The y-axis on the right shows the VOC level on its own scale because it has a significantly different magnitude than the other terms.

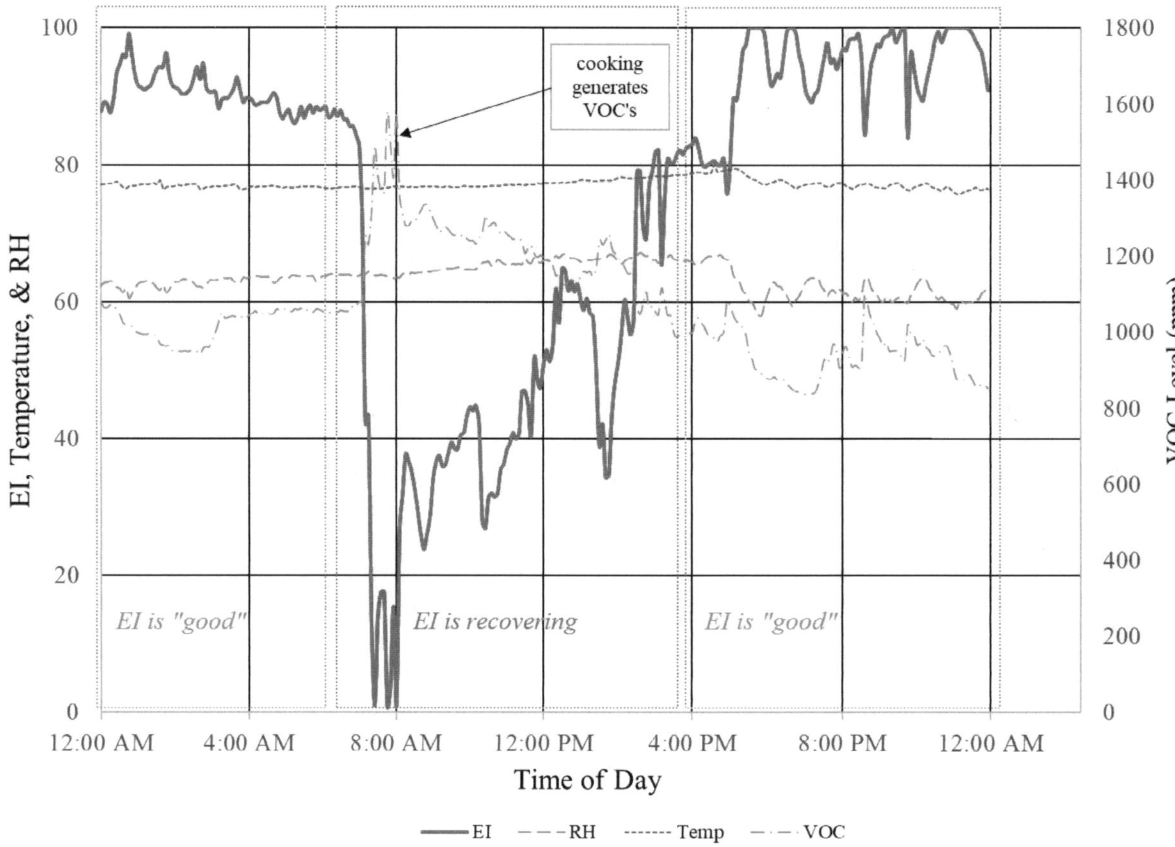

Figure 5. Environmental Index is impacted by real time conditions in the zone

Figure 5 shows that the temperature and humidity stayed relatively stable throughout the test because the HVAC system conditioned the air in the house constantly. The VOC level was initially less than 1000 ppm. With the combination of all three parameters, the computed Environmental Index started at a high value, over 90. However, when the time reached 6am, the VOC level spiked and the Environmental Index decreased accordingly. The VOC level reached its peak around 8am and the EI value dropped to the lowest value of the day, nearly 0. After noon, the VOC level dropped down as time passed and the EI value recovered to a healthier range. The possible explanation of this trend is the human activities that occur in the morning, such as cooking and showering. These activities release particles into the room and therefore increase the VOC level. Similar effects are expected for other HVAC-conditioned buildings, where temperature and humidity stay relatively stable while sudden changes in VOC levels had the biggest impact on EI.

6. CONCLUSION

Indoor Air Quality is important for maintaining health and comfort in the built environment. The Environmental Index was developed as a user-friendly indicator of IAQ that a homeowner could easily understand, interpret, and act upon. The EI is defined based on measurements of VOC's, temperature, and humidity that are compared to established benchmarks. The EI was evaluated in the context of a graphic user interface developed to track the performance of a Biowall, a plant-based filter for improving IAQ in homes. A preliminary test of the graphic user interface showed that it worked as expected. The VOC level had the strongest impact on EI computations because the real time VOC level can rapidly change based on conditions inside a home.

REFERENCES

Alraddadi, Osama Saad (2016), *"Investigating The Toluene Removal of a Botanical Air Filter With a Loose-Packed Growth Media for Growth Media for Potential Energy Savings ". Open Access Theses*. 1231. Purdue University https://docs.lib.purdue.edu/open_access_theses/1231

Allocca, D (2019). *Reducing Size and Cost of Biowall Control System.* Purdue School of Engineering Technology, West Lafayette. 10-20

Newkirk, D., W.; Hutzel, W. J., Dana, M., Qu, M., (2014). *Energy Modeling of a Botanical Air Filter. Paper presented at: International High Performance Buildings Conference*. West Lafayette, IN: Purdue University.

Parkinson, T., Parkinson, A., Dear, R. (2019). *Continuous IEQ monitoring system: Context and development.* Building and Environment, Published by Elsevier Ltd. https://doi.org/10.1016/j.buildenv.2018.12.010

Soreanu, G., Dixon, M., & Darlington, A. (2013*). Botanical Biofiltration of Indoor Gaseous Pollutants - a Mini-Review.* Chemical Engineering Journal, 229 (8), 585-594

Switzer, M. (2018). Applying *User Experience (UX) Principles to Net Zero Energy Buildings.* 5th International High Performance Building Conference at Purdue.

Tom, S. (2008). Manage Energy and Comfort. *ASHRAE Journal, June 2008*, 18-23.

Characteristics of Graphite and MnO_2 Enhanced Cement-Based Thermoelectric Composite for Low-Temperature Applications

Xiaoli LIU[1], Ming QU[2]*, Alan Phong Tran NGUYEN[3]

[1]Lyles School of Civil Engineering, Purdue University,
West Lafayette, IN 47907, U.S.A
liu2251@purdue.edu

[2]Lyles School of Civil Engineering, Purdue University,
West Lafayette, IN 47907, U.S.A
mqu@purdue.edu

[3]School of Mechanical Engineering, Purdue University,
West Lafayette, IN 47907, U.S.A
nguye218@purdue.edu

* Corresponding Author

ABSTRACT

Thermoelectric (TE) cement composite is a new type of TE material. Unlike ordinary cement, TE cement can mutually convert thermal energy to electrical energy due to the addition of carbon fibers, metal oxide nanoparticles, etc. In hot summer or cold winter, the significant temperature difference between indoor and outdoor can be used by TE cement to generate electricity. On the other hand, given power input, the same material can provide cooling/ heating to adjust room temperature. Therefore, TE cement has certain energy-saving potential in the application of building enclosures and energy systems. Its ability to convert different forms of energy and use low-grade energy is conducive to the operation of net-zero buildings. In this study, a novel TE cement composite, MnO_2 and graphite enhanced cement, was firstly fabricated. The surface morphology of the composites was analyzed by scanning electron microscopy. The performance indicators of TE materials include the power factor and dimensionless figure of merit ZT, which are calculated by Seebeck coefficient, thermal conductivity, and electrical conductivity. The values of these five TE properties were measured and calculated by a Physical Property Measurement System at different temperatures. Compared with the cement reinforced by graphite alone, it is confirmed that MnO_2 nanoparticles have a positive effect on the enhancement of the TE performance for cement composites. The highest ZT of 5wt.% graphite and 10wt.% MnO_2 enhanced cement composite prepared by the dry method is about 6.2×10^{-6} at 350 K.

1. INTRODUCTION

Energy crisis and environmental degradation make people pay more and more attention to the rational use of energy. Therefore, thermoelectric (TE) technology which can improve the utilization of clean and renewable energy and recycle low-grade and waste energy has become a promising research topic. TE technology can convert heat into electricity. In the hot summer, the exterior surface of the city (about 60 °C) will aggravate the urban heat island effect and have an essential impact on urban energy consumption. Therefore, researchers combine the traditional TE module with the pavement, which not only takes away the heat of the road to cool the surface, but also make use of low-grade energy to generate electricity (Datta et al., 2017; Hasebe et al., 2006; Jiang et al., 2017, 2018; Tahami et al., 2019). On the other hand, TE technology can directly convert electricity into heat absorption and dissipation. TE cooler/heater has been used in building envelope, to provide active cooling or heating for buildings (Ibañez-Puy et al., 2018; Irshad et al., 2015; Khire et al., 2005; Z. B. Liu et al., 2015; Zhongbing Liu et al., 2015; ZhongBing Liu et al., 2015). The system shows the ability of TE module to regulate room temperature and reduce heating or cooling loads. The energy conversion through the TE module in buildings has high flexibility, scalability, and reliability, and does not have harmful effects on the environment in operation (He et al., 2015). Thus, there are continuously increasing studies focused on the application of TE materials in pavement and building envelopes. However, the existing studies use commercial TE modules that need to be embedded inside or on the surface of the infrastructure. In order to improve the integration of the TE system and maximize the use of temperature differences for a better performance of heat transfer, the TE cement composite as building envelope materials has been proposed.

The key to improving the performance of TE cement-based enclosure is to develop cement with high TE performance. A unique property of TE materials is the Seebeck coefficient (S), which is the ratio of the output open-circuit voltage (ΔV) to the input temperature difference (ΔT). There are also two performance indicators, power factor (PF) and the dimensionless figure of merit (ZT), representing the overall performance of power conversion efficiency and the coefficient of performance of cooling/heating system. For TE cement, the choice of additives plays a vital role in its TE properties. As shown in Figure 1, researchers studied carbon fiber and steel fiber as additives from 1998 to 2008. Since 2013, more and more studies have been conducted on metal oxide powders such as Fe_2O_3, ZnO, and MnO_2. In recent years, research began to focus on graphite/graphene and carbon nanotubes as additives.

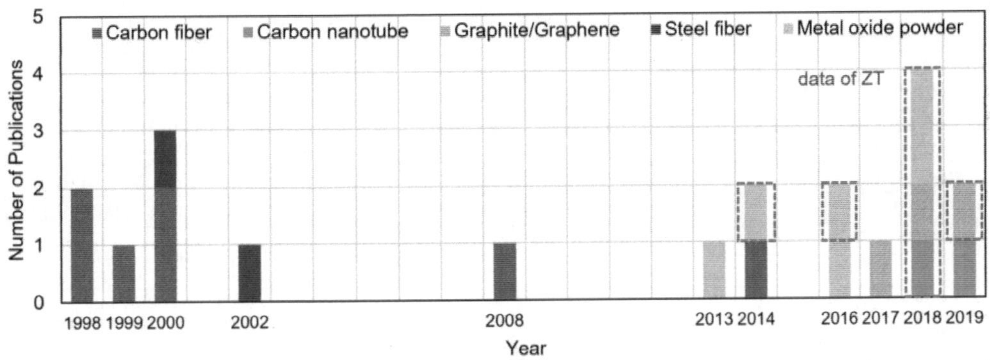

Figure 1: Number of publications related to TE performances of enhanced cement composites versus the year

The data of TE properties for different TE cement published in the literature from 1998 to 2020 are summarized in Table 1. It is found that the most potential additives for ZT enhancement are graphene and aluminum-doped zinc oxide (AZO). Graphene and expanded graphite help to improve conductivity, while metal oxides such as AZO and MnO_2 help to improve the Seebeck coefficient. Therefore, it is necessary to study the effect of graphite combined with MnO_2 on the TE performance. In terms of characterization, before 2014, most studies only included measurements of the Seebeck coefficient. However, the Seebeck coefficient should not be a single indicator of TE performance. The overall performance of TE materials needs a comprehensive analysis of thermal conductivity, electrical conductivity, Seebeck coefficient, etc.

This study firstly tried to produce high-ZT TE cement composites with graphite and MnO_2 as additives. A graphite enhanced cement composite and a new MnO_2 and graphite enhanced cement composite were prepared. Then the Seebeck coefficient, thermal conductivity, and conductivity of the samples were measured by DynaCool physical property measurement system (PPMS) to study the variation of TE properties at different temperatures. In this characterization, all three TE properties are measured almost simultaneously on the same sample. The paper is divided into four parts. The second section describes the detailed experimental methods from three aspects including material preparation, sample fabrication, and characterization techniques. Results of the Seebeck coefficient, thermal conductivity, conductivity, power factor, and are presented and analyzed in Section 3. Finally, the conclusions of the experiment are made in Section 4.

Table 1: The data of TE properties for different TE cement published in the literature from 1998 to 2020

Type	Additives	Maximum Seebeck μV/K	Electrical conductivity S/m	Thermal conductivity W/mK	ZT	Ref.
Carbon fiber	PAN-based	+16.67	-	-	-	(M. Sun et al., 1998a, 1998b)
	Isotropic pitch-based	+5.44	-	-	-	(Wen & Chung, 1999)
	Bromine-intercalated	+5.47	-	-	-	(Wen & Chung, 2000b)
	Isotropic pitch-based	+21.2	-	-	-	(Wen & Chung, 2000b)
	Carbon fiber	+127	-	-	-	(Bahar & Salih, 2008)
	PAN-based	+22.07	0.2008	0.22	1.33×10^{-7}	(Wei, Nie, et al., 2014)
Carbon nanotube	Ordinary CNT	+52.5	80	0.947	7×10^{-5}	(Wei, Fan, et al., 2018)
	p-doped CNT n-doped CNT	-100 -50	0.55 1.9	-	-	(Tzounis et al., 2019)
Graphite graphene	PAN + graphite	-10	0.078	-	-	(Wei et al., 2017)
	Expanded graphite	-20	2480	3.213	6.82×10^{-4}	(Wei, Zhao, et al., 2018)
	Graphene	34	1168	1.067	4.4×10^{-4}	(Ghosh et al., 2019)
	Graphite+AZO	141.5	1390	0.99	1.01×10^{-2}	(Ghosh et al., 2020)
Steel fiber	Stainless steel fiber	-63.9	3.13×10^{-3}	-	-	(Wen & Chung, 2000a)
	Stainless steel fiber	-59	2.08×10^{-3}	-	-	(Wen & Chung, 2002)
Metal oxide	$Ca_3Co_4O_9$ + PAN	+58.6	-	-	-	(Wei et al., 2013)
	Fe_2O_3/Bi_2O_3 + PAN	+100.28	-	-	-	(Wei, Hao, et al., 2014)
	Fe_2O_3 + PAN	+1200	5×10^{-3}	-	8×10^{-4}	(Wei et al., 2016)
	ZnO + Fe_2O_3	+3300	1.7×10^{-6}	0.75	7.41×10^{-9}	(Ji et al., 2016)
	MnO_2	-3085	1.88×10^{-4}	0.72	7.60×10^{-7}	(Ji et al., 2018)
	ZnO/AZO	+0.188	5.87×10^{-2}	0.6	1.03×10^{-12}	(Ghahari et al., 2017)

2. Methodology

2.1 Material Preparation

This study used Portland cement (Type III) from Buzzi Unicem, USA. The composition of Type III cement is very similar to that of type I cement, but the particle size of cement is smaller. The additive is surface-enhanced flake graphite (SEFG), purchased from Asbury Graphite Mills, USA. It is a very light material with a typical surface area as high as $27 m^2/g$. This flake graphite is manufactured by heating expandable graphite at high temperatures to expand its molecular structure, and then compress it into a flake shape. Another additive used is the manganese dioxide nanoparticles (MnO_2) with an average diameter of 50 nanometers, from US Research Nanomaterials, Inc in the United States. The purity is 98%, and the density is 5.02 g/cm^3.

2.2 Specimen Fabrication

Two samples are fabricated in this research. The first sample of cement composite (CC) contains SEFG and type III cement, and the mass ratio of SEFG to cement is 10%, marked as 10SEFG/CC. The second sample of cement composite (CC) contains SEFG and type III cement, and the mass ratio of SEFG to cement and MnO_2 are 10 and 5%, respectively, marked as 10SEFG-5MnO_2/CC. Schematic diagrams of the preparation process by the dry mixing and compressing method are illustrated in Figure 2. In the dry method, the cement powder and additives were mixed without the use of water and dispersant. Then, the homogeneous mixture was placed in a customized cylindrical stainless-steel mold with a metal strip and die block. A Mechanical Test System (MTS) applied the pressure at 50 MPa to the metal strip, which compressed the dry mixture into a cylinder with a diameter of 9.5 mm (3/8 inches) and a height of 5~10 mm. The compact cylinder sample was first put on a saturated (water) sponge for 24 hours to pre-cure, and it then was soaked in a tank full of water for three days to cure the sample at ambient atmosphere fully. After curing, the sample was stored in the lab under a constant temperature of 25 °C and relative humidity of 50%.

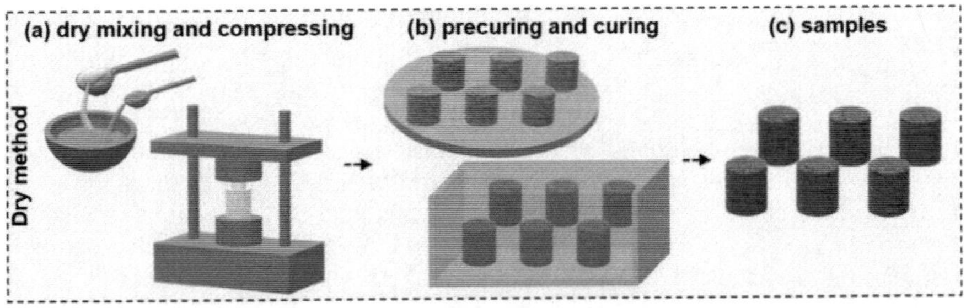

Figure 2: The schematic diagram of the procedures for sample fabrication via the dry method: (a) dry mixing and compressing, (b) pre-curing and curing, and (c) sample storage

2.3 Characterization Techniques of TE Properties

The Thermal Transport Option (TTO) module in DynaCool® Physical Property Measurement System (PPMS) was adopted to measure TE properties at low temperatures (300-350K), including the Seebeck coefficient, thermal conductivity, and electrical conductivity of the TE cement composite. Figure 3 shows the puck for sample mounting in the TTO module. Two copper plates were bonded at the ends of the cement sample by the silver epoxy EPO-TEK® H20E. The copper plates were tightly connected with two sensors, one heater, and a cold foot by screws. The top-right sensor is used to measure the voltage as well as the hot-side temperature. The sensor at the left bottom is for voltage and cold-side temperature measurement. The heat was imposed on the sample by a heater at the top-left side. It also helps to impose the current to the sample for resistance measurement.

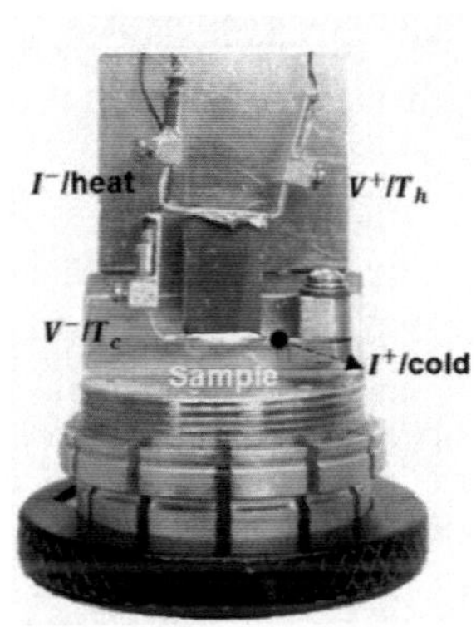

Figure 3: The diagram of the cement sample mounted at the testing puck

In this study, the PPMS-TTO measured the thermal conductivity by the steady-state heat conduction method. It monitored the temperature gradient $(T_h - T_c)$ across the sample given a known heat input through the sample. When the system achieved a steady state (assuming the ratio of temperature variation to the absolute temperature is lower than 0.1%), the thermal conductivity could be calculated based on the heat input, the size of the sample, and the temperature difference across the sample, according to Fourier's Law of heat conduction (equation (1)). The temperature gradient was determined by measuring the absolute temperature of Cernox resistive thermometers at the hot and cold end of the sample. The sample is relatively small to ensure the uniform heat is transferred through the sample vertically. The puck was thermally insulated in the chamber so that the heat loss was minimized. The light influence of convective and radiative heat loss was then considered in the correction procedure.

$$k = \frac{QL}{A\Delta T} \tag{1}$$

At the same time, due to the temperature gradient across the sample, the TE cement composite automatically generated a voltage output $(V^+ - V^-)$. The ratio of the voltage output to the temperature gradient, both measured in real-time, was the value of the Seebeck coefficient (equation (2)). The thermal voltage across the sample was measured simultaneously using a nano voltmeter. The low voltage signal was amplified by an ultra-low-noise amplifier.

$$S = \frac{\Delta V}{\Delta T} \quad (2)$$

After achieving the steady-state and measuring the corresponding Seebeck coefficient and thermal conductivity, the system switched to the electrical conductivity measurement mode. After removing the heat and allowing the sample to return to an isothermal state, an AC resistance bridge measured the two-probe electrical conductivity according to Ohm's Law.

The sample was initially stabilized at 300K before all measurements were taken. The operating temperature of the sample, determined by averaging T_h and T_c, was increased in increments of 10K up to 350K. Therefore, six tests were conducted in each sample measurement. Each test is initialized by bringing the PPMS chamber to a high-vacuum state.

2.4 Performance indicators

There are also two performance indicators calculated by three TE properties measured above, that is, the power factor (PF) and the dimensionless figure of merit (ZT). Power factor is the product of the square of Seebeck coefficient and electrical conductivity, as listed in equation (3). High power factor is the key to the actual capacity of power generation of the TE generator.

$$PF = \sigma S^2 \quad (3)$$

The figure of merit equals power factor divided by thermal conductivity, as listed in equation (4). High ZT is conducive to improving the coefficient of performance of TE cooler, that is, reducing the operating cost.

$$ZT = \sigma S^2 / k \quad (4)$$

3. Results and Analysis

In this section, the results of TE transport properties of SEFG-enhanced cement composites obtained from PPMS operating in TTO mode, including thermal conductivity, electrical conductivity, Seebeck coefficient, the dimensionless Figure of merit, as well as the power factor are compared and analyzed. The impact of MnO_2 on the TE performance of SEFG/CC is observed

3.1 Seebeck coefficient

Figure 4 shows the comparisons of the Seebeck coefficient of 10SEFG/CC and $5MnO_2$-10SEFG/CC. It can be found that by adding 5wt.% MnO_2 powder, the new cement composite exhibits a relatively higher Seebeck coefficient than 10SEFG/CC. The average values of the Seebeck coefficient of 10SEFG/CC and $5MnO_2$-10SEFG/CC were 16.74 and 14.99 μV/K, respectively. This may be due to the fact that adding MnO_2 brings more carrier scattering. And the variations of the Seebeck coefficient with respect to temperature for different materials are also different. The Seebeck coefficient of 10SEFG/CC decreases with higher temperature while that of $5MnO_2$-10SEFG/CC is more stable.

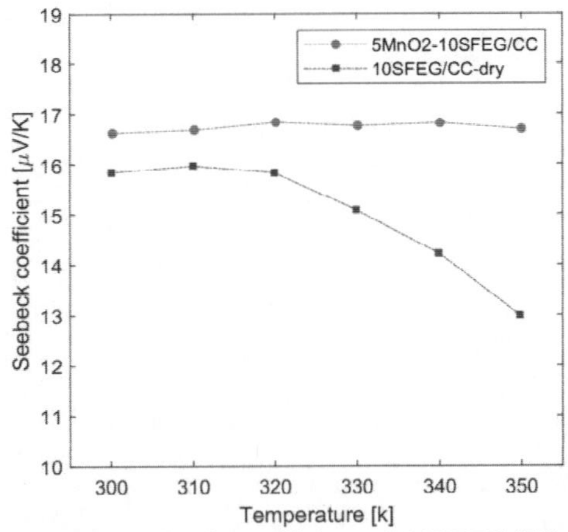

Figure 4: The comparisons of Seebeck coefficient of 10SEFG/CC and $5MnO_2$-10SEFG/CC

3.2 Thermal conductivity and electrical conductivity

Figure 5 shows the comparisons of thermal conductivity of 10SEFG/CC as well as the $5MnO_2$-10SEFG/CC. It can be found that by adding 5wt.% MnO_2 powder, the new cement composite exhibits relatively higher thermal conductivity than 10SEFG/CC. The average value of thermal conductivity of 10SEFG/CC was 1.78 W/mK. The average value of thermal conductivity of $5MnO_2$-10SEFG/CC was 2.16 W/mK. This is because introducing MnO_2 increases the conductive paths inside the cement matrix. This also explains the improvement of electrical conductivity by adding MnO_2 to the cement composite. The average values of electrical conductivity of 10SEFG/CC and $5MnO_2$-10SEFG/CC were 102.58 S/m and 135.83 S/m, respectively.

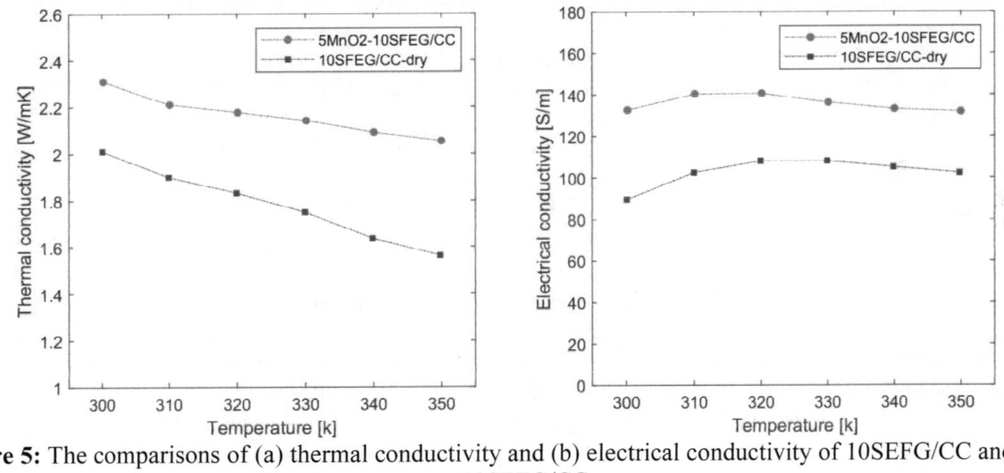

Figure 5: The comparisons of (a) thermal conductivity and (b) electrical conductivity of 10SEFG/CC and $5MnO_2$-10SEFG/CC

3.3 Power Factor and the figure of merit

Figure 6 shows the comparisons of power factor and figure of merit of 10SEFG/CC as well as the $5MnO_2$-10SEFG/CC. Even though the introduction of MnO_2 leads to higher thermal conductivity, which is not conducive to ZT, the increase in electrical conductivity and Seebeck coefficient offset the negative effects and as a result, the overall PF and ZT are improved. The average PF and ZT were increased by 64.43% and 36.21%, respectively. The highest PF was 0.0399 at 320K, and the highest ZT was 6.2×10^{-6} at 350K for $5MnO_2$-10SEFG/CC.

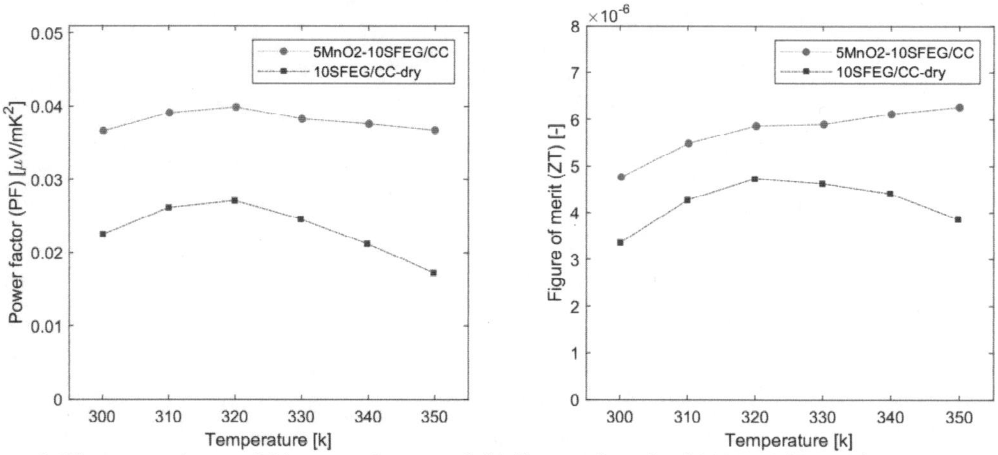

Figure 6: The comparisons of (a) power factor and (b) figure of merit of 10SEFG/CC and $5MnO_2$-10SEFG/CC

3.4 Discussion

Although Maxwell-Boltzman theory cannot be directly applied to the cementitious material, which is amorphous and non-crystal, it is still possible to use the band theory to explain the variation of TE properties of cement composites. The improvements of thermal conductivity and electrical conductivity of graphite-MnO_2 enhanced cement composite is mainly due to the increase of carrier concentration induced by MnO_2 nano powder. However, the higher carrier concentration usually leads to a lower Seebeck coefficient. Hence, a possible solution to the simultaneous increase of conductivity and Seebeck coefficient is that the effective mass of the composite has been enhanced by the MnO_2 nano powder and it counteracts the adverse effects of increased concentration on Seebeck (J. Sun et al., 2010). The introduction of nanoparticles increases the creation of energy barriers which prevent carriers from moving between grains. Carrier scattering and filtering helps to improve the electrochemical potential of the substance (Martin et al., 2009). As a result, the enlarged effective mass, coupled with the enhancement in carrier concentration and high-density defects interfaces, lead to the increases of all TE performance simultaneously.

4. CONCLUSIONS

In this study, the thermoelectric performance of the graphite and MnO_2 enhances cement composite was investigated experimentally. The physical property measurement system was used to simultaneously measure the Seebeck coefficient, thermal conductivity, and electrical conductivity of the samples to study the variation of the figure of merit and the power factor at different operating temperatures ranging from 300K to 350K.

The measured data showed that introducing MnO_2 nanoparticles to SEFG/CC can enhance TE behavior. By adding 5wt.% MnO_2 nanopowder, we observed that the brand-new cement composite exhibited a better TE performance than the 10SEFG/CC. The highest *ZT* of 6.2×10^{-6} was obtained at 350K, which was increased by 36.2% compared with the dry 10SEFG/CC.

NOMENCLATURE

CC	Cement Composite	(–)
k	Thermal conductivity	(W/mK)
MTS	Mechanical Test System	(–)
PF	Power Factor	(μV/mK^2)
PPMS	Physical Property Measurement System	(–)
SEFG	Surface-Enhanced Flake Graphite	(–)
SEM	Surface-Enhanced Flake Graphite	(–)
S	Seebeck coefficient	(μV/K)
TE	Thermoelectric	(–)
TTO	Thermal Transport Option	(–)

ZT	Figure of merit coefficient	(–)
σ	Electrical conductivity	(S/m)

REFERENCES

Bahar, D., & Salih, Y. (2008). Thermoelectric behavior of carbon fiber reinforced lightweight concrete with mineral admixtures. *New Carbon Materials*, *23*(1), 21–24. https://doi.org/10.1016/S1872-5805(08)60009-8

Datta, U., Dessouky, S., & Papagiannakis, A. T. (2017). Harvesting Thermoelectric Energy from Asphalt Pavements. *Transportation Research Record: Journal of the Transportation Research Board*. https://doi.org/10.3141/2628-02

Ghahari, S., Ghafari, E., & Lu, N. (2017). Effect of ZnO nanoparticles on thermoelectric properties of cement composite for waste heat harvesting. *Construction and Building Materials*, *146*, 755–763. https://doi.org/10.1016/J.CONBUILDMAT.2017.04.165

Ghosh, S., Harish, S., Ohtaki, M., & Saha, B. B. (2020). Enhanced figure of merit of cement composites with graphene and ZnO nanoinclusions for efficient energy harvesting in buildings. *Energy*, *198*, 117396. https://doi.org/10.1016/j.energy.2020.117396

Ghosh, S., Harish, S., Rocky, K. A., Ohtaki, M., & Saha, B. B. (2019). Graphene enhanced thermoelectric properties of cement based composites for building energy harvesting. *Energy and Buildings*, 109419. https://doi.org/10.1016/j.enbuild.2019.109419

Hasebe, M., Kamikawa, Y., & Meiarashi, S. (2006). Thermoelectric Generators using Solar Thermal Energy in Heated Road Pavement. *2006 25th International Conference on Thermoelectrics*, 697–700. https://doi.org/10.1109/ICT.2006.331237

He, W., Zhang, G., Zhang, X., Ji, J., Li, G., & Zhao, X. (2015). Recent development and application of thermoelectric generator and cooler. *Applied Energy*, *143*, 1–25. https://doi.org/10.1016/J.APENERGY.2014.12.075

Ibañez-Puy, M., Martín-Gómez, C., Bermejo-Busto, J., Sacristán, J. A., & Ibañez-Puy, E. (2018). Ventilated Active Thermoelectric Envelope (VATE): Analysis of its energy performance when integrated in a building. *Energy and Buildings*, *158*, 1586–1592. https://doi.org/10.1016/J.ENBUILD.2017.11.037

Irshad, K., Habib, K., Thirumalaiswamy, N., & Saha, B. B. (2015). Performance analysis of a thermoelectric air duct system for energy-efficient buildings. *Energy*, *91*, 1009–1017. https://doi.org/10.1016/J.ENERGY.2015.08.102

Ji, T., Zhang, X., & Li, W. (2016). Enhanced thermoelectric effect of cement composite by addition of metallic oxide nanopowders for energy harvesting in buildings. *Construction and Building Materials*, *115*, 576–581. https://doi.org/10.1016/j.conbuildmat.2016.04.035

Ji, T., Zhang, X., Zhang, X., Zhang, Y., & Li, W. (2018). Effect of Manganese Dioxide Nanorods on the Thermoelectric Properties of Cement Composites. *Journal of Materials in Civil Engineering*, *30*(9), 04018224. https://doi.org/10.1061/(ASCE)MT.1943-5533.0002401

Jiang, W., Xiao, J., Yuan, D., Lu, H., Xu, S., & Huang, Y. (2018). Design and experiment of thermoelectric asphalt pavements with power-generation and temperature-reduction functions. *Energy and Buildings*. https://doi.org/10.1016/j.enbuild.2018.03.049

Jiang, W., Yuan, D., Xu, S., Hu, H., Xiao, J., Sha, A., & Huang, Y. (2017). Energy harvesting from asphalt pavement using thermoelectric technology. *Applied Energy*. https://doi.org/10.1016/j.apenergy.2017.08.091

Khire, R. A., Messac, A., & Van Dessel, S. (2005). Design of thermoelectric heat pump unit for active building envelope systems. *International Journal of Heat and Mass Transfer*, *48*(19–20), 4028–4040. https://doi.org/10.1016/J.IJHEATMASSTRANSFER.2005.04.028

Liu, Z. B., Zhang, L., Gong, G., Luo, Y., & Meng, F. (2015). Experimental study and performance analysis of a solar thermoelectric air conditioner with hot water supply. *Energy and Buildings*, *86*, 619–625. https://doi.org/10.1016/J.ENBUILD.2014.10.053

Liu, ZhongBing, Zhang, L., Gong, G., & Han, T. (2015). Experimental evaluation of an active solar thermoelectric radiant wall system. *Energy Conversion and Management*, *94*, 253–260. https://doi.org/10.1016/J.ENCONMAN.2015.01.077

Liu, Zhongbing, Zhang, L., Gong, G., Luo, Y., & Meng, F. (2015). Evaluation of a prototype active solar thermoelectric radiant wall system in winter conditions. *Applied Thermal Engineering*, *89*, 36–43. https://doi.org/10.1016/J.APPLTHERMALENG.2015.05.076

Martin, J., Wang, L., Chen, L., & Nolas, G. S. (2009). Enhanced Seebeck coefficient through energy-barrier

scattering in PbTe nanocomposites. *Physical Review B - Condensed Matter and Materials Physics*, *79*(11), 115311. https://doi.org/10.1103/PhysRevB.79.115311
Sun, J., Yeh, M. L., Jung, B. J., Zhang, B., Feser, J., Majumdar, A., & Katz, H. E. (2010). Simultaneous increase in seebeck coefficient and conductivity in a doped poly(alkylthiophene) blend with defined density of states. *Macromolecules*, *43*(6), 2897–2903. https://doi.org/10.1021/ma902467k
Sun, M., Li, Z., Mao, Q., & Shen, D. (1998a). Study on the Hole Conduction Phenomenon in Carbon Fiber-Reinforced Concrete. *Cement and Concrete Research*, *28*(4), 549–554. https://doi.org/10.1016/S0008-8846(98)00011-8
Sun, M., Li, Z., Mao, Q., & Shen, D. (1998b). Thermoelectric percolation phenomena in carbon fiber-reinforced concrete. *Cement and Concrete Research*, *28*(12), 1707–1712. https://doi.org/10.1016/S0008-8846(98)00161-6
Tahami, S. A., Gholikhani, M., Nasouri, R., Dessouky, S., & Papagiannakis, A. T. (2019). Developing a new thermoelectric approach for energy harvesting from asphalt pavements. *Applied Energy*. https://doi.org/10.1016/j.apenergy.2019.01.152
Tzounis, L., Liebscher, M., Fuge, R., Leonhardt, A., & Mechtcherine, V. (2019). P- and n-type thermoelectric cement composites with CVD grown p- and n-doped carbon nanotubes: Demonstration of a structural thermoelectric generator. *Energy and Buildings*, *191*, 151–163. https://doi.org/10.1016/j.enbuild.2019.03.027
Wei, J., Fan, Y., Zhao, L., Xue, F., Hao, L., & Zhang, Q. (2018). Thermoelectric properties of carbon nanotube reinforced cement-based composites fabricated by compression shear. *Ceramics International*, *44*(6), 5829–5833. https://doi.org/10.1016/J.CERAMINT.2018.01.074
Wei, J., Hao, L., He, G. P., & Yang, C. L. (2013). Thermoelectric Power of Carbon Fiber Reinforced Cement Composites Enhanced by Ca3Co4O9. *Applied Mechanics and Materials*, *320*, 354–357. https://doi.org/10.4028/www.scientific.net/AMM.320.354
Wei, J., Hao, L., He, G., & Yang, C. (2014). Enhanced thermoelectric effect of carbon fiber reinforced cement composites by metallic oxide/cement interface. *Ceramics International*, *40*(6), 8261–8263. https://doi.org/10.1016/J.CERAMINT.2014.01.024
Wei, J., Nie, Z., He, G., Hao, L., Zhao, L., & Zhang, Q. (2014). Energy harvesting from solar irradiation in cities using the thermoelectric behavior of carbon fiber reinforced cement composites. *RSC Adv.*, *4*(89), 48128–48134. https://doi.org/10.1039/C4RA07864K
Wei, J., Zhang, Q., Zhao, L., Hao, L., & Nie, Z. (2017). Effect of moisture on the thermoelectric properties in expanded graphite/carbon fiber cement composites. *Ceramics International*, *43*(14), 10763–10769. https://doi.org/10.1016/J.CERAMINT.2017.05.088
Wei, J., Zhang, Q., Zhao, L., Hao, L., & Yang, C. (2016). Enhanced thermoelectric properties of carbon fiber reinforced cement composites. *Ceramics International*, *42*(10), 11568–11573. https://doi.org/10.1016/J.CERAMINT.2016.04.014
Wei, J., Zhao, L., Zhang, Q., Nie, Z., & Hao, L. (2018). Enhanced thermoelectric properties of cement-based composites with expanded graphite for climate adaptation and large-scale energy harvesting. *Energy and Buildings*, *159*, 66–74. https://doi.org/10.1016/J.ENBUILD.2017.10.032
Wen, S., & Chung, D. D. . (1999). Seebeck effect in carbon fiber-reinforced cement. *Cement and Concrete Research*, *29*(12), 1989–1993. https://doi.org/10.1016/S0008-8846(99)00185-4
Wen, S., & Chung, D. D. . (2000a). Seebeck effect in steel fiber reinforced cement. *Cement and Concrete Research*, *30*(4), 661–664. https://doi.org/10.1016/S0008-8846(00)00205-2
Wen, S., & Chung, D. D. . (2000b). Enhancing the Seebeck effect in carbon fiber-reinforced cement by using intercalated carbon fibers. *Cement and Concrete Research*, *30*(8), 1295–1298. https://doi.org/10.1016/S0008-8846(00)00341-0
Wen, S., & Chung, D. D. . D. L. (2002). Origin of the thermoelectric behavior of steel fiber cement paste. *Cement and Concrete Research*, *32*(5), 821–823. https://doi.org/10.1016/S0008-8846(01)00754-2

ACKNOWLEDGEMENT

This paper is based upon work supported by the National Science Foundation under grant no. CBET-1805818.

3D CFD Simulation of a New Ventilated Roof

Michele CALATI[1], Luca DORETTI[2], Cristiano GATTO[3], Simone MANCIN[1,*]

[1]University of Padova, Department of Management and Engineering,
Vicenza, 36100, Italy
michele.calati@phd.unipd.it, simone.mancin@unipd.it

[2]University of Padova, Department of Civil, Environmental and Architectural Engineering,
Padova, 35131, Italy
luca.doretti@unipd.it

[3]AERTetto s.r.l.,
Preganziol (TV), 31022, Italy
gatto.cristiano@aertetto.it

* Corresponding Author

ABSTRACT

In the last decades, energy management and saving have become challenging issues. Considering the building sector (residential or industrial), different technologies have been developed in order to realize tangible energy savings, such as: ventilated roof, double facades, glazed surfaces, etc. Nonetheless, it is important for these new technologies to contemporary assure the human thermal comfort. Passive cooling (or heating) technologies are of actual interest. Low or near-zero energy buildings can only be realized as a result of the good design of all their components; specifically, the roofs call for particular attention as they take large parts of a building's total surface area. This paper presents a comparison between an innovative ventilated roof, based on an original design of the support and a traditional one. A 3D numerical model is developed to analyze the air flow and to compute the achievable benefits in terms of reduction of the summer heating gains. The simulations were performed by varying the solar irradiance from 600 to 1000 W m^{-2}. The investigation is conducted comparing a ventilated roof assembly to the same traditional structure, assuming buoyancy-driven airflow. Two roof types are studied: an insulated roof and a non-insulated one. The results reveal that the ventilated roof leads to a great reduction of the total amount of solar heat gains for all the simulated scenarios.

1. INTRODUCTION

In the last decades, for engineers, architect, designers, indoor thermal comfort has become a significative challenge. Focusing on building air conditioning, Gagliano *et al.* (2012) reported that in the European Countries the total cooled of floor area is destined to grow up to 2000 million m^2 in 2020 (while it was 1000 million m^2 in 2012). It means that the energy consumption will reach more than 100000 GWh/year only for building cooling.

The building envelopes are in direct contact with the external air and they play a fundamental role in the thermal equilibrium of the buildings' indoor ambient. Therefore, different technologies have been proposed and analyzed to dissipate the heat coming from the outdoors and to prevent it to be transferred into the indoor environment. Ventilated roof is one the most interesting technology to limit the entrance of the thermal loads into a confined space.

The typical structure of the ventilated roof is the following: an upper slab, an air cavity and a lower slab. The upper slab, which can be composed of metal or tiles is in direct contact with the external environment. The lower slab, instead, separates the indoor space from the roof structure itself. If the roof is insulated, different insulation layers can usually compose the lower slab. The air flows in the cavity between the slabs. If the airflow is buoyancy driven, the ventilated roof is considered a passive technology for reducing heat loads.

In the open literature different experimental, theoretical and numerical studies on the ventilated roofs can be retrieved. Puangsombut *et al.* (2007) focused on an inclined roof cavity and studied the airflow rate and the convective heat transfer applying a constant heat flux on the upper cavity surface. They gave results in terms of Reynolds and Nusselt

numbers. The authors considered the presence of a radiant barrier, and they stated that the heat gains can be reduced up to 50%. Susanti *et al.* (2011) investigated a 4.8-m long inclined roof. They analyzed different opening ratio and heating conditions measuring the air velocity and temperature in correlation to the heat flux, opening ratio, cavity spacing and the structure inclination. Tong and Li (2014), using the circuit transformation theory, developed a theoretical model to estimate the amount of heat flux transferrable by a naturally ventilated roof. They obtained that the heat flux is influenced by the solar reflectance of exterior surface, the infrared emittance of cavity surface, the thermal resistance of lower and upper roof slabs, the roof inclination and cavity spacing. Ciampi *et al.* (2005) carried out an energy analysis of ventilated and micro-ventilated roofs. They verified that when the cavity is less than 0.06 m thick the airflow is laminar and the micro-ventilation occurs.
Gagliano *et al.* (2012) analysed three different typologies of ventilated roof, setting the thermal insulation layers in different locations, by means of 2D CFD simulations They found that the performance of the ventilated roof is related to the position of the insulation layer; the best results were achieved when it was inserted under the air layer near the cold surface. Bianco *et al.* (2018) conducted 2D analysis to investigate wall temperature profiles, air velocity and heat transfer coefficients in cavities to find out the optimal configuration. Ray *et al.* (2014) simulated a 3D full-scale naturally ventilated atrium using k-ε, RNG k-ε and LES turbulence models. Liberati *et al.* (2009) analyzed the 3D profile of air temperature in a duct of a ventilated roof for livestock housing. They evaluated the temperature in the house and obtained the optimal cavity thickness which minimized it. An unsteady 3D simulation of a naturally ventilated roof was also conducted by Li *et al.* (2016). In general, three-dimensional numerical analyses permit to obtain more accurate solutions and they should be preferred.
This paper presents a steady-state 3D CFD simulation in Ansys Fluent 18.2 of a new ventilated roof. In particular, the presence of a novel design of the support and its effect on the air thermal fluid dynamics is discussed. Significative reduction of the specific heat flux transferred into the room is obtained for two ventilated roof typologies (insulated and non-insulated) at three different solar thermal loads (I = 600, 800 and 1000 W/m^2).

2. PHYSICAL AND NUMERICAL MODELS

2.1 New Ventilated Roof Description

a) b)

Figure 1: Details of the ventilated roof structure a), of the support b)

In Figure 1a, a portion of the new ventilated roof is presented. It can be observed that the ventilated roof is composed by a series of convex tiles (forming the upper slab) which are simply placed on the plastic support (Figure 1b). They don't need to be fixed by other tools like screws, etc.
To create the air cavity, a concave tile is located onto two adjacent convex tiles, as it can be appreciated in Figure 1a. The main properties of the support and tiles are listed in Table 1.

Table 1: Material main properties

Material	λ [W m^{-1} K^{-1}]	ρ [kg m^{-3}]	C_p [J kg^{-1} K^{-1}]
Polipropylene (Support)	0.22	915	1925
Tile Covering	0.7	1600	1000

This new ventilated roof offers an easy installation which can be suitable for new buildings and retrofitting being an interesting tool for energy management and energy saving improvement.
For this purpose, two different roof typologies have been investigated: an insulated roof (U = 0.33 W m^{-2} K^{-1}) and a non-insulated one. The main properties of the materials use in the two roof types are listed in Table 2 and 3.

Table 2: Insulated Roof, Details of materials composing the lower slab

Material	λ [W m^{-1} K^{-1}]	**Thickness** [m]
Bituminous Roof Sealant	0.75	0.005
Rock Wool	0.035	0.1
Wooden Beads	0.17	0.025

Table 3: Non-insulated roof, Details of material composing the lower slab

Material	λ [W m^{-1} K^{-1}]	**Thickness** [m]
Concrete	1.6	0.1

2.2 Numerical Model

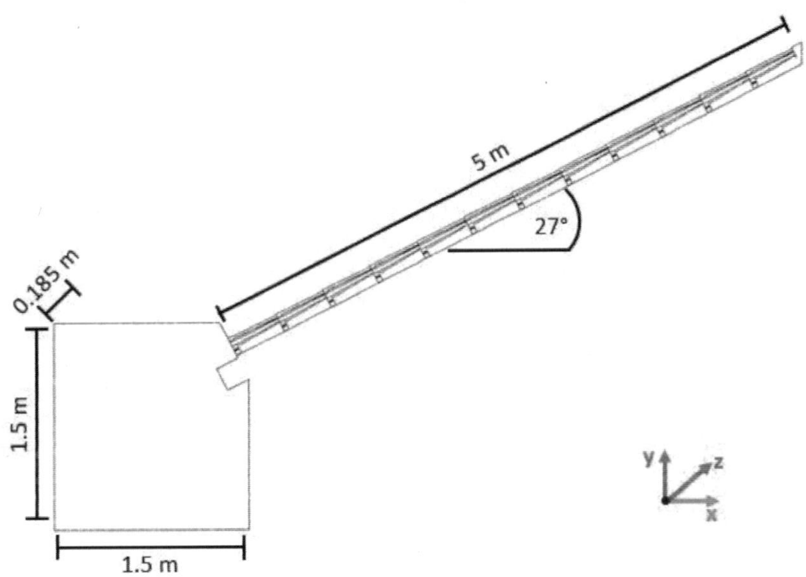

Figure 2: 2D Profile of the 3D simulated module

A 3D module of the ventilated roof has been developed consisting of two tiles. A portion of the entire roof has been obtained by repeating the module in the z-direction up to the desired length. To generate its complementary symmetric structure a mirroring along the y-axis has been done.
The roof is meant for residential buildings and it is inclined of 27°, as showed in Figure 2. An "air reservoir" has been designed, as proposed by Bianco *et al.* (2018) to simulate the ambient air, which feeds the ventilation cavity. An obstacle, i.e. a gutter, has also been inserted just below the inlet of the channel.
A heat transfer coefficient of 7 W m^{-2} K^{-1} has been set on the internal face in order to model the air heat transfer with the internal room, which has been considered at a constant temperature of 26 °C.
On the top surface of the tiles, the temperature sol-air model as proposed by Villi *et al.*, 2009 has been implemented to account for both convection and solar radiation between the tiles and the outdoor environment. The emissivity of the tiles has been set equal to 0.9. The global heat transfer coefficient h of 20 W/m^2K which takes into account the convective and radiative contributions has been considered. The sol-air model calculates a fictitious free stream temperature T_{sa} to simulate the heat transfer considering the solar irradiance, in accordance with the following equation:

$$T_{sa} = T_{ext} + \frac{I * a}{h} \tag{1}$$

Where:

I = solar irradiance [W m^{-2}]
a = absorptance [-]
T_{ext} = external air temperature [K]

A fixed value of 0.9 has been chosen for the tile covering absorptance.

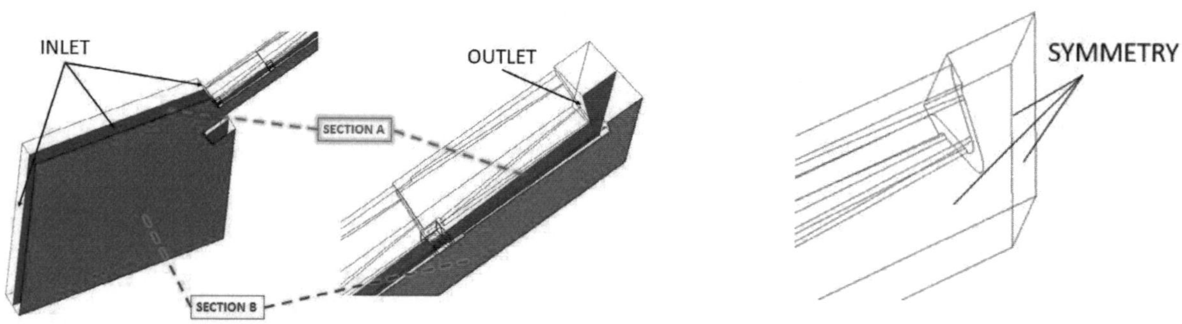

a) b)

Figure 3: Named selections

As reported in Figure 3a, a "pressure-inlet" boundary condition has been assigned to the inlet. Constant atmospheric pressure and temperature of 300 K have been fixed to simulate the outdoor environment. Similarly, "pressure-outlet" condition has been assigned to the outlet, as shown in Figure 3a. Besides, in order to investigate the thermal fluid dynamics inside the channel, two sections have been defined: section A (blue) that is located along the central axis of the channel and passes through the support, and section B (red) is the side symmetry plane. Figure 3b reports the ridge, which is located at the end of the ventilation channel and lets the air to exit. For the sake of clarity, sections A and B are not shown here. The sides of the 3D module have been modelled as "symmetry". All the other surfaces have been considered as "wall".
No wind effect has been considered. Besides, the tiles' slab has been supposed to be airtight. Therefore, the air could only enter the ventilation channel from the intake vent and exit from the exhaust one.

2.2.1 Governing Equations

To study the thermal fluid dynamics, the commercial CFD software "Ansys Fluent 18.2" has been used. The software solves the three Navier-Stokes conservation equations (mass, momentum, energy). For a steady-state solution and a compressible flow the following equations have been used (they are all reported for an i-th direction):

<u>*Mass Conservation*</u>:

$$\frac{\partial}{\partial t} \nabla \cdot (\rho \boldsymbol{v}) = 0 \tag{2}$$

Where:
ρ = air density
$\boldsymbol{v}$ = velocity vector

<u>*Momentum Conservation*</u>:

$$\frac{\partial}{\partial t}(\rho \boldsymbol{v}) + \nabla(\rho \boldsymbol{v}\boldsymbol{v}) = \nabla p + \nabla(\boldsymbol{\tau}) + \rho \boldsymbol{g} + \boldsymbol{F} \tag{3}$$

Where:
p = static pressure
$\boldsymbol{\tau}$ = shear stress tensor
$\rho \boldsymbol{g}$ = body force
$\boldsymbol{F}$ = external force

Energy Conservation:

$$\frac{\partial}{\partial t}(\rho E) + \nabla \cdot \left[\boldsymbol{v}(\rho E + p)\right] = \nabla \cdot \left[k_{eff}\nabla T - \sum_i h_i \boldsymbol{j}_i + (\boldsymbol{\tau}_{eff} \cdot \boldsymbol{v})\right] + S_h \tag{4}$$

Where the terms in square brackets in the right side of equation (4) are the conductive, convective, viscous dissipative energy exchanges.

S_h = heat produced by chemical reactions

Fluent needs two additional transport equations to solve the turbulence. To model the turbulence the standard k-ε model has been adopted in accordance with the literature. k is the turbulent kinetic energy while ε is the kinetic energy of turbulence dissipation. They are calculated using equations (5) and (6), respectively.

$$\frac{\partial}{\partial t}(\rho k) + \frac{\partial}{\partial x_i}(\rho k u_i) = \frac{\partial}{\partial x_j}\left[\left(\mu + \frac{\mu_t}{\sigma_k}\right)\frac{\partial k}{\partial x_j}\right] + G_k + G_b - \rho\varepsilon - Y_M + S_k \tag{5}$$

$$\frac{\partial}{\partial t}(\rho \varepsilon) + \frac{\partial}{\partial x_i}(\rho \varepsilon u_i) = \frac{\partial}{\partial x_j}\left[\left(\mu + \frac{\mu_t}{\sigma_\varepsilon}\right)\frac{\partial \varepsilon}{\partial x_j}\right] + C_{1\varepsilon}\frac{\varepsilon}{k}(G_k + C_{3\varepsilon}G_b) - C_{2\varepsilon}\rho\frac{\varepsilon^2}{k} + S_\varepsilon \tag{6}$$

With the turbulent viscosity calculated as follow:

$$\mu_t = \rho C_\mu \frac{k^2}{\varepsilon} \tag{7}$$

For the constants, the following values have been used as proposed by Bianco *et al.* (2018):

$C_{1\varepsilon}$ =1.44, $C_{2\varepsilon}$ =1.92, $C_{3\varepsilon}$ = 1.0, C_μ = 0.09, σ_k = 1.0, σ_ε = 1.3

The SIMPLE scheme has been selected to solve the pressure-velocity coupling.

2.3 Mesh Sensitivity

In order to evaluate the solution robustness, a mesh sensitivity analysis has been conducted. In particular, three different meshes have been tested varying the number of the elements, as reported in Table 4.

Table 4: Effects of three different meshes

Case	Number of Elements	Time to Convergence	P_{room} [W]
A	10494148	500 min	2.12
B	14592370	500 min	2.04
C	21355522	1300 min	2.17

The results refer to the ventilated insulated roof behaviour when the solar irradiance is equal to 600 W m^{-2}. The comparison of three different cases reveals that the solution is independent from the studied meshes. The solution convergence has been reached when the residuals have been lesser then 10^{-3} for the continuity, k and ε, while 10^{-6} for the energy and the velocity terms. As it can be seen in Table 4, the results showed that there is not any noticeable difference between the three studied meshes in terms of total power transferred to the internal room.

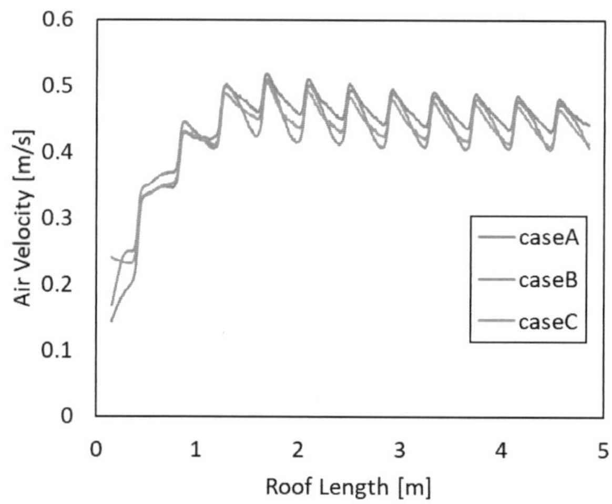

Figure 4: Mesh sensitivity analysis. Air velocity as function of the roof length.
(I = 600 W m^{-2}, ventilated insulated roof)

As showed in Figure 4, another parameter has been considered to choose the most suitable mesh; in particular, the air velocity profile along the axis of the thickest section of the ventilation cavity has been studied. Hence, considering the trade-off between time to convergence and accuracy of the solution, in terms of thermal power transferred to the room P_{room} and air velocity along the roof, the caseB has been selected. In fact, it allowed to obtain a reliable solution with a time to convergence that was almost a third of caseC.

a)

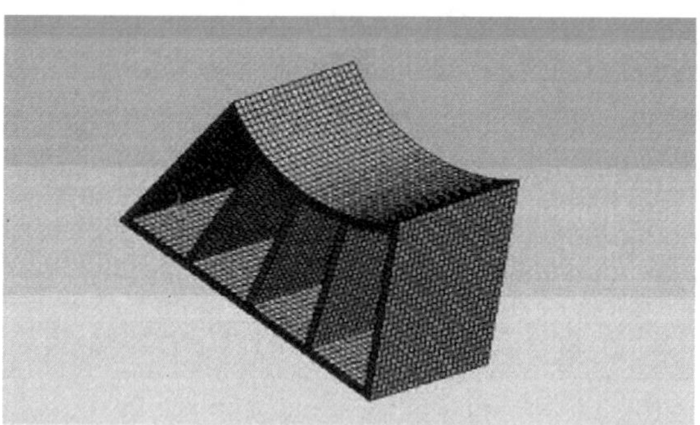

b)

Figure 5: Mesh details

3. RESULTS AND DISCUSSION

As it can be noticed from the sequence of images in Figures 6 (a-h), the proposed numerical model has successfully been implemented to simulate the thermal fluid dynamics of the new ventilated roof. For the sake of brevity, only the results for the insulated ventilated roof under a solar irradiance of 800 W m^{-2} are reported. Besides, similar results have been obtained for the other cases.

Figure 6a shows the pressure field at the intake vent. The pressure at the inlet of the channel assumes negative values because the air is sucked through the channel by the density gradient induced by the heat coming from the tiles. A buoyancy driven flow is therefore generated and fresh air moves upward.

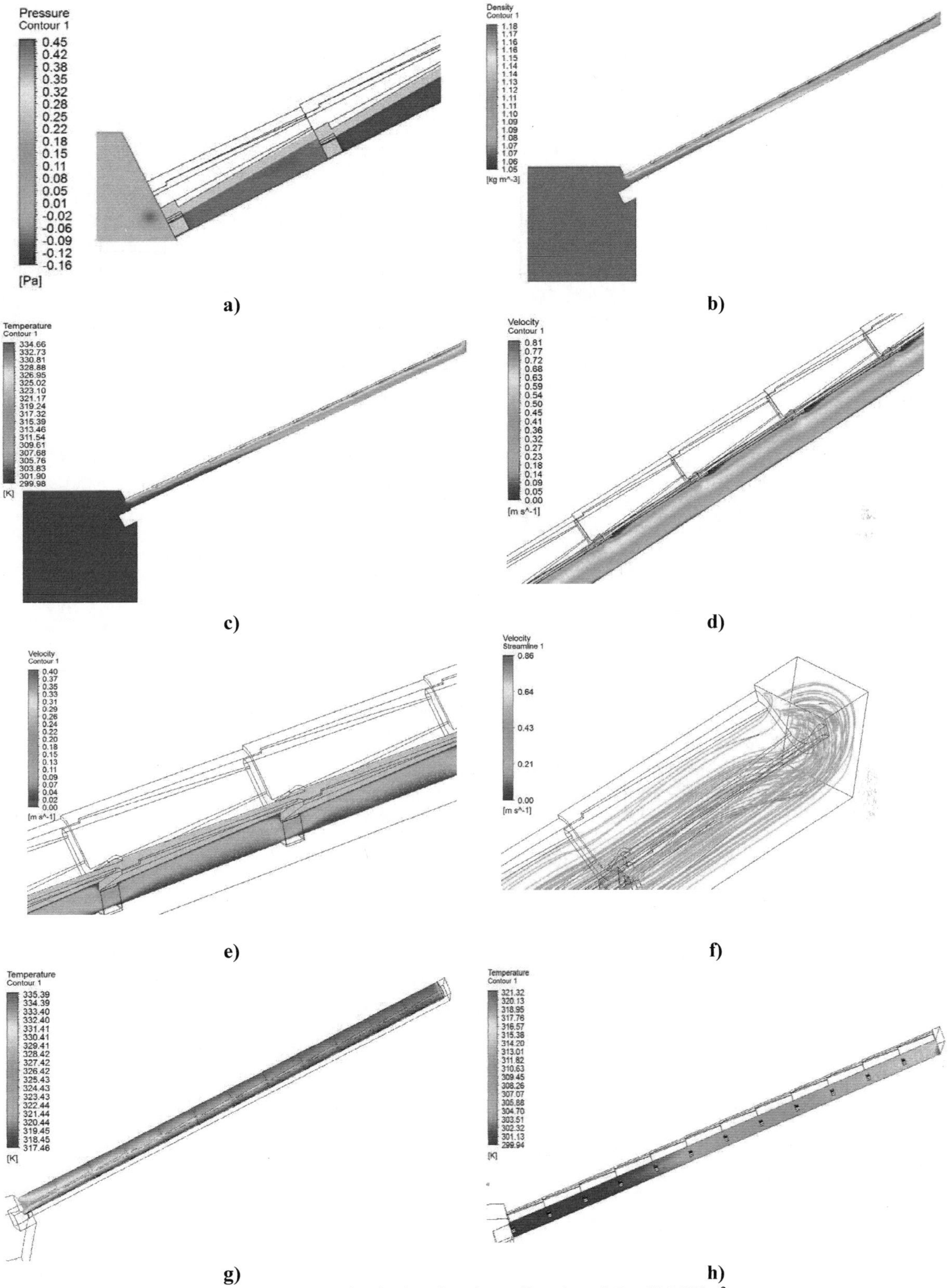

Figure 6: Results for insulated ventilated roof, I = 800 W/m^2

In Figure 6b, the density contours can be seen, the air is warmed up through the channel (Figure 6c) and its density decreases accordingly. The velocity contours of section B are reported in Figure 6d, the velocity ranges between 0.4 and 0.6 m/s. When considering section A, located in the middle of the support (Figure 6e), it is possible to highlight the great benefit of the new support: differently from the traditional one, the air can flow through and the average velocity is around 0.25 m/s. Figure 6f shows the streamlines at the outlet of the ventilated roof, the air is fairly homogenously distributed and leaves the channel from the ridge.
The tiles' temperature contours are shown in Figure 6g, while the temperature distribution on the wall facing the room can be found in Figure 6h. The tiles' surface warms up along the length and it reaches around 335 K, while the bottom surface does not go over 320 K.
The velocity profiles along the axis of section B are plotted in Figure 7. The curves permit to clearly show the effects of the supports; at all the investigated heat fluxes, the fluid flow develops in the first 1.5 m and then it stabilizes. The peaks and valleys are located in correspondence of the supports. At a solar irradiance of 1000 W/m^2, the two curves mostly overlap, showing the almost similar maximum and minimum values. When decreasing the solar irradiation, the two are slightly different and the peak-to-valley velocity difference decreases.

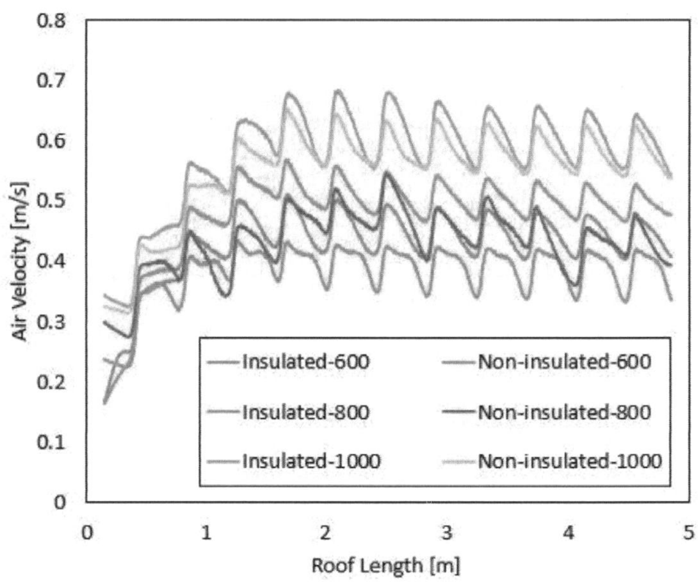

Figure 7: Air velocity profile along the axis of the thickest section of the cavity for insulated and non-insulated ventilated roof, I = 600, 800, 1000 W m^{-2}

Table 5: Comparison of Specific Thermal Power transmitted into the room for all I-values between insulated and non-insulated, ventilated and non-ventilated roof.

	Insulated Roof			Non-insulated Roof		
I	Ventilated [W m^{-2}]	Non-Ventilated [W m^{-2}]	Savings [%]	Ventilated [W m^{-2}]	Non-Ventilated [W m^{-2}]	Savings [%]
600 W m^{-2}	2.04	7.16	72	13.76	34.74	60
800 W m^{-2}	2.68	9.53	72	18.61	47.62	61
1000 W m^{-2}	2.82	11.9	76	22.38	60.37	63

As resumed in Table 5, for the insulated ventilated roof an average reduction of more than 70% of the specific thermal load to the indoor space was calculated as compared to the traditional non-ventilated roof; considering the non-insulated cases, a reduction of some 60% specific themal load was achieved. The ventilated roof allows for a huge reduction of the thermal loads to the indoor rooms. The results can be also showed in the two diagrams of Figure 8, the remarkable reduction of the specific heat flux is clearly noticeable between a ventilated roof and a non-ventilated one for all three solar irradiation values. In the studied scenarios, a great amount of heat is taken away by the air flowing in the channel.

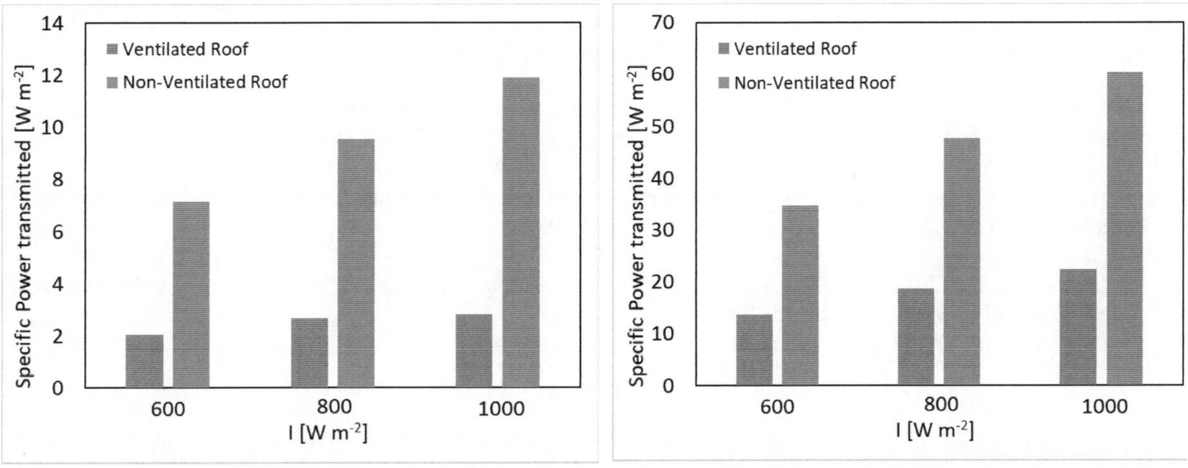

a) b)

Figure 8: Specific Thermal Power entering the room: insulated roof a), non-insulated roof b)

4. CONCLUSIONS

In this work, a 3D CFD numerical analysis of a new ventilated roof structure is proposed. Six different scenarios of a typical summer day have been simulated, varying the roof's typology and the solar irradiance intensity. In particular, insulated and non-insulated roofs have been studied under I = 600, 800, 1000 W m^{-2}, keeping constant at the outdoor temperature at 300 K. The tiles-supports structure and its effects on the fluid-dynamics of the airflow in the ventilation cavity and on the transmitted thermal power to the indoor space have been investigated.

The results reveal a remarkable reduction of the specific thermal loads transferred to the indoor space. For the insulated roof, an average savings of more than 70 % on the specific thermal load has been estimated while for a non-insulated one, the value has been around 60%. Finally, these 3D CFD simulations have confirmed the great benefits of a ventilated roof and, in particular of the novel support design, in promoting an effective mitigation of the thermal load during the summer.

REFERENCES

Bianco, V., Diana A., Manca O., Nardini S. (2018). Numerical investigation of an inclined rectangular cavity for ventilated roofs applications. *Thermal Science and Engineering Progress*, 6, 426-435.

Ciampi, M., Leccese, F., Tuoni, G. (2005). Energy analysis of ventilated and microventilated roofs. *Solar Energy*, 79, 183-192.

Gagliano, A., Patania, F., Nocera F., Ferlito A., Galesi A. (2012). Thermal performance of ventilated roofs during summer period. *Energy and Buildings*, 49, 611-618.

Li D., Zheng Y., Liu C., Qi H., Liu X. (2016). Numerical analysis on thermal performance of naturally ventilated roofs with different influencing parameters. *Sustainable Cities and Society*, 22, 86-93.

Liberati, P., Spiga, M., Zappavigna, P. (2009). Optimization of ventilated roofs for livestock housing. *International Communications in Heat and Mass Transfer*, 36, 432-437

Puangsombut W., Hirunlabh J., Khedari J., Zeghmati B., Win M. M. (2007). Enhancement of natural ventilation rate and attic heat gain reduction of roof solar collector using radiant barrier. *Building and Environment*, 42, 2218-2226.

Raya, S. D., Gong, N. W., Glicksman, L. R., Paradiso, J. A. (2014). Experimental characterization of full-scale naturally ventilated atrium and validation of CFD simulations. *Energy and Buildings*, 69, 285-291.

Susanti L., Homma H., Matsumoto H. (2011). A naturally ventilated cavity roof as potential benefits for improving thermal environment and cooling load of a factory building. *Energy and Buildings*, 46, 211-218.

Tong S., Li H. (2014). An efficient model development and experimental study for the heat transfer in naturally ventilated inclined roofs. *Building and Environment*, 81, 296-308.

Villi, G., Pasut, W., De Carli, M. (2009). CFD modelling and thermal performance analysis of a wooden ventilated roof structure. *Building Simulation*, 2, 215-228

ACKNOWLEDGEMENT

This research work was partially funded by AERTetto s.r.l., Preganziol (TV), 31022, Italy, with a research grant titled: "Analisi del comportamento termico e dei materiali del sistema denominato supporto per tegole" in collaboration with Department of Civil, Environmental and Architectural Engineering (DICEA) of University of Padova.

Design and Development of a Human Building Interaction Laboratory

Sourabh D. YADAV[1]*, Parveen DHILLON[1], Orkan KURTULUS[1], Charles BAXTER[1], Davide ZIVIANI[1], Travis HORTON[2], Panagiota KARAVA[2], James E. BRAUN[1],

[1]Ray W. Herrick Laboratories, School of Mechanical Engineering, Purdue University,
West Lafayette, IN, USA

[2]Lyles School of Civil Engineering, Purdue University,
West Lafayette, IN, USA

yadav35@purdue.edu; pdhillon@purdue.edu; orkan@purdue.edu;
cbaxter@purdue.edu; dziviani@purdue.edu; wthorton@purdue.edu; pkarava@purdue.edu;
jbraun@purdue.edu

* Corresponding Author

ABSTRACT

A future is envisioned where buildings are assembled on-site from factory manufactured modular elements that integrate the smart technology needed to enable scalable, cost-effective solutions with autonomous, occupant-responsive, healthy, and sustainable features. The use of modular elements would mean that buildings are assembled rather than constructed on-site with better quality control, less material waste, and more predictable schedules. The use of manufactured building elements can enable more cost-effective integration of new sensors, embedded intelligence, networking, adaptive interfaces, renewable energy, energy recovery, comfort delivery, and resiliency technologies, making high-performance buildings more affordable. To explore and evaluate these modular and intelligent comfort delivery concepts and advanced approaches for interaction with occupants, a new human-building interaction laboratory (HBIL) has been designed and is under development. The facility has a modular construction layout with thermally active panels. The interior surface temperature of each panel can be individually controlled using a hydronic system. Such configuration allows us to emulate different climate zones and building type conditions and perform studies such as the effect of different active building surfaces on thermal comfort, localized comfort delivery, and occupant comfort control, among others. Moreover, each panel is reconfigurable to allow investigating different interior surface treatments for different visual and acoustic comfort conditions. In this paper, the overall design approach of the facility is presented. Furthermore, a prototype panel has been constructed to validate the design and assess the dynamic and steady-state thermal performance. Test results for the prototype panel are also presented here with a discussion on their agreement with design phase modeling results.

1. INTRODUCTION

In an envisioned future, modular building elements can interact with occupants to optimize energy usage and provide localized thermal comfort. These human-building interaction concepts could involve smart sensors and voice assistants with advanced control systems and technologies. However, before these innovative building technologies are implemented in the field, they need to be engineered and evaluated, both technology-wise and occupant perception-wise. Radiant heating and cooling (radiant systems) is one of the ways to provide local thermal comfort with modular wall elements and thermo-active surfaces. Rhee & Woo (2015) reviewed the development in radiant heating and cooling systems over the past fifty years and suggested future research areas that would improve radiant heating and cooling technologies. These include designing a system that can provide heating, cooling with multi-zone control, developing practical and straightforward technology to sense surface temperature, optimizing hydronic equipment for radiant systems, and integrating radiant systems with building envelopes. The concept of a human-building interaction laboratory (HBIL) with thermo-active wall panels was conceived to study modular construction, localized thermal comfort delivery, development of radiant heating and cooling systems with multi-zone control, application of intelligent sensors and advanced control approaches, and different novel comfort delivery technologies embedded in walls such as micro heat pump systems, thermoelectric devices, among others. Figure 1 illustrates the HBIL concept where an occupant inside a space equipped with thermally active panels is communicating with an integrated smart voice assistant for control. The occupant provides a command stating that he/she is feeling hot. In turn, the voice assistant sends a command to the controller to provide localized space cooling around the occupant with thermo-active panels to provide the desired thermal environment.

Figure 1. Modular comfort delivery with voice control

The HBIL facility is a 6.1m × 3.65m × 3.04m (20' × 12' × 10') (length x width x height) reconfigurable test-space to facilitate laboratory research for residential and commercial building systems. It will be able to emulate conditions for one or two exterior walls with a glazing area on one wall to enable a realistic representation of a house or a perimeter office space. It will also allow the evaluation of localized thermal comfort using thermo-active surfaces and generate initial data on occupant interactions with modular comfort delivery systems using novel audio and visual interfaces. The main design features include modular construction, reconfigurable wall and interior surface finishes, thermally active wall panel elements, variable/localized thermal comfort delivery, reconfigurable lighting/visual comfort, and reconfigurable acoustic comfort.

The facility is designed to have the capability to individually control the interior surface temperature of each panel, thus allowing studies of the effect of surface temperature on thermal comfort, localized occupant comfort control, and so on. In addition, the facility will also have air comfort delivery capabilities. For this, slot diffusers are embedded in some of the ceiling panels, which will be connected to a main supply air duct from an air-handling unit (AHU). Figure 2 shows the exploded view of the final design of the thermo-active wall panel, and Figure 3 shows the isometric view of the overall facility design.

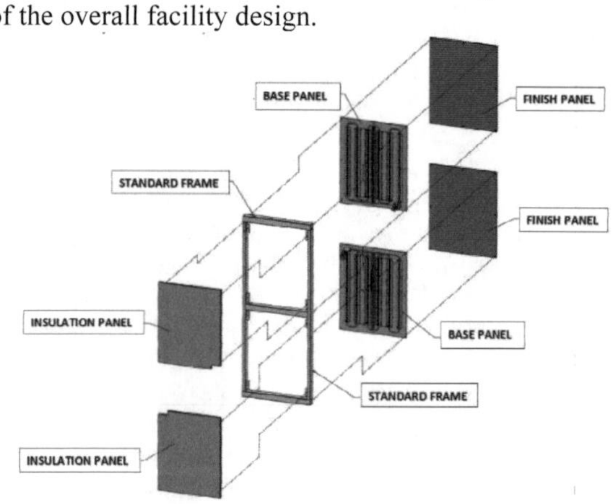

Figure 2. Panel exploded view

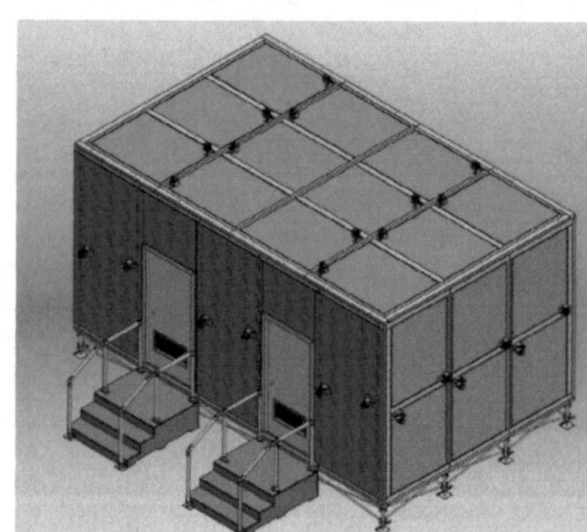

Figure 3. HBIL facility isometric view

This paper presents the HBIL facility design approach with detailed steps leading to the final design configuration. The design and testing of a prototype wall panel to study and validate the design is also presented. A 1-D transient numerical model of the panel and its validation based on test results are also presented in this paper.

2. FACILITY DESIGN APPROACH

To help in determining the range of internal surface temperatures required in the test facility, the DOE reference (Deru et al., 2011) small-office building model was used to simulate energy flows for Chicago. The purpose was to determine typical ranges for interior wall surface temperatures for both winter and summer conditions. Based on the simulation results and typical surface temperatures for floor/ceiling radiant heating and cooling systems, a surface temperature range of 15.5-26.7°C (60°F−80°F) was selected for the scope of research to be performed in this facility.

2.1 Load Estimation

To select an approach, and size equipment for controlling the surface temperature over the range of interest, the first step was to determine facility heating and cooling loads based on heat loss/gain estimates from interior surfaces for the most extreme temperature conditions. Thus, convective and radiative heat gains/losses were estimated for the room dimensions (surface area (A) of walls, floor, and ceiling), interior surface temperature ($T_{surface}$), and indoor air temperature (T_{air}) conditions that are shown in Table 1.

Table 1. Load estimation design parameters

Room Dimension: 20×12×10 ft	Heating Design Temperatures	Cooling Design Temperatures
A_{wall} = 640 ft^2 (59.2 m^2)	$T_{surface}$ = 80°F (26.7°C)	$T_{surface}$ = 60°F (15.5°C)
A_{floor} = A_{roof} = 240 ft^2 (22.3 m^2)	T_{air} = 70°F (21.1°C)	T_{air}= 74°F (23.3°C)

The convective heat transfer from the interior surfaces (walls and floor) was estimated using convective heat transfer correlations from Goldstein & Novoselca (2010) based on the designed comfort delivery system implementation with ceiling diffusers. The radiative heat transfer from the interior surfaces (walls and floor) was estimated using a radiative heat transfer correlation from Bergman et al. (2011) in which the long-wave radiation is linearized with a view factor of unity. Table 2 shows the load estimation results based on heat transfer from interior surfaces (walls and floor) for the heating and cooling case. The heating case represents the minimum heating input required to maintain a positive heat gain to the room from the interior surfaces at the specified conditions. The negative sign in the cooling case conveys that the heat transfer is into the surface from the indoor air.

Table 2. Load estimation for heating and cooling case

Operation Mode	Convection [W]	Radiation (*1) [W]	Total [W]	Heat Flux [W/m^2]
Heating	1592	2592	**4184**	**51**
Cooling	-2229	-3469	**-5698**	**-70**

*1 Radiation heat transfer assuming worst-case scenario for each surface, i.e., the surface in the calculation is at $T_{surface}$ whereas other walls, floor, and roof are at T_{air}, and assuming paint (ϵ=0.96)

2.2 Investigation of Design Approaches for Panel Surface Temperature Control System

Four different system design approaches were investigated to control panel surface temperature for heating and cooling. Each approach was evaluated based on its potential to provide both heating and cooling to satisfy the design conditions, ease of individual panel temperature control, and overall cost. Table 3 summarizes the different design approaches along with their pros and cons. Based on these considerations, a hydronic system that uses chilled and hot water loops with mixing valves was selected for controlling surface temperatures of individual panels. A detailed description of the panel design and performance is presented in Section 3.1.

Table 3. Design approaches for HBIL facility

Design Approach	Description	Remarks
Thermoelectric based system	Thermoelectric modules embedded in the test facility surfaces for both heating and cooling.	Has the capability to provide both heating and cooling and can meet the required heat flux. However, this system would have required many modules and control hardware due to the small size of each module, thereby leading to very high total cost.

Cold air and heating element based system	Cold air through ducts between the interior surface. Exterior insulation layer for cooling and resistive heating element for heating.	An airflow with a velocity of 1 m/s would not be sufficient to maintain the interior surface temperature at the lowest target setpoint. Higher air velocities could result in noise issues.
Chilled water and heating element based system	Chilled water through tube-heat spreader attached to the interior surface for cooling, resistive heating element for heating.	With this design configuration, the lower thermal resistance between the heating element and the chilled water loop compared to the thermal resistance between the heating element and the indoor air could lead to control issues and very large cooling requirements when the system is operating in the transition between cooling and heating mode.
Chilled and hot water-based system	Cold and hot water loop with a 3-way mixing valve to control the supply water temperature and subsequently control the panel surface temperature.	This system suits the design requirements for surface temperature target range and with a reasonable control approach without any issues mentioned above for other design approaches.

For the chosen system configuration, the panel interior surface temperatures are controlled based on mixing chilled and hot water for the desired target temperature. The idea is to have two parallel chilled and hot water loops and utilize a 3-way mixing valve to change the water inlet temperature to each panel to maintain the desired interior surface temperature. The water flow rate in each panel is kept constant to reduce the overall system cost and control complexity.

Preliminary steady-state simulations were performed in EES (Engineering Equation Solver) (Klein, 2021) using a finite volume model of the panel described in Section 3.2 to determine a water flow rate to achieve an acceptably uniform temperature distribution over the panel surface. In the simulation, the inside air temperature (T_{ID}) was assumed to be 21.11°C (70°F), and the outside air temperature was 23.33°C (74°F) (as the facility will be located in an indoor environment). Figure 4 shows the steady-state temperature distribution across the panel height for a cooling case, with a water inlet temperature of 12.77°C (55°F) and a flow rate of 0.95 GPM. Figure 5 shows simulation results with the same flow rate for a heating case, with a water inlet temperature of 29.44°C (85°F). It can be seen from both the plots that this water flow rate results in a temperature change in water (black line) across the panel height of around 0.6°C (1.2°F) and a temperature difference along the panel surface (green line) of around 0.7°C (1.4°F). These are within acceptable limits. Based on the analysis results and 3-way mixing valve selection constraints, the water flow rate for each cooling circuit of a panel was specified to be 1 GPM.

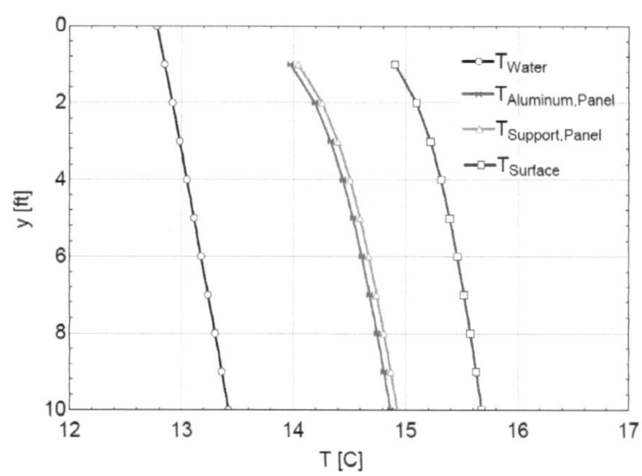

Figure 4. Panel steady-state simulation results for cooling

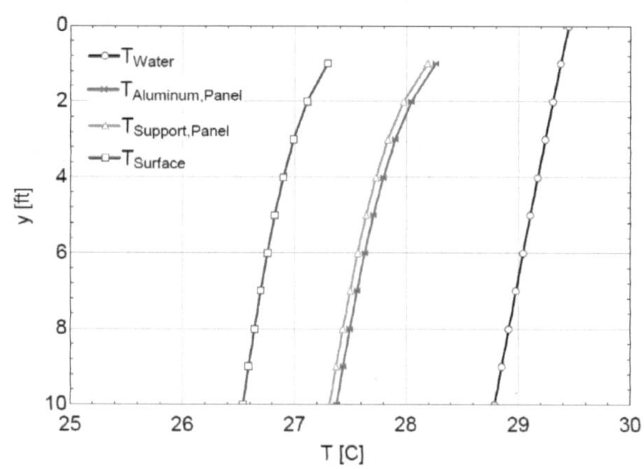

Figure 5. Panel steady-state simulation results for heating

The facility will be composed of 52 modular 1.22m × 1.52m (4'×5') (width × height) wall, ceiling, and floor panels attached to an overall frame. A four-pipe hydronic distribution system will provide hot and cold water to pairs of panels that are arranged in a series flow arrangement as depicted in Figure 6. The hot and cold water will be mixed using a 3-way valve to allow individual control of a temperature at the interior surface of each panel pair. There will also be a diverting valve at the outlet of each panel pair that will divide the mixed water into hot and cold water return loops in the same proportion as the mixing valve in order to maintain balanced flows. The hydronic system will continuously circulate cooling and heating water in the two separate loops to each 3-way mixing valve. A temperature sensor will be embedded in an appropriate location at the interior surface treatment of each 4'×5' panel and used to provide feedback control of local comfort conditions within the interior space. A water-to-water heat pump will provide heat transfer between chilled and hot water tanks that will be coupled to the chilled and hot water loops. In addition, there will be a resistance heating element in the hot water tank and an auxiliary hot water fan coil unit for heat rejection that will be used, when necessary, for load balancing. Pressure-regulating valves will be employed in both hot and cold water loops to maintain the required head pressure. The 3-way valves will be adjusted using feedback control based on surface temperature setpoints that can vary locally within the indoor space. Humidity control and ventilation requirements will be satisfied by dedicated air diffusers.

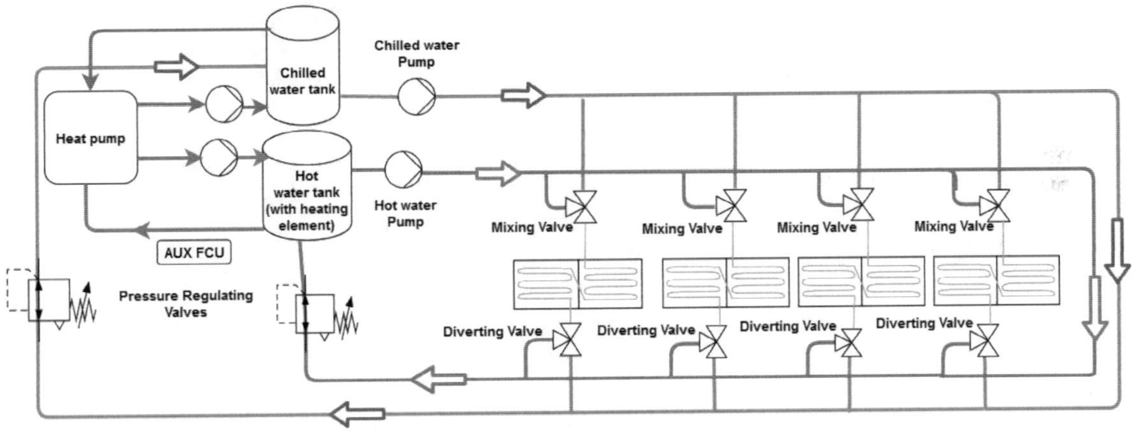

Figure 6. HBIL facility hydronic system

3. PROTOTYPE PANEL DESIGN, TESTING, AND NUMERICAL MODEL VALIDATION

A prototype panel was constructed to validate the design and assess the steady-state and dynamic thermal performance of the panel and it is shown in Figure 7. As mentioned before, the design goal for the panel was to provide uniform surface temperature from 15.5°C (60°F) to 26.7°C (80°F). Additional transient response criteria were established, so that panel surface temperature would be able to change from minimum to maximum design value in less than 50 minutes. A numerical model was developed for the panel that is described in this section, along with steady-state and transient response validation results. This numerical model will be useful as a tool to design alternative wall panels with different surface treatments in the future.

A test setup for a prototype panel testing was designed and constructed, as shown in Figure 8. Twelve thermocouple sensors were attached to measure the surface temperature, as shown in the front view. One thermocouple sensor was attached to the back panel to measure aluminum backplate temperature, as shown in the 'rear view' of Figure 8. Hot, cold, and mixed water temperatures were measured using in-line thermocouples inserted in the water stream. Hot and cold water bypass valves were used to maintain the constant head pressure on the hot and cold loop. Head pressures of hot, cold, and mixed water were also measured. Furthermore, a differential pressure sensor is installed between the inlet and outlet of the panel to estimate the pressure drop across the panel.

Figure 7. Prototype panel test setup

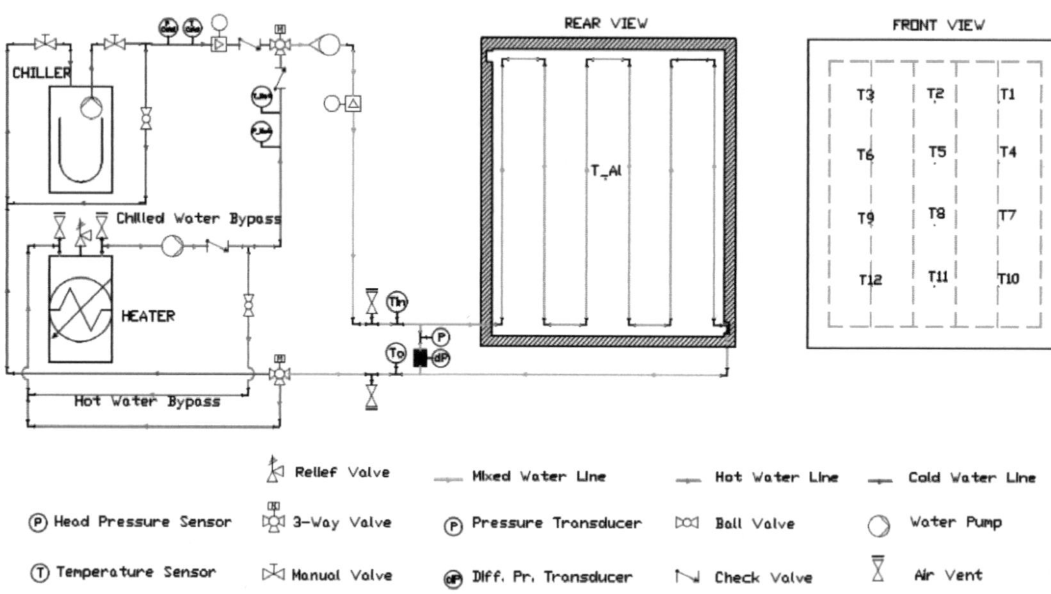

Figure 8. Prototype panel test setup schematic drawing

A temperature stratification was observed in the room air temperature from the ground level to 7 feet height. The air temperature at 1 foot above the floor was around 1°C lower than the temperature at a level of 7 feet above the floor level. Also, because all the panel layers were installed inside an aluminum frame, it was assumed that there would be heat transfer occurring through the edges of the panel, thereby resulting in sightly different surface temperature readings as compared to the central part. Keeping these factors in mind, the design goal was to have the surface temperature distribution across the twelve thermocouples to be within 1°C to 1.5°C. Different piping configurations were tested to achieve uniform surface temperature distribution across the panel surface along the same horizontal plane of a specific elevation from the floor. A series flow configuration was finalized because the parallel piping configurations showed non-uniform flow across different parallel passes, which led to a non-uniform surface temperature distribution. Six heat spreaders are used for the six straight sections of copper tubing. They have a semi-circular cavity slot to hold the copper tubes on one side and have a flat surface on the other side attached to the flat aluminum backplate surface. Based on experimental investigation, C-type heat spreaders showed better heat transfer capability than the U-type heat spreader. Figure 9 shows C-type and U-type heat spreaders.

C-type heat spreader U-type heat spreader

Figure 9: Heat Spreaders

3.1 Panel Performance Limit Assessment

Limits on the panel thermal performance were assessed for heating and cooling mode by evaluating the maximum and minimum surface temperature the panel can achieve with the given chiller and heater capacities for typical indoor conditions. The temperature on the panel surface was measured at twelve locations, as shown in the 'front view' of Figure 8. In a performance assessment test, the heating mode was initiated at first, and hot water at 45°C was circulated through the hydronic circuit at the panel back until the surface temperature reached a steady-state. After a 10-minute steady-state measurement, the system was switched to cooling mode, and chilled water at 4.5°C circulated through the panel until a steady-state was reached.

Figure 10 shows test results with PEX tubing in the hydronic circuit at the back of the panel. The design goal for the surface temperature in heating and cooling was achieved; however, the transient response was slow. It took around 90 minutes to reach the minimum surface temperature from the maximum, which did not meet the transient design goal. To improve the transient response of the panel, PEX tubes were replaced with better-conducting copper tubes. This modification resulted in a faster dynamic response in heating as well as in cooling mode, as shown in Figure 11. The time required to reach maximum heating or minimum cooling steady-state temperature was reduced by 40% compared to the PEX tubing case. In addition to the quicker response, there was around a 2°C rise in heating and a 1.5°C drop in cooling steady-state panel surface temperature compared to the test with PEX tubing. Further, the effect of rear insulation on the panel surface temperatures was also studied, as shown in Figure 11. Without rear insulation, the surface temperature dropped around 1°C due to heat losses from the panel back surface.

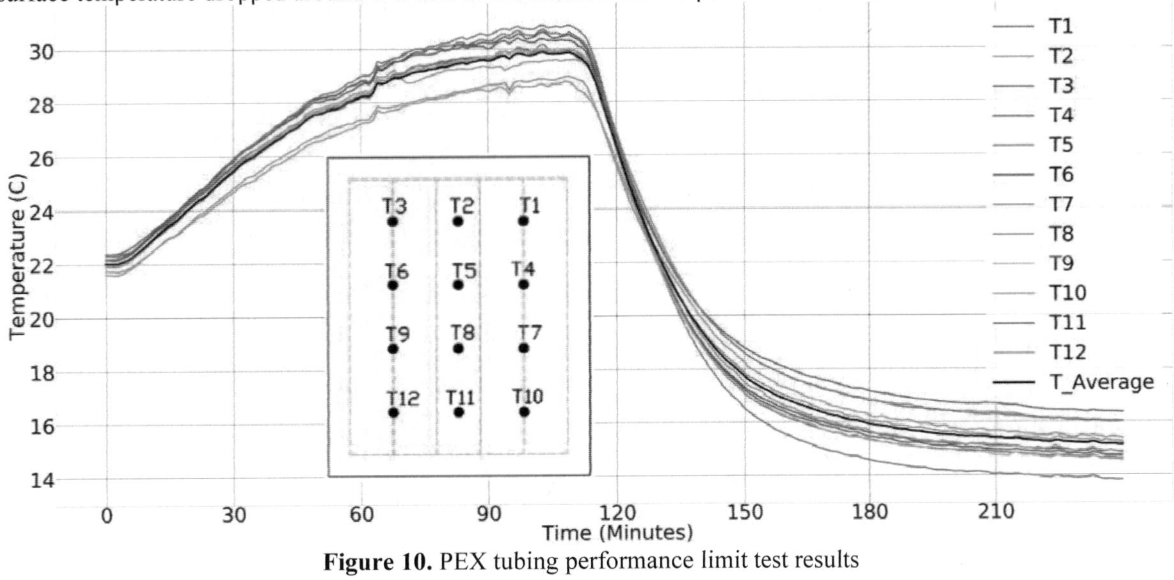

Figure 10. PEX tubing performance limit test results

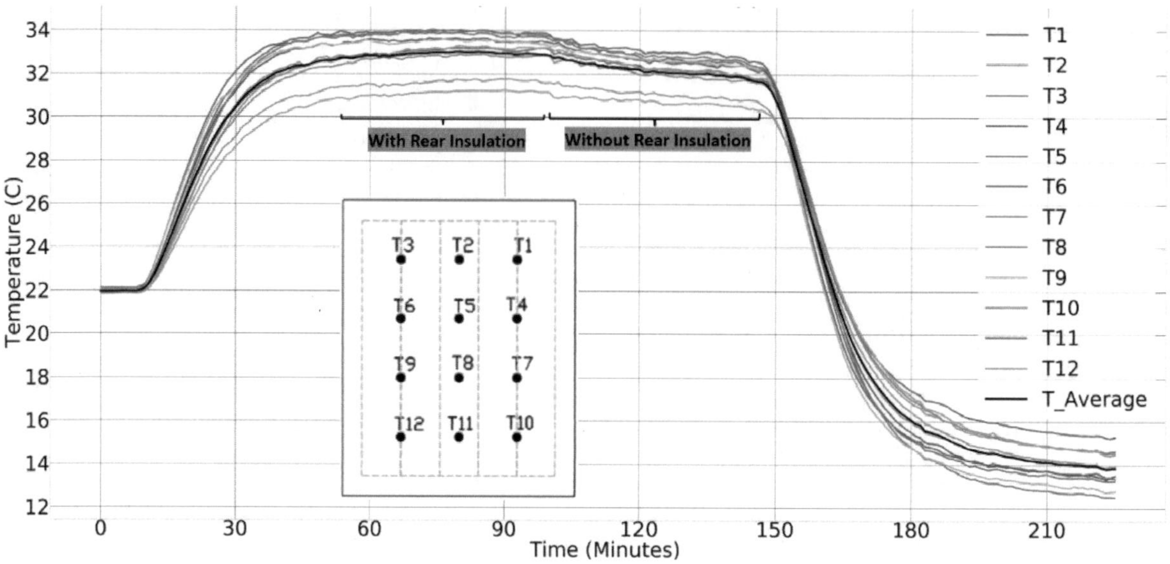

Figure 11. Copper tubing performance limit test results

Figure 12 and Figure 13 are panel surface thermal images during steady-state heating and cooling conditions with copper tubing, respectively. A significant heat loss was observed from the panel edges. Consequently, the region with the lowest temperature value was in the center of the panel. This temperature distribution aligns with the thermocouple readings plotted in Figure 11. It can also be observed that the temperature measured by thermocouples closer to the floor (T12, T11) was lower than the rest of the thermocouples, which aligns with the indoor air temperature stratification explained above. It should be noted that panel edge losses should be dramatically reduced when the panels are installed within the facility if adjacent panels are operating at similar conditions. It should also be noted that these experiments were extremely useful in finalizing the prototype panel design and validating the model, as described in the next section.

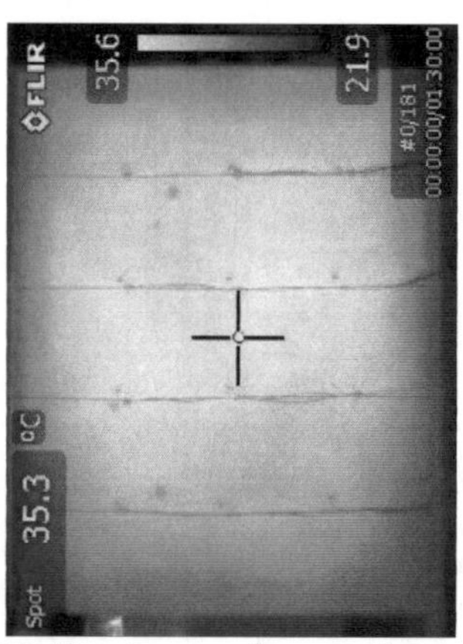

Figure 12. Heating steady-state thermal image

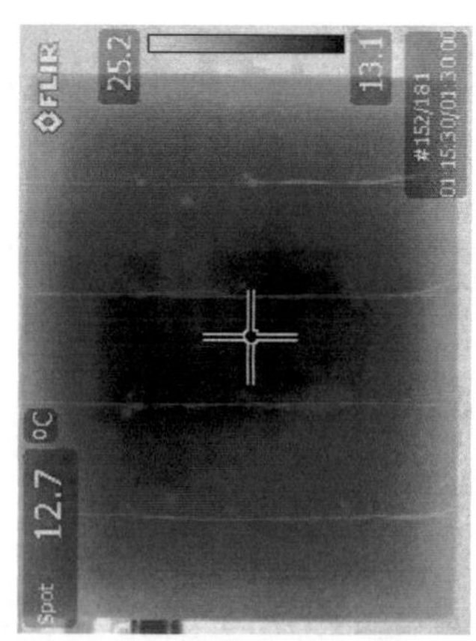

Figure 13. Cooling steady-state thermal image

3.2 Validation of Prototype Panel Heat Transfer Model Using Experimental Results

The prototype wall panel is essentially a composite wall with copper tubes embedded in heat spreaders on the backside and a gypsum layer on the front side. A 1-D transient thermal network model for this prototype wall panel was developed in EES (Klein, 2021). The critical inputs for the model are mixed water inlet temperature, outdoor temperature, and indoor air temperature values. The model calculates values of panel surface temperature, water outlet temperature, and aluminum backplate temperature. In addition, the heat transfer that occurs from water to the indoor air ($\dot{Q}_{front}$) and the heat lost to the outdoor air ($\dot{Q}_{loss}$) are computed. The model uses 'bond conductance' to account for the thermal resistance for heat transfer between water and the heat spreader. The model was tuned by varying the values of bond conductance, capacitances, and thermal and contact resistances. Figure 14 shows the thermal network diagram for the model with different layers of the prototype wall panel.

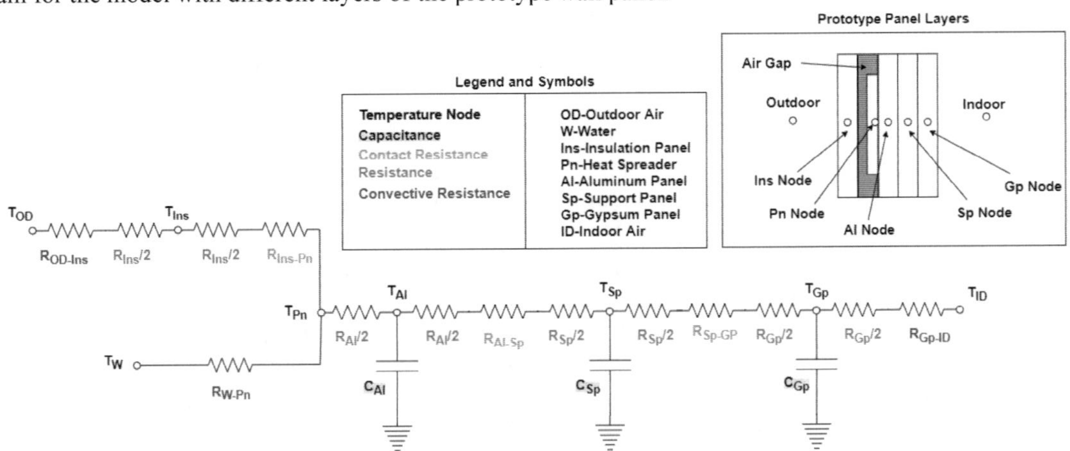

Figure 14: Thermal network model for prototype panel

The thermal circuit in Figure 14 leads to the following equations (equations 1-4) that represent the heat transfer rates. To model heat transfer from copper tubes to the heat spreader panel, equation 5 is used, which is analogous to a solar flat plate collector scheme (Kalogirou, 2009). The heat transfer coefficient between the water and tube inner surface (h_{fi}) is evaluated using an inbuilt EES function for internal pipe flow from Nellis and Klien (2009). Convective and long-wave radiative heat transfers to or from the panel surface are assumed to occur in parallel with exchange occurring with a common indoor condition. The long-wave radiation is linearized with a view factor of unity (Bergman et al., 2011) to the surrounding surfaces (equation 6). In contrast, a natural convection correlation from Nellis and Klein (2009) for a vertical wall (also an EES built-in function) is employed.

$$\dot{Q}_{in} = \dot{Q}_{front} + \dot{Q}_{loss} \tag{1}$$

$$\dot{Q}_{in} = (T_{W\,in} - T_{Pn})/R_{W-Pn} \tag{2}$$

$$\dot{Q}_{loss} = (T_{Pn} - T_{OD})/R_{Ins-Al} \tag{3}$$

$$\dot{Q}_{front} = (T_{Pn} - T_{Al})/(R_{Panel} + R_{Gp-ID}) \tag{4}$$

$$q'_u = \frac{(T_b - T_f)}{(1/C_b) + (1/\pi D_i h_{fi})} = q'_{fin} + q'_{tube} \tag{5}$$

$$h_r = \varepsilon_{gp} \times \sigma \times \left(T^2_{Surface} + T^2_{ID}\right) \times (T_{Surface} + T_{ID}) \tag{6}$$

where $\dot{Q}_{in}$ is the heat transfer rate from the water inside the tubes to the heat spreader node, $\dot{Q}_{front}$ is the heat transfer rate from the heat spreader node to indoor air through interior surface treatment and $\dot{Q}_{loss}$ is the heat transfer rate occurring from the heat spreader node to outdoor air through insulation panel. The thermal capacitances of the heat spreader (due to its small size and mass) and insulation panel (due to its low mass density) are neglected because they were found not to significantly contribute to the transient behavior of the prototype wall panel. Using the heat transfer rate ($\dot{Q}_{front}$), an energy balance was set up to calculate a time-dependent temperature gradient (dT_{Al}/dt) equation 7. This was integrated numerically over the total duration of the test (*Et*) to calculate aluminum node temperature (T_{Al}) (equation 8). A similar set of energy balance and integration equations were set up between the aluminum and support panel nodes and between the support and gypsum panel nodes to calculate panel surface temperature. The equation system is closed by an energy balance (equation 9) between the total heat transfer rate entering the system and energy gained or lost through interaction with the cooling or heating water, respectively.

$$\dot{Q}_{front} = \{Cap_{Al} \times (dT_{Al}/dt)\} + \frac{(T_{Al} - T_{Sp})}{(R_{Al}/2 + R_{Sp}/2 + R_{Al,Sp}/2)} \tag{7}$$

$$T_{Al} = T_{ID} \times \int_0^{Et} (dT_{Al}/dt)\ dtime \tag{8}$$

$$\dot{Q}_{in} = \dot{m}_w \times Cp_w \times (T_{w,out} - T_{w,in}) \tag{9}$$

The key outputs of the model are surface temperatures and aluminum backplate temperatures obtained for each time step (every minute). Figure 15 shows the panel surface temperature results from the tuned model (T_{Model}) and experiments (T_{Expt}). It can be seen that model results are within 0.5 °C of the experimental results throughout the test interval. Similarly, Figure 16 shows the aluminum backplate temperature results ($T_{Al,Back,Model}$) overlaid on experimental results ($T_{Al,Back,Expt}$). In this case, the differences in the model and experimental values are slightly higher, about 1°C, especially at steady-state. This is because the sensor measuring the aluminum backplate temperature was covered using insulation tape, thereby blocking the convective and radiative heat transfer with the air. This resulted in lower values for steady-state cooling and higher values steady-state heating cases. On the contrary, the temperature sensors on the panel surface had a transparent thin thermal tape allowing relatively more convective and radiation heat exchange with the indoor air. Additional validation tests were performed on the prototype at different operating conditions with similar results.

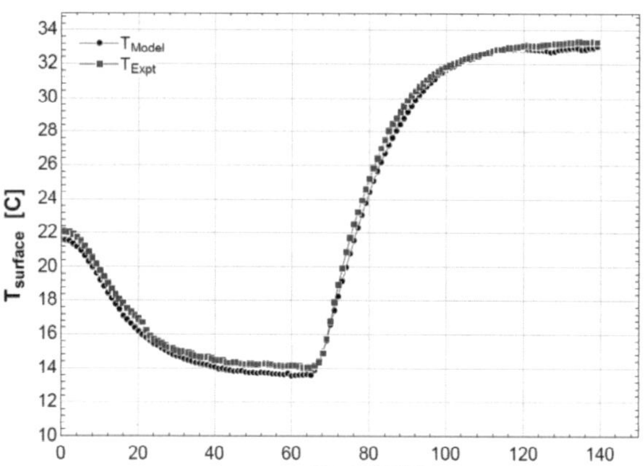

Figure 15. Comparisons of the model and experimental results (Panel surface temperature)

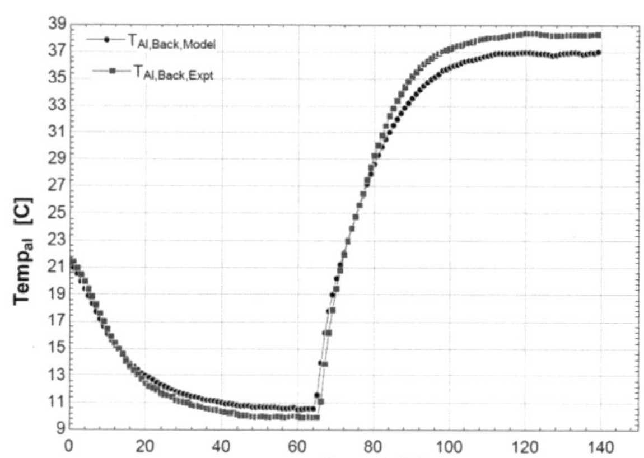

Figure 16. Comparison of the model and experimental results (Aluminum backplate temperature)

4. CONCLUSIONS

This paper presented an overview and design of a human-building interaction laboratory (HBIL) that will enable evaluation of next-generation concepts for comfort delivery and building interaction. Furthermore, a prototype thermally active panel for the facility was presented along with the assessment of its transient and steady-state thermal performance. The panel is capable of generating a wide range of interior surface temperatures that could be locally controlled to provide local heating or heating within the facility. Modeling results for the panel agree well with measurements, such that the model could be used in the future to develop alternative thermally active panels for installation in this modular facility.

The future steps involve the assembly of the HBIL facility, installation and commissioning of the facility's hydronic system, design, commissioning of control architecture, and voice interaction. Once fully commissioned, the facility will be used to emulate different climate zones and/or study localized thermal comfort delivery with human test subjects.

NOMENCLATURE

Symbols		Subscripts	
T:	Temperature	Pn:	Heat Spreader Panel
R:	Thermal Resistance	Al:	Aluminum Back Panel
R_C:	Thermal Contact Resistance	Sp:	Support Panel
$\dot{Q}_{loss}$:	Heat transfer rate from *Al* to *OD* through *Ins* (W)	Gp:	Gypsum Panel
$\dot{Q}_{in}$:	Heat transfer rate from *W* to *Al* through *Pn* (W)	Ins:	Insulation Panel
$\dot{Q}_{front}$:	Heat transfer rate from *Al* to the *Surface* (W)	W:	Water
D_i:	Tube Internal Diameter (m)	ID:	Indoor Air
q'_u:	Heat rate per unit length of the tube (W/m)	OD:	Outdoor Air
C_b:	Bond Conductance (W/m-K)	$Surface$:	Panel Surface
h_{fi}:	*W*-Tube inside wall heat transfer coefficient (W/m^2-K)		
h_c:	*Surface-ID* convective heat transfer coefficient (W/m^2K)		
h_r:	*Surface-ID* radiative heat transfer coefficient (W/m^2K)		
ΔT:	$T_{surface}$- T_{ID}		
ε_{gp}:	Emissivity for gypsum surface		
σ:	Stefan-Boltzmann constant (W/m^2K^4)		
T_b:	*Al* Temperature (K)		
T_f:	Water Inlet Temperature (K)		

REFERENCES

Bergman, T., Lavine, A., Incropera, F., & Dewitt, D. (2011). Radiation. In *Fundamentals of Heat and Mass Transfer* (7th ed., pp. 8–10). John Wiley & Sons.

Deru, M., Field, K., Studer, D., Benne, K., Griffith, B., Torcellini, P., Liu, B., Halverson, M., Winiarski, D., Rosenberg, M., Yazdanian, M., Huang, J., & Crawley, D. (2011). U.S. Department of Energy commercial reference building models of the national building stock. *Publications (E)*, *February 2011*, 1–118. http://digitalscholarship.unlv.edu/renew_pubs/44

Goldstein, K., & Novoselca, A. (2010). Convective Heat Transfer in Rooms with Ceiling Slot Diffusers. *HVAC&R RESEARCH*, *16*(5), 629–655. https://doi.org/http://dx.doi.org/10.1080/10789669.2010.10390925

Kalogirou, S. (2009). *Solar Energy Engineering* (an O. M. C. Safari (Ed.); 1st edition) [Book].

Klein, S. A. (2021). *EES-Engineering Equation Solver* (Academic Professional V-10.834-3D [2020-05-27]). www.fChartSoftware.com

Nellis, G. (2009). *Heat transfer* (Sanford A. Klein (Ed.)) [Book]. Cambridge University Press.

Rhee, K., & Woo, K. (2015). A 50 year review of basic and applied research in radiant heating and cooling systems for the built environment. *Building and Environment*, *91*, 166–190. https://doi.org/10.1016/j.buildenv.2015.03.040

ACKNOWLEDGEMENTS

Development of the HBIL was funded, in part, by the Center for High Performance Buildings at Purdue. The design and manufacturing of all the panels and base structure and its installation are being carried out by Bridgewater Studious. Their support and work are greatly appreciated. The essential equipment of the facility's hydronic system have been donated by WaterFurnace (Heat Pumps and Water tanks), Belimo (3-way mixing valves), and Johnson Controls (Auxiliary Fan Coil Unit). The HBIL team is very grateful for their interest in this project and the equipment donations. We also acknowledge Feng Wu, our fellow research student contributing to the voice-control system design and its integration with the HBIL facility and future HBIL testing. Lastly, the HBIL team is grateful to the Herrick shop staff for their technical support and guidance.

Performance evaluation of three virtual metering methods to estimate zone-level perimeter heater energy requirement

Darwish Darwazeh[1,*], Jean Duquette[2], and Burak Gunay[1]

[1] Carleton University, Department of Civil and Environmental Engineering, Ottawa, Ontario, Canada
[2]Carleton University, Department of Mechanical and Aerospace Engineering, Ottawa, Ontario, Canada
* Corresponding author e-mail address: darwish.darwazeh@carleton.ca

ABSTRACT

Virtual metering provides a cost-effective alternative to physical meters to monitor building energy performance and capture unmetered energy flows at the zone-level. Virtual metering accuracy depends on the modelling method and its ability to represent the heating and cooling processes at a building thermal zone. This paper employs three virtual metering methods to estimate the heating energy of zone-level perimeter heaters: a steady-state modelling method, a transient modelling method, and a load disaggregation modelling method. Inverse models representing these three virtual metering methods are trained using data obtained from seven perimeter offices in an academic building in Ottawa, Canada. Model parameters are identified using the genetic algorithm and used for creating virtual meters that estimate the energy requirement of zone-level perimeter heaters. The virtual meters' accuracy is assessed by comparing the results to measured heating energy obtained from physical meters installed in the seven offices. The three virtual metering methods' performance is evaluated through illustrative examples in terms of modelling assumptions, data requirements, and virtual metering accuracy. The results indicate that the three virtual metering methods can estimate the daily heating energy supplied by perimeter heaters at a normalized root-mean-square error between 13% and 23%.

1. INTRODUCTION

Metering of building heating and cooling energy at the zone-level is uncommon due to cost and practical issues, leaving critical quantities unmeasured such as the heat added by perimeter radiant heaters. Virtual metering provides a non-intrusive alternative to physical meters that can capture zone-level unmetered energy flows by utilizing available sensor and actuator measurements from a building automation system (BAS). Virtual meters (VMs) can be employed at a building zone-level to support operational decisions and improve energy performance and occupant comfort by allowing early detection of operational inefficiencies. However, virtual metering accuracy depends on the modelling method utilized and its ability to represent zone-level heating and cooling processes. These modelling methods involve assumptions regarding data processing and the degree to which the model agrees with the physical reality.

Selection of a suitable modelling method that can represent energy flows at a building thermal zone is essential for developing accurate zone-level VMs, describing building energy performance, identifying energy consumption patterns, developing model-based predictive control strategies, and detecting zone-level system faults (Bacher & Madsen, 2011; Gunay et al., 2016). The process of developing a zone model includes the selection of a model structure, the estimation of model parameters, and model validation (Madsen et al., 2016). The purpose of the zone model and the measured data available from a building thermal zone guide selecting a suitable model structure. Data collected from zone-level sensors, meters, and actuators are processed and used to train the model. Model parameters that represent unmeasured values are estimated using a model optimization algorithm that fits the model prediction to measured values (Balan et al., 2011). These estimated model parameters should be within a reasonable range of the underlying physical characteristics they represent; for example, an over-estimated value for the parameter representing the zone thermal resistance under ideal operating conditions indicates an unsuitable model structure.

VMs can be used to estimate unmetered heating energy of zone-level heating devices to understand its impact on the energy consumption of a building. Perimeter zones are subjected to internal heat gains from lights, occupants, and equipment, solar heat gains through windows, and heat gains by transmission through the envelope. The zone air temperature setpoint is maintained using supply air provided by an air handling unit (AHU) through a variable air volume (VAV) terminal box. These zones are often equipped with hydronic radiant heaters to satisfy the heating loads.

VMs can be employed to estimate the heat added by hydronic radiant heaters using the following three distinct modelling methods that emulate the heat transfer mechanisms within a given perimeter zone:

(1) *A steady-state inverse greybox modelling method*: this modelling method is useful for describing linear and stationary relations between model inputs and output (Madsen, 2008; Madsen et al., 2016) by assuming no energy is stored within the zone (Nassif et al., 2008a). The measurements' sampling time is usually averaged over a longer period since the system's dynamic behaviour is not described (Madsen et al., 2016; Rabl, 1988). Steady-state models are commonly used to evaluate energy efficiency measures and retrofit performance (Corrado & Fabrizio, 2007; Dall' o' et al., 2012; Firth & Lomas, 2009; Heo et al., 2012), and to find the effective thermal resistance of a building wall (Albatici et al., 2015; Nardi et al., 2015; Zheng, Cho et al., 2016).

(2) *A transient inverse greybox modelling method*: this modelling method describes the dynamic behaviour of the zone at different levels of complexity (Afram & Janabi-Sharifi, 2014b; Rabl, 1988). Heat transfer mechanisms are usually approximated by an equivalent thermal resistance-capacitance network (*i.e.*, RC network) (Gunay et al., 2017; Madsen et al., 2016). Transient models are typically used to evaluate building retrofits (Abushakra, 1999; Braun & Chaturvedi, 2002; Dong et al., 2005), to develop model-based predictive control strategies (Candanedo et al., 2013; Gunay et al., 2014; Ma et al., 2012; Široký et al., 2011), to detect HVAC system faults (Capozzoli et al., 2015; Ranade et al., 2020; Shi et al., 2016), and to characterize heating and cooling load patterns (Gunay et al., 2017; Wang et al., 2016).

(3) *A load disaggregation modelling method*: this modelling method breaks down the total load measured at the source into subsystem loads. The disaggregated loads provide a better understanding of the subsystems' energy performance (Yan et al., 2012). Load disaggregation models are commonly used in the literature to disaggregate the total electricity load across zone-level appliances in buildings (Dinesh et al., 2016; Patri et al., 2014; Valovage & Gini, 2017), and to disaggregate building total water consumption across zone-level water devices (Larson et al., 2012; Wang et al., 2018).

While models derived from these modelling methods are proposed in the literature for building energy management and building operation applications, to our knowledge, no comparisons have been made of these three modelling methods for virtual metering of heat added by perimeter radiant heaters. To this end, this paper analyzes the performance of the three modelling methods by developing VMs to estimate the heat added by zone-level perimeter heaters. Data obtained from seven perimeter offices in a highly instrumented academic building in Ottawa, Canada, are used to train the models. The accuracy of the VMs is assessed by comparing the results to measured heat obtained from physical meters installed in these seven offices. The modelling methods' performance is demonstrated through illustrative examples that compare modelling assumptions, data requirements, and virtual metering accuracy.

2. METHODOLOGY

As illustrated in Fig. 1, three distinct zone model structures are formulated via the application of a steady-state inverse greybox modelling method, a transient inverse greybox modelling method, and a load disaggregation modelling method. Data obtained from seven perimeter offices are used to train the models. Model parameters are identified using the genetic algorithm (GA) and used to develop VMs to estimate the heating energy of zone-level radiant heaters. VM results are compared with measured heat data obtained from physical meters installed at Zones (1) to (7). The root-mean-square error normalized by the range of daily measured heat at each radiant heater (NRMSE) is used to verify the accuracy of the VMs and evaluate the performance of the three modelling methods.

2.1 Data Collection and Processing

Data from seven perimeter offices in an academic building in Ottawa, Canada, are collected from November 2019 to January 2020 at 15-minute intervals. As illustrated in Fig. 2, the seven offices are served by two VAV terminal boxes VB_1 and VB_2 that receive supply air from an AHU. The supply air is reheated in the VBs and discharged into the offices via discharge air diffusers. These perimeter offices are provided with hydronic radiant heaters that use hot water from a steam/hot water heat exchanger. The radiant heaters are equipped with modulating valves to control the hot water flow rate into the offices. The seven offices are considered as seven thermal zones since each office uses an independent radiant heater to control the office temperature during the period of this study. The zone air temperature, $T_{z,1}$(°C) to $T_{z,7}$(°C), modulating valve state, X_1(%) to X_7(%), and the occupancy indicators, B_1 to B_7 , are measured

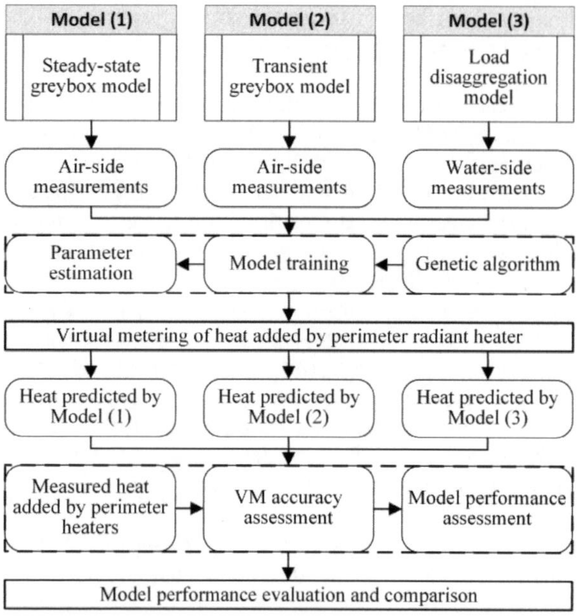

Figure 1: Schematic illustrating the use of three distinct modelling methods to develop zone-level radiant heater VMs. The accuracy of the VMs is used to evaluate the performance of the modelling methods

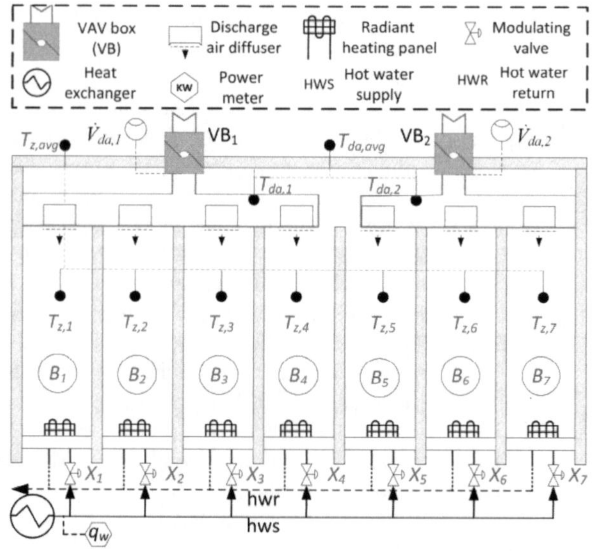

Figure 2: Data collected from seven perimeter zones in the academic building. $T_{z,i}$(°C), X_i(%), and B_i are zone air temperature, modulating valve state, and occupancy indicator in zone (i), respectively. $T_{da,j}$(°C) and $\dot{V}_{da,j}$(m^3/s) are the discharge air temperature and discharge airflow rate from VB_j, respectively. q_w(kW) represents the total hot water load.

at each zone, whereas the discharge air temperature, $T_{da,1}$(°C) and $T_{da,2}$(°C), and discharge airflow rate, $\dot{V}_{da,1}(m^3/s)$ and $\dot{V}_{da,2}(m^3/s)$, are measured at the two VAV terminal boxes. Additionally, the total hot water load, $q_w(kW)$, is measured at the heat exchanger hot water outlet. The zone air temperature and the discharge air temperature are modelled as single temperature nodes by averaging the measured data at every time step (Bleil et al., 2012; Platt et al., 2010). The total discharge airflow rate is obtained by summing up the discharge airflow rates from the two VAV terminal boxes VB_1 and VB_2. Finally, the binary occupancy signals (*i.e.,* 0 for non-occupied, 1 for occupied) are summed up for the seven zones. The processed dataset is used to train the three model structures developed in the following section.

2.2 Development of Zone Model Structures

Three distinct model structures are developed by applying a transient inverse greybox modelling method, a steady-state inverse greybox modelling method, and a load disaggregation modelling method. A description of each modelling method is provided in the following sections.

2.2.1 Transient inverse greybox modelling method

As shown in Fig. 3, the primary sources of heat gains and losses at a perimeter zone are internal heat gains from lights, occupants, and equipment, $q_{ig}(W)$, heat transfer through the envelope, $q_{env}(W)$, heat added or extracted through the discharge air diffuser, $q_{da}(W)$, solar heat gains, $q_{sol}(W)$, and heat added by the radiant panels, $q_{rad}(W)$. The rate of change in zone internal energy can be expressed as (Afram et al., 2014a; Nassif et al., 2008b):

$$c_a \cdot \rho_a \cdot V_z \cdot \frac{dT_z}{dt} = q_{da} + q_{env} + q_{rad} + q_{ig} + q_{sol} \tag{1}$$

where c_a is the specific heat of air (1006 $J/kg.$°C at 20 °C), ρ_a is the density of air (1.225 kg/m^3), T_z (°C) is the zone air temperature, and V_z (m^3) is the zone volume. The transient heat flow is approximated by an equivalent RC network with one resistance and one capacitance (*i.e.,* 1R1C network), as described in Fig. 4. The thermal capacitance, C_z (J/°C), represents the lumped thermal capacitance of a zone; and the thermal resistance, R_z (°C/W), represents the effective thermal resistance between zone indoors and outdoors, including heat transfer through the zone envelope and via air infiltration. Taking these terms into consideration, Equation (1) can be rewritten as (Gunay et al., 2016):

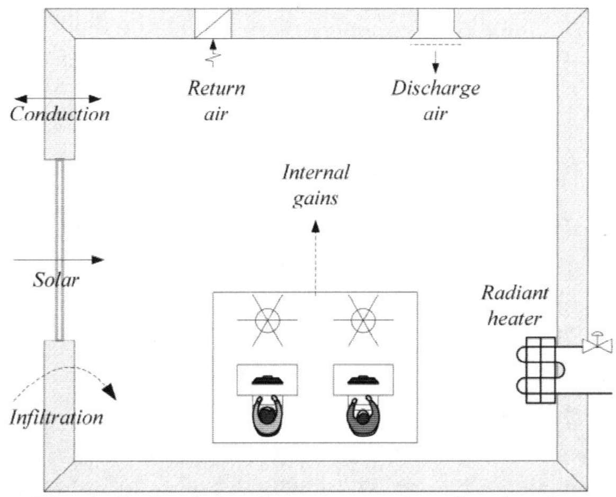

Figure 3: A schematic representation showing the primary sources of heat gains/losses in a single building VAV zone.

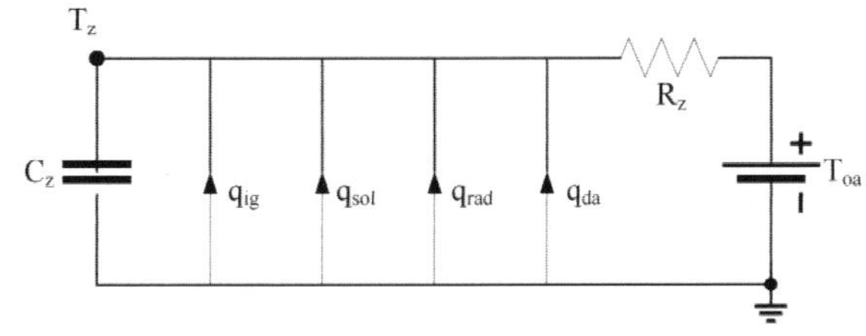

Figure 4: An approximation of transient heat flow with an equivalent thermal network comprising one resistance and one capacitance (*i.e.*, 1R1C network). Temperatures (°C) of zone air (T_z) and outdoor air (T_{oa}) are presented as electric voltages, whereas heat gains (W) from internal heat gains (q_{ig}), solar radiation (q_{sol}), radiant panels (q_{rad}), and discharge air (q_{da}) are presented as electric current. C_z ($J/$°C) and R_z (°C$/W$) represent the lumped thermal capacitance and resistance of the zone, respectively.

$$C_z \cdot \frac{dT_z}{dt} = \dot{V}_{da} \cdot c_a \cdot \rho_a \cdot (T_{da} - T_z) + \frac{(T_{oa} - T_z)}{R_z} + q_{rad} + q_{ig} + q_{sol} \quad (2)$$

where $\dot{V}_{da}$ (m^3/s) is the discharge airflow rate into the zone; and T_{da} (°C) and T_{oa} (°C) are the discharge air temperature and outdoor air temperature, respectively. A linear relationship is assumed between the hot water modulating valve state, X_{rad} (%), and the heat added by the radiant heater, q_{rad} (W), and between the binary occupancy indicator, B, and internal heat gains from lights, occupants, and equipment, q_{ig} (W). The impact of solar heat gains, q_{sol} (W), on the zone energy balance is minimized by only considering data points occurring after 4 pm and before 9 am. Under these assumptions, Equation (2) is further expanded to estimate the heat added or extracted through the discharge air diffuser as follows:

$$\dot{V}_{da} \cdot c_a \cdot \rho_a \cdot \left(T_z^{k-1} - T_{da}^{k-1}\right) = \frac{\left(T_{oa}^{k-1} - T_z^{k-1}\right)}{x_1} + x_2 \cdot B^{k-1} + d \cdot X^{k-1} - x_3 \cdot \frac{T_z^k - T_z^{k-1}}{\Delta t} \quad (3)$$

where the superscript k is the index number of measured data, and Δt (s) is the time step interval. Parameters x_1 to x_3 estimate the effect of the zone thermal resistance, internal heat gains, and the transient response of zone thermal capacitance, respectively, on the energy balance of the zone, and parameter d estimates the capacity of the radiant heater. Equation (3) represents a single-zone model that can be expanded to multi-zones by taking the average zone air temperature, $T_{z,avg}$ (°C), the average discharge air temperature, $T_{da,avg}$ (°C), the total airflow rate from the discharge air diffusers, $\sum \dot{V}_{da,i} (m^3/s)$, and the total occupancy indicator from the individual zones, $\sum B_i$. Hence, the transient inverse greybox model for the seven perimeter zones is expressed as:

$$c_a \cdot \rho_a \cdot \sum_{i=1}^{7} \dot{V}_{da,i}^{k-1} \cdot \left(T_{z,avg}^{k-1} - T_{da,avg}^{k-1}\right) = \frac{\left(T_{oa}^{k-1} - T_{z,avg}^{k-1}\right)}{x_1} + x_2 \cdot \sum_{i=1}^{7} B_i^{k-1} + \left(\sum_{i=1}^{7} d_i \cdot X_i^{k-1}\right) - \left(x_3 \cdot \frac{T_{z,avg}^k - T_{z,avg}^{k-1}}{\Delta t}\right) + e \quad (4)$$

Parameters d_1 to d_7 estimate the capacity of the radiant heaters in Zones (1) to (7), and parameter e is an error term that accounts for unmodelled heat flows within the zone.

2.2.2 Steady-state inverse greybox modelling method

The steady-state modelling method assumes no thermal storage in the zone and hence no change in the zone air temperature occurs over time (Nassif et al., 2008b). The steady-state model is formulated by omitting the transient term in Equation (4) (*i.e.*, $T_{z,avg}^k - T_{z,avg}^{k-1} = 0$) as follows:

$$c_a \cdot \rho_a \cdot \left(\sum_{i=1}^{7} \dot{V}_{da,i}^{k-1}\right) \cdot \left(T_{z,avg}^{k-1} - T_{da,avg}^{k-1}\right) = \left(\frac{\left(T_{oa}^{k-1} - T_{z,avg}^{k-1}\right)}{x_1}\right) + \left(x_2 \cdot \sum_{i=1}^{7} B_i^{k-1}\right) + \left(\sum_{i=1}^{7} d_i \cdot X_i^{k-1}\right) + e \tag{5}$$

Since the steady-state modelling method neglects the transiency of the heat and mass transfer processes, the measured data can be averaged over more extended sampling periods (*i.e.*, downsampling). The impact of temporal averaging of data on the steady-state model's performance is evaluated by averaging the data collected from Zones (1) to (7) over sampling periods ranging from 0.25 hours to 24 hours. VMs results corresponding to each sampling period are compared with measured heat to assess the impact of temporal averaging of data on the accuracy of the VMs.

2.2.3 Load disaggregation modelling method

The load disaggregation method breaks down the total energy supplied to a group of radiant heaters into individual radiant heater energy components based on the state of the modulating valve. As illustrated in Fig. 5, hot water is provided by a steam/hot water heat exchanger that distributes hot water to the radiant heaters in the seven perimeter zones. The total hot water load, q_w (kW), is disaggregated across the seven radiant heaters by utilizing the modulating valve analog signals, X_i (%) , that control the hot water flow rate into the zone-level radiant heaters as expressed in Equation (6).

$$q_w = \left(\sum_{i=1}^{7} d_i \cdot X_i\right) + e \tag{6}$$

Model parameters d_1 to d_7 correspond to individual radiant heater capacities at Zones (1) to (7), respectively, and the error term e accounts for errors in the measured data and any potential leakage.

2.3 Virtual Metering and Model Performance Evaluation

The dataset created for the seven perimeter zones is used to train the three model structures given by Equations (4) to (6). An optimization problem is formulated to minimize the root-mean-square error (RMSE) between the estimated and measured discharge air load, q_{da} (kW), for the transient and steady-state models, and between estimated and measured total hot water load, q_w (kW), for the load disaggregation model. The GA is selected to solve the optimization problem and search for optimal model parameters since the upper and lower bounds for the parameters can be identified from the physical quantities presented by the models. The estimated values of model parameters

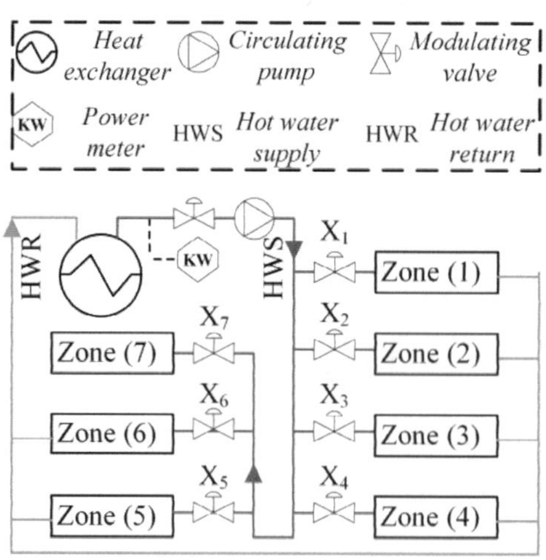

Figure 5: A schematic showing the disaggregation of total hot water load measured at the outlet of a steam/hot water heat exchanger across zone-level radiant heaters using the modulating valve state signals X_1 to X_7.

corresponding to each modelling method are used to create VMs to estimate the daily heat added by the radiant heaters, $\hat{E}_{rad,daily}$ (kWh/day), at each of the seven perimeter zones. The daily heat is obtained by summing the energy at 15-minute intervals for each zone as follows:

$$\hat{E}_{rad,daily} = \sum_{k=1}^{M} \hat{E}_{rad}^{k} = \sum_{k=1}^{M} (q_{rad}^{k} \cdot \Delta t) \tag{7}$$

where $\hat{E}_{rad}^{k}$ (kWh) is the estimated heat added during time step k, Δt (hrs) is the timestep interval, and M represents the number of time steps in one day. Measured data collected from physical meters are used to calculate the actual daily heat added by radiant heaters, $E_{rad,daily}$ (kWh/day) at Zones (1) to (7). The measured daily heat is utilized to verify the VM results using the RMSE (kWh/day) normalized by the range of daily measured heat at each zone (NRMSE), as follows:

$$NRMSE = \frac{\sqrt{\frac{1}{M} \cdot \sum_{k=1}^{M} (\hat{E}_{rad,daily}^{k} - E_{rad,daily}^{k})^2}}{(E_{rad,daily}^{max} - E_{rad,daily}^{min})} \tag{8}$$

where $E_{rad,daily}^{max}$ and $E_{rad,daily}^{min}$ represent the maximum and minimum values of measured daily heat. The accuracy of the VMs is used to analyze the performance of the steady-state, transient, and load disaggregation models. The three modelling methods are compared in terms of modelling assumptions, data requirements, and VMs accuracy.

3. RESULTS AND DISCUSSION

VM results corresponding to the three modelling methods are compared with measured heat collected from physical meters at Zones (1) to (7) to assess the accuracy of the VMs and evaluate the performance of the underlying modelling methods. As shown in Table 1, the three modelling methods can estimate the daily heat added by radiant heaters at an average NRMSE between 13% and 23%. The average RMSE from the seven zones for the transient modelling method, the steady-state modelling method, and the load disaggregation modelling method is 1.73 kWh/day, 2.67 kWh/day, and 1.63 kWh/day, respectively, as shown in Fig. 6. These results indicate better performance for the transient and load disaggregation modelling methods as compared to the steady-state modelling method. The decrease in accuracy for the VMs developed using a steady-state modelling method can be related to steady-state modelling assumptions and model inputs. For example, Fig. 7 shows the total estimated heat supplied by radiant heaters at Zones (1) to (7) from November 2019 to January 2020 using the three modelling methods. The total

Table 1: VM results assessment for the three modelling approaches

	Zone 1	Zone 2	Zone 3	Zone 4	Zone 5	Zone 6	Zone 7	Overall
Total measured heat (kWh)	334.61	231.05	274.83	405.67	335.29	194.02	357.63	2133.10
	Transient inverse greybox modelling approach							
Measured data required	$[T_{oa}(°C)]$, $[T_z(°C)]$, $[T_{da}(°C)]$, $[\dot{V}_z\ (m^3/s)]$, (B_z), $[X_z\ (\%)]$							
Total estimated heat (kWh)	346.23	216.06	337.19	460.02	357.57	177.06	359.34	2253.48
RMSE (kWh/day)	1.62	1.95	1.59	2.05	1.86	1.39	1.52	1.73
NRMSE (%)	19.29%	18.12%	23.19%	21.31%	15.36%	17.31%	13.04%	14.01%
	Steady-state inverse greybox modelling approach							
Measured data required	$[T_{oa}(°C)]$, $[T_z(°C)]$, $[T_{da}(°C)]$, $[\dot{V}_z\ (m^3/s)]$, (B_z), $[X_z\ (\%)]$							
Total estimated heat (kWh)	141.37	51.55	259.10	118.90	88.51	161.42	103.90	924.73
RMSE (kWh/day)	2.57	2.85	1.34	3.60	3.47	1.42	3.46	2.67
NRMSE (%)	30.66%	26.45%	19.52%	37.37%	28.62%	17.71%	29.65%	22.84%
	Load disaggregation modelling approach							
Measured data required	$[q_w\ (kW)]$, $[X_z\ (\%)]$							
Total estimated heat (kWh)	332.01	222.47	283.02	393.81	324.88	182.77	356.08	2095.03
RMSE (kWh/day)	1.58	1.95	1.36	1.74	1.82	1.39	1.52	1.63
NRMSE (%)	18.80%	18.10%	19.79%	18.09%	15.03%	17.26%	13.03%	13.26%

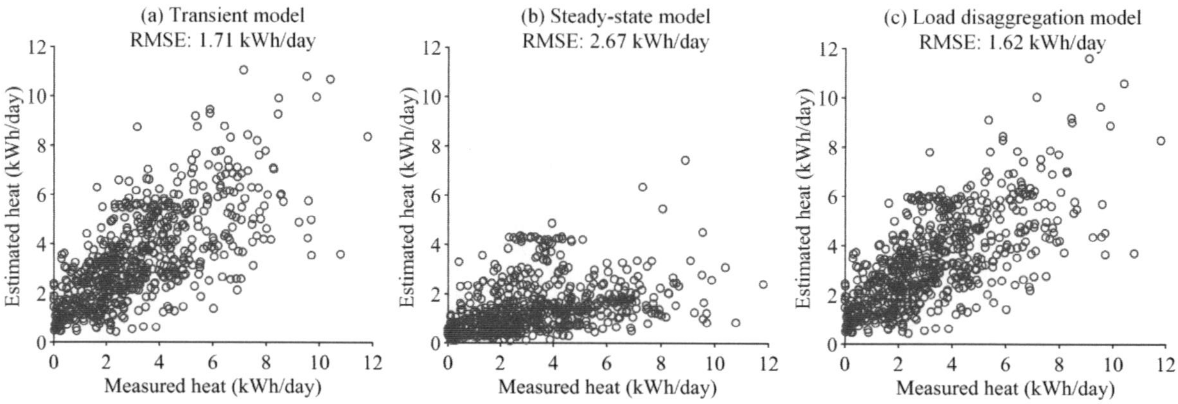

Figure 6: Predictive accuracy of VMs in the seven zones of the academic building developed by using (a) a transient modelling method, (b) a steady-state modelling method, and (c) a load disaggregation modelling method

Estimated heat for the seven zones when applying the steady-state modelling method is around 43% of the total measured heat. This substantial difference between the measured and estimated heat is believed to be related to the steady-state model structure, which assumes that no change in zone air temperature occurs over time. By checking the zone air temperature behaviour in Zone (3) for three days in January 2020, as shown in Fig. 8, it is noticed that the zone air temperature starts to decline when the VAV system shuts down at 22:45 hrs., and rises again upon VAV system start-up at 7:00 hrs. During the VAV system shut-down and start-up periods, the zone exhibits a dynamic behaviour not captured by the steady-state model. When data points corresponding to these periods are used as model inputs, the accuracy of the VMs declines, as indicated in Table 1.

The results presented in Table 1 are obtained using a 15-minute data sampling frequency for the three modelling methods. At this sampling frequency, a higher error is noticed for the steady-state modelling method than the transient and load disaggregation modelling methods (see Table 1). The impact of temporal averaging of measured

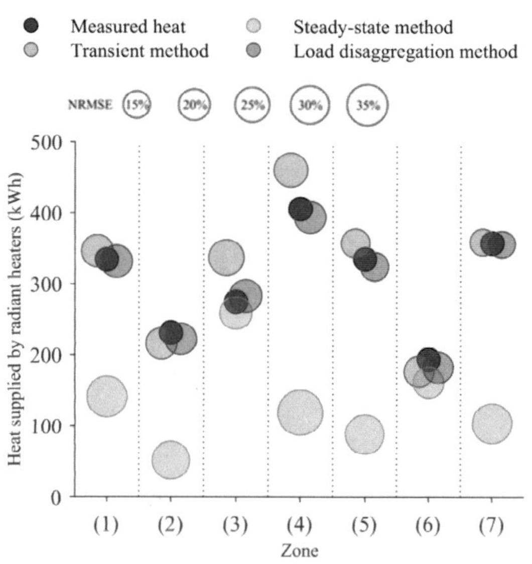

Figure 7: Estimated heat supplied by radiant heaters at Zones (1) to (7) from November 2019 to January 2020 using three modelling methods. The size of the bubble is proportional to the NRMSE obtained for each zone.

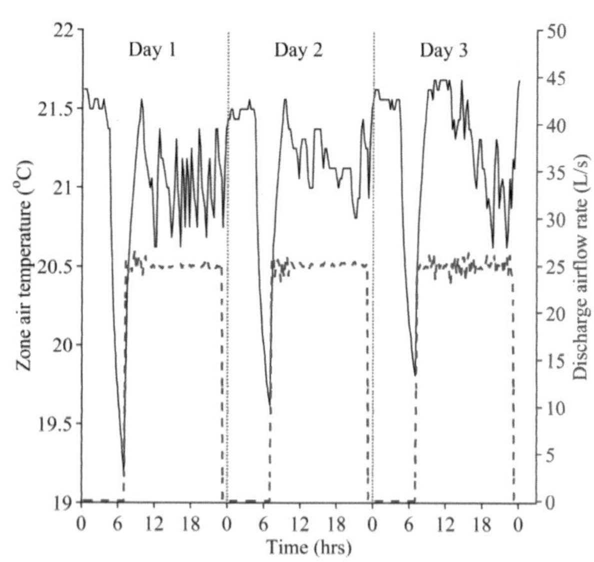

Figure 8: Zone air temperature in Zone (3) for three days in December 2019. During the start-up and shut-down of the VAV system, the zone's dynamic behaviour is not captured by the steady-state modelling method due to modelling assumptions.

data on the accuracy of the steady-state modelling method is examined by training the steady-state model using datasets averaged over sampling periods from 15 minutes to 24 hours. VMs are developed using the parameters identified from the trained models corresponding to each sampling period. As shown in Fig. 9, the average NRMSE of the VMs from the seven zones declines from 23% at a 15-minute sampling period to 13% when the sampling period is extended to 24 hours. The total predicted heat added by the radiant heaters in the seven zones compared to the measured heat indicates a better performance of the steady-state model at longer sampling periods. The decline in error at more extended sampling periods is attributed to minimizing the transient effect when the measured data is averaged over intervals with similar initial and final conditions. For example, the transient effect caused by starting-up and shutting-down the VAV system in Fig. 8 can be minimized by averaging the data over a 24-hour sampling period. Although using longer sampling periods with the steady-state modelling method improves the VM accuracy, additional data processing is required to average the data over longer sampling periods. In contrast, short sampling periods are appropriate for the transient and load disaggregation modelling methods.

Since the three modelling methods require different sorts of data, the availability of measured data from the BAS can be used as a means to select an appropriate modelling method. Measurements of outdoor air temperature, T_{oa}(°C), zone air temperature, $T_{z,i}$(°C), occupancy indicator, B_i, discharge airflow rate, $\dot{V}_{da,i}$, and radiant heater modulating valve position, X_i (%), are often available in the BAS of commercial and institutional buildings. However, measurements of discharge air temperature from the VAV system, $T_{da,i}$ (°C), and total hot water load from the steam/hot water heat exchanger, q_w (kW), are not always available. In cases when the total hot water load measurement is not available, the transient or the steady-state models should be selected, provided that the measurements required for these two modelling methods are available. Similarly, in cases when the measurement of discharge air temperature is not available, the load disaggregation modelling approach should be selected, provided that the total hot water load measurement is available. Hence, the availability of sensors and meters plays a critical role in the model selection process.

This study demonstrates the use of a steady-state modelling method, a transient modelling method, and a load disaggregation modelling method for creating VMs that estimate the heat added by zone-level perimeter heaters. The three modelling methods are compared in terms of modelling assumptions, data requirements, and virtual metering accuracy. The models developed upon applying these three modelling methods can be integrated into the BAS to estimate unmetered heat added by perimeter heaters. The estimated heating energy at a building zone-level provides critical information to building operators and facility managers to improve building energy performance and facilitate operational decisions.

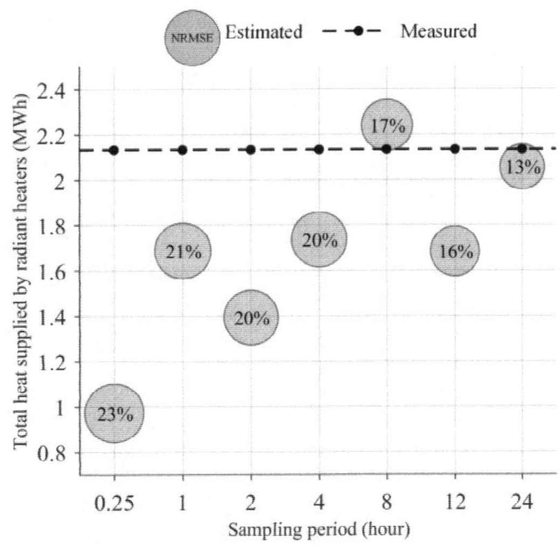

Figure 9: Total estimated heat supplied by the radiant heaters in Zones (1) to (7) compared to measured heat using the steady-state modelling method. Measurements are averaged over sampling periods from 15 minutes to 24 hours. The size of the bubble is proportional to the NRMSE obtained for each sampling period.

4. CONCLUSION

This paper evaluated the performance of three zone modelling methods used in developing VMs that can estimate the heat supplied by zone-level perimeter radiant heaters. Zone models were developed by applying a transient inverse greybox modelling method, a steady-state inverse greybox modelling method, and a load disaggregation modelling method. The models were trained using a dataset comprising three months of measured data collected from seven offices in an academic building in Ottawa, Canada. VMs were created by using model parameters that were identified using the genetic algorithm. The accuracy of the VMs was assessed using additional heat measurements collected from physical meters installed at the seven offices. The three modelling methods were compared in terms of modelling assumptions, model data requirements, and VMs accuracy.

The VMs' accuracy assessment showed that the three modelling methods could estimate the heat added by perimeter radiant heaters at an average NRMSE between 13% and 23%. The accuracy of the steady-state modelling method was further analyzed by checking the impact of temporal averaging of measured data over longer sampling periods. The results showed that the NRMSE improves by 10% when the sampling period increases from 15 minutes to 24 hours.

The model selection process was discussed by comparing the measurements required for the three modelling methods. Practical scenarios were presented, showing how the availability of measured data from the BAS impacts the selection of a suitable modelling method. Although an initial evaluation of the three modelling approaches was presented in this paper, there remain several unresolved issues for future research such as a) evaluating the performance of the three modelling methods as the number of zones increases, and b) developing visualization tools to present the VMs results to different building stakeholders.

ACKNOWLEDGEMENT

This research is supported by the research funding provided by the National Sciences and Engineering Research Council (NSERC) of Canada, Brookfield Global Integrated Solutions (BGIS), Public Services and Procurement Canada (PSPC), and CopperTree Analytics.

REFERENCES

Abushakra, B. (1999). An inverse model to predict and evaluate the energy performance of large commercial and institutional buildings. In *IBPSA Building Simulation* (Vol. 8, pp. 8–49). Retrieved from http://www.ibpsa.org/%5Cproceedings%5CBS1997%5CBS97_P183.pdf

Afram, A., & Janabi-Sharifi, F. (2014a). Review of modeling methods for HVAC systems. *Applied Thermal Engineering*. https://doi.org/10.1016/j.applthermaleng.2014.03.055

Afram, A., & Janabi-Sharifi, F. (2014b). Review of modeling methods for HVAC systems. *Applied Thermal Engineering*, *67*(1–2), 507–519. https://doi.org/10.1016/j.applthermaleng.2014.03.055

Albatici, R., Tonelli, A. M., & Chiogna, M. (2015). A comprehensive experimental approach for the validation of quantitative infrared thermography in the evaluation of building thermal transmittance. *Applied Energy*, *141*, 218–228. https://doi.org/10.1016/j.apenergy.2014.12.035

Bacher, P., & Madsen, H. (2011). Identifying suitable models for the heat dynamics of buildings. *Energy and Buildings*, *43*(7), 1511–1522. https://doi.org/10.1016/j.enbuild.2011.02.005

Balan, R., Cooper, J., Chao, K. M., Stan, S., & Donca, R. (2011). Parameter identification and model based predictive control of temperature inside a house. *Energy and Buildings*, *43*, 748–758. https://doi.org/10.1016/j.enbuild.2010.10.023

Bleil, C., Souza, D., & Alsaadani, S. (2012). Thermal zoning in speculative office buildings: discussing the connections between space layout and inside temperature control.

Braun, J. E., & Chaturvedi, N. (2002). An inverse gray-box model for transient building load prediction. *HVAC and R Research*, *8*(1), 73–99. https://doi.org/10.1080/10789669.2002.10391290

Candanedo, J. A., Dehkordi, V. R., & Lopez, P. (2013). A control-oriented simplified building modelling strategy. In *Proceedings of BS 2013: 13th Conference of the International Building Performance Simulation Association* (pp. 3682–3689).

Capozzoli, A., Lauro, F., & Khan, I. (2015). Fault detection analysis using data mining techniques for a cluster of smart office buildings. *Expert Systems with Applications*, *42*(9), 4324–4338. https://doi.org/10.1016/j.eswa.2015.01.010

Corrado, V., & Fabrizio, E. (2007). Assessment of building cooling energy need through a quasi-steady state model: Simplified correlation for gain-loss mismatch. *Energy and Buildings*, *39*(5), 569–579. https://doi.org/10.1016/j.enbuild.2006.09.012

Dall' o', G., Galante, A., & Torri, M. (2012). A methodology for the energy performance classification of residential building stock on an urban scale. *Energy and Buildings*, *48*, 211–219. https://doi.org/10.1016/j.enbuild.2012.01.034

Dinesh, C., Nettasinghe, B. W., Godaliyadda, R. I., Ekanayake, M. P. B., Ekanayake, J., & Wijayakulasooriya, J. V. (2016). Residential Appliance Identification Based on Spectral Information of Low Frequency Smart Meter Measurements. *IEEE Transactions on Smart Grid*, *7*(6), 2781–2792. https://doi.org/10.1109/TSG.2015.2484258

Dong, B., Cao, C., & Lee, S. E. (2005). Applying support vector machines to predict building energy consumption in tropical region. *Energy and Buildings*, *37*(5), 545–553. https://doi.org/10.1016/j.enbuild.2004.09.009

Firth, S. K., & Lomas, K. J. (2009). INVESTIGATING CO 2 EMISSION REDUCTIONS IN EXISTING URBAN HOUSING USING A COMMUNITY DOMESTIC ENERGY MODEL. *Building Simulation*.

Gunay, B., Shen, W., & Newsham, G. (2017). Inverse blackbox modeling of the heating and cooling load in office buildings. *Energy and Buildings*, *142*, 200–210. https://doi.org/10.1016/j.enbuild.2017.02.064

Gunay, H. B., Bursill, J., Huchuk, B., O'Brien, W., & Beausoleil-Morrison, I. (2014). Shortest-prediction-horizon model-based predictive control for individual offices. *Building and Environment*, *82*, 408–419. https://doi.org/10.1016/j.buildenv.2014.09.011

Gunay, H. B., O'Brien, W., & Beausoleil-Morrison, I. (2016). Control-oriented inverse modeling of the thermal characteristics in an office. *Science and Technology for the Built Environment*, *22*(5), 586–605. https://doi.org/10.1080/23744731.2016.1175893

Heo, Y., Choudhary, R., & Augenbroe, G. A. (2012). Calibration of building energy models for retrofit analysis under uncertainty. *Energy and Buildings*, *47*, 550–560. https://doi.org/10.1016/j.enbuild.2011.12.029

Homod, R. Z. (2013). Review on the HVAC System Modeling Types and the Shortcomings of Their Application. *Journal of Energy*, *2013*. https://doi.org/10.1155/2013/768632

Larson, E., Froehlich, J., Campbell, T., Haggerty, C., Atlas, L., Fogarty, J., & Patel, S. N. (2012). Disaggregated water sensing from a single, pressure-based sensor: An extended analysis of HydroSense using staged experiments. *Pervasive and Mobile Computing*, *8*(1), 82–102. https://doi.org/10.1016/j.pmcj.2010.08.008

Ma, J., Qin, J., Salsbury, T., & Xu, P. (2012). Demand reduction in building energy systems based on economic model predictive control. *Chemical Engineering Science*, *67*(1), 92–100. https://doi.org/10.1016/j.ces.2011.07.052

Madsen, H. (2008). Time Series Analysis. 1st.

Madsen, Henrik, Bacher, P., Bauwens, G., Deconinck, A.-H., Reynders, G., Roels, S., … Lethé, G. (2016). Thermal Performance Characterization using Time Series Data. *IEA EBC Annex 58*.

Mayhorn, E. T., Sullivan, G. P., Butner, R. S., Hao, H., & Baechler, M. C. (2015). *Characteristics and Performance of Existing Load Disaggregation Technologies*. *Pacific Northwest National Laboratory, Richland, WS* (Vol. null). Richland, WA (United States). https://doi.org/10.2172/1203912

Nardi, I., Ambrosini, D., De Rubeis, T., Sfarra, S., Perilli, S., & Pasqualoni, G. (2015). A comparison between thermographic and flow-meter methods for the evaluation of thermal transmittance of different wall constructions. *Journal of Physics: Conference Series*, *655*(1). https://doi.org/10.1088/1742-6596/655/1/012007

Nassif, N., Moujaes, S., & Zaheeruddin, M. (2008a). Self-tuning dynamic models of HVAC system components. *Energy and Buildings*, *40*(9), 1709–1720. https://doi.org/10.1016/j.enbuild.2008.02.026

Nassif, N., Moujaes, S., & Zaheeruddin, M. (2008b). Self-tuning dynamic models of HVAC system components. *Energy and Buildings*, *40*(9), 1709–1720. https://doi.org/10.1016/J.ENBUILD.2008.02.026

Patri, O. P., Panangadan, A. V., Chelmis, C., & Prasanna, V. K. (2014). Extracting discriminative features for event-based electricity disaggregation. In *2014 IEEE Conference on Technologies for Sustainability, SusTech 2014* (pp. 232–238). Institute of Electrical and Electronics Engineers Inc. https://doi.org/10.1109/SusTech.2014.7046249

Platt, G., Li, J., Li, R., Poulton, G., James, G., & Wall, J. (2010). Adaptive HVAC zone modeling for sustainable buildings. *Energy and Buildings*, *42*(4), 412–421. https://doi.org/10.1016/j.enbuild.2009.10.009

Rabl, A. (1988). Parameter estimation in buildings: Methods for dynamic analysis of measured energy use. *Journal of Solar Energy Engineering, Transactions of the ASME*, *110*(1), 52–66. https://doi.org/10.1115/1.3268237

Ranade, A., Provan, G., El-Din Mady, A., & O'Sullivan, D. (2020). A computationally efficient method for fault diagnosis of fan-coil unit terminals in building Heating Ventilation and Air Conditioning systems. *Journal of Building Engineering*, *27*, 100955. https://doi.org/10.1016/j.jobe.2019.100955

Shi, Z., O'Brien, W., & Gunay, B. (2016). Building Zone Fault Detection with Kalmnan Filter based Methods. In *eSim 2016*. Retrieved from https://www.researchgate.net/publication/306006612

Široký, J., Oldewurtel, F., Cigler, J., & Prívara, S. (2011). Experimental analysis of model predictive control for an energy efficient building heating system. *Applied Energy*, *88*(9), 3079–3087. https://doi.org/10.1016/j.apenergy.2011.03.009

Thavlov, A. ;, & Bindner, H. W. (2012). *Thermal Models for Intelligent Heating of Buildings*. APA.

Valovage, M., & Gini, M. (2017). Label correction and event detection for electricity disaggregation. In *Proceedings of the International Joint Conference on Autonomous Agents and Multiagent Systems, AAMAS* (Vol. 2, pp. 990–998). Retrieved from www.ifaamas.org

Wang, B., Chen, Z., Boedihardjo, A. P., & Lu, C. T. (2018). Virtual metering: An efficient water disaggregation algorithm via non-intrusive load monitoring. *ACM Transactions on Intelligent Systems and Technology*, *9*(4), 1–30. https://doi.org/10.1145/3141770

Wang, H., Xu, P., Lu, X., & Yuan, D. (2016). Methodology of comprehensive building energy performance diagnosis for large commercial buildings at multiple levels. *Applied Energy*, *169*, 14–27. https://doi.org/10.1016/j.apenergy.2016.01.054

Yan, C., Wang, S., & Xiao, F. (2012). A simplified energy performance assessment method for existing buildings based on energy bill disaggregation. *Energy and Buildings*, *55*, 563–574. https://doi.org/10.1016/j.enbuild.2012.09.043

Zheng, K., Cho, Y. K., Wang, C., & Li, H. (2016). Noninvasive Residential Building Envelope R-Value Measurement Method Based on Interfacial Thermal Resistance. *Journal of Architectural Engineering*, *22*(4). https://doi.org/10.1061/(ASCE)AE.1943-5568.0000182

Standard Testing Protocols for CO_2 Sensors and CO_2-based Demand Control Ventilation Systems

Antonea FRASIER[1], Frederick MEYERS[2], Derrick ROSS[2], Theresa PISTOCHINI[2], Matthew J. ELLIS[1*]

[1] Department of Chemical Engineering, University of California, Davis
Davis, CA 95616
(mjellis@ucdavis.edu)

[2] Western Cooling Efficiency Center, University of California, Davis
Davis, CA 95616
(tepistochini@ucdavis.edu)

* Corresponding Author

ABSTRACT

Carbon dioxide (CO_2)-based demand control ventilation (DCV) automatically adjusts building ventilation rates based on indoor CO_2 concentration. Since the indoor CO_2 concentration is directly related to the occupancy, the purpose of CO_2-based DCV is to conserve energy by reducing the ventilation rates during periods of low occupancy. In this work, two standard testing protocols for CO_2 sensors and CO_2-based DCV system controllers are developed and performed on several currently available CO_2 sensors and DCV system controllers. A few sample test results are provided.

1. INTRODUCTION

Proper building ventilation is crucial for achieving healthy indoor air quality since it maintains the concentrations of carbon dioxide (CO_2), volatile organic compounds (VOCs), and other indoor containments at or below their maximum healthy levels. ASHRAE Standard 62.1 specifies the ventilation requirements for acceptable indoor air quality (ASHRAE, 2010). Well-ventilated commercial buildings are associated with lower sick leave rates than poorly ventilated buildings (Emmerich & Persily, 2001). Under-ventilated buildings have been linked to a condition known as sick building syndrome (Emmerich & Persily, 2001). However, over ventilation increases energy consumption because fresh outdoor air must be conditioned. Recently, the focus on ventilation has been magnified by the SARS-CoV-2 pandemic (United States Centers for Disease Control, 2020) and wildfires in California and Oregon (California Division of Occupational Safety and Health, 2019).

Demand control ventilation (DCV) is an approach to building ventilation that automatically adjusts the building ventilation rate (i.e., the flow rate of outdoor air supplied to the building) in response to a sensor signal (Fisk & de Almeida, 1998). Most of the research and development on single-zone DCV occurred in the 1990s and early 2000s; refer to the review papers (Fisk & de Almeida, 1998; Emmerich & Persily, 2001; Apte, 2006) and the references therein. The more recent work on DCV has focused on DCV strategies for multi-zone configurations (Nassif, 2012; Lin & Lau, 2014; Liu et al., 2012; O'Neill et al., 2020). CO_2-based DCV system controllers manipulate the ventilation rate based on the indoor CO_2 concentration. Since CO_2 concentration is correlated with the occupancy amount, CO_2-based DCV system controllers have been demonstrated to conserve energy by reducing the outdoor airflow rate when spaces are not fully occupied (Schell & Inthout, 2001; Taylor, 2006). Giacomo (1999); Dougan and Damiano (2004); Mysen et al. (2005); Acker and Wymelenberg (2011) analyze the benefits and risks of DCV. While there has been some debate on the overall benefit of DCV, CO_2-based DCV for single-zone applications is now a well-established technology, and standard products exist from vendors (e.g., Honeywell Jade, Belimo Zip, and Johnson Controls Peak). Several simulations and field tests have been performed on DCV strategies (Emmerich & Persily, 2001) although most of these studies are now over twenty years old.

CO_2-based DCV system controllers require CO_2 sensors to function properly. A common type of low-cost CO_2 sensor used in heating, ventilation, and air conditioning (HVAC) applications is a non-dispersive infrared (NDIR) sensor (Shrestha, 2009). NDIR CO_2 sensors utilize an infrared radiation (IR) source. The IR passes through an air sample, and a detector (located on the opposite end from the IR source) measures the resulting intensity. A correlation is used to relate the measured intensity to the concentration as the intensity is proportional to the CO_2 concentration.

The effectiveness of CO_2-based DCV depends on the accuracy of CO_2 sensors. In (Jones et al., 1997), a study of 29 sensors was conducted. The testing procedure and setup were described where pure CO_2 was added over a period to obtain a certain concentration in a chamber housing the sensors. From the test results, absolute sensor errors of 100 ppm or more were possible. As pointed out in (Jones et al., 1997), measurement errors of 100 ppm could result in DCV control action errors up to 25 percent, compared to that produced with an accurate CO_2 sensor. This result further emphasizes the importance of accurate CO_2 sensors. In (Shrestha & Maxwell, 2009), a testing procedure was designed to test HVAC-grade CO_2 sensors to evaluate many factors including sensitivity to humidity, temperature, and pressure. The study used dry calibrated N_2/CO_2 gas mixtures. Water vapor was added to the dry gas mixture to achieve the desired humidity level. From the test results, the absolute error in sensor measurement was shown to generally increase with CO_2 concentration. Moreover, for several sensors considered in the study, absolute errors of over 100 ppm were identified.

Quantifying the overall performance of currently available DCV system controllers (i.e., the ability to maintain the CO_2 concentration at or below the standard limit defined in ASHRAE Standard 62.1 (ASHRAE, 2010)) is important given the renewed attention in building ventilation. As illustrated by previous studies, the accuracy of CO_2 sensors and the control performance of DCV system controllers should be evaluated. While previous work has tested various CO_2 sensors and DCV strategies under various conditions, the test results are now out-of-date, and may not reflect the performance of currently available products. Thus, evaluating the current products based on a standardized testing protocol will help inform building operators and managers.

In this work, two testing protocols for CO_2 sensors and CO_2-based DCV system controllers are developed and performed on several available sensors and DCV system controllers. The objective of the first protocol is to quantify the accuracy of CO_2 sensors. The objective of the second protocol is to assess the performance of CO_2-based DCV system controllers. To establish a performance baseline, the test results are compared to the expected results under an ideal DCV strategy. The protocols are described and results after executing each of the protocols are presented.

2. CO_2 SENSOR AND CO_2-BASED DCV SYSTEM CONTROLLER TESTING PROTOCOLS

In this section, the CO_2 sensor and CO_2-based DCV system controller testing protocols are described. To assess the performance of the DCV system controller, a baseline control strategy assuming ideal conditions is developed, and the performance under the ideal baseline control strategy is compared to the performance of the DCV system controller under test. The baseline control strategy, referred to as the ideal DCV strategy for the remainder, is also described in this section.

2.1 CO_2 Sensor Test Protocol

The objective of the CO_2 sensor test protocol is to quantify the accuracy of HVAC-grade wall-mount CO_2 sensors used for DCV system controllers under typical building environmental conditions. To evaluate sensor accuracy, sensors are placed in an enclosure that is tightly sealed and is continuously flushed with a calibrated CO_2/N_2 gas mixture. The steady-state sensor measurements obtained from the sensors are compared to the known concentration of the calibrated gas mixture reported by the manufacturer. During the test, environmental conditions in the enclosure are maintained at a temperature of 75°F, an absolute pressure of 101 kPa, and less than 5 percent relative humidity. The conditions are measured every 30 seconds. To define a successful test, the temperature and pressure must satisfy two tolerance types: a test operation tolerance and a test condition tolerance. The test operating tolerance is defined as the difference between the maximum and minimum measurement over the test duration for any operating condition and is 5°F and 3 kPa for the temperature and pressure, respectively. The test condition tolerance is defined as the difference between the test condition and the average value over the test duration for any operating condition and is ±3°F and ±1.5 kPa for the temperature and pressure, respectively. To prevent infiltration, positive differential pressure between the enclosure and surrounding of at least 5 Pa is maintained. The differential pressure must not exceed 100 Pa.

The temperature and pressure conditions are selected to mimic standard building conditions. Previous research has demonstrated that CO_2 sensor response is reasonably consistent over the temperature range allowed by the test (Shrestha, 2009). The absolute pressure condition is selected to be representative of the absolute pressure at sea level. Regarding the relative humidity condition, water vapor could be added to the dry gas mixture so that the relative humidity of the resulting CO_2, N_2 , and water vapor mixture reflects standard building relative humidity. However, lab-grade measure-

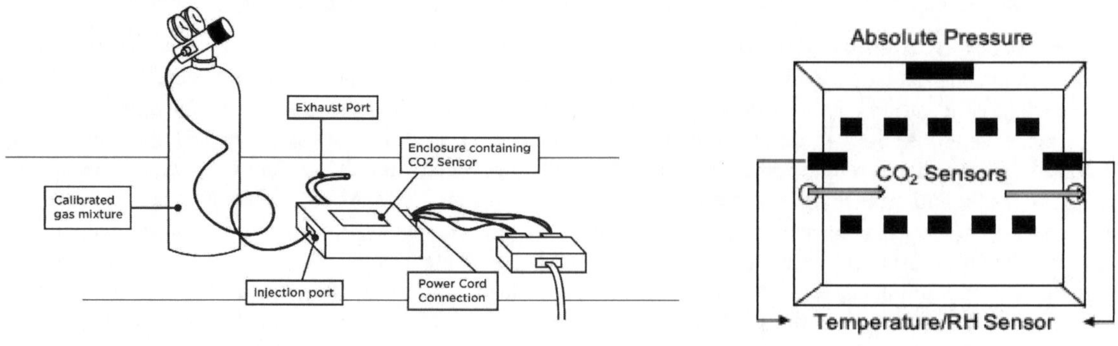

(a) **(b)**

Figure 1: (a) A diagram of the experimental setup and (b) a diagram of the enclosure with ten sensors.

ment equipment would be needed to measure the resulting CO_2 concentration, which is beyond the scope of the present study. Alternatively, when mixing the CO_2/N_2 with water vapor, the flow rates of the two streams could be tightly controlled so that the resulting concentration could be calculated, which was the approach employed in (Shrestha, 2009). Additional error from the flow controls and measurements would be introduced in the calculated concentration. Thus, dry gas mixtures are used in this study to leverage the fact that calibrated CO_2/N_2 gas mixtures are commercially available, which simplifies the test protocol.

The experimental setup includes a tank of calibrated CO_2/N_2 gas, a tightly sealed enclosure, a power supply, and a data acquisition system (Figure 1a). The test enclosure houses ten CO_2 sensors, a pressure gauge, and two temperature and relative humidity sensors (Figure 1b). The relative humidity and temperature sensor devices are located next to the injection and exhaust ports. Since some CO_2 sensors do not interface with the data acquisition system, the front of the enclosure is clear plastic so that the readings from the sensor display may be manually recorded. The enclosure has an injection port and exhaust port. Tubing is used to connect the tank regulator to the inlet port of the enclosure. Initially, the enclosure is filled with ambient air. During the test, the exhaust port allows for the calibrated gas mixture to displace the ambient air. The surrounding laboratory environment where the test is performed is temperature-controlled to satisfy the temperature test condition.

Three calibrated gas mixtures are used in the test: 425 ppm, 1100 ppm, and 1700 ppm. The lowest concentration is selected since it is approximately the concentration of ambient air. Many HVAC-grade CO_2 sensors are rated up to 2000 ppm, and saturate their reading at 2000 ppm if the measured concentration is greater than 2000 ppm. Therefore, the highest concentration is selected to be lower than 2000 ppm to test the full range of sensors while leaving room in case the sensor bias is substantial (i.e., greater than 100 ppm). Regarding the tolerance on the acceptable calibrated gas mixture, the actual concentration of the calibrated gas must be within 2 percent of the reported concentration (this is reported by the gas manufacturer), and the minimum and maximum concentration of the three gases are 400 and 450 ppm, 1000 and 1200 ppm and 1600 and 1800 ppm, respectively.

The test protocol is as follows. Ten CO_2 sensors are placed in an enclosure and powered with the manufacturer specified power requirements. The sensors are allowed to warm-up for at least the manufacturer-specified warm-up period (minimum on-time before obtaining data) in an environment where the absolute pressure requirement is met during the warm-up period. If no manufacturer-specified warm-up period is reported, then the warm-up period will be twenty-four hours. If the sensor utilizes an automatic background calibration (ABC) procedure, the CO_2 concentration of sensor environment must be maintained below 425 ppm for at least one hour during the warm-up period (the concentration during this period is determined by a CO_2 sensor without the ABC procedure calibrated using this procedure). After the warm-up period (if applicable), the test proceeds in order of increasing CO_2 concentration. The tubing between the enclosure and calibrated gas tank (425 ppm gas mixture) is connected. The regulator is adjusted to meet the absolute pressure and differential pressure requirements. Measurements from the CO_2 sensors and temperature, humidity, and pressure sensors are taken every 30 seconds.

When the CO_2 sensor measurements reach steady-state, 41 measurements are recorded in 30-second intervals over a total of 20 minutes. Steady-state is determined by fitting a time-constant for a first-order response to the data. The

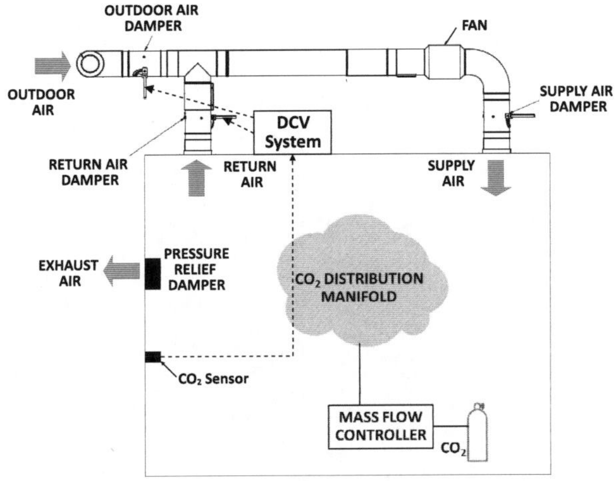

(a)

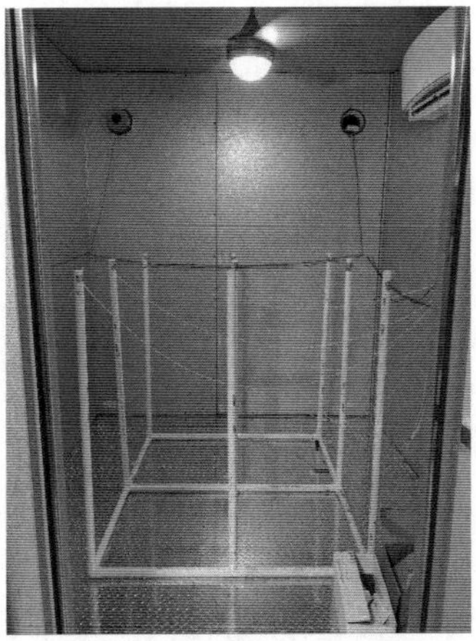

(b)

Figure 2: (a) A schematic of the test chamber and AHU and (b) a photo showing the CO_2 distribution system.

steady-state sensor data collection begins after waiting at least four time constants after the start of the test. The 20 minute period where measurements are collected is referred to as the recording period. After the recording period, the gas supply is turned off, and the next calibrated gas tank is connected to the injection tubing, and the procedure is repeated for the remaining two calibrated gas mixtures.

The average steady-state measurement for each sensor is determined based on measurements collected during the recording period. The sensor error, defined as the difference between the average steady-state measurement and the actual calibrated gas CO_2 concentration, is determined for all sensors and tests. The average and standard deviation of the sensor errors over all sensors is calculated. The minimum and maximum instantaneous measurement of any sensor over the recording period is determined. Finally, linear regression is used to calibrate the average sensor measurements to the actual concentrations for all sensors. The results are tabulated.

2.2 CO_2-based DCV System Controller Test Protocol

The objective of the CO_2-based DCV system controller test protocol is to assess the performance of DCV system controllers. The protocol features a laboratory procedure that tests the ability of the controller to maintain the indoor CO_2 concentration at a setpoint. The performance of the system is compared to the performance of an ideal controller, which will be discussed in detail in the next section. The protocol applies to DCV system controllers that receive a single CO_2 sensor input and modulate the outdoor and return air dampers for an HVAC system to maintain an indoor CO_2 setpoint. Parameters such as the ventilation rates, occupancy densities, and CO_2 generation rates are determined based on the 2019 California Building Energy Efficiency Standards (*Building energy efficiency standards for residential and nonresidential buildings*, 2018) and ASHRAE Standard 62.1 (ASHRAE, 2010) so the tests reflect realistic conditions.

To mimic an occupied building space where occupants exhale CO_2, a chamber equipped with a constant air volume air handling unit (AHU) and a CO_2 distribution system located inside the chamber is used to test the DCV system controllers. A schematic of the chamber and AHU are shown in Figure 2a, and a picture of the inside of the chamber is given in Figure 2b. The chamber has an interior height of 8 feet and a floor area of 56 square feet. The AHU is responsible for mixing the return air from the chamber with outdoor air and supplying the mixed air to the chamber. The fraction of outdoor air to return air is controlled by the DCV system controller, which modulates the outdoor air and return air dampers. The mixed air is supplied to the chamber at a constant flow rate using a constant speed supply fan. Additionally, a supply air damper controlled by a proportional-integral controller is used to tightly control the supply airflow rate. A relief damper is used to maintain a positive differential pressure in the chamber relative to the surrounding environment (1-10 Pa). The exhaust airflow rate is approximately equal to the outdoor airflow rate.

Within the chamber, a wall-mounted CO_2 sensor is placed four feet from the floor to measure the chamber CO_2 concen-

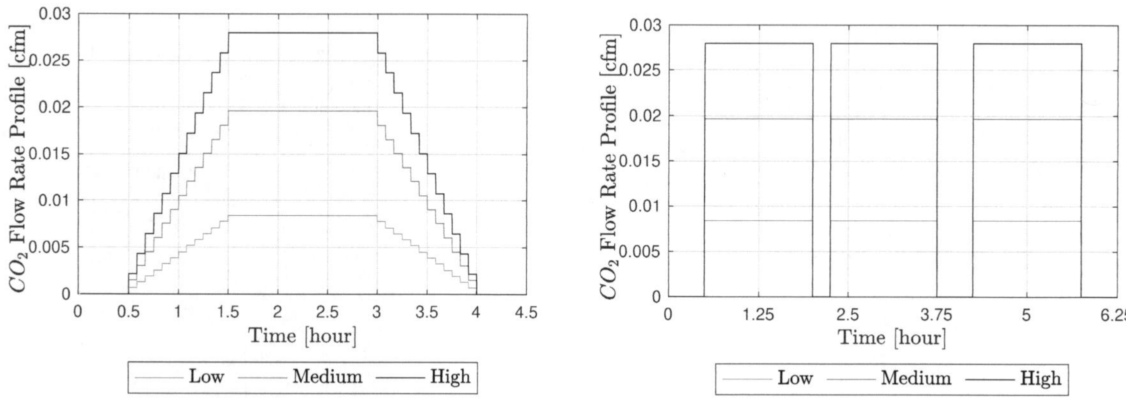

(a) (b)

Figure 3: CO_2 generation rate for (a) a gradually changing occupancy profile and (b) a series of step changes in the occupancy for three occupancy densities.

Table 1: Test chamber CO_2 generation, ventilation, and supply airflow rates for the 56 sq. ft. test chamber.

Occupant Density Category	Occupant Density (# people/1000 sq. ft.)	Occ_{max} (# people)	$G_{CO_2,max}$ ($\times 10^{-2}$ cfm)	$\dot{V}_{in,min}$ (cfm)	Supply Airflow Rate (cfm)
Low	15	0.84	0.84	8.4	56.0
Medium	35	1.96	1.96	8.4	56.0
High	50	2.80	2.80	8.4	56.0

tration and to report the value to the DCV system controllers. When compatible, the same (high accuracy HVAC-grade) sensor calibrated with the CO_2 sensor calibration procedure described in Section 2.1, referred to as the chamber CO_2 sensor, is used in all tests to send the CO_2 signal to the controller under test. The goal of using a calibrated CO_2 sensor is to isolate testing of the DCV system controller response characteristics. If the DCV manufacturer's CO_2 sensor must be used with the controller under test due to compatibility requirements, then all manufacturer recommendations for the operation of the CO_2 sensor are followed. In this case, the chamber CO_2 sensor is still used to record the CO_2 concentration during the tests for consistency. Additional calibrated CO_2 sensors are placed in the outdoor air duct and next to the relief damper to measure the exhaust air CO_2 concentration.

Pure CO_2 is released into the chamber at a controlled rate with a mass flow controller to simulate different occupant time profiles and densities. The CO_2 is dispersed through a tubing manifold with nine distribution locations at a height of four feet above the floor. Each tube has an identical length to ensure that the flow resistance of each tube is identical. In the chamber, a ceiling fan is used to promote the mixing of the supply air and the chamber air. A mini-split system and humidifier are used to regulate the chamber temperature and humidity to maintain the test conditions and tolerances specified below. Since the laboratory is located in a dry climate, dehumidification is not required.

A pre-specified CO_2 generation rate is used to mimic expected CO_2 generation rates of two occupant types: a profile with gradual changing occupancy and a profile with a series of step changes in the occupancy. For each profile, three maximum CO_2 generation rates are considered for a total of six tests. The generation rate as a function of time is computed based on each occupancy profile. Specifically, the occupancy profile (i.e., number of occupants) is computed by $\mathrm{Occ}(t) = \mathrm{Occ}_{frac}(t) \times \mathrm{Occ}_{max}$ where $\mathrm{Occ}(t)$ is the number of occupants at time t, $\mathrm{Occ}_{frac}(t)$ is the specified occupancy fraction at time t, and Occ_{max} is the maximum expected occupancy for each occupant density category. The generation rate is obtained from the occupancy profile and is given by $G_{\mathrm{CO_2}}(t) = \mathrm{Occ}(t) \times G_{\mathrm{CO_2,Occ}}$ where $G_{\mathrm{CO_2,Occ}}$ is the generation per occupant taken to be 0.01 cfm CO_2 per occupant and $G_{\mathrm{CO_2}}(t)$ is the generation rate in the chamber at time t, which has units of cfm CO_2. Similarly, the maximum CO_2 generation rate may be computed from Occ_{max} and is denoted by $G_{\mathrm{CO_2},max}$. The resulting generation rate profiles are shown in Figure 3 with parameters given in Table 1. For the remainder, the profiles shown in Figure 3a are referred to as the gradual profiles while the profiles shown

Table 2: Test conditions and test tolerances of the DCV system controller test.

Chamber property	Units	Test Condition	Test Operating Tolerance	Test Condition Tolerance
Absolute pressure	kPa	101	3	± 5
Dry-bulb temperature	°F	75	5	± 3
Relative humidity	%RH	40	20	± 10
CO_2 generation rate	SLPM	Figure 3	5% of test condition	$\pm 3\%$ of test condition
Outdoor air CO_2 concentration	ppm CO_2	≤ 425	$\leq 450^*$	$+10^*$

*Over the test, the average outdoor air CO_2 concentration must be maintained below 435 ppm. Additionally, the maximum outdoor air concentration over the test must be below 450 ppm.

in Figure 3b are referred to as the step profiles. The three occupancy densities are referred to as the low, medium, and high occupancy densities.

The minimum ventilation rate $\dot{V}_{in,min}$ is based on the 2019 California Building Energy Efficiency Standards Table 120.1-A (*Building energy efficiency standards for residential and nonresidential buildings*, 2018), which requires 0.15 cfm of outdoor air per square foot of floor area A, is

$$\dot{V}_{in,min} = (0.15\text{cfm/ft}^2)A \tag{1}$$

The minimum ventilation rate and the supply rates are given in Table 1. The supply airflow rate is based on a rule of thumb of 1 cfm per square foot of floor area.

The test protocol requires some preliminary setup tasks. The supply air fan speed is set to provide the required supply airflow rate for the test (see Table 1), which requires a one-time measurement of the supply airflow to ensure that it is within ± 10 percent of the desired flow rate. The minimum damper position is fixed to ensure that the minimum ventilation airflow rate is achieved for all tests (within ± 10 percent). The minimum damper position is determined using a tracer gas flow calibration. In the configuration of the DCV system controller for all tests, the maximum outdoor air damper position is set to its fully open position, which corresponds to a fully closed return damper position, so the AHU draws 100 percent outdoor air to supply the chamber when the DCV system controller commands the maximum ventilation rate. For consistency between all tests, the settings for the supply air fan, minimum damper position, and maximum damper position are recorded and used in all tests.

The first step in the protocol is to install and configure the DCV system controller with the manufacturer's CO_2 sensor (if required). The DCV controller is configured so that the command signal ranges between the minimum ventilation rate to 100 percent outdoor air, and the DCV system controller CO_2 concentration setpoint is set to 600 ppm above the outdoor air concentration. At the beginning of each test, the test chamber is flushed with outdoor air until the chamber CO_2 concentration is within the outdoor air concentration plus 30 ppm. The test chamber temperature, humidity, pressure, CO_2 generation rate, and outdoor air CO_2 concentration must be within the tolerances specified in Table 2. Once the preliminary setup tasks are completed, the main test protocol may be executed. CO_2 is added to the chamber through the CO_2 distribution system following one of the profiles shown in Figure 3 to simulate an occupancy pattern. As the chamber CO_2 concentration increases, the DCV system controller will start modulating the damper system to increase the ventilation rate. For all time-series data, measurements are sampled at 0.1 Hz. Throughout the test, the chamber temperature, humidity, pressure, supply airflow rate, and outdoor air CO_2 concentration are monitored. A successful test is one where the test is set up properly, the environmental variables are within the test tolerance (defined in Table 2), and the CO_2 generation rate follows one of the desired profiles (Figure 3). Upon completion of a test, the chamber is flushed with outdoor air to return the CO_2 concentration to within 30 ppm of the outdoor air concentration to begin the next test. The process is repeated until the tests for all six profiles are completed.

2.3 Ideal Baseline DCV Strategy

To help quantify the performance of each DCV system controller, the chamber CO_2 concentration obtained from the tests under each DCV system controller is compared to the expected concentration under an ideal DCV strategy. The ideal DCV strategy is a theoretical controller whose input is the outdoor CO_2 concentration, the generation rate, and the initial chamber CO_2 concentration recorded during the test and whose output is the optimal ventilation rate. The optimal ventilation rate maintains the minimum ventilation rate when the chamber CO_2 concentration is below the setpoint. For all other times, the optimal ventilation rate is the one that maintains the chamber CO_2 concentration at the setpoint. To generate the expected chamber CO_2 concentration, a closed-loop simulation is performed of the chamber under the ideal DCV strategy. Comparing the results obtained from the DCV system controller test protocol with that obtained from the closed-loop simulation under the ideal DCV strategy gives a measure of how close the DCV system controller performance is relative to the optimal performance.

The ideal DCV strategy is a model-based strategy, which uses a dynamic model to describe the chamber CO_2 concentration over time. Assuming that the chamber is well-mixed so that the contents are spatially uniform, an overall mass balance of CO_2 over the chamber and duct yields

$$\frac{dm_{\mathrm{CO_2}}(t)}{dt} = \dot{m}_{\mathrm{CO_2},in}(t) - \dot{m}_{\mathrm{CO_2},out}(t) + \bar{G}_{\mathrm{CO_2}}(t) \tag{2}$$

where $m_{\mathrm{CO_2}}(t)$ is the total mass of CO_2 in the chamber, $\dot{m}_{\mathrm{CO_2},in}(t)$ is the mass flow rate of CO_2 in the outdoor air stream, $\dot{m}_{\mathrm{CO_2},out}(t)$ is the mass flow rate of CO_2 in the chamber exhaust air stream, and $\bar{G}_{\mathrm{CO_2}}(t)$ is the mass generation rate of CO_2. Under the assumptions that the air density is constant and the outdoor airflow rate into the chamber is equal to the airflow rate leaving the chamber through the exhaust relief damper, (2) simplifies to

$$V\frac{dC_{\mathrm{CO_2}}(t)}{dt} = \dot{V}_{in}(t)(C_{\mathrm{CO_2},oa}(t) - C_{\mathrm{CO_2}}(t)) + G_{\mathrm{CO_2}}(t) \tag{3}$$

where $C_{\mathrm{CO_2}}(t)$ is the CO_2 concentration in the chamber, $\dot{V}_{in}(t)$ is the outdoor airflow rate (ventilation rate) that will be determined by the ideal DCV strategy, and $G_{\mathrm{CO_2}}(t)$ is the generation rate of the CO_2 in the chamber.

As described above, the ideal DCV strategy may be considered to be a feedforward controller, which utilizes perfect information of the CO_2 generation rate in the chamber to compute a ventilation rate that exactly rejects the effect of the disturbance. The CO_2 generation rate, which imitates occupancy in the chamber, is considered to be the disturbance. On the contrary, DCV system controllers are feedback controllers (i.e., reactive instead of proactive) since measuring the generation rate is not practical. The ideal DCV strategy is to maintain the minimum ventilation rate if the expected CO_2 concentration is at or below the maximum CO_2 concentration. Otherwise, the strategy selects the ventilation rate that exactly maintains the CO_2 concentration at its setpoint. Determining the ventilation rate that maintains the chamber concentration at exactly its setpoint, requires the solution of (3).

A simultaneous solution strategy is employed to determine the ventilation rate from the ideal DCV strategy and the solution of (3). Provided the input data including the chamber air volume, the outdoor air CO_2 concentration profile, the CO_2 generation rate profile, and an initial chamber CO_2 concentration, (3) may be numerically solved. For a fair comparison between the ideal DCV strategy and each DCV system controller test, the outdoor air CO_2 concentration profile and initial concentration are taken to be equal to the recorded data from each DCV system controller test. The explicit Euler method is employed to solve (3) with a sufficiently small integration time step, which gives

$$C_{\mathrm{CO_2}}(t+\Delta t) = C_{\mathrm{CO_2}}(t) + \frac{\Delta t}{V}\left(\dot{V}_{in}(t)(C_{\mathrm{CO_2},oa}(t) - C_{\mathrm{CO_2}}(t)) + G_{\mathrm{CO_2}}(t)\right) \tag{4}$$

where Δt is the integration time step, $C_{\mathrm{CO_2},oa}(t)$ is the CO_2 concentration recorded during the DCV system controller test and $C_{\mathrm{CO_2}}(0)$ is set to be equal to the initial chamber concentration at the beginning of the DCV system controller test. In (4), there is an implicit conversion of $G(t)$, which has units of cfm CO_2, to the units of (cfm air)/(ppm CO_2). The conversion has been omitted in the equation for simplicity of the presentation, but must be accounted for in the calculation. To determine $\dot{V}_{in}(t)$ under the ideal DCV strategy, first consider that $\dot{V}_{in}(t) = \dot{V}_{in,min}$. The expected CO_2 concentration at the next integration time is

$$C_{\mathrm{CO_2}}(t+\Delta t) = C_{\mathrm{CO_2}}(t) + \frac{\Delta t}{V}\left(\dot{V}_{in,min}(C_{\mathrm{CO_2},oa}(t) - C_{\mathrm{CO_2}}(t)) + G_{\mathrm{CO_2}}(t)\right)$$

If $C_{\mathrm{CO_2}}(t+\Delta t) \leq C_{\mathrm{CO_2},max}$, $\dot{V}_{in}(t)$ is set to $\dot{V}_{in,min}$ under the ideal DCV strategy. Else, the ventilation rate that keeps $C_{\mathrm{CO_2}}(t+\Delta t) = C_{\mathrm{CO_2},max}$ is computed by

$$\dot{V}_{in}(t) = \frac{V(C_{\mathrm{CO_2},max} - C_{\mathrm{CO_2}}(t)) - G(t)}{\Delta t(C_{\mathrm{CO_2},oa}(t) - C_{\mathrm{CO_2}}(t))}$$

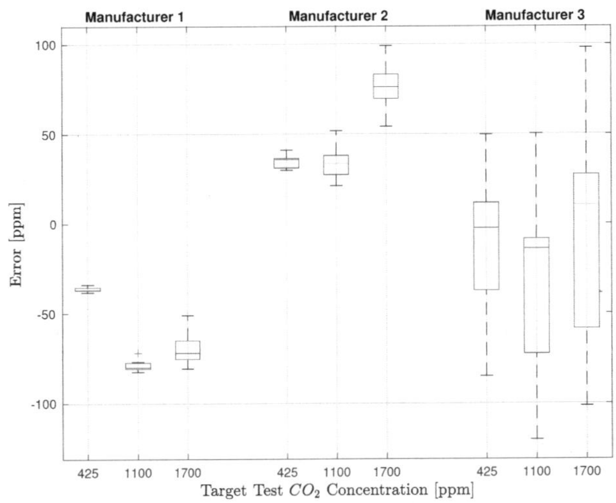

Figure 4: A box-and-whisker plot of the errors of each calibrated gas mixture. The box represents the second and third quartile, the line represents the full range (minimum to maximum), and the red line is the median error.

Table 3: CO_2 sensor test results. All units are ppm.

Manu-facturer	Value	425 ppm	1100 ppm	1700 ppm
1	Actual CO_2 Conc.	422	1098	1710
	Average	386	1019	1640
	Minimum	382	1013	1627
	Maximum	391	1028	1661
	Std. Dev.	1.6	3.2	8.6
	Error	-36.2	-79.2	-69.7
2	Actual CO_2 Conc.	425	1101	1706
	Average	460	1135	1782
	Minimum	447	1114	1750
	Maximum	474	1165	1823
	Std. Dev.	4.8	10.1	14.4
	Error	35.1	34.3	76.3
3	Actual CO_2 Conc.	422	1098	1710
	Average	410	1064	1702
	Minimum	289	942	1515
	Maximum	527	1215	1929
	Std. Dev.	38.4	49.6	59.2
	Error	-11.3	-33.7	-7.0

Therefore, the ideal DCV strategy is given by

$$\dot{V}_{in}(t) = \max\left\{\frac{V(C_{CO_2,max} - C_{CO_2}(t)) - G(t)}{\Delta t(C_{CO_2,oa}(t) - C_{CO_2}(t))}, \dot{V}_{in,min}\right\} \tag{5}$$

3. RESULTS AND DISCUSSION

Due to space limitations, only a few example results are given here. The complete results will be presented at the conference and will also be posted to `https://empowerprocurement.com` after the conference.

3.1 CO_2 Sensor Tests

The CO_2 sensor test protocol was executed on three sensor model types from three different manufacturers (labeled as Manufacturer 1, Manufacturer 2, and Manufacturer 3). Each test considered ten sensors of the same model. Figure 4 presents a box-and-whisker plot of the error for each sensor model and each calibrated gas test. The x-axis of Figure 4 is the target concentration of the test calibrated gas. Table 3 presents a summary of the key statistics collected from the test and includes the actual CO_2 concentration of the gas used in each test. In Table 3, the average, standard deviation, and error are taken over all average steady-state sensor measurements while the minimum and maximum are the minimum and maximum instantaneous measurement of any sensor over the recording period.

From Figure 4 and Table 3, the steady-state average measurements of the ten sensors of Manufacturer 1 and 2 were consistently less than and greater than, respectively, the actual concentration. Manufacturer 1 sensors had the smallest standard deviation compared to that of the other sensors. However, Manufacturer 2 sensors had a comparable standard deviation as that of Manufacturer 1 sensors. Manufacturer 3 sensors had a substantially larger standard deviation compared to that of the other two sensor types. Interestingly, the average error across all ten sensors was the smallest for Manufacturer 3 sensors. While the standard deviation increased with CO_2 concentration for all sensors, the sensor error did not monotonically increase or decrease with the CO_2 concentration.

3.2 DCV System Tests

The DCV system controller test protocol was executed on an example DCV system controller. Since the local outdoor air that supplies the chamber air handling unit was generally above 400 ppm and reached levels above 425 ppm at times, the CO_2 setpoint was set to 1025 ppm (425 ppm plus 600 ppm). The resulting chamber CO_2 concentration is

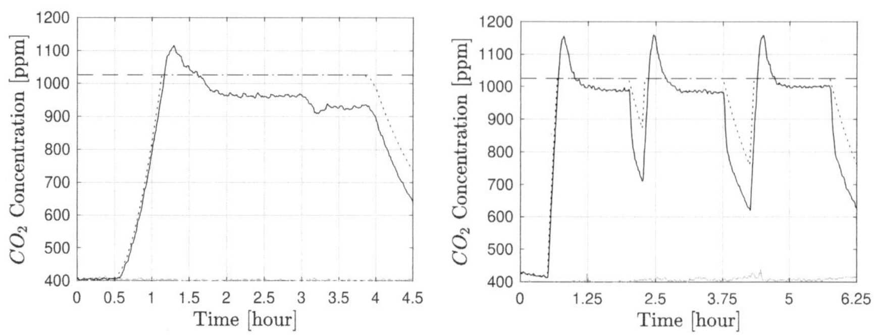

Figure 5: The chamber CO_2 concentration (solid), expected concentration under the ideal strategy (dotted), outdoor concentration (grey), and setpoint (dashed) for the high density (a) gradual profile and (b) step change profile tests.

Table 4: Time fraction within and outside the ideal ventilation target.

Test	Time Fraction Above 75 ppm Ideal Ventilation Target (%)	Time Fraction Below 75 ppm Ideal Ventilation Target (%)	Time Fraction within Ideal Ventilation Target (%)
Low step	0.0	33.1	66.9
Medium step	3.8	23.4	72.8
High step	7.1	21.9	71.0
Low gradual	0.0	31.2	68.8
Medium gradual	0.0	29.7	70.3
High gradual	1.5	31.6	66.9

shown in Figure 5. For performance comparison, the chamber was simulated with the ideal DCV strategy (5) with the initial chamber and the outdoor CO_2 concentrations recorded during the DCV test as inputs to the simulation. The expected chamber concentrations under the ideal DCV strategy are shown in Figure 5 (dotted black trajectory).

For the gradual profile tests (Figure 5a), the behavior under the DCV system controller is similar. Initially, the chamber CO_2 concentration is maintained at its ambient levels. After 30 minutes when CO_2 is introduced into the chamber through the CO_2 distribution system, the chamber CO_2 concentration starts to increase until the CO_2 concentration exceeds the setpoint. The DCV system controller responds by increasing the outdoor airflow rate. The chamber CO_2 concentration decreases below the setpoint where it is maintained at a constant level for a period. Finally, as the CO_2 generation rate decreases, the chamber CO_2 concentration decreases. For the step profile tests (Figure 5b), similar behaviors are observed as that of the gradual profile tests. The DCV system controller reactively increases the ventilation rate after the concentration exceeds the setpoint.

Comparing the responses under the DCV system controller with the expected responses under the ideal DCV strategy, there is general agreement between the two responses initially for all cases. The ideal DCV strategy can prevent the chamber CO_2 concentration from exceeding the setpoint owing to the feedforward nature of the ideal DCV strategy. Once the expected CO_2 concentration under the ideal DCV strategy reaches the setpoint, it is maintained there for a period. Once the generation rate decreases, the expected CO_2 concentration decreases at a comparable rate to that under the DCV system controller. The main differences between responses under the DCV system controller and the ideal DCV strategy occur when the concentration increases to the setpoint. The DCV system controller takes time to respond so the concentration increases past the setpoint (i.e., overshoots the setpoint). Eventually, the DCV system controller responds to decrease the concentration until the concentration decreases past the setpoint (over-ventilation). On the other hand, the ideal DCV strategy can anticipate the concentration reaching the setpoint and then, adjusts the ventilation rate to maintain the concentration at setpoint (i.e., prevents overshoot and then, over-ventilation).

To quantify the difference between the performance under the DCV system and that expected under the ideal DCV strategy, the difference between the measured CO_2 concentration under the DCV system and the expected CO_2 concentration under the ideal DCV stategy was computed. Table 4 tabulates the time fraction within and outside the ideal ventilation rate, which are defined as the time fraction that the difference between the two concentrations are within 75 ppm and the time fraction that the absolute value of the difference are greater than 75 ppm, respectively. Differences greater than 75 ppm represents under-ventilation while differences less than -75 ppm represents over-ventilation. From

Table 4, the chamber CO_2 concentration was within the target concentration for the majority of the time.

4. CONCLUSIONS

Two test protocols were presented: one for evaluating the accuracy of HVAC-grade CO_2 sensors used in CO_2-based DCV system controllers and the other for assessing the performance of CO_2-based DCV system controllers. The test protocols were performed on commercially available sensors and DCV system controllers. Of the three sensor models tested, one sensor model consistently measured the concentration less than the actual, one consistently measured the concentration greater than the actual, and the third was able to measure the CO_2 concentration with less than a 35 ppm absolute error on average, but had a large variance between the measurements of each sensors. From the DCV system controller test results, the main differences in the closed-loop performance of the DCV system controller test compared to that under an ideal DCV strategy is that the chamber CO_2 concentration under the DCV system showed considerable overshoot and over-ventilation not present in the expected CO_2 concentration under the ideal DCV strategy.

REFERENCES

Acker, B., & Wymelenberg, K. V. D. (2011). Demand control ventilation: Lessons from the field-How to avoid common problems. *ASHRAE Transactions*, *117*, 502–508.

Apte, M. G. (2006). *A review of demand control ventilation* (Tech. Rep.). Lawrence Berkely National Laboratory.

ASHRAE. (2010). ANSI/ASHRAE Standard 62.1-2010. Ventilation for acceptable indoor air quality [Computer software manual]. Atlanta, GA.

Building energy efficiency standards for residential and nonresidential buildings (Tech. Rep. No. CEC-400-2018-020-CMF). (2018, Dec.). California Energy Commission.

California Division of Occupational Safety and Health. (2019, Sept.). *Protecting indoor workplaces from wildfire smoke with building ventilation systems and other methods.* `https://www.dir.ca.gov/dosh/wildfire/Indoor-Protection-from-Wildfire-Smoke.html`.

Dougan, D. S., & Damiano, L. (2004). CO_2-based demand control ventilation: Do risks outweigh potential rewards? *ASHRAE Journal*, *46*, 47–54.

Emmerich, S. J., & Persily, A. K. (2001). *State-of-the-art review of CO_2 demand controlled ventilation technology and application* (Tech. Rep.). National Institute of Standards and Technology.

Fisk, W. J., & de Almeida, A. T. (1998). Sensor-based demand-controlled ventilation: A review. *Energy and Buildings*, *29*, 35–45.

Giacomo, S. M. D. (1999). Differential CO_2 based demand control ventilation (maximum energy savings & optimized IAQ): History, theory, and myths. *Energy Engineering*, *96*, 58–76.

Jones, J., Meyers, D., Singh, H., & Rojeski, P. (1997). Performance analysis for commercially available CO_2 sensors. *Journal of Architectural Engineering*, *3*, 25–31.

Lin, X., & Lau, J. (2014). Demand controlled ventilation for multiple zone HVAC systems: CO_2-based dynamic reset (RP 1547). *HVAC&R Research*, *20*, 875–888.

Liu, G., Zhang, J., & Dasu, A. (2012). *Review of literature on terminal box control, occupancy sensing technology and multi-zone demand control ventiliation (DCV)* (Tech. Rep.). Pacific Northwest National Laboratory.

Mysen, M., Berntsen, S., Nafstad, P., & Schild, P. G. (2005). Occupancy density and benefits of demand-controlled ventilation in Norwegian primary schools. *Energy and Buildings*, *37*, 1234–1240.

Nassif, N. (2012). A robust CO_2-based demand-controlled ventilation control strategy for multi-zone HVAC systems. *Energy and Buildings*, *45*, 72–81.

O'Neill, Z. D., Li, Y., Cheng, H. C., Zhou, X., & Taylor, S. T. (2020). Energy savings and ventilation performance from CO_2-based demand controlled ventilation: Simulation results from ASHRAE RP-1747 (ASHRAE RP-1747). *Science and Technology for the Built Environment*, *26*, 257–281.

Schell, M., & Inthout, D. (2001). Demand control ventilation using CO_2. *ASHRAE Journal*, 18–29.

Shrestha, S. S. (2009). *Performance evaluation of carbon-dioxide sensors used in building HVAC applications* (Unpublished doctoral dissertation). Iowa State University, Ames, Iowa.

Shrestha, S. S., & Maxwell, G. M. (2009). An experimental evaluation of HVAC-grade carbon dioxide sensors–Part I: Test and evaluation procedure. *ASHRAE Transactions*, *115*, 471–483.

Taylor, S. T. (2006). CO_2-based DCV using 62.1-2004. *ASHRAE Journal*, *48*, 67–75.

United States Centers for Disease Control. (2020, Oct.). *COVID-19 employer information for office buildings.* `https://www.cdc.gov/coronavirus/2019-ncov/community/office-buildings.html`.

Energy Modeling and Energy Efficiency Opportunities for a Public Library Building in the Upper Peninsula of Michigan

Amarnath WARTY[1], Sunil MEHENDALE[1]*

[1]Michigan Technological University

Houghton, MI, USA

awarty@mtu.edu, ssmehend@mtu.edu

* Corresponding Author

ABSTRACT

Energy modeling tools have been extensively used for analyzing building performance as well as for assessing energy efficiency opportunities. The present research has a twofold aim: (1) to model the natural gas consumption and the total electricity consumption of a 12600 sq. ft. public library building in Houghton, MI, and (2) to identify any opportunities to improve energy efficiency. This was accomplished by first developing and validating an eQUEST model for the library building in conjunction with a linear regression model correlating the natural gas consumption (during winter) with heating degree days and the electricity consumption (during summer) as a function of cooling degree days. The said library building, which is serviced by two rooftop furnaces, each with a DX coil, and a hot-water loop using two boilers, has been facing HVAC issues for a few years now, the most common complaint being that it gets too hot in the winter despite the thermostat being set to provide comfortable temperatures. This motivated us to model the building and try and discover the causes for complaints like the above-mentioned while keeping an eye on any energy saving opportunities. The eQUEST model Mean Base Error (MBE) is -3.60% and 2.48% for natural gas consumption and electricity consumption respectively. The coefficient of variation of Root Mean Squared Error - Cv (RMSE) is 7.33% and 4.14% for the natural gas consumption and for electricity consumption respectively. Having thus gained confidence in the ability of the model to provide reasonable predictions, the same was then exercised to understand the key factors responsible for energy consumption in the building and to check for energy efficiency opportunities. Preliminary results suggest that the principal factors affecting the building energy consumption are the lighting, HVAC loads, and occupancy and that the natural gas consumption of the building could be lowered by up to 20% using the furnace units alone (i.e., by dispensing with the boilers), while still meeting the building's heating requirements. One of the issues also identified in the course of the analysis was that the thermostat is located about 10 ft. away from the radiators and grills and hence it is possible that the HVAC system responds to a lower average temperature than that in the room.

1. INTRODUCTION

Energy modelling tools are used for predicting building energy consumption and verifying design parameters like HVAC system capacity and building schedules. In the design phase, results from accurately modeled buildings help engineers and architects to fine tune their design parameters and also ensure that their design is cost-effective. Energy modelling programs are also used for existing buildings to assess different energy efficiency improvement options and help consumers in cutting down energy consumption for heating, cooling, lighting, domestic hot water heating, etc.

eQUEST and EnergyPlus are two of the most commonly used open-source programs and there have been several studies comparing the accuracy and other capabilities of eQUEST and EnergyPlus. Rallapalli (2010) mentions in her whole building energy simulation study that EnergyPlus offers the advantages of sub-hourly time steps, independent radiation modelling, customizable HVAC systems, and integrated simulations for accurate results. eQUEST, on the other hand, is user-friendly and rapid (it gives results within minutes), but is limited to hourly time steps. Performance comparison showed that eQUEST energy consumption predictions were closer to the measured data than EnergyPlus results. Zerroug & Dzelzitis (2015) reiterated that eQUEST is easier to use compared to EnergyPlus. They also found eQUEST predictions to be much closer to the measured data than those of EnergyPlus. Zhu *et al.* (2013) compared EnergyPlus, DeST and DOE – 2.1E for their performance characteristics and found that DOE – 2.1E can provide relatively inaccurate results in double zone models (conditioned zone and adjacent non-conditioned zone) as DOE – 2.1E uses the previous hour's adjacent space temperature values for current calculations.

A public library building in Houghton, MI has been facing HVAC issues for a few years now. A common complaint from some of the library employees has been that it gets too hot in the winter despite the thermostat being set to a comfortable temperature. This motivated us to model the building to try and find the issues along with an examination of any energy efficiency opportunities. As previous studies have shown eQUEST to be user-friendly and since it had the relevant HVAC system (DX coil with furnace) in its library, we decided to use this software for modelling the library building. Before we modelled the building, it was important to understand prior research work using eQUEST and confirm its capability in modelling existing buildings, and the following studies have been summarized with this purpose in mind. Xing *et al.* (2015) modelled a hotel building in Tianjin, China using eQUEST and were able to successfully determine the factors affecting the accuracy of the baseline model. The study found that, as expected, increasing the space heating efficiency would significantly reduce the total energy consumed. Wang *et al* (2015) modelled a hotel building in Taiwan and their validated model indicated that energy savings of up to 10.5% per month could be obtained by optimizing the chiller capacity. Sobha *et al.* (2016) modelled a house in Missouri for demand side cooling management using eQUEST for energy consumption estimation. This work mainly focused on managing HVAC loads when the utility costs were low. Algarni *et al* (2017) used eQUEST to model a residential system in Jazan, Saudi Arabia. The air treatment process employed in the HVAC system leads to removal of condensate water which is usually wasted. In this study, the authors used this waste condensate water to clean the dusty roof surfaces (which increased building cooling loads) and observed up to 19% reduction in the annual building cooling energy consumption.

From the above studies, it is clear that eQUEST can be successfully used for modelling buildings and can be further employed to explore energy conservation opportunities. For the present study, the public library building (whose HVAC system includes two rooftop furnaces with DX coils and a hot-water loop using boilers) is modeled using eQUEST software and the model is validated using actual monthly utility bills for the building. The results show that by shutting off the hot water loop (boilers), energy savings of up to 20% could be realized, while still meeting the building's heating requirements. Details of the HVAC system and the eQUEST model for the same will be discussed in the following section.

2. ENERGY MODELING & VALIDATION

The eQUEST model was developed for the 12,600 sq. ft. (1170.6 m^2) library area (single floor) located in Houghton, Michigan (Figure 1). For modelling the HVAC system, HVAC and maintenance documents available with the library were used. The library has two roof-top units, each consisting of a furnace and a direct expansion (DX) coil cooling system. The total furnace output power rating is 600 kBtu/hr (175.8 kW). A hot water loop (comprising boilers and radiators) is also included in the library to provide supplemental heating in addition to the heating provided by the furnace. The loop has two boilers each having an output power rating of 134 kBtu/hr (39.3 kW) with 84% boiler efficiency, as well as a dedicated domestic hot water heating (storage) system with a tank capacity of 48 gallons (0.23 m^3) and an input power rating of 60 kBtu/hr (17.6 kW).

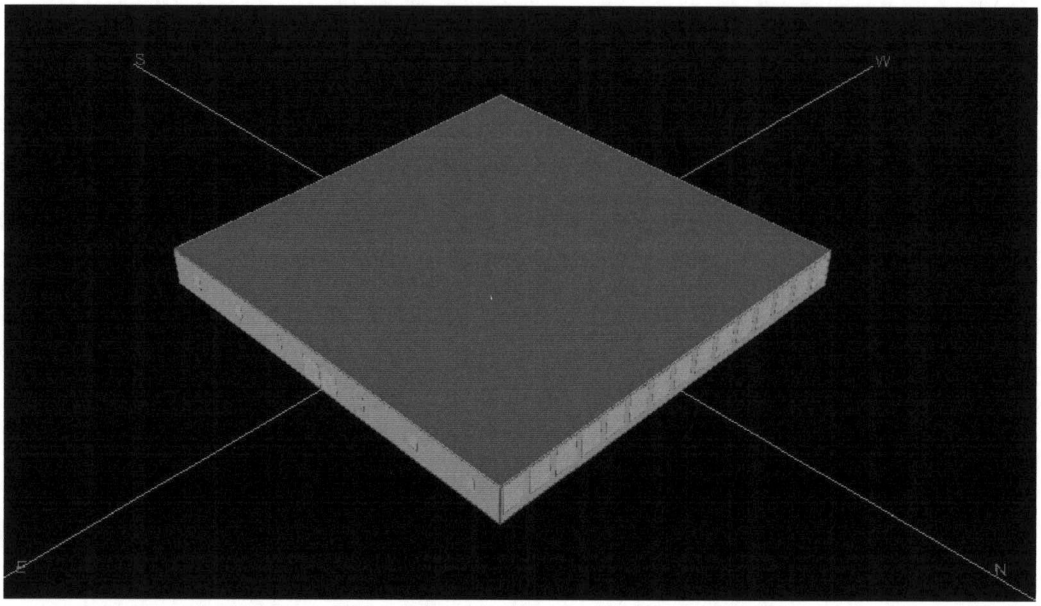

Figure 1: eQUEST model of the library building

Once the library building is modelled on eQUEST, the model needs to be validated to ensure its accuracy. The actual utility bills for the building can be used to verify the energy consumption predicted by eQUEST. An issue which arises here is that the weather profile used by eQUEST is not the same as the actual weather profile observed during the test period. Gould & Hawkins (2015) have suggested a simple linear regression model to overcome this issue. Natural gas consumption can be directly related to the Heating Degree Days (HDD) (during winter) and electricity consumption can be directly related to the Cooling Degree Days (CDD) (during summer). HDD are a measure of how low the temperature was on a given day or during a specified number of days compared to a standard temperature, usually 65°F (291.5 K) in the United States. Here, HDD is the difference between 65°F and the average daily temperature at the location of interest. Similarly, CDD are a measure of how high the temperature was on a given day or during a specified number of days compared to 65°F. Here, CDD is the difference between the average daily temperature at the location of interest and 65°F. An equation can be developed using simple linear regression to predict the natural gas consumption and electricity consumption respectively using HDD and CDD values and the actual utility bills. HDD and CDD can be calculated for the weather profile used by eQUEST and the regression equation generated from the actual utility bills can be used to generate the monthly electricity and natural gas consumption. These monthly values can then be used to validate the eQUEST model.

To generate the linear regression models, the utility bills for the years 2018-19 were collected from the library and the HDD and CDD were calculated for the bill periods. The HDD and CDD were calculated as suggested on BizEE Software and validated analytically. The HDD and CDD values were then divided by the relevant time period and the HDD/day and CDD/day values were obtained. Similarly, the natural gas consumption and electricity consumption obtained from the utility bills were divided by the time period to obtain therm/day (1.055×10^8 J/day) and kWh/day (3.6×10^{-6} J/day) values, respectively. Figure 2 shows the plot relating the natural gas consumed per day to the HDD per day for the winter of 2018-19 and Figure 3 shows the plot relating the electricity consumed per day to the CDD per day for the summer of 2018. It is noted that the plots have been generated based on the available library utility bills.

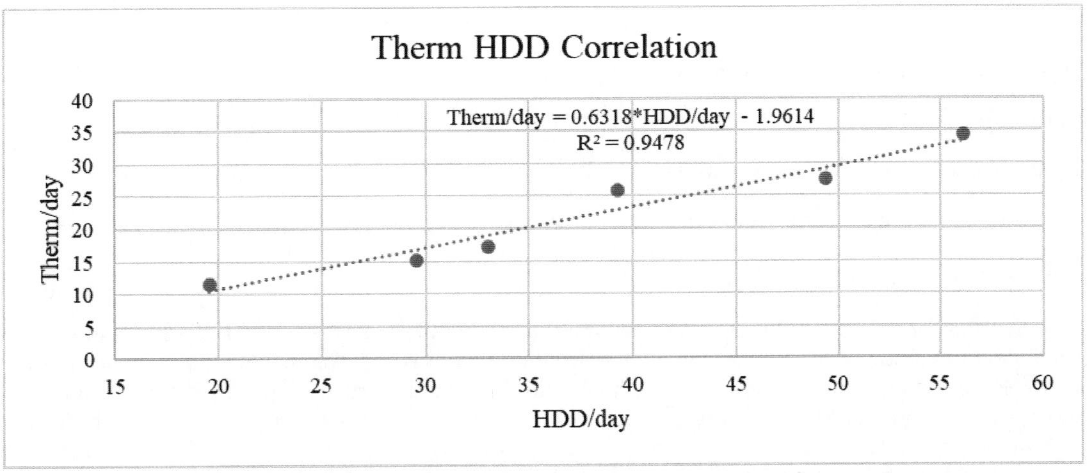

Figure 2: Therm/day (natural gas consumption) and HDD/day correlation (winter season)

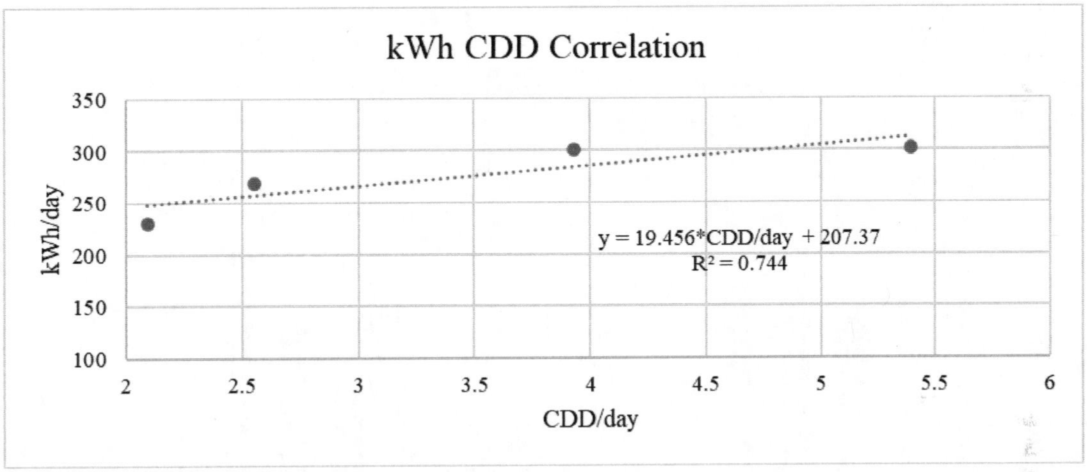

Figure 3: kWh/day (electricity consumption) and CDD/day correlation (summer season)

As observed in Figure 2, the R^2 value (0.9478) for the Therm/day – HDD/day, indicates a high positive correlation. The R^2 value for the kWh/day – CDD/day, 0.744, also shows a reasonable positive correlation. It should be noted that this latter value is not as high as the R^2 value for Therm/day – HDD/day, as electricity consumption is related not just to the cooling requirements but also to the lighting and miscellaneous equipment (such as computers, copiers, microwave ovens, etc.). If the HDD and CDD for eQUEST's weather profile are calculated and inserted in these equations, realistic utility bill values for the eQUEST model (with its associated weather profile) can be generated.

For the assessment of the eQUEST model and utility bill comparison, the statistical parameters of mean base error (MBE) and coefficient of variation of root mean squared error Cv (RMSE) were used as demonstrated by Xing *et al.* (2015). The parameters are defined as:

$$\text{MBE} = \frac{\sum_{\text{Period}} (\text{M-A})_{\text{Interval}}}{\sum_{\text{Period}} \text{A}_{\text{Interval}}} \tag{1}$$

$$\text{Cv (RMSE)} = \frac{\sqrt{\frac{1}{\text{N}_{\text{Interval}}} \sum_{\text{Period}} (\text{M-A})_{\text{Interval}}^{\;2}}}{\frac{1}{\text{N}_{\text{Interval}}} \sum_{\text{Period}} \text{A}_{\text{Interval}}} \tag{2}$$

where,

M: Energy consumption predicted by the model

A: Actual energy consumption

Period: Time period considered for the analysis

Interval: Time span diving the period into equal parts, i.e., 1 month

$N_{Interval}$: Total number of intervals in the period, i.e., 12 (months)

Figure 4 shows the comparison of linear regression adjusted utility bills and the eQUEST model results for natural gas consumption. As observed in Figure 4, the highest absolute error was observed in the months of April, May, and October, when the heating requirement was not as high as in the other months. The highest absolute error of 16.4% was observed in the month of May. This observation indicates that the linear regression model works best when the heating requirements are high, and the heating systems are being used frequently. In other words, the model works best when the HDDs are relatively high. The mean base error (MBE) for the natural gas consumption is -3.60% and the Cv (RMSE) is 7.33% which is well within acceptable limits of ± 5% and 15% respectively, as discussed by Xing *et al.* (2015).

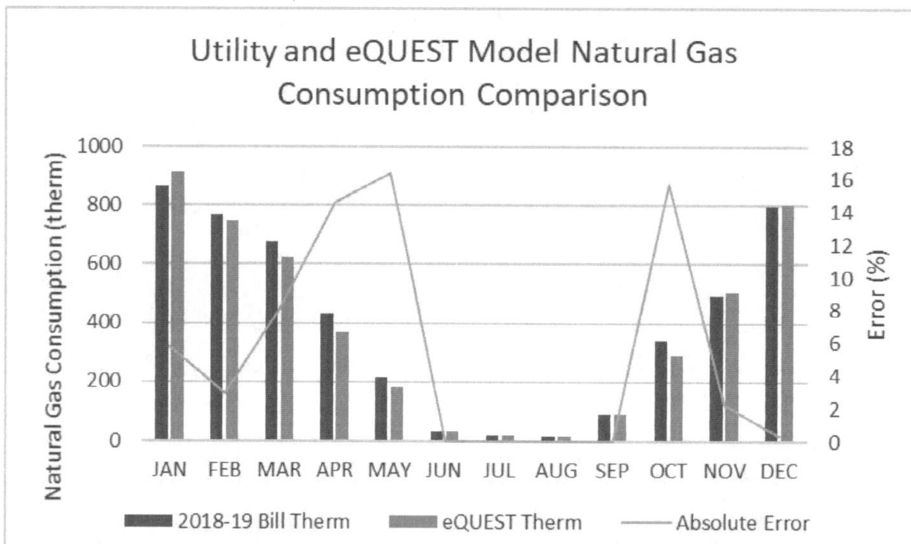

Figure 4: Utility and eQUEST model natural gas consumption

For electricity consumption, two models were used to predict the utility bills for the eQUEST weather profile. In the first model, the utility bill values for the cooling months (June through September) were calculated by directly using the equation obtained from the model shown in Figure 3 and for the other months, the utility bill values were obtained by extrapolating the equation and setting CDD equal to zero. In the second model, the equation from the model shown in Figure 3 was divided into two parts, a linearly increasing part and a constant part corresponding to the linear equation (kWh/day = m*CDD/day + c), where m is the slope of the line and c is the constant part of the equation. The constant part of the equation was calculated by averaging the utility bills from the winter months, October through May (when CDD = 0). Rebuilding the equation using the constant value c thus obtained, the utility bill values for electricity consumption can then be predicted.

The two models for kWh/day have been statistically compared in Table 1. The results indicate that model 2 provides more accurate expected electricity consumption values for the non-cooling months. For model 2, a maximum absolute error of 9.43% was observed in the month of May. This can be attributed to the fact that the month of May actually does consist of some days which require cooling. The month of May was not considered for deriving the equation in Figure 3 as it was considered to be part of the winter (heating) season.

Table 1: Statistical comparison of two models for kWh/day

Model	MBE	Cv (RMSE)	Avg. absolute error
kWh/day Model 1	-3.76%	4.85%	4.23%
kWh/day Model 2	2.48%	4.14%	3.33%

Figure 5 and Figure 6 show the comparison of the linear regression adjusted utility bills and eQUEST model results for electricity consumption.

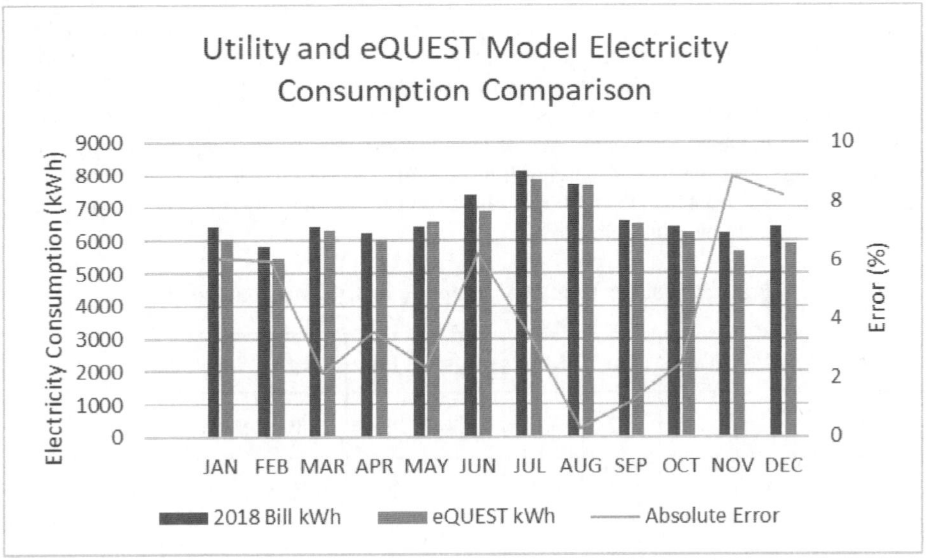

Figure 5: Utility and eQUEST model electricity consumption comparison

(from linear regression model – Model 1)

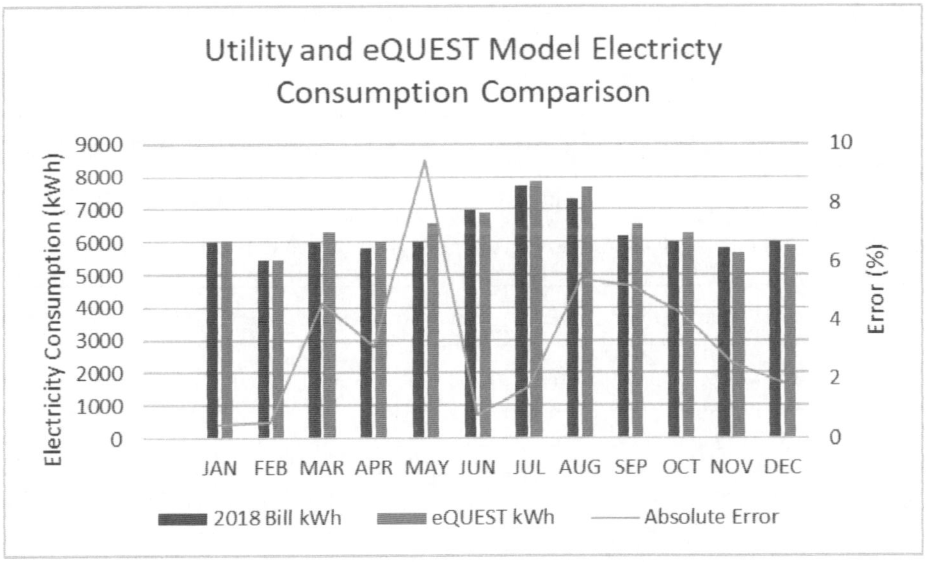

Figure 6: Utility and eQUEST model Electricity Consumption Comparison

(from modified method – Model 2)

This model discussed above has been used to identify the key factors affecting energy consumption and energy efficiency opportunities. The next section focuses on these aspects of this study.

3. RESULTS AND DISCUSSION

3.1 Model Results & Factors Affecting Energy Consumption

Annual electricity and natural gas consumption for the eQUEST library building model in 2018 is provided in Table 2 and Table 3 respectively. The principal factors affecting the building energy consumption are lighting, the HVAC loads and occupancy. 57.4% of the total electricity consumption in 2018 was used to provide lighting in the library. Space cooling accounted for 10.8% and other HVAC equipment (fans, pumps, etc.) accounted for 13.8% of the annual electricity consumption. In hot and humid locations, space cooling may account for the highest percentage share of the electricity consumption; however, as Houghton, MI does not experience such weather for most part of the year, the primary electricity usage turns out to be for lighting in this case. Natural gas is used to provide space heating and domestic hot water in the library. 97.4% of the total natural gas consumption in 2018 was used to provide space heating whereas domestic hot water accounted only for 2.6%.

Table 2: Library annual electricity consumption distribution in kWh (3.6 x 10^{-6} J)

	Jan	Feb	Mar	Apr	May	Jun	Jul	Aug	Sep	Oct	Nov	Dec	Total
Space Cooling	0	0	0	40	340	1830	2870	2330	940	10	0	0	8360
Ventilation Fans	620	530	580	560	620	930	860	940	530	620	580	560	7930
Pumps & Auxiliary Equipment	250	230	250	240	220	200	200	200	210	240	240	250	2730
Miscellaneous Equipment	1160	1050	1220	1160	1210	1170	1160	1240	1080	1210	1080	1140	13880
Task Lights	30	20	30	30	30	30	30	30	30	30	30	30	350
Area Lights	3980	3620	4210	3980	4160	2770	2740	2940	3740	4160	3740	3920	43960
Total	6040	5450	6290	6010	6580	6930	7860	7680	6530	6270	5670	5900	77210

Table 3: Library annual natural gas consumption distribution in therm (1.055 x 10^{8} J)

	Jan	Feb	Mar	Apr	May	Jun	Jul	Aug	Sep	Oct	Nov	Dec	Total
Space Heating	906.4	738.9	613.7	360.9	169.5	84.3	40.7	58.6	87.1	279.1	497.3	790.2	4626.7
Hot Water	11.2	10.5	12.1	11.4	11.1	10	9.4	9.6	8.5	9.7	9.3	10.4	123.2
Total	917.6	749.4	625.8	372.3	180.6	94.3	50.1	68.2	95.6	288.8	506.6	800.6	4749.9

It needs to be noted that higher lighting and occupancy help in reducing the HVAC loads in winter (which lowers the natural gas consumption) as these components add heat to the zone, which is the function of the HVAC system. In summer, these components increase the HVAC loads (which increases the electricity consumption). The results obtained were for a design occupancy of 50% (which is approximately 42 people in the 12600 sq. ft. library). For an occupancy of 75%, a decrease of 2.1% in the annual natural gas consumption was observed whereas an increase of 1.7% in the electricity consumption was observed. For an occupancy of 25%, an increase of 2.2% in the annual natural gas consumption was observed whereas a decrease of 1.7% in the electricity consumption was seen. The results are summarized in Table 4.

Table 4: Effect of occupancy on natural gas and electricity consumption

Occupancy (Occupancy for base model – 50%)	Change in Annual Natural Gas Consumption	Change in Annual Electricity Consumption
75%	-2.1%	+1.7%
25%	+2.2%	-1.7%

3.2 Energy Efficiency Opportunities

A maximum HVAC heating load of 330 kBtu/hr (96.7 kW) was observed in the month of January. The furnace installed in the library has a total output power rating of 600 kBtu/hr (175.8 kW). Also, as discussed earlier, one of the most common complaints among the library occupants/customers was that it got too warm in the winter months. These facts motivated us to check if the hot water loop for space heating is actually necessary. A parallel simulation was performed for the library building, with all the conditions remaining the same, but with only the furnace (i.e., with only the direct expansion coils) as the HVAC system. The results indicated that the furnace could still support the heating demand for the library. If the hot water loop were not used, the natural gas consumption would be lowered by 20% and the electricity consumption by 3.6% per year. In this study, it was also incidentally observed that the central thermostat was located about 10 ft. (3.05 m) away from the radiators and grills with a lot of the seating arrangement being around the radiators, which might have led to higher temperatures around the seating areas than the thermostat set temperature. Thus, as predicted by the validated model, turning off the hot water loop for space heating would be a significant energy saving opportunity for the library.

4. CONCLUSIONS

In this research, eQUEST was selected for modeling a public library building, due to its user-friendly interface and to the fact that it already included the HVAC system supporting the building's heating and cooling requirements in its system library,

After developing the eQUEST model for the building, it was essential to validate it with actual utility bills. Linear regression-based methodology was used to compare eQUEST results for natural gas and electricity consumption with the library's utility bills. The model to estimate natural gas consumption from HDDs indicated that it was more accurate for the months having higher HDDs, with a maximum absolute error of 16.4% observed in the month of May. The mean base error observed was -3.60% and the coefficient of variation of root mean squared error was 7.33%. Multiple models to estimate the electricity consumption from CDDs were developed and it was observed that the model using average electricity consumption values from the winter months for the constant part in the linear equation was more accurate than the model using the linear equation derived from the electricity consumption – CDD relation. For this model, the mean base error observed was 2.48% and the coefficient of variation of root mean squared error observed was 4.14%, with the highest absolute error being 9.43% for the month of May. Error analysis indicated that the errors observed in the models were within reasonable limits.

Based on the simulation, we found that the library could save up to 20% of its natural gas consumption and 3.6% of its electricity consumption per year by using the furnace units alone, while still comfortably meeting the library's heating requirements.

NOMENCLATURE

HDD	heating degree day	(HDD)
CDD	cooling degree day	(CDD)
HDD/day	average heating degree day in a given period	(HDD/day)
CDD/day	average cooling degree day in a given period	(CDD/day)
therm/day	average natural gas consumption in a given period	(therm/day)
kWh/day	average electricity consumption in a given period	(kWh/day)
R^2	linear regression coefficient of determination	(–)
MBE	mean base error	(%)
Cv(RMSE)	coefficient of variation of root mean squared error	(%)
m	line slope	(–)
c	constant part of linear equation	(–)

REFERENCES

Algarni S., Abdelmohimen M., & Saleel C. (2017), Potential of Energy and Water Sustainability from HVAC Unit Condensate Water in Hot-Humid Climate of Jazan, Saudi Arabia, *Proceedings of Second International Conference on Energy and Indoor Environment for Hot Climates, ASHRAE* (212-218)

BizEE Software. Heating & Cooling Degree Days - Free Worldwide Data Calculation. https://www.degreedays.net/

Gould S. & Hawkins M. (2015), Modeling Multifamily Buildings with eQuest: A Case Study of Prediction versus Reality, *Proceedings of 2015 ASHRAE Winter Conference* (1-8)

Rallapalli, H. (2010), *Whole Building Energy Simulation Results for a Medium Sized Office Building* [Master's thesis, Arizona State University]. ASU Repository.

Sobha P., Patne N. & Jadhav A. (2016), A Smart Demand Side Management Mechanism for Domestic Energy Consumers with Major HVAC Load, *Proceedings of 2016 International Conference on Electrical Power and Energy Systems (ICEPES), Maulana Azad National Institute of Technology, Bhopal, India*, (504-511)

Wang F., Lin H., Tu W., Wang Y. & Huang Y. (2015), *Energy Modeling and Chillers Sizing of HVAC System for a Hotel Building*, Procedia Engineering, vol. 121, 1812-1818

Xing J., Ren P. & Ling J. (2015), *Analysis of energy efficiency retrofit scheme for hotel buildings using eQuest software: A case study from Tianjin, China*, Energy and Buildings, vol. 87, 14-24

Zerroug A. & Dzelzitis E. (2015). Analysis of results of energy consumption simulation with eQUEST and EnergyPlus, *Proceedings of 5th International Scientific Conference, Latvia University of Agriculture* (102-107)

Zhu, D., Hong, T., Yan, D. & Wang, C. (2013), *A detailed loads comparison of three building energy modeling programs: EnergyPlus, DeST and DOE-2.1E*, Building Simulation, 6(3), 323-335.

Modelling the NIST IBAL

Adam KOPACH[1]*, Greg NELLIS[2], David BRADLEY[3], Doug REINDL[4], Amanda PERTZBORN[5]

[1]University of Wisconsin, Department of Mechanical Engineering,
Madison, WI, United States
akopach@wisc.edu

[2]University of Wisconsin, Department of Mechanical Engineering,
Madison, WI, United States
gfnellis@engr.wisc.edu

[4]Thermal Energy System Specialists, Principal Investigator
Madison, WI, United States
d.bradley@tess-inc.com

[3]University of Wisconsin, Department of Mechanical Engineering,
Madison, WI, United States
dreindl@wisc.edu

[5]NIST, Engineering Laboratory
Gaithersburg, MD, United States
amanda.pertzborn@nist.gov

* Corresponding Author

ABSTRACT

The Intelligent Building Agents Laboratory (IBAL) at the National Institute of Standards and Technology (NIST) will be used to experimentally study control strategies within the realm of commercial building air-conditioning systems. A simulation software (Klein, 2020) model of the IBAL has been created to help guide the experiments in the IBAL and to provide a software-based method of evaluating control strategies implemented in this facility. This simulation model is composed of the various components comprising the IBAL HVAC system such as heating coils, cooling coils, fans and dampers that can accurately simulate the actual component behavior under transient and off-design conditions. Additionally, pressure and airflow throughout the system are simulated to support associated controls for modulating damper positions and fan speeds. The resulting system simulation is flexible; it is possible to study the expected IBAL system response to forcing functions such as weather and loads under the influence of different control strategies. The initial controller included in the IBAL simulation model replicates the controls used in most commercial buildings. The resulting tool will be validated in the IBAL and allow NIST to test alternate, advanced control strategies quickly, providing guidance for the experimental work.

1. INTRODUCTION

According to the U.S. Energy Information Administration, the commercial buildings sector consumed over 106 quadrillion kJ, or roughly 18 % of primary energy, in the United States in 2018 (US EIA, 2018). Work towards a goal of achieving net zero energy use buildings will require substantial reductions in the energy consumption within the commercial sector. One proposed step towards attaining this goal is the use of intelligent agents for controlling heating, ventilation, and air conditioning (HVAC) systems used in commercial buildings. An intelligent agent is an autonomous entity that acquires information about the state of the system, makes an optimal or near-optimal control decision, and communicates that decision to another agent or to an actuator that executes the decision. Kelly and

Bushby at the National Institute of Standards and Technology (NIST) completed a feasibility study to determine if intelligent agent technology could lead to significant reductions in HVAC energy consumption. The results of this study demonstrated intelligent agents could produce a cost savings of 21 % over the course of a day (Kelly, 2012). To further test this concept, NIST created the Intelligent Building Agents Laboratory (IBAL) (Pertzborn,2016-2018). The laboratory is comprised of an HVAC system that emulates a small office building in terms of the complexity and the type of HVAC systems and equipment that would typically be present. It has an air side and water side; the water side contains the plant equipment (e.g., chillers) and the air side, which is the focus of this work, is shown in Figure 1. The airflow path begins with an Outdoor Air Unit (OAU) that emulates outdoor air conditions for the downstream system by preconditioning the actual outdoor air by use of a cooling coil, an electric heating coil, and a steam spray humidifier. After the air is preconditioned in the OAU, it flows into the two Air Handling Units (AHUs). Each AHU is composed of a preheat coil, a cooling coil, and a fan, all in series as shown in Figure 2. The air exiting each AHU flows to two Variable Air Volume (VAV) boxes. The VAV boxes are each composed of a damper blade and a reheat coil in series. Each VAV box delivers supply air to a single zone. The zone simulator includes an electric heater and a steam spray humidifier to emulate the sensible and latent loads of an occupied space. Upon leaving the zone simulators, a portion of the return air stream combines and mixes with the emulated "outdoor air" from the OAU, while the remainder exhausts outdoors. This project considers the various components that condition and control air being delivered to the zones, shown in Figure 2.

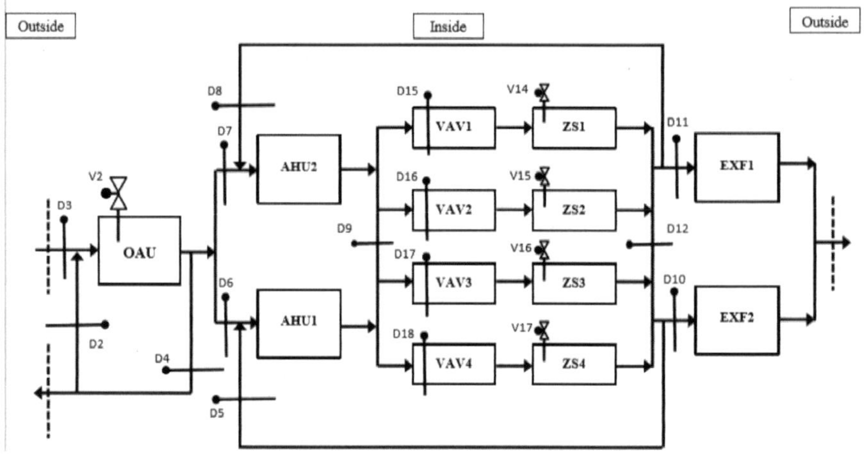

Figure 1: The IBAL Schematic (Pertzborn, 2018)

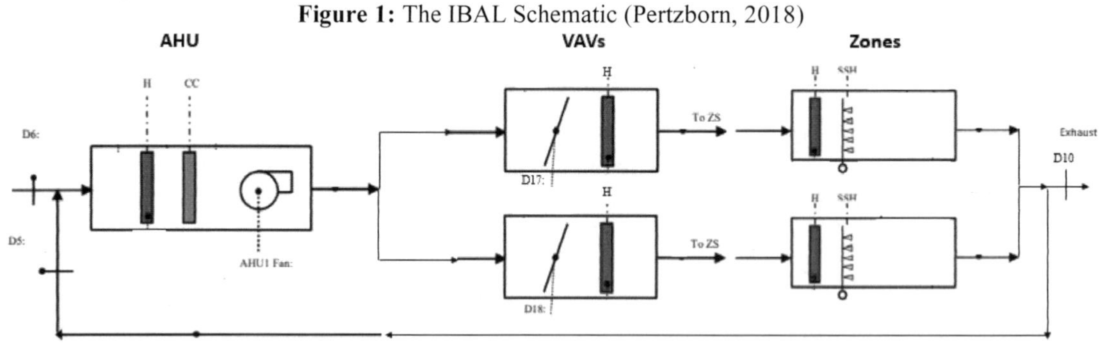

Figure 2: The HVAC system that is being controlled (CC = cooling coil, H = heater, and SSH = steam spray humidifier) (Pertzborn, 2018)

2. COMPONENTS

The simulation modelling focuses on the AHUs and VAV boxes. Each AHU is comprised of an electric heating coil, cooling coil, and draw-thru fan. The electric heating coil preheats mixed air to avoid freezing the glycol mixture flowing through the cooling coil. The cooling coil can provide both sensible and latent heat removal from the supply air stream. Finally, the fan ensures that sufficient airflow is provided to the rest of the system. The VAV boxes include a damper to adjust the flow of the air to the zone and a reheat coil to sensibly heat the air before delivery to the zone if required. The models developed in the context of this project differ in three ways from previously developed models of VAV air handlers. First, rather than treating the air handler as a whole, models are written for each component (heater, cooling coil, damper, etc.). Second, the models are not load based but are instead designed to compute air and water outlet conditions knowing only the inlet conditions and actuator settings. Third, they need to be able to predict behavior at a small time scale (on the order of seconds) and so are not quasi steady state models, but instead treat the transient behavior of each component.

2.1 Electric Heater

The electric heater model requires two parameters: heater thermal capacitance (C_h) and overall heat conductance (UA). The model also receives operating condition inputs including mass flow ($\dot{m}$), air temperature (T_{in}), relative humidity (RH_{in}), and heater electric power ($\dot{q}_{elec}$).

An energy balance on the heating coil provides:

$$C_h \frac{dT_h}{dt} = \dot{q}_{elec} - \dot{q}_{out} \tag{1}$$

where $\dot{q}_{out}$ is the heat transfer rate from the coil to the air as given by:

$$\dot{q}_{out} = \varepsilon \; \dot{m} \; c\left(T_h - T_{in}\right) \tag{2}$$

where c is the specific heat capacity of air. The parameter, T_h, is the instantaneous coil temperature and ε is the heating coil effectiveness:

$$\varepsilon = 1 - \exp\left(-\frac{UA}{\dot{m}\,c}\right) \tag{3}$$

The differential equation (1) is solved analytically during each time step. The outlet temperature of the air leaving the coil is given by:

$$T_{out} = T_{in} + \varepsilon\left(T_h - T_{in}\right) \tag{4}$$

The parameters C_h and UA are determined through a "calibration" process that involves tuning the model to match experimental data from the IBAL as it operates over a range of conditions. A number of tests were carried out to accomplish this tuning process including ones in which both airflow rate and heating coil power input are varied with time. In total, four tests were used with approximately 2 000 data points per test. An example of one of these tests is shown in Figure 3. Figure 4 shows the measured temperature rise as a function of predicted temperature rise for all the tests considered and indicates that the model can predict the behavior of the coil to within approximately 1 °C under most conditions. The VAV reheat coil uses the same model and the agreement with data from the IBAL system is similar after tuning the parameters.

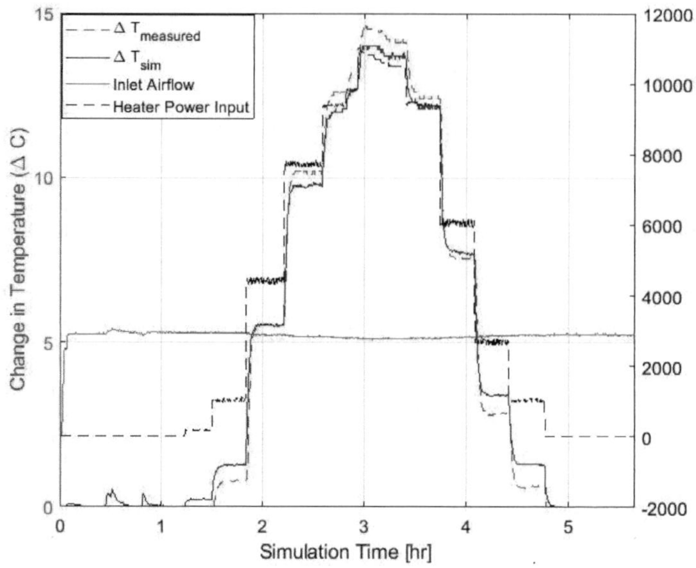

Figure 3: AHU Heater Test Case with Varying Heater Power Input

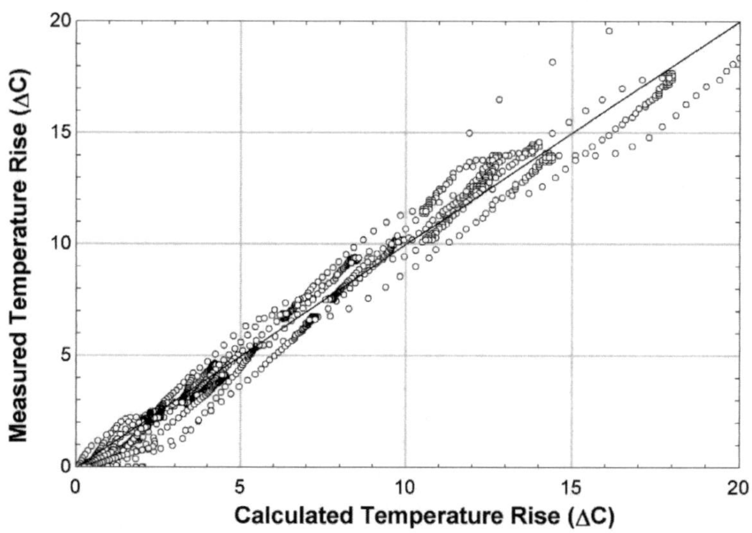

Figure 4: Compiled AHU Heater Data Compared to Simulated Model Output

2.2 Cooling Coil

Figure 5 shows a schematic of a "bypass model" (Stanke, 2000) used as the basis for simulating the cooling coil. The parameters required by this model include the cooling coil capacitance (C_c), the coolant-side conductance (UA_{coil}), and the air bypass fraction (ϕ). The model also requires, as inputs, the inlet air temperature, relative humidity, and air mass flow rate ($T_{a,in}$, RH_{in}, and $\dot{m}_a$, respectively) as well as the inlet secondary fluid temperature and mass flow rate ($T_{liquid,in}$ and $\dot{m}_{liquid}$, respectively).

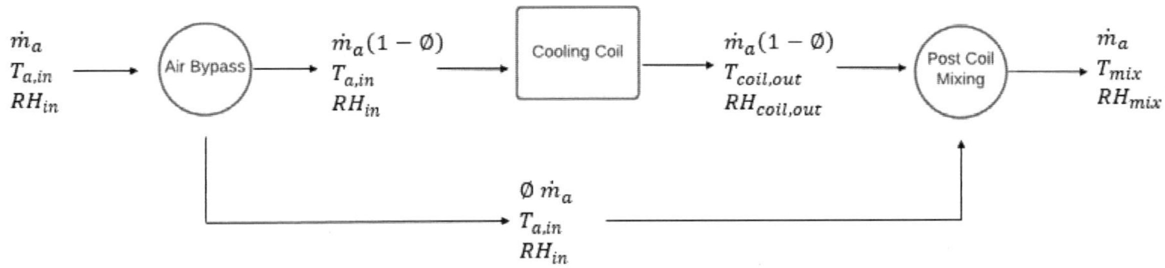

Figure 5: Schematic of the air side of the Cooling Coil bypass model

An energy balance on the cooling coil becomes:

$$\frac{dT_c}{dt}C_c = \dot{q}_{air} - \dot{q}_{liquid} \tag{5}$$

where the rate of heat transfer to the coolant is given by:

$$\dot{q}_{liquid} = \dot{m}_{liquid}\, c_{liquid}\, \varepsilon_{coolant}\left(T_c - T_{liquid,in}\right) \tag{6}$$

where $\varepsilon_{coolant}$ is the effectiveness of the cooling coil based on the coolant side flow:

$$\varepsilon_{coolant} = 1 - \exp\left(-\frac{UA_{coil}}{\dot{m}_{liquid}\, c_{liquid}}\right) \tag{7}$$

The rate of heat transfer from the air is given by:

$$\dot{q}_{air} = \dot{m}_a\left(1-\phi\right)\left(h_{air,in} - h_{coil,out}\right) \tag{8}$$

where $h_{air,in}$ and $h_{coil,out}$ are the moist-air specific enthalpies of the air entering the cooling coil and leaving the cooling coil, respectively. The bypass model assumes that a fraction of the entering air, ϕ, bypasses the cooling coil and the remaining fraction of entering air, (1-ϕ), passes over the cooling coil and leaves the cooling coil at the cooling coil temperature, T_c. If this temperature is below the dew point temperature, then the air is saturated and some condensation has occurred. If the cooling coil temperature is above the entering air dew point temperature, then the air's humidity ratio is unchanged and no condensation occurs.

Within each time step, the simulation numerically solves the ordinary differential equations (ODEs) associated with these equations using an Euler technique to predict the outlet air conditions and the outlet coolant temperature. The outlet coolant temperature is computed by:

$$T_{liquid,out} = T_{liquid,in} + \varepsilon_{coolant}\left(T_c - T_{liquid,in}\right) \tag{9}$$

The bypass airflow mixes with the coil airflow and leaves the coil at temperature T_{mix} and relative humidity RH_{mix}, which are determined by an energy balance and water mass balance on this mixing process.

$$h_{air,mix} = \phi\, h_{air,in} + \left(1-\phi\right)h_{air,out} \tag{10}$$

$$\omega_{air,mix} = \phi\, \omega_{air,in} + \left(1-\phi\right)\omega_{air,out} \tag{11}$$

The parameters C_c, UA_c, and ϕ are tuned to allow the model to predict the behavior of the cooling coil installed in the IBAL. Figure 6 illustrates the results of one test where the coolant flow is varied over time and the predicted vs measured air-side temperature change compared.

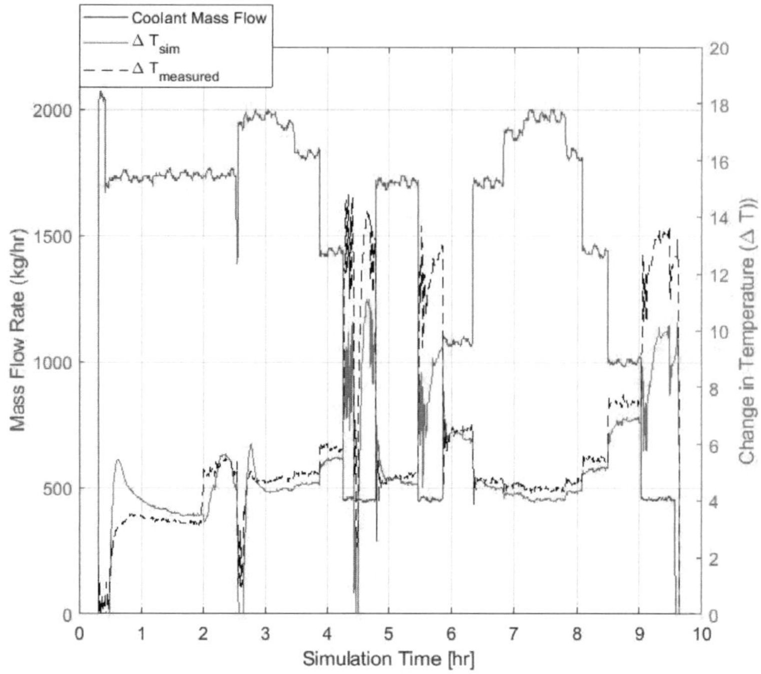

Figure 6: Cooling Coil model validation

Figure 6 shows that the model tracks the data under periods of high coolant flow, but has difficulty predicting the performance when there is extremely low coolant flow (essentially zero volumetric flow). This is an issue that might be addressed by modeling the coil conductance as a function of coolant flow.

2.3 Pressure/Flow Component

Two of the primary mechanisms used to control the HVAC operation are the AHU fan speed and the VAV damper position. The damper position is adjusted to maintain an airflow setpoint, which is determined by the difference between the zone temperature and the zone setpoint. The fan speed is adjusted to maintain a static pressure setpoint, which is reset periodically so that the dampers will be more open. The goal is to reduce fan power consumption by matching the fan speed to the airflow demand. The control system senses pressure as one of the monitoring signals. Therefore, it is necessary for the simulation model to predict how air pressure and airflow within the system change over time as the damper positions and fan speed change. This is accomplished in the Pressure/Flow component, which is a single component that simulates the flow and pressure that exists within one AHU serving two VAV/zones with the associated return air branch (D5), as shown in Figure 7. The model only uses the fan and damper curves to calculate pressure; it does not explicitly model the pressure drop across individual components such as a cooling coil.

The Pressure/Flow component requires the characteristics of the dampers and fans as inputs; these include fan curves (pressure rise vs flow) over a range of speeds and damper resistance curves (pressure drop vs flow) over a range of damper positions. The fan model was derived from the fan curve data provided by the manufacturer for the installed fans. The fan curves were normalized by the maximum pressure rise and maximum flow for each speed (189, 283, 377) rad/s ((1800, 2700, and 3600) rpm) to create the curve fit shown in Figure 8(a). The damper model is obtained from the damper curves provided by the manufacturer. The curves for each damper position, expressed as a % of wide open, %WO, were normalized by the maximum pressure rise and maximum flow for each damper position (45, 50, 60, 70, 80, 90) %WO to generate the curve fit shown in Figure 8(b). The pressure/flow component requires inputs of fan speed and position of all dampers as well as the upstream pressure and the downstream air pressures.

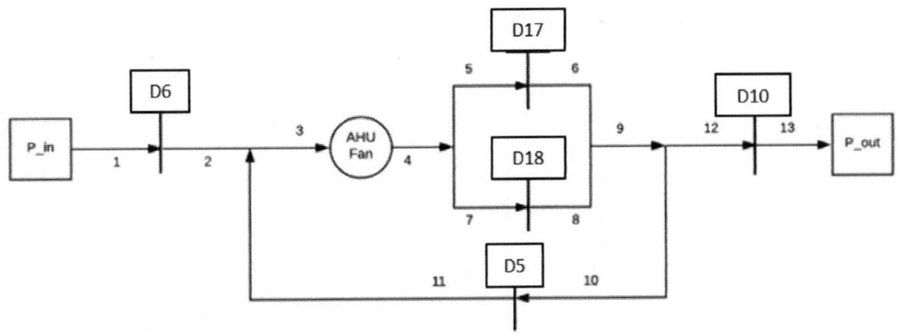

Figure 7: Pressure/flow component, which consists of a single AHU serving two VAV/zone branches and the return air duct. Note that D = damper and the numbered points indicate locations of distinct pressure and/or flow.

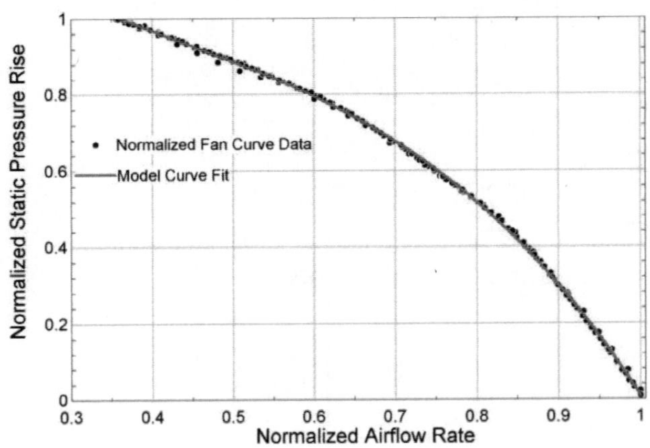

a)

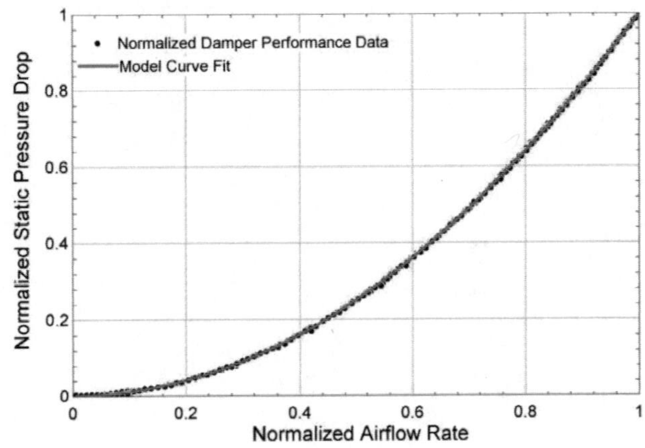

b)

Figure 8: a) Normalized fan pressure rise as a function of normalized flow for various fan speeds, also shown is the curve fit used by the model, b) Normalized damper pressure drop as a function of normalized airflow rate for various damper positions, also shown is the curve fit used by the model.

The model enforces continuity of flow and pressure at all nodal intersection points. The process is depicted in Figure 9. At each damper (D), the pressure drop is calculated based on the position of the damper and airflow across the damper. The system is assumed to respond instantaneously for the time scales of interest to changes in fan speed or damper position. The values of flow at node 1 and node 11 are varied until the pressure error terms are negligible, which indicates convergence. To verify the model is working correctly, a few test conditions were simulated. Figure 10 shows the pressures and flow rates at two locations (nodes 4 and 11 in Figure 7) for a case where all damper positions are held constant while the fan speed increases from 189 rad/s (1800 rpm) to 377 rad/s (3600 rpm). Referring to Figure 7, locations 4 and 11 correspond to the location downstream of the AHU fan and the return air branch downstream of Damper 2. Note that downstream of the AHU (at node 4), both the flow and pressure increase quickly as the fan speed increases. The flow and pressure at location 11 are also affected by this change in fan speed; the pressure goes down slightly and the flow increases. The future work associated with this component involves tuning its function to match data from the IBAL.

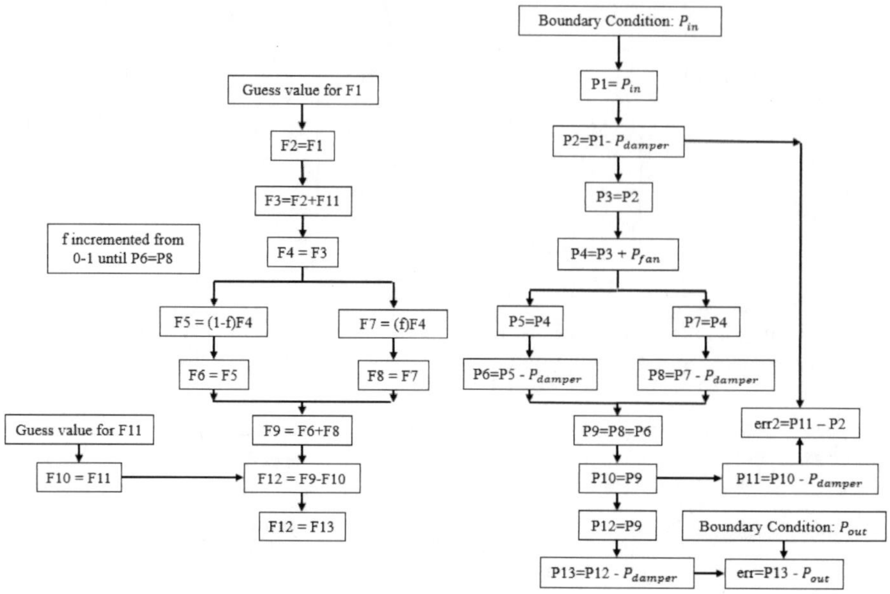

Figure 9: Airflow calculation flow chart

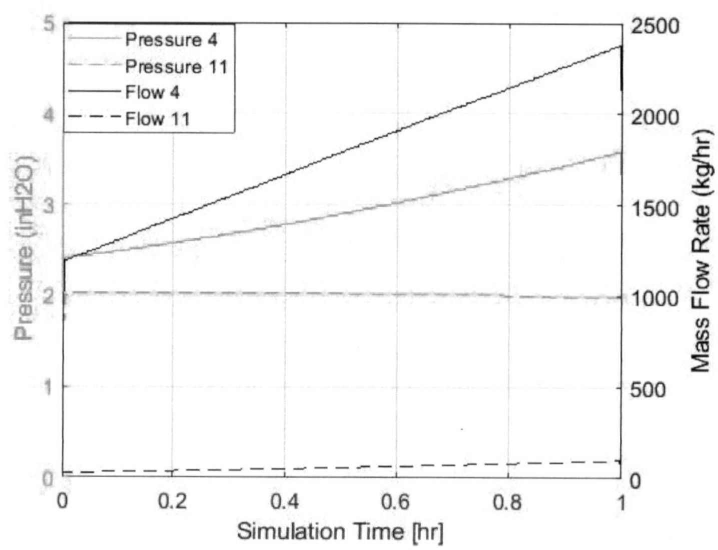

Figure 10: Pressure and flow rate downstream of the AHU (location 4) and in the return branch (location 11) as the fan speed increases.

3. CONTROLLERS

The IBAL simulation model discussed in this paper is flexible with regard to the control system. One of the primary purposes for developing a model is to allow for the simulation of a variety of advanced control strategies prior to deploying them in the IBAL. Initial demonstrations of control function were carried out using a conventional approach that consists of three separate controllers. The first controller is responsible for adjusting the damper in the VAV boxes and activating the reheat coil based on a comparison of the zone temperature to its setpoint. The second controller is responsible for adjusting the fan speed to maintain a setpoint pressure in the supply air duct downstream of the fan and upstream of the VAVs. The final controller is responsible for adjusting the coolant flow through the cooling coil based on the difference between the temperature of the air downstream of the cooling coil (supply air temperature) and its setpoint temperature.

The controllers were tested using a single day that was designed to require both heating and cooling logic. Figure 11(a) shows the outside temperature, the zone setpoint temperature, and the simulated temperature of the zone for this day. For this test, the model inputs and setpoints were set to emphasize the behavior of each controller, which means that the system was generally unable to meet the setpoint. This is not a reflection of what would be expected in a real building, but rather a construct for testing the simulated controllers. During the first 8 hours of the test, the outside temperature drives the zone temperature below the setpoint. Over the next 8 hours, the outside temperature rises, which drives the zone temperature above the setpoint temperature. In the final 8 hours, the outside temperature becomes cool again, which drives the zone temperature below the setpoint.

Figure 11(b) shows the behavior of the VAV controller in response to the zone temperature. During the first 8 hours, the damper position is held at the minimum position while the reheat signal is based on the difference between the zone temperature and the setpoint temperature. Over the next 8 hours, the reheat is turned off as heating is no longer needed. Instead, the controller actuates the damper to open based on the difference in the temperature of the zone compared to the setpoint. In the final 8 hours, the controller reverts to reheat mode. Figure 11(c) displays the cooling coil controller response. Over the first 8 hours and final 8 hours, the mass flow through the coil is at its lowest value because cooling is not needed. In the middle of the simulation, the mass flow increases in response to the rising air temperature downstream of the coil. Figure 11(d) highlights the response of the fan controller. The fan speed remains constant when the damper position is at its minimum value. As the damper opens in the middle portion of the test, the pressure within the system drops, causing the fan speed to increase, which provides a larger volume of cool air during the timeframe when the zone temperature is above the setpoint.

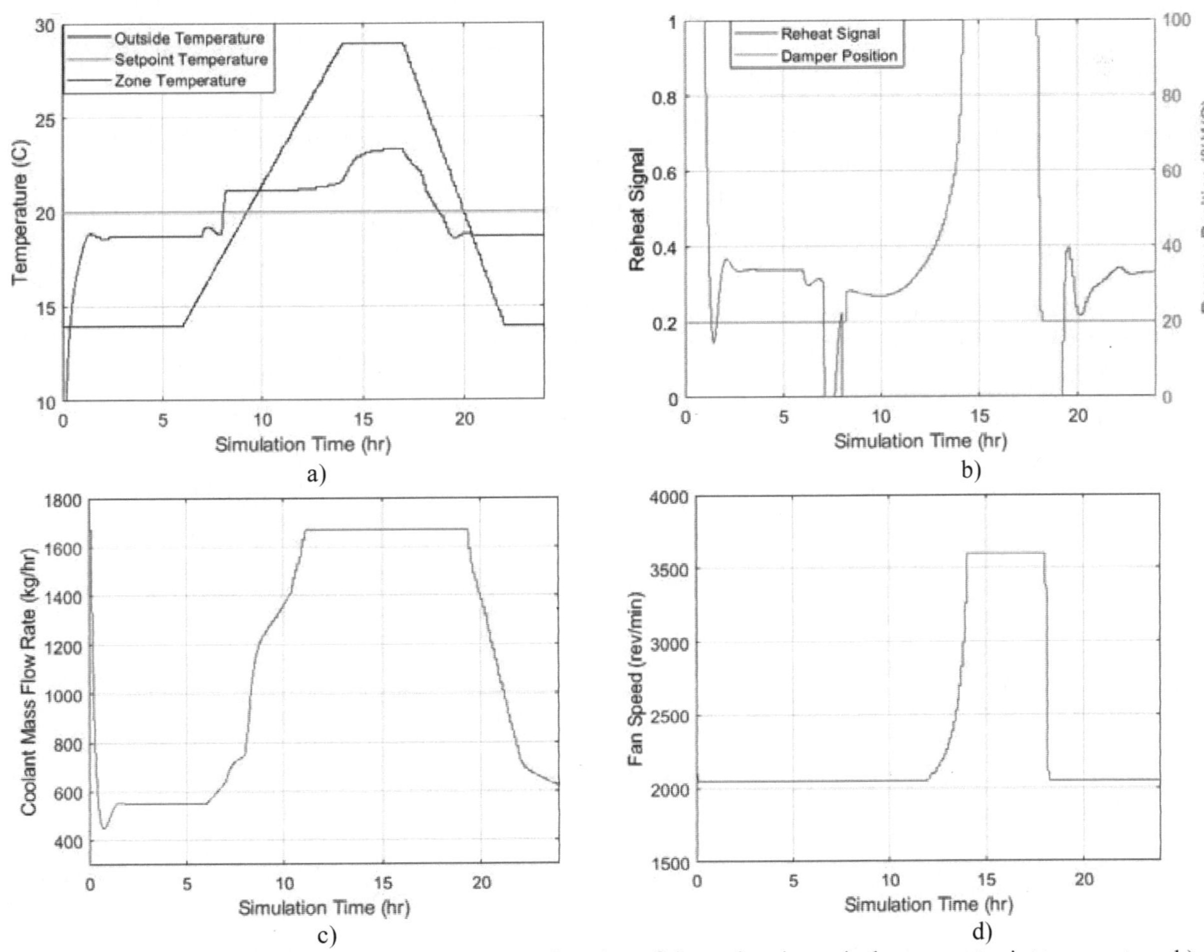

Figure 11: a) Outside air and zone temperature as a function of time, also shown is the zone setpoint temperature, b) VAV controller response including damper position and reheat signal, c) Cooling coil controller response showing coolant flow rate, and d) Fan controller response showing fan speed.

4. CONCLUSION

Once all components have been tuned and validated, the simulation model can be used to test arbitrary controller architectures quickly. The program is designed so that NIST can change between different control strategies seamlessly. One of the proposed control strategies is the use of variable setpoint temperatures throughout the system. This proposed method of control involves the use of predicted weather data to adjust the setpoint temperatures to the various controllers to provide a more optimal control that reduces energy use.

REFERENCES

Kelly, G. & Bushby, S. (2012). Are intelligent agents the key to optimizing building HVAC system performance? *HVAC R Res.*, *18*(4), 750–759.

Klein, S.A. et al. (2020). *TRNSYS* (v18) [A Transient Simulation Program]. WI Solar Energy Laboratory, University of Wisconsin.

Pertzborn, A. J. (2016). *Intelligent Building Agents Laboratory: Hydronic System Design* (NIST Technical Note 1933). National Institute of Standards and Technology.

Pertzborn, A. J. (2017). *Measurement Uncertainty in the Hydronic System in the IBAL* (NIST Technical Note 1970). National Institute of Standards and Technology.

Pertzborn, A. J. & Veronica D. A. (2018). *Intelligent Building Agents Laboratory: Air System Design* (NIST Technical Note 2025). National Institute of Standards and Technology.

Stanke, D. (2000). Dehumidify with Constant-Volume Systems: It May Take More than You Think. *Trane Engineers Newsletter*, *29*(4), 4-5.

US Energy Information Administration. (2018). *U.S. energy flow*. https://www.eia.gov/totalenergy/data/monthly/pdf/flow/total_energy.pdf

ACKNOWLEDGEMENT

This research would not be possible without the grant funded by NIST (NIST Award # 70NANB18H203).

Elementary School Renovation to Achieve Net Zero Energy

Songhao Wu[1]*, Brian Tedeschi[2], Zach Schreiber[3], William Hutzel[4], Luciana Debs[5]

[1]Purdue University, Lyles School of Civil Engineering,
West Lafayette, IN, United States
songhaowuu@gmail.edu

[2]Purdue University, Department of Mechanical Engineering Technology,
West Lafayette, IN, United States
Briantedeschi29@gmail.com

[3]Purdue University, Department of Mechanical Engineering Technology,
West Lafayette, IN, United States
zschreiber1@yahoo.com

[4]Purdue University, Department of Mechanical Engineering Technology,
West Lafayette, IN, United States
hutzelw@purdue.edu

[5]Purdue University, School of Construction Management
West Lafayette, IN, United States
ldecresc@purdue.edu

*Corresponding Author

ABSTRACT

This paper presents a comprehensive plan for renovating a thirty-year old elementary school located in the Midwest United States to achieve net zero energy. This preliminary design earned a 2nd place finish in the Elementary School Division of the U.S. Department of Energy's 2020 Solar Decathlon Design Challenge. In addition to adding solar energy, the proposed renovations included modifications to the building envelope, mechanical systems, lighting, plug loads, and building automation systems. Additionally, a living wall of plants called a Biowall is recommended for improving Indoor Environmental Quality. A corresponding economic analysis was conducted on each proposed renovation to identify which features should be implemented immediately and which features should be accomplished over time. The resulting plan for Phased Net Zero (PNZ) could become a template for similar elementary schools to save energy, reduce operating costs, and improve the student learning environment. This strategy for improving existing educational facilities is becoming important because many public schools are facing the challenges of declining budgets, aging facilities, and competition in the form of alternative forms of education.

1. INTRODUCTION

The traditional model of a public school is under significant pressure for change. Numerous reports identify concerns about school funding, safety, and overall quality of instruction. This has led to competition in the form of online education and charter schools, which siphon away tax dollars traditionally used by public schools. In states like Indiana, a growing school voucher program also allows parents to send their children to private schools of their choice, which diverts even more money away from public schools. The situation for traditional public schools is not all negative, because they are quickly evolving to meet society's expectations as a place where children go to learn, grow, and excel. One competitive advantage for public schools is their buildings. More than just physical structures, well-designed school buildings can create an exceptional learning environment by maintaining high levels of Indoor Environmental Quality and including sustainability features that save energy and reduce operating costs.

1.1 Net Zero Energy Elementary Schools

Net zero energy buildings have become a trend in the building industry. By 2018, nearly 500 zero energy buildings have been built across the U.S. and Canada, which is a 700% increase since 2012 (*U.S., Canada Home*, 2018). As an important sector of public buildings, net zero energy schools have gained much attention. High performance buildings not only promote energy saving but also contribute to teaching, learning, and student success as they often incorporate mindful designs. Because of this increase in interest, in 2018 ASHRAE published *the Advanced Energy Design Guide (AEDG) for K-12 School Buildings: Achieving Zero Energy* to provide strategies and recommendations for achieving a zero energy school.

When considering net zero energy buildings, it is important to distinguish between source energy and site energy. Source energy includes energy losses in the generation and transmission of energy, while site energy only refers to energy at its point of use. Some net zero energy buildings are defined based on site energy. However, calculations using source energy provide a more holistic view of energy performance by accounting for all forms of energy being used. For calculations, site energy can be converted to source energy by using accepted energy conversion factors. For example, a multiplier of 2.8 is used to convert electricity from the electric grid to its equivalent source energy. (*The Difference between Source and Site Energy*, 2020)

There are many net zero energy schools across the U.S. The 97,600-ft^2 (9,067 m^2) Discovery Elementary School in Arlington, VA is the largest net zero energy school building in the nation, setting the bar for energy efficiency and IEQ (Beaufait, 2018). Discovery Elementary School introduced a solar laboratory and dashboard system to integrate solar energy as part of its educational programs, which inspired the renovation for this project to connect with student learning. For example, the this project design includes plans to link science learning objectives for students with training modules based on the operation of its solar array.

In Indiana, many schools are considering on-site renewable energy due to the rising cost of electricity and the decreasing cost of solar energy installations. For commercial photovoltaic systems, installation costs have decreased 66% between 2010 and 2018, shortening financing payback (Fu et al., 2018). These school improvements are typically done by companies that offer renewable energy solutions along with upgrades to improve energy efficiency inside the school buildings. For example, a large Energy Services Corporation (ESCO) recently guided Hamilton Southeastern Schools in Indiana through the installation of 1,848 kW solar array at three schools in its district (*Construction begins*, 2019). In 2018, a single company helped five Indiana schools install solar panels (*2018 School Projects,* 2019) and it is expected that more schools will embrace renewable energy soon.

1.2 Solar Decathlon Design Competition

The Solar Decathlon Design Challenge is an annual international competition hosted by the U.S. Department of Energy (DOE) that encourages university students to develop innovative strategies for achieving net zero energy buildings. Student teams engage with the National Renewable Energy Laboratory (NREL) located in Golden, CO over an academic year. The work initially includes training in building sciences for all students participating in the competition. Student groups submit an initial proposal, a project update, and a detailed final report that are evaluated by an expert panel of judges. Under normal circumstances, the competition culminates with in-person presentations to the judges at NREL late Spring semester. However, due to COVID-19, the finalist presentations for the 2020 competition were delivered as a virtual event in April of 2020.

Traditionally the Solar Decathlon Design Challenge has focused on new construction, but a pilot project for the 2020 competition encouraged student teams to partner with real clients, allowing Purdue University students to partner with the nearby Tippecanoe School Corporation (TSC) to evaluate and propose improvements at an existing school. TSC is the fourth largest school district in the State of Indiana, both in terms of geographic size and number of students and is looking for good ideas to counter its increasing operating costs. The school district already has twenty schools and more schools are being planned to serve rapidly expanding communities surrounding Lafayette, IN.

TSC recognizes the potential of net zero energy schools as one way to counter ever-increasing energy expenditures. The school corporation is facing a 15% increase in electricity rates for 2020 with more rate increases expected in the future. TSC schools tend to be located in rural locations that have abundant space and TSC has considered installing wind turbines in the past. However, due to expectations for population growth, new county ordinances essentially prohibit the installation of new wind turbines. TSC's is interested in reviewing the potential for adding solar photovoltaics as a way of reducing long-term operating expenses.

1.3 James Cole Elementary School

James Cole Elementary (JCE) was the school targeted for renovations during the 2020 Solar Decathlon Design Challenge. It is located in rural Stockwell, IN and is one of 20 school buildings for the TSC. The school was built in 1988 with energy efficiency at its core, integrating an innovative earth berm design (Figure 1), daylighting, and geothermal heating/cooling. With time, the focus on sustainability has faded, and in the competition, students focused on realigning the building with its original concept. JCE could be the first in TSC to pursue net zero energy, making it a model for other renovations and future school designs to follow in this rapidly growing school district.

Figure 1: JCE's earth berm feature saves energy

Climatic conditions had a strong impact on the initial review of JCE's design and operation. Located in climate zone 5A, JCE experiences all four seasons, receiving roughly 190 sunny days per year, 15 days less than the U.S. average. The maximum temperature in Stockwell, IN is 84 °F (29 °C) and the minimum is 17 °F (-8 °C). The average rainfall is 38 inches (965 mm) and snowfall is 18 inches (457 mm). A properly designed building envelope is particularly important because the school uses significant amounts of energy for heating in the winter and cooling in the summer.

1.4 Design Goals

Much research has been done on building renovations for schools. Two core issues are energy efficiency and cost. In most cases, renovation projects are initiated to improve building performance and reduce energy costs. Research shows that cost-optimal solutions reduce primary energy consumption between 68% and 95 % compared with the reference scenario (Baglivo et al., 2015; Congedo et al., 2015; Zacà et al., 2015)

Congedo et al. (2016) proposed a recast of the Energy Performance of Buildings Directive (EPBD) as a comparative methodological framework to promote energy efficiency and establish minimum energy performance requirements in buildings at the lowest costs. They argued that cost-optimal methodology is useful in searching for the most appropriate retrofit measures for the renovation of existing public buildings. For school buildings, it is crucial to find the best balance between cost reduction and high comfort to optimize the environment for students and staff.

Hamdy et al., (2013) proposed a multi-stage optimization method for cost-optimal and nearly-zero-energy building solutions. It includes 1) an efficient algorithm to find cost-effective energy-efficient combinations of design variants like building-envelop and HVAC systems, 2) assess energy and life-cycle cost of the combinations, 3) improve financial/environmental viabilities of the optimal.

Wyon (2005) pointed out that poor indoor air quality can reduce the performance of office work by 6–9%. A similar influence is also true for a school setting. Experiments conducted by Wargocki & Wyon (2006) demonstrated that air quality and temperatures in classrooms are important factors in the learning process. Many works suggest that mental tasks that require concentration, memory, and original thought could be affected by the indoor environment (Lan et al., 2011; Tham, 2004; Wyon, 2005). Building on this existing research, the goals of this paper are to identify:

- Infrastructure upgrades to improve energy efficiency and achieve net zero energy at JCE
- Economic analysis to prioritize projects according to the budget limitations at JCE
- A renovation template for achieving Phased Net Zero at other similar elementary schools

2. Engineering and Operations

JCE had an energy-efficient design when constructed in 1988. However, innovations in technology have created opportunities to improve the building operations and energy efficiency. A comprehensive analysis of its existing engineering and operation features led to energy conservation measures that could be implemented.

2.1 Architecture

JCE has a compact one-story central core that centralizes the gym, library, cafeteria, and other common areas as shown in Figure 2. The high ceilings and large clerestory windows make the school welcoming and spacious, while its east-west orientation allows for many south-facing classrooms, taking advantage of solar heating and daylighting. The double-stacked corridors with exterior walls and roof exposure reduce thermal reservoirs and the school's earth berm design geothermally regulates heat loss.

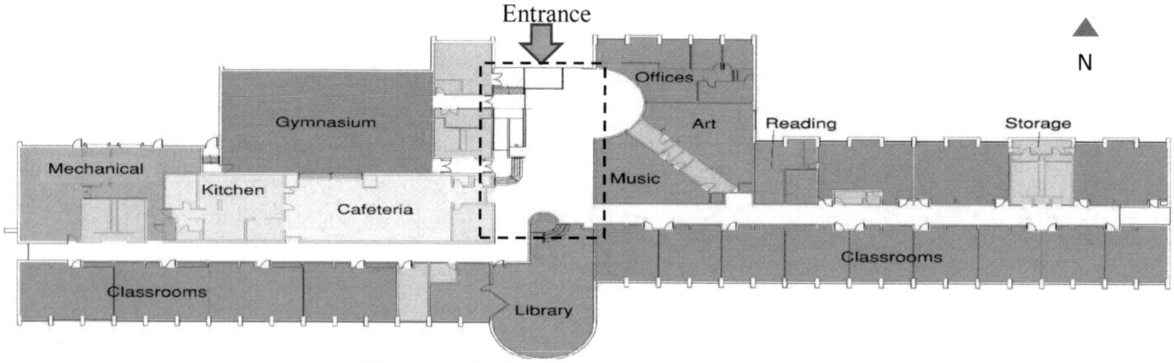

Figure 2: JCE's compact central core layout

2.2 Building Envelope

JCE has multiple building control layers for effectively regulating the flow of heat, water, air, and vapor. Its R-16 ($ft^2 \cdot °F \cdot h/Btu$) (R-2.82 $m^2 \cdot K/W$) metal roof is well below modern building insulation standards (greater than R-40 (R-7.05)). This contributes to a large amount of heat loss from the roof of the building. The various compositions of the walls include brick masonry, metal studs and paneling, concrete, insulation, and earth berm. The masonry walls with metal paneling and studs achieve R-values between 12 (2.11 $m^2 \cdot K/W$) and 13 (2.29 $m^2 \cdot K/W$). Three to five-foot earth berm wall sections increase the R-value to 15 (2.64 $m^2 \cdot K/W$) and 17 (3.00 $m^2 \cdot K/W$), respectively. Windows allow for great visibility and solar daylighting; however, the current U-0.58 [$Btu/(ft^2 \cdot h \cdot °F)$] [U-3.29 $W/(m^2 \cdot K)$] windows are below today's standards of less than U-0.25 (U-1.42). This contributes to large amounts of heat loss. In 2016, JCE received brick masonry replacement and air sealing to improve the building envelope and air sealing.

2.3 Mechanical System

The mechanical systems in JCE were designed to be efficient and comply with all building standards then. The HVAC system at JCE complies with ASHRAE 55-2004 (Thermal Environmental Conditions for Human Occupancy) and ASHRAE 90.1-2001 (Energy Standard for Buildings Except Low-Rise Residential Buildings) as currently written. The school uses a primary ground source heat pump, a central makeup air unit, and individual zone heat pumps throughout JCE for thermal conditioning.

A closed geothermal well loop is used for heating and cooling throughout the school as well as individual zone heat pumps. The heat pumps are connected to the well loop for zone heating or cooling based on demand and well field temperatures. The intelligent engineering design of zone heat pumps allows for individual zone conditioning, heating northern classrooms while cooling southern classrooms on a sunny day in spring.

The 8000-cfm (3.8 m^3/s) air handler supplies fresh air at a constant speed following ASHRAE 62.1-2016 (Ventilation for Acceptable Indoor Air Quality) guidelines for elementary schools. The 13 cfm per person rule of thumb in a 35-student classroom requires 3,900 cfm (1.8 m^3/s) every day for 300 students and staff and 250,000 Btu/h (73 kW) on a design winter day in Stockwell, IN. The built-in heat recovery system conditions 70% of the outside air, saving 70 MWh of electricity annually. Additional conditioning is accomplished by the zone heat pumps.

JCE has a 250,000 Btu/h (73 kW) water-heater located in the mechanical room on the west side, supplying all hot water for the school. Most hot water is used for cooking purposes in the kitchen located next to the mechanical room, minimizing distribution losses. The energy losses for distributing hot water to the rest of the school are reduced by the Building Automation System (BAS), circulating hot water only when school is in session.

2.4 Lighting

JCE benefits from a substantial amount of daylight from its east-west orientation and clerestory windows, which means that the lighting system doesn't need to run at full capacity all the time. Its light system was completely fluorescent but JCE has started LED upgrade projects to save energy and operation cost.

2.5 Plug loads

There are many appliances throughout the school that are old and inefficient. For example, a dairy cooler in the cafeteria uses far more energy than it should. It is old and its compressor runs continuously. There will be significant energy savings when this and other similar equipment are replaced.

2.6 Operations and BAS

JCE operations are controlled by a BAS to operate, monitor, and maintain systems efficiently within the school. The lighting, HVAC, and hot water distribution are scheduled on during the school day and off outside school activities, saving energy. It also monitors pumps, fans, heat pumps, temperatures, humidity, CO_2 levels, and equipment status. The BAS is web-enabled and can be remotely accessed for monitoring alarms and equipment faults.

3. Renovation Opportunities

Energy Use Intensity (EUI) expresses a building's energy use as a function of its size or other characteristics. The Solar Decathlon Competition uses the total source energy consumed annually divided by the gross floor area (kBtu/ft^2 or kWh/m^2) to express EUI as shown in Equation (1) (*2020 Design Challenge Rules*, 2020). EUI is used to quantify improvements in building energy performance or benchmark performance as compared to similar buildings.

$$EUI = \frac{Total\ Annual\ Energy\ Consumption}{Built\ Area} \quad (1)$$

Based on the existing engineering and architecture traits of JCE, an energy audit was conducted and an energy model was created to identify improvements for each area, evaluate energy savings, and certify compliance with building energy codes. Figure 3 compares the 2018 EUI at JCE to the EUI after energy efficiency upgrades are made (labeled "Retrofit"). After achieving all conservation measures, a solar PV array could be installed to achieve net zero energy (labeled "NZEB").

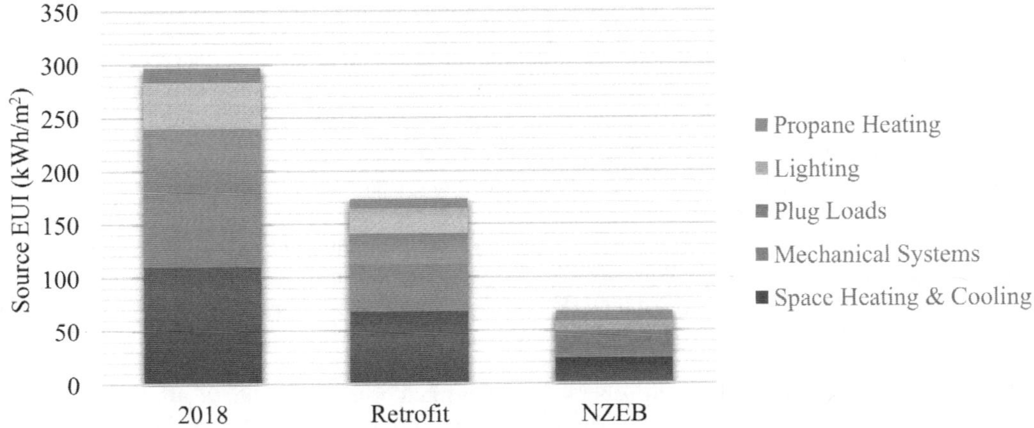

Figure 3: Retrofit and Solar to achieve net zero

The colored bars in Figure 3 identify how energy savings are accomplished. The Space Heating & Cooling (blue) represents the HVAC systems. The Mechanical Systems (orange) includes larger equipment like the cafeteria refrigerator. Plug loads (gray) include smaller appliances computers. Lighting (yellow) deals with electricity consumption from the lighting system in the school. Finally, propane heating (green) represents the energy used for the water heater. In 2018 the overall EUI for JCE was a 297 kWh/m^2 (94.1 kBtu/ft^2), but after efficiency improvements the EUI will drop to 174 (55.1). On-site renewable energy will bring the number further down to 68 (21.6).

Figure 4 shows the composition of retrofit energy savings. The improvements were initially identified by energy modeling and were later verified by building professionals who provided performance specifications and budget cost estimates. Implementing these renovations could result in significant savings.

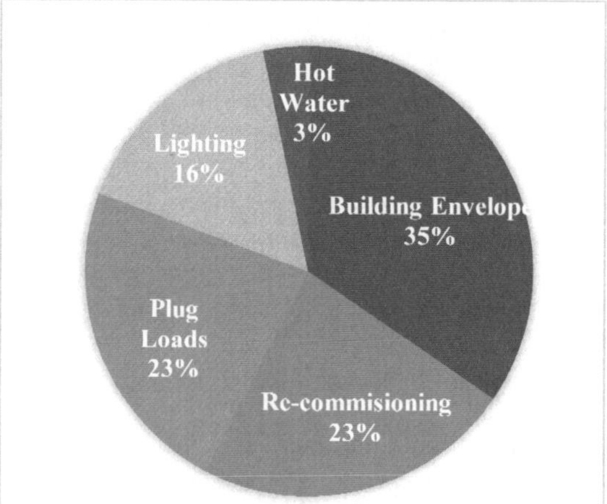

Figure 4: Retrofit targets for reducing energy use

3.1 Building Envelope

Building Envelope upgrades will provide 40% of the total energy savings by increasing the roof insulation from R-16 (ft^2·°F·h/Btu) (R-2.82 m^2·K/W) to R-45 (R-7.93), adding R-30 (R-5.29) fluted ISO insulation on top of the existing roof. Replacing the U-0.58 [Btu/(ft^2·h·°F)] [U-3.29 W/(m^2·K)] windows with new double-glazed U-0.3 (U-1.70) based on feasible renovation building specs will also contribute to the energy savings. Air sealing will be included with both renovations to limit infiltration and heat loss.

3.2 Mechanical System

Replacing the hot water boiler with a distributed tankless water heater system will provide 4% of the total energy savings, improving efficiency and on-demand hot water while reducing heat losses. The existing HVAC system is relatively efficient and upgrading will not be cost-effective now due to complex equipment replacement. The pumps and fans in the school are single speed and substantial energy savings can be realized by making this upgrade.

3.3 Recommissioning

Recommissioning of the building systems and staff training will provide 23% of energy savings. An investigation of systems, energy bills, and the BAS showed that the propane boiler switched on 24/7, consuming twice as much energy as normal. This is one example that identified an opportunity for a recommissioning of the building systems as well as additional training for facilities staff. Energy savings will be accomplished by:

- Improving the performance of the geothermal heat pumps by monitoring loop temperatures
- Improving the performance of the air handling systems by monitoring CO_2 levels
- Developing training programs for the facility engineers and staff operating the school

The use of Key Performance Indicators (KPI's) in Table 1 simplifies monitoring for facilities staff. In a large building, many systems must be monitored and maintained to ensure proper performance, efficiency, and conditioning. KPI's give specific details and instructions for monitoring crucial building systems to quickly identify problems and deploy maintenance. Targeting a few key metrics on a regular basis can yield substantial energy savings.

Table 1: Key Performance Indicators (KPI) help focus maintenance activity

Focus Area	Key Performance Indicator
Geothermal Heat Pump	Loop Temperatures between 40 °F (4.4 °C) (winter) and 90 °F (32.2 °C) (summer)
Air Handling Unit	Runs according to a timed schedule
Hot Water	Temperature and run time within specified limits
Indoor Environmental Quality	CO_2 levels within the recommended threshold for school
Solar	Array efficiency within 10% of expected performance

3.4 Lighting

A lighting retrofit will account for 16% of the total savings by replacing the remaining fluorescent lights in the school with dimmable LED with occupancy sensors. This will require a fixture replacement and allow for adjustable lighting as needed. The office, atrium, library, and gymnasium have already been converted to LED's, but there is plenty of work left. The existing T-8 170 W fluorescent lights in the cafeteria, corridors, and classrooms will be converted to LED lighting, and motion detectors will be added to each classroom as part of the upgrade.

3.5 Plug Loads

Reducing plug loads throughout the school could produce 23% of the total energy savings by replacing old and inefficient equipment as well as reducing equipment use as described by Table 2. The greatest energy savings can be obtained by replacing the walk-in refrigerator that uses obsolete R-22 refrigerant. Other replacements and reduction will also contribute to energy savings within the school.

Table 2: Estimated energy savings for reducing plug loads

Energy Saving Target	Annual Savings (kWh)
Replace walk-in refrigerator/freezer	38,000
Replace dairy cooler	4,100
Unplug water fountains	2,100
Replace refrigerator	1,800
Miscellaneous	1,300
Reduce use of carpet dryers	740

3.6 Biowall

Although not directly saving energy, a Biowall could be introduced into JCE to improve indoor air quality and mindfulness. A Biowall consists of a tray of plants grown in a loose-packed growth media. Small fans on the back of the Biowall pull air through the growth media where the Volatile Organic Compounds (VOCs) removal takes place. The cleaned air is re-introduced to space. The Biowall design proposed for JCE is based on U.S. patent 10,477,779 that was issued to Purdue University in November of 2019. In addition to the oxygen supplied by plants through photosynthesis, Purdue's Biowall improves IEQ by leveraging the ability of microbes surrounding the root zone of plants to metabolize VOCs (Alraddadi et al., 2016).

3.7 Renewable Energy

After implementing energy efficiency measures, net zero energy can be achieved by adding renewable energy. Solar array size in kW is computed by dividing the daily energy use in kWh by the sun hours at a given location along with a factor of safety (FS). Since both energy use and sun hours change by season, this computation is typically made in different months to get an average that accounts for the variation. PVWatts, a web-based tool supplied by the National Renewable Energy Lab, was also used to automate this process over an entire calendar year.

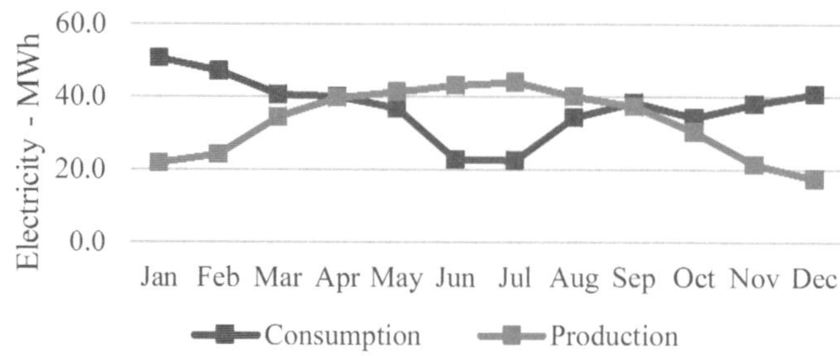

Figure 5: Matching electricity production with consumption

The results of the renewable energy analysis show that a 215-kW array is needed to make JCE Net Zero assuming that all energy efficiency measures are installed. However, a larger 300-kW solar array is recommended as it may not be cost-effective to implement all conservation measures. This array will produce 400 kWh of electricity on an annual basis. Figure 5 shows the relationship between energy consumption and solar energy production over a calendar year.

4. Financial Feasibility

Financial feasibility and market potential are key aspects of the JCE Project. This project has multiple stakeholders with different needs and wants from JCE. For this to be a financially responsible investment, it must offer clear benefits to the school and its direct users. In conjunction, the potential for repeatability across all TSC school buildings is important to create a template for future PNZ renovations.

4.1 Financial Analysis

Table 3 shows the parameters that went into the financial analysis for each system. The proposed lighting system parameters are given as an example. The table is color-coded based on the aspects of the calculation. The orange represents direct costs for the system including initial investment costs and annual maintenance. Green represents the first-year energy savings calculated from energy reductions in section 3. The blue represents human aspects including the number of students at JCE (300) and the state annual funding allotted for educational use per student ($6,500). The yellow represents percentages and rates for all calculations over a 20-year lifecycle. As time passes, fees and costs will rise, thus energy, repair, and student fee escalation rates were factored in. These values were constant and estimated considering the current economic state. The discount rate represents the method at which TSC or JCE will pay for these upgrades, either with current assets on hand or using borrowed funds. The change in student outcomes is meant to show that energy systems have a human impact and can allow for better use of each dollar spent.

Table 3: Financial calculation parameters example for lighting system

The life cycle of investment (years)	**20**
Installation cost ($)	-42,250
First-year energy savings ($)	1,750
Annual maintenance/repair costs ($)	-1000
Annual state allocation per student ($)	6,500
# of students	300
Student state fee escalation rate (%)	0.5
Energy escalation rate (%)	3
Repair/maintenance escalation rate (%)	1
Discount rate (%)	0
Change in student outcomes (%)	0.25

4.2 Results

Table 4 summarizes the system level economic analysis conducted for JCE. Eight topics are listed in order based on their potential for a financial return on investment. The installation cost, year 1 energy savings, impact on student/staff productivity, financial breakeven year, and 20-year net present value were all included in Table 4 to accurately depict the financial situation for implementation.

Table 4: Economic analysis for the proposed upgrades at JCE

	Solar	*Re-commissioning*	*Plug Loads*	*Lighting*	*Biowall*	*Hot Water*	*Windows*	*Roof*
Installation Cost ($)	569,000	21,800	3,500	42,250	33,900	46,360	300,000	850,000
Y1 Energy Savings ($)	44,200	4,464	4,464	1,785	0	821	4,500	4,464
Productivity Impact (%)	0	0	0	0.25	0.25	0	0	0
Breakeven Year	15	3	3	9	10	N/A	N/A	N/A
20-Year NPV ($)	210,000	120,000	63,800	43,900	27,600	-39,000	-226,600	-715,800
Priority	Do Now	Do Now	Do Now	Do Now	Do Now	End of Life	End of Life	End of Life

Table 4 shows that solar panels, recommissioning, plug loads, and lighting systems all have a positive Net Present Value (NPV) over the 20-year lifecycle, while the hot water, windows, and roofing systems all have negative values. This means that after 20 years, the system will have yielded either a savings or a cost to JCE and TSC. All the installation costs were gathered using quotes and outreach from professional contractors and businesses within the central Indiana region. The savings on energy cost were derived from the percent energy reduction from old to new systems. A productivity increase of 0.25% was taken into account for the Biowall and lighting systems. These systems deal with air quality and lighting intensity, two factors directly correlated with human performance and productivity.

4.3 Implementation

Implementation of these school upgrades to achieve net zero should be achieved through phased project planning to limit disruption of school operations for TSC. Table 4 shows the systems with a positive and negative NPV over the 20-year lifecycle. The plan for PNZ recommends implementing systems based on ease of implementation, lowest first cost, and breakeven year. Some of these improvements can be funded with the existing TSC maintenance budget. Other projects, such as the solar array, will likely require a loan in the form of a school bond. Projects like replacing the windows, roof, and hot water system all have value towards achieving net zero but are not financially feasible over 20 years. These systems should be implemented once the existing systems are at the end of their lifecycle.

5. Market Potential

JCE has the potential to make an impact in the local community and serve as a template for future PNZ projects of similar scope. Consulting with local contractors provided detailed quotes. Additionally, utilizing local businesses creates a community stake, helping support the local economy and supply chains. The PNZ template provides a methodology for analysis and feasibility that can be repeated and adapted to other building renovation projects.

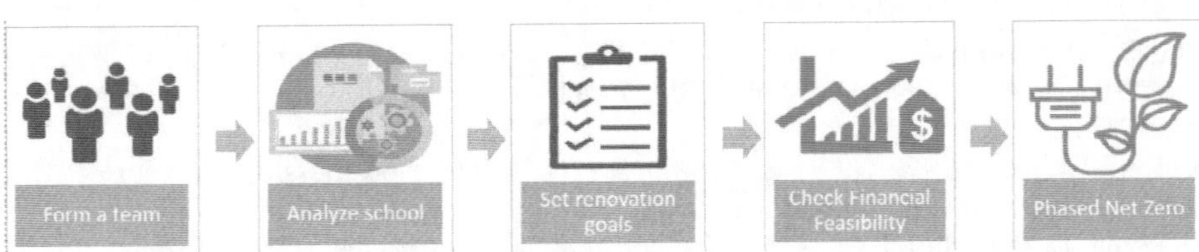

Figure 6: A template for a project plan for the Phased Net Zero Project.

Figure 6 outlines a potential plan to achieve a PNZ project. Selecting appropriate team members is important to involve all stakeholders from different backgrounds and roles for the project and create a diverse team. Analyzing and surveying the existing structures and systems identify opportunities for improvements. Setting renovation goals defines realistic and measurable investments and outcomes that give direction for a successful project. All investments need a comprehensive financial analysis to ensure a positive return at the end of a given time period. These investments should be implemented in phases around academic calendars to prevent disruption of operations and education. A PNZ plan will allow for a financial budget to be maintained, while also working around the schedule of a school. By offering distinct savings and benefit to the occupants, the concept of PNZ school renovations demonstrates opportunity and sustainability for the future.

6. CONCLUSIONS

A comprehensive analysis of James Cole Elementary (JCE) located in Stockwell, IN delivered a Phased Net Zero (PNZ) renovation plan for achieving net zero energy that met the fiscal and operational constraints of this active school. This comprehensive project earned a second-place finish for Purdue University students in the Elementary School Division of the 2020 Solar Decathlon Design Challenge. This project shows that improving energy efficiency and achieving net zero through phased and prioritized renovations is attainable and financially feasible for existing schools. The analysis of JCE has provided methodologies to create, analyze, and implement a PNZ energy plan while maintaining budgets, working within schedules, and collaborating with local suppliers to improve the economy. Education will always continue to be the main priority within the public-school system. By using a PNZ plan and moving towards net-zero energy, we are creating more sustainable infrastructure and shaping the design for future spaces. School buildings are important structures in our society and they must be maintained at the highest standard.

REFERENCES

2018 School Projects, Producing Energy in 2019. (2019). https://johnsonmellohsolutions.com/jms-2018-school-projects/

2020 Design Challenge Rules (Issue January). (2020).

Alraddadi, O., Leuner, H., Boor, B., Rajkhowa, B., Hutzel, W., Dana, M., & Lafayette, W. (2016). Air Cleaning Performance of a Biowall for Residential Applications. *International Compressor Engineering, Refrigeration and Air Conditioning, and High Performance Buildings Conferences*, 1–10.

Baglivo, C., Congedo, P. M., D'Agostino, D., & Zacà, I. (2015). Cost-optimal analysis and technical comparison between standard and high efficient mono-residential buildings in a warm climate. *Energy*. https://doi.org/10.1016/j.energy.2015.02.062

Beaufait, R. (2018). Paving a path for zero energy schools. *ASHRAE Journal*, *60*(5), 36–42.

Congedo, P. M., Baglivo, C., Zacà, I., & D'Agostino, D. (2015). High performance solutions and data for nZEBs offices located in warm climates. *Data in Brief*. https://doi.org/10.1016/j.dib.2015.09.041

Congedo, P. M., D'Agostino, D., Baglivo, C., Tornese, G., & Zacà, I. (2016). Efficient solutions and cost-optimal analysis for existing school buildings. *Energies*, *9*(10), 1–24. https://doi.org/10.3390/en9100851

Construction begins for solar project at HSE Schools. (2019). Current Publishing. http://youarecurrent.com/2019/07/26/construction-begins-for-solar-project-at-hse-schools/

Fu, R., Feldman, D., & Margolis, R. (2018). U.S. Solar Photovoltaic System Cost Benchmark: Q1 2018. *Nrel, Novmnber*, 1–47. https://doi.org/10.7799/1325002

Hamdy, M., Hasan, A., & Siren, K. (2013). A multi-stage optimization method for cost-optimal and nearly-zero-energy building solutions in line with the EPBD-recast 2010. *Energy and Buildings*. https://doi.org/10.1016/j.enbuild.2012.08.023

Lan, L., Wargocki, P., Wyon, D. P., & Lian, Z. (2011). Effects of thermal discomfort in an office on perceived air quality, SBS symptoms, physiological responses, and human performance. *Indoor Air*, *21*(5), 376–390. https://doi.org/10.1111/j.1600-0668.2011.00714.x

Tham, K. W. (2004). Effects of temperature and outdoor air supply rate on the performance of call center operators in the tropics. *Indoor Air, Supplement*, *14*(SUPPL. 7), 119–125. https://doi.org/10.1111/j.1600-0668.2004.00280.x

U.S., Canada Home To Nearly 500 Zero Energy Buildings. (2018). https://businessfacilities.com/2018/03/u-s-canada-home-to-nearly-500-zero-energy-buildings/

Wargocki, P., & Wyon, D. P. (2006). Effects of HVAC on student performance. *ASHRAE Journal*, *48*(12), 12.

The Difference between Source and Site Energy. (2020). ENERGY STAR Buildings and Plants | ENERGY STAR, www.energystar.gov/buildings/facility-owners-and-managers/existing-buildings/use-portfolio-manager/understand-metrics/difference.

Wyon, D. P. (2005). The effects of indoor air quality on performance, behaviour and productivity. *Pollution Atmospherique*, *14*(SPECIAL ISSUE), 35–41.

Zacà, I., D'Agostino, D., Congedo, P. M., & Baglivo, C. (2015). Data of cost-optimality and technical solutions for high energy performance buildings in warm climate. *Data in Brief*. https://doi.org/10.1016/j.dib.2015.05.015

The Role of Occupant Behavior Models in Multi-Objective Optimization Analysis

Julian Donges[1]*, Alessandro Prada[2], Francesca Cappelletti[3], Andrea Gasparella[1]

[1]Free University of Bozen-Bolzano, Faculty of Science and Technology,
Bozen-Bolzano, Italy
Tel.: +39 0471017890, Email: julian.donges@natec.unibz.it

[2]University of Trento, Dep.of Civil Environmental and Mechanical Engineering, Trento, Italy

[3]University IUAV of Venice, Dep. of Architecture and Arts,
Venice, Italy

* Corresponding Author

ABSTRACT

Building retrofit design aims at achieving a certain low energy target at a minimum cost. However, these buildings tend to be less comfortable than expected, prone to overheating, poor air quality, and less resilient towards different user profiles. Even when more accurate simulation models are used to calculate the energy demand, occupant behavior is usually oversimplified as a static schedule or rule-based model which often does not depend on comfort conditions and does not represent the actual occupants' reactions to manage indoor comfort. This can cause a significant gap between the simulated and the measured building performance. To address this gap, we have compared the performance indicators of optimal retrofit solutions obtained through a multi-objective optimization of a reference case and recalculated using different occupant-behavior models for the daily building operation – i.e. opening/closing windows. Recalculations of the optimal solutions have been performed with the generally used static schedules, a rule-based adaptive model, and an innovative probabilistic approach. The results have been analyzed through Pareto difference metrics to quantify the influence of behavioral models on energy consumption, cost, and comfort. Two referent scenarios – in a heating and cooling dominated climate – have been tested to observe the results under different boundary conditions. The findings demonstrate that the performance indicators vary strongly with each behavioral model severely compromising on the competing objectives of energy demand, i.e. cost, indoor air quality, and thermal comfort. The importance of realistic user behavior modeling is highlighted to prevent misleading conclusions on optimal solutions in the assessment of energy efficiency measures. It is pointed out that probabilistic behavior models are highly sensitive to variations of operating conditions, even leading to a positive rebound effect for certain retrofit strategies.

1. INTRODUCTION

Improvements in the performance of buildings can be done by a wide choice of energy efficiency measures. The most rational approach is to find an optimal solution representing the best trade-off energy and cost-wise. Since 2010 the European Committee has promoted the cost-optimal methodology to guide the selection of the best retrofit solutions for existing buildings, described for the first time by the Commission Delegated Regulation EU (European Comission, 2012). In this framework practitioners generally rely on Building Energy Simulation combined with Multi-Objective Optimization to find the optimal energy saving measures, but they usually refer to technical standard procedures mainly developed for energy certification which do not take into account realistic usage and actions to actively manage discomfort. The performance simulation of occupied buildings is always driven by assumptions made for occupant behavior. Especially, assumptions about building controls operated by occupants such as windows, lighting, shading, heating set-points or the use of fans need to be made before estimating the building's performance. If this is neglected, it can lead to a significant gap between the performance of the optimally designed solution and its actual performance, because people's interactions may affect energy consumption or indoor conditions severely. This not only hampers the indoor environmental livability but could as well raise complaints about the renovation quality or lead occupants to interact inappropriately (unnecessary opening of windows or operation of shadings) to improve their conditions. In international building codes, occupant-related approaches still vary considerably and need some further standardization (O'Brien et al., 2020), because they do not consider people as active. In recent years behavioral models relying on probabilities rather than schedules have gained more and more importance for modeling occupant behavior. This trend led to the development of several works on stochastic models to predict interactions with windows reviewed by Roetzel et al. (2010), Fabi et al. (2012) and Fabi et al. (2013). As a consensus in most research papers in the field, logistic regression models are calibrated using previously obtain experimental data (Andersen et al., 2016; Fabi et al., 2013, 2015; Haldi et al., 2017; Page et al., 2008). The use of building controls and especially windows is triggered by several physical conditions with varying statistical significance. Considering the most relevant predictors, neglecting the ones with minor importance, and assigning weighting factors as interpolation parameters to them, can reproduce experimental data for an examined reference case.

In the here presented study we investigate to which extend detailed user-behavior models for the daily building operation of windows impact on the performance of the optimal retrofit design configuration of a residential building. The design configuration has been optimized considering a standard user-behavior model and two optimization objectives (i.e. energy and cost). In a second step, two probabilistic occupant models and a rule-based adaptive model have been considered to recalculate by simulation the optimal solutions obtained focusing on changes in energy demand, total cost, and weighted discomfort time (WDT) for thermal comfort and indoor air quality (IAQ). For the current investigations, a minimalistic probabilistic approach developed by Nicol (2001) is considered taking solely outdoor temperature into account. As a second model, a more complex approach by Andersen et al. (2013) is chosen to account for the variety of environmental conditions. Here, the probability of action is predicted by indoor and outdoor temperature, CO_2 concentration, solar radiation, and relative humidity. Both models are originally developed and fitted to residential apartments. The last model, the rule-based adaptive model (Penna et al., 2016), allows for user dynamic interactions with windows and shadings triggered by thresholds for the indoor temperature and the beam solar radiation respectively.

2. METHOD

2.1 Case study

As a case study, a simplified semi-detached apartment building in a shoebox-like configuration has been considered. The apartment module has a square floor of 100m^2 and an internal height of 3m, which results in a shape factor of 0.97 m^{-1}. The floor area of the apartment is chosen according to the weighted average surface area for European residential buildings (UN Economic Commission for Europe, 2006). The window to floor ratio equals 14.4 %. The thermal transmittance is set to 1.03 W m^{-2} K^{-1} as a typical value between the first (Italian parliament, 1976) and the second energy legislation (Italian parliament, 1991) to represent the most common Italian building structures. The infiltration rate is set to 0.2 ACH. To quantify the influence of different climatic conditions, Milan and Messina are tested as representative Italian heating and cooling dominated climates. The national test reference years have been used (CTI 2016). In Table 1 the main building characteristics are summarized. More detailed information can be found in Penna et al. (2016). For each climate, the best retrofit strategies have been found through a multi-objective optimization approach considering energy efficiency and global cost as contrasting objectives and a standard occupational profile like in a typical asset rating procedure for energy labeling. For the optimization a genetic algorithm is used which refers to a selection of energy efficiency measures further explained in section 2.2. that are roof, floor and wall insulation, glazing, and boiler replacements, and the implementation of a mechanical ventilation system (MVS) with a heat recovery unit.

Table 1 – Characteristics of the reference building

Characterization		**Values**
Construction Period	1979 - 1991	U_{OPAQUE} =0.49 W m^{-2} K^{-1}
		$U_{GLAZING}$ =3.20 W m^{-2} K^{-1}
Glazing	Window Orientation: South	A_{GLAZED}=14.4 m^2
Climates tested	Milan	HDD_{20}= 2404 K d – Cfa, Köppen Classification
	Messina	HDD_{20}= 707 K d – Csa, Köppen Classification

2.2 Energy Efficiency Measures (EEMs)

The Multi-Objective Optimization systematically tests different combinations of Energy Efficiency Measures to define the optimal retrofit strategy for the building. The considered Energy Efficiency Measures are:

a) External wall insulation with incremental thicknesses increases of 1cm, from 1 to 20 cm
b) Roof external insulation with incremental thicknesses increases of 1cm, from 1 to 20 cm
c) Improved choice of glazing systems as stated in Table 2 including the replacement of the window frames with improved aluminum ones with thermal break and $U_{FRAME} = 1.2$ W m^{-2} K^{-1}
d) Replacement of old boiler with modulating or condensing boilers with a climatic control system
e) Installation of a mechanical ventilation system with heat recovery

It is to mention that some of these measures result in energy performance increases without causing additional costs, which are further specified in Penna et al. (2016).

Table 2 – Technical specifications of the Energy Efficiency Measures considered in the analysis

Thermal characteristics of external insulation: polystyrene EPS		
Thermal Conductivity λ (W m^{-1} K^{-1})	0.04	
Specific heat c (J Kg^{-1} K^{-1})	1470	
Density ρ (kg m^{-3})	40	
Thermal characteristics of glazing system	**U (W m^{-2} K^{-1})**	**SHGC**
DH – Double, high SHGC (4/9/4, krypton, low-e)	1.140	0.608
DL – Double, low SHGC (6/16/6, krypton, low-e)	1.099	0.352
TH – Triple, high SHGC (6/12/6/12/6 krypton, low-e)	0.613	0.575
TL – Triple, low SHGC (6/14/4/14/6 argon, low-e)	0.602	0.343
Nominal efficiency of the heating system based on LHV		
Standard (STD)	89 %	
Modulating (MD)	96 %	
Condensing (CD)	101 %	
Technical characteristics of the mechanical ventilation system		
Ventilation Rate (m^{3}h^{-1})	150.0	
Fan Power (W)	**59.7**	

2.3 Building Energy Simulation Modelling

The building energy simulation model is set up in TRNSYS (Solar Energy Laboratory, 2017). To perform the dynamic balance of energy, mass of vapor and carbon dioxide in the building the TRNSYS Multi-zone building subroutine Type 56 (Solar Energy Laboratory, 2012b) is implemented and equipped with Type 869 (Haller et al., 2011b, 2011a) as heating system. This basic building model is completed through the addition of an occupant behavior subroutine using excel which is able to represent the presented behavioral models. By coupling excel and TRNSYS the direct effects of occupant's interactions on the mentioned performance indicators can be simulated dynamically. The simulation time step is 10 min to balance the level of detail and computational time.

2.3.1 Standard behavioral Model (SM):

Optimization of retrofit solutions has been carried out applying a standard occupant behavior model. This model – the Standard Model – assumes a standardized static schedule provided by technical standards for window opening/closing and shading operations. The thermostat is set to an interval from 20 to 22°C to turn on and off the heating system. When replacing the boiler, the water supply temperature is assumed to be adjusted by an outdoor temperature reset control. The internal gains through the heating system are modeled half radiative and half convective according to UNI/TS 11300-1:2014 (UNI, 2014). The air change rate reaches 0.5 ACH during occupancy time (17-7 on weekdays; 15-10 on weekends). If mechanical ventilation is applied, a heat exchanger is used to recover exhaust heat. In summer the mechanical ventilation system is used to reduce overheating. Especially if indoor temperatures exceed comfort conditions and outside temperatures are cooler. The airflow rate of the mechanical ventilation system - if in use - is set to 0.5 ACH as recommended by EN 16798-7:2017 (CEN, 2017).

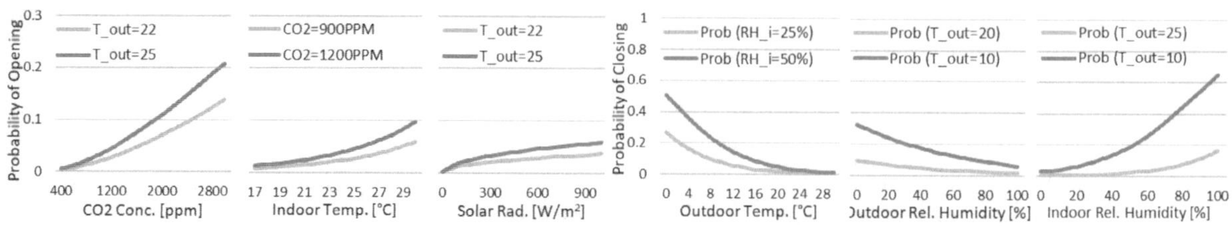

Figure 1 - Opening/Closing Probabilities for Andersen et al. (2013)

2.3.2 Probabilistic Model (PM):

The Probabilistic Model (PM) focuses on a logistic regression predicting interactions with windows. The regression parameters β_k and predictors x_i are taken from two research advancements (Andersen et al., 2013; Nicol, 2001). These models interpolate behavioral data previously obtained by measurements in residential buildings. In the next step, both models have been applied to the presented case study for the entire reference year to recalculate the energy demand, the total cost, and the weighted discomfort time of the optimal solutions obtained with the standard model. The PM is not based on fixed thresholds, but rather on probabilities, which are compared to a random number. Therefore, each decision, even for the same conditions at the same time, is ever changing. In the first examined regression model from Nicol (2001), survey data obtained in three different countries is interpolated using the mean values of all surveys throughout Europe. A best-fit line is obtained, showing the proportion of opened windows for a given outdoor temperature. The outdoor temperature is found as the most relevant predictor. Linking outdoor temperature and probability of opening/closing windows, is not creating a loop of action and considered as more robust and useful (Nicol, 2001). Nicol's work explicitly discusses the application of the model for building simulation and suggests considering other factors to complete the full analysis of stochastic processes involved and their effect on thermal comfort and IAQ. Therefore, a logistic regression analysis developed by Andersen et al. (2013) has been chosen as a second probabilistic model. It uses measurements in 15 residential buildings in Denmark, clusters them into similar groups and provides a set of regression parameters for each group. The model shows a higher level of complexity by defining different regression parameters depending on the time of the day and distinguishing between predictors for opening and closing behavior. For opening probabilities indoor temperature T_i, outdoor temperature T_O and indoor CO_2 concentration and solar radiation are considered. For closing probabilities indoor temperature T_i, outdoor temperature T_O, indoor CO_2 Concentration, indoor relative humidity RH_i and outdoor relative humidity RH_O are considered. Both models depend on the current state of the window but use different environmental conditions as predictors. Probabilities p for occupant actions are calculated as a logistic model (Eq. 1) that depends on a set of predictors x_i, e.g. the mentioned relevant environmental parameters, and regression parameters β_k obtained from the experimental data based on their statistical significance.

$$\text{logit}(p) = log\left(\frac{p}{1-p}\right) = \beta_0 + \sum_{k=1}^{n} \beta_k x_k \quad (1)$$

With the obtained probability a new state for windows and their corresponding opening angles are evaluated for each time interval of 10 min. The outcomes are discrete numbers (1 for open, 0 for closed). For the Andersen model, a continuous opening angle based on the probability p_O is calculated (Eq. 2) whereas in the Nicol model opening angles are assumed to be 90°.

$$\gamma = p_O * 90° \quad (2)$$

The logistic regression models are implemented in EXCEL and outcomes are coupled with TRNSYS Type 62 to vary the ventilation rate used for the dynamic building energy simulation.

2.3.3 Adaptive Model (AM):

The Adaptive Model is based on a previous research paper by Penna et al. (2016). Control actions regarding blind and window opening have been set as common-sense reactions to discomfort conditions. As stated by Nicol and Humphreys (2002) and Mahdavi (2011), it is reasonable to assume that people operate actively on buildings proportionally to their thermal perception of discomfort and that they behave differently in summer and winter conditions. The AM correlates occupant reactions to the ambient conditions using temperature thresholds. These temperature bounds are set to represent the range of acceptable comfort conditions according to comfort category II of the Standard EN 16798-

1:2019 (CEN 2019). According to the standard EN 16798-1 (CEN, 2019) the temperature thresholds are constant in the heating season and set to 20°C and 25°C respectively. In summer, these values vary according to the running mean outside temperature θ_{rm} as shown in equations (3a) and (3b).

$$\theta_{o,limit,upper} = 0.33\ \theta rm + 18.8 + 3 \quad (3a)$$

$$\theta_{o,limit,lower} = 0.33\ \theta rm + 18.8 - 3 \quad (3b)$$

The heating season is from 15th October to 15th April in Milan and from 1st December to 31st March in Messina according to the D.P.R. 74/2013 (Italian Parliament, 2013).

Depending on the thermal sensation of the occupants, actions that reduce thermal discomfort are taken: windows are opened or closed if the indoor temperature is outside or respectively inside of the defined boundaries, e.g. causing too cold or too hot sensations for the occupant. During the summer period occupants are considered as actively operating shadings to adapt to irradiation or increase the ventilation rate by opening a window. During not occupied times the shades are considered as closed with a shading factor of 0.8, else they are only closed if the beam solar radiation exceeds 150 W m^{-2}. For the windows, a similar strategy is set up. During the entire year, they are opened if inside temperatures rise above the upper comfort threshold and outside temperatures are lower. Windows are closed if indoor temperatures fall below the lower comfort threshold. The corresponding ventilation rate has been modelled according to EN 16798-7 (CEN, 2017), which considers wind speed, temperature difference between inside and outside and window opening angle. The opening angle has been set to 90° (fully opened) during summer and 5° (partially opened) during winter. Windows are then closed when the operative temperature decreases below the lower comfort bound.

2.4 Multi-Objective Optimization

To achieve several competing retrofit goals, a multi-objective optimization analysis is performed to optimize the initial reference building. This approach aims to minimize the target functions of energy demand and global cost equally. Optimizing these competing targets results in a trade-off solution, which is called the "Pareto front". Regarding the above-mentioned, the Pareto front will represent solutions with the lowest energy consumption (EP_H) at any given cost (NPV). The optimization is applied for a standard occupational behavior (SM).

To optimize the competing targets, a genetic algorithm (GA) is used where different combinations of EEM's are treated as genes. The most suitable among the population of genes are selected and recombined as the next generation for the following optimization run. This is achieved through an NSGA II - a Non-dominated Sorting Genetic Algorithm (Deb et al., 2002). The initial population is composed of 128 individuals chosen by the Sobol's Method, which refers to a random process selecting compositions uniformly distributed through the range of the problem. According to Saltelli et al. (2004), this provides higher efficiency in gene development towards the optimal ones. The individuals are chosen in fractions of 0.5 tournament selection, 0.8 of arithmetic crossover, and a mutation rate of 0.1. The Optimization is run using the *gamultiobj*-function in Matlab and the Optimization Toolbox (Matlab R2020b, The MathWorks, Natick, MA, USA).

2.5 Recalculation of Performance Indicators

Once the Pareto front for the SM is obtained, the optimal solutions are recalculated assuming the adaptive and probabilistic behavioral model. The indicators should allow characterizing the behavioral model regarding the occupant's well-being, resilience towards energy demand, and running costs. Therefore, the performance indicators chosen are the following:

- Energy Performance for Heating (EP_H) is the primary energy used per heated floor area is represented in the EP_H. Primary energy is needed to maintain the set temperature conditions during the year (ISO, 2017) and considers the boiler consumption of natural gas and electricity needed by pumps or mechanical ventilation systems.
- Net Present Value (NPV) represents the cash flow generated over the building's life span of 30 years. Referring to EU 244/2012 (European Commission, 2012), the following costs are taken into consideration:
  - Initial investment costs (IC) for the construction or renovation of the building are defined according to regional prices and reported in detail in Penna et al. (2016), Table 3.
  - Annual running costs are the sum of the annual energy cost for heating (EC) and the maintenance costs (MC) to ensure the functionality of the building. Energy cost increases are shown in Penna et al. (2016), Table 4.
    - Replacement costs (RC) for spare and wearing parts needed for buildings maintenance.
    - Residual values (RV) for equipment used longer than 30 years according to EN 15459 (UNI, 2018).

- Temperature Weighted Discomfort Time (WDT_t) is measured in degree-hours of discomfort (CEN, 2005). This parameter indicates how long (τ in h) and to what extent (wf in K) the temperature is outside the comfort range. It can be calculated using the Equations (5) and (6).

$$WDT_t = \sum wf * \tau \quad (5)$$

$$\text{With } wf = abs(\Theta_I - \Theta_{I,LowerLimit}) \text{ if } \Theta_I < \Theta_{I,LowerLimit} \text{ or} \quad (6a)$$

$$wf = \Theta_I - \Theta_{I,UpperLimit} \; if \; \Theta_I > \Theta_{I,UpperLimit} \quad (6b)$$

- CO_2 Weighted Discomfort Time (WDT_{CO2}) is measured in PPM-hours of discomfort. The parameter indicates the duration in hours weighted by the concentration above the specified threshold as indicated in formulas (7) and (8). Assuming a normal level of expectation (Category II according to ISO norms) the level of comfort for CO_2 concentration according to EN 16798-1 (CEN, 2019) discomfort occurs if the indoor CO_2 level raises 800ppm above outdoor air. Outdoor air is assumed to have a CO_2 concentration of 300ppm as according to ASHRAE 62.1 p.38 (Hedrick et al., 2013) generally acceptable outdoor CO_2 concentration ranges between 300-500ppm.

$$WDT_{CO2} = \sum wf * \tau \quad (7)$$

$$\text{With } wf = CO_2 - CO_{2,UpperLimit} \; if \; CO_2 > CO_{2,UpperLimit} \quad (8)$$

- Relative Humidity Weighted Discomfort Time (WDT_{RH}) is similarly defined considering discomfort outside of the foreseen comfort band between 25-65% as described in EN 16798:2019 (CEN, 2019).

3. RESULTS AND DISCUSSION

To analyze the impact of different behavioral models on the performance of the optimal solutions previously obtained with the standard model, three different steps have been conducted. First, the cost and energy optimal solutions have been recalculated after applying the presented occupant behavior models. Second, Pareto fronts have been compared as shown in Figure 2. In a last step, the comfort sensation for each occupational model has been studied regarding thermal discomfort and indoor air quality. Probabilistic models aim to grasp behavior through multiple predictors resulting in a probability of action to reproduce what occupants would actually do in order to restore their comfort according to an adaptive approach (Humphreys and Nicol, 1998). In this respect, results obtained are not always consistent with expectations showing an increase in discomfort, especially for naturally ventilated dwellings. This is due to a reduced estimation of window openings and ventilation in probabilistic models with respect to the high nominal ventilation rate that is considered in the standard approach, compliant with the guidelines, but not necessarily with reality in naturally ventilated buildings. This is also the reason for a better energy performance in the case of naturally ventilated buildings, or for a clear performance gap when configurations equipped with MVS are considered. As concerns the latter ones, window openings are adding ventilation to the already well-ventilated building. This results in higher energy demand in winter, but a comfort increase compared to the standard model, in summer. These diverse results are somehow consistent with previous studies on the performance gap by Shi et al. (2019), who reviewed papers on residential and office buildings, showing both positive and negative rebound effects. Turner & Frankel (2008) analysed 121 new LEED certified constructions showing that both performance gap and performance increase are commonly found.

In more detail, the general trend of the performance indicators can be clustered in different groups: Heating or Cooling Dominated Climates (HDC vs CDC), with or without MVS. This leaves us with four categories which generally show different outcomes regarding the role of behavioral models:

- *HDC with MVS* (empty markers in fig. 2, Milan): The Andersen Model does not affect Eph, NPV and WDT_t. Due to more frequent window openings, the Nicol model causes an increase in energy demand of about 35%, but effectively decreases thermal discomfort. The idealized adaptive approach strongly improves all WDT_t with similar NPV and Eph.
- *HDC without MVS* (solid markers in fig. 2, Milan): Both probabilistic approaches reduce energy and costs, because of the reduced ventilation with respect to the standard evaluation. However, only the Nicol model can reduce WDT_t

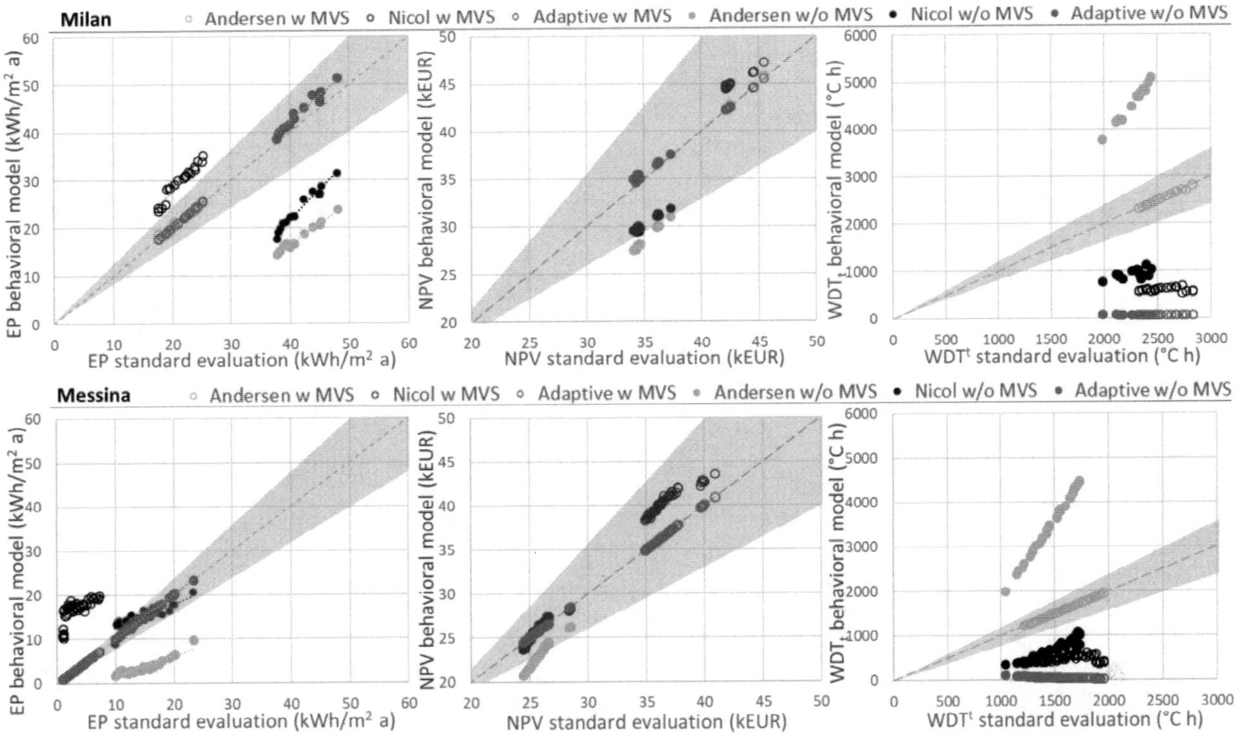

Figure 2 - EP, NPV and WDT_t for each occupational model and the reference climates of Milan and Messina

to only 1/3 of its original value. Unfortunately, IAQ (reported in fig. 3) worsens a lot, since the minimum ventilation rate cannot be maintained in the wintertime, which results in CO_2-Concentrations beyond the defined comfort range. The adaptive model shows an averaged increase in EPh of about 7 % and similar NPV, but is able to almost eliminate WDT_t.

- *CDC with MVS* (empty markers in fig. 2, Messina): In cooling dominated climates the performance indicators are trending similarly as in heating-dominated climates on EPh, NPV, and WDT_t. With the Andersen model, IAQ deteriorates more than with Nicol's due to fewer window openings. Again, the adaptive approach is able to perform as the standard model in terms of EPh and NPV but improves a lot WDT_t.
- *CDC without MVS (*solid markers in fig. 2, Messina): In this case EPh and NPV are less affected by the probabilistic models. In particular Nicol's model is not able to improve systematically the energy and cost performance with respect to the standard approach. As concerns thermal discomfort, trends are similar to the ones in the HDC. Again Andersen's model predictions result in the highest CO_2 discomfort of all scenarios.

Focusing more on the comfort aspects, some more details can be provided as follows:

- *WDT_t:* Assuming a static scheduled ventilation rate, will cause thermal discomfort, given that comfort is not restored actively. The two probabilistic models are considered to adjust temperature through window openings, but neglect the ventilation requisites from the regulations. Nicol's model foresees higher opening frequency and appears quite effective in reducing thermal discomfort, even if not sufficiently in winter. During summer times and milder periods, frequent opening events prevent the building from overheating and WDT_t is more than halved for solutions without MVS and reduced more than 75% for solutions with MVS. It is noted that, even if the model is less complex than Andersen, it performs better regarding all analyzed comfort aspects – especially thermal comfort. Conversely, applying the Anderson behavioral model, excess heat is not removed in summer times resulting in the highest thermal discomfort of all tested models. In Detail, windows are opened sufficiently, but opening angles are too small to ensure enough ventilation, especially if MVS is not present. In winter, thermal comfort is being maintained well but the model cuts on IAQ, resulting in high CO_2-Discomfort.
- *WDT_{CO2}:* Assuming a static scheduled ventilation rate provided by technical standards, will ensure sufficient ventilation, and provides a suitable indoor air quality. Given that adaptive actions can only increase ventilation above this minimum, air quality is similar also for the ideal adaptive model. Contrarily, for probabilistic occupant behav-

Milan						Messina					
Cost-Optimal Solution (MVS=0)						Cost-Optimal Solution (MVS=0)					
Model	Eph [%]	NPV [%]	WDT_T [°C h]	WDT_{CO2} [PPI	WDT_{RH} [h]	Model	Eph [%]	NPV [%]	WDT_T [°C h]	WDT_{CO2} [PPI	WDT_{RH} [h]
Std	0%	0%	1981	0	10	Std	0%	0%	1037	0	21
Nicol	-34%	-13%	772	1222391	20	Nicol	-11%	-3%	351	565313	67
Andersen	-50%	-19%	3772	2970167	36	Andersen	-58%	-15%	2002	3159896	125
Adaptive	7%	3%	72	0	13	Adaptive	0%	0%	118	0	71
Energy-Optimal Solution (MVS=1)						Energy-Optimal Solution (MVS=1)					
Model	Eph [%]	NPV [%]	WDT_T [°C h]	WDT_{CO2} [PPI	WDT_{RH} [h]	Model	Eph [%]	NPV [%]	WDT_T [°C h]	WDT_{CO2} [PPI	WDT_{RH} [h]
Std	0%	0%	2826	0	12	Std	0%	0%	1953	0	9
Nicol	38%	4%	577	0	13	Nicol	1010%	7%	420	0	53
Andersen	7%	1%	2822	0	12	Andersen	0%	0%	1953	0	9
Adaptive	1%	0%	66	0	13	Adaptive	7%	0%	33	0	64

Figure 3 – Comfort analysis (WDT_t [°C h], WDT_{CO2} [PPM h], WDT_{RH} [h]) for each occupational model

-ior without MVS, if probabilities of action are not reaching relevant levels, airchange relies only on infiltration (0.1 ACH). Therefore, in winter both Andersen and Nicol models do not ensure enough ventilation, which results in high CO_2 discomfort. In the Andersen model, probabilities for opening and closing are treated differently as described in section 2.3.2. Especially, complexity is increased using CO_2 as an additional predictor with the initial thought to restore not only thermal comfort but also improve CO_2 discomfort. Contrary to expectations, IAQ could not be improved. In the summertime, the model is quite sensitive to rising CO_2 discomfort. Whereas in wintertime enough ventilation (>0.5 ACH on average) can not be ensured if relying solely on occupants.

- *WDT_{RH}*: When analyzing discomfort regarding the time quantity of relative humidity values outside the foreseen comfort band, dissimilarities between the models are less evident. It can be noted that humidity discomfort is on average halved if MVS is applied given a sensible heat exchange and more efficient humidity removal. Furthermore, humidity comfort has a lower comfort limit as well, which if overcome results in a too dry indoor air.

4. CONCLUSIONS

In this paper, the influence of occupant behavior on the retrofit solutions of a specific reference building is presented using common practice optimization approaches in BPS. In the search for more suitable occupant behavior models to react to the performance gap in building simulation, four different approaches have been chosen and tested: scheduled ventilation (UNI, 2014), probabilistic actions on windows using one predictor (Nicol 2001) or multiple predictors (Andersen et al. 2013) and ideal adaptive actions on windows (Penna et al. 2015). A detailed comparison of the influence of behavioral modeling on predicted energy demand and costs has been conducted and comfort aspects are analyzed. The two probabilistic models try to grasp the complexity of the problem and reproduce what the occupant actually would do to restore comfort conditions. However, when the optimized retrofit configurations without MVS are simulated with the probabilistic behavioral models, energy needs and global costs are generally reduced while thermal discomfort improves only with one of the two models, i.e. Nicol's, because of the unrealistically high ventilation rate assumed by the standard model. Indoor air quality and thermal comfort cannot be maintained for renovation strategies without MVS, meaning that renovated buildings may be prone to overheating and poor indoor air quality if not properly equipped and operated. Indeed the ideal adaptive model shows the feasibility of controlling discomfort while maintaining low energy demand and cost predicted by the standard approach. As a future development, building management strategies based on the ideal adaptive model could be further enhanced and applied as an automated building operation system, providing higher efficiency and more resilience to occupational behavior. Moreover, it seems worth investigating, to implement resilience towards occupants as an optimization criterion to maintain adequate comfort conditions without increasing energy demand even if accounting for the individualism of each occupant.

NOMENCLATURE

U_{OPAQUE}	thermal transmittance opaque surface	$W\ m^{-2}\ K^{-1}$
$U_{GLAZING}$	thermal transmittance opaque surface	$W\ m^{-2}\ K^{-1}$
A_{GLAZED}	area of glazed surface	m^2
HDD_{20}	heating degree days above 20 °C	K d
λ	thermal conductivity	$W\ m^{-1}\ K^{-1}$
c	specific heat	$J\ Kg^{-1}\ K^{-1}$
ρ	density	$kg\ m^{-3}$

SHGC	solar heat gain coefficient	
$\theta_{o,limit,lower}$	temperature lower limit of comfort range	°C
$\theta_{o,limit,upper}$	temperature upper limit of comfort range	°C
θ_{rm}	running mean outside temperature	°C
α	weighing coefficient for calculation of θ_{rm}	
β_k	regression parameters	$[x_i^{-1}]$
x_i	predictors	°C ; % ; W m^{-2}
γ	opening angle window	deg
p_O	probability for opening a window	
p_C	probability for closing a window	
wf	degree difference to comfort range	K
τ	timeperiod in which discomfort occurred	h

REFERENCES

Andersen, R., Fabi, V., Toftum, J., Corgnati, S. P., & Olesen, B. W. (2013). Window opening behaviour modelled from measurements in Danish dwellings. *Building and Environment*, *69*, 101–113. https://doi.org/10.1016/j.buildenv.2013.07.005

Andersen, R. K., Fabi, V., & Corgnati, S. P. (2016). Predicted and actual indoor environmental quality : Verification of occupants ' behaviour models in residential buildings. *Energy & Buildings*, *127*, 105–115. https://doi.org/10.1016/j.enbuild.2016.05.074

CEN. (2019). EN 16798 - Energy performance of buildings - Ventilation for buildings - *Part 1*. Bruxelles, Belgium

CEN. (2017). EN 16798 - Energy performance of buildings - Ventilation for buildings - Part 7: Calculation methods for the determination of air flow rates in buildings including infiltration *(Modules M5-5)*. Bruxelles, Belgium

CEN. (2008). EN 15251:2008 - Indoor environmental input parameters for design and assessment of energy performance of buildings addressing indoor air quality, thermal environment, lighting and acoustics. Bruxelles, Belgium

CEN. (2018). EN 15459-1:2018 - Energy performance of buildings - Economic evaluation procedure for energy systems in buildings - Part 1: Calculation procedures, *Module M1-14*. Bruxelles, Belgium

CEN. (2005). EN 7730:2005 Ergonomics of the thermal environment – Analytical de-termination and interpretation of thermal comfort using calculation of the PMV and PPD indices and local thermal comfort criteria. Bruxelles, Belgium

Comission, E. (2012). COMMISSION DELEGATED REGULATION (EU) No 244/2012 of 16 January 2012 supplementing Directive 2010/31/EU of the European Parliament and of the Council on the energy performance of buildings.

Deb, K., Pratap, A., Agarwal, S., & Meyarivan, T. (2002). A fast and elitist multiobjective genetic algorithm: NSGA-II. *IEEE Transactions on Evolutionary Computation*, *6*(2), 182–197. https://doi.org/10.1109/4235.996017

Fabi, V., Andersen, R. K., & Corgnati, S. (2015). Verification of stochastic behavioural models of occupants' interactions with windows in residential buildings. *Building and Environment*, *94*(P1), 371–383. https://doi.org/10.1016/j.buildenv.2015.08.016

Fabi, V., Andersen, R. V., Corgnati, S. P., & Olesen, B. W. (2013). A methodology for modelling energy-related human behaviour: Application to window opening behaviour in residential buildings. *Building Simulation*, *6*(4), 415–427. https://doi.org/10.1007/s12273-013-0119-6

Fabi, V., Vinther, R., Corgnati, S., & Olesen, B. W. (2012). Occupants ' window opening behaviour : A literature review of factors in fl uencing occupant behaviour and models. 58, 188–198. https://doi.org/10.1016/j.buildenv.2012.07.009

Haldi, F., Calì, D., Andersen, R. K., Wesseling, M., & Müller, D. (2017). Modelling diversity in building occupant behaviour: a novel statistical approach. *Journal of Building Performance Simulation*, *10*(5–6), 527–544. https://doi.org/10.1080/19401493.2016.1269245

Haller, M. Y., Paavilainen, J., Konersmann, L., Haberl, R., Dröscher, A., Frank, E., Bales, C., & Streicher, W. (2011a). A unified model for the simulation of oil, gas and biomass space heating boilers for energy estimating purposes. Part I: Model development. *Journal of Building Performance Simulation*, *4*(1), 1–18. https://doi.org/10.1080/19401491003671944

Haller, M. Y., Paavilainen, J., Konersmann, L., Haberl, R., Dröscher, A., Frank, E., Bales, C., & Streicher, W. (2011b). A unified model for the simulation of oil, gas and biomass space heating boilers for energy estimating purposes. Part II: Parameterization and comparison with measurements. *Journal of Building*

Performance Simulation, *4*(1), 19–36. https://doi.org/10.1080/19401491003653629
Hedrick, R. L., Thomann, W. R., Mcfarland, J. K., Alevantis, L. E., Berlin, G. L., Brunner, G., Buttner, M. P., Chisholm, J. K., Damiano, L. A., Darwich, A. K. H., Fisher, F. J., Green, D. I., Herrmann, D. C., Howard, E. P., Howard, R. L., Lawton, W. M., Mcclendon, J. P., Mcguire, M. E., Rogers, L. J., … Bruning, S. F. (2013). ANSI/ASHRAE Standard 62.1-2013 - Ventilation for Acceptable Indoor Air Quality *(Vol. 2013)*.
Herring, H. (2006). Energy efficiency - A critical view. *Energy*, *31*(1 SPEC. ISS.), 10–20. https://doi.org/10.1016/j.energy.2004.04.055
Humphreys M.A., Nicol J.F. (1998). Understanding the adaptive approach to thermal comfort. *ASHRAE Trans, 104 (1998), pp. 991-1004*
Italian Parliament. (1976). Law 373 Norme per il contenimento del consumo energetico per usi termici negli edifici.
Italian Parliament. (1991). Law 10, Norme per l'attuazione del piano energetico nazionale in materia di uso nazionale dell'energia, di risparmio energetico e di sviluppo delle fonti rinnovabili di energia.
Italian Parliament, (2013). D.P.R. n.74/2013, Regolamento recante definizione dei criteri generali in materia di esercizio, conduzione, controllo, manutenzione e ispezione degli impianti termici per la climatizzazione invernale ad estiva degli edifici e per la preparazione dell'acqua calda per usi igienici sanitari
Mahdavi A., (2011). People in building performance simulation*, in Hensen J.L.M. & Lamberts R., Building performance simulation for design and operation, Spon Press,* New York, USA, p.56-83
MATLAB, R2020b. Natick, Massachusetts: The MathWorks Inc., 2020.
McLeod, R. S., Hopfe, C. J., & Kwan, A. (2013). An investigation into future performance and overheating risks in Passivhaus dwellings. *Building and Environment*, *70*, 189–209. https://doi.org/10.1016/j.buildenv.2013.08.024
Mlakar, J., & Štrancar, J. (2011). Overheating in residential passive house: Solution strategies revealed and confirmed through data analysis and simulations. *Energy and Buildings*, *43*(6), 1443–1451. https://doi.org/10.1016/j.enbuild.2011.02.008
Nicol, J. F. (2001). Characterising Occupant Behaviour in Buildings: Towards a Stochastic Model of Occupant Use of Windows, Lights, Blinds, Heaters and Fans.
Nicol J.F., Humphreys M.A. (2002). Adaptive thermal comfort and sustainable thermal standards for buildings. *Energy and Buildings*, 34: 563-72
O'Brien, W., Tahmasebi, F., Andersen, R. K., Azar, E., Barthelmes, V., Belafi, Z. D., Berger, C., Chen, D., De Simone, M., Simona d'Oca, Hong, T., Jin, Q., Khovalyg, D., Lamberts, R., Novakovic, V., Park, J. Y., Plagmann, M., Rajus, V. S., Vellei, M., … Zhou, J. (2020). An international review of occupant-related aspects of building energy codes and standards. *Building and Environment*, *179*(April). https://doi.org/10.1016/j.buildenv.2020.106906
Page, J., Robinson, D., Morel, N., & Scartezzini, J. L. (2008). A generalised stochastic model for the simulation of occupant presence. *Energy and Buildings*, *40*(2), 83–98. https://doi.org/10.1016/j.enbuild.2007.01.018
Penna, P., Prada, A., Cappelletti, F., & Gasparella, A. (2015). Multi-objectives optimization of Energy Efficiency Measures in existing buildings. *Energy & Buildings*, *95*, 57–69. https://doi.org/10.1016/j.enbuild.2014.11.003
Penna, P., Prada, A., Cappelletti, F., & Gasparella, A. (2016). Influence Of Building Management On Cost Optimality Definition In Residential Buildings Retrofitting.
Roetzel, A., Tsangrassoulis, A., Dietrich, U., & Busching, S. (2010). A review of occupant control on natural ventilation. *14*, 1001–1013. https://doi.org/10.1016/j.rser.2009.11.005
Saltelli A., Tarantola S., Campolongo F., Ratto M. (2004). Sample generation, in: *Sensitivity Analysis in Practice. A Guide to Assessing Scientific Models*, John Wiley & Sons, Chichester, UK, pp. 193–204
SdS srl, soluzionisds.com. (n.d.). CTI comitato termotecnico italiano. *Sito del Comitato Termotecnico Italiano. Retrieved December 17, 2020, from https://www.cti2000.it/*
Shi, X., Si, B., Zhao, J., Tian, Z., Wang, C., & Jin, X. (2019). Magnitude , Causes , and Solutions of the Performance Gap of Buildings : A Review. 1–21. https://doi.org/10.3390/su11030937
Solar Energy Laboratory. (2017). *Trnsys 18*. *3*, 49–50. http://www.trnsys.com/
Sorrell, S., Dimitropoulos, J., & Sommerville, M. (2009). Empirical estimates of the direct rebound effect: A review. *Energy Policy*, *37*(4), 1356–1371. https://doi.org/10.1016/j.enpol.2008.11.026
Tronchin, L., & Fabbri, K. (2010). A Round Robin Test for buildings energy performance in Italy. *Energy and Buildings*, *42*(10), 1862–1877. https://doi.org/10.1016/j.enbuild.2010.05.022
Turner, C., & Frankel, M. (2008). Energy Performance of LEED ® for New Construction Buildings. *New Buildings Institute*, 1–46.
UN Economic Commission for Europe. (2006). *UN Economic Commission for Europe* (Issue 1000, pp. 1–4).
UNI (Italian Organization for Standardization). (2014). UNI/TS 11300-1:2014, Energy performance of buildings – Part 1: Evaluation of energy need for space heating and cooling.

Design of Net Zero Energy Environmental Engineering Building In Peru

William HUTZEL[1]*, Pedro FLORES-LARICO[2], Travis HORTON[3], Mark ZIMPFER[4]

[1]Purdue University, Department of Mechanical Engineering Technology,
West Lafayette, IN, United States
765-494-7528, hutzelw@purdue.edu

[2]Universidad Nacional de San Agustin de Arequipa, Mechanical Engineering,
Arequipa, Peru
054-271951, pfloresl@unsa.edu.pe

[3]Purdue University, Lyles School of Civil Engineering,
West Lafayette, IN, United States
765-494-6098, wthorton@purdue.edu

[4]Purdue University, Department of Construction Management Technology
West Lafayette, IN, United States
765-494-7528, hutzelw@purdue.edu

* Corresponding Author

ABSTRACT

This paper describes a Net Zero Energy Building (NZEB) being constructed on the engineering campus of the Universidad Nacional de San Augustín (UNSA) in Arequipa, Peru. The project is one part of a larger collaboration between UNSA and Purdue University called the Arequipa Nexus Institute for Sustainable Food, Energy, Water and the Environment (The Nexus Institute) that started in March of 2018. As UNSA seeks to become a leader in environmental engineering, there is an increasing need for a new building that has dedicated resources for teaching and research on sustainable design. Solar panels to achieve net zero energy were an important late addition to the building design. The Arequipa region of Peru is particularly well-suited for solar energy because it enjoys an abundance of sunshine throughout the year. The mild climate is also helpful because buildings do not require extensive HVAC systems that can consume significant amounts of energy.

1. INTRODUCTION

The Universidad Nacional de San Agustín de Arequipa (UNSA) in Arequipa, Peru and Purdue University have partnered to create a new research, education, and innovation institute to address key challenges for a sustainable future for the peoples of Arequipa. The Arequipa Nexus Institute for Sustainable Food, Energy, Water and the Environment (The Nexus Institute) launched its multiphase collaboration in March, 2018. The Arequipa Nexus Institute leverages the institution-level faculty, student, and infrastructure commitment of UNSA and the combined strengths of research faculty and infrastructure from Purdue University's Colleges of Agriculture, Engineering, Health and Human Science, Liberal Arts, Science, Purdue Polytechnic Institute, and Purdue libraries. A variety of faculty-led research collaborations are developing methodologies to improve air quality, water quality, and soil quality primarily for agricultural applications. Energy is also an important focus.

As UNSA also seeks to become a leader in environmental engineering, a new state-of-art building with modern resources for teaching and research in this area has been planned for some time. In addition to classrooms, the building itself will serve as a living laboratory for hundreds of university students who are interested in sustainable design. The building will be four-stories tall and provide approximately 3,716 m^2 (40,000 ft^2) of space. It's layout and construction are very similar to a new Architecture building that was recently completed on the UNSA campus.

Since energy is also part of the Arequipa Nexus Institute, a new goal for net zero energy was established late in the conceptual design of the Environmental Engineering building. In collaboration with faculty from Purdue University, solar photovoltaic panels were added to the design so that its renewable energy can be an example for future sustainable communities in Peru. The solar panels will be placed on the roof to generate electricity. This paper documents the solar analysis and design to make this Net Zero Energy Building (NZEB) a reality.

NZEB are becoming more common in the United States as a way to decrease environmental impacts and reduce energy costs. However, a NZEB in Arequipa, Peru is a more interesting challenge because solar panels are not as widely deployed in buildings. Overall there is a lack 1) of up to date solar maps for estimating the solar resource, 2) nationally recognized standards for net metering, and 3) standardized specifications for solar photovoltaic installations. Despite these challenges, a detailed design has been developed and work is underway to make this facility a reality.

Figure 1 is an early conceptual drawing when UNSA first decided to add solar panels to the environmental engineering building. This drawing was used to get a preliminary estimate of the number of solar panels that would fit on the available roof area. Solar panels are shown on the low roof to the left and a higher roof in the back corner. Overall this initial layout was not ideal due to the potential for shading the panels at certain times of the day.

Figure 1: Initial concept for UNSA Environmental Engineering building

2. AREQUIPA CLIMATE

One significant advantage for building design in Arequipa is that the region enjoys mild weather and abundant sunshine year-round. Figure 2 shows two views of the Edificio de Ingenería de Materiales (Materials Science Building), a typical school building at UNSA that takes full advantage of the mild climate. The building has three floors and an "open" design, with classrooms and offices directly connected to the outdoors. The windows and doors to the building are operable and are frequently left open to allow fresh air to enter the building. The image to the left is the front view of the building facing north. The image to the right is the rear of the building facing south. The new Environmental Engineering building is being constructed in the open lot in the picture to the right.

Figure 2: Sensor locations for environmental monitoring in the Material Science building at UNSA

To begin quantifying the weather-related impacts on building operations in Arequipa more accurately, data loggers were installed in the Material Science Building at UNSA. This data will be used to calibrate and improve the results of future energy modeling work. The data loggers were installed at the four locations circled in red in Figure 2. Two small data loggers for temperature and humidity were installed outdoors on the shaded South side of the building. Two more data loggers for temperature and humidity were installed indoors, on the North side of the building, and directly facing the sun. The only difference between the two sets of sensors is that one set is rated for indoor use and the other set is rated for outdoor use. All the sensors measure and record data at 15 minute intervals and store more than one year of data. The sensor data is manually downloaded to a smartphone app when it is placed near the sensor and transmitted to a computer in a spreadsheet format for further analysis.

Figure 3 is a brief summary of the data collected over an eight month period in 2018-2019, highlighting the mild climate of Arequipa. The blue bar is the monthly average temperature, which is close to 20 °C every month. The fluctuation between day and night temperatures is typically less than 3 °C. The orange bar is the average daily relative humidity. During the "dry" season from May to October the relative humidity is on the order of 30%, but in the "wet" season from November to April the relative humidity is closer to 50% on average. Overall these results reinforce the idea that the mild climate in Arequipa has allowed "open" building designs, where the indoor and outdoor climate are close to the same.

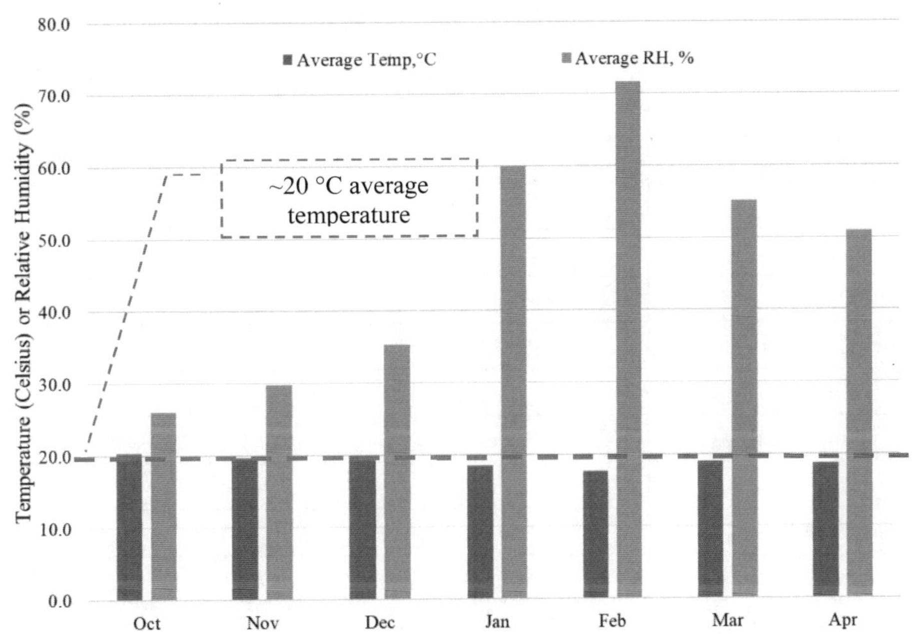

Figure 3: Daily temperature and relative humidity trends in Arequipa

3. BUILDING ENERGY ANALYSIS

Estimating the overall energy use in the Environmental Engineering at UNSA is a straightforward computation based on knowing 1) energy use of building systems and 2) the amount of time they are used. The building is all electric. Due to the mild climate and typical construction used in Peru, the building does not have a central HVAC system. The rooms in the building open directly to an outside corridor and ventilation is from operable windows. The building follows an operating schedule similar to many other universities, with two full semesters per academic year plus a limited summer calendar. Campus buildings are busy during the week-days with limited operation on weekends.

3.1 Building Energy Use

Table 1 shows the peak electric loads provided by the electrical engineers in Arequipa responsible for the building's design. The table is broken down into standard room categories like classrooms, which have similar electrical requirements. The electric loads in Table 1 are further broken down into general lighting and plug loads. For example, the building has 10 laboratories with a total of 5 kW of lighting and 3 kW of plug loads. All electric loads are added at the bottom of Table 1 to show 27.1 kW of peak electric use.

Table 1: Electric loads in Environmental Engineering building

Category	Peak Demand (kW)	
	General Lighting	Plug Loads
Laboratory (10)	5.0	3.0
Classroom (5)	4.5	0.3
Workshop (2)	1.8	0.2
Computer Lab	0.7	2.3
Library	1.1	0.2
Multipurpose Room	0.5	0.1
Office	2.1	2.3
Recirculating Pump	-	3.0
Elevator	-	-
Subtotal	15.7	11.4
Total	27.1	

There are several reasons why the electric loads in Table 1 are significantly less than what would be expected for a comparably sized building in the United States:

- Several large electricity consumers were consciously omitted from the list to avoid skewing the results. As an example, the "elevator" listed in Table 1 uses up to 8 kW power but is not frequently used and was left out.
- Abundant use of daylighting keeps the electricity for lighting to a minimum. The building is designed around a central open courtyard that receives sunshine every day and that is where people congregate.
- The overall building design is relatively simple, without things like vending machines, so plug loads are kept to a minimum.

3.2 Comparison to U.S. Buildings

Equation (1) shows how the daily energy use in kWh was estimated for the Environmental Engineering building. *Peak Demand* is 27.1 kW as shown in Table 1. The *Daily Hours of Operation* is 12 based on UNSA's classroom scheduling from 8 AM to 8 PM daily. A *Diversity Factor* of 0.6 is applied to account for times during the day that electrical equipment is not being used. The three terms multiplied together predict that the building will use approximately 195 kWh on a typical day.

$$\text{Daily Energy Use (kWh)} = \text{Peak Demand x Daily Hours of Operation x Diversity Factor} \quad (1)$$
$$195\text{ kWh} = 27.1\text{ kW} \times 12\text{ hours/day} \times 0.6$$

Equation (2) illustrates the assumptions made for estimating the annual energy use of the Environmental Engineering building. *Daily Energy Use* is 195 kWh from Equation (1). It was assumed that the building is used for 50 weeks in a year and 6 days a week, a total of 300 days. The three terms multiplied together predict that the Environmental Engineering building will use approximately 58,500 kWh annually.

$$\text{Annual Energy Use (kWh)} = \text{Daily Energy Use x Weeks in a Year x Days in a Week} \quad (2)$$
$$58{,}500\text{ kWh} = 195\text{ kWh} \times 50\text{ weeks/year} \times 6\text{ days/week}$$

Energy Use Intensity (EUI) is a common metric used for evaluating and comparing energy use for commercial buildings. The EUI for the Environmental Engineering buildings was computed from equation (3). EUI is the *Annual Energy Use* from equation (2) divided by the *Gross Floor Area* of the building. Dividing the Annual Energy Use (58,500 kWh) by the gross floor area (3,716 m^2) yields an EUI of 16.

$$\text{Site EUI} = \frac{\text{Annual Energy Use (kWh)}}{\text{Gross Floor Area (m}^2\text{)}} \quad (3)$$
$$16 = \frac{58{,}500\text{ kWh}}{3{,}716\text{ m}^2}$$

Table 2 illustrates the significant differences in EUI for a typical higher education building in the U.S. as compared to the new Environmental Engineering building in Peru. The U.S. value of 265 is a median value for an adult higher education building as reported by the U.S. Environmental Protection Agency's Portfolio Manager program. This value was converted from its English value of 84.3 kBtu/ft^2 to its equivalent SI value of 265 kWh/m^2. The more than ten-fold difference highlights significant differences in climatic conditions and building operations. The larger EUI in the U.S. is not "better", but is indicative of significantly higher energy use in all areas of a building in the U.S.

Table 2: Comparison of Energy Use Intensity (EUI) for Higher Education

Country	**Site EUI for Higher Education Building** (kWh/m^2)
United States	265
Peru	16

4. SOLAR ENERGY ANALYSIS

The addition of solar panels to achieve net zero energy came about fairly late in the design of the Environmental Engineering building. One of the key constraints was the amount of space on the roof for locating the panels. The design fills all of the available roof space and will come close to achieving net zero energy

4.1 Solar Array Sizing

Equation (4) is the fundamental computation for achieving net zero with a solar photovoltaic array. The array size in kW is computed by dividing the daily energy use in kWh by the sun hours at a given location along with a correction to account for energy losses in the system. Since both energy use and sun hours change by season, this computation was made in different months to get an average that accounts for the variation. A free web-based tool from the U.S. National Renewable Energy Laboratory (NREL), was used to make this computation over an entire calendar year. It was a little surprising that NREL had solar mapping information for this relatively remote region of Peru.

A sample of the calculations for computing the *Array Size* in kW for an average day is presented here. The *Daily Energy Use* of 195 kWh from equation (2) is divided by 6 *sun hours* and a *correction factor* of 0.8 to get an array size of 41 kW. The initial estimate of 41 kW is done for different months so that an overall average can be computed.

$$\text{Array Size (kW)} = \frac{\text{Daily Energy Use (kWh)}}{\text{sun hours} \times \text{correction}} \tag{4}$$

$$41\ \text{kW} \quad = \quad \frac{195\ \text{kWh}}{6\ \text{h} \times 0.8}$$

Equation (5) estimates the number of solar panels needed. The *Array Size* of 195 kWh from equation (4) is divided by the *Rated Power of One Panel.* The calculation shows approximately 128 solar panels, each rated at 320 Watts, are needed to achieve a net-zero-energy building. The roof will accommodate on the order of 120 panels, so the building has a good chance of providing most of its own power from the sun.

$$\#\ \text{of panels} = \frac{\text{Array Size (kW)}}{\text{Rated Power Of One Panel}} \tag{5}$$

$$128\ \text{panels} = \frac{41\ \text{kW}}{0.320\ \text{kW}}$$

The orientation and mounting angle of the solar panels is determined by geographic location. Arequipa, Peru is located in the southern hemisphere at a latitude of 16 °S. To optimize performance, a rule of thumb is to mount solar panels at an angle roughly equal to their latitude. Therefore, the panels in the Environmental Engineering building should be mounted facing north at an angle close to 16 ° with respect to horizontal.

4.2 Grid Interaction

The initial design of the photovoltaic array included batteries to store electricity on sunny days when the energy production of the array exceeded energy use in the building. This would have allowed the Environmental Engineering building to operate on its own power even at night, adding resilience to the building operation. Upon further investigation, the battery installation for achieving grid independence was too expensive and was dropped from the final design.

A less expensive grid interactive design was pursued using rules for distributed generation that are regulated by the Ministry of Energy and Mines in Peru. The Peru Distributed Generation Code states that to promote distributed generation 1) excess power can be delivered to the grid and 2) there is a maximum amount of power allowable for each renewable technology. The methodology for using solar panels is the same as what is used in the U.S.

DC power from the solar panels is routed to an inverter for conversion to AC. The AC power is first used within the Environmental Engineering building. Excess power is then used by nearby buildings on the UNSA campus. The amount of excess energy is small enough that it will always be used within the confines of the UNSA electrical infrastructure. The amount of solar electricity is relatively small compared to the overall campus power consumption.

The rules for net metering are not fully defined in Peru. In other words, it has not been clearly established whether local electric utilities provide reimbursement for the solar electricity provided to the grid. Without these rules in place it is more cost effective for UNSA to use the electricity within its local network.

4.3 Operation & Monitoring

Purdue provided assistance to UNSA engineers on safety, operations, and long-term monitoring of the photovoltaic array in the form of a sample Request for Proposal (RFP). The RFP was adopted from a template published by the National Renewable Energy Laboratory in the U.S. to promote best practices for solar array operation (Kiatreungwattana, 2018). This document provides guidance on applicable codes, design requirements, equipment specifications, commissioning, and long term monitoring. The RFP was provided in English with key parts translated to Spanish.

Possibly the most important piece of information in the RFP was detailed provisions for web-based monitoring of the photovoltaic array. The monitoring parameters included DC energy, AC energy, solar irradiance, energy use by the building, and equipment status. The data should be available in real-time and archived in 5-minute averages. The purpose of the monitoring is two-fold. First, the monitoring is important for students and faculty who will use the new Environmental Engineering building as a living laboratory. Secondly, the monitoring is crucial to ensure that the solar array is performing as intended. Arequipa is an arid climate that does not receive much rain and dirt tends to accumulate on solar panels. The web-based monitoring platform will detect the resulting performance degradation, indicating that the solar panels need to be cleaned.

5. ONGOING WORK

Figure 4 is the most recent architectural renderings showing the final plans for the Environmental Engineering building at UNSA. Although substantial changes were made to the building layout, the overall roof space available for solar photovoltaic panels is about the same. Figure 4 shows that the panels are located at both upper and lower levels of the building. Even with the re-design, the overall building should be very close to achieving net zero energy based on the original estimates of energy use discussed in this paper. Construction started in October of 2020 and the building should be fully operational in less than two years.

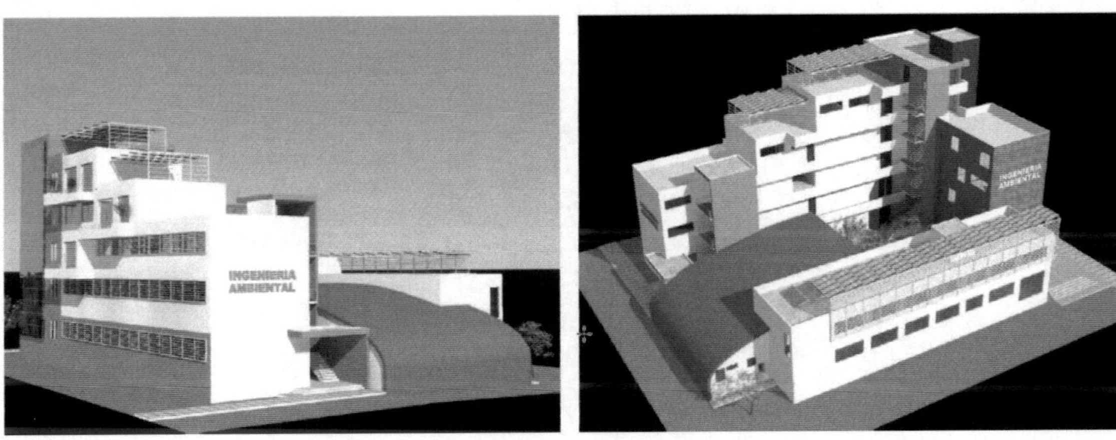

Figure 4: Final design of UNSA Environmental Engineering building

There are a number of opportunities for using the new Environmental Engineering building at UNSA as a platform for future teaching and research:

BIM Modeling - Figure 5 is a Building Information Model (BIM) for the exterior of the Environmental Engineering building using two dimensional CAD drawings provided by UNSA. Even though the layout of the building changed prior to construction, this model is still an example of the software tools that can be used for teaching design to undergraduate students in UNSA's engineering and architecture programs. This BIM model will be modified to include final details of the overall layout, mechanical systems, electrical systems, and other equipment that can be used for record keeping and scheduling maintenance.

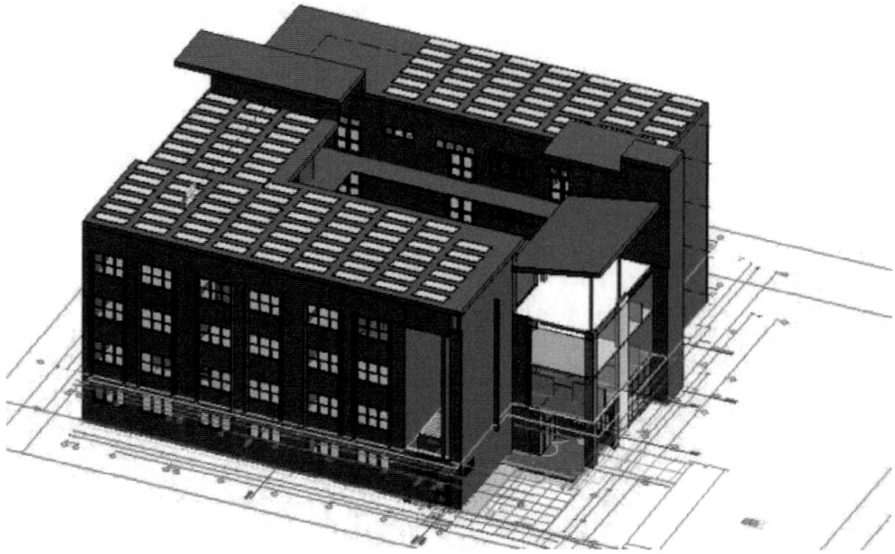

Figure 5: BIM model of UNSA Environmental Engineering building

Energy Modeling - Figure 6 is an early layout for the Environmental Engineering building that was the basis for additional energy modeling. This model was used to estimate energy use and energy generation from solar photovoltaics. Energy models are increasingly used for life cycle evaluations of carbon emissions to provide a more holistic view of overall sustainability. As stated earlier, this model will be updated according to the final design and used for teaching undergraduate students in engineering and architecture.

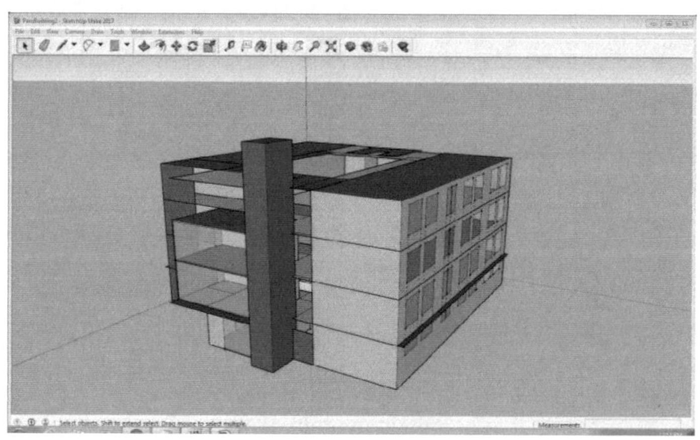

Figure 6: Basic 3-D model used for energy modeling

6. CONCLUSIONS

An international university partnership between the Universidad Nacional de San Agustín (UNSA) in Arequipa, Peru and Purdue in the U.S. is focusing on several dimensions of sustainability. One outcome of this collaboration is a net zero energy Environmental Engineering building for the UNSA campus. Solar panels on the roof of the building will provide all of its energy on an annual basis. Solar photovoltaics is a good renewable energy solution for Arequipa due to the abundant solar resource and the relatively low energy used by the building. Long term monitoring of this installation is needed for teaching, research, and to ensure that the solar panels are achieving their full potential.

NOMENCLATURE

AC	alternating current	(Watts)
DC	direct current	(Watts)
EUI	energy use intensity	(kWh/m^2)

REFERENCES

Kiatreungwattana, Kosol. (2018). RFP Template for Grid-Tied Solar PV Systems for State/City. Golden, CO: National Renewable Energy Laboratory. NREL/TP-7A40-71868.

US Energy Use Intensity by Property Type. (2018, August). Retrieved April 7, 2020, from https://portfoliomanager.energystar.gov/pdf/reference/US National Median Table.pdf

Peru Distributed Generation Code. (2018) April 7, 2020 from http://www3.vivienda.gob.pe/dnc/archivos/Estudios_Normalizacion/Normalizacion/normas/EM.80_INSTALACIONES.pdf

PVWatts. (2017, August). Retrieved April 7, 2020, from https://pvwatts.nrel.gov/pvwatts.php

ACKNOWLEDGEMENT

The authors acknowledge the Facilities Engineering department at UNSA and the architect for the overall building design, including the drawings of the Environmental Engineering building used in this paper.

A Data-driven Model Development for Generalized Building Energy Predictions

Tao Cao[1]*, Lei Gao[1], Vikrant Aute[1], Yunho Hwang[1]

[1] 4164 Glenn Martin Hall Bldg., Center for Environmental Energy Engineering
Department of Mechanical Engineering, University of Maryland,
College Park, MD., 20742, United States

*: Corresponding author: Tel: (301) 405-8672, email: taocao@umd.edu

ABSTRACT

Building energy predictions are in critical need in many fields. The conventional physic model-based approach (via EnergyPlus or similar tools) does decent work to predict energy consumptions. However, it is limited to single predefined building analysis and requires an extensive amount of time and labor to build models. Nevertheless, decision-makers usually need to quantify the energy savings of large building clusters within a short time. The thriving of big data and machine learning techniques enables predicting energy consumptions accurately for different applications within reasonable time frames. This study aims at developing data-driven models for generalized building energy predictions. The models can be used for establishing counter-factual baselines to validate the efficacy of energy-saving measures and energy production and usage planning. The former is usually for medium to long-term durations, while the latter is for short-term durations. We used real-world open data sets from ASHRAE, which covers energy consumptions of about 1,500 buildings for two years. We then preprocessed the data following the industry's standard practice. Multiple approaches of missing values imputations, outlier detections, and feature engineering were explored, based on which the best methods are suggested for building energy predictions. Gradient boosting (GB) based model has been developed for medium to long-term predictions, while the long short-term memory (LSTM) based model has been developed for short term predictions. Hyperparameter tuning was performed on model structures and parameters. We used root mean squared error (RMSE) between the predicted and true energy consumptions to evaluate performances. The results show that the GB based model achieves RMSE of 0.49 for electricity, 1.10 for chilled water, 1.25 for steam, and 1.32 for hot water. The LSTM model performs better with shorter prediction days and longer input days. However, further increasing input days beyond a week does not increase the performance. The LSTM model has about 38% lower prediction errors than the baselines, which are averages of energy consumptions from similar historical days. The study demonstrates the development process of data-driven models for general purpose building energy predictions, from data preparations, model selections, development, and evaluations.

Keywords: Building Energy Prediction, Gradient Boosting, Long Short Term Memory, Root Mean Squared Error

1. INTRODUCTION

Energy consumption of buildings, including commercial and residential sectors, keeps around 38 Quads annually, which takes nearly 40% of the total energy consumption in the past two decades (*LLNL Flow Charts*, n.d.). The energy-saving is urgent, and a large portion of consumption would dramatically improve the total energy efficiency. Therefore, high energy-efficient buildings are critical to the efficiency and sustainability of society. High energy-efficient buildings rely heavily on optimum design and control of energy conversion systems, which are both rooted in predicted energy consumption. Therefore, it is essential to build an accurate energy prediction model to capture consumption changes.

The models for building energy prediction can be categorized into physical models, data-driven models, and hybrid models that combine the previous two models. Physical models are established based on modularized building sectors and heat and mass transfer mechanisms. Nevertheless, these models become incredibly complex when considering the complicated mechanism and the coupling characteristics of every single module in buildings. Only by simulating or simplifying every subsystem can the whole system performance be evaluated. In this genre, many commercial software packages have been developed to assess the energy consumption of buildings, such as DOE-2, EnergyPlus, and BLAST. Although these tools are well-developed, they require detailed building information and environmental parameters before being fed into the simulation system. Consequently, predicting the energy consumption of any building would require an excessive amount of time, labor resources, and knowledge from experienced experts. These requirements make a comprehensive evaluation of building energy consumption cumbersome. On the contrary to the physical models, the data-driven models are sometimes deemed as black-box. Therefore, they have no specific physical meaning, such as machine learning (ML) approaches, which only implicitly extract features from data. These methods have advantages of robustness, flexibility, and rapidity when applying to prediction tasks compared with physical models. Concisely, the building energy prediction task is a type of supervised regression in the viewpoint of the ML field. 'Supervised' means the data used for constructing a prediction model has labels, and 'regression' means the prediction values are continuous. In this genre, any algorithms that are used to deal with supervised regression in the machine learning area have the potential to be shifted to building energy prediction tasks.

With the development of ML and data science, opportunities to leverage data for building energy prediction are manifold. Among all the paradigms of ML, neural networks (NN), support vector regression, decision tree, and Gaussian process regression are the most represented ones. These techniques and algorithms have all been intensively applied in this prediction task since 2000 (Zhong et al., 2019). Moreover, recent years have witnessed a similar migration of research interest as computer science marched in terms of ML algorithms. The trend has been alternated since the enlarged power of GPU makes the rapid auto gradient of a broader and deeper NN possible in the big data era. Taking enormous advantages of that, NN based ML algorithm deep learning (DL) has been reported to continuously outperform humans in many areas. DL's capability of dealing with the high-dimensional and large-scale data structure facilitated researchers probing into energy prediction tasks by this algorithm. Consequently, many investigations drew the support of various DL based algorithms for building energy prediction (Park et al., 2016). Among all the DL algorithms, long short-term memory (LSTM) showed its capability in building energy prediction tasks due to temporal characteristics along with the NN layers (Xu, 2019).

That being said, most studies are limited to developing models based on available data and particular buildings. There have been few investigations of generalized building energy prediction using different models based on various real-world settings. For instance, quantifying energy-saving measures' efficacy requires accurate and reliable predictions of baseline energy consumptions over medium to long time durations. The energy production and usage planning, however, focuses more on accurate predictions for short time durations. One needs to develop and optimize data-driven models for different applications, given the various available algorithms. For example, The LSTM may work well with short-term predictions, while the gradient boosting (GB) framework may work better with long-term predictions.

Our objective is to develop generalized building energy prediction methods for two applications mentioned above: short-term and medium- to long-term predictions. Our models were built upon the ASHARE dataset (*ASHRAE - Great Energy Predictor III*, n.d.). The paper is organized as follows. In section 2, we present the data preparations. In particular, we investigated multiple methods for missing value imputation, outlier detection, and feature engineering for building energy predictions. In section 3, we present the two models, the GB based model for medium- to long-term predictions and LSTM based model for short-term predictions. In section 4, we present each model's results, with a focus on their performance using root mean squared error (RMSE).

2. Data Preparation

2.1 Methodology

We followed the cross industry standard process for data mining (CRISP-DM) ("Cross-Industry Standard Process for Data Mining," 2019). CRISP-DM (*CRISP-DM Help Overview*, 2014) defines the process of data mining into six major phases: business understanding, data understanding, data preparation, modeling, evaluation, and deployment. This section presents our method of exploring the raw data and preparing them for model development.

2.2 Data Description

We used the open datasets released by ASHRAE (*ASHRAE - Great Energy Predictor III*, n.d.). The dataset covers about 15,00 buildings' energy consumptions over two years. Four types of energy consumption are covered: electricity, chilled water, steam, and hot water. Not all types of energy consumption are available to all buildings. All energy consumptions are metered real-world data. The datasets are organized into three sets. The first set provides building id (a unique integer number linked to each building), meter type (0 for electricity, 1 for chilled water, 2 for steam, 3 for hot water), time step, and meter readings. The second set contains building information, including site id (a unique number for a geographic or climate zone), building's primary usage (such as education and office), building's gross floor area, the year the building was opened, and the number of floors of the building. The last set provides weather data. For each site id, the set contains air temperature, could coverage, dew temperature, precipitation depth, sea level pressure, wind direction, and wind speed at each time step. The dataset is the most comprehensive real-world metered data on building energy consumptions we have seen so far. While the site location and building locations are not provided (possibly due to privacy concerns), it does not affect the model development. The total amount of data is about 4 gigabytes.

2.3 Data Processing

We first combined three datasets into one by joining the dataset according to building identifier and climate zone identifier. The aggregated dataset has more than ten features and 80 million entries. Next, we checked the missing values. Table 1 shows the missing counts and percentages for the ten features. Based on the feature types and missing percentages, two approaches are adopted. For those features with high missing percentages (>50%) or without significant impacts on the building energy consumptions (e.g., cloud coverage and precipitation depth), we dropped the entire feature column. For those features with low missing percentages, we used either median or linear interpolations to fill in the missing entries. The interpolations methods are suitable as all features are weather-related. They are either constant over long periods (e.g., sea level pressure) or changing linearly (e.g., temperatures).

Table 1: Missing counts and percentages of all features in the raw dataset

Feature	Missing Total	Missing Percentage (-)
Floor count	16,709,167	0.83
Year built	12,127,645	0.60
Cloud coverage	8,825,365	0.44
Precip. depth (1 hr)	3,749,023	0.19
Wind direction	1,449,048	0.07
Sea level pressure	1,231,669	0.06
Wind speed	143,676	0.01
Dew temperature	100,140	0.005
Air temperature	96,658	0.005
Meter reading	0	0

We then investigated outliers by investigating energy consumptions. Based on preliminary exploratory data analysis, there are two categories of outliers: extremely high energy readings and extensive long periods of zeros. For the first category, we found out that one building has several orders of magnitudes metered readings higher than all other buildings. As the building's features and energy consumptions are not representatives, it is removed from the dataset. For the zeros, we filtered out about 10% of them by determining whether they are true or false zeros. False zeros were when meters were not working correctly, or the facility was shut down for extensive periods (such as summer break for educational facilities). The false zeros were determined by comparing them against similar buildings at the same time.

Lastly, we performed feature engineering. Log transformation was applied to scale down metered energy readings. Further, new time features were added. We added months of the year, days of the week, hours of the day, and whether it is a holiday or not. Aside from this, we encoded cyclic features of hours, using a sine and cosine transformation, as shown in Eq. (1). Such feature transformations are widely used in time series regression problems.

$$
\begin{aligned}
x_{sin} &= \sin\frac{2\pi x}{x_{max}} \\
x_{cos} &= \cos\frac{2\pi x}{x_{max}}
\end{aligned} \tag{1}
$$

3. MODELS

3.1 Model Overview

The model development lies in two focuses. First, we built a model for medium to long term predictions. The model is meant to establish baselines to evaluate the efficacy of energy-saving measures. The model is built upon the GB framework, using Light Gradient Boosting Machine (LightGBM) (*Welcome to LightGBM's Documentation! — LightGBM 3.1.1.99 Documentation*, n.d.). Hereafter, we refer to it as the GB model. The second model is meant to be used for energy predictions within short terms. Similarly, as the model is LSTM based, we refer to it as the LSTM model hereafter.

3.2 GB Model

As mentioned before, quantifying the efficacy of energy-saving measures requires accurate baselines. Counter-fractural models are needed to establish such baselines by assuming nothing changed in the building. These models were used for energy predictions of long periods, typically ranging from several weeks to years. The model takes building and weather information as inputs and outputs energy consumptions at corresponding timestamps. The data was first normalized by removing the mean value of each feature, then scale it by dividing non-constant features by their standard deviations. The data was then fed into the GB model. We trained one model for one type of energy consumption. For each model, we used five k-fold cross-validations (*3.1. Cross-Validation: Evaluating Estimator Performance — Scikit-Learn 0.23.2 Documentation*, n.d.) by splitting data into 80% testing set and 20% validation set. K-fold cross-validation was acceptable here as we did not shuffle data. Further, the model did not use metered energy consumptions as inputs. The final prediction results are averages of 5 models from the cross-validations.

Table 2 shows the parameter settings of the GB model. RMSE was used to evaluate the model's performance. It evaluates the overall differences between true and predicted energy consumptions. We set the early stopping criteria as no improvements of RMSE on the validation set in 20 rounds. Manual parameter tunings were performed to improve the performance over default settings.

Table 2: Parameter setting of the GB model

Parameter	Value
Boosting type	gbdt
Objective	regression
Metric	rmse
Learning rate	0.25
Feature fraction	0.9
Bagging fraction	0.8
Bagging frequency	5
Number of boost rounds	2000
Early stopping rounds	20

3.3 LSTM Model

The LSTM model aims for short-term energy predictions, contrary to GB models. It is meant to be used for energy production and usage planning within short-term, from hours to days at most. The LSTM network has been widely used for similar time-series predictions ("Long Short-Term Memory," 2020).

Figure 1 shows the overall flow chart of the LSTM model. The LSTM model takes historical energy consumptions and other features as input and output energy consumptions of the immediate future time steps. Hence, historical energy consumptions are critical inputs. For buildings without historical energy consumptions (such as new constructions), we proposed constructing a historical profile based on the most similar buildings from the 1500 building pool. The similarities are measured in Gower distance (*Gower.Dist Function | R Documentation*, n.d.), using building information including footprint, building types, year built, and floor count.

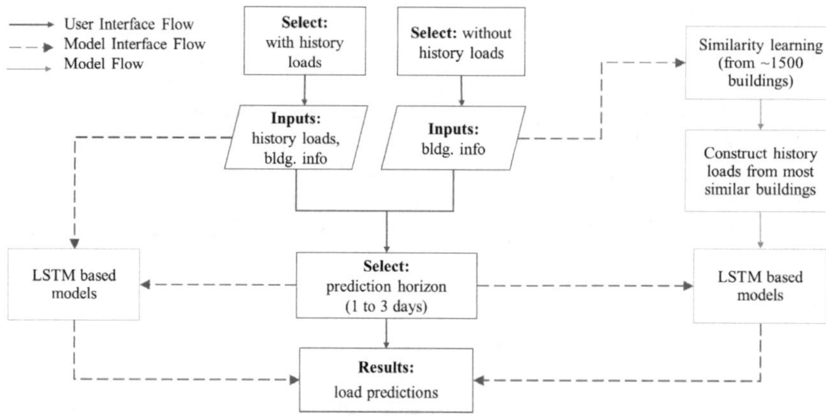

Figure 1: Flow chart of the LSTM model

The LSTM model is unique to each building. For each building, we used the first ten months of data as the training set and the remaining two months of data as the test set. The model accepts various historical hours to define the training set and different numbers of future hours to define the test set. For example, an LSTM model of 24 hours of inputs and 3 hours of outputs indicates the following. The inputs are energy consumptions and all other features (such as weather) in the past 24 hours, while the outputs are energy consumptions in the next three hours.

We conducted the hyperparameter tuning of the LSTM model using Keras Tuner (*Keras Tuner*, n.d.). Table 3 shows the parameter settings. The LSTM model structure is flexible with various layer combinations, optimizer learning rates, and early stopping rounds. The max epochs, batch size, and optimizer is fixed. We used the default loss function of mean squared error (MSE).

Table 3: Parameter setting of the LSTM model tuner

Parameter	Value (Range)
Layer	4~8
LSTM layer units	32 to 128 with a step of 16
Dropout rate	0.2 to 0.5 with a step of 0.1
Max epochs	1000
Batch size	32
Optimizer	Adam
Optimizer learning rate	1e-2, 1e-3, 1e-4
Loss function	mse
Early stopping rounds	20, 40, 60

Note that the Keras Tuner only performs preliminary training and validation. It behaves in such a manner to speed up the parameter tuning process. Therefore, we continued to train the model with the parameters selected by the Keras Tuner. The models were further trained with Keras from Tensorflow (Team, n.d.).

4. RESULTS

4.1 GB Model Results

The GB model was trained on the python 3.7 environments, on a windows machine with eight cores and 16 gigabytes of memory. The training process took about 30 minutes. Figure 2 shows the comparison between the true and predicted energy consumptions of four models for four meter types, with 0 for electricity, 1 for chilled water, 2 for steam, and 3 for hot water. The duration is an entire year. The energy readings are log-transformed. It is observed that the predictions agreed with the true values very well. The predictions have smoother trends than true values, especially when there are abrupt jumps. There could be several real-life operation mechanisms, which are difficult for the model to learn and capture. The power meter might have reading jumps when there were unstable voltage supplies. Also, the building may have occasional significant occupant schedule changes. These might be some of the reasons contributing to deviations when there were abrupt jumps.

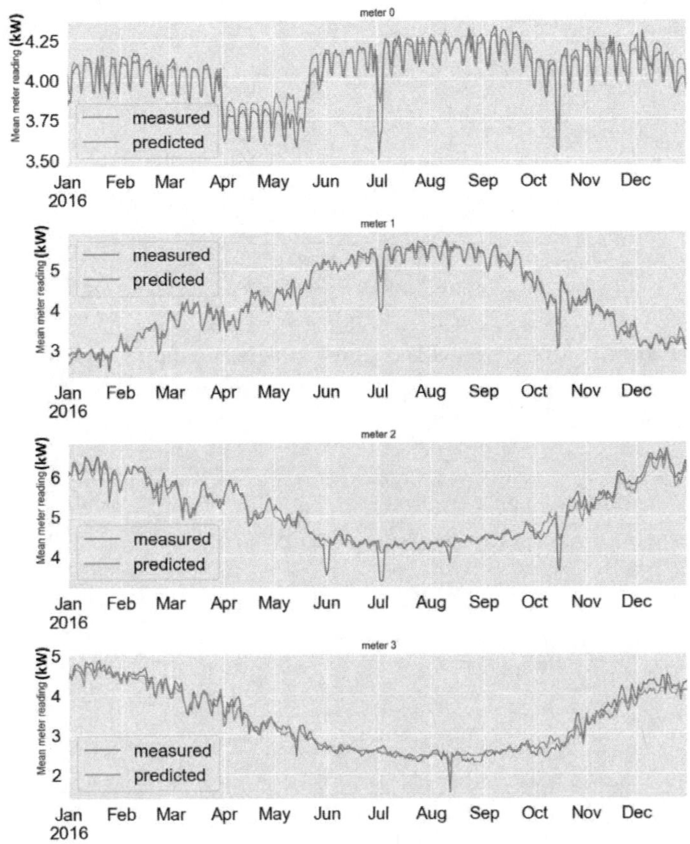

Figure 2: GB model predictions of four energy types

As mentioned in section 3, we used RMSE to evaluate model performance. Based on Figure 2, the models have an RMSE of 0.49 for electricity, 1.10 for chilled water, 1.25 for steam, and 1.32 for hot water. With more than 50% of entries are for electricity, we expect a significantly lower RMSE for electricity predictions.

4.2 LSTM Model Results

For the LSTM model development, we focused on the electricity data of several representative buildings for demonstration purposes. The model is trained on the python 3.7 environments, on a windows machine with eight cores and 16 gigabytes of memory. The training process took about 30 minutes for one building, including about 15 minutes for hyperparameter tuning. We monitored the losses of the training and validation set and determined the early stopping criteria to prevent overfitting. As an example (as shown in Figure 3), we stopped training until the validation set losses were not decreasing for minimal (e.g., 20) epochs. The model which presented minimal loss in the last 20 rounds was adopted. From Figure 3, epoch 87 was selected.

As mentioned in section 3.3, each model has the flexibility of various input and output periods. The results are presented by taking a look at the impacts of both variations. Figure 4 shows the prediction results of various prediction periods (1 to 3 days), with the same input periods (5 days). The red dots show the LSTM model's predictions, while the blue dots represent the baseline's predictions. The baseline is a simple average of similar three historical days of the week. We have observed that the LSTM model not only outperforms the baseline consistently but also agrees with the true values very well. The model's prediction starts to deviate more from the true further down the prediction period. It is expected as the model has less true historical information to rely on further into the future.

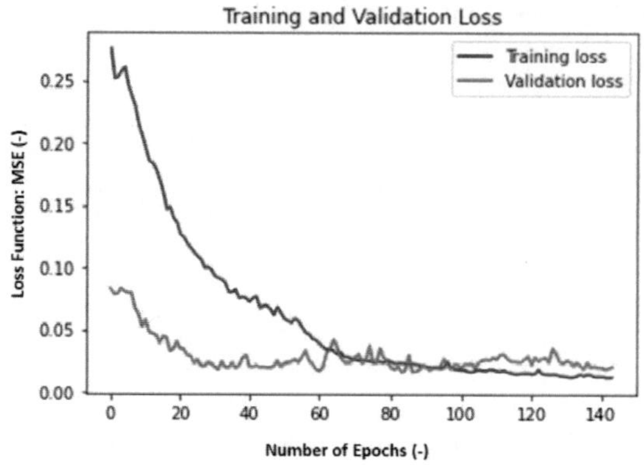

Figure 3: LSTM model loss function versus training epochs

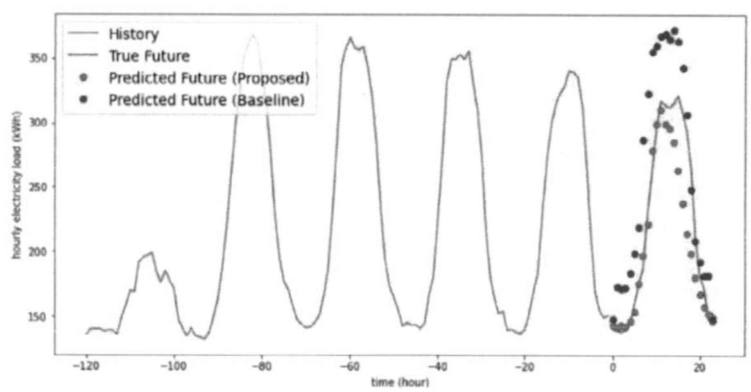

(a) 5 days to 1 day

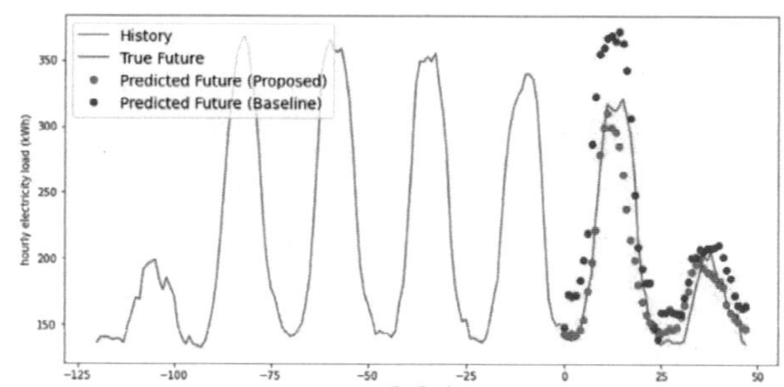

(a) 5 days to 2 days

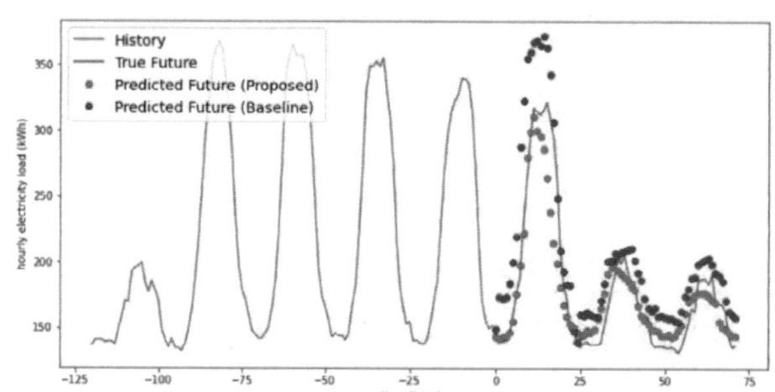

(a) 5 days to 3 days

Figure 4: LSTM model results with various prediction periods

The model achieved RMSE of 30.5, 23.5, and 20.5 for 1 day, 2 days, and 3 days prediction, respectively. The baseline has RMSE of 48.6, 37.7 and 33.2 for similar prediction days. Hence the LSTM model has about 38% lower prediction errors than the baseline. The lower values for the longer day predictions are due to how RMSE is calculated. Deviations at higher absolute values contribute more to higher RMSE.

Figure 5 presents the prediction results of various input periods (3 to 7 days), with the same prediction periods (2 days). The LSTM model performs better with increasing input periods, as observed from better agreements between the red dots and green line. Also, the model generally performs better than the baselines, with similar performances under three input days. The LSTM model achieved RMSE of 33.1, 23.5, 22.6 for 3 days, 5 days, and 7 input days. It is worth noting that increasing input days does not necessarily result in better model performance. We further increased the input days up to 14 days but did not observe increased model performance. The LSTM model performed worse when the input days approached 14 days. LSTM has been known to struggle to work with extensive long sequences.

It is still an open research topic (Trinh et al., 2018). Regarding the LSTM model's structure, we have found 4 to 6 layers are enough for all cases. The LSTM layer's units usually range from 64 to 128. The drop out rate is mostly 0.2.

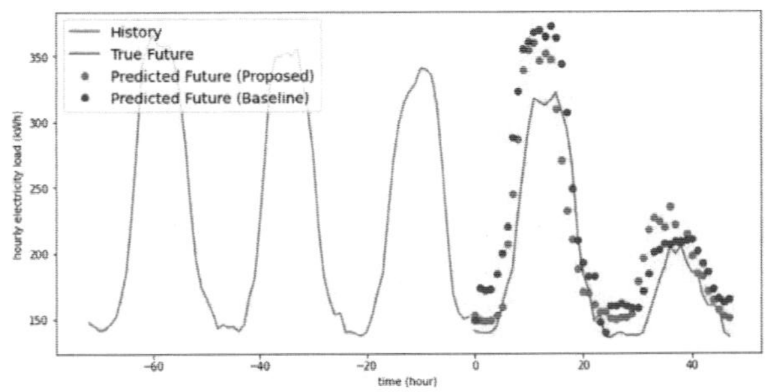

(a) 3 days to 2 days

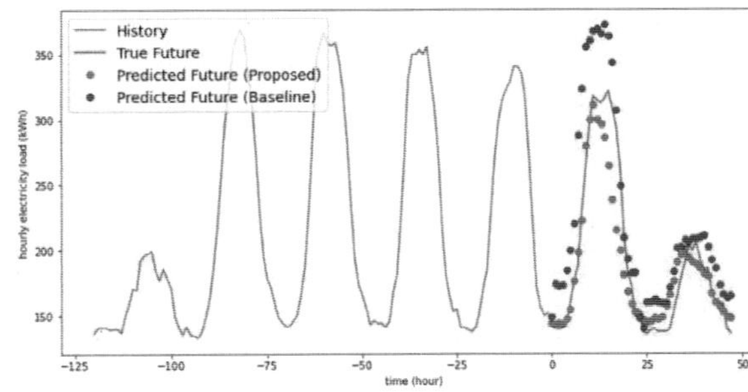

(a) 5 days to 2 days

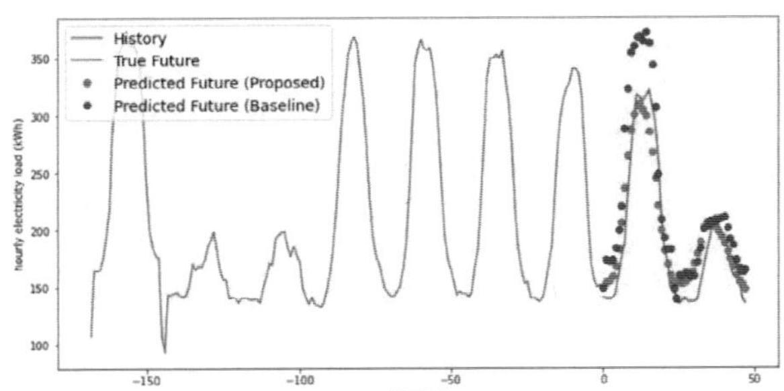

(a) 7 days to 2 days

Figure 5: LSTM model results with various historical periods

4. CONCLUSIONS

We have developed data-driven models for general building energy predictions using metered energy consumptions of about 1,500 buildings for two years. The data covers eight typical climate regions and a dozen types of buildings. The data is preprocessed accordingly to the industry's standard practice. In particular, we performed outlier detections to handle excessive false zeros in energy consumptions. New features were added to handle time series related predictions. We have developed two models for two types of applications. The GB model is suited for medium- to long-term predictions, used to establish counter-factual baselines to evaluate energy-saving measures' efficacy. The LSTM model is meant for short-term predictions of energy production and usage planning. RMSE is used to evaluate models' performances. For the GB model, we performed parameter tuning and used 5 fold cross-validations. The model achieved an RMSE of 0.49 for electricity, 1.10 for chilled water, 1.25 for steam, and 1.32 for hot water predictions (after log-transformed) over an entire year. For the LSTM model, we have developed the model to be flexible for buildings with or without historical metered energy consumptions. The model is capable of generating historical profiles from similar buildings. Further, the model is flexible with various input days and prediction days. We have further tuned the model structure and parameters. Results show that the LSTM model performs better with shorter prediction days and longer input days. Though further increasing input days beyond a week does not increase the performance. The LSTM model has about 38% lower prediction errors than the baselines, which are averages of energy consumptions from similar historical days.

NOMENCLATURE

ASHRAE	American Society of Heating, Refrigerating and Air-Conditioning Engineers
CRISP-DM	cross industry standard process for data mining
DL	deep learning
GB	gradient boosting
LightGBM	Light Gradient Boosting Machine

LSTM	long short term memory
ML	machine learning
MSE	mean squared error
NN	neural network
RMSE	root mean squared error

REFERENCES

3.1. Cross-validation: Evaluating estimator performance—Scikit-learn 0.23.2 documentation. (n.d.). Retrieved December 15, 2020, from https://scikit-learn.org/stable/modules/cross_validation.html

ASHRAE - Great Energy Predictor III. (n.d.). Retrieved March 31, 2020, from https://kaggle.com/c/ashrae-energy-prediction

CRISP-DM Help Overview. (2014, October 24). www.ibm.com/support/knowledgecenter/ss3ra7_15.0.0/com.ibm.spss.crispdm.help/crisp_overview.htm

Cross-industry standard process for data mining. (2019). In *Wikipedia.* https://en.wikipedia.org/w/index.php?title=Cross-industry_standard_process_for_data_mining&oldid=930958276

Gower.dist function | R Documentation. (n.d.). Retrieved December 15, 2020, from https://www.rdocumentation.org/packages/StatMatch/versions/1.2.0/topics/gower.dist

Keras Tuner. (n.d.). Retrieved December 15, 2020, from https://keras-team.github.io/keras-tuner/

LLNL Flow Charts. (n.d.). Retrieved December 16, 2020, from https://flowcharts.llnl.gov/

Long short-term memory. (2020). In *Wikipedia.* https://en.wikipedia.org/w/index.php?title=Long_short-term_memory&oldid=993897583

Park, H. S., Lee, M., Kang, H., Hong, T., & Jeong, J. (2016). Development of a new energy benchmark for improving the operational rating system of office buildings using various data-mining techniques. *Applied Energy*, *173*, 225–237. https://doi.org/10.1016/j.apenergy.2016.04.035

Team, K. (n.d.). *Keras documentation: About Keras.* Retrieved December 15, 2020, from https://keras.io/about/

Trinh, T. H., Dai, A. M., Luong, M.-T., & Le, Q. V. (2018). Learning Longer-term Dependencies in RNNs with Auxiliary Losses. *ArXiv:1803.00144 [Cs, Stat].* http://arxiv.org/abs/1803.00144

Welcome to LightGBM's documentation! —LightGBM 3.1.1.99 documentation. (n.d.). Retrieved December 15, 2020, from https://lightgbm.readthedocs.io/en/latest/index.html

Xu, L. (2019). Probabilistic load forecasting for buildings considering weather forecasting uncertainty and uncertain peak load. *Applied Energy*, 16.

Zhong, H., Wang, J., Jia, H., Mu, Y., & Lv, S. (2019). Vector field-based support vector regression for building energy consumption prediction. *Applied Energy*, *242*, 403–414. https://doi.org/10.1016/j.apenergy.2019.03.078

ACKNOWLEDGEMENT

We gratefully acknowledge the support of the Center for Environmental Energy Engineering (CEEE) at the University of Maryland.

A Whole Building Life-Cycle Assessment Approach to Support Decision-Making for Sustainable Buildings

Hao ZHANG[1], Jie CAI[1]*, James E. BRAUN[2]

[1]School of Aerospace and Mechanical Engineering, University of Oklahoma,
Norman, OK, USA
zhang906@ou.edu, jcai@ou.edu

[2]School of Mechanical Engineering, Purdue University
West Lafayette, IN, USA
jbraun@purdue.edu

* Corresponding Author

ABSTRACT

Buildings account for 39% of the CO_2 emissions in the U.S. Although emerging building energy-efficient technologies lead to improved energy efficiency and lower environmental impact in the operational phase, not much attention has been paid to the environmental impact associated with the materials and manufacturing of the building mechanical systems. This paper presents an integrated life-cycle assessment method for buildings that accounts for the embodied and use phase carbon impacts of the building and its mechanical systems. The proposed methodology relies on EnergyPlus to generate the use-phase energy consumption for any given building. A material 'grabber' routine was developed to automatically extract building envelope material information from EnergyPlus models, which is then used for envelope embodied carbon analysis. For the mechanical equipment, embodied carbon accounting was performed for two representative air-conditioning and heat pump units: a 4-ton packaged unit and a 4-ton split heat pump. The different elements were incorporated in the overall life-cycle assessment tool along with lighting embodied data to allow generation of a whole building environmental performance report. Five Department of Energy commercial building prototypes were used as case studies and analysis results for seven U.S. climate locations are presented in this paper. The results show that for the investigated prototypical buildings, the use phase energy consumption has a dominant impact on the overall building environmental performance: the embodied carbon contribution is less than 9% for all considered cases. However, the tradeoffs could change dramatically as the U.S. moves towards net-zero buildings and the tool presented in this paper could be used to consider these tradeoffs.

1. INTRODUCTION

The building sector accounts for 39% of total CO_2 emissions in U.S. (EIA 2018). The life cycle assessment (LCA) approach has been applied by researchers to investigate the environmental performance of buildings since 1990s (Wang et al., 2011). However, most of the previous work focused on building construction materials, e.g., (Su and Zhang, 2010); (Monteiro and Freire, 2012); (Lemay, 2011). Very limited research can be found that performed LCA studies on a whole building level due to difficulties in integrating the various construction, mechanical and electrical elements within a building. In addition, energy efficient technologies, such as light-emitting diode (LED) lights, solar- or geothermal-assisted heating/cooling equipment, are widely implemented in buildings with the purpose to reduce energy usage and the associated environmental impacts (Bribian et al., 2009). However, the embodied carbon emission of building systems could be increased by using these new technologies. Thus, there is a need to evaluate different energy efficient technologies in terms of environmental impact considering both manufacturing/construction and operation phases.

This paper fills a research gap by presenting a whole building LCA methodology that covers both embodied and use phases and the integration of building envelope and mechanical system design. The method utilizes EnergyPlus (National Renewable Energy Laboratory, 2017), a flagship building energy simulation software developed and maintained by the U.S. Department of Energy (DOE), to generate building energy uses in the operation phase. LCA, e.g., BEES,(National Institute of Standards and Technology, 2016), U.S.LCI, (National Renewable Energy Labratory, 2010), etc., databases are employed to estimate the building environmental impacts associated with the raw material acquisition, building material production and operation stages. Five DOE prototypical commercial buildings were considered as case studies to demonstrate the analysis approach. In-depth analysis was carried out for three types of lighting technologies: incandescent, compact fluorescent (CFL) and LED lights. Detailed embodied carbon analysis was carried out for a representative packaged rooftop air-conditioning unit and a split heat pump. Parametric analysis was performed for the different lighting options, building types and climate locations. As building owners or contractors are increasingly aspiring towards healthier and more environment-friendly products in the building design and retrofit phases, results presented in this paper can be used as a reference to facilitate decision making from an environmental perspective.

2. METHODOLOGY

Figure 1 describes the methodology proposed for whole-building LCA analysis. The overall process starts with a building information model (BIM), which contains detailed building construction information. The BIM is fed to a whole building energy simulation tool to generate building energy uses in the operation phase. OpenStudio (Guglielmetti et al., 2011), developed and maintained by the National Renewable Energy Laboratory, is an upfront graphical user interface for EnergyPlus. OpenStudio takes the BIM for a given building and calls EnergyPlus to execute whole building energy simulations. A "Material Grabber" routine was developed that can take an OpenStudio/EnergyPlus model and automatically generate the bill of materials for a given building. Based on the collected material information, the embodied carbon associated with the construction materials are estimated via queries to LCA databases. The U.S. LCI and ecoinvent databases (Wernet et al., 2016) were used as main LCA data sources for the material life-cycle stages. Building-specific databases, BEES and BIRDS (Lippiatt et al., 2013), were also used when pertinent data was not available in the U.S. LCI or ecoinvent databases. With the simulated operation phase energy uses and estimated embodied carbon, whole building LCA analysis is performed. Detailed LCA analysis for HVAC equipment and lighting systems relied on both laboratory collected data or publicly available literature. As summarized in Table 1, embodied carbon and carbon emissions associated with building operation were estimated in this study. Carbon emissions for the transportation and building construction were not considered as these effects are believed to be small relative to the other impacts.

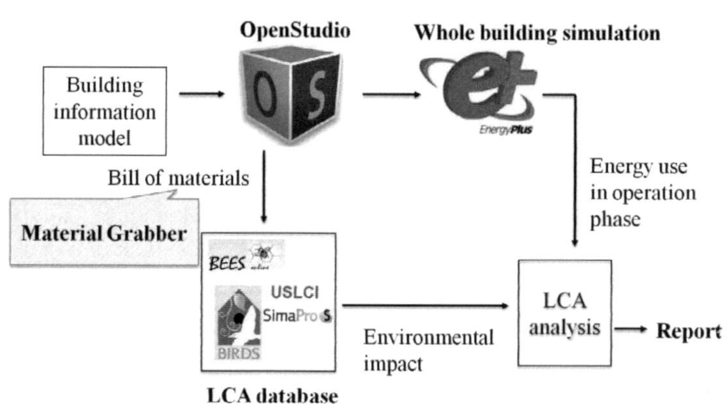

Figure 1. Proposed building LCA methodology.

Table 1. System boundary

Life cycle phases	Main assumptions for calculation
Embodied	Building envelope HVAC system Lighting system
Use phase	Operation (electricity and nature gas consumption)

Five DOE prototypical commercial buildings were used (small office, medium office, sit-down restaurant, standalone retail, and primary school) as case studies to demonstrate the processes in the proposed LCA approach. Detailed descriptions of these prototypical buildings can be found on the DOE website (U.S. Department of Energy, 2016). The LCA case studies considered seven locations with one representative location from each of the seven climate zones in the US: Miami, FL (1A), Houston, TX (2A), San Francisco, CA (3C), Washington, MD (4A), Chicago, IL (5A), Burlington, VT (6A), and Duluth, MN (7A). Parametric analysis covering different climate locations was performed to investigate the impact of climate conditions on the building environmental performance.

3. EMBODIED CARBON ANALYSIS

3.1 Building Envelope Embodied Carbon Analysis

Construction material carbon analysis for a medium office case study is presented in this section as an example to illustrate the process. Construction materials vary with respect to building location and the presented example is associated with San Francisco. The bill of materials for this prototype building was obtained using the "Material Grabber" routine. The global warming potential (GWP) per unit of construction material were mainly obtained from ecoinvent and U.S.LCI databases. The GWP values of storefront window materials were downloaded from a major manufacturer' website, which is 128 kg CO_2 eq. per m2 of glazing and 49.5 kg CO_2 eq. per m2 of frame (Kawneer Company, n.d.).The GWP of wall insulation (fiberglass) was calculated based on constituents and manufacturing energy requirements from BEES. Table 2 gives the collected material information and the corresponding carbon densities.

Table 2. Material inputs and corresponding GWP for the medium office at San Francisco

Building product	**Mass (or area)**	**Material**	**GWP (kg CO_2 eq.) /kg material**
½ IN Gypsum	13.2 Tonnes	Gypsum fiber	0.292
100 mm Normal weight concrete floor	3920 Tonnes	Concrete	0.112
1 IN Stucco	62.3 Tonnes	Stucco	0.074
8 IN Concrete block basement wall	784 Tonnes	Concrete block	0.123
8 IN HW Concrete	603 Tonnes	Concrete block	0.123
13 mm Gypsum board	52.1 Tonnes	Gypsum fiber	0.292
Metal roof surface	10.4 Tonnes	Metal panel	2.722
Roof membrane - highly reflective	17.7 Tonnes	Roofing membrane	0.488
Windows	652.83 m^2	Fixed storefront window	-
Wall insulation (2.54 cm thickness)	0.408 Tonnes	Fiberglass	0.806

Figure 2 shows the estimated embodied GWP associated with all the materials in a medium office across different locations. It can be seen that the material "100 mm Normal weight concrete floor" has the largest impact on the global warming compared with other materials. The medium office located in San Francisco (climate zone 3C) uses "8 IN Concrete block basement wall", while a medium office located in other climate zones uses "6 IN Normal weight concrete floor". That is one main difference in construction materials between different locations. The insulation level also differs significantly from one location to another, but the relative contribution of wall insulation to the total building envelope embodied carbon is very small, which makes it difficult to see the insulation variation for different locations. Windows contribute a significant portion of the total embodied carbon in the case study building. Materials associated with interior furnishings were not considered in this embodied carbon analysis. Similar patterns can be observed across all the other building types:

- Concrete makes the most significant contribution to the total embodied carbon;
- The impact coming from wall insulation is very small and probably can be neglected in the overall building LCA analysis;

- Windows contribute a noticeable portion of the total embodied carbon for all considered building types.

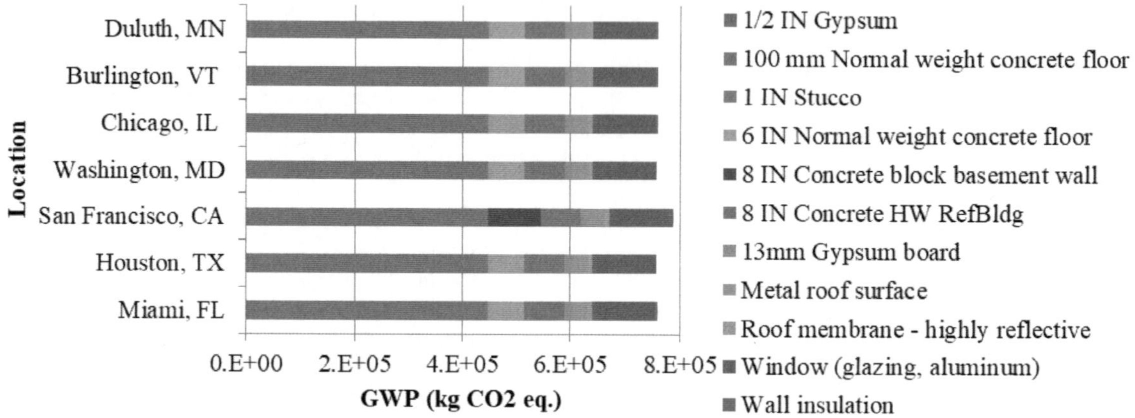

Figure 2. GWP by embodied materials for a medium office.

3.2 Lighting Embodied Carbon Analysis

Embodied carbon for three types of commonly used lighting systems (incandescent, CFL and LED) was analyzed as part of the overall building analysis. Carbon emissions associated with raw material extraction and manufacturing were considered. Material inputs for 12.5-W LED, 14-W CFL, and 60-W incandescent light bulbs were collected from (Scholand and Dillon, 2012), (Tuenge et al., 2013), (Zhang et al., 2017), respectively. Table 3 lists all the major material inputs for the three types of light bulbs. Since the specific manufacturing processes for metal components are not completely known, the generic "metal product manufacturing, average metal working/RER S" in the ecoinvent database was used for all metal components.

Table 3. Material inputs for lighting systems

Component	**Mass (g)**	**Material**
12.5-W LED		
Metal	68.2	Aluminum
Lamp holder	15.3	Copper
Plastic	16.1	Plastic
LED aluminum board	6.9	Aluminum
Power supply devices	43.7	Printed circuit board, capacitors, resistors, transistors
Rubber	25.2	Rubber
14-W CFL		
Power supply devices	14.1	Printed circuit board, capacitors, resistors, transistors
Bulb glass	18.2	Glass
Lamp holder	5.3	Copper
Plastic	11.9	Plastic
60-W Incandescent		
Lamp holder	7.5	Copper
Wire	3.2	Copper
Metal base	1.5	Tin plate
Filament	0.02	Tungsten
Bulb glass	16.4	Glass

Based on the collected raw material inputs shown in Table 3 and manufacturing process information, the total GWPs per lamp were calculated and the results are given in Table 4. It can be seen that the embodied carbon for a LED bulb is the highest while an incandescent bulb has the lowest embodied carbon. However, incandescent lights have much shorter life time and lower efficacy; thus, more light bulbs are needed for a given operation period.

Table 4. Characteristics of the three lighting options

Lamp type	**Lumens**	**Lifespan (hrs)**	**GWP (kg CO_2 eq.) per lamp**
60 W-Incandescent	860	1,000	0.726
14 W-CFL	900	10,000	2.29
12.5 W-LED	850	25,000	9.81

3.3 HVAC System Embodied Carbon Analysis

A 4-ton packaged gas heat & electric cooling unit and a 4-ton split heat pump unit with similar rated efficiency were identified and detailed embodied carbon analysis was carried out for these two representative HVAC units based on laboratory collected data and publicly available literature(Carrier Enterprise, 2018). The estimated GWPs for the two units are very close and only the results for the packaged unit are presented in this paper due to space limitations. Table 5 shows the collected material information and the corresponding carbon densities collected from the various LCA databases.

Table 5. Material inputs and corresponding GWP for the packaged unit

Components	**Mass (kg)**	**Materials**	**GWP (kg CO_2 eq.) per kg material**
Casing	53.66	Steel	1.81
Condenser coil	23.47	Aluminum	8.94
	5.89	Copper	1.99
Blower motor	9.43	NdFeB magnet motor (Navarro et al., 2014)	6.79
Blower wheel	2.99	Galvanized steel (American Galvanizers Association, 2017)	1.79
Evaporator coil	3.79	Copper	1.99
	8.37	Aluminum	8.94
Condenser motor	9.14	NdFeB magnet motor	6.79
Condenser fan	5.28	Galvanized steel	1.79
Heat exchanger	16.53	Galvanized steel	1.79
Compressor	3.18	Galvanized steel	1.79
	1.42	Copper	1.99
	0.71	Aluminum	8.94
	29.72	Cast iron	1.52

Figure 3 shows the GWP for the packaged system by materials. The results show that aluminum has the largest impact (46%) on the overall unit environmental impact, due to the high carbon density of aluminum and the significant amount of aluminum used in the fin-tube evaporator and condenser coils.

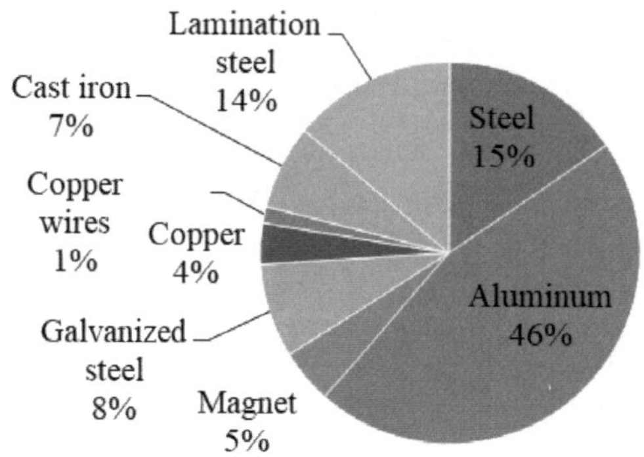

Figure 3. Packaged unit GWP (kg CO_2 eq.) by materials.

3.4 Use Phase Analysis: Simulation Results

In this section, simulation results in the use phase analysis are presented for the different building types and locations. The incandescent, CFL, and LED lights have quite different characteristics as shown in Table 4. Incandescent has the lowest efficacy (lumens/watt) of about 14 lumens/watt, while the efficacy of LED is the highest (68 lumens/watt) among the three lighting options. In the whole building simulations, a lighting requirement of 500 lumens/m2 was assumed for all building spaces. Based on lighting efficacies, the lighting power densities (LPD) for the different lighting options were configured accordingly in EnergyPlus.

EnergyPlus simulations were carried out to generate annual electricity and natural gas uses for all considered building types, lighting options and climate locations. Multiple 4-ton packaged units were assumed to serve all the building types and the numbers of units required were selected automatically in EnergyPlus based on the annual peak cooling demands. All building prototypes use gas furnaces for space heating except for the small and medium offices. The small office prototype is served by packaged air-source heat pumps with gas furnaces as back up. Although the analyzed packaged unit is not a heat pump, the embodied carbon of a heat pump should be similar and the estimated embodied carbon for the packaged unit was used directly for heat pump equipment in the prototype buildings. In the medium office, electrical heating is only used for variable air-volume (VAV) terminal boxes while the packaged unit uses gas as the primary heating source. Embodied carbon for VAV boxes and electrical heaters were not studied in this paper as they only show up in the medium office case studies and the impact should be small.

The different lighting systems (LED, incandescent, CFL) were applied to all the prototype buildings by setting the lighting power densities appropriately in EnergyPlus. As an example, Figure 4 shows the simulated annual energy end uses for the medium office. Among the three lighting options, incandescent has the highest lighting energy consumption due to the low efficacy. LED lights, which have highest efficacy, consume the least energy. The heat released from light bulbs also affects the energy consumption of HVAC systems. It can be observed that less efficient lights, e.g., incandescent bulbs, lead to noticeable reduction in heating energy uses for cold climate locations such as Duluth and Minneapolis; these lights result in higher cooling energy use for hot climate locations such as Miami and Houston. So applications of energy efficient lighting technologies could provide both lighting and HVAC energy savings for hot climates. Although LED and CFL lights lead to increased heating energy use, the lighting energy savings is much greater than the marginal heating energy increase leading to reduced total energy use.

Energy use patterns are very different from one building type to another. For small and medium offices, interior lighting and HVAC equipment are responsible for a significant portion of the total energy use. For a sit-down restaurant, the lighting energy contribution is very small while miscellaneous and HVAC loads are more significant. The distribution of energy use for the primary school and retail store have a stronger dependence on location and lighting type than for the other buildings. However, both HVAC and lighting energy uses are significant for these two building types.

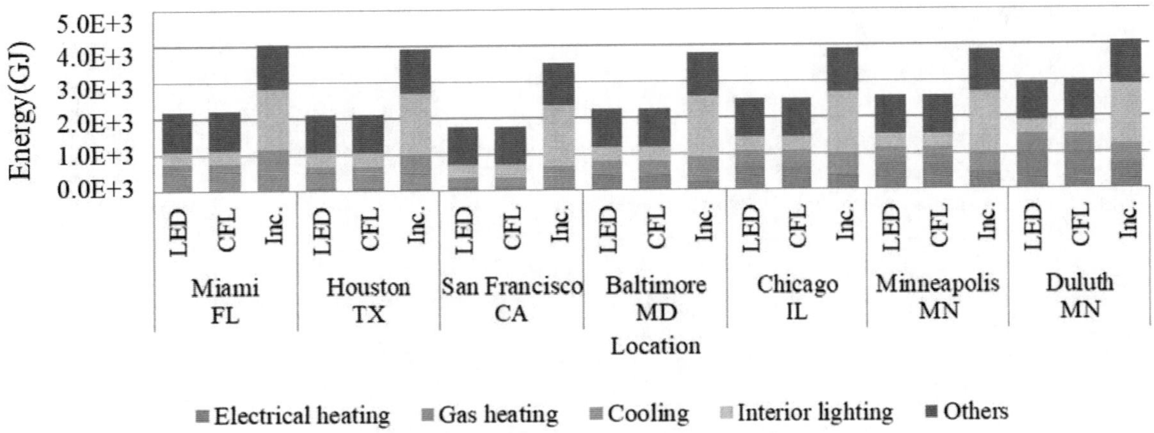

Figure 41. Annual energy uses for medium office.

3.5 Life-Cycle Environmental Analysis

3.5.1 Embodied GWP (50 years)

This study assumed a building life of 50 years for all case study buildings and life cycle assessment was carried out for the same period of time. Again, a lighting requirement of 500 lumens/m2 was assumed for all building spaces. The number of light bulbs required for each building type was calculated based on the design lighting power, efficacies, and the floor area. Different light bulbs have different lifespans (also listed in Table 4). LED bulbs have much longer life time and thus, need less frequent replacements. To calculate the total number of light bulbs needed for the analysis period, lighting schedules were extracted from the EnergyPlus prototypical building models and annual lit hours were calculated. In EnergyPlus, HVAC systems were auto-sized based on the peak cooling/heating demand during design days. All buildings were assumed to be served by identical 4-ton units and the number of units needed for each building was determined by the auto-sized cooling capacity. The lifespan of a HVAC system was assumed to be 10 years. Thus, HVAC equipment needs to be replaced 5 times during a building's life cycle.

The embodied GWP for building envelope, lighting, and HVAC equipment are compared for all considered building types. Figure 5 shows an example of the embodied carbon analysis for the medium office building. It can be clearly seen that the building envelope has the most significant contribution to the total embodied GWP. The impact from lighting systems is also significant and cannot be neglected. It should be highlighted that the embodied carbon per lamp is lowest for incandescent lights. But the total embodied carbon for incandescent lights is highest. That is because an incandescent bulb has a much shorter life time compared to LED and CFL and more bulbs are needed during a building's life cycle. HVAC equipment makes a noticeable contribution to the embodied carbon for restaurants (10%~18%) and retail stores (7%~12%) (results not presented in this paper), mainly due to the high cooling/heating densities; but the HVAC embodied contribution is negligible for the other building types.

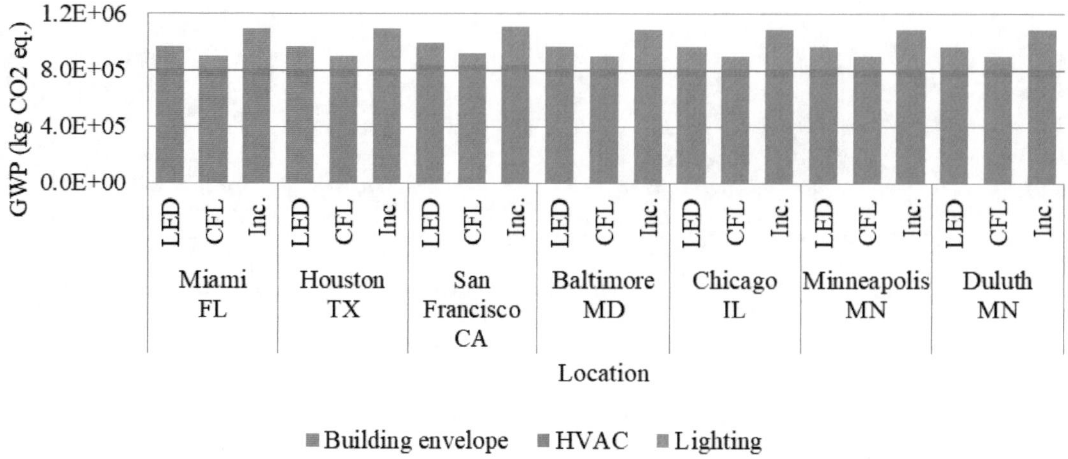

Figure 5. Embodied GWP for medium office.

3.5.2 Embodied and use phase GWP (50 years)

The whole building life-cycle analysis was performed for a 50-year building life time considering embodied and use phase GWP. Figure 6 shows example LCA results for the medium office buildings for different locations and lighting options.

The results show that the embodied carbon contribution to the life-cycle GWP is in general very small and can be neglected (2.5~8.2% for small office, 2.3%~6.4% for medium office, 1.3%~4.0% for standalone retail, 0.6%~1.4% for sit-down restaurant, and 1.4%~4.4% for primary school). However, the results were associated with typical building descriptions in 2010. As more energy efficient technologies are adopted and net-zero buildings are developed, the tradeoffs could change dramatically and the proposed methodology will be able to capture the tradeoffs.

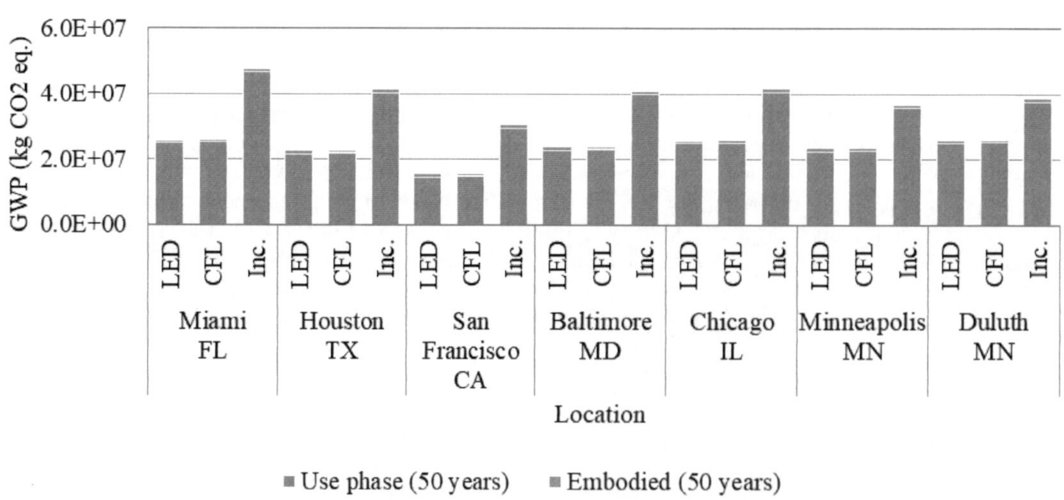

Figure 6. Embodied and use phase (50 years) global warming potential for medium office.

4. CONCLUSIONS

In this paper, a holistic building LCA methodology was developed and applied to 5 prototypical commercial buildings for different climate locations in the U.S. The overall methodology integrates whole building energy simulation and LCA together, providing a streamlined process for building life-cycle environmental analysis. In this paper, EnergyPlus was used to generate building energy end uses in the operation phase. A variety of LCA databases were utilized to estimate the building environmental impact in the construction phase. Then an overall analysis was carried out based on the collected information for the use and construction phases.

High-granularity environmental data are not available for HVAC systems in the existing LCA databases. To fill the gap, detailed embodied carbon analysis was performed for two representative HVAC systems, based on laboratory collected material data and manufacturer literature. In-depth analysis was carried out for three types of lights that are mostly used: incandescent, CFL, and LED. Parametric analysis was performed for the different lighting options, building types and climate locations. Based on the analysis results, the following conclusions can be drawn:

- Wall insulation differs significantly for buildings in different climate locations; but its embodied carbon contribution is negligible compared to the other construction materials;
- Windows make a noticeable contribution to the total building envelope embodied carbon;
- Lighting makes a significant contribution to the total building embodied carbon for all considered cases;
- Incandescent lights have lower embodied carbon per lamp; but their long term embodied carbon is highest due to the short lifespan and more bulb replacements;
- HVAC embodied carbon constitutes a small fraction of the overall building embodied carbon, except for restaurants (10%~18%) and retail stores (7%~12%) which have higher cooling/heating densities;
- Efficient lights such as CFL and LED lead to reduced cooling energy use for hot climate locations (further reducing use phase environmental impact); but they cause higher heating energy use for cold climate locations (lighting electricity savings are greater than heating energy increases resulting in reduced use phase environmental impact);
- HVAC energy use contributes small environmental impacts for cold climate locations and for buildings with natural gas as the primary heating source, as natural gas has much smaller environmental impact compared to electricity;
- Building use phase GWP are dominant in the total life-cycle GWP; the tradeoffs could change dramatically as building energy efficiency increases and as the U.S. moves towards net-zero buildings and renewable power generation.

REFERENCES

American Galvanizers Association, 2017. Hot-Dip Galvanizing for Sustainable Design.

Bribian, I.Z., Uson, A.A., Scarpellini, S., 2009. Life cycle assessment in buildings: State-of-the-art and simplified LCA methodology as a complement for building certification. Building and Environment 44, 2510–2520.

Carrier Enterprise, 2018. Carrier® PerformanceTM - 4 Ton, 15 SEER, 90,000 Btuh, Residential Packaged Gas Heat & Electric Cooling Unit, 2-Stage.

Guglielmetti, R., Macumber, D., Long, N., 2011. OpenStudio: an open source integrated analysis platform. National Renewable Energy Lab.(NREL), Golden, CO (United States).

Kawneer Company, n.d. Aluminum Storefront Framing Systems.

Lemay, L., 2011. Life Cycle Assessment of Concrete Buildings, Concrete Sustainability Report, CSR04. National Ready Mixed Concrete Association.

Lippiatt, B.C., Kneifel, J.D., Lavappa, P.D., Suh, S., Greig, A.L., 2013. Building Industry Reporting and Design for Sustainability (BIRDS) Technical Manual and User Guide.

Monteiro, H., Freire, F., 2012. Life-cycle assessment of a house with alternative exterior walls: Comparison of three impact assessment methods. Energy and Buildings 47, 572–583.

National Institute of Standards and Technology, 2016. BEES.

National Renewable Energy Laboratory, 2017. EnergyPlus.

National Renewable Energy Labratory, 2010. Life-Cycle Inventory Database (USLCI).

Navarro, J., Zhao, F., Sutherland, J., 2014. Comparative LCA of NdFeB and ferrite motors used in the microfabrication industry, in: 9th International Workshop on Microfactories. Retrieved from Http://Conf. Papercept. Net/Images/Temp/IWMF/Media/Files/0036. Pdf.

Scholand, M., Dillon, H.E., 2012. Life-cycle assessment of energy and environmental impacts of LED lighting products part 2: LED manufacturing and performance. Pacific Northwest National Lab.(PNNL), Richland, WA (United States).

Su, X., Zhang, X., 2010. Environmental performance optimization of window–wall ratio for different window type in hot summer and cold winter zone in China based on life cycle assessment. Energy and buildings 42, 198–202.

Tuenge, J.R., Hollomon, B., Dillon, H.E., Snowden-Swan, L.J., 2013. Life-cycle assessment of energy and environmental impacts of LED lighting products, part 3: LED environmental testing.
U.S. Department of Energy, 2016. Commercial Prototype Building Models.
Wang, E., Shen, Z., Barryman, C., 2011. A building LCA case study using Autodesk Ecotect and BIM model.
Wernet, G., Bauer, C., Steubing, B., Reinhard, J., Moreno-Ruiz, E., Weidema, B., 2016. The ecoinvent database version 3 (part I): overview and methodology. The International Journal of Life Cycle Assessment 21, 1218–1230.
Zhang, H., Burr, J., Zhao, F., 2017. A comparative life cycle assessment (LCA) of lighting technologies for greenhouse crop production. Journal of cleaner production 140, 705–713.

ACKNOWLEDGEMENT

This work was supported by the Purdue Center for High Performance Buildings.

Development of Climate Classification through Hierarchical Clustering for Building Energy Simulation

Giovanni PERNIGOTTO[1]*, Andrea GASPARELLA[2] and Jan L. M. HENSEN[3]

[1]Free University of Bozen-Bolzano, Faculty of Science and Technology,
Bolzano, Italy
Tel: +39 0471 017632, Fax: +39 0471 017009, email: giovanni.pernigotto@unibz.it

[2]Free University of Bozen-Bolzano, Faculty of Science and Technology,
Bolzano, Italy
Tel: +39 0471 017200, Fax: +39 0471 017009, email: andrea.gasparella@unibz.it

[3]Eindhoven University of Technology, Department of the Built Environment,
Eindhoven, The Netherlands
Tel: +31 (0)40 2472988, Fax: +31 (0) 40 2438595, email: j.hensen@tue.nl

* Corresponding Author

ABSTRACT

Climate classification plays an important role for the identification of homogeneous groups of climates, from which representative locations can be extracted and used for building energy simulation analyses. Nevertheless, according to the current state-of-the-art, the main reference systems consider just a fraction of those weather quantities which are relevant in the building energy balance, i.e., ambient temperature and humidity and solar radiation. To overcome this issue, in previous researches a new methodology was defined, based on monthly series of weather quantities, statistical analyses and data-mining techniques for climate clustering. In this work, with the aim of further developing such approach, a shorter time-discretization of weather quantities, i.e., a weekly discretization, was tested, alongside additional variables describing the daily range of ambient temperature and humidity. In order to investigate the potential of those modifications, a dataset with more than 300 European reference climates was analyzed and subdivided into climate classes according to the proposed clustering procedure.

1. INTRODUCTION

An accurate climate zoning is particular important for policy makers, when it comes to define effective strategies for building energy retrofitting or minimum energy performance requirements for new buildings. Indeed, a poor climate classification can undermine the efficacy of energy savings strategies or bring to requirements difficult to comply with by professionals, companies and public administrations operating in the building sector. Typical classifications adopted in several countries worldwide are based on heating – and sometimes cooling – degree-days or temperature-based quantities (Walsh *et al.*, 2018 and 2019). For example, popular climate classifications are the Köppen-Geiger system (Peel *et al.*, 2007) and the one based on the ANSI/ASHRAE 169 (ASHRAE, 2013), which account also for precipitation in order to group the different climates. Other weather variables, like humidity and solar irradiance, are rarely included in climate zoning, even though they play a significant role on building energy balance and performance.

In order to discuss the suitability of the climate zoning methods according to the state of the art and overcome their limitations, a new weather-based classification approach was proposed in previous works (Pernigotto and Gasparella, 2018; Pernigotto *et al.*, 2019). Specifically, all weather variables relevant for building energy balance were accounted for, i.e., dry bulb temperature, water vapour partial pressure, solar global horizontal irradiation, and processed by

means of Hierarchical Clustering. Involved quantities were analyzed on a monthly basis and characterized through annual averages and spreads of monthly series, in a fashion similar to the Köppen-Geiger system. Although achieved results were found promising, the adopted time discretization brought to loss of information which may be significant for the design of high-performance buildings, such as the variability of weather quantities within each day.
With the aim of further assessing the proposed climate classification and possibly adding improvements, on one hand a weekly time discretization of variables was tested while, on the other hand, statistics able to describe also the daily ranges of the weather quantities were considered as inputs for the Hierarchical Clustering.

2. METHODOLOGY

2.1 Dataset of climates

The analysis performed in this work exploited the same dataset of climates considered in (Pernigotto *et al.*, 2019). Specifically, all European weather files available in EnergyPlus online weather database (https://energyplus.net/weather) were considered except for the Italian climates. Indeed most of Italian weather files included in the EnergyPlus online database refer to the IGDG series, i.e., they were developed from multi-year series collected in the period 1951-1970 and are now outdated. Consequently, the most recent weather files, published by the Italian CTI (Comitato Termotecnico Italiano, https://try.cti2000.it/) and in agreement with the Italian technical standard UNI 10349-1:2016 on weather data for energy calculations, were preferred. For the other European countries, when more than a weather source was found for the same locality, the International Weather for Energy Calculations IWEC files were selected.
As a whole, 318 typical or reference years were included in the analysis. Table 1 and Figure 1 show, respectively, the distribution of the considered locations according to 11 different Köppen-Geiger climatic classes and their position on a map of Europe. *Dfb* (cold climate without dry season and with warm summer, in central and eastern Europe), *Csa* (temperate climate with dry and hot summer, in the Mediterranean areas), and *Cfb* (temperate climate without dry season and with warm summer, in western Europe) are the three most populated climate classes, with respectively 29.2 %, 23.6 %, and 17.9 % of the dataset. For further details regarding homogeneity and overlapping of the Köppen-Geiger climatic classification applied to this sample, see (Pernigotto *et al.*, 2019).

Table 1: Distribution of the dataset of climates according to the Köppen-Geiger classes

Köppen-Geiger class	Description	Number of locations	Fraction of the sample
BSk	Arid cold steppe climate	14	4.4 %
BWk	Arid desert cold climate	1	0.3 %
Csa	Temperate climate with dry and hot summer	75	23.6 %
Csb	Temperate climate with dry and warm summer	12	3.8 %
Cfa	Temperate climate without dry season and with hot summer	49	15.4 %
Cfb	Temperate climate without dry season and with warm summer	57	17.9 %
Dsb	Cold climate with dry and warm summer	2	0.6 %
Dfa	Cold climate without dry season and with hot summer	3	0.9 %
Dfb	Cold climate without dry season and with warm summer	93	29.2 %
Dfc	Cold climate without dry season and with cold summer	11	3.5 %
ET	Polar and tundra climate	1	0.3 %
Total		318	100 %

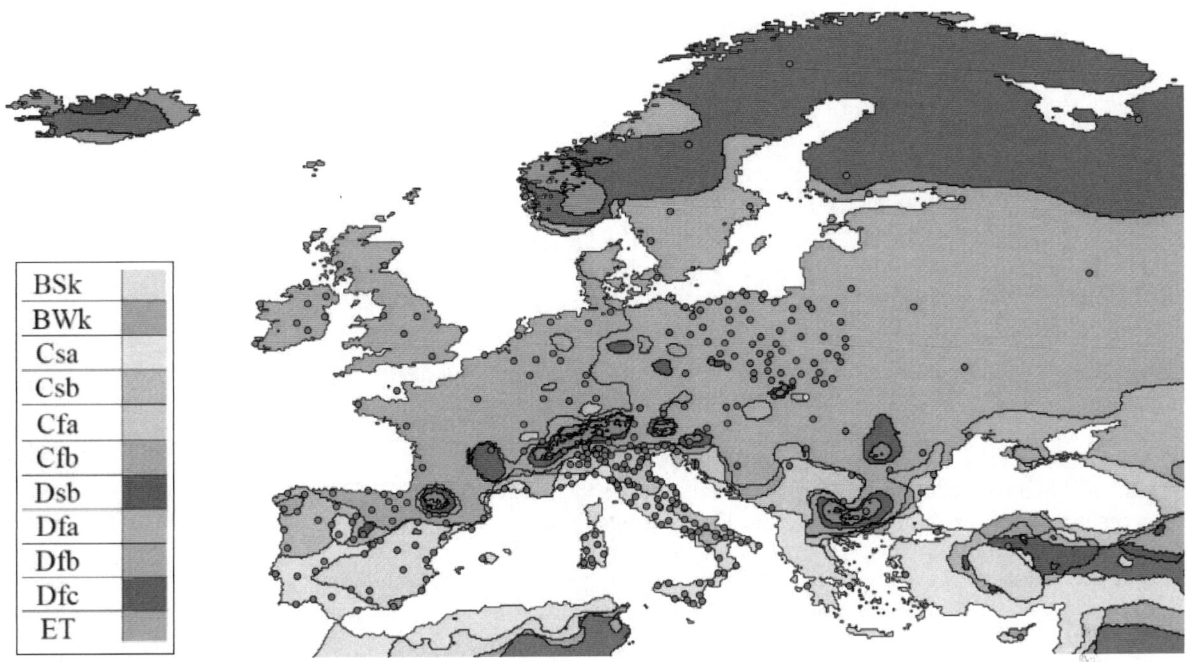

Figure 1: European map showing the position of the considered locations in the Köppen-Geiger climate classes (Pernigotto *et al.*, 2019; prepared with QGIS v. 3.4.2 based on the Köppen-Geiger GIS climate map by NASA ORNL DAAC)

Adopting the same approach as in (Pernigotto and Gasparella, 2018; Pernigotto *et al.*, 2019), the focus was put on those weather variables considered the most significant in the building energy balance, i.e., ambient temperature and humidity, and solar radiation. Respectively, ambient temperature was expressed as dry bulb temperature *DBT*, air humidity as water vapour partial pressure *WVP*, and solar radiation as global horizontal solar radiation *GHI*. While *DBT* and *WVP* were defined as average values, *GHI* was integrated over the considered time period, i.e., with monthly discretization. As explained in previous researches, wind speed was neglected due to representativeness issues incompatible with the goals of this analysis. Furthermore, considering the different number of weather files available for the different European countries in the dataset (see Figure 1), in particular with the majority located in Italy (110 locations equal to 34.6 % of the sample), in Poland (61 equal to 19.2 %) and in Spain (46 equal to 14.5 %), geographical distances and altitudes were not considered as variables for clustering. In such a way, uneven distributions of localities did not affect the classification.

2.2 Climate clustering

In previous researches (Pernigotto and Gasparella, 2018; Pernigotto *et al.*, 2019), preliminary analyses required for climate clustering were performed for each location, considering a series of 12 monthly values. These series were used to determine *annual average* and *spread* of monthly values for each weather quantity, as in Figure 2 (left). Then, these statistics were normalized and used as inputs for a Hierarchical Clustering, chosen for its ability to allow for a climate classification without the need of a predefined number of groups, as it is, for instance, in the *k*-means approach. Classes were identified with the goal of enhancing group homogeneity and avoiding overlapping, assessed respectively by means of *standard deviations* and *mean values* of the annual averages and spreads of *DBT*, *WVP* and *GHI*.

Even if the clustering can be performed for each weather quantity at a time (e.g., for dry bulb temperature), a more meaningful classification was found including all of them in the procedure. Moreover, each quantity was given the same relative importance, without any predetermined hierarchy. Although clustering can be repeated in multiple steps to run sub-classification, the study of the dendrogram characterizing the Hierarchical Clustering was recommended to target potential subdivisions when useful to increase group homogeneity and reduce the risk of overlapping.

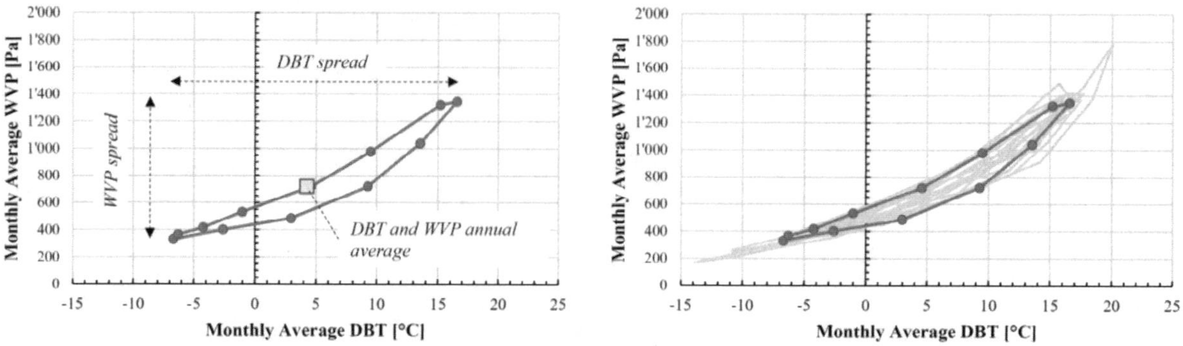

Figure 2: Annual average and spread for *DBT* and *WVP* in a given location (**left**); example of a class of climates to identify its homogeneity (**right**)

2.3 Modifications to the proposed methodology

As explained in the introduction, this work wants to further improve the proposed procedure for climate classification to get distinct groups with homogenous weather conditions, suitable to identify representative locations and to run building energy simulation studies to ease the definition of regional or national energy policies. With that goal in mind, it was decided to discuss if (1) a different time discretization of the input data series can improve clusters' uniformity, and if (2) an alternative definition of weather quantities, based on metrics assessing daily ranges, can lead to a different classification.

In order to investigate the first goal, instead of a series of 12 monthly values, 52 weekly values were used as inputs. It was assumed that each typical year begins conventionally with the first day of the week, Monday, and the last day of the year was included in the calculation of the previous week average value. The procedure was either repeated as in Section 2.2, i.e., with (A) annual *averages* and *spreads* of weekly weather quantities, and considering also different statistics. Specifically, in order to account for the fact that weekly series are more sensitive to extreme or anomalous input data, which can affect the calculation of overall statistics like spreads as well, alternatives were tested. In details, the clustering procedure was applied also working with (B) annual *averages* and *standard deviations* of weekly weather quantities, and with (C) annual *averages* coupled with *minimum* and *maximum* values in the weekly series weather quantities. All three approaches share a statistical variable, i.e., the *average*, which is clearly the same as in the case of monthly *DBT* and *WVP* series due to the properties of the arithmetic mean. To determine if a shorter time-discretization of the input data could be beneficial for the procedure, homogeneity of climate classes was expressed in terms of standard deviations of weather statistics and compared with those found in (Pernigotto *et al.*, 2019). Both partial clusterings, with all considered statistics for a single weather quantity at a time, and global clusterings were analysed for a more comprehensive understanding of the findings.

As regards the second goal, the original methodology with monthly series of weather quantities was applied. Nevertheless, along with monthly averages of *DBT* and *WVP*, also those of daily ranges of the same quantities, $\widehat{DBT}$ and $\widehat{WVP}$, were included. As regards the solar radiation, the monthly *GHI* integrals were analysed without any additional metrics. For all considered weather quantities annual *averages* and *spreads* were calculated, normalized and used as inputs for the Hierarchical Clustering. Besides the qualitative comparison of the obtained climate classes with those in (Pernigotto *et al.*, 2019), standard deviations were discussed as well. Again, for sake of completeness, both partial and global clusterings were performed.

3. RESULTS AND DISCUSSION

3.1 Impact of time-discretization of input weather data series

To discuss the impact of time-discretization of the weather inputs on the results of the climate classification, both partial and global clustering were performed. As indicated in Section 2.3, three alternative sets of statistics were considered (i.e., A – *averages* and *spreads*, B – *averages* and *standard deviations*, C – *averages*, *maximum* and *minimum* values). The number of groups was fixed to 7, as in (Pernigotto *et al.*, 2019), for both partial and global clusterings. Very different dendrograms were obtained in partial clusterings, highlighting the importance of the chosen weather statistics, as it can be seen for instance in Figure 3 for the global clusterings. Nevertheless, for both (A) and (B) 7 groups were identified at a height in the dendrogram around 0.5 for partial clustering and at a height of 1 for the global one, similarly to what found in the previous studies. Regarding (C), instead, slightly higher heights were observed, respectively around 0.7 and 1.2, for partial and global clustering. This can be associated to the higher number of variables included in the Hierarchical Clustering, i.e., 3 and 9 instead of 2 and 6.

The selection of weather statistics influenced also the distribution of the locations in the different classes. For instance, partial clustering (A) led to groups ranging from 4 to 91 elements for *DBT*, from 8 to 98 for *WVP*, and from 15 to 71 for *GHI*. The size of groups from partial clustering (B), instead, is between 3 or 4 and 130 elements for *DBT*, 6 or 7 and 138 for *WVP*, and 11 and 80 for *GHI*. Finally, partial clustering (C) generated groups with from 3 to 101 locations for *DBT*, from 19 to 104 for *WVP*, and from 7 to 58 for *GHI*. In case of global clustering, the minimum group size is 12, 24, and 15, and the maximum one 90, 94, and 68, respectively for (A), (B), and (C).

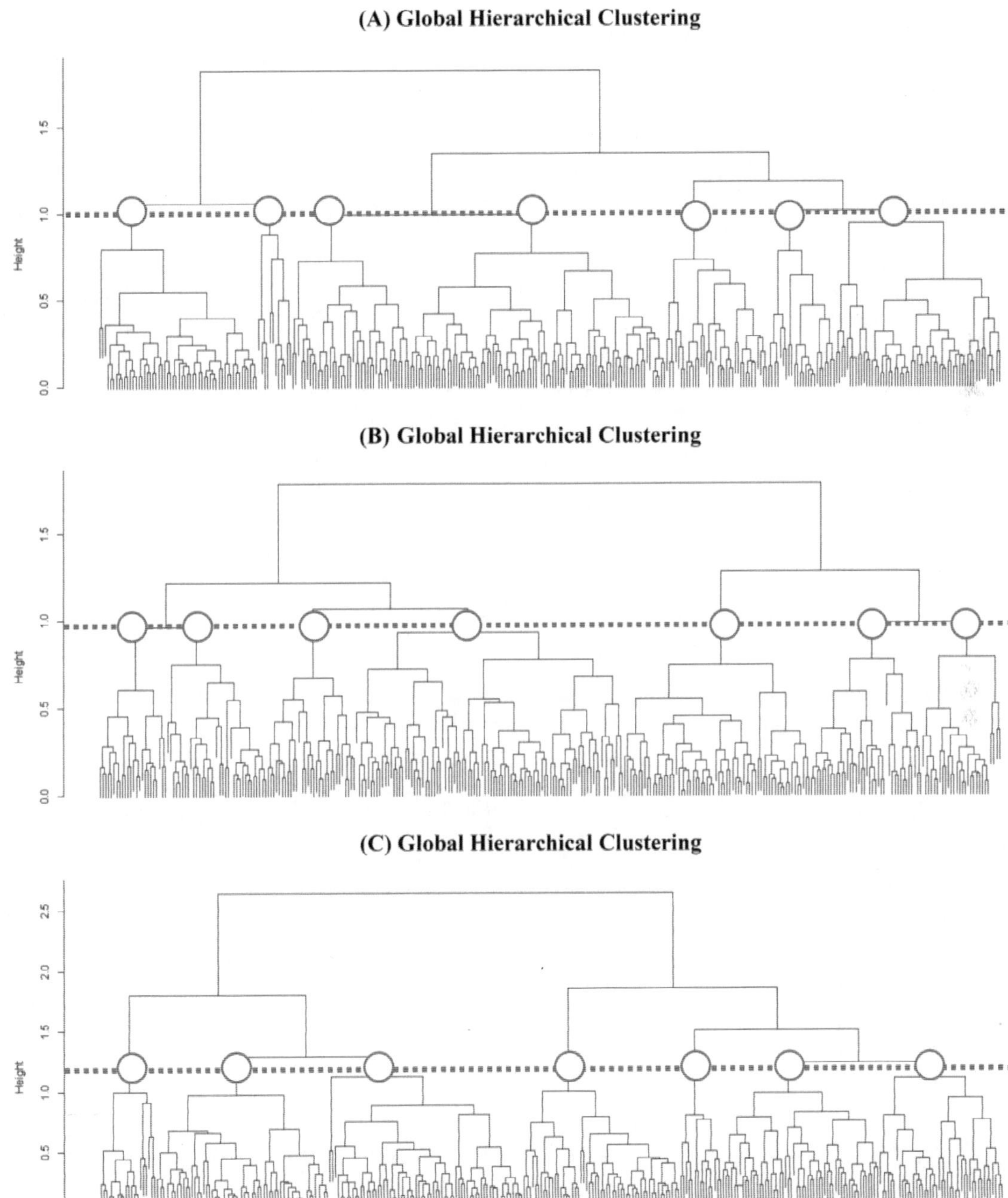

Figure 3: Global hierarchical clusterings according to different groups of weather statistics (A, B, C). The red line represents the height chosen for determining the number of clusters, identified with the circles.

As it can be seen in Tables 2-4, in some cases partial clustering gave groups with very close annual averages, as for instance classes **4**^{DBT} and **5**^{DBT}, and classes **6**^{DBT} and **7**^{DBT} for *DBT* in clustering (C) in Table 2, classes **1**^{WVP} and **2**^{WVP}, **3**^{WVP} and **4**^{WVP}, **5**^{WVP} and **6**^{WVP} for *WVP* in clustering (B) in Table 3, classes **1**^{GHI}, **2**^{GHI} and **3**^{GHI} for *GHI* in clustering (C) in Table 4. As a whole, the adoption of annual *spreads* as in clustering (A) seems to reduce the risk of overlapping with respect to the other tested alternatives. Indeed, as confirmed also in the comparison of global clusterings in Table 5, while just two classes show close annual averages in clustering (A) (specifically, classes **3** and **4** for *WVP*), this occurs more frequently for clustering (B) (see, for instance, groups **2** and **3** and groups **5** and **6** for *WVP*) and (C) (e.g., groups **1** and **2** and groups **4** and **5** for *GHI*).

Table 2: *DBT* Partial Hierarchical Clustering with weekly series of weather quantities: mean values and standard deviations for each class and annual averages and spreads

***DBT* Hierarchical Clustering**

			Classes and fraction of sample						
A	**Annual**	**Quantities**	**1 (1.3%)**	**2 (9.4 %)**	**3 (28.6%)**	**4 (6.9%)**	**5 (18.6%)**	**6 (23.3%)**	**7 (11.9%)**
	Average	*DBT* [°C]	2.7±3.0	6.8±2.3	9.1±1.5	9.7±1.4	12.6±1.1	15.3±1.4	16.5±1.6
		WVP [Pa]	694±21%	860±13%	940±8%	1005±10%	1131±10%	1215±13%	1357±9%
		GHI [kWh m^{-2} w^{-1}]	17.6±18%	18.7±12%	20.5±20%	18.9±13%	25.8±7%	29.2±8%	29.8±12%
	Spread	*DBT* [°C]	37.3±2.0	28.9±2.3	24.3±2.7	15.4±2.1	27.1±2.0	22.4±1.7	17.2±2.0
		WVP [Pa]	1554±23%	1434±14%	1326±19%	1052±10%	1709±15%	1443±24%	1441±23%
		GHI [kWh m^{-2} w^{-1}]	42±5%	41.7±9%	41.9±11%	40.2±9%	46.8±8%	45.9±8%	43.3±8%
B	**Annual**	**Quantities**	**1 (1.2%)**	**2 (0.9%)**	**3 (18.5%)**	**4 (6.2%)**	**5 (28.6%)**	**6 (40.8%)**	**7 (3.4%)**
	Average	*DBT* [°C]	0.8±1.9	3.9±2.0	8.6±1.7	9.5±1.3	10.0±2.1	15.0±1.8	15.7±1.6
		WVP [Pa]	572±12%	759±10%	933±9%	1004±10%	995±11%	1230±13%	1300±9%
		GHI [kWh m^{-2} w^{-1}]	16.0±8%	18.6±15%	20.8±17%	18.3±9%	21.5±20%	28.6±10%	29.5±10%
	Spread	*DBT* [°C]	27.1±6.2	38.2±0.5	27.0±3.3	15.2±2.2	24.6±3.4	22.1±3.4	20.1±5.1
		WVP [Pa]	1040±8%	1728±4%	1447±15%	1064±9%	1403±20%	1499±24%	1535±20%
		GHI [kWh m^{-2} w^{-1}]	39.3±13%	42.2±5%	42.3±9%	39.5±8%	43.1±10%	45.3±8%	47.1±7%
C	**Annual**	**Quantities**	**1 (0.9%)**	**2 (4.7%)**	**3 (11.3%)**	**4 (9.4%)**	**5 (24.5%)**	**6 (31.7%)**	**7 (17.2%)**
	Average	*DBT* [°C]	1.3±2.1	6.4±1.2	7.8±1.9	9.81±1.8	10.0±2.0	14.7±2.1	14.8±1.8
		WVP [Pa]	595±14%	825±6%	923±9%	1012±13%	1007±10%	1186±16%	1244±10%
		GHI [kWh m^{-2} w^{-1}]	15.7±8%	18.8±7%	18.1±8%	19.1±15%	21.8±17%	29.0±10%	27.8±8%
	Spread	*DBT* [°C]	34.4±4.1	29.3±4.3	26.3±2.7	17.7±4.2	25.9±3.1	21.2±4.0	23.1±2.8
		WVP [Pa]	1262±25%	1418±15%	1431±10%	1196±24%	1484±15%	1387±30%	1584±16%
		GHI [kWh m^{-2} w^{-1}]	41.9±6%	43.9±5%	39.6±9%	40.7±9%	42.7±10%	46.1±7%	45.3±8%

Table 3: *WVP* Partial Hierarchical Clustering with weekly series of weather quantities: mean values and standard deviations for each class and annual averages and spreads

***WVP* Hierarchical Clustering**

			Classes and fraction of sample						
A	**Annual**	**Quantities**	**1 (3.7%)**	**2 (2.5%)**	**3 (30.8%)**	**4 (20.7%)**	**5 (11.9%)**	**6 (17.2%)**	**7 (12.8%)**
	Average	*DBT* [°C]	6.3±4.7	12.5±2.1	9.7±2.3	9.6±2.6	14.1±2.1	14.0±1.6	16.8±1.2
		WVP [Pa]	725.±17%	873±10%	951±7%	982±9%	1172±5%	1254±6%	1396±7%
		GHI [kWh m^{-2} w^{-1}]	21.4±31%	30.7±6%	20.7±21%	21.7±17%	27.0±16%	26.9±7%	30.2±9%
	Spread	*DBT* [°C]	22.9±5.2	20.6±0.8	23.5±4.5	27.7±3.3	19.2±4.1	24.3±3.2	20.0±2.6
		WVP [Pa]	972.±9%	651±13%	1254±13%	1591±6%	1182±14%	1850±8%	1559±14%
		GHI [kWh m^{-2} w^{-1}]	43.9±12%	48.1±4%	42.1±11%	43.2±9%	44.1±8%	46.4±7%	43.7±9%
B	**Annual**	**Quantities**	**1 (1.8%)**	**2 (2.2%)**	**3 (43.3%)**	**4 (19.8%)**	**5 (6.6%)**	**6 (22.3%)**	**7 (3.7%)**
	Average	*DBT* [°C]	4.9±3.7	5.8±5.0	10.3±2.5	10.1±2.6	14.7±1.2	15.5±1.7	17.0±1.4
		WVP [Pa]	706±18%	712±15%	989±9%	991±10%	1293±8%	1299±7%	1414±9%
		GHI [kWh m^{-2} w^{-1}]	18.0±30%	22.3±31%	22.0±21%	22.6±21%	27.5±9%	28.8±9%	30.9±8%
	Spread	*DBT* [°C]	25.1±6.8	25.9±8.7	23.8±4.7	25.4±4.6	22.5±3.9	21.6±3.4	19.4±2.7
		WVP [Pa]	1003±8%	1106±40%	1338±18%	1411±24%	1721±21%	1575±17%	1678±18%
		GHI [kWh m^{-2} w^{-1}]	44.0±9%	42.8±14%	43.3±10%	42.6±11%	46.5±08%	44.9±8%	43.6±6%
C	**Annual**	**Quantities**	**1 (6.6%)**	**2 (12.2%)**	**3 (24.8%)**	**4 (32.7%)**	**5 (9.1%)**	**6 (8.4%)**	**7 (5.9%)**
	Average	*DBT* [°C]	5.9±2.4	8.94±2.9	11.5±3.0	13.1±3.0	13.5±2.3	13.0±4.4	14.0±1.5
		WVP [Pa]	786±13%	899±11%	1069±14%	1143±15%	1174±8%	1178±20%	1248±11%
		GHI [kWh m^{-2} w^{-1}]	18.4±17%	21.8±25%	22.9±23%	25.9±16%	26.4±14%	25.9±25%	26.8±7%
	Spread	*DBT* [°C]	28.3±6.0	24.0±3.6	21.7±5.0	23.8±3.7	22.3±5.1	23.4±4.7	24.0±3.1
		WVP [Pa]	1302±19%	1223±26%	1339±21%	1504±18%	1507±23%	1511±19%	1770±15%
		GHI [kWh m^{-2} w^{-1}]	43.2±7%	42.0±12%	43.5±11%	44.2±9%	43.5±9%	43.9±10%	46.5±8%

Table 4: *GHI* Partial Hierarchical Clustering with weekly series of weather quantities: mean values and standard deviations for each class and annual averages and spreads

GHI Hierarchical Clustering			Classes and fraction of sample						
A	**Annual**	**Quantities**	**1 (14.4%)**	**2 (12.2%)**	**3 (15.0%)**	**4 (21.3%)**	**5 (22.3%)**	**6 (9.7%)**	**7 (4.7%)**
	Average	*DBT* [°C]	7.9±2.2	7.6±2.2	10.2±2.4	12.6±2.2	13.9±1.9	15.9±1.7	17.4±1.2
		WVP [Pa]	922±11%	894±13%	1037±13%	1124±13%	1136±16%	1312±11%	1310±12%
		GHI [kWh m^{-2} w^{-1}]	17.2±6%	18.0±6%	21.4±9%	25.2±8%	29.1±6%	29.2±6%	33.8±2%
	Spread	*DBT* [°C]	23.0±5.6	25.6±5.5	24.2±5.5	25.1±3.4	22.6±3.5	20.6±2.8	19.3±2.3
		WVP [Pa]	1289±15%	1329±15%	1473±15%	1607±18%	1385±31%	1524±16%	1246±24%
		GHI [kWh m^{-2} w^{-1}]	36.8±4%	44.1±5%	40.2±5%	46.8±3%	48.4±4%	42.2±4%	42.4±5%
B	**Annual**	**Quantities**	**1 (12.2%)**	**2 (24.2%)**	**3 (25.1%)**	**4 (11.3%)**	**5 (10.0%)**	**6 (13.5%)**	**7 (3.4%)**
	Average	*DBT* [°C]	7.1±2.0	8.6±1.9	12.5±2.2	13.6±1.9	14.1±1.7	15.7±2.1	17.4±1.1
		WVP [Pa]	883±12%	947±10%	1109±15%	1182±11%	1219±11%	1212±18%	1335±8%
		GHI [kWh m^{-2} w^{-1}]	17.8±7%	18.4±8%	25.9±9%	26.2±8%	27.0±4%	31.7±5%	32.1±3%
	Spread	*DBT* [°C]	26.8±3.1	23.2±5.8	23.7±5.3	24.6±2.7	23.5±2.5	20.3±2.6	20.1±2.3
		WVP [Pa]	1400±11%	1312±15%	1496±24%	1621±17%	1642±16%	1242±31%	1435±17%
		GHI [kWh m^{-2} w^{-1}]	40.2±9%	40.8±9%	45.5±7%	46.2±9%	45.6±9%	45.8±6%	42.1±6%
C	**Annual**	**Quantities**	**1 (6.2%)**	**2 (17.9%)**	**3 (14.7%)**	**4 (22.6%)**	**5 (18.2%)**	**6 (17.9%)**	**7 (2.2%)**
	Average	*DBT* [°C]	8.2±2.0	7.8±1.7	8.6±2.4	14.0±2.3	13.2±2.2	15.0±2.0	15.5±2.4
		WVP [Pa]	935.±9%	914±10%	941±13%	1208±12%	1098±17%	1223±13%	1289±17%
		GHI [kWh m^{-2} w^{-1}]	18.2±9%	18.3±9%	18.8±9%	27.0±10%	27.6±9%	29.1±11%	30.8±12%
	Spread	*DBT* [°C]	26.7±3.6	25.8±3.8	22.4±6.7	23.6±3.4	22.2±4.5	21.9±4.1	22.7±4.9
		WVP [Pa]	1424±10%	1392±13%	1290±17%	1598±15%	1334±31%	1460±27%	1620±12%
		GHI [kWh m^{-2} w^{-1}]	40.1±9%	41.5±10%	39.8±6%	44.8±9%	46.1±7%	46.3±6%	47.3±8%

Table 5: Global Hierarchical Clustering with weekly series of weather quantities: mean values and standard deviations for each class and annual averages and spreads

Global Hierarchical Clustering			Classes and fraction of sample						
A	**Annual**	**Quantities**	**1 (3.7%)**	**2 (17.6%)**	**3 (18.5%)**	**4 (7.5%)**	**5 (11.3%)**	**6 (28.3%)**	**7 (12.8%)**
	Average	*DBT* [°C]	4.0±2.7	7.9±1.3	10.0±2.2	10.6±1.8	13.2±2.0	13.9±1.7	16.7±1.4
		WVP [Pa]	722±16%	918±6%	1016±12%	1064±10%	982±15%	1204±10%	1345±9%
		GHI [kWh m^{-2} w^{-1}]	17.4±11%	18.0±9%	21.2±13%	19.4±12%	29.4±08%	27.3±6%	30.8±8%
	Spread	*DBT* [°C]	30.8±5.9	26.1±3.0	25.4±2.9	14.9±2.2	20.3±2.8	24.7±3.1	19.7±2.3
		WVP [Pa]	1310±24%	1420±10%	1502±13%	1106±11%	911±20%	1707±13%	1428±15%
		GHI [kWh m^{-2} w^{-1}]	43.6±6%	38.2±6%	43.2±7%	39.2±8%	47.5±3%	47.9±4%	42.3±5%
B	**Annual**	**Quantities**	**1 (8.8%)**	**2 (12.5%)**	**3 (20.7%)**	**4 (29.5%)**	**5 (7.5%)**	**6 (11.0%)**	**7 (9.7%)**
	Average	*DBT* [°C]	6.2±2.8	8.7±1.7	8.9±1.2	13.1±1.6	14.6±1.5	15.2±2.0	16.9±1.2
		WVP [Pa]	833±16%	950±8%	963±7%	1099±13%	1269±11%	1253±12%	1354±10%
		GHI [kWh m^{-2} w^{-1}]	17.3±7%	20.4±13%	18.6±9%	27.0±10%	28.2±8%	28.3±8%	31.4±7%
	Spread	*DBT* [°C]	27.3±3.6	26.9±4.3	22.6±5.4	23.4±4.2	22.0±4.0	22.3±2.6	19.7±2.8
		WVP [Pa]	1350±13%	1486±12%	1304±15%	1408±29%	1615±22%	1563±19%	1495±20%
		GHI [kWh m^{-2} w^{-1}]	40.8±10%	41.5±10%	40.8±8%	46.3±8%	46.6±6%	45.1±8%	44.0±6%
C	**Annual**	**Quantities**	**1 (6.2%)**	**2 (17.2%)**	**3 (18.8%)**	**4 (21.3%)**	**5 (13.8%)**	**6 (17.6%)**	**7 (4.7%)**
	Average	*DBT* [°C]	5.6±2.3	7.7±1.5	9.9±1.2	14.1±2.3	14.2±1.7	14.2±2.3	16.3±1.6
		WVP [Pa]	784±13%	919±8%	1005±7%	1211±12%	1197±12%	1124±18%	1332±14%
		GHI [kWh m^{-2} w^{-1}]	17.9±11%	18.1±8%	19.8±11%	27.3±8%	27.3±9%	29.4±10%	30.8±9%
	Spread	*DBT* [°C]	28.9±5.8	26.3±2.5	21.7±5.1	24.0±3.5	23.8±4.0	20.2±3.7	21.0±3.9
		WVP [Pa]	1344±19%	1428±9%	1313±16%	1601±16%	1616±21%	1216±33%	1541±22%
		GHI [kWh m^{-2} w^{-1}]	43.4±6%	39.6±9%	40.9±8%	45.3±8%	47.1±6%	45.4±7%	47.0±7%

In order to discuss the groups' homogeneity in the different clusterings, the maximum standard deviations found for annual *averages* and *spreads* were compared to those determined in (Pernigotto *et al.*, 2019, Table 2 and Figure 6). Clusters with less than 10 climates were discarded in this analysis and the largest standard deviations observed in the current research compared to those detected in previous analyses. Regarding clustering (A), (1) in *DBT* partial clustering, slightly larger standard deviations were found for *DBT*, both *average* and *spreads*, and for *WVP averages*, while slightly lower ones were observed for *WVP spreads* and *GHI*; (2) in *WVP* partial clustering, improvements were recorded just for *WVP* and *GHI spreads*; finally, in (3) *GHI* partial clustering limited reductions of the maximum standard deviations were detected for *DBT averages* and *GHI*. As far as clustering (B) is concerned, (1) in *DBT* partial clustering significant worsening of standard deviations was seen for *DBT spreads*, alongside negligible variations for the other statistics; (2) in *WVP* partial clustering larger standard deviations were found for *DBT* and *WVP spreads*, while reductions were noticed for *DBT* and *WVP averages* and for *GHI*; in (3) *GHI* partial clustering, standard deviations of all the weather statistics but *DBT averages* increased. Clustering (C) gave the worst performance in terms of groups' homogeneity: limited improvements were registered for *GHI* (in *DBT* partial clustering and, just for *GHI spreads*, in *WVP* partial clustering) and for *DBT averages* (in *GHI* partial clustering); worsening in homogeneity was observed in all partial clustering for the remaining weather statistics, particularly for *DBT spreads*. Global clusterings do not differ too much each other in terms of homogeneity: a significant worsening is observed for *DBT spread*, a general worsening for all variables except for *GHI*, which was found slightly improved.

In conclusion of this section, the adoption of weekly series of weather quantities brought mixed changes with respect to the original method based on monthly series, with improvements in climate clusters' homogeneity limited to the quantity characterizing solar radiation but with significant worsening in case of the ambient temperature. As a consequence, it should be observed that the calculation of shorter data series has not practical benefits in the framework of the proposed methodology.

3.2 Daily range weather quantities

Again, in this second analysis 7 climate classes were identified. This time, due to higher number of variables involved in clustering, i.e., 4 in each partial clustering and 10 in the global one, the 7 groups were determined at heights in the dendrograms equal to 0.8 and 1.3, respectively. Groups' sizes varied from 3 to 94 (*DBT* partial clustering), from 1 to 172 (*WVP* partial clustering), and from 4 to 130 (global clustering). As a whole, with respect to the classification in Section 3.1 and to previous researches (Pernigotto *et al.*, 2019), dominant larger groups emerged.

Although partial clusterings with *DBT* and *WVP* (Table 6) revealed some overlapping between classes considering the annual *averages* of weather statistics, e.g., *DBT* or *WVP*, those groups were found differentiated by $\widehat{DBT}$ or $\widehat{WVP}$ (see for instance in *DBT* partial clustering, classes $\mathbf{4^{DBT}}$, $\mathbf{5^{DBT}}$ and $\mathbf{6^{DBT}}$ as regards *DBT* and $\widehat{DBT}$ values). The same is confirmed in global clustering (Table 7): for example, while groups **2** and **3** have similar *DBT averages*, they have very different $\widehat{DBT}$ *averages*.

As expected, due to the presence of one or more larger classes, homogeneity is lower compared to the monthly-based clustering performed in (Pernigotto *et al.*, 2019). For both partial and global clusterings, a general worsening of homogeneity was detected, in particular for *DBT*; limited or negligible improvements were observed just for *GHI spreads*.

Table 6: Partial Hierarchical Clustering with additional weather quantities: mean values and standard deviations for each class and annual averages and spreads

***DBT* Hierarchical Clustering**

		Classes and fraction of sample						
Annual	**Quantities**	**1 (2.5%)**	**2 (0.9%)**	**3 (29.5%)**	**4 (19.4%)**	**5 (11.0%)**	**6 (26.4%)**	**7 (10.0%)**
Average	*DBT* [°C]	4.0±2.4	10.3±5.5	11.0±3.3	11.9±3.1	12.2±3.2	12.5±3.4	13.6±3.9
	$\widehat{DBT}$ [°C]	6.4±1.0	1.7±0.1	7.2±1.2	5.3±1.1	2.9±0.9	6.3±1.0	2.9±0.9
	WVP [Pa]	734±14%	1039±34%	1040±17%	1077±14%	1096±16%	1122±16%	1198±18%
	$\widehat{WVP}$ [Pa]	320±22%	156±53%	385±51%	411±40%	310±70%	301±18%	363±74%
	GHI [kWh m^{-2} m^{-1}]	17.7±12%	21.7±26%	23.7±21%	24.2±24%	23.4±25%	25.1±16%	26.5±22%
Spread	*DBT* [°C]	33.9±3.9	20.2±5.2	24.9±3.9	22.1±4.3	18.5±4.7	24.3±3.0	22.8±4.3
	WVP [Pa]	1459±19%	1518±35%	1461±22%	1309±24%	1247±26%	1529±16%	1517±21%
	GHI [kWh m^{-2} m^{-1}]	42.9±4%	40.5±12%	43.5±10%	43.4±10%	42.6±11%	44.5±8%	44.7±11%

***WVP* Hierarchical Clustering**

		Classes and fraction of sample						
Annual	**Quantities**	**1 (0.3%)**	**2 (8.4%)**	**3 (54.0%)**	**4 (12.2%)**	**5 (12.5%)**	**6 (4.7%)**	**7 (7.5%)**
Average	*DBT* [°C]	16.2	11.0±3.8	11.6±3.4	11.6±4.7	11.5±2.8	11.8±5.3	14.4±1.7
	$\widehat{DBT}$ [°C]	2.9	5.9±2.1	6.1±1.4	4.7±2.2	5.2±2.0	5.9±2.5	3.98±2.3
	WVP [Pa]	1022	1041±19%	1069±16%	1078±23%	1096±14%	1131±27%	1192±12%
	$\widehat{WVP}$ [Pa]	1112	619±26%	348±26%	545±17%	260±41%	131±59%	49.1±47%
	GHI [kWh m^{-2} m^{-1}]	32.0	23.4±29%	24.2±20%	24.7±24%	22.4±25%	24.4±20%	27.4±8%
Spread	*DBT* [°C]	21.9	23.8±3.9	24.5±4.1	22.5±4.7	18.9±5.0	24.0±5.1	24.6±3.8
	WVP [Pa]	582	1345±21%	1449±19%	1389±26%	1268±24%	1615±25%	1677±15%
	GHI [kWh m^{-2} m^{-1}]	49.4	42.4±10%	43.8±9%	42.4±10%	42.4±10%	44.5±9%	48.6±5%

Table 7: Global Hierarchical Clustering with additional weather quantities: mean values and standard deviations for each class and annual averages and spreads

		Global Hierarchical Clustering						
		Classes and fraction of sample						
Annual	**Quantities**	**1 (1.2%)**	**2 (40.8%)**	**3 (6.2%)**	**4 (29.8%)**	**5 (8.4%)**	**6 (10.0%)**	**7 (3.1%)**
Average	*DBT* [°C]	9.5±4.7	10.9±3.5	11.2±2.9	11.9±4.0	13.3±3.9	13.7±1.5	14.7±1.6
	$\overline{DBT}$ [°C]	2.2±1.0	6.3±1.6	2.9±0.9	6.1±1.1	3.1±1.1	6.9±1.0	2.1±0.6
	WVP [Pa]	988±31%	1027±16%	1081±13%	1096±19%	1153±19%	1173±14%	1259±10%
	$\overline{WVP}$ [Pa]	172±43%	459±27%	151±47%	290±18%	594±30%	156±74%	52.4±45%
	GHI [kWh m^{-2} m^{-1}]	19.8±30%	23.6±24%	21.0±24%	24.2±19%	26.4±23%	26.8±8%	27.5±9%
Spread	*DBT* [°C]	20.4±4.2	24.0±4.4	16.6±4.5	24.3±4.3	22.5±4.5	24.5±3.2	22.6±4.9
	WVP [Pa]	1373±38%	1362±22%	1211±22%	1491±18%	1423±26%	1621±19%	1670±18%
	GHI [kWh m^{-2} m^{-1}]	42.1±12%	42.8±10%	41.1±11%	44.2±9%	43.1±10%	47.0±6%	48.9±5%

Considering the geographical distribution of the obtained groups (Figure 4):

- Class **1** (grey) includes two alpine climates (Kasprowy Wierch and Sniezka), at the border between Poland and Czech Republic, Reykjavik, Iceland, and Bergen, Norway;
- Class **2** (yellow) is the largest and most varied one, with 130 climates located mainly in Italy, in the Alpin region, and in the Balkans;
- Class **3** (blue) is composed by 20 localities in the British Islands and in the French and Spanish regions surrounding the Bay of Biscay;
- Class **4** (green) includes 95 climates, located mostly in Scandinavia, central and eastern Europe and Russia;
- Class **5** (purple) has 27 coastal localities, mainly in Italy;
- Class **6** (light blue) has 32 locations, distributed mostly in inland Spain;
- Class **7** (red) includes 10 cities along the coast of the Iberian peninsula.

Although different from the classification developed in (Pernigotto *et al.*, 2019), it can be noticed a good degree of land and geographical continuity in the clusters. According to the dendrogram, further subdivisions show the northern Scandinavian and the Russian locations separated from the rest of class **4**, and class **2** divided into 3 sub-groups, distinguishing mountain, inland and coastal climates, in particular for the Italian peninsula.

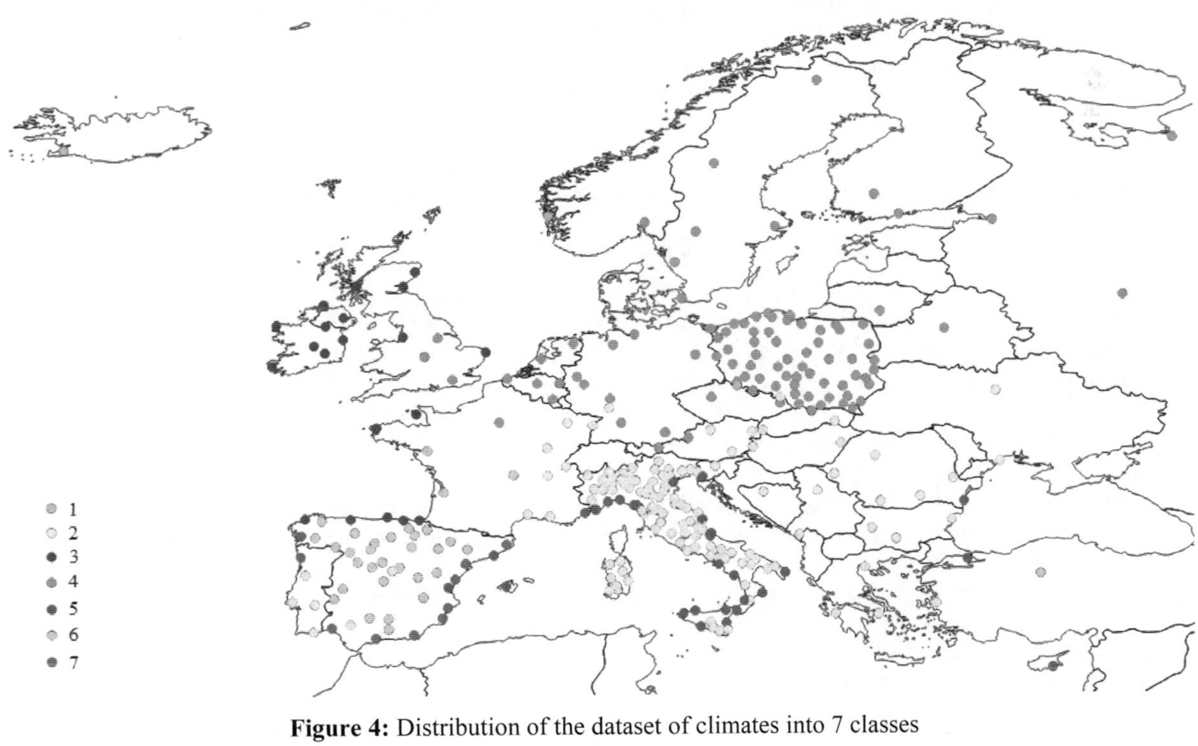

Figure 4: Distribution of the dataset of climates into 7 classes

4. CONCLUSIONS

This work tested two possible improvements to a methodology previously proposed by the Authors to perform climate clustering with the aim of identifying groups of locations sufficiently homogeneous for a robust selection of representative climates to use in building energy simulation analyses. According to the presented approach, typical or reference years are used as input to calculate monthly series of weather quantities which play a primary role on the building energy balance, i.e., the ambient temperature and humidity and the solar radiation. The series are then characterized by means of some weather statistics, e.g., annual average and spread of input series, and processed by means of a Hierarchical Clustering. The first goal of this research was to understand if a monthly time-discretization was adequate or some improvements were achievable using a shorter one, i.e., working with weekly series of weather quantities. The second goal, instead, aimed at discussing the impact of additional quantities describing the daily range of weather data, specifically for ambient temperature and humidity.
In order to answer to those research questions, the same dataset of more than 300 typical European climates already analyzed in previous studies by the Authors, was considered. Features of the generated climate classes were discussed and compared, considering also additional statistics besides average and spreads. In particular, the presence of distinct groups and their level of homogeneity were studied through mean and standard deviation values of weather statistics. We found that:

1. The adoption of weekly series of weather quantities can lead to groups different from those obtained with monthly series. Furthermore, significantly different results can be observed in Hierarchical Clustering, depending on the employed statistics. Among the tested alternatives, the combination of annual averages and spreads as statistical indicators seems the one giving the better results in terms of limiting interclass overlapping and enhancing uniformity.
2. As a whole, the improvements achieved by adopting shorter time-discretization of the inputs are limited and often related just to solar radiation. Furthermore, they are largely counterbalanced by important reductions in homogeneity for the other weather quantities, in particular for the ambient temperature. Consequently, the adoption of weekly series of inputs does not bring advantages in the framework of the proposed methodology.
3. As regards additional weather quantities descriptive of daily range of temperature and humidity, their inclusion in the clustering procedure led to a geographically meaningful classification with good degree of land continuity. Nevertheless, homogeneity was found decreased compared to previous analyses, requiring a further refinement of the largest classes. In general terms, the increase of the number of groups and a slightly loss of homogeneity look reasonable, especially for complex building systems for which daily variations of weather solicitation can have large effects on the energy performance.

ACKNOWLEDGEMENT

Funded by the project "Klimahouse and Energy Production", in the framework of the programmatic-financial agreement with the Autonomous Province of Bozen-Bolzano of Research Capacity Building.

REFERENCES

ASHRAE. (2013). ANSI / ASHRAE Standard 169-2013, Climatic Data for Building Design.

Peel, M.C., Finlayson, B.L., & McMahon, T.A. (2007). Updated world map of the Köppen-Geiger climate classification. *Hydrology and Earth System Sciences*, 11, 1633–1644.

Pernigotto, G., & Gasparella, A. (2018). Classification of European Climates for Building Energy Simulation Analyses. *Proceedings of the V International High Performance Buildings Conference at Purdue*. West Lafayette (Indiana, U.S.), 9-12 July 2018.

Pernigotto, G., Walsh, A., Gasparella, A., & Hensen, J.L.M. (2019). Clustering of European Climates and Representative Climate Identification for Building Energy Simulation Analyses. *Proceedings of Building Simulation 2019*. Rome (Italy), 2-4 September 2019.
Available at: http://www.ibpsa.org/proceedings/BS2019/BS2019_210938.pdf

Walsh, A., Cóstola, D., & Labaki, L.C. (2017). Review of methods for climatic zoning for building energy efficiency programs. *Building and Environment*, 112, 337-350.

Walsh, A., Cóstola, D., & Labaki, L.C. (2018). Performance-based validation of climatic zoning for building energy efficiency applications, *Applied Energy*, 212, 416-427.

Developing a Framework for Determining Optimum Dispatch of Energy to a Building from Conventional and Renewable Sources

Sultan AHMAD[1]*, W. Travis HORTON[2]

[1]Purdue University, Lyles School of Civil Engineering,
West Lafayette, IN, USA
+1 (765) 237-7802, sultan.ahmadsid@gmail.com

[2]Purdue University, Lyles School of Civil Engineering,
West Lafayette, IN, USA
+1 (765) 494-6098, wthorton@purdue.edu

* Corresponding Author

ABSTRACT

In an optimally designed grid-connected system with distributed energy resources where the grid plays the role of an energy buffer, it is interesting to analyze the economic feasibility to employ energy storage systems. A grid-connected system synchronizes with the power fluctuations, lowering the costs of energy compared to the cost of using conventional energy storage systems. An adaptive code is developed using computer programming that is used for lifetime simulation of the energy dispatch system with a specified time step and for optimization algorithm with comprehensive reliability/cost assessment. The results can be extended to a long period considering various economic factors. The programming code can be integrated with any system model, which can be flexibly implemented to any number of applications. In the present work, a strategic framework is developed for determining the optimal energy technology allocation to a typically selected commercial building located in the United States. The optimum design and management strategy of grid connected renewable generating systems composed of energy conversion units is considered. The provision of a hybrid system of energy storage is investigated. A genetic algorithm optimization-based approach is adopted for carrying out the optimization. The optimization of the set problem consisted of the minimization of the total lifecycle costs considered as the objective function, whereas the fulfillment of the users demand for energy was considered as the key constraint. The most suitable systems with an operation on hourly basis and the best strategy for the storage of energy were considered to generate the optimization results providing the optimal size and total cost of the system components. Furthermore, the possibility of using alternative energy dispatch systems was explored that might reduce the total lifetime costs below the cost of a benchmark case in which the entire demand for electricity is fulfilled from the grid. Four scenarios were analyzed to measure the impact of planning and operating the distributed energy resources: typical, off grid, on grid, feed-in-tariffs.

1. INTRODUCTION

In the recent times, increased energy demands have led to a dramatic rise in the consumption of fossil fuels and this in turn has led to highly raised energy prices and environmental pollution issues. Considering the effect of using fossil fuels on the environment, the necessity for using renewable energy (RE) to meet the increasing energy demand has been developed. The main challenges of using renewable energy include the associated high costs and the unpredictability of the renewable energy technologies. In this context, a promising scenario would be to overcome the challenges associated with renewable energy by integrating the renewable energy sources in order to meet the demand of energy in each area (Sharafi & ELMekkawy, 2014).

With the help of a widespread literature search, some implications are drawn regarding the Distributed Energy Resources [DER] systems and about its modeling research. It becomes evident that almost all DER systems and their modeling research originates from Asian and European regions, not dealing explicitly with the US regions energy policy. There has been no research which investigates a large range of energy technologies considering simultaneously- renewable (e.g. wind, solar), modern (e.g. geothermal pumps, different type of fuel cells), conventional (e.g. boilers, engines) and storage (e.g. pumped storage, batteries). Very few researches study technical, economic and environmental aspects all clubbed together (Mallikarjun & Lewis, 2014).

The design optimization of a DER system is considered critical, particularly when the net present cost of a DER system is high and therefore, the adaptation of a substandard optimal design would have a negative impact on the economics of the DER system significantly over a long run. In addition, the Kyoto protocol, implemented in the year 1997, obligates that the countries, which were industrialized should reduce their greenhouse gas emissions. Therefore, there is a need for research that would help in minimizing the lifetime costs as well as emissions from fossil fuel (Sharafi & ELMekkawy, 2014).

In the present study, an attempt is made to develop a multi-objective strategic framework for a typically selected commercial building with a main objective to determine the optimal size of energy components allocated for the DER system. The proposed framework considers economic, technical, and environmental concerns simultaneously. The minimization of the total annual energy costs and CO_2 emissions is considered as the problem's optimization target, while the main constraint is the fulfillment of the users' requirement for the electricity. An optimization tool is developed simulating an optimization algorithm. The heating, cooling, and electricity load profile on hourly basis are defined for the building in a reference year. The input data to run the model include hourly weather data of the year and the actual cost of the technologies. The objective functions are optimized by using the input data into the energy balance equations and relevant constraints. To achieve this goal, the research is extended to the following four sub-objectives:

1. Carry out review of the relevant literature on studies pertaining to existing modelling and simulation of various DER systems and applications of different techniques for optimization of the DER systems.
2. Formulate models to calculate the annual load requirements and CO_2 emissions by the DER system.
3. Implement optimization techniques in order to ensure that the desired energy efficiency is achieved without exceeding the level of CO_2 emissions beyond maximum permissible limits.
4. Conduct case studies on the optimization of the selected DER systems.

Different groups, who can be benefitted from the framework proposed through the present research, are listed below:

1. Occupants: People would be aware of the energy systems that might match with their requirements in a better way.
2. Policymakers: They will be able to assess the effect of subsidy schemes, such as steering taxes, net-metering or feed-in-tariffs, and carbon tax.
3. Companies: Information on sizing the system such that it is appealing to the users.
4. Researchers: People would be able to study novel technologies, which can be integrated with the DER.

2. RESEARCH METHODOLOGY

2.1 Problem Description

The trigeneration system is composed of four sources of energy, four transformation components, two storage components, and six end-uses. The four energy sources are solar, wind, electricity grid and natural gas grid. Energy conversion unit is composed of a photovoltaic (PV) panel, a wind turbine, an electrolyzer and a cogeneration fuel cell. The two types of storages are batteries and hydrogen tank. The energy flux of the proposed system framework is presented with a schematic diagram in Figure 1. Surplus energy is stored in respective storage components when the production of energy is more than the required load. Batteries are used to store the excess energy, which is produced for electricity. A part of excess energy, generated for electricity demand, is stored into batteries and the remaining part of the energy is put into the electrolyzer to convert into hydrogen. Electrolyzer produced hydrogen is transported and stored in the hydrogen tank (H_2-tank). Based on the H_2 tank capacity and the state of charge present in batteries, the fuel cell and batteries can provide the intended power to meet the load requirements when the energy produced by the electricity type does not meet the load. Electricity from the grid can be used as a source of an emergency power supply when the fuel cell and batteries are not able to satisfy the energy deficit.

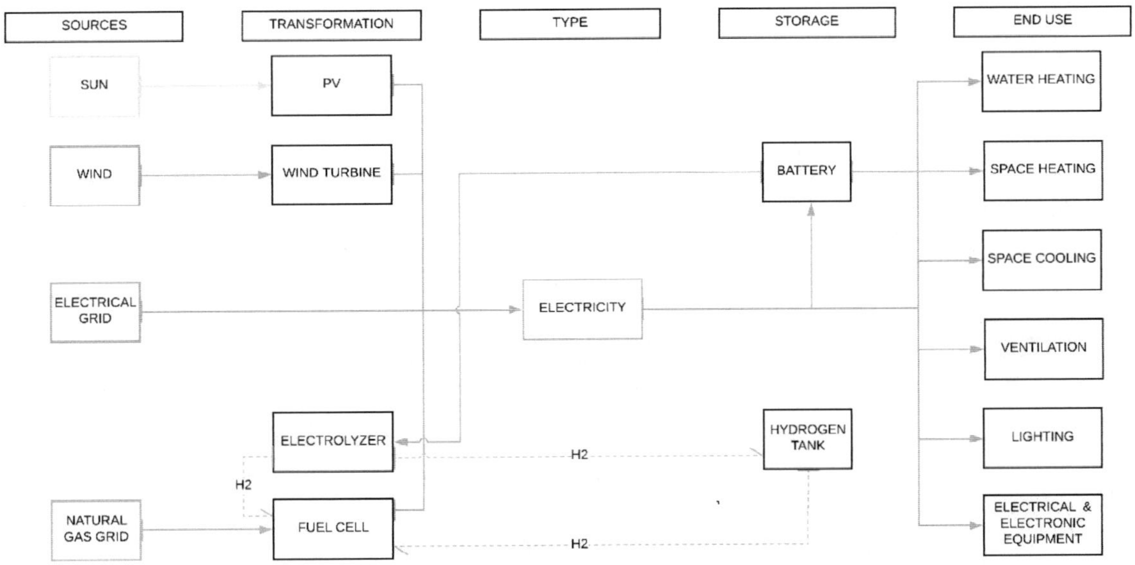

Figure 1: Energy flux of the proposed system framework

2.1.1 End Use: The reference building is a medium office building located in Albuquerque, New Mexico, United States. The building has a total rooftop area of 4982 m^2. Main constraint is the fulfillment of the occupant's load demand that includes electricity, space heating, space cooling and hot water. Electricity grid, PV panels, wind turbines, and electrolyzer produce electricity to meet the energy requirement of the users. Figure 2 shows the average monthly energy use of the building per type of electric load. The total energy load of the building is 4.25E+05 kWh.

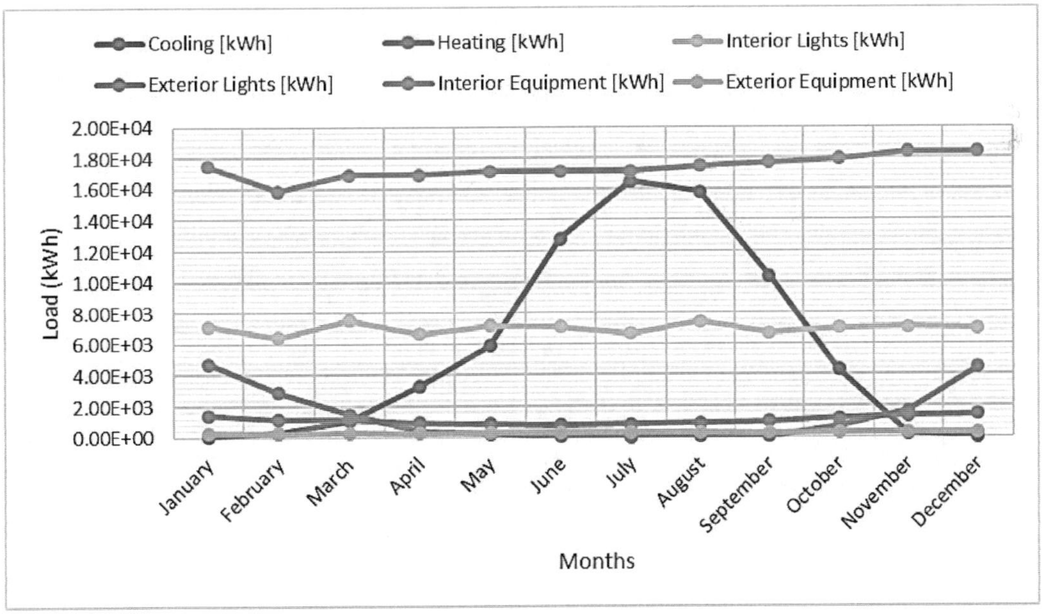

Figure 2: Monthly energy consumption by the building for different electric load type

The energy load requirement of the building is fulfilled by electricity. Energy load, E_{Load} can be classified as a combination of the Heating Load (L_H), Cooling Load (L_C), Electricity Load (L_{EL}), Hot Water Load (L_{HW}). All energy loads are provided in kWh, as shown in Figure 2. Based on the energy usage, electricity seems to be the primary source of heating.

3. MATHEMATICAL MODELS

The mathematical models, utilized in the simulation of conversion and storage components, are summarized below.

3.1 PV Panel

Photovoltaic cells or solar panels are devices that transform solar energy into electricity. Solar irradiation, wind velocity and air temperature data are used in the model in order to estimate the performance of the PV system. This data is obtained either from real-time measures or from online database and are location specific. The amount of energy supplied for each hour by the solar panel is measured using this data. A fixed tilt angle of 55° is used for the selected building location. The energy produced by PV panels can be calculated using Equation (1) (Ekren & Ekren, 2010).

$$E_{PV}(t) = (\eta_{PV} A_{PV}) I_T(t) \tag{1}$$

where, η_{pv} is the module efficiency of PVs, A_{PV} is the solar panel area (m^2), and I_T (t) is total solar radiation (hourly) received by the tilted surface. A constant η_{pv} equal to 15% is assumed in this study. The module efficiency value accounts for loss of power in solar panels due to losses in inverter, shade, temperature change, dirt, etc. The main constraint governing the conversion of solar energy into electricity is the direct dependence on the incoming radiation. The device may generate more power than its nominal output if the radiation is higher than that at the normal rating conditions (1000 Wm^{-2}). The rated panel power output is 6.2 kW.

3.2 Wind Turbine

In wind turbines, conversion of wind kinetic energy into electrical energy results in generation of wind power. The energy generated from a wind turbine can be calculated using Equation (2) (Ekren & Ekren, 2010).

$$E_{WT}(t) = \begin{cases} 0, & v < V_c \\ \frac{1}{2} C_p \rho A_{WT} v^3(t) \Delta t, & V_c < v < V_r \\ 0, & v > V_o \\ P_{WG,r}, & V_c < v < V_r \end{cases} \tag{2}$$

V(t) is an input variable and is defined as the hourly wind speed. C_p is the coefficient of performance, which is calculated as the ratio of the power output to maximum power and is obtained from the manufacturer's handbook. A_{WT} denotes to rotor swept area and ρ refers to the air density. V_c and V_r are the cut-in and rated wind velocity, respectively. V_c is taken as 4 m/s while V_r is taken as 14 m/s. V_o is the cut-off wind speed and is fixed at 20 m/s. $P_{WG,r}$ refers to the rated power of wind turbine which is 6.5 kw (Dufo-López & Bernal-Agustín, 2008).

3.3 Battery

In the battery-based renewable energy system, the battery size continuously changes due to the intermittent supply of electricity from solar panels and wind turbines. The state of charge (SOC) of the battery at any time period is determined using Equation (3):

$$SOC(t) = SOC(t-1) \pm \frac{E_{bat}(t)}{P_{bat}} \cdot 100 \tag{3}$$

where, SOC (t) and SOC (t-1) are the state of charge of battery in time interval t and t-1, respectively. $E_{bat}(t)$ is energy charged or discharged by the battery during hour t. P_{bat} is 10 kW which is the nominal capacity of battery. The sign is positive while charging and negative when it is discharging. Until it exceeds the lower limit of SOC_{min}, which is 30%, the battery will provide electricity to the system. In addition, the battery can be charged until 100% of SOC_{max} is achieved. (Abedi et al., 2012). The battery is in a charging state when the combined output of PV panels and wind turbines is greater than the load requirement. The battery is in discharge status when the combined output of PV panels and wind turbines is smaller than the demand for load. The charge quantity of the battery at time t can be obtained by Equation (4).

$$E_{bat}(t) = E_{bat}(t-1) \times (1-\sigma) \pm \left[E_{PV}(t) + E_{WT}(t) - \frac{E_{Load}(t)}{\eta_{inv}} \right] \times \eta_{bat} \tag{4}$$

where E_{bat} (t) and E_{bat} (t-1) are the battery charge at time t and t-1, σ = 5%, η_{inv} = 90 % represents the inverter efficiency, E_{load} refers to the load requirement, and η_{bat} = 80% is the battery efficiency in charging state. For this study, the discharge efficiency of battery is taken as η_{bat} = 100%.

3.4 Electrolyzer

Oxygen and hydrogen are extracted from water through electrical energy with the use of an electrolyzer. $Elec_{EL}$ is the electrical consumption capacity of the electrolyzer and is modeled as a function of nominal hydrogen mass flow rate (Q_{n-H_2}) (kg/h), and actual hydrogen mass flow rate (Q_{H_2}= 0.9 Q_{n-H_2}) (kg/h), as given in Equation (5). In this study, heating value, HHV_{H2} is considered as 38.7 kwh/kg, β_E = 40 kwh/kg and α_E = 20 kwh/kg, (Dufo-López & Bernal-Agustín, 2008)..

$$Elec_{EL} = \alpha_E Q_{n-H_2} + \beta_E Q_{H_2} \tag{5}$$

where, α_E, β_E are defined as the coefficients of electricity utilization curve. The efficiency of the electrolyzer is expressed as the produced HHV_{H2} divided by the consumption of electricity, as shown in Equation (6) (Dufo-López & Bernal-Agustín, 2008).

$$\eta_{EL} = \frac{Q_{H_2} \times HHV_{H_2}}{Elec_{EL}} \tag{6}$$

3.5 Fuel Cell

Fuel cells (FC) are selected as backup generator as they convert hydrogen and oxidants chemical energy into electrical energy. The total power output for a FC is defined through a function ($H_{2,cons-FC}$) (kg/h) which is the hydrogen consumption of the fuel cell and it is shown in Equation (7) (Dufo-López & Bernal-Agustín, 2008).

$$H_{2_{cons-FC}} = (\alpha_{FC} \times P_{n-FC}) + (\beta_{FC} \times P_{a-FC}) \tag{7}$$

where α_{FC}, β_{FC} are defined as the coefficients of hydrogen consumption curve and is an input of the model. Also, α_{FC} = 0.005 kg/kwh, β_{FC} = 0.06 kg/kwh. P_{n-FC} (kW) is defined as the nominal output power while P_{a-FC} (kW) is defined as the actual power of fuel cell. The manufacturers' manual recommends the maximum output power to be 90% of the nominal power. The energy efficiency is η_{FC} and it is calculated using Equation (8), where, LHV_{H2} = 33.3 kWh/kg.

$$\eta_{FC} = \frac{P_{a-FC}}{H_{2_{cons-FC}} \times LHV_{H_2}} \tag{8}$$

The energy produced by fuel cell during hour (t) is obtained by Equation (9):

$$E_{FC}(t) = H_{2_{cons-FC}}(t) \times LHV_{H_2} \tag{9}$$

3.6 Hydrogen Tank

To estimate the efficiency of the hydrogen tank, the charging efficiency of the electrolyzer and the discharge efficiency of the FC is taken into consideration. The electrolyzer would be used to fill the hydrogen tanks if the power produced from the PV/wind system is greater than the load requirement at time step t. Hydrogen level of the hydrogen tank at time t ($H_{2,level}$ (t)) is based upon hydrogen level at time t-1 (H_{2level}(t-1)), hydrogen consumption by fuel cells at time t,($H_{2,cons-FC}$ (t)), and hydrogen production by the electrolyzer at time t (Q_{H2}), Equation (10).

$$H_{2_{level}}(t) = H_{2_{level}}(t-1) + Q_{H_2}(t) - \frac{H_{2_{cons-FC}}(t)}{\eta_{H_2-tank}} \tag{10}$$

where, $\eta_{H2-tank}$ is defined as the efficiency of the hydrogen tank storage which specifies losses associated with pumping and leakage. $\eta_{H2-tank}$ is assumed 95% for this study. Additionally, the hydrogen level is demarcated with upper limit defined as the tank's nominal power while the lower limit is set as 5% of rated capacity (Kashefi et al, 2009). The capacity of hydrogen tank at time step t is obtained by Equation (11):

$$P_{H_{2_{tank}}}(t) = H_{2_{level}}(t) \times HHV_{H_2} \tag{11}$$

4. OPTIMIZATION MODELLING

A novel approach is presented for optimizing the size of the DER system. The genetic algorithm (GA) is utilized to minimize the objectives. GA is a method based on a natural selection process that mimics biological evolution to solve both constrained and unconstrained optimization problems. The algorithm modifies a population of individual solutions repeatedly. The GA randomly selects individuals from the current population at each stage and uses them as parents to create the children for the next generation. The population "evolves" toward an optimum solution over successive generations. An adaptive code is developed using multi-paradigm programming language that is used for the optimization algorithm.

4.1 Objective Functions

The objective functions to be minimized are:

- The total cost over the lifetime: Cost ($)
- The CO_2 emissions: CO_2 (kg/year)

The complete lifetime of a system is considered to be 25 years, similar to the life of solar panels, as they are considered to be the components having a better lifetime (Dufo-López & Bernal-Agustín, 2008).

4.1.1 System Installation Costs: The system installation cost consists fuel cost, investment cost, replacement cost, operation and maintenance cost, spread over the project lifetime Equation (12).

$$Installation\ Cost = \sum_j \left(\left[C_{I,x} + C_{O\&M,x} \times \frac{1}{CRF(i,T)} + C_{rep,x} \times K_x\right] \times P_x\right) + C_{fuel} \times fuel_{cons,yr} \times \frac{1}{CRF(i,T)} \quad (12)$$

where, x is system type, C_{Ix} is the capital cost per unit ($/unit), $C_{rep,x}$ is the replacement cost per unit ($/unit), $C_{O\&M,j}$ is the operation and maintenance cost per unit ($/unit), P_x is the size of each system. C_{fuel} is the fuel cost per unit ($/lit) and $fuel_{cons,yr}$ is the consumption of fuel per year (lit/year). K_x is net present cost and CRF is the capital recovery factor, and are represented by Equation (13, 14, 15) (Abedi et al., 2012; Dufo-López & Bernal-Agustín, 2008; Kashefi et al., 2009).

$$K_j = \sum_{n=1}^{R} \frac{1}{(1+i)^{L\times n}}, \quad \begin{cases} R = \left[\frac{T}{L}\right] - 1 & T\%L = 0 \\ R = \left[\frac{T}{L}\right] & T\%L \neq 0 \end{cases}, \quad CRF = \frac{i(1+i)^T}{i(1+i)^T - 1} \quad (13, 14, 15)$$

where, the interest rate is i, L and R are defined as lifetime and number of component replacements, respectively. T is the lifetime of the project that is assumed to be 25 years in this analysis.

4.1.2 Net Present Value of System Savings: NPV is presented in dollars which is estimated using sum of the total future cash flows over the investment lifetime. It is a common metric where the present value of future cash flows is calculated using the discount rate. For DER, the resulting energy savings are termed as future cash flows (House, 2017). NPV of each system (S_j) is calculated using the Equation (16):

$$NPV(S_j) = \sum_j \sum_{n=0}^{T} \frac{Cash\ Flow_n}{(1+d)^n} \quad (16)$$

Where, T is the lifetime of the project, n is a year within the lifetime, cash flow is the system cost for years n = 0 through N and d is the discount rate. Table 1 provides the data required for the calculation of the system used:

Table 1: Input Data for Net Present Value

Parameter	System	Parameter	System
Electricity Cost per kWh	14%	Fuel Inflation Rate	2%
Discount Rate	6%	Annual Degradation	0.6%

4.1.3 CO_2 Emission Cost: In this research, an overall environmental profile of different energy generation technologies has been analyzed via the concept of carbon footprint. The purpose of such analysis is to evaluate the complete life cycle of the energy producing technology, ranging from resource and fuel mining through construction to operation and waste management (Edenhofer et al., 2011). The cumulative ton of CO_2 released by the system over a period of 1 year is considered to be the right estimate of pollutant emissions. A CO_2 cost is an economic metric that lets consumers determine whether to stop, reduce or continue polluting and pay for it. In this way, the overall environmental goal is accomplished in the most adaptable and least expensive way to society and, therefore, it can be used as the objective to be minimized. CO_2 cost is calculated by using Equation (17).

$$CO_2\ emission\ cost = \sum_{t=1}^{8760} P_j(t) \times \Delta t \times EF \times CO_{2\,\mathrm{price}} \quad (17)$$

where P_j is the capacity of the technologies. $CO_{2,price}$ is $ 0.055 per kg of CO_2 produced per unit of electrical energy generated (Brooks, n.d.). EF is known as the emission factor and it depends on the type of fuel or technology used. The emission factor for different renewable energy technologies is summarized in Table 2 (Milousi et al., 2019).

Table 2: Life cycle CO_2 emission factors of the analyzed DER systems

Type	Emission Factor (g CO_2/kWh)	Type	Emission Factor (g CO_2/kWh)
PV	26-60	**Hydrogen tank**	10-24
Wind Turbine	9-35	**Electrolyzer**	26
Battery	35	**Fuel Cell**	26

4.1.4 Total System Costs: The total system cost is estimated by adding CO_2 cost to the difference between the cost of system installation and cumulative future cash flows over the lifetime. It is presented using Equation (18):

$$Total\ Cost = [\text{Installation Cost} - \text{NPV}(S_j) + CO_2\ emission\ cost] \quad (18)$$

4.2 Decision Variables

Decision variables are number of components in each technology and capacities of the technology. The following vector summarizes the capacities of the systems: $P = \{P_{PV}, P_{WT}, P_{Bat}, P_{Tank}, P_{El}, P_{FC}\}$, where, P_{PV} is defined as the capacity of PV panels (kW), P_{WT} refers to the capacity of the wind turbine (kW), P_{Bat} is the capacity of battery banks (kWh), P_{Tank} links to the capacity of $H_{2\text{-tank}}$ (kWh), P_{EL} refers to the capacity of electrolyzer (kW), P_{FC} is known as the capacity of fuel cell (kW).

4.3 Constraints

Operation constraints are applied on the component's energy and storage levels. The energy flux in each time step ($E_j(t)$) must be less than the component capacity as shown in Equation (19).

$$E_j(t) \leq P_j(t) \times \Delta t \quad (19)$$

where, Δt is defined as time interval of 1 h. There is a constraint on the area usable for the installation of solar panels on the building roof, as shown in Equation (20).

$$A_{PV} \leq A_{max} \quad (20)$$

where, A_{PV} is the area for installing PV panels. A_{max} is the maximum area, which is the total roof area of 4982 m^2 for the selected building. Constraint is put on the rotor swept area of wind turbines and is shown in Equation (21):

$$A_{WT,min} \leq A_{WT} \leq A_{WT,max}[\text{m}^2] \quad (21)$$

where, A_{WT} is the area of wind turbines. $A_{WT,min}$ and $A_{WT,max}$ is the minimum and maximum area of wind turbine respectively, considered as: $A_{WT,min} = 5$ m^2 and $A_{WT,max} = 8$ m^2. As mentioned before, the batteries can provide energy up to SOC_{min} lower limit and can be charged before SOC_{max} is achieved, Equation (22).

$$SOC_{min} \leq SOC(t) \leq SOC_{max} \quad (22)$$

where, SOC_{min} = 30% and SOC_{max} = 100% (Abedi et al., 2012). The demarcation of hydrogen level is shown in Equation (23).

$$H_{2_{level-min}} \leq H_{2_{level}}(t) \leq H_{2_{level-max}} \quad (23)$$

where, $H_{2,level-min}$ is the nominal capacity of H_2-tank (P_{Tank}), and $H_{2,level-max}$ is considered 5% of the rated capacity. There are nonnegativity constraints for decision variables and energy flux Equation (24), (25).

$$0 \leq P_j(t) \leq P_{j_{max}} \quad (24)$$

$$E_j(t) \geq 0 \quad (25)$$

The maximum unmet load (UL) permitted is an input of the proposed design tool. The UL (kWh/year) is specified as the amount of energy that could not be provided by the DER system over the lifetime. The unmet load is defined as the difference between total annual load demand and energy provided by DER system. The constraint on the unmet load is provided in Equation (26).

$$UL\ (\%) \leq 0\% \quad (26)$$

5. RESULTS AND DISCUSSION

The present study explores how the usage of renewable energy resources can result in the reduction of total system costs and carbon dioxide emissions relative to a base case scenario where all the energy used by the selected building is provided through the electricity grid. Four scenarios are analyzed to measure the impact of optimizing the DER systems to design and operate residential or commercial energy systems: typical, off grid, on grid and feed-in-tariffs. Case 1: 'Typical' pertains to the fulfillment of the demand for electricity from the grid. Case 2: 'Off grid' relates to a consumer that does not take electricity from the grid but can sell electricity to grid. Case 3: 'On grid' refers to a consumer that has provision to satisfy the demand for electricity from the grid as well as from all possible renewable sources. However, excess electricity produced cannot be exported to the grid. Case 4: 'Feed-in-tariffs' corresponds to the exchange (export and import) of electricity at the same price. Figure 3 shows the yearly energy produced by solar panels and wind turbines in each case scenario. Case 1 is the reference case where all the energy requirement is provided by the electricity grid. Case 2 has almost equal contribution by solar panels and wind turbines. The yearly energy reduction by solar panels is 50% whereas by wind turbines is 41%. Cases 3 & 4 mostly favor the installation of solar panels. Energy reduction by solar panels and wind turbines is 74% and 16% respectively.

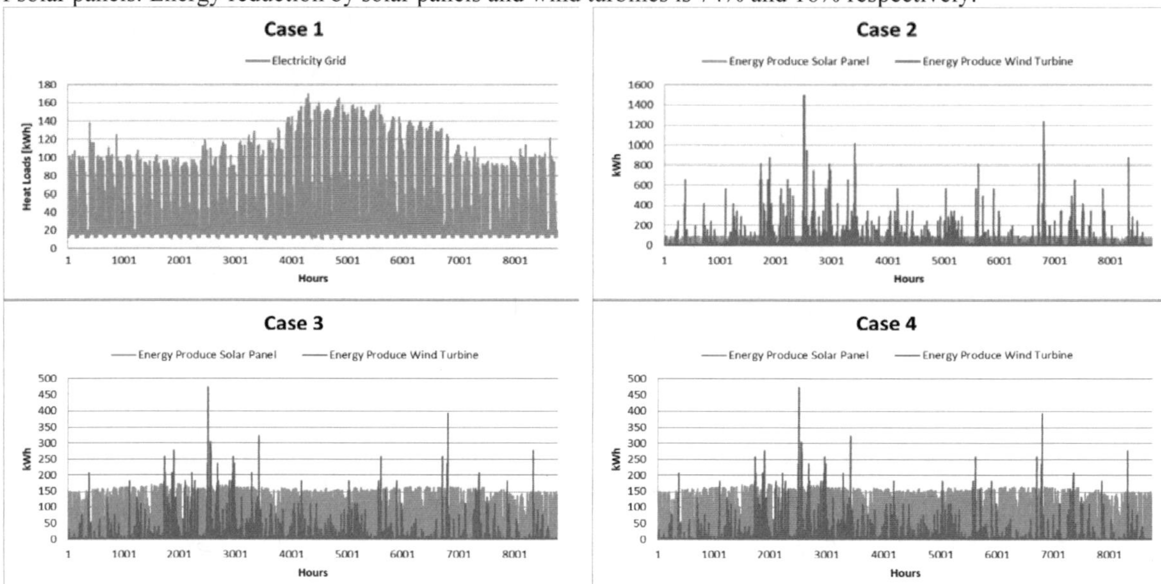

Figure 3: Energy produced by the system components

Figure 4 shows the optimal size of the system component for each case scenario. There is a 50% increase in the system total capacity for 'on grid and 'feed-in-tariffs' compared to that of the 'off grid' case scenario. The reason for the increase of the battery capacity by almost four times is that now the system cannot sell the excess energy to the grid and to accommodate any unexpected peak load requirements, the system needs to be always well equipped. This can be achieved through an increase in the storage size (battery capacity), an increase in the fuel cell capacity, or a combination of both. Indeed, 'on grid' case observes a growth in the capacity of fuel cell from 16.42 to 30.59 kW and rise in the solar panel capacity from 65 to 122 kW. Wind capacity is however decreased due to its increased installation costs. The most extreme change in the capacity is that of the battery, due to two key factors. Firstly, the solar panels and the fuel cell generate ample energy to sustain a battery charge state high enough to cover any unpredicted variation in the load demands. Secondly, the life cycle cost of increasing the capacity of battery is almost three times lesser compared to the life cycle cost of increasing the capacity of the fuel cell.

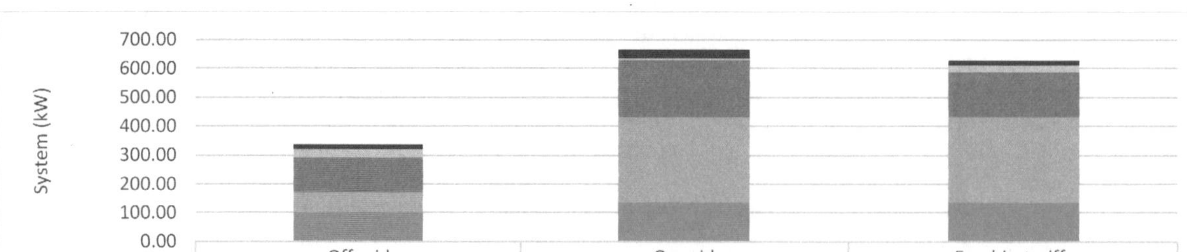

	Off grid	On grid	Feed-in-tariffs
Fuel cell	16.42	30.59	15.42
Electrolyzer	28.27	5.23	23.67
Hydrogen Tank	120.68	197.44	154.72
Battery	68.53	297.49	298.99
Wind Turbine	37.53	11.92	11.92
Solar Panel	65.72	122.81	122.81

Figure 4: System component size

Figure 5 shows the system component cost and savings in its lifetime of 25 years. The cost of electricity provided by the grid is $919,192. Both solar panels and wind turbines provide energy cost savings, which is higher than their installation costs. Therefore, it has net present savings after the studied period of 25 years. Off grid system has higher savings due to wind turbines compared to solar panels. However, both on grid and feed-in-tariffs systems have a higher savings due to solar panels. On grid storage systems have a lower life cycle costs compared to the other two cases. Interestingly, within the considered lifetime of 25 years, there is total life cycle savings of $267,815 only by the off grid scenario. There is no savings by the other two cases in the first 25 years. However, there is almost 50% decrease in the life cycle costs by the feed-in-tariffs system compared to on grid system.

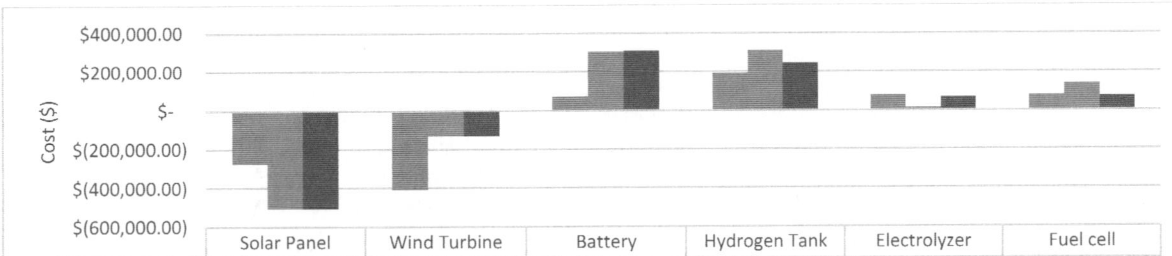

	Solar Panel	Wind Turbine	Battery	Hydrogen Tank	Electrolyzer	Fuel cell
Off grid	$(271,350.97)	$(404,039.28)	$70,331.36	$188,571.74	$76,084.72	$73,483.99
On grid	$(507,035.76)	$(128,329.78)	$305,334.55	$308,516.52	$14,073.84	$136,907.31
Feed-in-tariffs	$(507,035.76)	$(128,329.78)	$306,874.08	$241,766.75	$63,721.82	$69,007.77

Figure 5: System component life cycle cost

Table 3 provides a summary of the results obtained through the optimization. According to the results, off grid system seems like the best option as it has the lowest levelized cost of energy of 0.053 $/kWh. In addition, it has a simple payback period of 16 years, which is better than the higher payback period of on grid and feed-in-tariffs which gives a return after additional 10 years.

Table 3: Summary of results

	Off grid	**On grid**	**Feed-in-tariffs**
Area of solar panels (m^2)	529.86	990.07	990.07
Number of wind turbines	5.77	1.83	1.83
Number of batteries	6.85	29.75	29.9
Levelized cost of energy ($/kWh)	0.053	0.090	0.082
Simple payback (years)	16.95	28.38	26.5

Even though there are no added incentives that promote the use of renewable energy, it is possible to minimize lifecycle costs and CO_2 emissions by combining today's available technologies. Uncertainty in energy production and requirement is not considered while obtaining the results. Overall, the developed tool proves to be instrumental in determining investment decisions that are resilient in terms of uncertainty in weather parameters.

6. CONCLUSION

An innovative and novel method is presented in this paper that can be used to optimize the size of a distributed energy resource system. The genetic algorithm method is utilized to minimize the objectives that include the total lifecycle cost of the system and CO_2 emission. In order to solve the multi-objective optimization problem, the developed tool used a simulation-based method. One of the key advantages of the proposed method is its easy and effortlessness execution, which results in computational efficiency. The proposed study is evaluated in a case study that includes solar panels, wind turbines, batteries, hydrogen tank, electrolyzer and fuel cell. Four scenarios were analyzed to measure the impact of planning and operating the distributed energy resources: typical, off grid, on grid, feed-in-tariffs. By comparing the four scenarios, it was concluded that the total cost was improved in all the cases, with additional cost savings in one of the them. The proposed tool can be utilized in research studies and the design of a DER system. The approach can easily be extended to heating/cooling loads as well as domestic hot water loads. The framework is developed in such a way that any number of conventional and renewable energy resources can be added to a selected building to obtain an optimized system with minimum cost and CO_2 emissions.

For future research, a sensitivity analysis can be done in order to see the effect of a certain technology based on its economic or environmental impact under different climatic conditions. Another sensitivity analysis can be performed on the developed model to analyze the sensibility of the input parameters. The analysis would be used to predict the outcome if the efficiency and cost of the system components are changed. Furthermore, a comprehensive analysis on the load, that separately takes into account the heating/cooling and electricity, can be done to generate an optimum system configuration that matches the load. Lastly, uncertainty on the availability of intermittent energy sources can be studied as well.

REFERENCES

Abedi, S., Alimardani, A., Gharehpetian, G. B., Riahy, G. H., & Hosseinian, S. H. (2012). A comprehensive method for optimal power management and design of hybrid RES-based autonomous energy systems. *Renewable and Sustainable Energy Reviews*, *16*(3), 1577–1587. https://doi.org/10.1016/j.rser.2011.11.030

Brooks, S. (n.d.). The true cost of carbon pollution. Retrieved from Environmental Defense Fund website: https://www.edf.org/true-cost-carbon-pollution

Dufo-López, R., & Bernal-Agustín, J. L. (2008). Multi-objective design of PV-wind-diesel-hydrogen-battery systems. *Renewable Energy*, *33*(12), 2559–2572. https://doi.org/10.1016/j.renene.2008.02.027

Edenhofer, O., Pichs-Madruga, R., Sokona, Y., Seyboth, K., Eickemeier, P., Matschoss, P., … Stechow, C. Von. (2011). IPCC, 2011: Summary for Policymakers. In: IPCC Special Report on Renewable Energy Sources and Climate Change Mitigation. In *Cambridge University Press*. https://doi.org/10.5860/CHOICE.49-6309

Ekren, O., & Ekren, B. Y. (2010). Size optimization of a PV/wind hybrid energy conversion system with battery storage using simulated annealing. *Applied Energy*, *87*(2), 592–598. https://doi.org/10.1016/J.APENERGY.2009.05.022

House, T. (2017). Quantifying the Value of a Solar Installation: Some Helpful Metrics. Retrieved from AuroraSolar website: https://blog.aurorasolar.com/quantifying-the-value-of-a-solar-installation/

Kashefi Kaviani, A., Riahy, G. H., & Kouhsari, S. M. (2009). Optimal design of a reliable hydrogen-based stand-alone wind/PV generating system, considering component outages. *Renewable Energy*, *34*(11), 2380–2390. https://doi.org/10.1016/j.renene.2009.03.020

Mallikarjun, S., & Lewis, H. F. (2014). Energy technology allocation for distributed energy resources: A strategic technology-policy framework. *Energy*, *72*, 783–799. https://doi.org/10.1016/j.energy.2014.05.113

Milousi, M., Souliotis, M., Arampatzis, G., Papaefthimiou, S., Intergovernmental Panel on Climate Change, Intergovernmental Panel on Climate Change, … Shahsavar, A. (2019). Multi-criteria decision support system of the photovoltaic and solar thermal energy systems using the multi-objective optimization algorithm. *Energy*, *11*(March), 1–10. https://doi.org/10.3390/su11092539

Sharafi, M., & ELMekkawy, T. Y. (2014). Multi-objective optimal design of hybrid renewable energy systems using PSO-simulation based approach. *Renewable Energy*, *68*, 67–79. https://doi.org/10.1016/j.renene.2014.01.011

ACKNOWLEDGEMENT

We would like to acknowledge the support received from the Lyles School of Civil Engineering at Purdue University, West Lafayette, USA, in conducting the presented research work.

Development of a shoeboxing approach for Urban Building Energy Modeling

Federico BATTINI[1]*, Giovanni PERNIGOTTO[2], Andrea GASPARELLA[3]

[1]Free University of Bozen-Bolzano, Faculty of Science and Technology,
Bolzano, Italy
Tel: +39 0471 017888, Fax: +39 0471 017009, email: federico.battini@natec.unibz.it

[2]Free University of Bozen-Bolzano, Faculty of Science and Technology,
Bolzano, Italy
Tel: +39 0471 017632, Fax: +39 0471 017009, email: giovanni.pernigotto@unibz.it

[3]Free University of Bozen-Bolzano, Faculty of Science and Technology,
Bolzano, Italy
Tel: +39 0471 017200, Fax: +39 0471 017009, email: andrea.gasparella@unibz.it

* Corresponding Author

ABSTRACT

Urban Building Energy Modeling aims at assessing the building energy performance at city scale with as little computational effort as possible. Thus, different methods have been developed in the last years to reduce the required calculation time by simplifying the modeling approach, selecting only representative buildings, or minimizing the building description. Starting from the latter ones, this work proposes a novel algorithm capable of abstracting a randomly shaped building into a representative shoebox. The presented shoebox generation algorithm is based on a preliminary sensitivity screening analysis on a set of reference parallelepiped-shaped thermal zones. This allowed the identification of the most significant geometry indicators influencing the building's performance. Based on this, more complex geometries have been simplified to the shoebox with the same indicators and the accuracy of the algorithm has been evaluated comparing the simulated performance of simplified and original buildings. The approach includes the definition of equivalent shading surfaces, to account for self-shading elements in the original building geometry. The algorithm has shown good accuracy not only on the hourly thermal loads, but also the zones' hourly temperature profiles, reducing to one third the energy simulation time with respect to the detailed building model. Although not as fast as other urban modelling approaches in the literature, it can retain accurate results at a finer time scale, i.e., on hourly basis, which is necessary in applications such as district heating and energy networks.

1. INTRODUCTION

Cities account for 60 to 80 % of the world's energy consumption and generate as much as 70 % of human-induced greenhouse gas emissions (United Nations Habitat, 2019). The building sector, in particular, is responsible for a large share of the overall energy use, corresponding for instance to 40 % in the European Union (European Parliament and Council of the European Union, 2010). Consequently, decreasing buildings' energy demand is essential to tackle the climate crisis. In the European Union, the current annual renovation rate of the building stock varies between 0.4 and 1.2 %, depending on the Member State, and it is largely insufficient for a fast reduction of the global building energy demand. By means of the European Green Deal, the European Commission aims at least to double the current renovation rate to meet its long-term climate objectives (European Commission, 2019). To achieve these goals, however, it is necessary to adopt new approaches in order to estimate the building energy demand at large spatial scales, focusing on groups of buildings and districts. Among the available alternatives, the most known is the Urban Building Energy Modeling *UBEM*, which allows for a bottom-up city-scale energy modeling using dedicated Building Performance Simulation codes (Johari *et al.*, 2020) specifically introduced to overcome the limitations of traditional energy modeling at single building level. Indeed, the same workflow employed for characterizing the energy performance of a single building cannot be directly applied to large groups, since it would require too much information and would be too computationally demanding (Reinhart and Davila, 2015). Therefore, to guarantee time and efficiency, *UBEM* relies on domain and model simplifications (Abbasabadi and Ashayeri, 2019), such as (i) reduction of building definition complexity, (ii) introduction of simplification algorithms, and (iii) simulation of representative buildings out of the entire building stock.

The first option pertains thermal model definition and simulation choice, which can considerably reduce the simulation time compared to detailed models. For instance, building energy balance can be solved as a steady state heat balance for each single thermal zone, including the surrounding context (Nouvel *et al.*, 2013) or not (Dall'O *et al.*, 2012). Furthermore, the number of thermal zones can be reduced as well, ranging from multi-zone (Mastrucci *et al.*, 2014) to single-zone models (Caputo *et al.*, 2013).

The second possibility regards the introduction of less resource intensive procedures. Such approaches are very different from one another and include, for example, low order thermal network models (Lauster *et al.*, 2014), abstracting and arbitrary building massing into a meaningful group of thermal shoebox models (Dogan and Reinhart, 2013), and reorganizing the building geometries with functional clustering and radiation analysis scaling (İşeri and Dino, 2020).

The third alternative goes beyond the usual archetype definition, often employed mainly for defining buildings' non-geometrical features. Representative buildings are found by means of a set of descriptive indicators to capture the energy-relevant features such as physical, operational and contextual characteristics. In this way, a more generic approach is developed for stock classification considering the dynamic nature of the urban context (Ghiassi *et al.*, 2017).

Despite all solutions listed above, the optimization of *UBEM* computational efficiency is still an open issue (Hong *et al.*, 2019). In this work, taking advantage of previous studies showing the potential of shoeboxes in urban simulation (Dogan and Reinhart, 2017), a new simplification algorithm is proposed. The algorithm is meant to simplify each building in the urban model into an equivalent representative shoebox, taking into account buildings' adjacencies and context obstructions. With respect to similar approaches in the literature, the outlined algorithm aims to improve the accuracy of energy demand prediction when dealing with complex building geometries, even at fine time scales such as hourly, while reducing the computational cost of the simulation.

2. METHODOLOGY

The workflow to develop a shoeboxing approach for *UBEM* consists of three parts: (a) identification and selection of geometry indicators for building simplification through sensitivity analysis, (b) shoeboxing algorithm development, and (c) extensive test of algorithm effectiveness.

2.1 Geometry indicators identification and selection

To simplify a randomly shaped building into a meaningful shoebox, the first step is to identify a set of geometry indicators properly describing a building. In this work, geometrical indices employed in the literature have been considered and selected through a multivariate sensitivity analysis performed on annual (1) heating needs, (2) cooling needs, (3) heating peak loads, and (4) cooling peak loads.

2.1.1 Indicators identification: Geometrical indicators can describe either the building's footprint or the entire massing. As regards the former one, the polygon shape indices used by SAGA-GIS have been considered (Conrad, 2020). Concerning the latter one, this study has adopted the same indices used by Kodors (2017) for geometric feature selection at the urban level and by Ghiassi and Mahdavi (2017) for multivariate cluster analysis in the development of a reductive bottom-up urban energy model. Finally, the shape factor has been also considered, given its influence on the final energy demand (Danielski *et al.*, 2012). As a whole, 13 indicators have been used and tested.

2.1.2 Indicators selection: To select the most relevant indices affecting the thermal demand, a sensitivity analysis has been performed on a parallelepiped-shaped thermal zone, with a fixed height and length and depth varied between 4 and 12 m according to a parametric plan, as in previous studies (Pernigotto *et al.*, 2014). A portion of each envelope surface has been modelled as adiabatic to simulate adjacencies to thermal zones at the same conditions. The adiabatic fraction of each surface has been changed independently and can vary between 0 and 1, i.e., each surface can be totally exposed to the external environment or completely adiabatic.

Table 1: Spearman's rank correlation coefficients color-coded in grey scale to highlight correlations' strength (the darker the cell, the stronger the correlation).

Geometry index	Heating		Cooling	
	Annual	Peak	Annual	Peak
Plan area	0.43	0.55	0.18	0.37
Plan perimeter	0.43	0.55	0.18	0.37
Edge ratio	-0.43	-0.54	-0.18	-0.37
Elongation1	-0.10	-0.13	-0.05	-0.09
Elongation2	0.10	0.13	0.05	0.09
Eccentricity	-0.10	-0.13	-0.05	-0.09
Form factor	0.10	0.13	0.05	0.09
Compactness	-0.10	-0.13	-0.05	-0.09
Volume	0.43	0.55	0.18	0.37
Surface area	0.43	0.55	0.18	0.37
Shape factor	-0.43	-0.54	-0.18	-0.37
Non-adiabatic surface area	0.44	0.52	0.13	0.34
Non-adiabatic area shape factor	0.00	-0.04	-0.07	-0.04

Windows have been modelled as portions of the externally exposed vertical walls by a window-to-wall ratio ranging between 0 and 0.9. The geometrical definition of the model has been completed by adding vertical and horizontal overhangs whose depth can vary between 0 and 2.8 m. The opaque envelope has been characterized by an insulating and a massive layer, for which only the thickness has been considered as variable. The same wall composition, characterized by a thermal transmittance varying between 0.17 and 3.28 W m^{-2} K^{-1}, has been applied to all opaque surfaces. Windows have been defined by means of the type of glass, filling gas and number of gaps, giving a total of 17 combinations with transmittance and solar heat gain coefficient ranging from 0.41 W m^{-2} K^{-1} and 0.49 to 5.75 W m^{-2} K^{-1} and 0.84, respectively. For the remaining heat balance contributions and boundary conditions time-constant values have been assigned. The ventilation rate has been expressed as air changes per hour ranging between 0 and 1. Internal gains varying from 0 to 8 W m^{-2} have been included. Finally, the HVAC system has been considered as always active. Moreover, constant heating and cooling setpoints have been varied between 18 and 22 °C and from 22 to 30 °C, respectively.

As sampling technique, the Latin hypercube sampling has been used, and the number of variants has been progressively increased until convergence of the employed statistics has been reached. An overall number of 20000 different thermal zone configurations have been simulated with EnergyPlus, assuming ideal heating and cooling systems in the climate of Denver, Colorado. The Spearman's rank correlation coefficient ρ has been adopted. This statistic evaluates how well the relationship between two variables can be described by a monotonic function. It can be either positive or negative, i.e., $-1 \leq \rho \leq 1$. A value of $|\rho|$ larger than 0.7 points out a strong correlation, a value between 0.3 and 0.7 a moderate correlation, while a value lower than 0.3 a weak or even null correlation. ρ for each of the four simulation outputs has been computed for every investigated indicator. The outcome of the sensitivity analysis is reported in Table 1. Since a shoebox can be defined by three dimensions, it has been decided to select three indicators, in order to solve a system of three equations with three unknowns. Indicators have been selected considering the correlation strength, as well as their linear correlations. Furthermore, more complex indicators have been preferred to the simplest ones (e.g., although characterized by similar ρ, the edge ratio has been selected instead of area and perimeter). As shown in Table 1 and Equation (1), the selected indices are edge ratio, shape factor and the first elongation definition:

$$\begin{cases} \text{edge ratio} = \dfrac{\text{perimeter}}{\text{area}} \\ \text{elongation} = \dfrac{\text{minor length}}{\text{major length}} \\ \text{shape factor} = \dfrac{\text{surface area}}{\text{volume}} \end{cases} \quad (1)$$

2.2 Shoeboxing algorithm development

Once the most meaningful indicators have been selected, it has been possible to proceed with the development of the simplification algorithm. The algorithm has been generated in the framework of Rhinoceros as CAD modeling platform, its plug-in Grasshopper, and EnergyPlus as Building Performance Simulation tool. In details, the workflow has been built in Grasshopper by means of customized components written in Python programming language. Geometrical modelling tasks have been implemented using RhinoCommon API (Robert McNell & Associates, 2020), while the conversion from massing model to thermal zones performed by using the Ladybug Tools core SDK (Ladybug Tools LLC, 2020). First, each building has been converted into a representative shoebox; then, the three selected indicators have been computed, based on its geometrical characteristics. The calculated indicators have been employed to estimate the representative shoebox's dimensions (x, y and z, respectively for width, depth and height) by solving a system of non-linear equations, as reported in Equation (2).

$$\begin{cases} \text{edge ratio} = \dfrac{2\,(x+y)}{x \cdot y} \\ \text{elongation} = \dfrac{x}{y} \\ \text{shape factor} = \dfrac{2(x \cdot z) + 2(y \cdot z) + 2(x \cdot y)}{x \cdot y \cdot z} \end{cases} \tag{2}$$

The resulting shoebox has been characterized by the same window-to-wall ratio as the starting building and the same floor height. Also, the thermophysical properties have been assigned accordingly and the building's self-shading included. Specifically, the impact of the building's self-shading has been taken into account through equivalent shading surfaces obstructing a portion of the windows. Those shading elements have been determined by means of a radiation analysis performed on both starting geometry and resulting shoebox according to the all-weather Perez sky model. For sake of simplicity, for the starting geometry the radiation analysis focused only on the windows' areas, while the entire geometry has been accounted for in the shoebox. Since considering only windows surfaces has considerably reduced the area to analyse, a calculation grid of 0.2 m has been employed. The average annual entering radiation has been obtained for each cardinal direction and floor for both models, and the ratios between starting geometry and shoebox results computed. Then, an equivalent obstruction ratio has been calculated for each façade and floor:

$$\text{obstruction ratio} = 1 - \frac{\text{starting geometry entering radiation}}{\text{shoebox entering radiation}} \tag{3}$$

For every window in a given floor and façade of the simplified shoebox model, the corresponding ratio has been used to calculate the equivalent shading area out of the window's area. The modelled shadings have the same height of the window, while their length has been computed to match the equivalent shading area calculated before. The equivalent shading surfaces have been placed 1 cm outside the window to fully obstruct all incident radiation. In this research, the starting building and the corresponding shoebox have been modelled as stand-alone since the focus has been put on the determination of equivalent shading surfaces to account for self-shading. Nevertheless, this approach can be generalized to account for the urban context by modelling also the building's surroundings. Moreover, also adjacencies can be taken into account by means of adiabatic elements in the facades, whose ratios can be computed independently for each cardinal direction by analyzing the façades of the starting building and considering a ± 45° tolerance range. As a final step, detailed and simplified models have been automatically exported as idf files and edited by means of Python scripts with the aid of the library eppy (Philip, 2020), to interface EnergyPlus with Python language.

2.3 Extensive test

The driving purpose of the proposed algorithm is the improvement in building's performance prediction when dealing with complex massing shapes. In the literature, various works studying the urban context relied on the creation from scratch of sample neighborhoods (Lyu *et al.*, 2019), urban layouts (Xu *et al.*, 2019) or building grid models (Natanian *et al.*, 2019). However, in such configurations buildings are simple in shape, mainly with rectangular footprints. Furthermore, the same pattern is repeated to parametrically generate buildings' arrangement.

Table 2: Number of polyominoes depending on the number of cells

Number of cells	1	2	3	4	5	6	7	8	9	10
Number of polyominoes	1	1	2	5	12	35	108	369	1285	4655
Polyominoes with holes	0	0	0	0	0	0	1	6	37	195
Polyominoes without holes	1	1	2	5	12	35	197	363	1248	4460

Differently, Javanroodi *et al.* (2018) employed a series of form generation rules based on a basic module to generate buildings' forms based on urban density ranges. In this work a similar approach has been adopted, defining buildings' shapes out of polyominoes (Golomb, 1994) in order to guarantee complex and various shapes. A polyomino is created out of square cells by joining one or more squares edge to edge; their number increases according to the number of cells following the integer sequence A000105 (Sloane, 2020), as reported in Table 2. To keep the size of the analysis manageable, the number of cells utilized to create polyominoes has been varied between two and six. Among them, rectangular in shape geometries have been discarded, resulting in 48 unused forms. Figure 1(a) reports all possible 55 shapes, which have been used as base for the building geometries. As reported in Table 2, up to six cells no polyomino has holes. Therefore, courtyard buildings are not considered in this analysis unlike other studies (Javanroodi *et al.*, 2019). Nevertheless, the variety of shapes considered, even without courtyard, has been considered adequate to guarantee properly testing of the algorithm. To be consistent with the sensitivity analysis previously performed, polyominoes have been generated from a starting square cell with an edge of 4 m. To explore the space of solutions as much as possible, each resulting shape has been: (i) rotated according to the four cardinal points, (ii) scaled by a factor equal to one, two, three or four, and (iii) extruded by a number of floors equal to one, two, three or four, with each floor 3 m tall. Thus, from each polyomino, a total of 58 building shapes has been generated, as shown for instance in Figure 1(b). The total amount of generated buildings to be simulated has been 3072.

Once all building shapes have been generated, each building has been converted into its representative shoebox according to the simplification algorithm. At this point, the same thermophysical and operational features have been assigned to every simulation file (Table 3). The envelope definition is the same for every external opaque surface, with an external insulating and an internal massive layer, i.e., polystyrene and concrete respectively. As for the envelope, one definition has been adopted for the windows as well. For sake of simplicity, the frame has been neglected and a triple glass window with air filling (3 mm clear glass panels, with the innermost panel with low-e treatment, and 6 mm air gaps) has been considered. Finally, to test the effectiveness of the algorithm under different boundary conditions, the buildings have been simulated in three climates - Bolzano and Messina, Italy, and Denver, U.S., selected to include both a heating- (Bolzano) and a cooling-dominated (Messina) climate, as well as a climate characterized by large daily temperature variations (Denver).

The investigated outcomes of this research have been both temperature and energy needs. The temperature output has been evaluated on hourly basis, considering just the hours for which the temperature is larger than 20 °C and lower than 26 °C, and calculating also the annual positive and negative maximum differences. To understand how well the simplification algorithm performs overall, the share of hours in which the hourly deviation lays within ± 0.5 °C has been calculated. The energy needs have been analyzed at annual and hourly scale. Annual absolute and relative differences between detailed and simplified models have been computed and the Root Mean Square Error *RMSE* calculated. Furthermore, the percentage of hours of the entire year in which the hourly difference between the models lays within specific intervals (i.e., ± 5 %, ± 10 %, ± 20 %) has been computed to assess hourly deviations. Such evaluations have been carried out at building and floor level, since every floor has been modelled as a thermal zone.

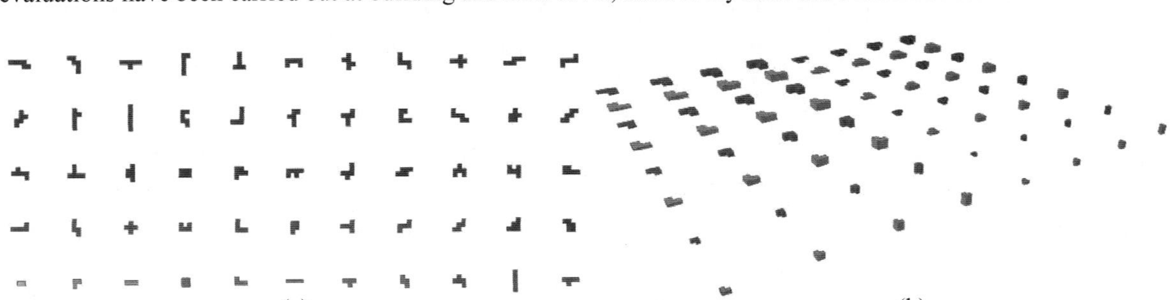

(a) (b)

Figure 1: (a) Possible polyominoes obtained by combining up to 6 square cells and **(b)** all buildings' combinations resulting from one polyomino's footprint after rotation, scaling and extrusion

Table 3: Energy model settings

OPAQUE ENVELOPE					
Layer	**Thickness [m]**	**Thermal conductivity [W m^{-1} K^{-1}]**	**Specific heat capacity [J kg^{-1} K^{-1}]**	**Density [kg m^{-3}]**	**U-Value no film [W m^{-2} K^{-1}]**
Polystyrene	0.1	0.05	1470	40	0.329
Concrete	0.2	0.37	840	1190	

FENESTRATION			
Window-to-wall ratio	**Glass U-Value [W m^{-2} K^{-1}]**	**Glass SHGC**	**Glass Visible Transmittance**
40 %	1.772	0.578	0.698

HEATING SYSTEM			COOLING SYSTEM		
Setpoint	**Schedule**	**Type**	**Setpoint**	**Schedule**	**Type**
20 °C	Always ON	Ideal	26 °C	Always ON	Ideal

VENTILATION			INTERNAL GAINS		
Type	**Rate [ach]**	**Schedule**	**Type**	**Power [W m^{-1}]**	**Schedule**
Infiltration	0.3	Always ON	Other equipment	0.3	Always ON

3. RESULTS AND DISCUSSION

3.1 Overview

Temperature and energy needs differences between the detailed and simplified models are presented by means of boxplots and histograms. Regardless the magnitude of the deviation occurring, the shoeboxing algorithm reduced the time required for the thermal simulation to one third for the simplified models with respect to the detailed ones for all the considered cases.

3.2 Temperature

Figure 2 reports the boxplots regarding the share of hours for which the temperature deviation in each floor lays within ± 0.5 °C for the three climates. Bolzano and Messina show a similar behavior: for both climates the temperature deviations lay within ± 0.5 °C for more than 95 % of time for all simulated floors, excluding the outliers, and the temperature prediction for Bolzano is the most accurate overall. On the contrary, the outputs obtained for Denver are less accurate: indeed, excluding the outliers, temperature deviations stay within the ± 0.5 °C interval for only at least 75 % of time, depending on the considered floor.

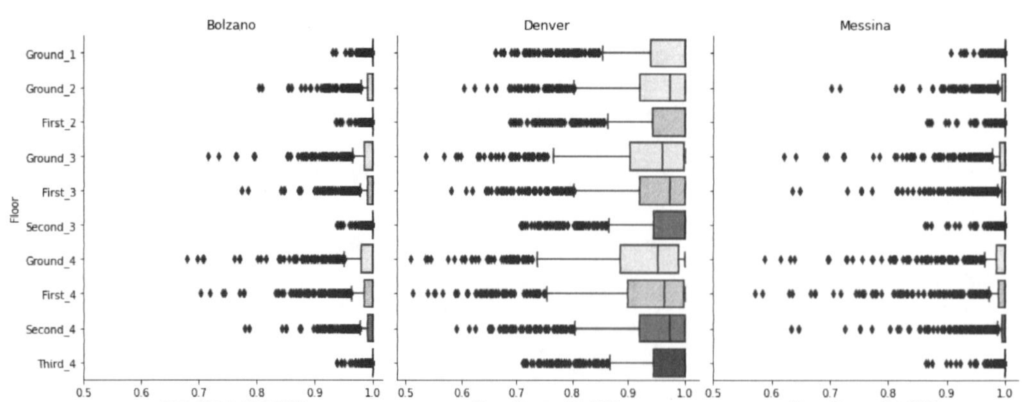

Figure 2: Resulting boxplots of the share of hours within ± 0.5 °C temperature deviation for each floor. Outliers are plotted outside the boxes' whiskers. The floors, color-coded in grey scale, are indicated with the floor name and the number of floors present in the building of which the floor is part of.

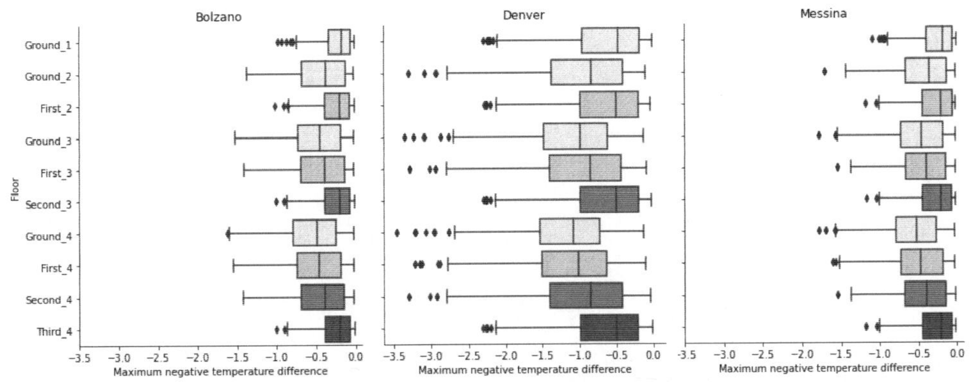

Figure 3: Resulting boxplots of the largest negative temperature difference

The most shaded floors are characterized by the largest temperature deviations. This has been observed in all the considered climates and particularly in Denver. Moreover, the higher the number of floors, the lower the accuracy. Finally, the maximum positive and negative temperature deviations are often included within the ± 0.5 °C interval for the climates of Bolzano and Messina, while larger differences occur for the climate of Denver. Such behaviour is shown by Figure 3 in which the boxplot regarding the largest negative temperature differences are reported. The largest positive temperature differences, even though not reported, experience lower differences.

3.3 Heating and cooling needs

The energy need has been investigated on annual and hourly basis. Figure 4 and Figure 5 depict with boxplots the relative differences of heating and cooling needs, as well as the *RMSEs*. As for the temperatures, it is possible to see that the climate of Bolzano is the one with the best predictions: indeed, the heating need stays within ± 5 % for all buildings and the cooling need within ± 10 % for the most of them. The accuracy of this prediction is confirmed also by the *RMSE* values. The climate of Messina may seem the one with the worst predictions, in particular regarding the heating need. However, from Figure 6 it is clear that the large relative deviations are mainly due to the very low magnitude of the heating needs. As it can be noticed in Figure 6 and Figure 7, the results obtained in the climate of Denver are overall the least accurate. Nonetheless, Figure 4 shows that the model annual need predictions do not exceed ± 10 % and ± 20 % deviation, respectively for cooling and heating and for most of buildings. Figure 6 and Figure 7 also highlight how, for the considered climates, the heating demand is underestimated when not properly predicted. On the contrary, the cooling demand is generally overestimated by the simplified model. This behavior can be explained considering that the simplified shoeboxes are smaller than the starting geometry. Even if the obstruction is properly shading the windows and the solar heat flux entering in the zone is the same for both models, since the volume of air is lower in the simplified model, so is its capacitance – consequently, a slightly different impact of solar gains is obtained. Overall, it is possible to see that the deviations increase with the number of floors, as a consequence of larger deviations occurring at the lower floors as observed also in the analysis of temperature deviations.

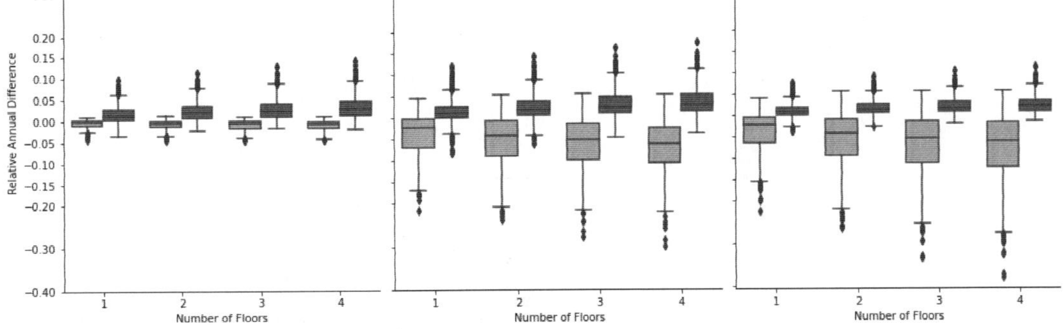

Figure 4: Boxplots for annual heating and cooling needs relative differences in function of buildings' number of floors

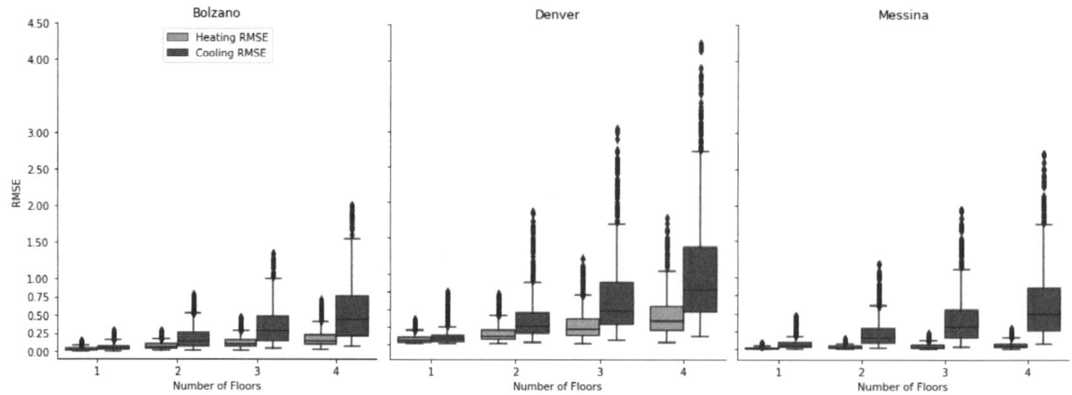

Figure 5: Boxplots for annual heating and cooling *RMSE* in function of buildings' number of floors

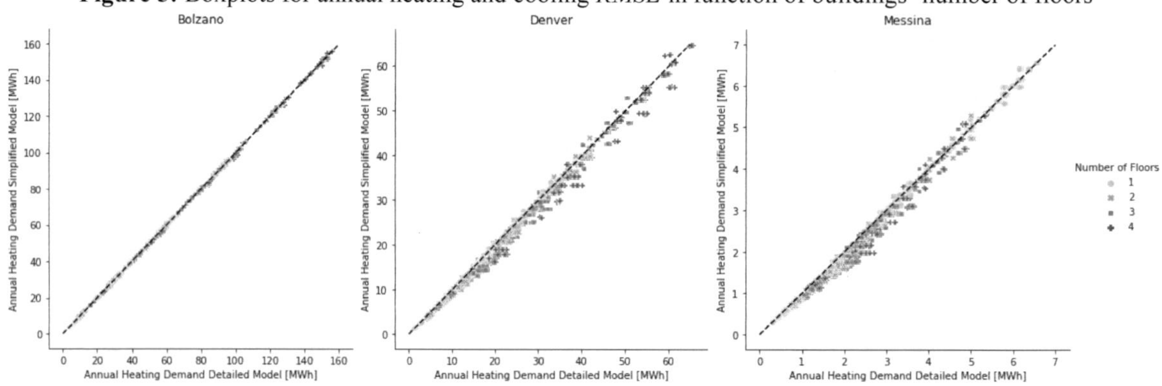

Figure 6: Simplified and detailed model heating annual outcomes in function of buildings' number of floors

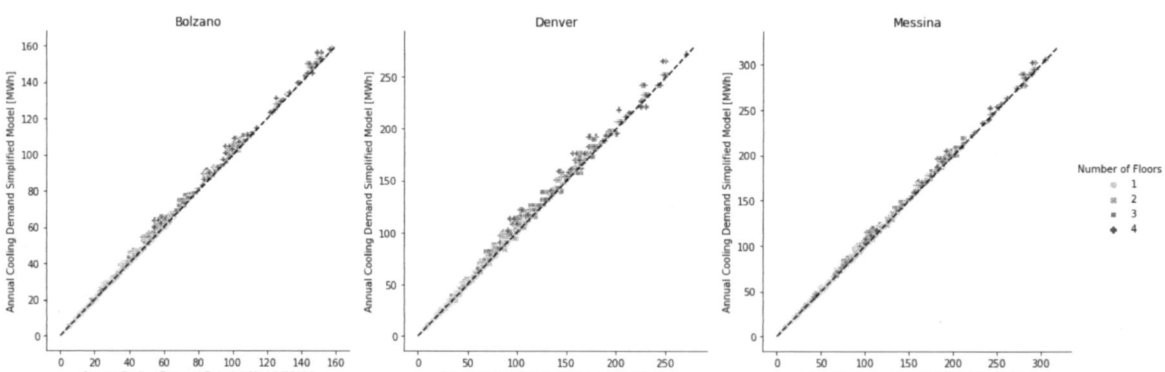

Figure 7: Simplified and detailed model cooling annual outcomes in function of buildings' number of floors

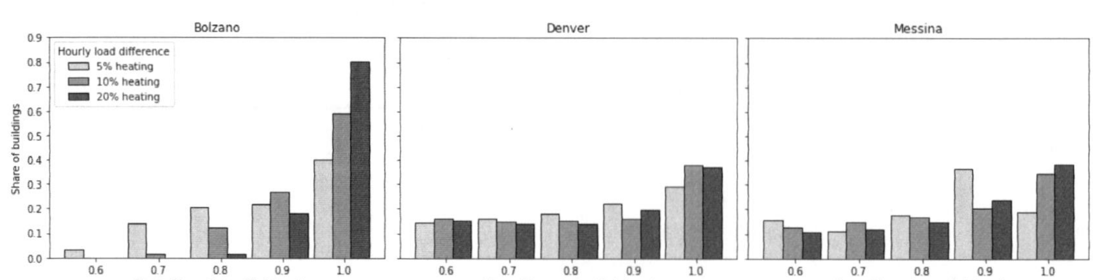

Figure 8: Share of buildings within a specific hourly heating range for a certain number of hours along the year

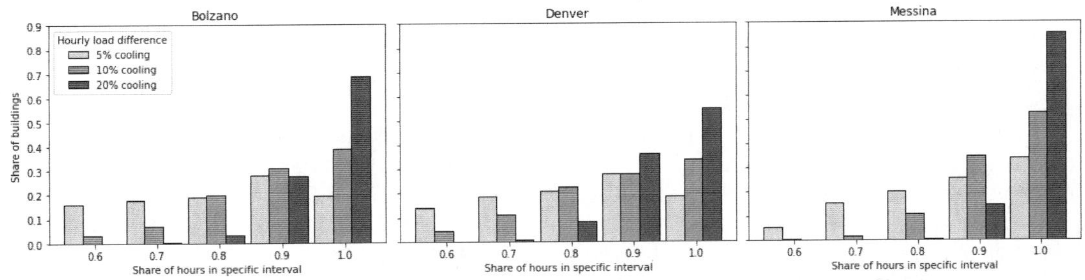

Figure 9: Share of buildings within a specific hourly cooling range for a certain number of hours along the year

Regarding the hourly energy needs, Figure 8 and Figure 9 report the share of buildings for which the relative hourly deviations lay within three deviation intervals (5 %, 10 % and 20 %) for a certain time of the year.
As for the annual needs, the outputs of Bolzano are those with the best prediction for the heating: up to 70 % of buildings have more than 90 % of hours along the year with a maximum relative deviation between the detailed and simplified model of ± 20 %. Furthermore, all buildings are in this interval if the minimum percentage of hours considered is 70 %. Regarding the cooling needs, hourly accuracy is lower. The hourly results for the climates of Denver and Messina are aligned to the annual ones. Messina shows lower accuracy for the heating demand, but, as seen for the annual results, it is related to the very low magnitude of the needs. Instead, the cooling hourly prediction is the best among the three cases. Denver shows higher accuracy in cooling estimation with respect to heating. This is not only due to the capability of the algorithm but also to the lower magnitude of the heating needs with respect to the cooling ones.
Although floor level results have not been reported, they show the same behavior of the building level outcomes and they underline how the prediction for the least shaded floors is the most precise.
Finally, even though it seems that on the hourly basis the need prediction is less accurate, a closer look at the data reveals that larger deviations occur in hours characterized by lower energy needs. For what concerns the heating demand, peak loads and those periods of the year with higher demands are well predicted.

4. CONCLUSIONS

In this work, a novel simplification algorithm to ease *UBEM* computing resources has been presented. The algorithm aims at simplifying complex building shapes into representative shoeboxes, in order to reduce the required simulation time without compromising modeling accuracy and resolution. The automated simplification procedure creates the simplified model from a set of geometry indicators, includes equivalent shading surfaces to account for self-shading elements and defines a thermal zone for each of the building floors. To test the proposed approach, 3072 buildings have been generated and simplified into representative shoeboxes. Both detailed and simplified models have been simulated in three different climatic conditions, and hourly temperatures and heating and cooling needs analyzed.
The algorithm has shown a good level of accuracy in estimating both hourly and annual needs, as well as temperature profiles for each floor, reducing the time required for the simulation to one third for the simplified models with respect to the detailed ones. However, larger discrepancies have been observed in climates characterized by quick temperature variations and large daily temperature differences.
In order to further improve the proposed algorithm, a thorough analysis of outliers will be performed in future developments, with the aim to optimize the shoebox definition and to increase the prediction accuracy. Moreover, the shoeboxing procedure efficacy will be also tested by considering the context and adjacent buildings, paving the way to the implementation in a urban model made of building shoebox metamodels, allowing to increase considerably the computational efficiency.

REFERENCES

Aasabadi, N., & Ashayeri, M. (2019). Urban energy use modeling methods and tools: A review and an outlook. *Building and environment 161*.

Caputo, P., Costa, G., & Ferrari, S. (2013). A supporting method for defining energy strategies in the building sector at urban scale. *Energy Policy*, 261-270.

Conrad, O. (2020, August 6th). *Tool Polygon Shape Indices*. Retrieved from SAGA-GIS Tool Library Documentation (v7.1.1): http://www.saga-gis.org/saga_tool_doc/7.1.1/shapes_polygons_7.html

Dall'O, G., Galante, A., & Torri, M. (2012). A methodology for the energy performance classification of residential building stock on an urban scale. *Energy and Buildings*, 211-219.
Danielski, I., Fröling, M., & Joelsson, A. (2012). The Impact of the Shape Factor on Final Energy Demand in Residential Buildings in Nordic Climates. *WREF - The World Renewable Energy Forum.* Denver.
Dogan, T., & Reinhart, C. (2013). Automated conversion of architectural massing models into thermal 'shoebox' models. *Proceedings of BS2013: 13th Conference of International Building Performance Simulation Association, Chambéry, France, August 26-28*, (pp. 3745-3752). Chambéry.
Dogan, T., & Reinhart, C. (2017). Shoeboxer: An algorithm for abstracted rapid multi-zone building energy model generation and simulation. *Energy and Buildings, 140*, 140-153.
European Commission. (2019, 12 11). The European Green Deal. Brussels.
European Parliament and Council of the European Union. (2010, May 19). Directive 2010/31/EU of the European Parliament and of the Council on the energy performance of buildings. Official Journal of the European Union.
Ghiassi, N., & Mahadavi, A. (2017). Reductive bottom-up urban energy computing supported by multivariate cluster analysis. *Energy and Buildings*, 372-386.
Ghiassi, N., Tahmasebi, F., & Mahdavi, A. (2017). Harnessing buildings' operational diversity in a computational framework for high-resolution urban energy modeling. *Building Simulation*, 1005-1021.
Golomb, S. W. (1994). *Polyominoes: Puzzles, Patterns, Problems, and Packings.* Princeton, New Jersey: Princeton University Press.
Hong, T., Chen, Y., Luo, X., Luo, N., & Lee, S. (2019). Ten questions on urban building energy modeling. *Building and Environment, 168*.
İşeri, O. K., & Dino, İ. G. (2020). An Algorithm for Efficient Urban Building Energy Modeling and Simulation. *2020 Proceedings of the Symposium on Simulation for Architecture and Urban Design*, (pp. 455-462).
Javanroodi, K., Mahdavinejad, M., & Nik, V. M. (2018). Impacts of urban morphology on reducing cooling load and increasing ventilation potential in hot-arid climate. *Applied Energy*, 714-746.
Javanroodi, K., Mahdavinejad, M., & Nik, V. M. (2019). A novel design-based optimization framework for enhancing the energy efficiency of high-rise office buildings in urban areas. *Sustainable Cities and Society*.
Johari, F., Peronato, G., Sadeghian, P., Zhao, X., & Widén, J. (2020). Urban building energy modeling: State of the art and future prospects. *Renewable and Sustainable Energy Reviews*.
Kodors, S. (2017). Geometric Feature Selection of Building Shape for Urban Classification. *Environment Technology Resources Proceedings of the International Scientific and Practical Conference*, (pp. 78-83). Rezekne.
Ladybug Tools LLC. (2020, October 7). Retrieved from Ladybug Tools: www.ladybug.tools
Lauster, M., Teichmann, J., Fuchs, M., Streblow, R., & Mueller, D. (2014). Low order thermal network models for dynamic simulations of buildings on city district scale. *Building and Environment*, 223-231.
Lyu, T., Buccolieri, R., & Gao, Z. (2019). A Numerical Study on the Correlation between Sky View Factor and Summer Microclimate of Local Climate Zones. *Atmosphere*.
Mastrucci, A., Baume, O., Stazi, F., Salvucci, S., & Leopold, U. (2014). A GIS-based approach to estimate energy savings and indoor thermal comfort for urban housing stock retrofitting. *BauSIM 2014*, (pp. 190-197). Aachen.
Natanian, J., Aleksandrowicz, O., & Auer, T. (2019). A parametric approach to optimizing urban form, energy balance and environmental quality: The case of Mediterranean districts. *Applied Energy*.
Nouvel, R., Schulte, C., Eicker, U., Pietruschka, D., & Coors, V. (2013). CityGML-based 3D city model for energy diagnostics and urban energy policy support. *Proceedings of BS2013: 13th Conference of International Building Performance Simulation Association, Chambéry, France, August 26-28*, (pp. 218-225). Chambéry.
Pernigotto, G., Prada, A., Gasparella, A., & Hensen, J. (2014). Development Of Sets Of Simplified Building Models For Building Simulation. *International High Performance Buildings Conference.* Purdue.
Philip, S. (2020). *scripting language for E+, Energyplus*. Retrieved from https://github.com/santoshphilip/eppy/
Reinhart, C., & Davila, C. (2015, December 2). Urban building energy modeling - A review of a nascent field. *Building and Environment, 97*, 196-202.
Robert McNell & Associates. (2020, August 6). *RhinoCommon API.* Retrieved from Rhino Developer Docs: https://developer.rhino3d.com/api/RhinoCommon/html/R_Project_RhinoCommon.htm
Sloane, N. A. (2020). *A000105 Number of free polyominoes (or square animals) with n cells.* Retrieved from The on-line encyclopedia of integer sequences: https://oeis.org/A000105
United Nations Habitat. (2019). The strategic plan 2020-2023.
Xu, X., Liu, Y., Wang, W., Xu, N., Liu, K., & Yu, G. (2019). Urban Layout Optimization Based on Genetic Algorithm for Microclimate Performance in the Cold Region of China. *Applied sciences*.

Field Evaluation of Solar Thermal Air Collectors

Zachary SCHREIBER[1]*, James RIESER[2], William Hutzel[3]

[1]Purdue University, School of Engineering Technology,
West Lafayette, IN, United States
219-775-6489, zschreiber1@yahoo.com

[2]Purdue University, School of Engineering Technology,
West Lafayette, IN, United States
765-412-5406, jrieser@purdue.edu

[3]Purdue University, School of Engineering Technology,
West Lafayette, IN, United States
765-494-7528, hutzelw@purdue.edu

ABSTRACT

This paper documents the performance of three different active flat plate solar thermal air collectors that were designed, built, and installed on the roof of a university building. The three collectors circulate air and use small fans for capturing and delivering supplemental heat for the building. These collectors are part of a larger solar energy installation used in a university setting for teaching and research about renewable energy. Each collector has a different configuration to illustrate how a solar collector's design impacts its ability to capture thermal energy. A web-based building automation system provides real time data on solar collector performance that can be used by students and researchers. The three new collectors achieved 20% to 25% energy conversion efficiency when analyzed according to ASHRAE Standard 93 "Methods Of Testing To Determine The Thermal Performance Of Solar Collectors". Testing also showed that the amount of air flowing through the collectors had a significant impact on overall performance. The solar collectors also achieved an energy factor (the ratio of heat collected to fan power) of approximately 5, effectively demonstrating the viability of solar thermal technologies.

1. INTRODUCTION

Solar thermal systems are one of the oldest renewable energy technologies. They capture the sun's energy and convert it to heat for use in various residential and commercial applications. This energy can be collected using passive or active collection techniques. Passive techniques require no additional energy or system inputs such as heating through windows and skylights. Active techniques use additional systems and equipment to continually capture energy. This equipment can include motors, pumps, and more to track the sun, move fluids, or operate other forms of input to improve performance.

Many researchers have analyzed solar thermal air collector designs to understand the parameters that improve performance. Raam Dheep and Sreekumar (2020) looked at the effects of absorber design and varying flow rates through active loop collectors, finding lower flow rates are desirable to achieve higher efficiencies. Khanlari et al (2020) analyzed collector designs finding that more absorber surface area achieves higher thermal efficiencies. Saffarian et al (2020) analyzed the use of nanofluids and different flow directions in solar collectors to increase convective heat transfer. Abuşka et al (2019) performed an experimental analysis of the heat storage ability of a phase change material to continue collector heating after dusk. These studies show improvements in design and methodology that identified the key opportunities for designing new solar thermal air collectors for use in a university setting.

All three forms of heat transfer are important for capturing heat and improving performance in a solar thermal collector. *Radiant* energy is the mechanism that delivers solar energy to the collector. *Conduction* is what allows energy to enter the collector and plays a significant role in heat loss during cold weather. *Convection* is important in an active loop collector for heating a secondary fluid. If the secondary fluid is air, optimal convection heat transfer is needed to transfer heat from the collector surface and to the air being conveyed through the collector.

Standards provide methods for testing and evaluating collectors for uniform comparison. ASHRAE Standard 93-2010 defines methods for testing and evaluating air or liquid solar collector efficiency and thermal performance. Section 8.5 of ASHRAE 93 specifies measurements and processes for evaluating solar collector performance to provide uniform, state of the art comparisons between collectors and quantify how different geometries and flow rates affect the energy captured. Work done by the Florida Solar Energy Center (FSEC) and the Solar Rating & Certification Corporation (SRCC) was also investigated for additional insight into practical considerations for testing and evaluation of solar thermal panels.

This field evaluation of solar thermal air collectors deviated from ASHRAE 93 in one important aspect. This was a cold-weather test, where the outdoor temperature ranged from 0 to 5 °C. Most test standards require testing when outdoor air temperature and solar intensity are above 15 °C and 600 W/m^2, respectively. The cold weather likely reduced the performance of the solar collectors due to conduction and convection losses, but the cold weather data is more realistic. It reflects a time of year when the heat from the solar air collectors is most beneficial for a building and its occupants.

Solar thermal air collectors can save energy and money. Many locations worldwide experience days below 18 °C, where building heating is needed. Solar energy is a particularly good option when natural gas or propane are unavailable. By using renewable technologies like solar air collectors, heat can be captured using a small amount of electricity for a fan, reducing the use of fossil fuels and saving money. Furthermore, these systems can be utilized for drying food (Khanlari et al, 2020) and other applications requiring heat, saving on energy costs.

2. DESIGN

The Applied Energy Laboratory (AEL) at Purdue University is used for teaching and research into high performance buildings. The laboratory equipment mimics what can be found in a modern net zero energy building and includes HVAC equipment, building automation systems, and a variety of solar thermal and solar photovoltaic systems. These systems provide renewable energy for the laboratory and supplemental energy for the entire building.

This research project replaced three active loop solar air collectors in the Applied Energy Laboratory that were built more than 30 years ago. The new collectors were design to fit within an existing rack containing other solar collectors that were already installed on the roof of the building. The cost was kept to a minimum by using commercially available materials and fabrication techniques available in the university's sheet metal shop.

2.1 Materials

The solar air collectors are a six sided box, where the one side facing the sun is a glass surface. Aluminum was chosen as the framing material for five of the six sides because of its durability, low cost, light weight, and corrosion resistance. The glass absorber surface is a double paned U-0.5 glass patio door with a glazed surface and low iron composition. Low iron glass permits more light penetration allowing for slightly higher efficiency and performance.

2.2 Heat Loss

A steady state conduction heat transfer analysis was performed using equation 1 to determine the amount of insulation to manage the heat loss from the collector during cold weather. A is the surface area, ΔT is the temperature differential inside and outside the solar collector, and R is the overall R-value of the insulated collector box. The sensitivity analysis was conducted to determine the amount of insulation (R) needed for limiting the total heat loss (Q) for a range of temperature differentials (ΔT).

$$Q = \frac{A * \Delta T}{R} \tag{1}$$

Figure 1 is the results of the sensitivity analysis, showing the R-value of insulation (R_IN) to maintain heat losses at a fixed percentage of the total heat collected for various temperature differentials. The solid, dashed, and dotted lines represent 5%, 10%, and 15% total heat loss, respectively. Figure 1 clearly shows that he overall amount of heat loss becomes unacceptably large during extremely cold weather, which justifies the need to select a specific design range that reflects the most common outdoor conditions during cold weather. The selected range for this analysis is between a ΔT of 30 and 50 °C and is shown as a dashed box in Figure 1.

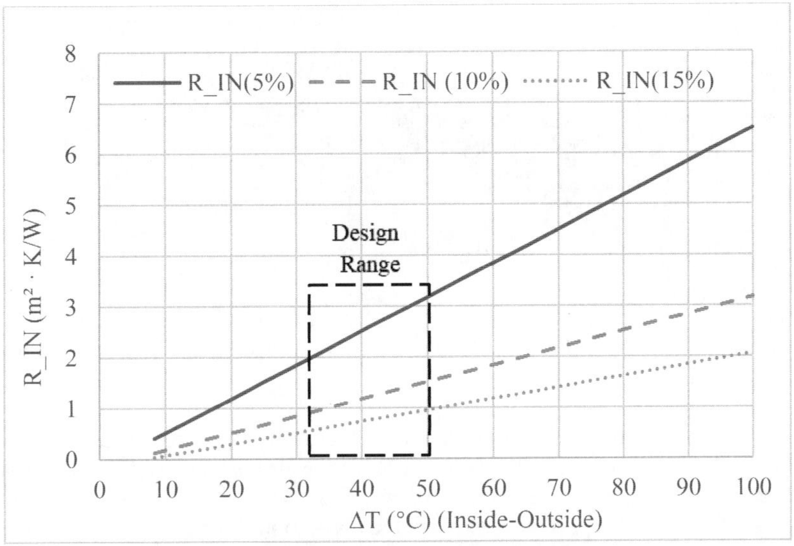

Figure 1: Required insulation to reduce heat loss

The results of this analysis show that heat loss can be kept to an acceptable level by using the correct amount of insulation. Using an insulation thickness of 25.4 mm (1 inch), the best material found was R_{SI}-1 ($R_V = 6$) insulation. This insulation limited heat loss to 15% of the total heat collected in a worst-case scenario design day (ΔT = 50°C). As the inside to outside temperature differential decreases, the % of heat loss also decreases with a fixed insulation.

2.3 Designs

Figure 2 shows the three new solar air collectors (6, 7, & 8) that were designed, fabricated, and installed. All three collectors have a centered absorber plate that allows air to travel on both sides. Air collector 6 is a simple flat absorber plate with a flat black color. The second air collector (7) uses a similar flat black plate, but with additional X-shaped perforated baffles, similar to the design presented by Khanlari et al. The third collector (8) has a wavy, corrugated center plate. These were all mounted facing south at a 53-degree angle to take full advantage of winter sun.

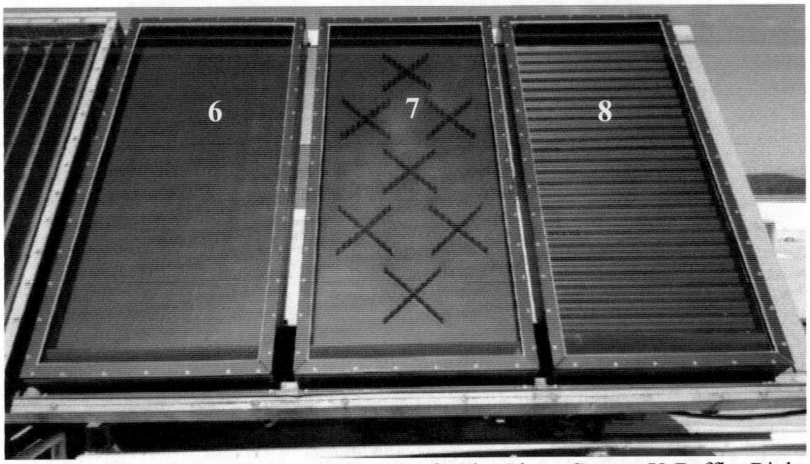

Figure 2: Three solar thermal air collector designs (Left: Flat Plate; Center: X-Baffle; Right: Corrugated)

Figure 3 provides further detail on how the absorber surface is centered in the frame to allow air travel on either side of the absorber plate. The cross wing design is shown in Figure 3, but the same concept is common across all three solar collectors. Air is forced through the bottom of the collector by a fan inside a fan box housing on the back of each collector. There are two curved manifolds (highlighted in blue) within the collector that direct air flow to both sides. These manifolds have different size and number of holes to encourage air flow in one direction. The lower manifold has small diameter holes while the upper manifold has large diameter holes.

Figure 3: X-Baffle flat plate collector CAD rendering showcasing frame and dual side air flow

2.2 Fabrication

The collector frames are made of 3 mm (1/8 inch) aluminum that is welded and riveted for rigidity, with silicone sealing for air tightness. All three frames are the same with only difference being the internal absorber plates. All collectors have 25.4 mm (1-inch) insulation on 5 of the 6 sides of the panels that has an insulating value of R_{SI}-1 (R_V-6). Both the interior and exterior of the collectors were painted black to improve their ability to absorb radiant energy. The absorber surface on all three collectors is a double-paned glass patio door made of low-iron glass to improve its heat transfer characteristics. The three new collectors cost $14,200 to design, fabricate, and install.

3. OPERATIONS

The solar collectors are monitored and controlled using a web-based Building Automation System (BAS). Figure 4 is a graphic interface showing the temperature differential for all three collectors. All three collectors typically achieve approximately the same temperature differential for air leaving versus the air entering.

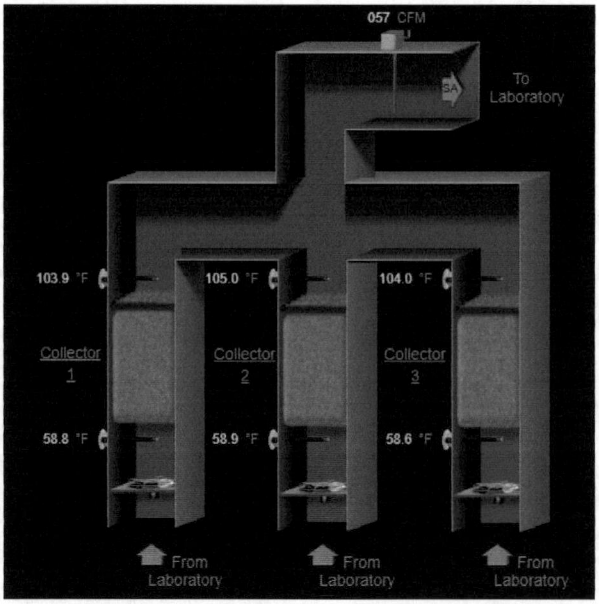

Figure 4: Graphic interface recording parameters for all three collectors

Figure 5 is a detailed graphic interface for one individual collector, demonstrating further parameters that are recorded every 15 minutes. A solar pyranometer measures the solar intensity (radiant energy) as a basis for comparison to the heat energy captured. Fans for the solar thermal system are programmed to run when the solar pyranometer reads a solar intensity higher than 100 W/m^2 . These parameters are recorded at 15 minute intervals and are used to analyze and compare the collected and potential energy of each collector.

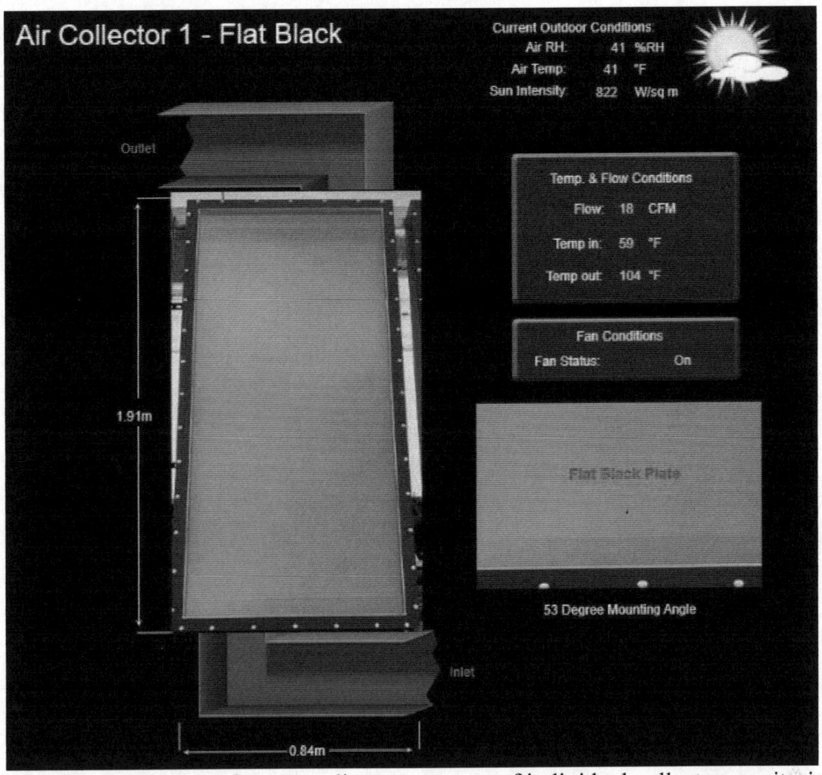

Figure 5: Graphic interface recording parameters of individual collector monitoring

3.1 Performance Data

Roughly 4 months of data were recorded for this project. From that large data set, many sunny and nearly cloudless days were chosen to analyze performance. A sample of the data acquired is given in Table 2. These parameters are recorded every 15 minutes into the online BAS data base. Data for two different air flow rates, 50 CFM (86 m^3/hr) and 20 CFM (34 m^3/hr), were collected to assess the impact of airflow on overall collector performance.

Table 2: Sample Performance Data

Date & Time	AV Solar Intensity (W/m^2)	Source Energy (W)	Outside Humidity (%)	Outside Temp (°F)	Return Temp (°F)	Supply Temp (°F)	Energy Captured (BTU/HR)
2/16/2020 09:45:00 AM EST	667	173	55	21	85	57	584
2/16/2020 10:00:00 AM EST	732	173	52	22	92	58	686

The Energy Captured from Table 2 was calculated within the BAS using the parameters shown in equation 2. This is a common equation used for computing the sensible energy for air at atmospheric conditions using English units. The results of equation 2 were converted to Watts for performance comparisons in the results section of this paper. The Source Energy in Table 2 is the electricity in Watts supplied to all three fans for the collectors, so the true energy for one fan is the source energy divided by three.

$$Energy\ Captured = 1.1 * CFM * (T_{out} - T_{in}) \tag{2}$$

4. RESULTS

Solar air collectors perform the best on a clear sunny day with few clouds. Figure 6 displays the energy captured by a solar air collector as well as solar intensity on an ideal sunny day. The energy captured tends to be lower in the morning because the collector is heating up. The peak energy capture occurs at mid-day when the solar intensity is highest. The energy capture decreases along with the solar intensity after mid-day.

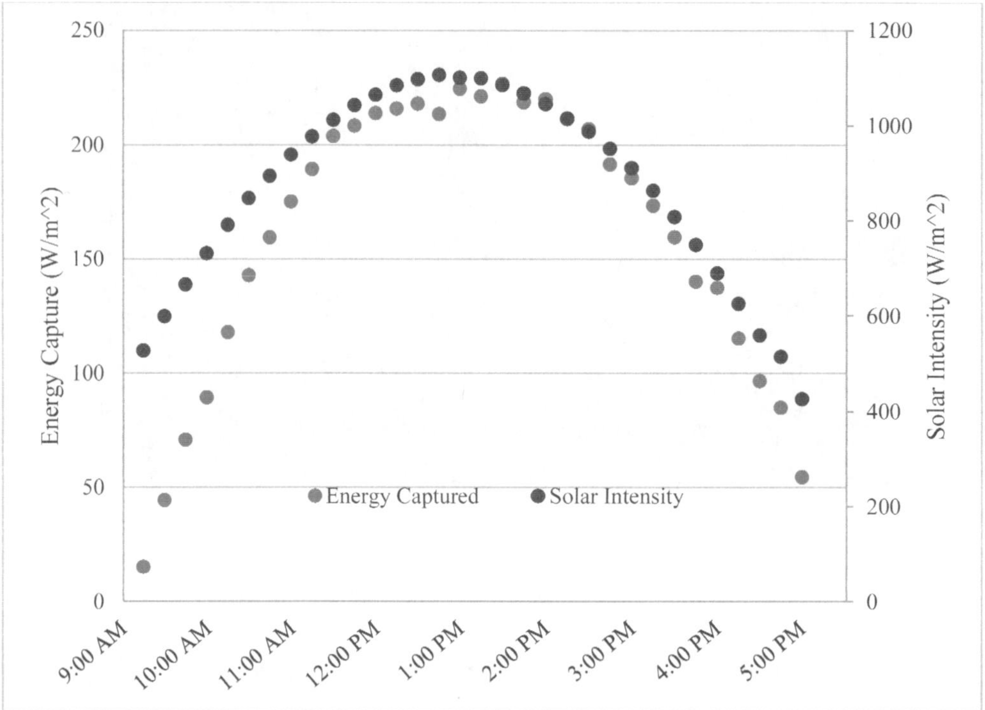

Figure 6: Energy and efficiency changing over a day

ASHRAE 93 was used to analyze performance data, calculate results, and compare the collectors. Section 8.5, equation 8.22 of ASHRAE 93 was specifically used to evaluate solar collector performance to provide uniform comparisons between all three collectors and analyze how different geometries and flow rates affect the energy captured by the collectors. Fan source energy is factored into efficiency calculations to determine true efficiency as shown by equation 3, but it does not have a large influence on the overall efficiency calculations.

$$Efficiency = \frac{Energy\ Captured - Source\ Energy}{Solar\ Intensity} \tag{3}$$

Figure 7 illustrates the impact of both outdoor temperature and air flow rate on solar collector efficiency. An analysis was conducted on days with different outdoor temperatures and air flows of either 34 m^3/hr (20 cfm) or 86 m^3/hr (50 cfm). Figure 7 shows that solar collector efficiency was higher at warmer outdoor temperatures, which is not surprising since the unwanted heat loss from the collector to the outdoor environment would be less during warmer weather. Figure 7 also shows higher efficiency was achieved at lower air flow rates. All three collectors displayed similar trends. These overall findings are supported by the research of Raam Dheep, and Sreekumar.

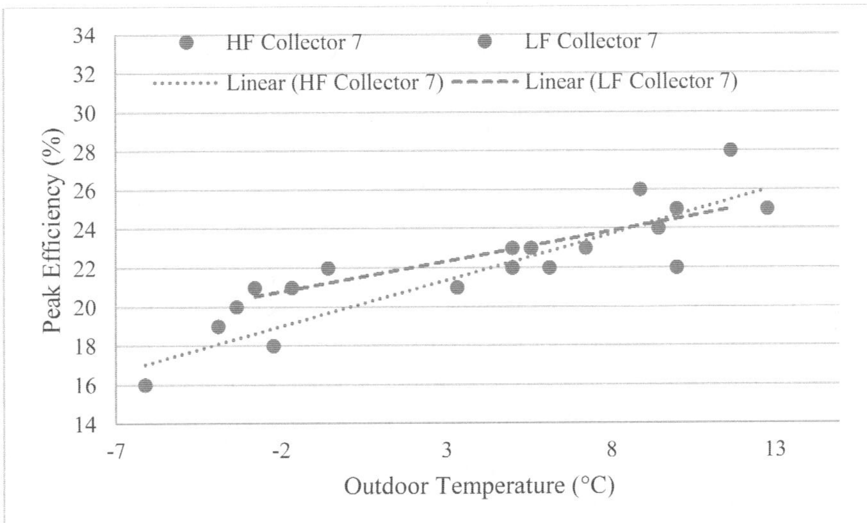

Figure 7: Efficiency improves with higher temperatures and lower flow rates

Figure 8 compares the efficiency of the three different collector designs at a low flowrate. The results are displayed over a range of outdoor temperatures. The x-wing (7) and corrugated (8) collecotors performed so close to each other that their data was superimposed in Figure 8. Although the efficiency of all three collectors was similar, the trendlines show that collectors 7 and 8 had slightly higher efficiencies than collector 6. This is not surprising since collectors 7 and 8 have slightly more absorber surface area, plus these designs help disrupt and distribute flow more effectively throughout the collector.

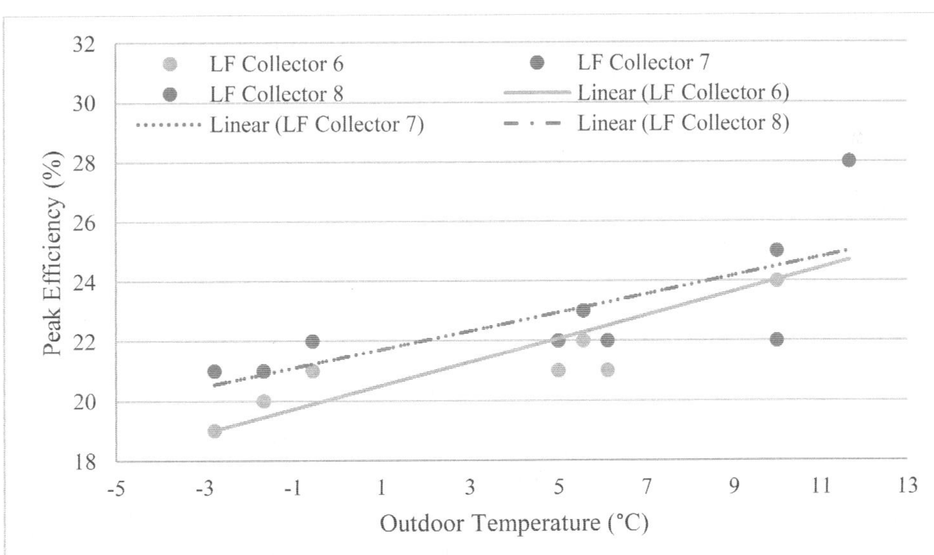

Figure 8: Collector absorber design impacts efficiency slightly

The Solar Energy Factor (SEF) shown in equation 4 is another dimensionless parameter that helps quantify the performance of solar thermal collectors. The numerator is the total heat energy captured for one day. The denominator is the total source energy is the electricity used to drive the fans for one day. The SEF should be greater than one to justify operation of the solar collectors.

$$Solar\ Energy\ Factor = \frac{Total\ Energy\ Captured}{Total\ Source\ Energy} \quad (4)$$

Equations 5 and 6 show how the two terms of the SEF were computed using the trapezoidal rule for integrating data collected at discrete time intervals t and t_i. Equation 5 quantifies the heat energy captured. Equation 6 quantifies the source electricity used by the fan, which includes a site to source multiplier of 1.0 because the electricity to run the fan comes from photovoltaic panels located on site. Equations 5 & 6 were computed from data that was collected at 15 minute intervals over one day.

$$Total\ Energy\ Captured\ (Wh) = \int_{t}^{t_i} \frac{Energy\ Capture_t + Energy\ Capture_{t_i}}{2}\, dt \tag{5}$$

$$Total\ Source\ Energy\ (Wh) = \int_{t}^{t_i} \frac{Source\ Energy_t + Source\ Energy_{t_i}}{2}\, dt \tag{6}$$

Figure 9 shows the results of the SEF computations for several days that span a range of outdoor temperatures. All the data was collected for low flow rate, which is labelled "LF" on the figure. The SEF ranged from 4.5 in cold weather to 5.5 in warmer weather. The larger SEF in warmer weather is mostly likely because heat loss is reduced from the solar collectors during warmer weather Figure 9 also shows that collector 8 had the highest SEF of all three collectors, further illustrating that absorber design can impact solar collector performance.

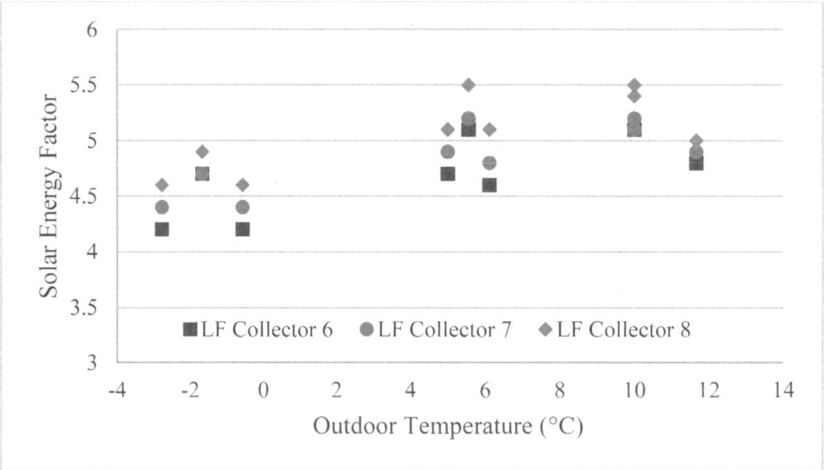

Figure 9: Collector absorber design impacts SEF

4.1 Environmental and Economic Impacts

Table 3 estimates the annual energy heating energy provided when all three solar thermal air collectors were used for supplemental heating at their location in northwestern Indiana. The estimate was directly measured based on 4 months of cold weather performance and extrapolated to encompass a 5 month heating season. The far left column is the amount of heat provided by all three solar thermal collectors, in kilowatt-hours, over a five month heating season. The remaining columns show fossil fuel equivalencies for electricity, natural gas, propane, and heating oil. The solar thermal collectors used about 200 kWh of electricity during the 5 month heating season to power the small fans that move the air through the collectors and that electricity was provided by a solar photovoltaic array on the same roof.

Table 3: Annual Equivalent Fuel Savings

Solar Energy	Fuel Equivalencies			
Solar Collector (kWh)	Electricity Saved (kWh)	Natural Gas Saved (m^3)	Propane Saved (liter)	Heating Oil Saved (liter)
1300	1300	120	180	120

The economic payback for the three new solar collectors in terms of their ability to provide "free" heating is not particularly attractive. The three custom-made solar thermal air collectors cost on the order of $15,000 to design, build, and install on the roof of the building. This cost is significantly inflated because extra sensors and other equipment were included to support teaching and research. The value of the annual energy from Table 3 (1,300 kWh) is on the order of $200 assuming an electricity rate of $0.15/kWh. Comparing the $15,000 first cost to the $200 annual savings suggests that this is not a viable installation for a typical homeowner. However, a better payback could be achieved by aggressively pursuing less expensive fabrication and installation techniques commonly used in residential construction. In this situation, the extra cost is justifiable because the main purpose of these solar thermal air collectors is for teaching and getting students interested in renewable energy technologies.

5. CONCLUSIONS

Three different flat plate solar thermal air collectors were designed, built, and tested according to ASHRAE Standard 93. They achieved a cold weather efficiency in the range of 20 to 25% and had a Solar Energy Factor of approximately 5. A comparison of the three collectors showed that increasing the complexity of the absorber surface using either a finned or corrugated surface increased the total heat transfer, efficiency, and solar energy factor. Reducing the flow rate of the air also improved the total heat transfer, efficiency, and solar energy factor of the solar collectors. Based on this data, solar thermal air collectors are a viable option for supplemental heating in locations where there is limited access to traditional fossil fuels. These solar collectors are also environmentally friendly as they provide heat without the need to consume fossil fuels.

NOMENCLATURE

ASHRAE	American Society of Heating, Refrigeration, and Air-Conditioning Engineers
ISO	International Standards Organization
FSEC	Florida Solar Energy Center
SRCC	Solar Rating & Certification Corporation
AEL	Applied Energy Lab
LF	Low Flow
HF	High Flow
BAS	Building Automation System
BTU	British Thermal Unit
CFM	Cubic Feet per Minute
CAD	Computer Aided Design
Q	Heat Loss
A	Area
T	Temperature Difference
R	Insulation

REFERENCES

Abuşka, M., Şevik, S., & Kayapunar, A. (2019). Experimental analysis of solar air collector with PCM-honeycomb combination under the natural convection. *Solar Energy Materials and Solar Cells*, 195, 299-308. doi:10.1016/j.solmat.2019.02.040

Khanlari, A., Güler, H. Ö., Tuncer, A. D., Şirin, C., Bilge, Y. C., Yılmaz, Y., & Güngör, A. (2020). Experimental and numerical study of the effect of integrating plus-shaped perforated baffles to solar air collector in drying application. *Renewable Energy*, 145, 1677-1692. doi:10.1016/j.renene.2019.07.076

Raam Dheep, G., & Sreekumar, A. (2020). Experimental studies on energy and exergy analysis of a single-pass parallel flow solar air heater. *Journal of Solar Energy Engineering, Transactions of the ASME*, 142(1) doi:10.1115/1.4044127

Saffarian, M. R., Moravej, M., & Doranehgard, M. H. (2020). Heat transfer enhancement in a flat plate solar collector with different flow path shapes using nanofluid. *Renewable Energy*, 146, 2316-2329. doi:10.1016/j.renene.2019.08.081

Applied Energy Lab Building Automation System. WebCTRL: Solar PV Monitor. Retrieved from http://technrg1.ecn.purdue.edu/

ASHRAE Standard 93. (2014). Methods of Testing to Determine the Thermal Performance of Solar Collectors. Retrieved from https://subscriptions.techstreet.com/subscriptions/index

Florida Solar Energy Center (n.d) Florida Solar Standards. Retrieved from http://www.fsec.ucf.edu/en/certification-testing/solarstandards/index.htm

International Code Council, Inc. (2015, August). ICC 901/SRCC 100 – 2015 Solar Thermal Collector Standard (ICC 901 – 2015). Retrieved from https://codes.iccsafe.org/content/ICC9012015

National Renewable Energy Laboratory (2000). Transpired Solar Collectors. Retrieved from https://www.nrel.gov/docs/fy00osti/23667.pdf

Technical and Economic Analysis of a Residential Heat Pump with Photovoltaic Solar Panels in Self-Consumption for Space Heating, Cooling and DHW

Pedro JORQUERA FERRAT[1]*, Michèle MONDOT[1]

[1]CETIAT, Centre Technique des Industries Aérauliques et Thermiques, Villeurbanne, France.

*Corresponding author e-mail: pedro.jorquera-ferrat@cetiat.fr

ABSTRACT

Two different thermodynamic systems for space heating, cooling, and domestic hot water production coupled to photovoltaic panels have been studied. The aim is to assess the PV electricity self-consumption variation with the heat storage volume and the applied control strategy. Each system has been studied using a 1D unsteady state modelisation software to simulate its performance in a house complying with the French building code (RT2012) having a DHW demand according to ErP's load profile M and with Paris weather data. The main parameters studied are the electricity consumption of the thermodynamic system, the seasonal performance factor (SPF), the self-consumption ratio, the self-production ratio, the coverage ratio and the Return on Investment. The results show that the seasonal efficiency can be doubled thanks to the optimization of the PV electricity self-consumption, which results in a return on investment lower than 15 years.

1. INTRODUCTION

The necessity to reduce both the energy consumption of residential heating appliances and the electrical network peak load (i.e. through aggregation scenarios) as well as the technical evolution of some heating or hot water production appliances (mainly thanks to new control strategies) allow us to consider possible the photovoltaic (PV) self-consumption in a detached house in order to feed a heat pump (HP) for space heating and domestic hot water (DHW) production.

Unsteady state, 1D simulations and calculations are widely used to study the PV self-consumption and the controls of the system. Pichler *et al.* (2016) showed that for a detached house with a HP+PV system for heating and DHW production, the use of a predictive control model (PCM) has the potential to increase the self-consumption ratio from 20% (base case scenario) to 50%. Haller *et al.* (2017) proved that the self-consumption ratio can be increased by 50% with a simple control strategy consisting of overheating the storage tank for the DHW production. However, Baeten (2017) indicated that large storage volumes may penalize the HP system performance. Thür *et al.* (2018) established that the implementation of a smart control strategy is far more efficient for the PV self-consumption than the increase of the thermal storage volume in a ground source HP+PV system for heating and DHW production. On the other hand, Jorquera (2018) showed that whilst thermal storage allows an increase of PV self-consumption of around 50% compared to a system without thermal storage, the use of electrical batteries increases the PV self-consumption up to 80%, but is not profitable as the costs of replacing the batteries are far too great.

Scientific publications show that studies generally focus on the controls of the system for one or two services at a time (heating + DHW production, cooling + heating, …). The objective of this study is to enlarge the knowledge of a thermodynamic system with PV panels. A techno-economic assessment has been carried out to optimize the utilization of PV electricity in order to minimize the electricity bill of a detached house.

The paper is divided in three parts:

- Description of the two systems studied
- Results for system #1
- Results for system #2

2. DESCRIPTION OF THE SYSTEMS

The thermodynamic systems have been studied for the two installation schemes represented in figures 1 and 2. The PV electricity produced by the photovoltaic panels installed on the roof of the house is either consumed by the house or exported to the electrical network, depending on the type of contract for the sale of PV electricity. In the first case, PV electricity not consumed by the thermodynamic system can be consumed by other usages in the house and the electricity not consumed by these is exported to the grid. In the second case, all produced PV is exported. Each component of the system and the associated simulation model are described in this chapter.

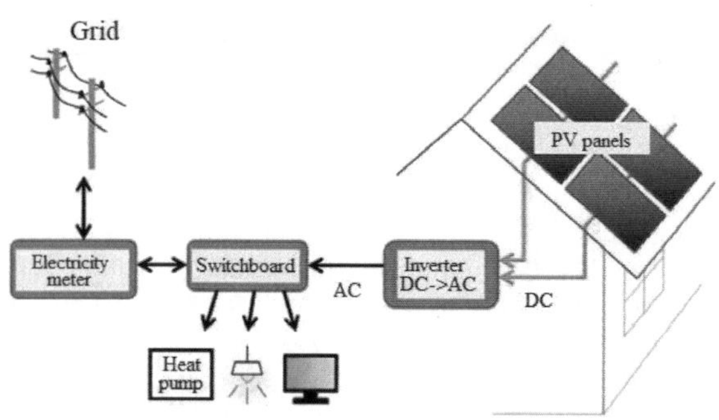

Figure 1: HP + PV system diagram with a contract of self-consumption of PV electricity

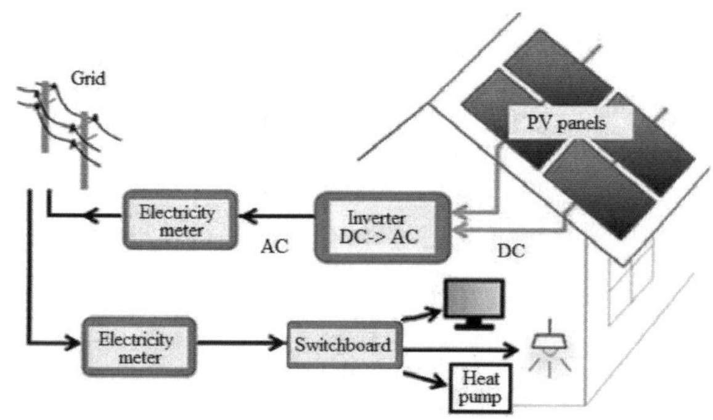

Figure 2: HP + PV system diagram with a contract of total PV electricity exportation

2.1 The Detached House

A detached house of medium inertia with a surface heat loss coefficient Ubat of 0.4 $W{\cdot}m^{-2}{\cdot}K^{-1}$ was considered. The heated surface area is 111.7 m^2, the ceiling height of 2.5 m and the glazed area facing South / East is 5 m^2. The ventilation air renewal is 140 m^3/h. The simulations were carried out over the period from January 1st to December 31st (365 days) with Paris weather data. The modelling of the building corresponds to a 5R/3C model inspired from NF EN ISO 13790:2013, as presented by Noël (2016). It takes into account the outdoor temperature, the solar gains and walls characteristics to calculate the thermal energy load as well as the air and walls temperature evolution over time. The 5R3C developed model has been compared to the BESTEST benchmark, a benchmark in the field of building thermal simulation presented by Judkoff and Neymark (1995). The comparison between model results and reference results is excellent. The electricity consumption profile is defined according to Bouvenot (2015) findings.

2.2Thermodynamic System

The thermodynamic system for heating, cooling and DHW production is described for the two following assemblies. In system #1 an air-to-water heat pump (HP) ensures the space heating and domestic hot water through a buffer tank and a DHW storage tank, respectively (see figure 3). It is associated with an air-to-air air conditioning (AC) unit for the cooling needs of the house (see figure 4). In system #2 an air-to-air reversible heat pump (RHP) provides both heating and cooling for the house (see figure 5). It is associated with a heat pump water heater (HPWH) for DHW production. Data used for the simulations is listed in table 1.

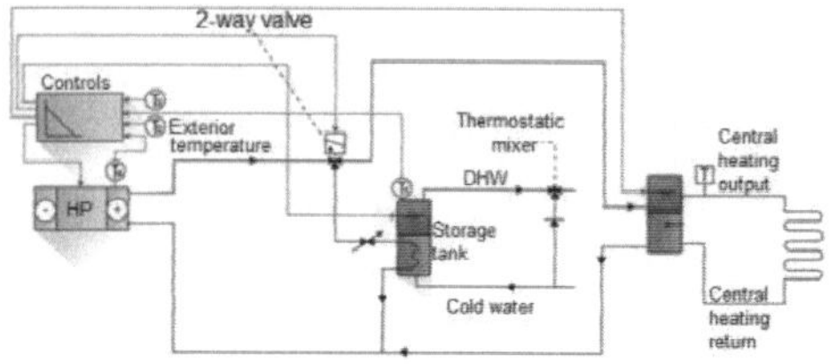

Figure 3: Hydraulic diagram of the HP for heating and DHW

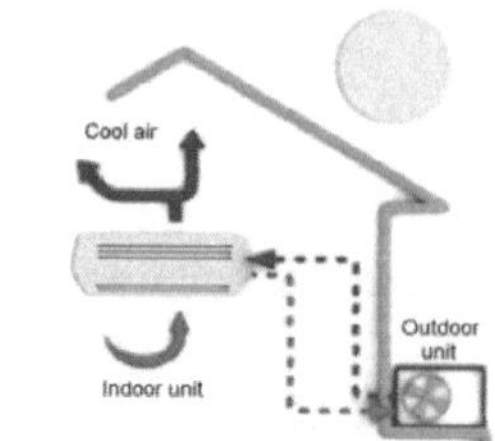

Figure 4: AC unit

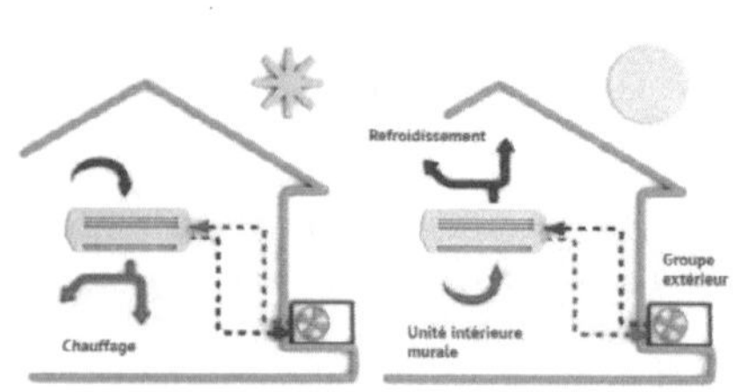

Figure 5: Representation of the reversible heat pump for heating and cooling

Table 1: Performances of the thermodynamic systems

Air-to-water heat pump (HP)	Air-to-air reversible heat pump (RHP)
Variable speed compressor	Variable speed compressor
Rated heating capacity @7/30-35 = 6 kW	Rated heating capacity @7/20 and cooling capacity @35/27 = 6 kW
COP @7/30-35 = 4,16	COP @7/20 = 4,16 & EER @35/27 = 4
heating capacity at LRContMin[(1)] = 2,4 kW	Heating (cooling) capacity at LRContMin = 2,4 kW
COP at LRContMin = 4,16	COP (EER) at LRContMin = 4,16 (4)

[(1)] LRContMin: switch point to On / Off operation mode of the heat pump or minimum heating capacity in continuous operation.

The HPWH and the DHW function of the air-to-water HP are characterized by a COP equal to 3.37 for a load profile L, corresponding to a total tapped energy of 11.655 kWh, i.e. 0.2 m^3 of hot water at 60°C per day. The performance of the thermodynamic system (HP, RHP, HPWH and AC unit) is calculated using a performance matrix established on the basis of the rated COP and the temperature ranges of the heat source (outdoor air) and the heat sink (recycled air or water). The model has a simple dynamic (small volume heated or cooled). The delivered capacity depends on a set point and the correlation laws in the form of a matrix. The modelling also takes into account the part load operation of the HP. The energy balance (energy requirement, energy supplied and energy consumed) is calculated by the model with a time step equal to 5 seconds and then integrated over the total simulated duration (in this case 1 year). The seasonal performance factor (SPF) is calculated using equation (1). The energies involved are calculated by the integration over time of the corresponding capacities and powers obtained by the simulation model.

$$SPF = \frac{E_{cooling} + E_{heating} + E_{DHW} + E_{backup}}{E_{consumed\ total} - E_{PV\ self-consumed}} \tag{1}$$

2.3 Control Strategies

A base case has been defined by means of a standard control strategy, without self-consumption improvement techniques. This base control strategy is defined as follows:

- In heating mode
  - Room set temperature (Ts_room) = 19°C (hysteresis ±0,5 K);
  - In system #1: heating buffer tank set temperature = 45°C (hysteresis – 5 K);
  - Heating season: between October 17th and April 1st;
- In DHW mode
  - Hot water storage tank set temperature = 45°C (hysteresis -5 K);
  - DHW set temperature = 39°C;
- In cooling mode
  - Room set temperature (Ts_room) = 24°C (hysteresis ±0,5 K);
  - Cooling season: between May 2nd and September 11th.

In order to increase PV electricity self-consumption several "enhanced" control strategies have been studied. This will allow us to determine their impact on system performance and overall economics. These control strategies are described below:

Control strategy 1:

- Heating mode:
  - If PV production over 500 W then room set temperature (Ts_room) and buffer set temperature (Ts_buffer) increased by +2 K and +10 K, respectively;
  - Heating reduction periods (00h-6h, 8h-10h and 22h-00h): Ts_room -3 K and Ts_buffer -5 K;
- DHW mode
  - If PV production over 500 W then storage set temperature (Ts_storage) increased to 55°C;
  - Ts_storage reduced by -10 K between 23h and 6h;
- Cooling mode
  - If PV production over 500 W then Ts_room = 21°C;

- Between heating and cooling seasons, the AC unit is allowed to start if set temperature is not met when PV production > 500 W.

Control strategy 2: Control strategy 1 and:

- Heating / DHW / air cooling, if PV production > 500 W:
  - Heat pump capacity adjustment to PV power production.
    The modulation of the power input of the heat pump is obtained by calculating the output set temperature of the heat pump according to the available PV power and the operating conditions (water flow, return temperature) as well as the COP of the heat pump;
    In heating and DHW the set temperature is limited between 40°C and 70 ° C.

Control strategy 3: Control strategy 2 and:

- Heating / DHW / cooling, if PV production > 500 W:
  - For buffer and storage tanks: The hysteresis is reduced from 5 K to 1 K in order to restart the system quickly and to consume the photovoltaic production;
  - For the building: The air conditioning start hysteresis is reduced to 0 K and the stop hysteresis is increased to 2 K.

Control strategy 4: Control strategy 3 and:

- Heating: If PV production then heating of the buffer tank up to 60°C;
- DHW: If PV production then heating of the storage tank up to 60°C.

2.4 PV Panels

In this study, a power of 3.0 kWp was considered, i.e. 10 photovoltaic panels (~ 15 m^2) of 300 Wp each. The model is based on TRNSYS 16, Type 180. Three parameters are calculated in order to characterize the use of photovoltaic electricity. The self-consumption rate indicates the share of photovoltaic production consumed directly by the house. The self-production rate indicates the share of total electricity consumption supplied by photovoltaic panels. The coverage rate makes it possible to establish an energy balance between annual production and annual electricity consumption. Two types of electricity contracts have been taken into account: Sale of the total PV production and Self-consumption with sale of PV production excess (Gabriele, 2019).

2.5 Water Storage Tanks

For system #1, three storage tank volumes for heating and DHW are simulated, as shown in Table 2 with their heat losses. For system #2, only DHW tank is simulated. A multi-layer model is taken into account with a defined number of layers (~ 100) of constant thickness. A convection / diffusion equation is completed by an algorithm for suppressing inverse gradients. The conductivity of the water and the metal of the wall are taken into account for the thermal diffusion in standby, as described by Noël *et al.* (2010).

Table 2: Characteristics of water storage tanks

Mode	Volume (l)	Thermal losses (W)
Heating	100	42.4
	200	52.6
	500	71.4
DHW	150	48.0
	300	60.1
	600	76.0

2.6 Return on Investment (ROI)

The data used to calculate the ROI for the two types of PV contract are shown in Table 3. A calculation of ROI based on the savings generated thanks to the PV, compared to the annual electricity bill without PV, was considered. The annual electricity bill without PV for year 1 is calculated according to equation (2). To calculate the electricity bill for the following years the price of the imported kWh and the cost of the subscription are updated. To simplify the calculations, it was considered that all costs (subscription, TURPE, kWh imported and exported) change at the same rate, obtaining equation (3). Finally, the electricity bill over n years is calculated using equation (4).

$$F_0 = C_{tot} * p_{imp} + A \tag{2}$$

$$F_i = C_{tot} * p_{imp} * (1+e)^i + A * (1+e)^i \tag{3}$$

$$F_n = \sum_{i=0}^{n} F_i = \sum_{i=0}^{n} (C_{tot} * p_{imp} + A) * (1+e)^i \quad (4)$$

Similarly, the electricity bill over n years for the system with PV panels for the excess PV production sale contract is calculated using equation (5). The electricity bill over n years for the system with PV panels for the total PV production sale contract is calculated with equation (6).

$$F_n^{PV} = \left(C_{imp} * p_{imp} + A + TURPE_{sp}\right) * \frac{(1+e)^n - 1}{e} - C_{ex} * p_{ex} * n \quad (5)$$

$$F_n^{PV} = \left(C_{tot} * p_{imp} + A + TURPE_{vs}\right) * \frac{(1+e)^n - 1}{e} - C_{ex} * p_{ex,100\%} * \frac{(1+g)^n - 1}{g} \quad (6)$$

The ROI corresponds to the number of years (n) in which the savings generated by the PV panels (F_n-F_{nPV}) equalize the initial investment and, if higher than 10 years, the replacement of the PV panels inverter.

Table 3: Economic data for the calculation of the ROI

Parameter	**Value**	
PV	7500 €	
Replacement of the PV panels inverter after 10 years	1000 €	
Thermal storage	Buffer tank 100 litres : 300 € Buffer tank 250 litres : 400 € Buffer tank 500 litres : 600 €	Storage tank 150 litres: 300 € Storage tank 300 litres: 400 € Storage tank 600 litres: 600 €
Electricity imported from the grid	15,6 c€/kWh +2,5%/year	
Electricity exported to the grid	Total PV production sale: 18,73 c€/kWh -2,5%/year capped at 10 c€/kWh Sell of excess: 10 c€/kWh + bonus of 1200 €	
Electricity meter Linky	Total PV production sale: 500 €	
Electricity subscription	150,93 €/year +2,5%/year	
TURPE	Total PV production sale: 41,6 €/year +2,5%/year Sell of excess: 10,8 €/year +2,5%/year	

3. RESULTS FOR SYSTEM #1: AIR-TO-AIR HEAT PUMP WITH AC UNIT

3.1 Electricity consumption and SPF

The annual electricity consumption shown in Figure 6 includes the electricity consumption of the HP, the electrical backup and the AC unit. The results are presented according to the storage volume (heating buffer volume – DHW storage tank volume) and the control strategy (1 to 4). These results show that a greater heat storage capacity implies a higher electricity consumption (Figure 6). With control strategies to increase PV self-consumption, system electricity consumption is increased in most cases from 2.5% to 10%. This increase is induced by greater thermal losses from the storage tanks heated up to a higher temperature. It is interesting to note that the difference in electricity consumption between the smallest volume and the largest volume increases with a strategy to optimize self-consumption of electricity. The results show a significant growth in SPF with the first control strategy compared to base case (Figure 7). This strategy allows to almost double the SPF of the system. The second control strategy (control strategy 1 with modulation of the heat pump capacity) brings little compared to the control strategy 1. The third control strategy (control strategy 2 with reduction of the hysteresis) is much more effective than the control strategy 2 and an optimum storage volume is found at 250-300 litres. The integration of the fourth strategy (control strategy 3 with +10 K water tank overheating) has a negative impact on the overall efficiency of the system and the SPF is reduced compared to control strategy 3 (water tank overheating at higher set temperature). This indicates that, for a thermodynamic system with heat storage for both heating and DHW, a compromise between self-consumption and performance must be found depending on the storage volume and the maximum heating temperature of the water storage tank. With any of the studied control strategies, the SPF of system #1 can therefore be more than doubled compared to the system with a basic control strategy.

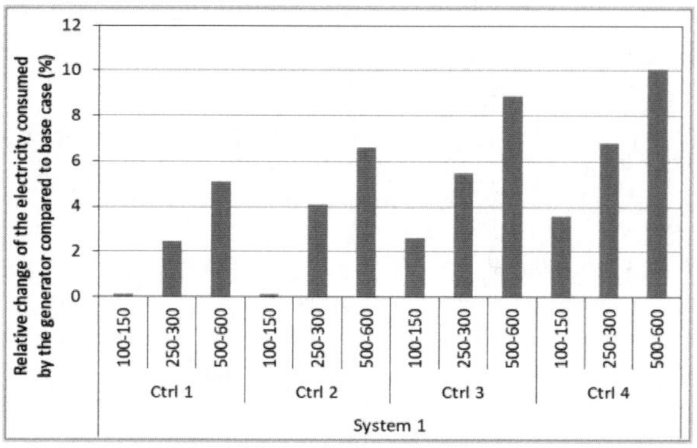

Figure 6: Electricity consumption of system #1

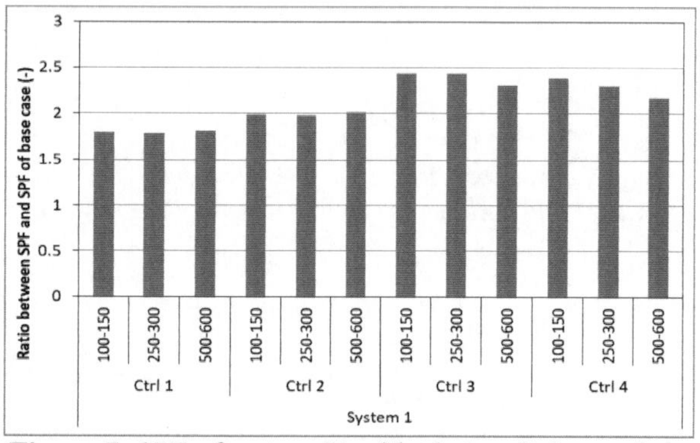

Figure 7: SPF of system #1 with photovoltaic panels in self-consumption

3.2 Self-consumption, self-production and coverage ratio

The results show that the implemented control strategies are very effective in increasing the self-consumption of the thermodynamic system. However, this increase leads to a decrease in the total self-consumption of the house (Figure 8). One can see that each additional control strategy to increase self-consumption is less and less effective, reaching a plateau at around 63% of self-consumption by the thermodynamic system alone (bottom bars) and 90% of the overall self-consumption for the house (bottom plus top bars). Going beyond this mark would require setting up a control strategy taking into account the consumption of other appliances in the house, the remaining 10% being to the other uses in the home (TV equipment, hi-fi, lighting, household appliances, etc.). The self-production rate results, in Figure 9, show that the control strategies are very effective, making it possible to triple the share of PV electricity consumed on site without any excessive increase of the electricty consumption. An optimum is found with an intermediate storage volume and the control strategy 3.

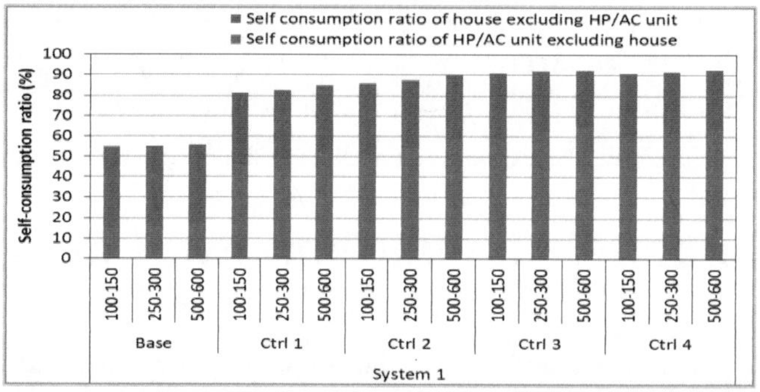

Figure 8: Self-consumption ratio with system #1

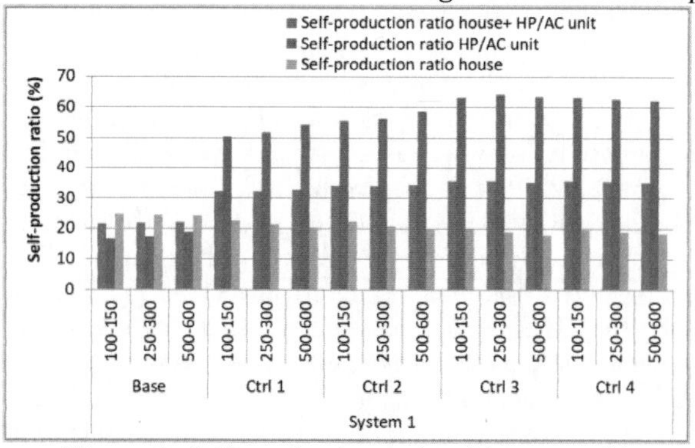

Figure 9: Self-production ratio with system #1

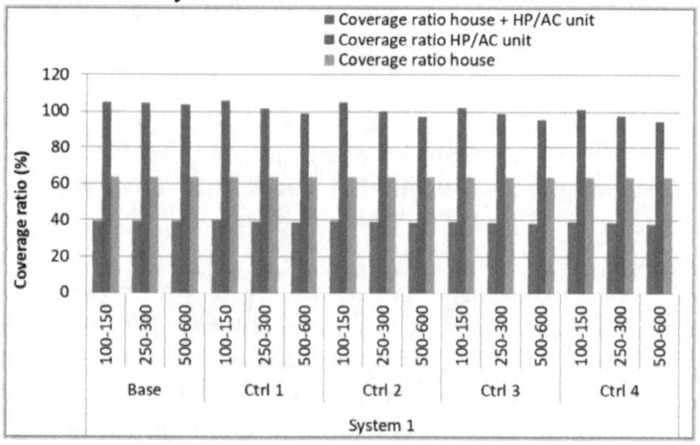

Figure 10: Coverage ratio with system #1

The coverage rate shown in Figure 10 is relatively constant. Indeed, it depends on the consumption of the thermodynamic system, the house and the photovoltaic production. However, the consumption of the system barely varies with the studied parameters (control strategy, storage volume) and both the photovoltaic production and the consumption of the house are kept constant.

3.3 Return on Investment

The initial investment costs include purchase and installation of the photovoltaic panels with the peripherals, and costs associated with thermal storage. The costs of the thermodynamic system are not included. The ROI then varies depending on the ratio of exported and self-consumed electricity as well as on the volume of storage tanks. The ROI with the total PV production sale contract does not depend on the self-consumption control strategy because with this contract all the PV production is exported and sold to the grid. As can be observed in Figure 11, the ROI with this contract increases with the storage volume, from 22 to almost 24 years. This is because larger storage volume requires a higher initial investment cost and induces higher thermal losses, whereas the PV production (then the PV exportation) is constant.

On the other hand, with an excess PV sales contract, the ROI varies according to the self-consumed and exported parts of PV production. Greater self-consumption reduces the import of electricity from the grid (at a price of around 15 c € / kWh) but also the sale of electricity exported to the network (at the price of 10 c € / kWh). With the considered costs, the ROI only becomes lower than 15 years when applying control strategies that lead to improving the self-consumption of the thermodynamic system. Thus, a system with control strategies to maximize self-consumption of PV electricity leads to a ROI of around 13 years, which is 2 to 3 years less than with a base control strategy. Regarding storage volumes, bests results are obtained with a medium to small storage volume.

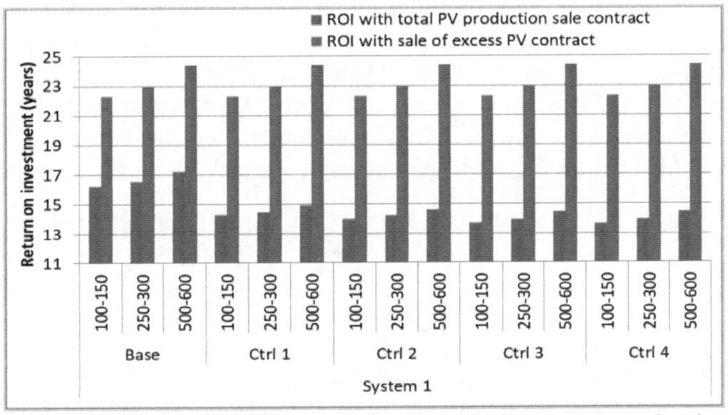

Figure 11: Return on investment with system #1 depending on PV production sale contract

4. RESULTS FOR SYSTEM #2: REVERSIBLE AIR-TO-AIR HEAT PUMP WITH HEAT PUMP WATER HEATER

4.1 Electricity consumption and SPF

The annual electricity consumption includes the electricity consumption of the reversible air-air heat pump, the heat pump water heater and an electrical back-up heater. The results in Figure 12 show that the electricity consumption of system #2 is weakly influenced by the control strategy implemented or by the storage volume. Control strategies aimed at increasing self-consumption of PV electricity have led to a reduction in the total consumption of the thermodynamic system by around 3.5%. This is because increase of non-heating and non-cooling periods in fact induced lower energy consumption (while maintaining comfort criteria). A small increase in SPF is observed with control strategies 1 to 4 (Figure 13), compared to base case. These results suggest that, compared to system #1, the lack of heat storage for heating (buffer tank) is greatly limiting the amount of PV electricity that can be self-consumed by the system. A control strategy for increasing self-consumption of PV electricity does not allow an increase in SPF of more than 26%.

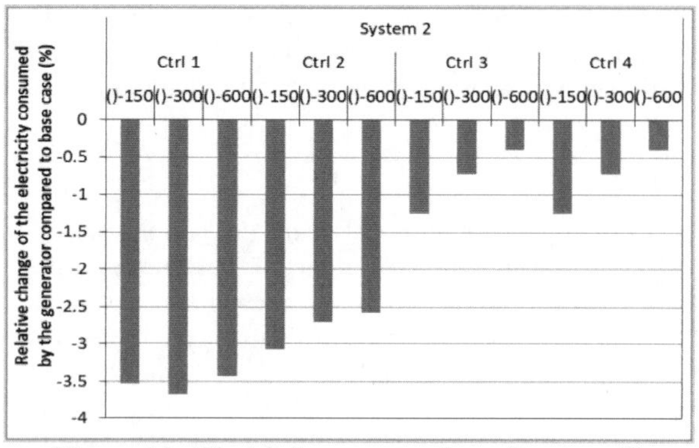

Figure 12: Electricity consumption of system #2

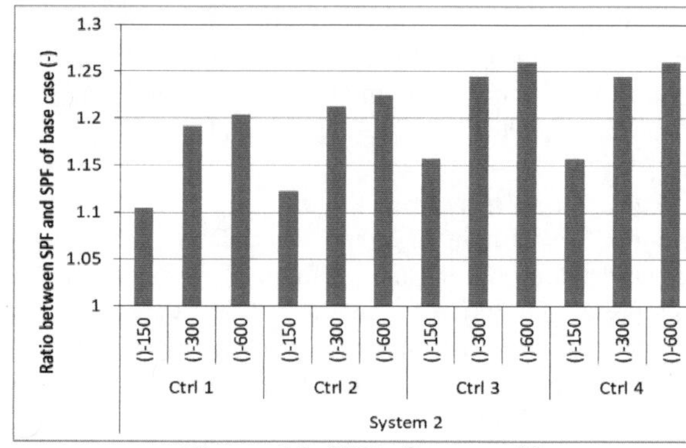

Figure 13: SPF of system #2

4.2 Self-consumption, self-production and coverage ratio

The control strategies introduced for system #2 lead to self-consumption rates of up to 70% (house + thermodynamic system), as shown in Figure 14. Compared to system #1, the base control strategy results in the same self-consumption rate of around 55%. However, with the optimized control strategies (1 to 4), the levels of self-consumption achieved with system #2 are much lower than with system #1. The self-production rate is around 23% with the base control strategy (Figure 15). With enhanced control strategies, the self-production rate only reaches 29%. Thermal storage for heating is therefore recommended in order to maximize self-consumption of PV electricity. The impact of the DHW water storage volume is noted with these enhanced control strategies, making it possible to gain around 2% of self-production with a larger storage volume. Other control strategies must be developed in order to maximize the consumption of this system during periods of PV production. The thermodynamic system coverage rate is greater than 100% (Figure 16) but the system's self-production rate is limited to 40%. This implies that the system consumption from PV is limited to 40% but could be completely covered as the system's coverage rate is over 100%. These results confirm a limited flexibility of system #2 to increase PV self-consumption due to studied control strategies or the lack of heat storage possibilities.

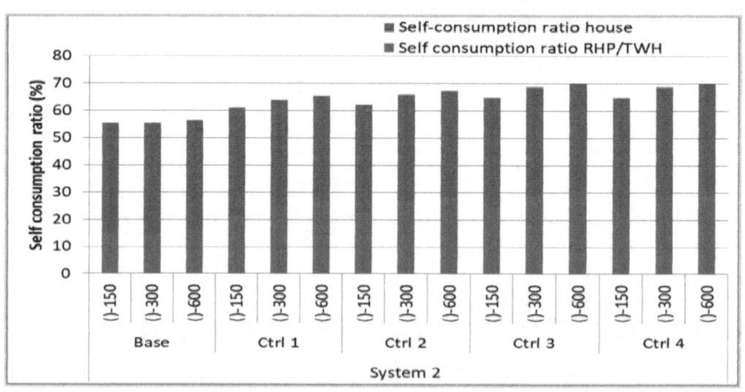

Figure 14: Self-consumption ratio with system # 2

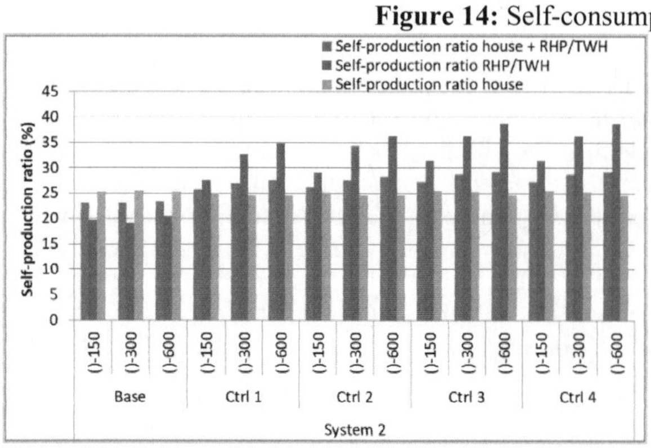

Figure 15: Self-production factor with system # 2

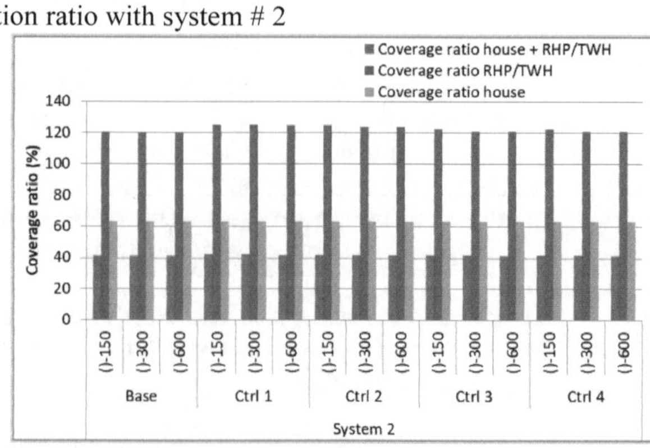

Figure 16: Coverage ratio with system # 2

4.3 Return on Investment

The payback time for system # 2 is shown in Figure 17. The ROI with the total PV sale contract is independent of the control strategy applied to the thermodynamic system, as explained in 3.3. This ROI is too high (around 22 years) and is negatively impacted by the increased cost of thermal storage. The ROI with an excess PV production sale contract is around 16 years with the base control strategy. It is reduced to around 15 years with the studied control strategies, thus having very little impact of ROI.

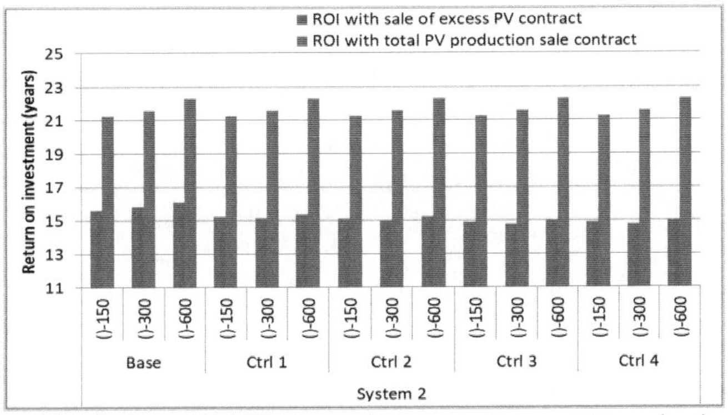

Figure 17: Return on investment with system # 2 depending on PV production sale contract

5. CONCLUSIONS

In this study, two types of thermodynamic systems for heating, cooling and DHW production associated with PV panels in self-consumption were simulated with different storage volumes and different control strategies:

- System # 1: air-to-water heat pump with heating storage and DHW storage, and air-to-air air conditioning unit;
- System # 2: reversible air-to-air heat pump and heat pump water heater.

The results showed that:

- The electricity consumption of system # 1 increases between +2.5 to +10% with the studied control strategies and the storage volumes;
- The electricity consumption of system # 2 can be reduced by 3.5% with control strategies;
- The SPF of system # 1 can be doubled thanks to optimized control strategies while, for system # 2, the SPF increases little (+ 26%);
- A self-consumption rate of 90% is achieved with system # 1. However, an increase in the system's self-consumption is obtained at the expense of the self-consumption of the house other uses. To further increase the total self-consumption rate, it would therefore be necessary to integrate a control strategy taking into account the consumption of the whole house as well as the state of charge of the storage;
- System # 2, because of no thermal storage for heating, leads to much lower self-consumption rates than system # 1. The control strategies studied are therefore less appropriate or effective for this system;
- System # 1 with an excess PV sale contract and enhanced control strategies lead to an ROI of less than 15 years;
- System # 1 shows a much greater ability than system # 2 for self-consumption and PV production management.

By comparing the results obtained with the two systems, it emerges that thermal storage for heating is recommended in order to maximize self-consumption of PV electricity, which makes it possible to achieve an SPF in the order of 7 to 9. The results also showed that, for a thermodynamic system with heat storage for heating and DHW, a compromise between self-consumption and performance must be found depending on the storage volume and its set temperature. From this study, it can be concluded that, to improve the performance and self-consumption of the HP + PV system, it is first necessary to prioritize the development effort over the control strategy put in place, which allows self-consumption to be achieved.

NOMENCLATURE

A	Electricity contract price per year	(€)
E	Energy	(kWh)
e	Annual increase in the price of electric kWh	(-)
F	Electrify bill	(€)
f	Annual increase in the cost of the electricity contract subscription	(-)
g	Evolution of PV price exported to grid per year with total PV production sale contract	(-)
n	Number of years	(-)
p	Price	(€)
T	Temperature	(°C)
Ubat	surface heat loss coefficient	($W \cdot m^{-2} \cdot K^{-1}$)

Abbreviations

AC	Air conditioning
COP	Coefficient of performance
DHW	Domestic hot water
ErP	Energy related products
HP	Heat pump
HPWH	Heat pump water heater
PCM	Predictive control model
PV	Photovoltaic
RHP	Reversible heat pump
ROI	Return on Investment
SPF	Seasonal performance factor
TURPE	Tariff of Use of Public Electricity Networks

Subscript

100%	Total PV production sale contract
ex	Exported
imp	Imported
p	Peak
s	Set
sp	Excess PV sale contract
tot	Total

REFERENCES

Baeten, B. (2017). Residential Heating Using Heat Pumps and Hot Water Storage Tanks - Tank Sizing to Minimize Environmental Impact in a Renewable Energy Context [Doctoral dissertation, Faculty of Engineering Science, Belgium].

Bouvenot, J. P. (2015). Etudes expérimentales et numériques de systèmes de micro cogénération couplés aux bâtiments d'habitation et au réseau électrique. Energie électrique. [Doctoral dissertation, University of Strasbourg].

Gabriele (2019, July 25). Panneau Solaire Photovoltaïque. Le Guide de Référence 2019. Medium. https://www.insunwetrust.solar/blog/le-solaire-et-vous/prix-panneaux-photovoltaiques/

Haller, M., Battaglia, M., Haberl, R., Reber, A., Bamberger, E., Borner, M. (2017). Steigerung des Photovoltaik-Eigenverbrauchs durch intelligente Wärmepumpen. *Tagung des BFE-Forschungsprogramms Wärmepumpen und Kälte, June 14th, 2017*, HTI Burgdorf, Switzerland.

Jorquera, P. (2018). Analyse technico-économique d'un système PAC+PV en auto-consommation. NT 2018/047, Villeurbanne, France: CETIAT.

Judkoff, R., & Neymark, J. International Energy Agency building energy simulation test (BESTEST) and diagnostic method. United States. doi:10.2172/90674.

NF EN ISO 13790, September 2013. Energy Performance of Buildings – Calculation of Energy Use for Space Heating and Cooling.

Noël, J., Heintz, J., Albaric, M. (2010). Modélisation, implémentation et validation d'un modèle général de ballon ECS. *Conference IBPSA France 2010.*

Noël, J. (2016). Modélisation "5R3C" d'un bâtiment - Module BOOST "BAT16". NT2016/006. Villeurbanne, France: CETIAT.

Pichler, M. F., Heinz, A., Rieberer, R. (2017). Model predictive heat pump and building control to maximize PV-power on site use. *Proceedings of the 12th HEA Heat Pump Conference.*

Thür. A., Calabresen T., Streicher, W. (2018). Smart grid and PV driven ground heat pump as thermal battery in small buildings for optimized electricity consumption. *Solar Energy 174* (2018) 273–285.

Demand Side Management and Battery Storage Utilization to Increase PV Self-consumption of a Modulating Heat Pump

Maria PINAMONTI[1]*, Alessandro PRADA[2], Paolo BAGGIO[2]

[1]Free University of Bolzano, Faculty of Science and Technology,
Bolzano, Italy
maria.pinamonti@natec.unibz.it

[2]University of Trento, Department or Equivalent,
Trento, Italy

* Corresponding Author

ABSTRACT

The combination of photovoltaic (PV) systems and heat pumps for heating and cooling of buildings is a promising solution to increase the share of renewable energy in the residential sector. The interaction between the system components is fundamental to assure a high performance of the system. The level of PV energy self-consumption is strictly dependent on the control strategy applied to the system. The solar source is intermittent and it does not always match the building loads for heating and cooling.

Furthermore, even the heating and cooling demands are strongly time-dependent in high performance buildings. For these reasons, an efficient control system is essential to ensure the high performance. Several papers in the literature have proposed advanced control techniques based on the model predictive control (MPC). However, their implementation in residential buildings is often limited due to high device costs.

This paper proposes a rule-based control strategy for a modulating air-source heat pump coupled with a PV plant, which provide space heating, space cooling and domestic hot water in a residential building. The proposed control strategy can be easily implemented in residential buildings by using low-cost board shields. The heat pump is modulated and optimized depending on the instantaneous PV production, to maximize the direct use of solar energy onsite. When an overproduction of PV energy occurs, the heat pump operates to store the solar energy as thermal energy, exploiting thermal storage tanks and the building thermal capacitance (aka virtual battery). The heat pump is controlled by varying its compressor rotational speed. The compressor is regulated to operate at the maximum capacity level compatible with the supplied PV power. The control strategy is evaluated in combination with a electric storage system. The efficacy of the control strategy is assessed by means of dynamic energy simulations. The simulations are run for the whole year. A parametric analysis is carried out by considering different PV and battery size, to understand the impact of the system component size on the results.

1. INTRODUCTION

Around 40% of the European Union (EU) energy consumption is related to the building use. The total energy demand of buildings is around 192 Mtoe, which is divided in 79% related to space heating (SH), 15% to domestic hot water (DHW), and 6% to space cooling (SC). Additionally, the use of space cooling in residential buildings is reported to be increasing in the last decade, because of several factors such as higher required comfort levels, climate change and the resulting increase in external air temperature. As reported by Eurostat, only the 19% of the energy consumed for heating and cooling in the European countries comes from renewable energy sources (RES) (Eurostat, 2018). A larger use of renewable energy sources is one of the main challenges to decarbonize the building energy sector.

The use of heat pump units is one of the most valuable solutions to reduce the primary energy demand and to integrate RES in buildings. Specifically, the combination of photovoltaic systems and heat pumps for heating and cooling of buildings is a promising solution to increase the use of renewable energy produced on-site. However,

several factors, such as the time gap between the energy availability and the building loads, are a limit for the system to reach high levels of renewable energy self-consumption. Considering the strong time dependency of the solar source and the building demand, the control strategy applied to the system is fundamental to increase the level of PV self-consumption. The interaction between the system components is fundamental to assure a high performance level of the system. Load-control strategies exploit the ability of the building system devices to work during preset operation periods to match the energy demand and the production. These strategies are commonly defined as Demand Side Management (DSM), and they can be applied to a wide range of electric applications. Heat pumps are characterized by a great potential for load flexibility, hence they are a promising tool for DSM applications. The control strategies can be divided in two main categories, depending on their conceptual structure: rule based (RB) controls and model predictive controls (MPC). The first group can be implemented in the system with the use of simple low-cost controllers, while the second group involves more complex solutions that involve models for weather forecast and predictions about the building/system behavior. Several papers in the literature have shown the efficacy of advanced control techniques based on the MPC (Fischer and Madani, 2017; Péan, Salom and Costa-Castelló, 2019). However, their implementation in residential buildings is often limited due to high device costs.
RBC are designed to shape the electric load with non-predictive approaches. Different strategies can be implemented depending on the trigger parameters that are used to control the system. Arteconi et al. (Arteconi, Hewitt and Polonara, 2013) studied a control strategy to shift the heat pump operation out of the peak hours (from 16:00h to 19:00h). They found a benefit of shifting of the HP operation in terms of a reduction of the energy costs. Tatjewski et al. (Tatjewski *et al.*, 2016) designed a control algorithm to increase the performance of the HP, reaching an increase of the COP around 10%. Dentel and Betzold (Dentel and Betzold, 2017) proposed a control strategy based on the instantaneous PV production. The results showed an increase of self-consumption by 21%. RBC can be integrated with no additional costs into the controllers of modern HP units, which are already provided of internet connectivity. The efficacy of these solutions needs to be studied further, to better understand their potential in comparison to more sophisticated solutions (i.e. MPC). In most cases, the studies about MPC compared the results to a poorly design controller as a reference case. This assumption leads to an overestimation of the MPC benefits, and the potential of well-designed RBC can be analyzed more in detail (Fischer *et al.*, 2017).
This paper presents a RBC for a modulating air-source heat pump, which provide SH, SC and DHW for a residential building. The system is coupled with a rooftop PV system. The proposed control strategy can be easily implemented in residential buildings by using low-cost board shields The heat pump is modulated and optimized depending on the instantaneous PV production, to maximize the direct use of solar energy onsite. A similar system has been analyzed in a previous study (Pinamonti, Prada and Baggio, 2020), and the results showed the efficacy of the proposed control strategy to increase the level of self-consumption of the system, and to decrease the energy purchased from the grid. The results proved the efficacy of the control strategy under different boundary conditions. In this paper, the integration of a battery system is evaluated, to point out the effect of the control strategy over the optimal size of the electric storage. The efficacy of storing the PV energy using the thermal capacitance of the buildings is analyzed considering different battery storage sizes.

2. METHODS

This study proposes a control strategy to increase the PV self-consumption of a HP system in a residential building. The heat pump has a variable-speed compressor, and it works to provide space heating, space cooling and domestic hot water to the building. When an overproduction of PV energy occurs, the heat pump operates to store the solar energy as thermal energy, exploiting thermal storage tanks and the building thermal capacitance (aka virtual battery). The heat pump is controlled varying its compressor rotational speed. The compressor is regulated to operate at the maximum capacity level compatible with the supplied PV power. The efficacy of the control strategy is assessed by means of dynamic energy simulations. The simulations are run for 1 year.

2.1 Case study

The case study consists of a reference building located in Bolzano, northern Italy. The building is developed on two floors above ground, with a total net area of 140 m^2 and a ratio S/V of 0.59. The heated volume is divided into 4 thermal zones. The air change rate of the building is set to 0.5 ACH (European Committee for Standardization-CEN, 2017), and it is increased to 1.5 ACH during the summer period to represent the window opening by the users. The DHW demand is 186 l/day (UNI - Ente Italiano di Normazione, 2014). The DHW demand profile is defined as prescribed by the European Standards (European Committee for Standardization-CEN, 2016).

The building system consists of an air-source HP, which provides SH, SC and DHW. Two water tanks are integrated in the system, one dedicated to space heating and space cooling, the other one for DHW. The tanks are filled with water, with a volume of 250 liters (SH/SC tank) and 150 liters (DHW tank). The tanks are subjected to thermal stratification. The HP is controlled to maintain the temperature in the tanks at the set-point level, with priority to the DHW tank. The set-point for the DHW tank is set to 50 °C, while the SH/SC tank set-point varies following a outside temperature reset (OTR) curve. In heating mode, the set-point varies from 40°C to 20°C in relation to the outdoor air temperature going from -5°C ad 20°C respectively. For cooling mode, the set-point goes from 12°C to 20°C with the external air temperature varying from 35°C to 26°C. A radiant floor system is connected to the SH/SC tank, and a regulation system controls the water flows for the 4 zones separately. The indoor air temperature is maintained in between 20°C and 26°C ±1 °C. A rooftop PV plant is installed on the building toward South, with a slope of 45°. The nominal power of the system is around 3.20 kW. The heat pump is modeled as a variable-speed compressor unit, based on a performance map that describes the operating behavior of the unit. The performance map was defined starting from detailed steady-state measurements supplied by the manufacturer for a new generation heat pump. The independent variable for the HP model are the inlet air temperature, the outlet water temperature and the compressor speed (i.e. inverter frequency). The dependent variable are the heating rate capacity and the electric input of the compressor. The electric input (Y_{el}) is defined with a polynomial equation (Eq. 1) obtained from the manufacturer data for given sink and source temperatures.

$$Y_{el} = b_0 + b_1 \cdot T_{cond.out}{}^2 + b_2 \cdot T_{ev.in} + b_3 \cdot T_{ev,in}{}^2 \quad (1)$$

The compressor speed is regulated depending on the set-point temperature and the inlet flow temperature in the HP. For DHW production, the HP is working at its maximum capacity. To improve the HP working conditions, the heat pump model is regulate with a minimum stop and running time, and imposing a maximum frequency variation of the inverter of 5 Hz/min.

The proposed control strategy is implemented with a simple control algorithm. Its operation requires common-use sensors for temperature (external air, water flow, thermal storage) and power measurements (domestic appliances, heat pump consumption and photovoltaic production). The control strategy is based on the study presented by Pinamonti et al. (Pinamonti, Prada and Baggio, 2020).

The PV surplus ($P_{PV,production}$) is assessed excluding the appliances' electric load, as in Eq. 2.

$$P_{PV,surplus} = P_{PV,production} \cdot 0.9 - P_{appliances} \quad (2)$$

A reduction factor of 10% is applied to the $P_{PV,production}$ in the formula. The percentage of electric input ($Y_{el,\%,surplus}$) is defined as the ratio between the available PV surplus and the electric input (Yel) at the actual operating conditions of the heat pump calculated as in Eq. (19). The $Y_{el,\%,surplus}$ is calculated as in Eq. 3.

$$Y_{el,\%,surplus} = \frac{P_{PV,surplus}}{Y_{el}} \quad (3)$$

Then, this value is used to define the maximum frequency for the HP operation allowed by the PV availability, using Eq. 4.

$$f = -0.0042 \cdot Y_{el,\%}{}^2 + 1.3861 \cdot Y_{el,\%} - 1.3342 \quad (4)$$

This level of frequency is used in the model to control the heat pump operation during surplus periods.

The PV energy is stored by the system using 3 different solutions: by overheating (or overcooling during summer) the water tanks for SH/SC and DHW, by changing the indoor air temperature set-point by ±2 °C to exploit the building thermal mass as virtual battery, and by charging a Li-ion battery system. Two different strategies are proposed for the energy storage in the water tanks, considering the prioritization of the DHW tank over the SH/SC tank (CS1), and vice versa (CS2). For the battery storage, a battery size varying from 0 to 19.2 kWh is analyzed. The battery has a charging efficiency of 90% and the low limit for the fractional state of charge (FSOC) is set to 0.2. The regulator efficiency is set to 78%, and the inverter efficiency to 96%. Two different charging modes (CM) are proposed. In a first case, the HP is controlled as described above for CS1 and CS2, and any further PV overproduction is stored in the battery system. This charging mode is defined as CM1. In a second case (CM2), the PV surplus is sent with priority to the battery, and, when the battery is totally charged, the PV energy is used to operate the HP following the different CSs. The battery charge is controlled through the fractional state of charge (FSOC).

The system operation is evaluated by means of the self-consumption ($SC_{\%}$) and the self-sufficiency ($SS_{\%}$), as in Eq. 5 and Eq. 6.

$$SC_{\%} = \frac{Self\ consumption}{PV\ production} \quad (5)$$

$$SS_{\%} = \frac{Self\ consumption}{Total\ loads} \quad (6)$$

The efficacy of the thermal and electric storage is assessed considering different combination of the proposed solutions. In particular, 3 main cases are analyzed:

- Battery storage only (CM1 and CM2);
- Battery storage + overheating of the water tanks (CS1 and CS2);
- Battery storage + overheating of the storage tanks + building thermal mass activation (CS1+ and CS2+).

3. RESULTS

The system performance is evaluated for the standard control strategies, and the proposed control strategies with different energy storage solutions (electric storage, thermal storage within the water tanks and within the building thermal inertia). The analysis is carried out using a dynamic energy simulation software. The simulations are run for 1 year, using a 1-minute time step.

In a first analysis, the proposed control strategies are combined with a battery system, considering an electric storage size varying from 1.2 to 19.2 kWh.

The self-consumption level of the system is analyzed with and without battery integration. The results related to the different control strategies and storage capacities are shown in Figure 1 for the CM1 and CM2.

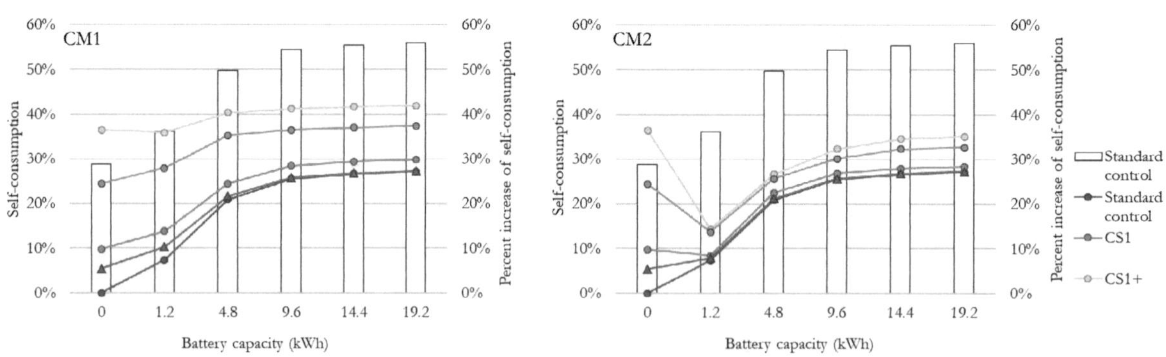

Figure 1 - Level of self-consumption (%) for the different control strategies in relation to the battery capacity (kWh) with the CM1 (left) and CM2 (right)

The first graph shows that, for the standard control strategy, a maximum increase of self-consumption of 27% is achievable with the largest battery capacity (19.2 kWh). In this case, increasing the battery size above 9.6 kWh leads to a minimal benefits to the self-consumption of the system. The increase of $SC_{\%}$ related to the battery integration is reduced for the system working with to the CSs characterized by higher levels of self-consumption. For instance, the application of CS1 leads to an increase of self-consumption of 37% without the battery integration. For the same control strategy, the maximum increase of $SC_{\%}$ related to the battery integration is limited to 5%. The second graph shows the results for the application of CM2. For all the cases, the self-consumption levels are reduced in comparison to CM1. Moreover, the impact of the battery integration leads to a reduction of self-consumption in relation to CS1+ and CS2+, for small battery sizes. For CS2+, the benefit of the thermal mass activation over the SC is compensate with a battery storage capacity of 4.8 kWh. A noticeable difference between the two charging modes

(CM1 and CM2) is that the impact of the storage sizes is reduced for the CSs with the highest levels of self-consumption in the case of CM1. For CM2, the battery size impacts the results until a capacity level around 9.6-14.4 kWh, depending on the control strategy.

Then, the annual amount of energy required from the grid for each solution is analyzed. The results are shown in Figure 2 for the CM1 and CM2.

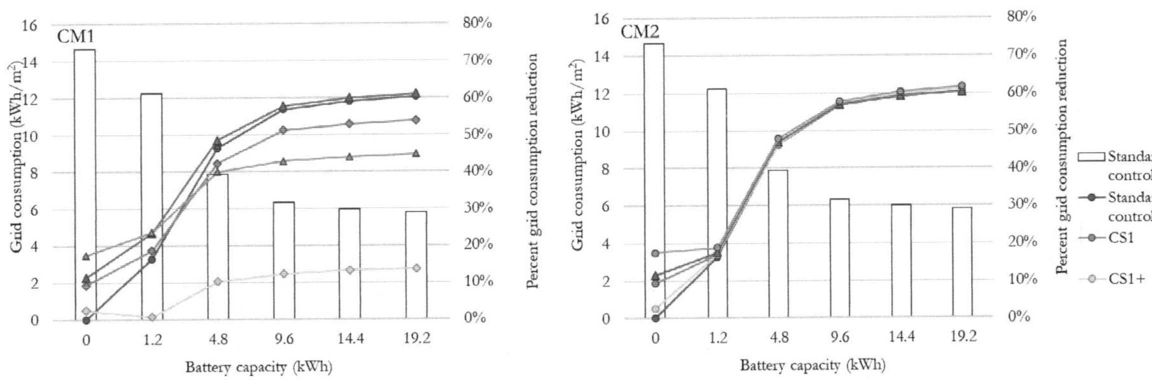

Figure 2 - Grid energy consumption (kWh/m^{2}y) in relation to the battery capacity for the standard control strategy and the percent consumption achievable with the application of the CS2 for different battery sizes with CM1 (left) and CM2 (right)

For CM1 with the standard control strategy, the maximum energy reduction level achievable with the battery integration is 60%. The control strategy CS1+, which leads to the highest level of self-consumption, is characterized by the lowest level of grid consumption reduction (14%). The highest level of energy reduction is achievable with the application of CS2 (61%), with a small increase in comparison to the standard control. The results for the standard control move towards the results of CS2 with increasing battery capacities. Contrarily to the results for the system without battery storage, CS2+ leads to a limited level of grid consumption reduction (45%) in this case. Considering the charging mode CM2, the impact of the different control strategies over the grid consumption is minimized in relation to the majority of the analyzed storage capacity levels. In this case, a slightly higher level of energy reduction is achievable with CS2+ (62%) in comparison to the other CSs. The grid reduction achievable with the CS2+ is only 2%, but an increase of self-consumption around 5% is achievable. It is worth noticing that similar levels of grid energy reduction can be achieved by applying the CS2+ or with the integration of 1.2 kWh electric storage capacity in the system with the standard control strategy. For the self-consumption level, a battery of 9.6 kWh is required with the standard control strategy to reach the same $SC_{\%}$ achievable with the CS2+.

Moreover, the impact of the battery integration over the grid demand magnitude is analyzed and the results are shown in Figure 3 and Figure 4.

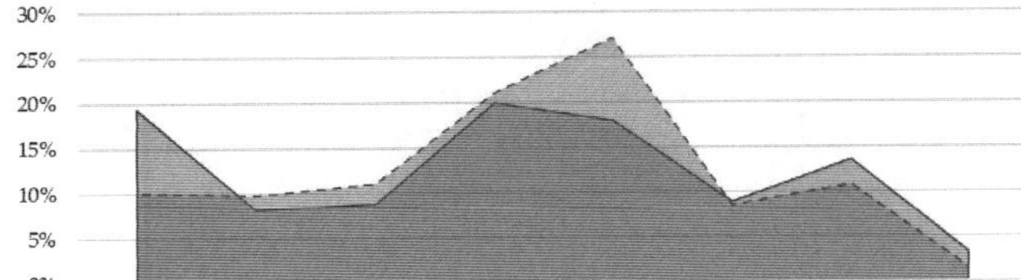

(kJ/h)	0-1000	1000-2000	2000-3000	3000-4000	4000-5000	5000-6000	6000-7000	7000-8000
Standard control	19%	8%	9%	20%	18%	9%	14%	3%
Standard control Battery 9.6 kWh	10%	10%	11%	21%	27%	9%	11%	2%

Figure 3 - Frequency distribution of the grid demand magnitude (kJ/h) of the system throughout the year for the standard control with and without battery integration

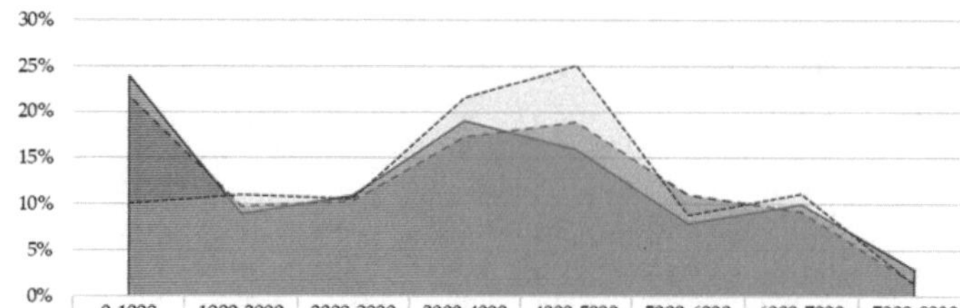

	0-1000	1000-2000	2000-3000	3000-4000	4000-5000	5000-6000	6000-7000	7000-8000 (kJ/h)
CS2+	24%	9%	11%	19%	16%	8%	10%	3%
CM1_CS2+ Battery 9.8 kWh	22%	10%	10%	17%	19%	11%	9%	1%
CM2_CS2+ Battery 9.6 kWh	10%	11%	11%	22%	25%	9%	11%	1%

Figure 4 - Frequency distribution of the grid demand magnitude (kJ/h) of the system throughout the year for the CS2+, with and without battery, considering the two charging modes CM1 and CM2

The first graph shows that the integration of the battery leads to an increase of the medium-magnitude grid withdrawals (+9%). At the same time, the use of the battery decreases the purchase of energy at low and high load levels (-9% and -3%). The second graph shows the distribution of the grid demand magnitude in combination with the CS2+ without battery, and with battery for the two charging modes CM1 and CM2. As shown before, the CS2+ alone is able to slightly reduce the demand peak load periods by 4%, increasing the frequency of smaller grid withdrawals. Similar results are found with the integration of the battery and selecting the CM1, with a slight reduction of low and high load peaks, and with a shift of the peak demand to slightly higher load levels (4000-5000 kJ/h). With the application of CM2, the results are similar to the standard control case. In this case, the battery integration leads to a decrease of low-load levels (-14%), but it does not significantly affect the high loads (-2%), which are already decreased by the application of the control strategy. The difference for the battery integration with and without control strategy is found in relation to a slight reduction of the medium- and high-load withdrawals (1-2%).

A parametric analysis is carried out by varying the PV size and the battery size. The size of the PV plant is defined by the number of panels. Starting from 12 (reference case), the number of PV panels is varied from 6 to 18. The results are analyzed in terms of grid consumption, self-consumption and self-sufficiency, as shown in Figure 5 and Figure 6. Considering the grid consumption (left column), the integration of the battery reduces the impact of the control strategies with increasing storage sizes as seen before. Specifically, when installing a battery size of 1.2 kWh, the control strategy is able to reduce the grid consumption up to 2.0 kWh/m^2, in case of charging mode CM1 and the building thermal mass activation with a 30-m^2 PV installation. Nonetheless, with the integration of a 1.2-kWh electric storage, the thermal mass activation (CS2+) shows higher grid consumption than the other solutions in combination with small PV areas. Considering larger battery sizes, the impact of the different control strategies is neutralized, with the exception of the CS2+ CM1 that shows higher grid consumption. This result shows the inability of this solution to exploit the battery storage. Looking at the self-consumption rate, the solution CS2+ with CM1 shows the highest level of self-consumption in all the analyzed cases. In all the cases, the thermal mass activation (CS2+ with both CM1 and CM2) is able to increase the self-consumption rate for every size of PV and battery. Moreover, the application of CS2+ with CM1 leads to higher levels of self-sufficiency with small battery size (1.2 kWh). With a battery storage of 4.8 kWh, CS2+ leads to higher level of self-sufficiency only in relation to large PV areas (20 m2). Considering battery sizes larger than 4.8 kWh, the solution that leads to the maximum level of self-sufficiency is CS2+ with CM2.

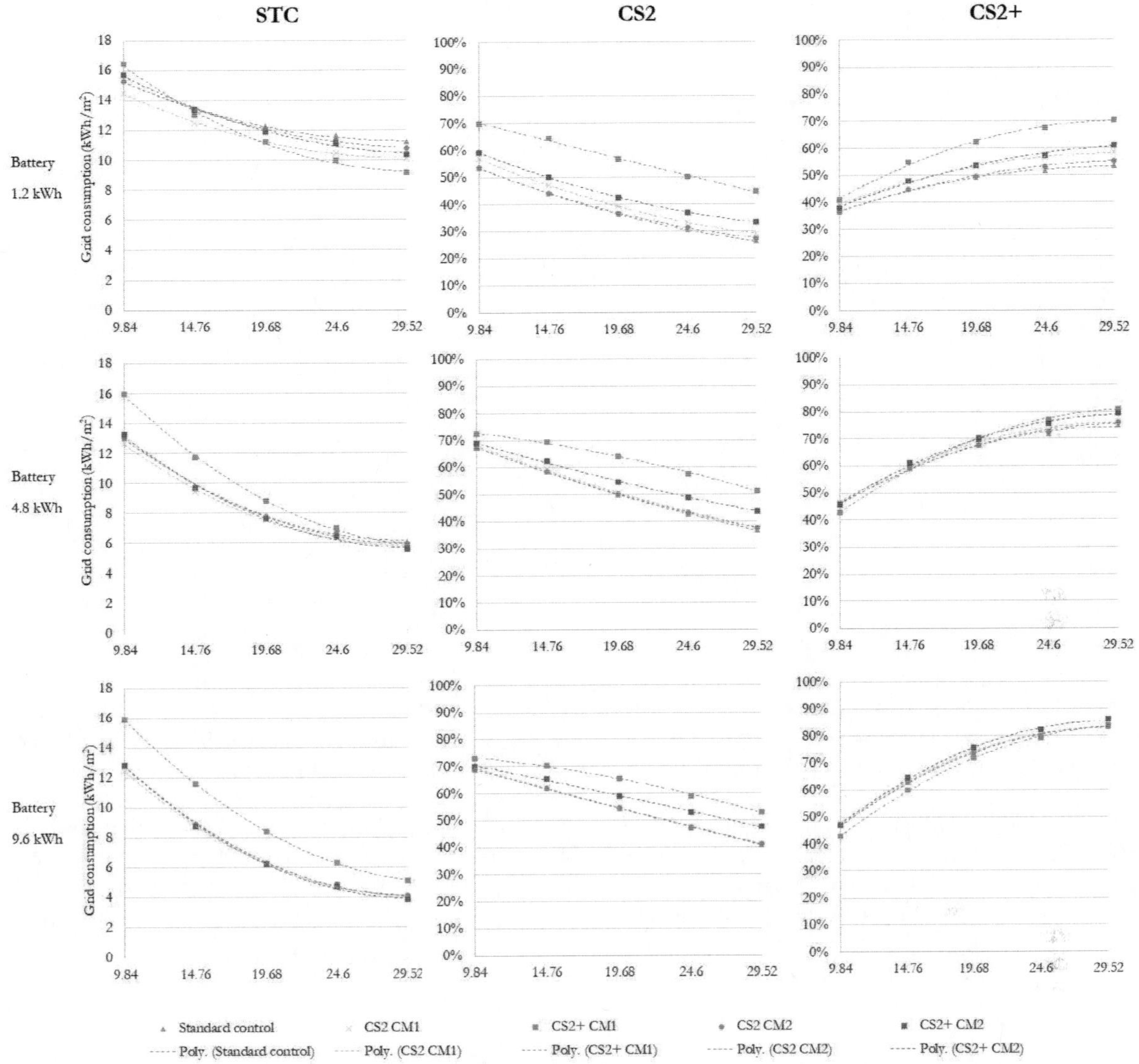

Figure 5 - Relation between the main performance indicators of the system (grid consumption, self-consumption, and self-sufficiency) and the PV area, considering different battery size (from 1.2 to 9.6 kWh) and control strategy

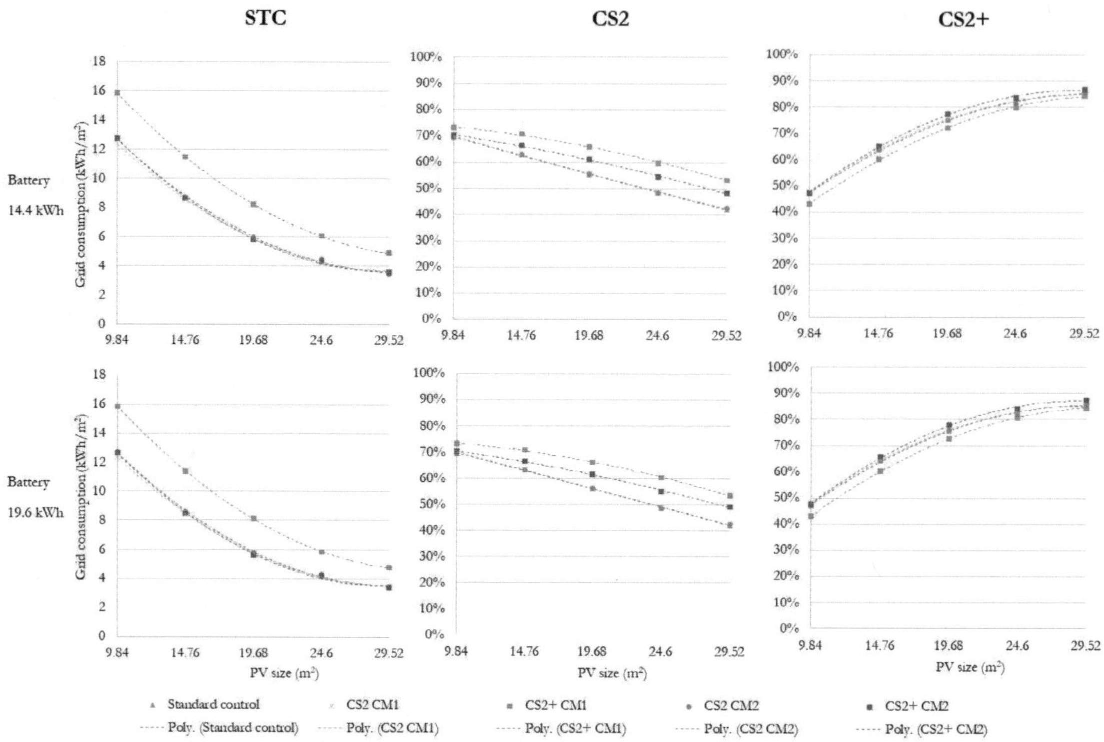

Figure 6 - Relation between the main performance indicator of the system (grid consumption, self-consumption, and self-sufficiency) and the PV area, considering different battery size (from 1.2 to 9.6 kWh) and control strategies

Lastly, the relation between the self-consumption and the grid consumption for different PV and battery sizes is shown in Figure 7. The best solutions are characterized by low grid consumption and high self-consumption levels (upper left part of the graph). The thermal mass activation (triangles) shows again its potential to increase the level of self-consumption while reducing the grid consumption.

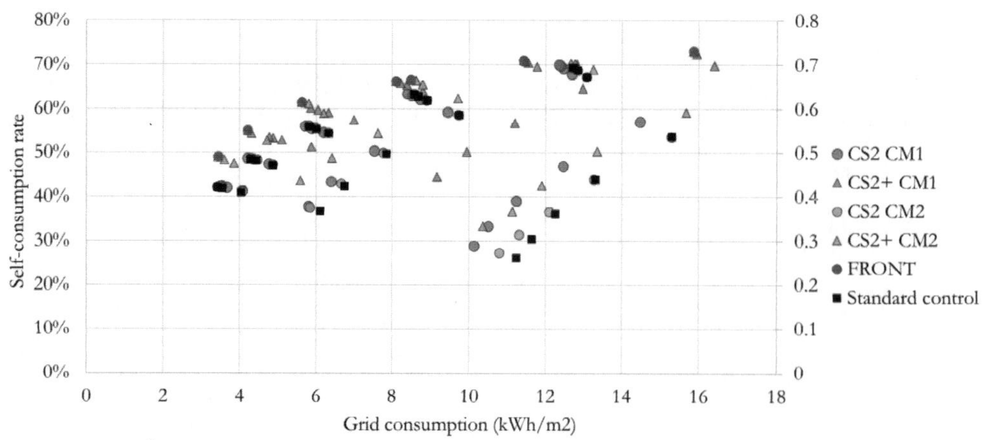

Figure 7 - Self-consumption and grid consumption relation for the different control strategies with battery integration and variable PV size, identification of the Pareto front

The graph shows the identification of the Pareto front, which identifies the solutions that are able to maximize the level of self-consumption.

4. CONCLUSIONS

This study shows the efficacy of a rule-based control strategy to increase the self-consumption of a PV and HP system for heating and cooling supplies, in combination with a an electric storage system. The results showed a constant increase of self-consumption in relation to different battery sizes. Nevertheless, considering the annual grid consumption, the benefits of the control strategy are neutralized with the integration of a battery storage larger than 1.2 kWh. Specifically, the application of the proposed CS2+ without battery and the installation of a 1.2 kWh reach similar results for the analyzed cases. A parametric analysis of the PV and battery size showed that, in combination to small battery size and large PV areas (30 m^2), the thermal mass activation (CS2+) is able to achieve a grid consumption reduction of 2.0 kWh/m^2. Nonetheless, with small PV areas, the same control strategy leads to an increase of grid consumption, in comparison to the other solutions. The thermal mass activation showed to be able to increase the self-consumption of the system for all the analyzed battery and PV sizes. Specifically, the increase of self-consumption increases with increasing PV areas.

NOMENCLATURE

CM	charging mode
CS	control strategy
DHW	domestic hot water
FSOC	fractional state of charge
HP	heat pump
MPC	model predictive control
RES	renewable energy source
SC	space cooling
$SC_{\%}$	self-consumption rate
SH	space heating
$SS_{\%}$	self-sufficiency rate
PV	photovoltaic

REFERENCES

Arteconi, A., Hewitt, N. J. and Polonara, F. (2013) 'Domestic demand-side management (DSM): Role of heat pumps and thermal energy storage (TES) systems', *Applied Thermal Engineering*. Elsevier Ltd, 51(1–2), pp. 155–165. doi: 10.1016/j.applthermaleng.2012.09.023.

Dentel, A. and Betzold, C. (2017) 'Control Strategies for Geothermal Heat Pump Systems in Combination with Thermal and Electrical Storage Units', in *Building Simulation*, pp. 292–298. doi: 10.26868/25222708.2017.081.

European Committee for Standardization-CEN (2016) 'EN12381-3 Energy performance of buildings - method for calculation of design heat load - Part 3: Domestic hot water systems heat load and characterisation of needs, Module M8-2, M8-3'.

European Committee for Standardization-CEN (2017) 'EN 12831-3 Energy Performance of buildings - method for calculation of the design heat load - Part 3: Domestic hot water systems heat load and characterisation of needs, Module M8-2, M8-3'.

Eurostat (2018) *Energy statistics - Share of energy from renewable energy sources*. Available at: https://ec.europa.eu/eurostat/data/database.

Fischer, D. *et al.* (2017) 'Comparison of control approaches for variable speed air source heat pumps considering time variable electricity prices and PV', *Applied Energy*. Elsevier Ltd, 204, pp. 93–105. doi: 10.1016/j.apenergy.2017.06.110.

Fischer, D. and Madani, H. (2017) 'On heat pumps in smart grids: A review', *Renewable and Sustainable Energy Reviews*, 70(October), pp. 342–357. doi: 10.1016/j.rser.2016.11.182.

Péan, T. Q., Salom, J. and Costa-Castelló, R. (2019) 'Review of control strategies for improving the energy flexibility provided by heat pump systems in buildings', *Journal of Process Control*. Elsevier Ltd, 74, pp. 35–49. doi: 10.1016/j.jprocont.2018.03.006.
Pinamonti, M., Prada, A. and Baggio, P. (2020) 'Rule-Based Control Strategy to Increase Photovoltaic Self-Consumption of a Modulating Heat Pump Using Water Storages and Building Mass Activation', *Energies*, 13(23), p. 6282. doi: 10.3390/en13236282.
Tatjewski, P. *et al.* (2016) 'Design and implementation of the air/water heat pump controller with increased coefficient of performance', in *21st International Conference on Methods and Models in Automation and Robotics, MMAR*. IEEE, pp. 959–964. doi: 10.1109/MMAR.2016.7575267.
UNI - Ente Italiano di Normazione (2014) 'Prestazioni energetiche degli edifici Parte 2 : Determinazione del fabbisogno di energia primaria e dei rendimenti per la climatizzazione invernale, per la produzione di acqua calda sanitaria, per la ventilazione e per l'illuminazione in edifici non resid'.

Pathways to Decarbonization of Residential Heating: Potential for Thermally-Driven Heat Pumps

Aleksandr FRIDLYAND[1]*, Paul GLANVILLE[1], Michael GARRABRANT[2]

[1]Gas Technology Institute

Des Plaines, IL 60018

(847) 768-0500; afridlyand@gti.energy

[2]Stone Mountain Technologies Inc.

Johnson City, TN 37604

* Corresponding Author

ABSTRACT

Space and water heating are the largest consumers of energy in residential buildings, up to 85% in cold climates. In addition to conservation measures, improving the building thermal envelope and optimizing the heat and hot water distribution, improving the performance of the heating technology is critical to provide a pathway to significantly reduce carbon emissions, with a focus on existing buildings. Conventional fuel-fired heating systems that use existing infrastructure have approached their thermodynamic limit over the past 15-20 years. While the performance of electrically-driven heat pumps, that would use electricity produced at the margin in a fuel-switching scenario, have improved, their performance in cold climates is highly dependent on regional and seasonal characteristics of the supplied electricity and the capacity of the heat pump and the grid itself to support this peak winter demand. The Gas Absorption Heat Pump (GAHP) is a thermally driven technology that can serve as the next step in fuel-fired heating with Coefficient of Performance (COP) value of 1.4 at design conditions with sufficient capacity to meet peak heating loads. The GAHP would serve as a drop-in replacement for furnaces, boilers and other conventional technologies, significantly reduce emissions, require no updates to existing infrastructure, and provide optionality as delivered decarbonized fuels become more widely available. This study looked at several locations in North America and compared potential emission and economic savings versus several baseline technologies, including furnaces, boilers, electric heat pumps, and various water heating options. The results show that the GAHP provided the lowest operating cost and highest emissions reduction (CO2e). Based on the results, the use and pursuit of GAHPs as a pathway to decarbonization is necessary to meet climate goals.

1. INTRODUCTION

Representing a combined 4.3 quads per year, the natural gas and propane consumed by U.S. homes is a significant energy expenditure and, as a result, source of greenhouse gas (GHG) emissions. The majority of this expenditure is used to provide heat and provide hot water, 3.7 TCF of natural gas and 3.4 billion gallons of propane to single-family and multifamily buildings, which predominantly use these delivered fuels for space and water heating during the heating season (EIA RECS, 2015), with >45% of the natural gas being used just between December and February (EIA, 2019). The challenge, as policymakers, builders, utilities, and end-users seek to reduce the greenhouse gas (GHG) impact of our homes, is that the installed base of space and water heating equipment is low-efficiency. For fuel-fired water heating, majority of which are storage-type (>90%), only 5% of the storage water heaters are 0.67 Uniform Energy Factor (UEF) or greater, qualifying for EnergyStar with the rest near or at the minimum allowable efficiency (Ryan, 2016). For central warm-air furnaces, which are in 85% of homes heated with natural gas or propane, less than half have a rated efficiency of 90% or greater, as an Annual Fuel Utilization Efficiency (AFUE) (DOE SNOPR, 2016). Further penetration of higher-efficiency conventional heating equipment, generally understood to have UEF > 0.90 and AFUE > 90%, is hampered by persistently low natural gas prices, the challenges with cost-effectiveness of infrastructure upgrades in retrofits (e.g. replacing appliance venting), and the absence of strong regulatory drivers to limit GHG emissions (e.g. carbon pricing), among other trends. Thus, the potential remains to mitigate the GHG impact of how we heat our homes, a particular challenge in existing and cold climate buildings.

To address this challenge, in this paper the authors examined the potential of a non-incremental solution for high-efficiency heating in residences, the gas absorption heat pump (GAHP). Applied as a combined space and water heating solution, the GAHP has a potential for a 40% reduction in fuel use for both loads and proportional reduction in emissions, while providing equal or better thermal comfort cost-effectively, compared to baseline equipment. This potential was explored via a techno-economic analysis, using open-source building energy simulation (BES) tools that are calibrated to experimental and field-derived GAHP datasets. This study looked at several locations in North America and compared potential emission and economic savings versus several baseline technologies, including furnaces, boilers, electric heat pumps, and various water heating options.

2. DESCRIBING THE GAS ABSORPTION HEAT PUMP

In prior efforts (Glanville, 2019 and Garrabrant, 2020), the authors have described the development and early demonstration of a pre-commercial gas-fired absorption heat pump (GAHP) for residential and light commercial applications (see Figure 1), based on the single-effect ammonia-water vapor absorption cycle in a heating-only mode. The GAHP is designed with a focus on easily manufactured components, at a projected equipment price of approximately $5,000, 30-50% that of GAHPs currently available on an approximately equal sizing basis. The absorption cycle is driven by a variable, 16 kW burner required to drive the refrigerant vapor from its absorbed state in the desorber (or “generator”). The unitary air-to-brine GAHP is installed outdoors and connected to indoor loads via a hydronic loop, with a nominal output of 23 kW, 4:1 system modulation, and operation without backup or supplemental heating.

Figure 1: Pre-commercial GAHPs operating at WI demonstration sites

During initial proof-of-concept development, a team including an absorption technology development company, a research and development (R&D) institute, and industry partners demonstrated several early generation prototypes in a laboratory environment. With a design described in prior reporting (Garrabrant, 2016), this GAHP was based on a simplified single-effect absorption cycle design, with a “thermal compressor” intended for mass manufacture. This “thermal compressor”, comprised of an absorber, desorber, solution pump, solution heat exchanger, and rectifier, is integrated with the remainder of the heat pump components, including the condenser, expansion valve, evaporator, and refrigerant heat exchanger. A direct-fired combustion system drives the cycle, from which approximately 40% of the GAHP heat output to the hydronic loop is drawn from the evaporator load. The remaining 60% of heat output is heat recovery from (a) the hot weak solution and its heat of sorption within the absorber and (b) the warm flue gases exiting the direct-fired desorber at ~149°C via a flue-to-hydronic condensing heat exchanger (CHX). The operating efficiency, as a Coefficient of Performance on a gas-input basis (COP_{Gas}) ranges from 1.40-1.90, depending on ambient and loop temperatures and modulation stage, with “condensing” level combustion efficiency by virtue of the CHX. These GAHPs demonstrated a nominal COP_{Gas} value of 1.65 at 8°C ambient and 38°C return and an extreme cold condition performance of 1.2 at -25°C with a return of 35°C, via 3rd party testing and on a higher heating value basis, with a more complete performance map shown in Figure 2 below. Peak power demand of early prototypes was up to 600 W, driving the solution pump, evaporator fan, combustion blower, and system controls. Subsequent generations of GAHPs were characterized, both in a) further laboratory testing, including testing to the ANSI Z21.40.4 standard methodology, resulting in an AFUE of 139% for U.S. Climate Region IV, and certification to the Ultra-Low NO_x emission standard of < 14 ng/J output, and b) the installation of eight GAHPs in residential field sites, in TN, IL, WI, and ON, demonstrating more than 10,000 operating hours, up to 53% fuel savings for space/water heating versus a measured baseline, and operation at down to -34°C outdoor conditions. Additionally, improvements in effective cycling, managing simultaneous space and water heating demands, end-user comfort, and power consumption were demonstrated, with several sites continuing to operate at the time of writing.

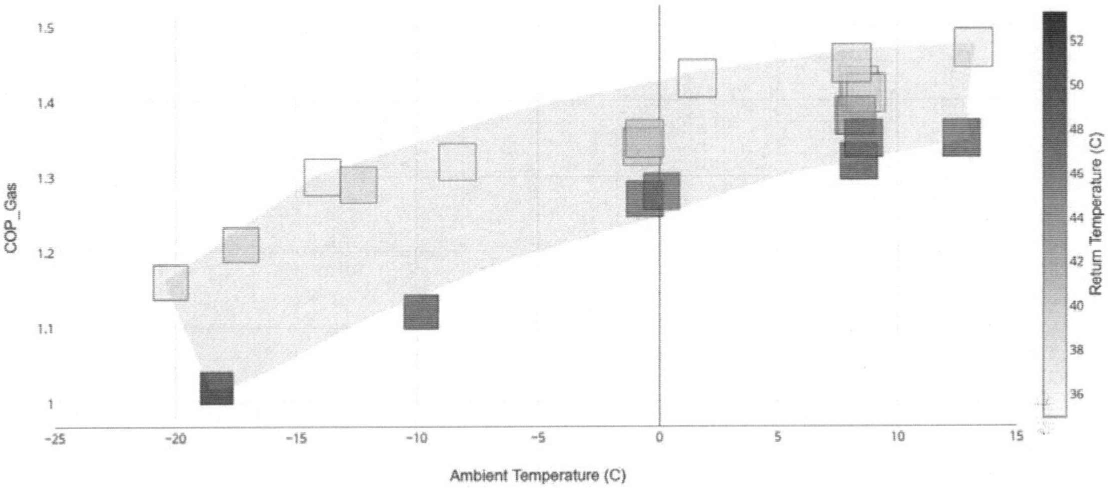

Figure 2: Steady-state efficiency of pre-commercial GAHP in 3rd party laboratory testing

3. MODELING METHODOLOGY

The objective of this analysis was to estimate the energy and cost savings potential of the GAHP combined space and water heating system ("combi" system) in cold climate retrofit applications. The analysis was performed using a combination of open-source BES tools maintained by the U.S. Department of Energy, specifically BEopt and EnergyPlus, and a custom GAHP combi model that uses a performance map based aforementioned laboratory and field-derived datasets. BEopt was used to develop prototypical residential building models and as a source of performance data of electrically-driven heat pumps, to be used for comparison. Three different home sizes were simulated, in nine different locations, and six different HVAC scenarios. Table 1 summarizes the high-level building characteristics used in the analysis, where each home was assumed to be two-story, with four bedrooms, three bathrooms, and a two-car attached garage. The building models were built to approximately International Energy Conservation Code (IECC) 2006 building code. Per the table, three building sizes are considered, though not in all locations/climate zones, reflecting expected nominal capacities of the GAHP at commercialization of (e.g., 35, 29, 23, 18 and 12 kW). Concerning the assumptions regarding the building construction, default assumptions within BEopt from the Building America House Simulation Protocols were used (Wilson, 2014) were applied, where not prescribed in the IECC 2006 codes. Typical home construction varies significantly by vintage, location, and extent to which energy efficiency retrofits have been performed, making it difficult to state definitively what is a "typical" retrofit scenario in each location. The IECC 2006 code was therefore chosen as the common reference point between the newest construction and older construction homes. Further details on the building construction are provided in detailed reporting (Garrabrant, 2020).

Table 1: Locations and buildings used in simulation

<table>
<tr><th>Building Location</th><th>Climate Zone</th><th>Building 1 Size (Large)</th><th>Building 2 Size (Mid)</th><th>Building 3 Size (Small)</th><th>Foundation</th></tr>
<tr><td>Fargo, ND</td><td>7-dry</td><td rowspan="9">278 m²</td><td rowspan="5">223 m²</td><td rowspan="2">N/A</td><td>Unfinished Basement</td></tr>
<tr><td>Rochester, NY</td><td>6-moist</td><td>Unfinished Basement</td></tr>
<tr><td>Minneapolis, MN</td><td>6-moist</td><td rowspan="2">167 m²</td><td>Unfinished Basement</td></tr>
<tr><td>Chicago, IL</td><td>5-moist</td><td>Unfinished Basement</td></tr>
<tr><td>Philadelphia, PA</td><td>5-moist</td><td rowspan="5">N/A</td><td>Unfinished Basement</td></tr>
<tr><td>Denver, CO</td><td>5-dry</td><td rowspan="4">N/A</td><td>Unfinished Basement</td></tr>
<tr><td>Portland, OR</td><td>4-marine</td><td>Slab</td></tr>
<tr><td>Louisville, KY</td><td>4-moist</td><td>Unfinished Basement</td></tr>
<tr><td>San Francisco, CA</td><td>3-marine</td><td>Slab</td></tr>
</table>

3.1 Modeling of HVAC and Water Heating Equipment

The table below summarizes the HVAC scenarios considered in the analysis. All HVAC system models were taken directly from BEopt with minor modifications and the baseline storage water heater (0.62 EF) model is a custom option added to BEopt to reflect the current federal minimum. The variable speed heat pump option was modified to size the equipment based on the maximum load (as opposed to cooling load only by default), to better reflect how a "cold-climate" heat pump would be sized. The furnace models were modified to include part-load efficiency curves, including equation 1, for condensing furnace, equation 2, for non-condensing furnace, with the part-load ratio (PLR) defined as in equation 3. These modifications were made to better account for cycling efficiency losses of typical furnaces. These curves were added directly to the EnergyPlus models generated by BEopt and simulated separately.

$$Normalized\ Efficiency = 0.9 + 0.1 * PLR \tag{1}$$

$$Normalized\ Efficiency = 0.8 + 0.2 * PLR \tag{2}$$

$$PLR = load/capacity \tag{3}$$

Table 2: Equipment Scenarios

Equipment Case	Space Heating Details	HVAC Equipment Sizing	Water Heating Details
Baseline / Gas	80% AFUE Furnace	Auto-sized for peak heating load	0.62 EF, 180 L Storage
Better / Gas	95% AFUE Furnace	Auto-sized for peak heating load	0.62 EF, 180 L Storage
Best / Gas	95% AFUE Furnace	Auto-sized for peak heating load	0.96 EF, 58 kW Tankless
GAHP Combi	13 kW output minimum	Auto-sized for peak heating load	246 L IST
Baseline / Electric	7.7 HSPF Heat Pump	Auto-sized for peak cooling load	0.92 EF, 225 L Storage
Best / Electric	10.0 HSPF Variable Speed Heat Pump	Auto-sized for peak heating load	0.92 EF, 225 L Storage

The equipment cases represented were meant to represent "likely" retrofit scenarios to be encountered. For this reason, electric heat pump water heaters were not considered given their low market penetration and the unlikely scenario that both the central HP and the electric water heater would be swapped out for much more expensive options. Similarly, the "Best / Gas" scenario is also unlikely given the long payback periods of tankless water heaters in retrofit scenarios (Fridlyand, 2019). However, it was left for comparison given its comparable performance to tankless water heater-based combis, which could be more cost effective in retrofit scenarios (Fridlyand, 2021).

3.1 Modeling of the GAHP Combi System

The GAHP combi was simulated by modifying the "Best Gas" scenario BEopt model. The indoor furnace was replaced with a hot water heating coil to represent the hydronic air handler. The storage water heater was replaced with a stratified 9-node storage tank model, indirectly heated by the GAHP, and with heat loss characteristics similar to the tanks used in the field (~44 W in standby). The GAHP itself was modeled as a user-defined plant component and a custom Energy Runtime Language (ERL) script within the EnergyPlus input file. The "Best Gas" scenario was chosen as the basis to accurately capture the reduced infiltration of closed combustion heating systems. The duct heat-loss model as implemented in BEopt was preserved, overall space heating demand was not impacted.

The GAHP plant model acted as a simple heating device responding to demand from the space and water heating branches. The maximum heating capacity and the efficiency of the GAHP was determined using six unique performance curves, following a similar approach used in EnergyPlus for electric heat pumps (US DOE, 2018). The overall heating capacity of the GAHP was determined from the rated heating capacity and a function (CAPFT) of outdoor dry bulb temperature ($Tamb$) and hydronic return temperature ($Tret$), as shown in equations 4 and 5, with coefficients defined in Table 3. Knowing the maximum heating capacity of the GAHP for a given condition, the PLR was determined according to equation 6.

$$GAHP\ Heating\ Capacity = Rated\ Heating\ Capacity * CAPFT \tag{4}$$

$$CAPFT = \ a1 * Tamb + b1 * Tamb^2 + c1 * Tret + d1 * Tret^2 + e1 * Tret * Tamb + f1 \tag{5}$$

$$Part\ Load\ Ratio\ (PLR) = \frac{load}{capacity}\ for\ 0.2 \leq PLR \leq 1 \tag{6}$$

Table 3: Equation 5 and 8 Coefficients

	a	b	c	d	e	f
Eq. 5	0.00428	-8.6E-05	0.004093	-0.00014	2.26E-06	1.011452
Eq. 8	-0.00318	6.6E-05	0.011763	-6.1E-05	-4.8E-05	0.382999

The minimum modulation level of the GAHP is 25%, with a minimum cycle time of 45 minutes, resulting in a minimum PLR of ~0.2. If the requested load was below the minimum PLR, the GAHP would not activate. At other conditions, the GAHP would meet the heating load and consume gas according to the following equations, where, $EIRFT$ is the energy input ratio function of outdoor dry bulb and hydronic return water temperatures, with coefficients defined in Table 3. In equation 9, $EIRFPLR$ is the energy input ratio function of the part load ratio. To account for the defrost performance penalty (up to 4% near -2.8°C), an additional energy input ratio function of outdoor dry bulb temperature was used defined in equation 10. Between the minimum modulation rate at $PLRmin = 0.25$ and the minimum cycle time at $PLR = 0.2$, the GAHP is expected to cycle to meet load. An additional cycling penalty function was defined to account for increased energy use in this range, defined in equation 11. In one example, Figure 3 compares the predicted versus measured efficiency as a function of outdoor dry bulb temperature and part load ratio.

$$Gas\ Use = \frac{Load * EIRFT * EIRFPLR * EIRDEFROST}{CRF} \tag{7}$$

$$EIRFT = \ a2 * Tamb + b2 * Tamb^2 + c2 * Tret + d2 * Tret^2 + e2 * Tret * Tamb + f2 \tag{8}$$

$$EIRFPLR = \ 0.0864 * PLR^2 \ - \ 0.0681 * PLR \ + \ 0.9814\, for\ 0.25 \leq PLR \leq 1 \tag{9}$$

$$EIRDEFROST = \ -0.0011 * Tamb^2 \ - \ 0.006 * Tamb \ + \ 1.0317\, for - 8.89°C \leq Tamb \leq 3.333 \tag{10}$$

$$CRF = 0.4167 * CR + 0.5833; \quad CR = \frac{PLR}{PLRmin}\ for\ 0.2 \leq PLR \leq 0.25 \tag{11}$$

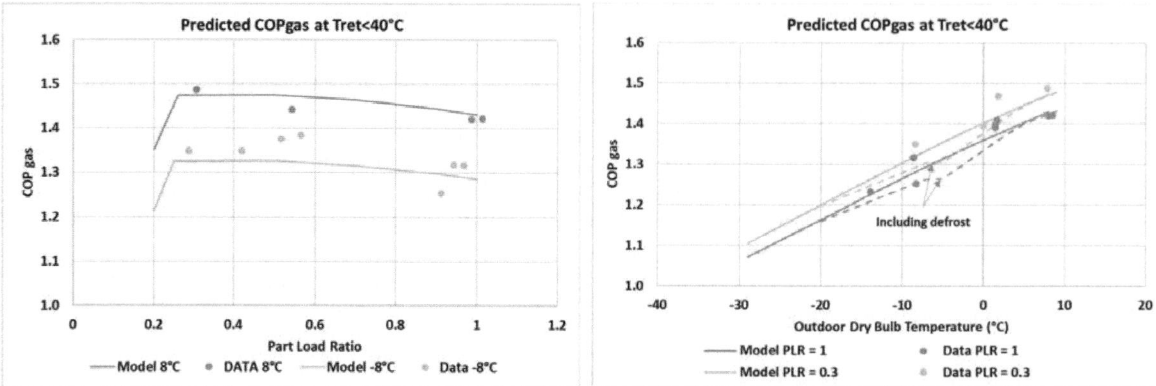

Figure 3: Predicted COP_{Gas} as function of PLR (Left) and ambient temperature (Right) for T_{Return} = 37°C

The electricity use of the GAHP was modeled using a simple approach, defining the consumption of the outdoor unit as 2.7% of the heat delivered and the pump energy as 1% of the heat delivered, both consistent with field measurements. The supply temperature setpoint for the GAHP followed an outdoor air temperature reset for space heating, summarized in Table 4. The supply water temperature setpoint was fixed at 54.4°C for hot water recovery, to meet the mid-tank setpoint target of 51.7°C. The balance of the EnergyPlus model, e.g., tank and water heating coil heat transfer effectiveness, was tuned to achieve a typical return water temperature of ~40.5°C, as observed at the field demonstration sites. For each location and building type, the rated heating capacity of the GAHP was auto-sized to better match the other HVAC scenarios which were all auto-sized by BEopt and to ensure that the GAHP could meet the peak heating load over the course of the year. This capacity was determined by dividing the maximum observed hourly heating load by the output of the *CAPFT* function at the corresponding outdoor dry bulb temperature and an assumed return water temperature of 37°C. The minimum rated capacity of the GAHP was fixed at 13.2 kW to ensure that there was enough capacity to meet the DHW demand. Economic and environmental assumptions are shown in Table 5 (SEEAT, 2019).

Table 4: Outdoor air temperature reset strategy

Ambient Dry Bulb Temperature, °C	Supply Water Temperature, °C
12.7 or above	43.33
12.7 to -20.6	Linear curve fit based on end points
-20.6 or below	51.67

Table 5: Economic and environmental assumptions by region, based on EIA and US EPA Data (SEEAT, 2019)

Location	Gas Winter Price ($/Therm)	Annual Average Electricity Price ($/kWh)	Electric Grid CO2e* (kg/MWh)	Gas CO2e Emissions (kg/MWh)
Chicago	$ 0.76	$ 0.125	285	
Denver	$ 0.69	$ 0.121	267	
Fargo	$ 0.67	$ 0.102	281	
Louisville	$ 0.98	$ 0.105	266	
Minneapolis	$ 0.77	$ 0.127	281	230
Philadelphia	$ 0.98	$ 0.139	225	
Portland	$ 1.10	$ 0.107	240	
Rochester	$ 1.05	$ 0.176	175	
San Francisco	$ 1.14	$ 0.174	165	
*Non-baseload (marginal) power plants				

4. RESULTS AND DISCUSSION

An important feature of modeling exercises of this nature is the assumed sizing of the equipment, as over-sized equipment can suffer poor performance at part-load and otherwise due to short-cycling and conversely, while uncommon, under-sized equipment can excessively rely on lower efficiency auxiliary/backup heating (when applicable). Figure 4 highlights for Building 1, the modeled peak heating demand versus the calculated GAHP capacity. In Figure 5, annual performance predicted for the GAHP Combi in each location for Building 1 is shown, defined on a fuel-only basis (COP_{Gas}) and total fuel and electricity input basis (COP_{Total}). The results are comparable to field observations and align with expected performance in cold climates. Figure 6 plots the variation of the annual COP_{gas} by location and building size. The hot water use for each building stays the same and the primary difference between the buildings is the size and therefore the overall heating load. Figure 7 compares operating cost savings.

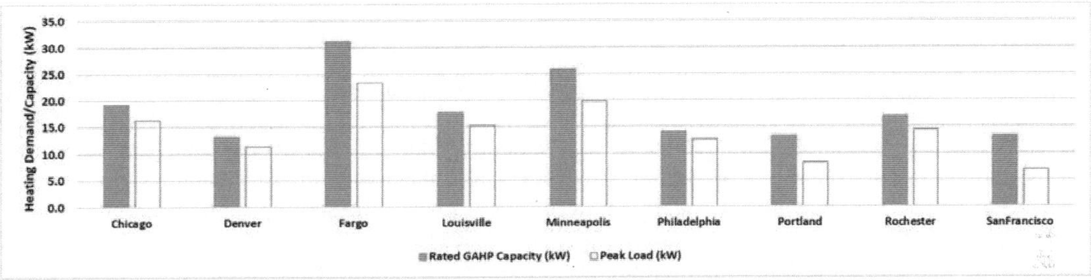

Figure 4: Building 1 peak heating demand and resultant GAHP sizing

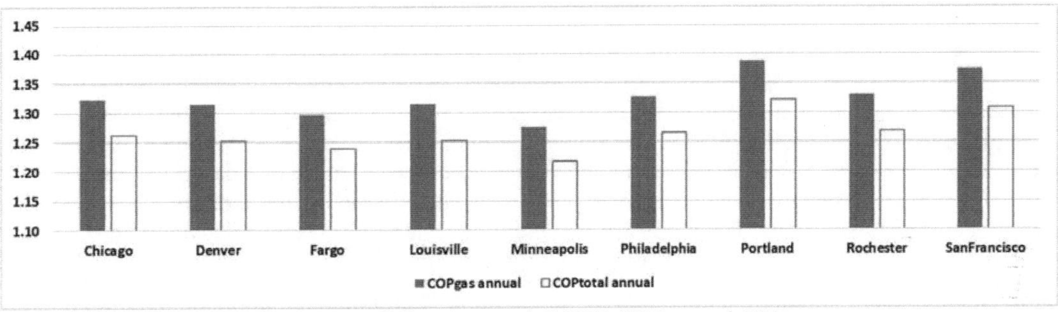

Figure 5: Building 1 annual GAHP COP_{Gas} and COP_{total}

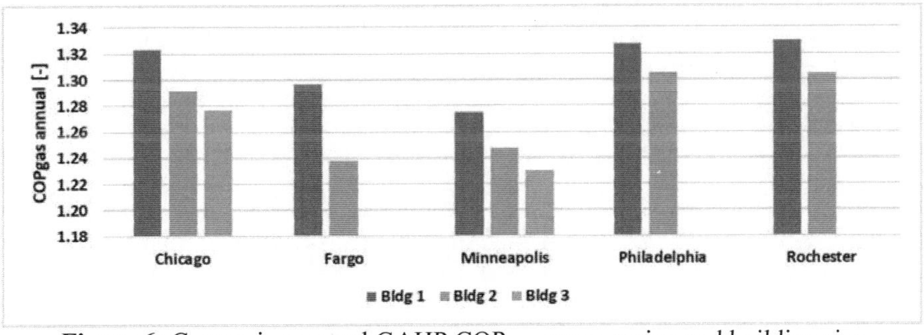

Figure 6: Comparing annual GAHP COP_{Gas} across region and building size

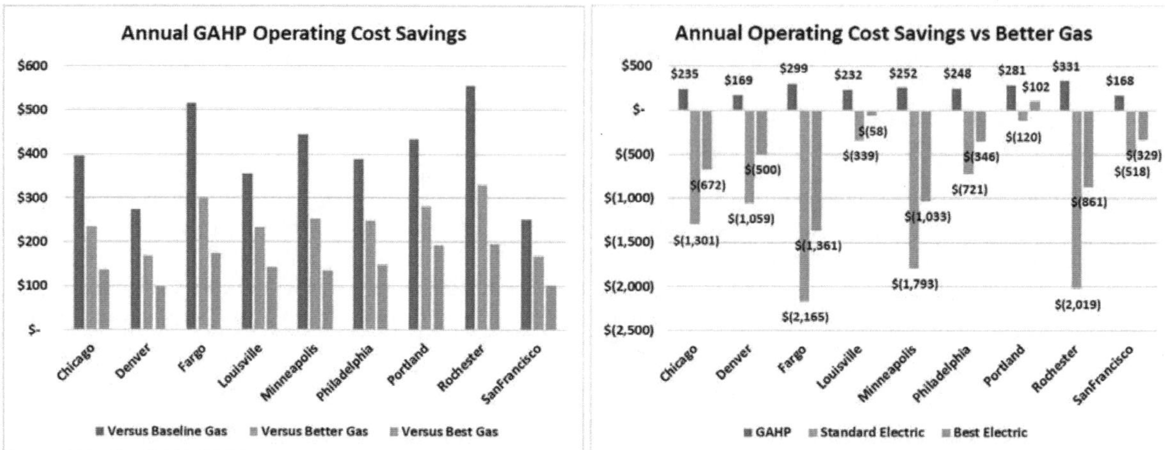

Figure 7: Comparing annual HVAC operating cost savings for gas options (L) and electric vs GAHP options (R)

The projected GAHP energy and cost savings are the highest compared to the baseline gas case which uses both non-condensing space and water heating technologies. The savings are the lowest when compared to the "Best Gas" scenario, condensing furnace and tankless water heating. However, due to the tough economics for tankless water heaters, e.g., 20+ year payback periods, this is not a likely retrofit scenario to occur (Fridlyand, 2019). With increasing prevalence of condensing furnaces and low penetration of high-efficiency water heating gas technologies, the "Better Gas" scenario is the more likely retrofit scenario to be encountered in the near future, against which a GAHP combi system delivers improved performance.

Figure 8 highlights the CO_2e emission savings, for Building 1, for the GAHP combi system versus all other equipment scenarios, highlighting the attractive emission savings in all regions. Similarly, the GAHP combi system provides operating cost and energy consumption savings in all cases as well. It is important to highlight that the marginal electricity grid emission were considered here. Per U.S. Environmental Protection Agency (EPA), renewable efficiency measures typically do not impact baseload power generation. The increased or decreased demand for electricity in buildings is managed on the margin (US EPA, 2020), which for the foreseeable future will still be dominated by fuel-fired power plants. Other scenarios including on-site renewable electricity generation, increases in renewable gas use, or changes to the grid generation mix were not considered due to the speculative nature of any future projections and to limit the scope of the analysis.

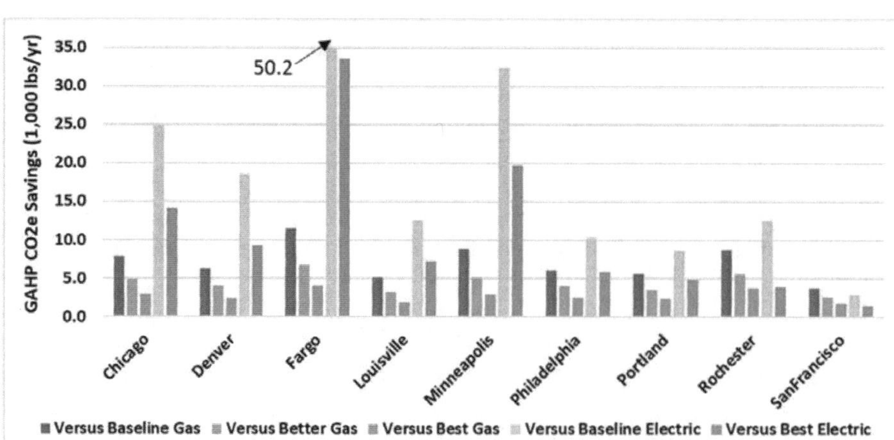

Figure 8: Annual CO_2e savings for GAHP combi versus other equipment scenarios for building 1

When compared to fuel-fired equipment scenarios, the most likely GAHP retrofit scenario, efficiency benefits are coupled with the elimination of negative interactive effects from low-efficiency gas water heaters on HVAC (combustion/dilution air requirements). This impact is significant, however it is somewhat blunted by the GAHP system's increased power demand. For example, when comparing the Chicago-area Building #1 from GAHP system to baseline gas, the annual fuel savings are significant, at 48%, however operating costs and CO_2 reductions are not as significant, $395/year and 35% respectively, due to the incremental power requirements of the GAHP and the hydronic heating system, increasing electricity consumption by 83% or 824.1 kWh/year.

6. CONCLUSIONS

This study built on the results of prior successful laboratory and field evaluations of a single-effect Gas Absorption Heat Pump (GAHP), to estimate the potential for energy and greenhouse gas emissions (GHG) savings in different locations and retrofit scenarios in North America. Using performance data measured as a function of outdoor and return water temperature to the GAHP, a combined space and water heating system was simulated in nine different locations, three different building sizes, and compared to five other gas and electric HVAC scenarios. In all cases considered, the GAHP system offered the greatest GHG emissions savings and the lowest annual operating costs, based on the current average retail energy prices and marginal electric grid GHG emission factors. While the electric grid power generation mix will continue to change, energy prices will vary, and the economics may change, presently, the GAHP combined space and water heating system in cold climates is predicted to be best and most cost-effective option for reducing GHG emissions in residential buildings.

REFERENCES

EnergyStar Program, via the U.S. Dept. of Energy (2010). *Energy Star® Water Heater Market Profile*, report prepared under contract BOA#4200000327.

Fridlyand, A., Liszka, M. (2019). *Lab Testing of Tankless Water Heater Systems*, NEEA Report E19-306, https://neea.org/resources/lab-testing-of-tankless-water-heater-systems

Fridlyand, A., Baez Guada, A., Kingston, T., Glanville, P. (2021). *Modeling Modern, Residential, Combined Space and Water Heating Systems Using EnergyPlus*, ASHRAE Transactions, (in press).

Garrabrant, M. Stout, R. Keinath, C. and Glanville, P. (2016). *Experimental Evaluation of Low-Cost Gas Heat Pump Prototypes for Building Space Heating*, International Refrigeration and Air Conditioning Conference Paper 1573.

Garrabrant, M., and Keinath, C. (2020). *Pre-Commercial Scale-Up of a Gas-Fired Absorption Heat Pump*, Report DOE-SMTI-8219-1 prepared for the U.S. Dept. of Energy under contract DE- EE0008219.

Glanville, P, Fridlyand, A., Keinath, C., and Garrabrant, M. (2019). *Demonstration and Simulation of Gas Heat Pump-Driven Residential Combination Space and Water Heating System Performance*, ASHRAE Transactions; Atlanta Vol. 125, pp. 264-272.

Raustad, R., (2011). *Modeling Variable Refrigerant Flow Heat Pump and Heat Recovery Equipment in EnergyPlus*, FSEC-CR-1960-11.

Ryan, S. (2016). *Making ENERGY STAR® Water Heaters a National Early Replacement Priority*, presentation made at the ACEEE Hot Water Forum, Portland, OR.

Source Energy and Emissions Analysis Tool (SEEAT), (Accessed 2019). http://seeatcalc.gastechnology.org/.

U.S. Census. (2019 ed.), *American Housing Survey*.

U.S. Dept. of Energy (DOE), (2016). *Supplemental Notice of Proposed Rulemaking (SNOPR) 2016. Technical Support Document: Residential Furnaces*, Prepared for the DOE by Navigant Consulting, Inc. and Lawrence Berkeley National Laboratory.

U.S. Dept. of Energy (DOE), (2018). *EnergyPlus Version 8.9 Documentation: Engineering Reference*, https://energyplus.net/sites/all/modules/custom/nrel_custom/pdfs/pdfs_v8.9.0/EngineeringReference.pdf

U.S. Energy Information Administration (EIA), (2009-2015). *Residential Energy Consumption Survey (RECS).* www.eia.gov/consumption/residential/

U.S. Energy Information Administration (EIA), (Accessed 2019). *Natural Gas Monthly Report.*

U.S. Environmental Protection Agency (EPA), (Accessed 2021). *Greenhouse Gases Equivalencies Calculator - Calculations and References.* https://www.epa.gov/energy/greenhouse-gases-equivalencies-calculator-calculations-and-references

Wilson, E., Horowitz, S. (2014). *Building America Housing Simulation Protocols*, NREL/TP-5500-60988.

ACKNOWLEDGEMENT

This material is based upon work supported by the U.S. Department of Energy's Office of Energy Efficiency and Renewable Energy (EERE) under the Buildings Energy Efficiency Frontiers & Innovation Technologies (BENEFIT) 2017, Award Number DE-0008219."

Disclaimer: "This report was prepared as an account of work sponsored by an agency of the United States Government. Neither the United States Government nor any agency thereof, nor any of their employees, makes any warranty, express or implied, or assumes any legal liability or responsibility for the accuracy, completeness, or usefulness of any information, apparatus, product, or process disclosed, or represents that its use would not infringe privately owned rights. Reference herein to any specific commercial product, process, or service by trade name, trademark, manufacturer, or otherwise does not necessarily constitute or imply its endorsement, recommendation, or favoring by the United States Government or any agency thereof. The views and opinions of authors expressed herein do not necessarily state or reflect those of the United States Government or any agency thereof."

Additional support was provided by Utilization Technology Development under projects 1.13.F and 1.16.H.

Material Selection and Sizing of a Thermoelectric Generator (TEG) for Power Generation in a Self-Powered Heating System

Tim J. LACLAIR[1]*, Kyle R. GLUESENKAMP[1], Ahmad ABU-HEIBA[1], Praveen CHEEKATAMARLA[1], Hsin WANG[1], Sandeep ALAVANDI[2], Hamid ABBASI[2], David CYGAN[2], Alexander KIRK[2], and Uttam GHOSHAL[3]

[1]Oak Ridge National Laboratory
Oak Ridge, Tennessee, USA
Contact: laclairtj@ornl.gov I 865-341-1305

[2]Gas Technology Institute
Des Plaines, Illinois, USA
Contact: salavandi@gti.energy I 847-768-0571

[3]Sheetak Inc.,
Austin, Texas, USA
Contact: ghoshal@sheetak.com I 512-382-0744

* Corresponding Author

ABSTRACT

By employing the high temperature heat source to directly generate the electricity needed to power auxiliary systems in a natural gas furnace, boiler or hot water heater, a "self-powered" heating system can provide several benefits. Compared with a traditional furnace, boiler or hot water heater, when overall fuel utilization is kept constant, the self-powered system will have a higher primary energy efficiency, lower operating costs, and dramatically improved building safety and resilience during electric grid outages. Furthermore, a self-powered heating system only has a single utility connection – natural gas, without an electric connection – thus simplifying installation. A thermoelectric generator can be used for direct energy conversion of thermal energy to electricity with no moving parts, which offers a very simple means to provide power for the self-powered heating system, and the operation is without noise or vibration and can thus provide a very long system life. This paper provides an analysis focused on materials selection and the thermal power requirements for a thermoelectric generator (TEG) for use in a self-powered heating system. The dimensionless figure of merit for thermoelectric materials, zT, is used to estimate the optimal efficiency that can be achieved with a TEG to produce the electric power required in such an application. Comparisons of the predicted efficiency, the required heat transfer rate to the TEG and the heat transfer area needed for sustained operation under thermal conditions relevant to the self-powered heating application are made for several potential thermoelectric materials. This analysis was used to develop system requirements for a self-powered hot water heater using a TEG for electric power generation.

1. INTRODUCTION

Conventional natural gas heating systems employed in buildings (furnaces, boilers and water heaters) require an electrical supply to power auxiliary equipment such as fans, ignitor and electronic controllers, requiring a second utility connection to the heating system even though electricity is not employed for the heating. These heating devices cannot function during power outages, which can eliminate the ability to provide heat when it is most needed, for

Notice: This manuscript has been authored by UT-Battelle, LLC under Contract No. DE-AC05-00OR22725 with the U.S. Department of Energy. The United States Government retains and the publisher, by accepting the article for publication, acknowledges that the United States Government retains a non-exclusive, paid-up, irrevocable, world-wide license to publish or reproduce the published form of this manuscript, or allow others to do so, for United States Government purposes. The Department of Energy will provide public access to these results of federally sponsored research in accordance with the DOE Public Access Plan (http://energy.gov/downloads/doe-public-access-plan).

example following severe storms when temperature extremes may be experienced for extended periods. Therefore, a self-powered heating system—one that generates its own electrical power as opposed to being connected to the electric grid—can improve building safety and resilience during electric grid outages. Additionally, a self-powered device simplifies installation since only a single utility connection is required, and there are potential benefits in primary energy usage when the power is generated locally (Gluesenkamp, 2019). Self-powered heating systems, including furnaces, boilers and hot water heaters, have been investigated in previous studies, and the literature shows examples of systems utilizing several possible power cycles for the electrical generation. A review of power conversion options for self-powered furnaces is provided in Abu-Heiba *et al.* (2021). Some specific applications are described in the following. Qiu and Hayden (2008) reported development of a self-powered residential hydronic heating system for which a PbSnTe thermoelectric generator (TEG) was used as the electricity generation device. The system generated 553.9 W of electric power at a hot surface temperature of 637°C and a cold surface temperature of 85°C. Butcher *et al.* (2011) built an oil-fired self-powered boiler that employed a thermophotovoltaic (TPV) module for the power generation, with a targeted electrical output of 100 W. Various emitter material systems and GaSb TPV cell configurations were evaluated for the impact on the heat transfer efficiency to the emitter. Qiu and Hayden (2014) integrated a GaSb TPV generator into a residential combi boiler. A maximum power of 246.4 W at an emitter temperature of 1265°C was generated with a fuel input of 12.3 kW. Alptekin *et al.* (2017) developed a self-powered condensing combi boiler based on a TEG for electrical generation, and the system was evaluated experimentally at several different firing and water flow rates.

This paper focuses on TEG materials for electric power generation in self-powered heating systems. A methodology for TEG material evaluation is used to characterize the performance of several TEG materials for use in natural gas applications, and associated thermal power requirements are evaluated.

2. MATERIAL PROPERTY CONSIDERATIONS FOR THERMOELECTRIC DEVICES

2.1 Thermoelectric Material Figure of Merit

The dimensionless thermoelectric (TE) figure of merit, zT, for a single p-type or n-type material is given by Equation (1),

$$zT = \frac{S^2}{\rho k} T, \tag{1}$$

where S is the Seebeck coefficient, ρ is the electrical resistivity, k is the thermal conductivity, and T is the absolute temperature (Rowe, 2006). It is noted that a TEG requires a combination of both p- and n-type materials, while the figure of merit zT is defined for individual materials. Nonetheless, this property provides a convenient and useful metric of the TE performance of individual materials for use in TE devices and is frequently reported as a function of temperature in the literature. In this report, zT data was obtained for 16 materials (Bjork, 2015) covering the feasible temperature range of each material. The data was digitized so detailed numerical analysis could be conducted for material selection and sizing for the TEG design. Figure 1 shows the zT data for these materials.

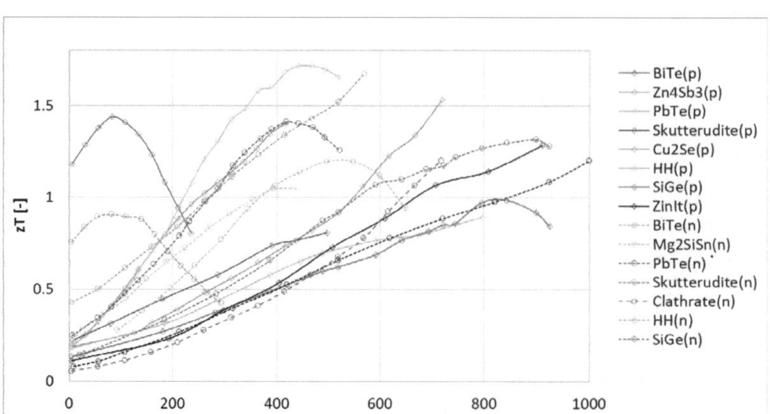

Figure 1: Dimensionless material-specific figure of merit, zT, for p- and n-type thermoelectric materials. [Data from Bjork (2015)]

A TEG consists of p- and n-type TE materials that are connected electrically in series and thermally in parallel, as shown in Figure 2.

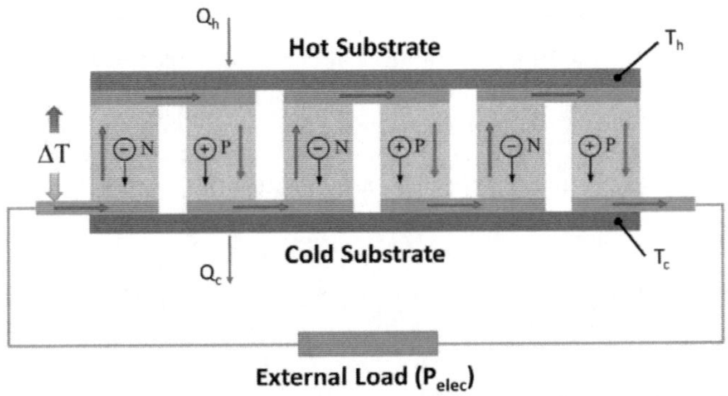

Figure 2: TEG operation: heat is transferred from an external source at the rate Q_h, and removed at the rate Q_c. Electrical power, P_{elec}, is generated by the TEG and provided to an external load.

2.2 TEG Device Figure of Merit

The device figure of merit, *Z*, for a TEG is given by

$$Z = \frac{(S_p - S_n)^2}{\left[(\rho_p k_p)^{1/2} + (\rho_n k_n)^{1/2}\right]^2}, \qquad (2)$$

where the subscripts correspond to the p- and n-type materials used for each couple in the TEG. Under some conditions, the value of *Z* can be reasonably approximated as the mean of the material figures of merit, z_p and z_n. The TEG efficiency, η, is defined as the electrical output divided by the high temperature thermal input ($\eta = P_{elec}/Q_h$). Once a value of *Z* is determined, the TEG maximum efficiency, η, can be calculated according to Equation (3):

$$\eta = \Delta T/T_h \;\left(\sqrt{(1+ZT)} - 1\right) / \left(\frac{\sqrt{(1+ZT)}+T_c}{T_h}\right), \qquad (3)$$

where T_c and T_h are the cold- and hot-side substrate temperatures of the TEG, and ΔT is the temperature difference T_h-T_c across the TEG (Bjork, 2015). For the analysis in this work, the value of *ZT* was evaluated at the average operating temperature of the TEG, $T_{avg}=(T_c+T_h)/2$, and *Z* was approximated as the mean of z_n and z_p at T_{avg}, as described previously. For comparison purposes, the *ZT* determined from average *zT* values are shown in Figure 3 as a function of average TEG temperature for combinations of p- and n-type TE materials based on the same un-doped material.

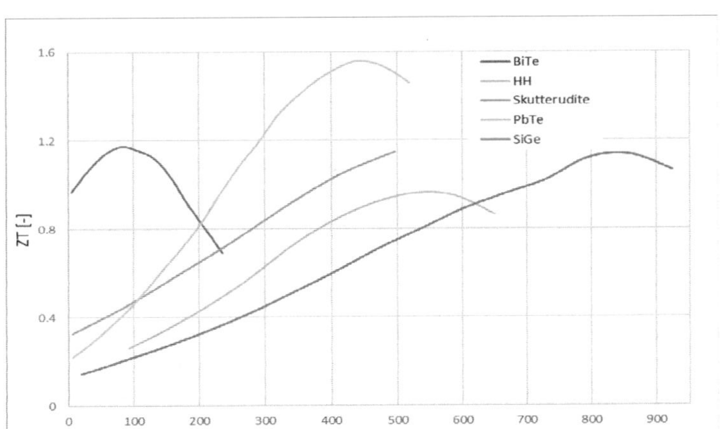

Figure 3: Dimensionless device figure of merit, *ZT*, as a function of average temperature for TEGs based on p- and n-type materials derived from BiTe, Half-Heusler, Skutterudite, PbTe and SiGe.

3. TEG PEAK EFFICIENCY, REQUIRED HEAT TRANSFER AND DEVICE AREA FOR DIFFERENT MATERIALS

Equation (3) for the TEG efficiency is used to determine and compare device efficiencies, which are subsequently used to calculate thermal power requirements and required TEG sizing based on different materials. These results are presented individually for the set of materials shown in Figure 3 at different operating conditions/boundary conditions relevant to the hot water heater operation. Then, plots are presented for each material at individual operating conditions to allow direct comparisons of the performance expected based on material selection.

3.1 Bismuth Telluride, BiTe

The calculated efficiency as a function of T_h for a TEG composed of n- and p-type bismuth telluride (BiTe) is shown in Figure 4 for three levels of T_c (50, 100 and 200°C), which are used throughout the analysis.

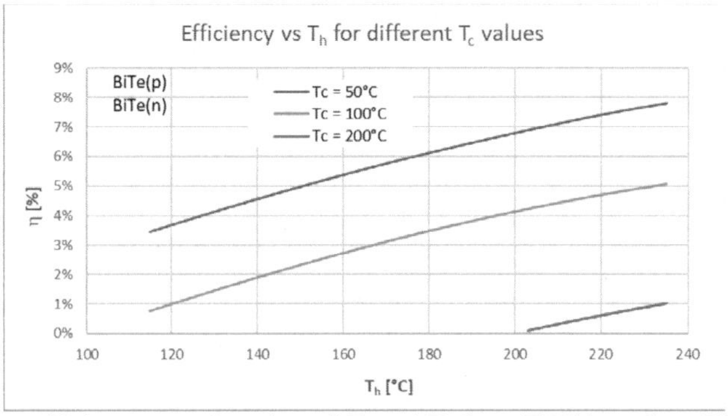

Figure 4: Maximum efficiency as a function of T_h and T_c for a TEG based on BiTe(p) and BiTe(n).

Since the TEG's conversion efficiency is the ratio of the electric power output to the thermal power input to the device, the rate of heat transfer, Q_h, that must be provided to the TEG from the heat source can be calculated from the efficiency and the required electrical power output, P_{elec}, as shown in Equation (4).

$$Q_h = \frac{P_{elec}}{\eta} \tag{4}$$

Figure 5 shows the thermal power required for the BiTe(p)-BiTe(n) TEG, per 100 W_{elec} of required electric power.

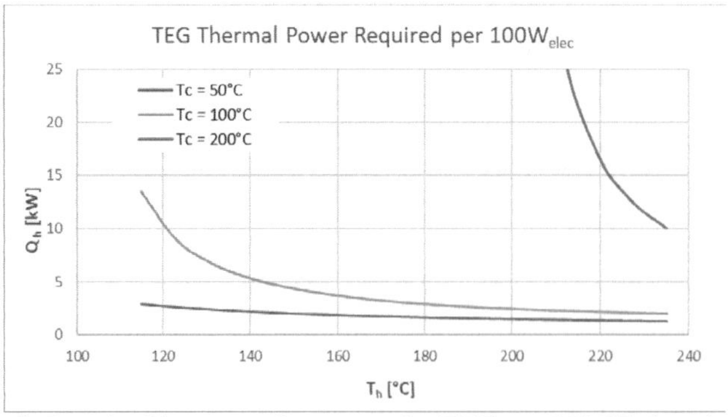

Figure 5: Required heat transfer rate to the TEG per 100 W electric power as a function of T_h and T_c for a TEG based on BiTe (p) and BiTe (n).

The heat transfer to the TEG hot side comes from an external heat source (e.g. a combustion gas stream), and heat must also be removed from the TEG cold side. For each side (hot and cold), there is an overall heat transfer coefficient that couples the external source and sink temperatures ($T_{h,\infty}$ and $T_{c,\infty}$) to the hot and cold substrate temperatures T_h and T_c, respectively. The heat transfer area, A, of each surface and the overall heat transfer coefficient, U, determine the heat transfer rate to/from the TEG. It should be noted that an array of TEG's may be required to achieve the desired power output for practical applications. Furthermore, the power output of the materials could be reduced due to parasitic electrical losses or Joule heating, and heat transfer losses should be considered when estimating the overall heat transfer coefficient to the TEG since they could potentially lower the electrical power output.

For the hot side, the heat transfer rate is given by

$$Q_h = U_h A_h \left(T_{h,\infty} - T_h\right) = U_h A_h \Delta T_h \,, \tag{5}$$

and a similar relation governs the heat removal from the cold side:

$$Q_c = U_c A_c \left(T_c - T_{c,\infty}\right) = U_c A_c \, \Delta T_c \,. \tag{6}$$

By inspection of Equation (6), it is clear that T_c will decrease with increasing $U_c A_c$ and will increase with decreasing $U_c A_c$. Therefore, without precisely calculating all parameters on the cold side, we can parameterize our analysis very simply by considering a few assumed T_c values that would be expected for $T_{c,\infty}$=20°C. We considered three cases for T_c: 50°C, 100°C and 200°C for high, medium and low $U_c A_c$ cases, respectively.

Returning to our analysis of the hot side heat transfer and sizing, Equation (4) provides Q_h as a function of T_h for the three considered levels of T_c (see Figure 5). Then $U_h A_h$ is uniquely determined when $T_{h,\infty}$ is specified, according to Equation (5). Depending on where the TEG is placed in the hot water heater, we can assume a value for $T_{h,\infty}$ ranging up to about 1400°C (if using the flame or combustion gases as the source). The overall heat transfer coefficient U_h will depend on how the TEG is thermally coupled to the burner/HX, but it can be reasonably approximated. For a typical forced convection heat transfer coefficient to air, we can expect a h value on the order of 100 W/m^2-K, but considering the additional resistances through a heat exchanger and to the TEG from the combustion gases will result in a lower overall heat transfer coefficient. We assume a value of U_h=50 W/m^2-K for the remainder of this analysis.

Our conclusion resulting from this reasoning is that the heat transfer area A_h is also determined as a function of T_h for each of the $U_c A_c$ levels considered (high, medium, low). The resulting A_h is the minimum surface area required at the hot side of the TEG to provide the heat transfer rate necessary for the TEG to sustain the specified electric power output P_{elec}. The proper equation is obtained by combining Equations (4) and (5). The final result is

$$A_h = \frac{P_{elec}}{U_h(T_{h,\infty} - T_h)\,\eta(T_h; T_c)}. \tag{7}$$

Considering the effect of the hot-side heat transfer coefficient on the required heat transfer area, from Equation (7) it can be noted that A_h is inversely proportional to U_h, independent of the other parameters used in the analysis. This indicates that the required heat transfer area will decrease by the same proportion as any increase in the overall heat transfer coefficient, emphasizing the importance of maximizing the coefficient between the heat source and the TEG.

Figure 6 shows the hot side area required per 100 W of electric power generation for the BiTe-based TEG with an assumed $T_{h,\infty}$ = 1400°C. Note that this source temperature significantly exceeds the temperature limit for the TEG, so it is assumed that the heat exchanger will maintain the hot substrate at a feasible temperature and that sufficient cooling at the cold side is always available to maintain the hot-side temperature of the TEG below its upper limit of operation.

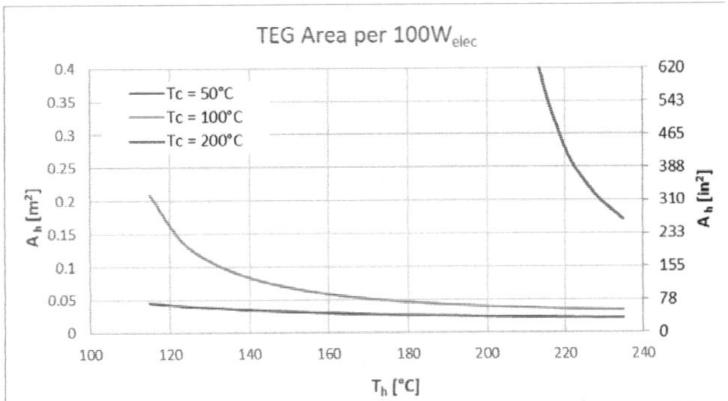

Figure 6: Required hot side heat transfer area, per 100 W electric power, as a function of T_h and T_c for a TEG based on BiTe (p) and BiTe (n).

3.2 Lead Telluride, PbTe

Using the equations for the efficiency, thermal power requirement and required heat transfer area as a function of the TEG hot- and cold-side temperatures, we compare results for the different TE materials. The same assumptions for U_h and $T_{h,\infty}$ are used in all cases to provide a "clean" comparison. Figure 7 shows the result of calculations for PbTe(p) and PbTe(n) materials. The ability to operate at higher temperatures and the higher zT values for this material result in considerably higher maximum efficiency than the BiTe material considered previously. Although for the same T_h operating range, the BiTe actually had higher efficiency, the PbTe can function at temperatures over 500°C (vs. only 235°C for BiTe), which enables a larger temperature difference and a greater maximum efficiency. For use in the hot water heater application, using the natural gas flame with temperatures of approximately 1400°C as the TEG's heat source, it makes sense to take advantage of the material enabling the higher temperature operation. We see at the highest hot side temperatures, the required thermal power and area are quite insensitive to the cold side temperature.

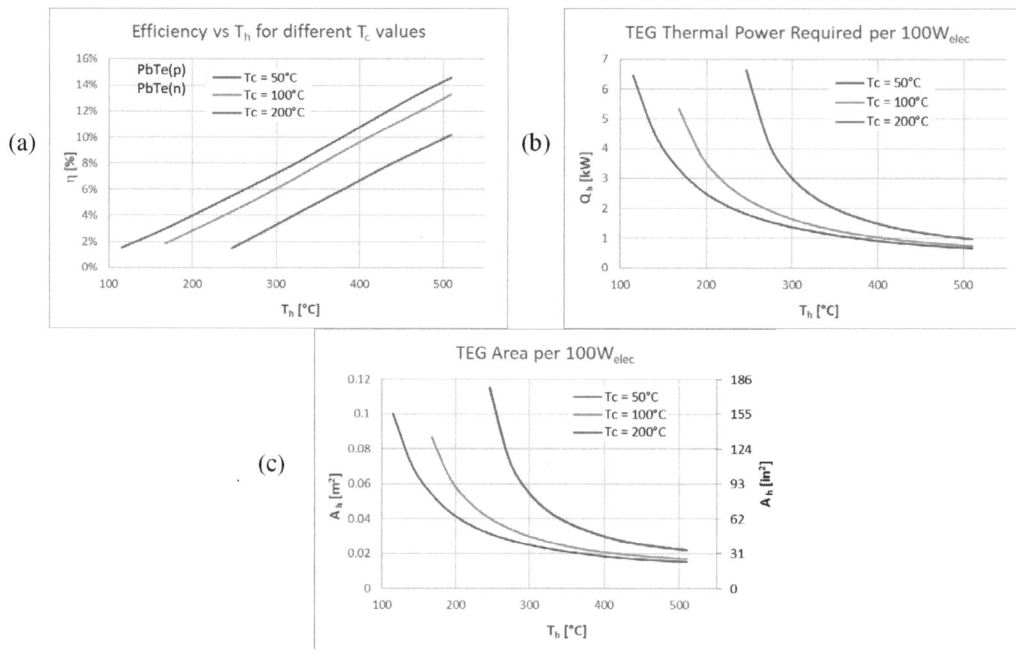

Figure 7: (a) TEG Efficiency, (b) Thermal power requirement per 100W electric power, and (c) required TEG heat transfer area for a TEG based on PbTe (p) and PbTe (n).

3.3 Half-Heusler, HH

Figure 8 shows results for a TEG based on Half-Heusler (HH) p- and n-type materials. The peak TEG efficiency is less for this material than for the lower temperature PbTe materials since the ZT value is considerably lower for the HH.

3.4 Skutterudite

Figure 9 shows the results for a TEG based on Skutterudite p- and n-type materials. The results appear fairly similar to those for the PbTe p- and n-type TEG, except that the maximum temperature is a little lower for the Skutterudite and its maximum efficiency at peak operating temperature is reduced considerably due to the lower ZT of Skutterudite material at temperatures approaching the limits of its operating range.

3.5 Silicon Germanium, SiGe

Figure 10 shows the results for the TEG based on SiGe(p) and SiGe(n) materials. The area requirement begins to increase very slightly for T_h>800°C. Although the device efficiency increases up to the maximum temperature, due to the decrease in the driving temperature difference ΔT_h as the hot side temperature is increased, the area requirement increases with rising T_h, so there is no benefit in terms of sizing obtained from the increased efficiency. For this material, if the source temperature could be increased to $T_{h,\infty}$ = 1800°C, the minimum heat transfer area required decreases to approximately 0.017 m^2 (26 in^2) per 100W electric power produced, but at the flame temperatures typical in a hot water heater, the performance does not reach the same level as for any of the previously considered materials.

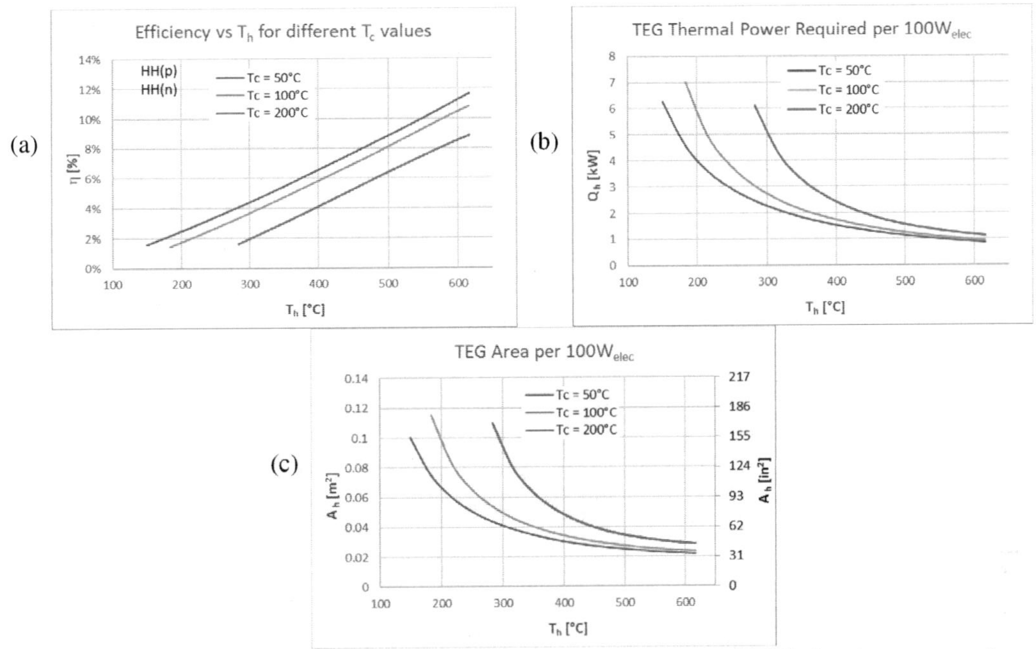

Figure 8: (a) TEG Efficiency, (b) Thermal power requirement per 100W electric power, and (c) required TEG heat transfer area for a TEG based on HH (p) and HH (n).

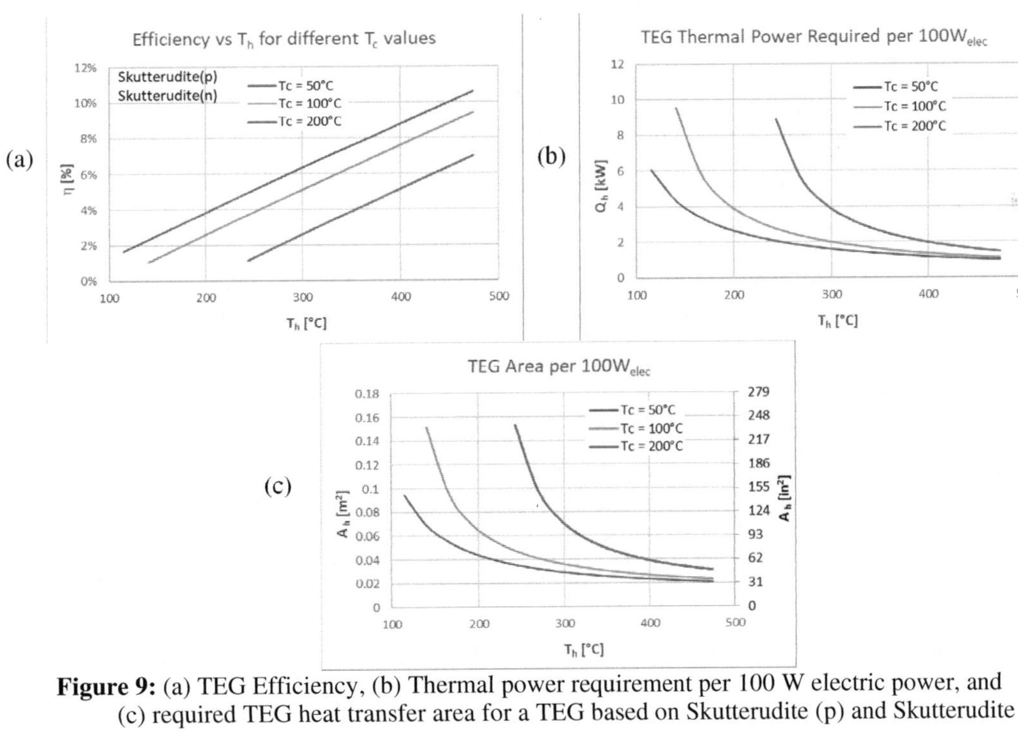

Figure 9: (a) TEG Efficiency, (b) Thermal power requirement per 100 W electric power, and (c) required TEG heat transfer area for a TEG based on Skutterudite (p) and Skutterudite (n).

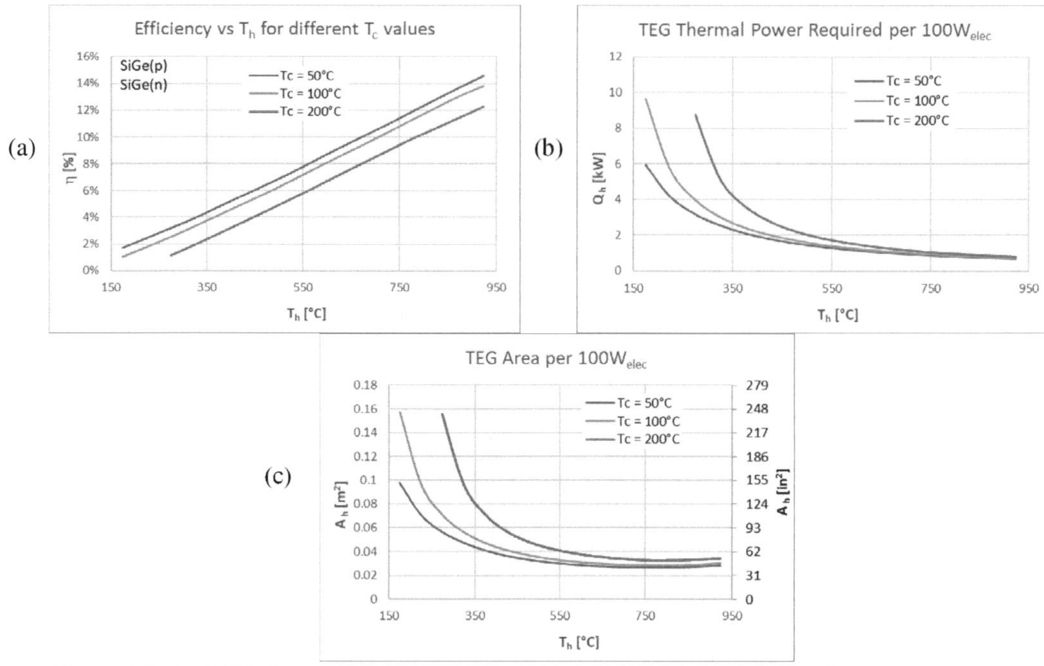

Figure 10: (a) TEG Efficiency, (b) Thermal power requirement per 100 W electric power, and (c) required TEG heat transfer area for a TEG based on SiGe(p) and SiGe(n).

4. PERFORMANCE COMPARISON PLOTS BETWEEN TE MATERIALS

To provide a direct comparison of results for different material selections, the Q_h and A_h curves are shown in Figure 11 for T_c=50°C and Figure 12 for T_c=200°C, for the five primary material combinations considered above (p- and n-type materials derived from the same undoped materials). The results in Figure 12 do not include BiTe, since for this material the upper operating temperature is only about 235°C, and the high T_c value would allow only a very narrow band of operation.

The results from this analysis show that TEG material with the lowest required heat transfer area depends on the cold side substrate temperature, T_c. For T_c=50°C (Figure 11), the lowest heat transfer area is achieved with PbTe, although BiTe can provide the same power with a very similar area while operating at a much lower hot-side temperature. SiGe can provide the highest power output for a given temperature difference and it can handle much higher temperatures, although the area requirement is higher due to a lower efficiency. For T_c=200°C (Figure 12), the smallest area can be achieved with a TEG utilizing PbTe. Further improvements in efficiency can be achieved and the required heat transfer area can be reduced if higher source temperature can be used, and also if the thermal coupling of the TEG to the heat source is increased. Improved coupling to the cold sink improves the performance of the TEG, but the performance is much more sensitive to $U_h A_h$ than to $U_c A_c$. In the results for PbTe presented in Figure 7, at T_h=500°C, increasing the value of U_h would result in a *proportional* increase in the heat transfer Q_h to the TEG and a corresponding decrease in required heat transfer area. On the other hand, doubling U_c (i.e. reducing T_c) produces only about a 5% reduction in area at the most effective operating condition.

5. CONCLUSIONS

Parametric analyses were conducted to evaluate the impacts of different operating conditions, including the hot- and cold-side temperatures of the TEG substrate, and the temperature of the heat source. These parameters are convenient for the TEG analysis and can be determined from the specific design configuration of the TEG in the self-powered hot water heater. Based on expected operating temperatures for our application, the model predicts that the p- and n-type PbTe material can provide the required power using the smallest size TEG device. Although SiGe allows higher operating temperatures and the greatest power output for a given temperature difference, its efficiency remains lower until much higher substrate temperatures are reached, and a larger heat transfer area is required. Bismuth telluride,

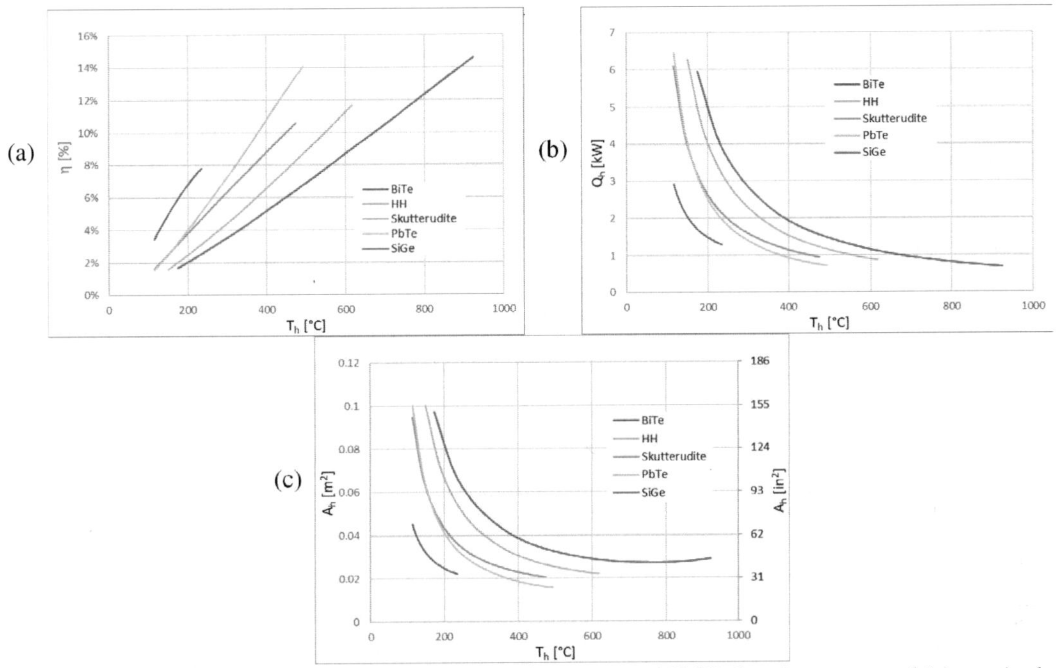

Figure 11: (a) Efficiency, (b) thermal power requirement per 100 W electric power, and (c) required TEG heat transfer area for the primary materials considered above. Results correspond to the lowest heat transfer performance on the cold side, i.e. T_c=50°C.

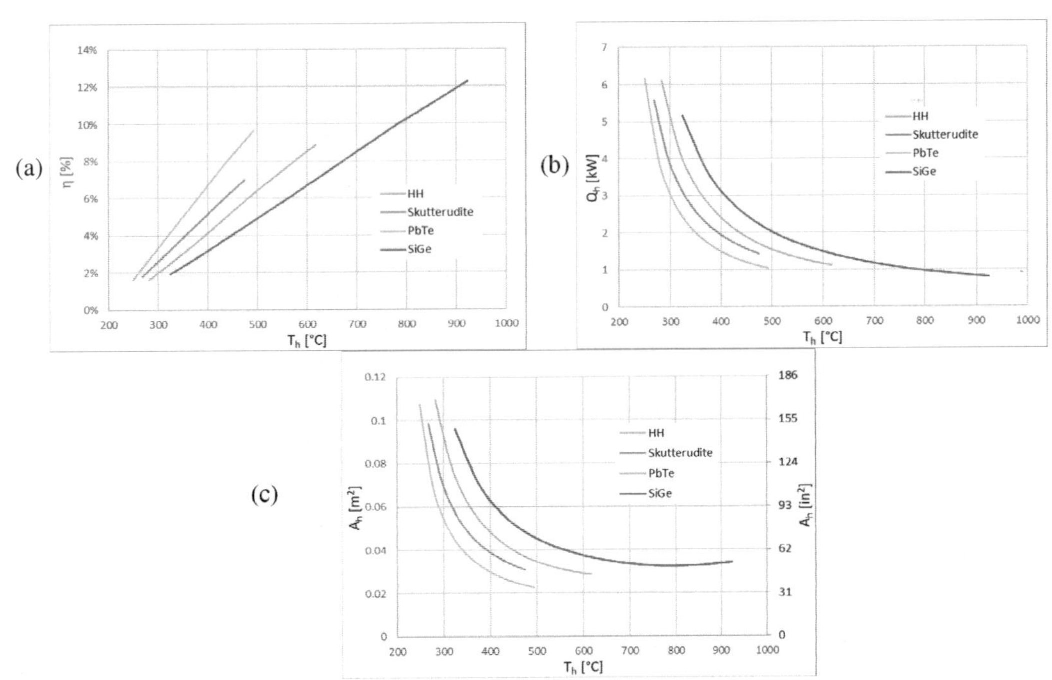

Figure 12: (a) Efficiency, (b) thermal power requirement per 100 W electric, and (c) required TEG heat transfer area for the primary materials. Results are for the lowest heat transfer performance on the cold side, i.e. T_c=200°C.

which is readily available on the commercial market, is limited to temperatures below 250°C, which limits the efficiency that can be achieved in the hot water heater application, unless the cold substrate temperature is kept quite cool (for example, 50°C). Skutterudite and Half-Heusler materials operate in a similar temperature range as the lead telluride, but the efficiency is somewhat reduced and a larger TEG would be required to generate the required power. We note that, other than bismuth telluride, there are few commercially available high temperature modules on the market, and development of customized TEGs would likely be required for other materials, although there have been recent product additions. Furthermore, in practical systems the efficiency will be lower than the ideal conditions due to thermal and electrical interface losses and long-term degradation.

NOMENCLATURE

A	area
k	thermal conductivity
P	power
Q	heat transfer rate
S	Seebeck coefficient
T	absolute temperature
ΔT	temperature difference across a TEG
U	overall heat transfer coefficient
z	material figure of merit for a TEG
Z	device figure of merit for a TEG
η	efficiency
ρ	electrical resistivity

Subscript

avg	average
c	cold side
elec	electrical
h	hot side
n	n-type material used for a couple in a TEG
p	p-type material used for a couple in a TEG
∞	source or sink

REFERENCES

Abu-Heiba, A., Gluesenkamp, K. R., LaClair, T. J., Cheekatamarla, P., Munk, J., Thomas, J., Boudreaux, P. (2021). Analysis of power conversion technology options for a self-powered furnace. *Appl. Therm. Eng.* (Accepted for publication).

Alptekin, M., Calisir, T., & Baskaya, S. (2017). Design and experimental investigation of a thermoelectric self-powered heating system. *Energy Conversion and Management, 146*, 244-252.

Bjork, R. (2015). Universal influence of contact resistance on the efficiency of a thermoelectric generator. *J. Electron. Mater., 44*(8), 2869-2876.

Butcher, T. A., Hammonds, J. S., Horne, E., Kamath, B., Carpenter, J., & Woods, D. R. (2011). Heat transfer and thermophotovoltaic power generation in oil-fired heating systems. *Applied Energy, 88*(5), 1543-1548.

Gluesenkamp, Kyle R. (2019). Drop-in, retrofit furnace with maximum efficiency – self powered system." DOE Building Technologies Office 7th Annual Peer Review, April 15–10, 2019, Crystal City, Virginia. Available at https://www.energy.gov/sites/prod/files/2019/05/f62/bto-peer%E2%80%932019-ornl-drop-in-retrofit-furnace-max-eff.pdf

Qiu, K. & Hayden, A. C. S. (2008). Development of a thermoelectric self-powered residential heating system. *Journal of Power Sources*, 180(2), 884-889.

Qiu, K. & Hayden, A. C. S. (2014). Implementation of a TPV integrated boiler for micro-CHP in residential buildings. *Applied Energy*, *134*, 143-149.

Rowe, D. M. (2006). *Thermoelectrics Handbook—Macro to Nano*. Boca Raton, FL: Taylor and Francis Group.

The Case for DC: Motivation of Modern Topologies, DC-Powered Solutions, and Applications within Residential Environments

Jonathan ORE[1]*, Fatih MERAL[1], Davide ZIVIANI[1], Eckhard GROLL[1]

[1]Purdue University, Mechanical Engineering,
West Lafayette, IN, USA
jore@purdue.edu

* Corresponding Author

ABSTRACT

The more than a century-old debate between AC and DC has its roots in an electrical distribution challenge, tackled by rival inventors Nikola Tesla and Thomas Edison during the late 1800s. Although originally collaborating on the improvements of Edison's work, the pair eventually parted ways due to conflicts in their personalities and business pursuits. Edison's direct current (DC) system leveraged a constant voltage and current to supply electricity, which was initially sufficient for small locales and geographical regions. However, DC encountered a major obstacle when longer-range transmission was required; there was simply no way to easily convert it between higher and lower voltages. These step-up and step-down conversions were critical for transmission, as power line losses are reduced significantly when proportionally increasing voltage levels. Tesla's alternating current (AC), on the other hand, was readily compatible with the newly christened *transformer*, a device which possessed the ability to effectively adjust AC voltages on demand. With more and more entities investing into the AC-based distribution scheme, it seemed that the war of currents had been firmly decided in favor of Tesla's solution. However, a century later has revealed an outmoded and fragile electrical ecosystem, with new energy sources and infrastructures that reposition DC as a primary contender for distribution and consumption. This paper will outline these current challenges, and explore the implementation of practical DC solutions across both the larger power grid and within residential applications.

1. INTRODUCTION

The U.S. power grid is one of the most complex engineering systems in the world, involving highly convoluted nonlinear behaviors, tens of thousands of individual nodes, and scores of intricate power electronics throughout. Its construction and historical progression hold far-reaching implications on the advancement of modern society, and ultimately the sustainability of the global environment. Insights into transportation and distribution mechanisms, the electrical needs of contemporary appliances and devices, and the revolution of renewable energy resources are paramount for discerning the path forward of electricity and the next generation of its development. More than a century ago, engineering heavyweights Nikola Tesla, George Westinghouse, and Thomas Edison battled over what would become the foundation for present-day electricity distribution. A major asset of Tesla's Alternating Current (AC) at the time was its ability to be converted to higher and lower voltages, giving it a critical advantage over Edison's Direct Current (DC). The battle was clinched during the Chicago World's Fair in 1893, during which Westinghouse successfully underbid Edison to supply power to the Fair. This demonstration of AC power consequently resulted in the Niagara Falls Power Company awarding Westinghouse a major contract to construct a power generator for the falls. In 1896, the Niagara Fall's hydroelectric power plant was successfully launched, providing power to the Buffalo, NY area. Following this success, General Electric and a multitude of other firms shifted their investment into AC power, securing AC as the predominant means of power distribution in the U.S (U.S. Department of Energy, 2014).

The war between AC and DC ultimately concluded in favor of AC, owing to its capability to be transformed to higher voltages and thus minimize power losses over longer transmission distances. This feature is a direct corollary of Ohm's law and the Joule-Lenz law, in which electrical power lost decreases quadratically with respect to current. Since by

Ohm's law, current can be reduced by a corresponding increase in voltage over a constant resistance, the flexibility of AC voltage to be modulated by the newly christened transformer of the late 1800s gave it a distinct advantage over the competing DC option (Gómez-Expósito *et al.*, 2018).

In parallel fashion to the U.S.'s geographical development, the overall power grid, or *macrogrid*, has evolved in a bottom-up regional fashion across the country, resulting in three primary interconnections defined by the North American Electric Reliability Corporation (NERC): The Eastern Interconnection, Western Interconnection, and Texas Interconnection (NERC, 2019). Each system maintains its own AC frequency and experiences unique load shapes, power flows, and stability challenges. The macrogrid is a composite of transmission and distribution networks, which directly contribute to the control and stability challenges surrounding it. Increasing energy demands and greater dependence on the macrogrid have fostered a delicate balance between satisfaction of consumer needs and reliability of operations. The spatial layout of the power grid in North America makes it especially vulnerable to cascading failures, with the Federal Energy Regulatory Commission (FERC) explaining that a loss of merely nine key substations out of 55,000 could result in a country-wide blackout (Smith, 2014). Compounding this, in 2017 the American Society of Civil Engineers (ASCE) scored the U.S. energy system with a grade of a *D+* for overall reliability, citing aging distribution lines, capacity bottlenecks, and climate impacts as leading factors in their evaluation (ASCE, 2017). In short, the urgency for a plausible strategy toward improvement and upgrade is clear, but to better appreciate the direction it should take, an understanding of the requirements at the end point of use must be considered.

2. REEVALUATING THE GRID

2.1 Power Transmission

According to the U.S. Energy Information Administration (EIA), approximately 5% of electricity transmitted is lost due to grid inefficiencies. Between 2000 and 2015, more than 172 quads of electricity were transferred through the U.S. electrical grid, equivalent to approximately 50 trillion kWh. From distribution losses alone, the amount of energy dissipated would be sufficient to power 306,000 houses over the same time period, assuming an average consumption of 914 kWh per house per month (EIA, 2019). An individual study in New York identified utility transmission losses of up to 5.8% and distribution losses of up to 4.6%; however, these values were obtained *after* utilities had enacted a variety of improvements to reduce losses in the distribution system. Ultimately, AC transmission and distribution schemes must contend with an array of loss-mechanisms, including but not limited to (a) ohmic losses, (b) corona losses, and (c) other distribution losses (Jackson *et al.*, 2015).

A little over a century from its induction, DC no longer faces the same challenges with respect to voltage transformation. In fact, similar to high voltage AC, high voltage DC (HVDC) can also be employed to transfer power over long distances while minimizing electrical losses. The first HVDC transmission lines were enacted in the 1950's in both Sweden and Italy, with dozens of new projects presently under construction or completed (Arrillaga, 1998). In 2019, China demonstrated an HVDC link using a 1,100 kV transmission line over a span of 3,300 km, supporting a maximum bulk power transfer of 12 GW (Ying *et al.*, 2019). With installations such as these, both intra- and intercontinental networks are feasible, with a host of benefits in contrast to AC equivalent systems.

Costs for transmission lines are associated with a variety of parameters, including occupied space for transmission line towers (referred to as right-of-way (ROW)), physical cost of towers, conduction line costs, electrical equipment and terminators, and other necessary power electronics. DC transmission has the immediate benefit over AC via its conductor real-estate needs; while AC requires three conductors to carry power (hot, line, and neutral), DC only requires two (positive and negative). Since high voltage transmission lines require a minimum amount of spacing between conductors to avoid ground-faults and arcing, HVDC transmission also benefits from requiring a smaller ROW through spacing management between its two conductors as opposed to three, as illustrated in Figure 1. The reduced amount of conduction material also affords smaller conductor losses, such as ohmic losses discussed in the previous section, and the avoidance of other AC-specific losses, such as the skin effect (Rashid, 2011).

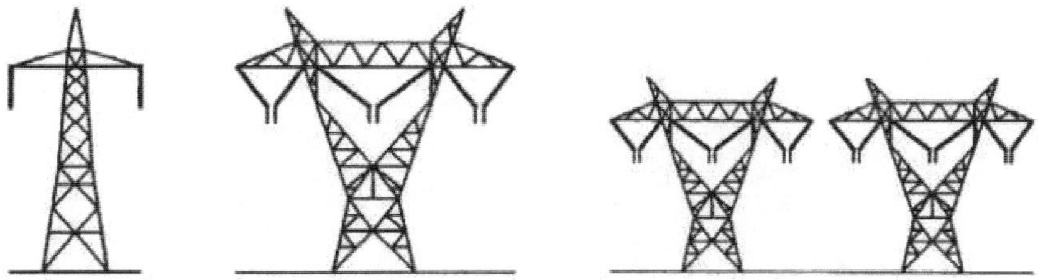

Figure 1: Equivalent DC vs AC ROW for long distance power transmission (Rashid, 2011)

An economic analysis of high voltage AC and DC transmission can be performed in consideration of the specific costs associated with each system configuration. A break-even distance occurs where the capital investment for an HVDC system is more cost-effective than the corresponding AC system. Several studies have determined this distance to be approximately 500 km, but there are some caveats with this calculation (Rashid, 2011). When directly compared to HVDC, more AC transmission lines are often required to provide sufficient stability between endpoints, thereby increasing overall cost. In addition, AC transmission lines must contend with complex power consumption and generation, and therefore require switching stations along a transmission path to appropriately manage power distribution (Bahrman, 2008). As a result, the true benefit of utilizing HVDC is underappreciated when performing a pure comparison in transmission and equipment costs. From a top-down perspective of the macrogrid, HVDC affords more opportunities for efficient energy transmission, practical long-distance transportation, and distinct economic and structural benefits as compared to conventional AC transmission.

2.2 Characterization of Energy Consumption

In a survey of 24 countries representing 92% of energy consumed worldwide during 2018, the International Energy Agency (IEA) determined that residential consumers represented up to 20% of the end-use of energy, as illustrated in Figure 2. Of this sector, space heating and cooling accounted for more than half of energy consumed as shown in Figure 3, positioning HVAC systems as key points of interest for analysis. Appliance energy consumption followed in second, accounting for another one-fifth of residential net energy consumed.

Owing in part to their significant partition of total energy consumption and homogeneity of specific load types, residential spaces occupy a pivotal juncture in the path toward the future architecture of the electrical grid. Many studies have evaluated the individual characteristics of residential loads, such as Anzar *et al.* (2017), Chauhan *et al.* (2017), Gerber *et al.* (2019), and Luo *et al.* (2019), for the purposes of assessing both demand side management and the opportunity for retrofit suitable with DC power. In particular, Gerber *et al.* (2019), classified common residential electrical loads as either (a) lighting, (b) electronics, (c) heating elements, or (d) motor loads. Of these, all but the heating elements were surmised to benefit from a direct-DC supply or suitable DC-retrofit compared to the baseline AC versions. In order to distribute DC power to these endpoints in such a setting, however, innovative and modernized electrical architectures must first be formulated, prefixed with the SI names corresponding to their relative size. These are known as the *microgrid*, *nanogrid*, *picogrid*, and so on.

2.3 State of the Art Topologies

As defined by the Lawrence Berkeley National Laboratory (LBNL), a microgrid is represented by a group of loads and distributed energy resources (DERs), which have a defined electrical boundary from other entities, can be controlled as a single body, and maintain the ability to either interact with the grid, or disconnect and operate independently in an islanding mode (Black, 2021). The International Council on Large Electric Systems (CIGRÉ, 2010) provides a similar interpretation, further specifying that DERs include all manner of energy generation in the microgrid (e.g., fossil fuels, combined heat and power (CHP), photovoltaics, wind, etc.), and that storage devices can possess a diverse collection of implementations (e.g., electrical, mechanical, gravitational, thermal, chemical, etc.). To this end, a microgrid is a unit which can coexist alongside conventional electrical distribution mechanisms, but affords the capability to operate on its own if necessary. Furthermore, the microgrid can distribute energy back to the macrogrid if required, offering an additional advantage to stability.

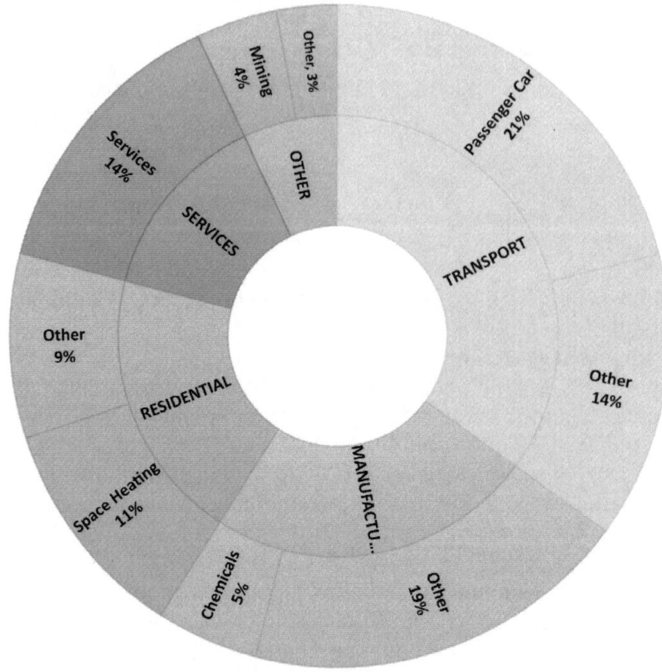

Figure 2: End-use energy consumption by sector during the 2018 year (IEA, 2018).

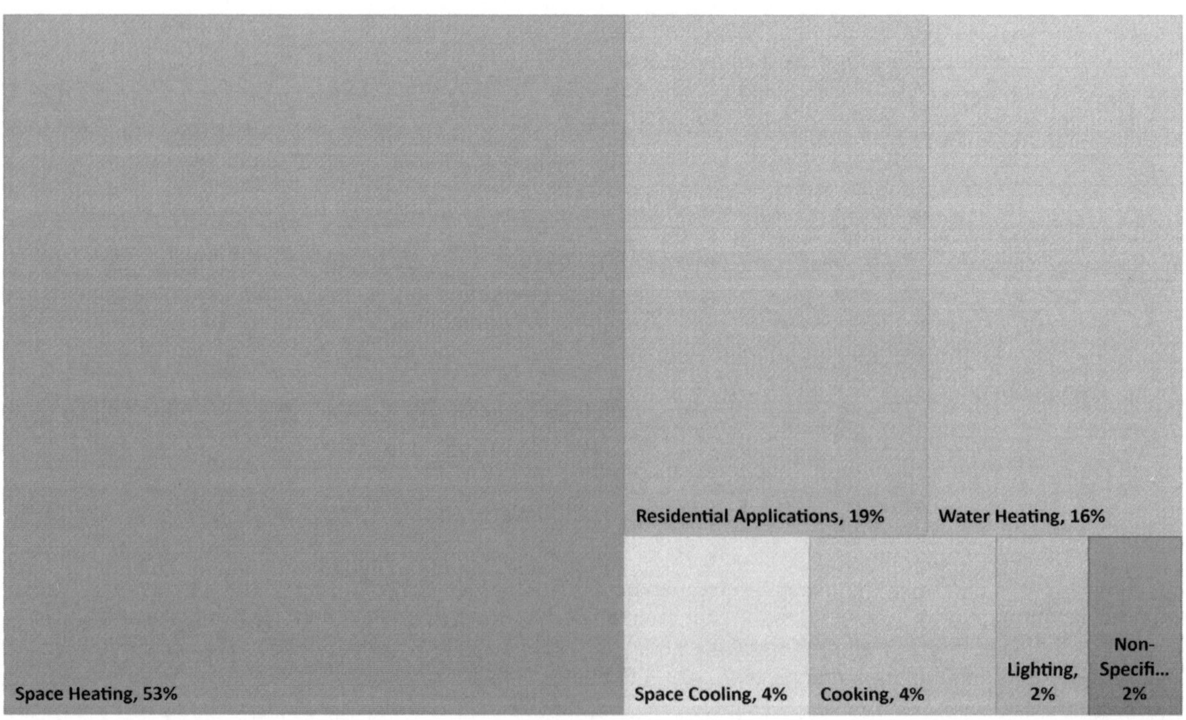

Figure 3: End-use energy consumption breakdown in residential environments during the 2018 year (IEA, 2018)

Similar to microgrids, nanogrids offer much of the same benefits on a smaller scale. While a microgrid might encompass an entire neighborhood of interconnected houses, a power plant, and local energy storage, a nanogrid could be comprised of a single home with a solar installation. Nordman (2010) defines a nanogrid to represent a single controllable entity with at least one load, and at least one connection to external grids (e.g., a larger microgrid, the overall macrogrid, etc.). A critical difference from the microgrid, however, is the requirement for storage; a nanogrid may or may not have energy storage integrated into its design. As a result, by its formal definition a nanogrid is not required to support islanding operation.

Microgrids and nanogrids are not required to stipulate a specific voltage type for distribution, although DC and other hybrid combinations are common. These systems benefit from their flexibility to integrate into a panoply of applications, ranging from industrial facilities, commercial buildings, cul-de-sacs, individual homes, and many other structures. Microgrids and nanogrids offer a bottom-up solution to transmission and distribution challenges, requiring minimal coordination and the flexibility to function alongside the existing AC infrastructure. With integrated battery storage, these topologies offer resiliency to grid disruptions, load-balancing opportunities, and ready-compatibility with DC-power producing renewable energy sources. Nanogrids advocate these benefits another step further, yielding configurations with reduced conversions between distribution and devices, and increasing the potential for energy savings. Combining these systems with DC-compatible devices and furthering retrofit-research into additional appliances capable of supporting DC could render AC the minority in power distribution, rather than the predominant entity.

Two possible microgrid topologies are demonstrated in Figure 4, with the left side indicating an AC/DC hybrid structure, and the right side presenting a pure DC structure. In these diagrams, darkened circles represent a switch, open circles represent power electronics, and arrows represent the direction of power flow. Under both schemes, the electrical storage could be removed without affecting the overarching integrity of the design. The hybrid structure affords some of the benefits of the microgrid architecture without fundamentally altering the common distribution mechanism. The right style, however, yields an ideal configuration for a DC-based solution. In this case, all conversions from AC to DC have been eliminated (with the exception of the primary grid-tie), and renewable energy generation, energy storage, and loads can benefit from direct-DC supplies with minimal and highly efficient DC-DC conversions. These topologies illuminate one of the most critical elements for the motivation of innovative grid layouts: the integration of DERs.

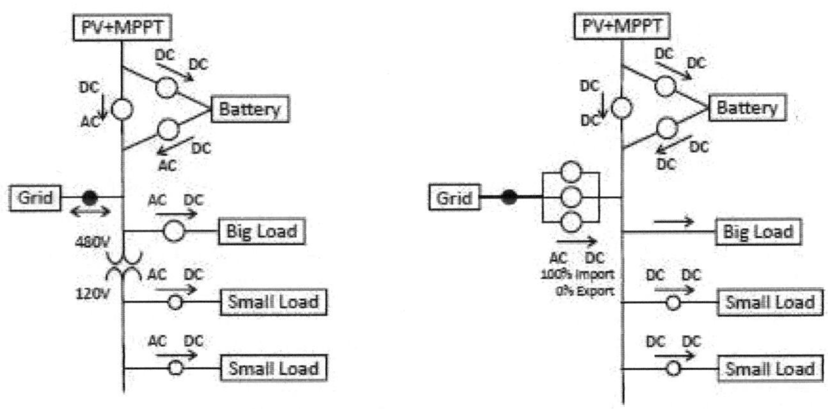

Figure 4: Sample AC and DC microgrid/nanogrid topology reference layouts (Backhaus *et al.*, 2015)

2.4 Distributed Energy Resources

As the macrogrid increases in size and throughput, so do its stability challenges and supply requirements. In addition, higher instantaneous power demands can lead to cascading failures during disruptions, as evidenced by the 2003 blackout resulting in over 50,000,000 impacted customers, and many other similar backout events during the past decade (Minkel, 2008). These obstacles can be overcome with the help of a newly emerging asset – DERs.

According to NERC, a DER is, "any resource on the distribution system that produces electricity and is not otherwise included in the formal NERC definition of the Bulk Electric System (BES)" (NERC, 2017, p. 1). As a result, DERs occupy a broad range of resources, including energy storage, renewable energy generation, electric vehicle (EV) charging stations, back-up generators, and even microgrids themselves. Fundamentally, to be classified formally as a DER an entity must be capable of producing electricity, thus supporting the inclusion of energy storage systems.

Interestingly, individual equipment and devices which maintain their own storage may also function as a DER (such as an EV) if the electrical connection is bidirectional (NERC, 2017).
With the acceleration in deployment of renewable energy generation, especially that of solar photovoltaics (PV) and wind power, the macrogrid is presented with an increasingly diverse supply of energy resources. Unlike traditional base load power plants, such as coal and nuclear, renewable resources are frequently volatile and therefore unsuitable for satisfaction of base load power demands. Instead, these systems are far more effective for responding to transient shifts in demand, such as intermediate loads arising during the daytime in winter and summer seasons. If these sources are coupled with energy storage systems, such as compressed air energy storage (CAES), pumped heat energy storage, or other conventional battery storage systems, the variability of energy generation can be significantly mitigated. In addition, storage systems within the macrogrid provide a relaxation to the demands on other generation processes, and bolster the stability of the neighboring grid elements.

Opponents to the integration of renewable energy sources and the larger composite of DERs have argued that due to their inherently unstable nature, these resources are unreliable and thus incapable of replacing established fossil-fuel based power generation sources. Contrary to this suggestion, multiple studies have recognized the considerable benefit of incorporating sources such as wind and solar PV, as Miller *et al.* (2014) confirmed in their multiphase Western Wind and Solar Integration Study (WWSIS). The WWSIS report sought to understand whether the macrogrid (specifically the Western Interconnection) could withstand the inclusion of extensive amounts of wind and solar energy generation without inducing instabilities and resulting in undesirable strain to the overall system. Far from any unfavorable impact, the WWSIS's first phase determined that up to 35% of the region's power production could be substituted by wind and power generation without requiring significant restructuring of the grid. Furthermore, the same modification would also provide an equivalent benefit to the environment roughly similar to removing up to 36 million cars off the road (compared to the existing system operation). Utility and operation costs were also discovered to decrease under the addition of these renewable resources, and transmission segments could provide better utilization due to the locality of energy generation. Finally, distributing the points of energy generation geographically was also recognized to reduce variability in production, as the prevalence of wind and solar conditions becomes proportionally more consistent as the spatial size of the region considered increases (Miller *et al.*, 2014).

3. RESIDENTIAL APPLICATIONS

3.1 The Residential Nanogrid

Owing to the explosion of growth in renewable energy generation in residential locations, and the aforementioned benefits from the inclusion of DERs within the existing macrogrid infrastructure, the application of DC-based architectures within residential spaces has become a compelling topic. In its *2020 World Energy Outlook* publication, the IEA (2020a) states that, "for projects with low-cost financing that tap high-quality resources, solar PV is now the cheapest source of electricity in history" (para. 5). Even while global electricity demand curtailed due to the COVID-19 pandemic, renewables maintained a year-over-year growth of nearly 7% during the 2020 year (IEA, 2020b). According to Solar Energy Industry Associates (SEIA), this international advancement was fueled in-part by a 43% increase in new electricity production of solar PV within the U.S. in the same year, and incidentally was also the single largest increase of that category during the previous decade. Contributions to new production included a 14% increase in residential solar installations between the second and third quarters of 2020 within the U.S., and new residential solar capacity additions of approximately 3 GWdc during the same year (SEIA, 2020).

From the perspective of consumption, the majority of modern devices and appliances consume DC power either directly or indirectly, yielding further credence to DC-based topologies. In a residential setting where DERs are present, maintaining conventional AC distribution results in a multitude of potentially dissipative AC to DC and DC to AC conversions, which diminish the capacity for micro/nanogrid benefits. As a result, a centralized DC-distribution architecture affords the capability to mitigate these impacts, reducing the need for complex and expensive power electronics, and offering viable high efficiency DC-DC conversions. Hybrid structures of the layouts previously described in Figure 4 can also be attainable, where DERs and DC-compatible loads can share a common DC bus, while conventional AC loads and DC-indifferent loads can remain on a traditional AC infrastructure. As Backhaus *et al.* (2015) enumerates, there are nine key performance indicators (KPIs) that can be applied to analyze an electrical architecture's benefit:

1. Safety and Protection
2. Reliability
3. Capital Costs
4. Energy Efficiency
5. Operating Costs
6. Engineering Costs
7. Environmental Impact
8. Power Quality
9. Resilience (p. 2).

After reviewing numerous studies for evidence and implementation details, including Vossos *et al.* (2014), Wang and Jain (2010), Burmester *et al.* (2017), Garbesi *et al.* (2011), and many others, these nine KPIs were summarized over various AC and DC topologies to reveal distinct advantages of DC designs for categories (3), (4), (6), and (8), and neutral or marginal benefits for the remaining categories (Backhaus *et al.*, 2015). As a result, architectures including significant DC composition certainly merit additional investigation.

3.1.1 Conversion Efficiencies

The losses associated with various AC and DC conversion and inversion processes is a prime mover for the topic of modern electrical topologies, but continues to persist as a significant topic of contention surrounding the explicit degree of benefit. Challenges to clarifying and simplifying these issues include the vast array of converter technologies and types, continuing developments from novel research efforts, manufacturer specifications versus actual performance, and the use-cases and scenarios under which findings are reported. As an example, Backhaus *et al.* (2015) reported peak load conversion efficiencies from high voltage AC to low voltage DC of approximately 95%. However, several common household devices studied by Santos *et al.* (2018) described conversion efficiencies for devices such as a laptop, LED, phone charger, fan, and 1 kW inverter as low as 87%, 49%, 71%, 60%, and 65%, respectively. Similarly, Burmester *et al.* (2017) indicated typical AC to DC conversion efficiencies of common appliances, including refrigerators, computers, televisions, lighting, and water heaters of approximately 87%, 80%, 85%, 82%, and 88%, respectively. As a result, although peak conversion efficiency studies frequently record values in the high 90 percentile range for all manner of conversion combinations (i.e., AC-AC, AC-DC, DC-AC, DC-DC), ordinary devices usually fall significantly short of this range.

An intrinsic aspect of the various conversion mechanisms and their impact on the overall electrical architecture's benefit encompasses the specific location where the conversion needs to occur. Figure 5 provides an illustration of common conversion placements and values in both an AC and DC-based micro/nanogrid layout employing local energy generation and storage elements. Under an AC scheme, energy generated from renewable sources is fundamentally DC in nature (even from wind turbines) but must be converted to AC to be distributed within the home. Multiple studies have indicated potential improvements through the use of a DC-based architecture ranging from 14% (Vossos *et al.*, 2014), 18% (Gerber *et al.*, 2018), and even as high as 30% (Hofer *et al.*, 2017) over AC-equivalent counterparts. These achievements are made feasible in part by (a) isolating primary AC-DC inversion processes to one central, highly specialized and efficient module, (b) instituting multiple high-efficiency DC-DC conversion devices driven by a principal DC bus, and (c) eliminating multiple conversion losses from energy generation and storage elements. In the case of an AC-based micro/nanogrid, the potential utilization path for wind power generation could include as many as six or more conversions (e.g., AC-DC (from generation), DC-AC (for distribution), AC-DC (for initial storage), DC-AC (for subsequent distribution), AC-DC (for appliance rectification stages), and DC-AC (for motor consumption)).

3.1.2 The DC House Project

The Purdue DC Nanogrid House project is a research effort located within a residential home in West Lafayette, IN, with the ultimate objective to retrofit the entire building and all appliances and devices from AC to DC power. The home is a two-story 1920's era structure, which functions as a legitimate *living-laboratory*, and houses three graduate students who live and work within the dwelling. Total floor space of the building is approximately 208 m^2, the exterior of which includes a detached garage and a property lot occupying 595 m^2. Each of the house's 32 individual circuits have been instrumented with current transformers (CTs), voltage clamps, and Wi-Fi circuit breakers to monitor historical and real-time energy and power consumption, and to serve as a baseline of comparison against the planned

DC retrofits. Currently, the nanogrid architecture outlined by Ore and Groll (2020) is under construction, with an intent to commission early in the 2021 year. Included in this design is a 14.3 kW rooftop solar installation, 20 kWh LiFePO4 battery system, and novel Energy Management System (EMS) and Building Management System (BMS) for optimal control of the nanogrid and individual loads. Following successful energization of the completed nanogrid, the first DC loads to be investigated include a retrofitted-hybrid heat pump system, capable of operating on either 380 VDC or 230 VAC, and a PoE-based LED lighting system. During the life of the system, additional loads will be added and compared against the previous years of AC baseline data to thoroughly report on observed energy efficiency improvement, system benefits and challenges, and potential for reproducibility in other locations.

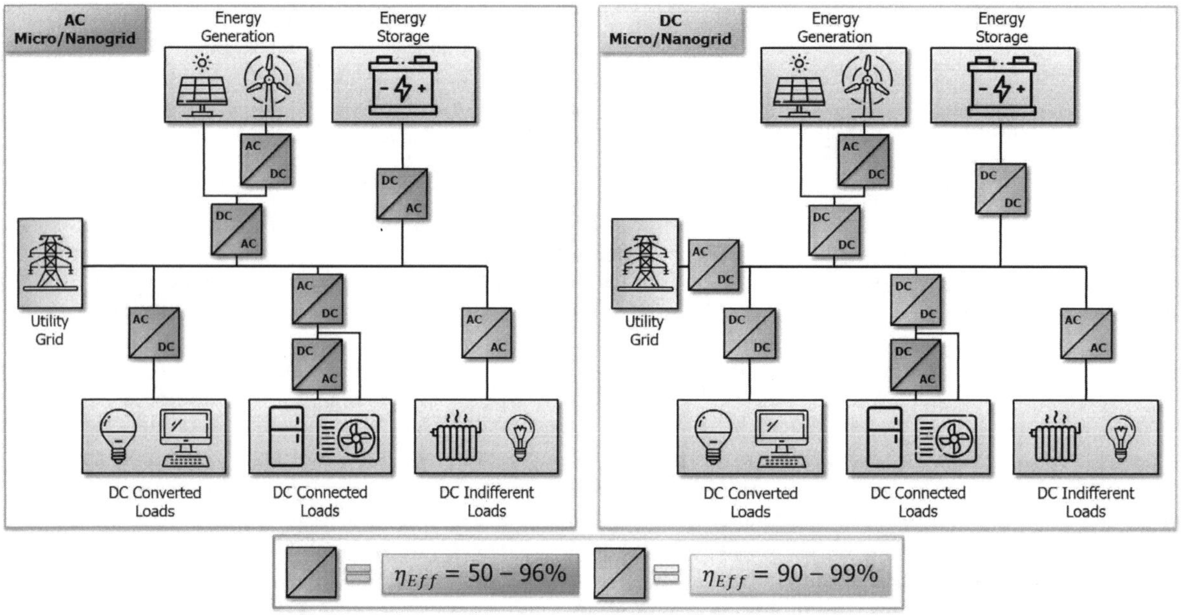

Figure 5: DC vs AC micro/nanogrid architectural depiction highlighting relative conversion efficiencies

4. CONCLUSIONS

A wealth of previous studies and research endeavors have reiterated the pressing necessity to evaluate DC and hybrid AC/DC based electrical topologies to satisfy the ever-increasing demands of the modern power grid. Though many articles frame this issue as a dilemma of proper choice between AC or DC, or even Tesla versus Edison, the reality is more fundamental to simply *using the right tool for the job*. In the same way obsolete wooden waterwheels are viewed today, doubtlessly too will fossil fuels and other similar power sources seem primitive to those of the coming centuries. However, the present time must commit to continual examination of constructive and propitious distribution strategies, independent from political or commercial inclinations, and remain focused on the ultimate goal of preserving the environment and promoting sustainability. As the well-known anthropologist Joseph Tainter (1988) wrote in his book, *The Collapse of Complex Societies*, "sociopolitical systems require energy for their maintenance, [and] increased complexity carries with it increased costs per capita…a new energy subsidy is necessary if a declining standard of living and a future global collapse are to be averted." (pp. 194–215). This admonition has held true through the ages, and continues through today; for the survival of both our individual societies and world as a whole, we must ceaselessly seek practical and effective means of leveraging our available resources to supply our civilization.

REFERENCES

Anzar, M., Ejaz, S., Iqra, R., Alvarez-Alvarado, M. S., Kousar, A., & Zafar, A. K. (2017). Optimization of Home Energy Management System in Smart Grid for Effective Demand Side Management. *International Renewable and Sustainable Energy Conference*, 1–6.

Arrillaga, J. (1998). *High Voltage Direct Current Transmission*. (2nd ed.). Stevenage, UK: The Institution of Engineering and Technology.

ASCE. (2017). *2017 Infrastructure Report Card*. https://www.infrastructurereportcard.org/making-the-grade/

Backhaus, S. N., Swift, G. W., Chatzivasileiadis, S., Tschudi, W., Glover, S., Starke, M., & Hammerstrom, D. (2015). *DC microgrids scoping study. Estimate of technical and economic benefits* (No. LA-UR-15-22097). Los Alamos National Lab. (LANL), Los Alamos, NM (United States).

Bahrman, M. P. (2008). HVDC transmission overview. *IEEE PES Transmission and Distribution Conference and Exposition*. 1–7.

Black, D. (2021). Microgrids and Vehicle-Grid Integration. Grid Integration Group. https://gridintegration.lbl.gov/microgrids-vehicle-grid-integration

Burmester, D., Rayudu, R., Seah, W., & Akinyele, D. (2017). A review of nanogrid topologies and technologies. *Renewable and Sustainable Energy Reviews*, *67*, 760–775.

Chauhan, R. K., Phurailatpam, C., Rajpurohit, B. S., Gonzalez-Longatt, F. M., & Singh, S. N. (2017). Demand-Side Management System for Autonomous DC Microgrid for Building. *Technology and Economics of Smart Grids and Sustainable Energy*, *2*(1), 1–6.

CIGRÉ. (2010). Microgrids 1: Engineering Economics & Experience forthcoming, CIGRE. Working Group C6.22 Terms of Reference.

EIA. (2019). *How much electricity is lost in electricity transmission and distribution in the United States?* https://www.eia.gov/tools/faqs/faq.php?id=105&t=3

Garbesi, K., Vossos, V., Sanstad, A. H., & Burch, G. (2011). *Optimizing Energy Savings from Direct-DC in U.S. Residential Buildings*. (No. LBNL-5193E). Lawrence Berkeley National Laboratory (LBNL), Berkeley, CA.

Gerber, D. L., Vossos, V., Feng, W., Marnay, C., Nordman, B., & Brown, R. (2018). A simulation-based efficiency comparison of AC and DC power distribution networks in commercial buildings. *Applied Energy*, *210*, 1167–1187.

Gerber, D. L., Liou, R., & Brown, R. (2019). Energy-saving opportunities of direct-DC loads in buildings. *Applied Energy*, *248*, 274–287.

Gómez-Expósito, A., Conejo, A. J., & Cañizares, C. A. (2018). Electric Energy Systems: An Overview. *Electric Energy Systems: Analysis and Operation* (1–54). Boca Raton, FL: CRC Press.

Hofer, J., Svetozarevic, B., & Schlueter, A. (2017). Hybrid AC/DC building microgrid for solar PV and battery storage integration. *2017 IEEE Second International Conference on DC Microgrids (ICDCM)*, 188–191.

IEA. (2018). *Largest end-uses of energy by sector in selected IEA countries*. Independent Statistics & Analysis. https://www.iea.org/data-and-statistics/charts/largest-end-uses-of-energy-by-sector-in-selected-iea-countries-2018

IEA. (2020a). *World Energy Outlook 2020*. IEA. https://www.iea.org/reports/world-energy-outlook-2020

IEA. (2020b). *Renewables 2020*. IEA. https://www.iea.org/reports/renewables-2020

Jackson, R., Onar, O. C., Kirkham, H., Fisher, E., Burkes, K., Starke, M., Mohammed, O., & Weeks, G. (2015). Opportunities for Energy Efficiency Improvements in the U.S. Electricity Transmission and Distribution System. *Oak Ridge National Laboratory*.

Luo, F., Ranzi, G., Wang, S., & Dong, Z. Y. (2019). Hierarchical Energy Management System for Home Microgrids. *IEEE Transactions on Smart Grid, 10*(5), 5536–5545.

Miller, N. W., Shao, M., Pajic, S., & D'Aquila, R. (2014). *Western Wind and Solar Integration Study Phase 3 – Frequency Response and Transient Stability*. (No. KAGN-4-23474-01). National Renewable Energy Laboratory. (NREL), Golden, CO.

Minkel, J. R. (2008). The 2003 Northeast Blackout--Five Years Later. *Scientific American*, 13.

NERC. (2017). *Distributed Energy Resources: Connection Modeling and Reliability Considerations* [PDF file]. https://www.nerc.com/comm/other/essntlrlbltysrvcstskfrcdl/distributed_energy_resources_report.pdf

NERC. (2019). NERC Interconnections. https://www.nerc.com/AboutNERC/keyplayers/Pages/default.aspx

Nordman, B. (2010). Nanogrids Evolving our electricity systems from the bottom up. *LBNL*.

Ore, J. P., & Groll, E. A. (2020). Analysis of a Residential House for the Design and Implementation of a DC Nanogrid. *2020 IEEE PES Innovative Smart Grid Technologies Europe (ISGT-Europe)*, 749–753.

Rashid, M. H. (2011). *Power Electronics Handbook: Devices, Circuits, and Applications*. (3rd ed.). Oxford, UK: Elsevier.

Santos, A. F. B., Duggan, G. P., Lute, C. D., & Zimmerle, D. J. (2018). An efficiency comparison study for small appliances operating in DC and AC in minigrids. *2018 IEEE Global Humanitarian Technology Conference (GHTC)*, 1–2.

SEIA. (2020). *Solar Market Insight Report 2020 Q4*. https://www.seia.org/about-solar-market-insight-report

Smith, R. (2014). *U.S. Risks National Blackout From Small-Scale Attack*. The Wall Street Journal. https://www.wsj.com/articles/u-s-risks-national-blackout-from-small-scale-attack-1394664965

Tainter, J. A. (1988). *The collapse of Complex Societies*. Cambridge University Press.

U.S. Department of Energy. (2014). The War of the Currents: AC vs. DC Power. https://www.energy.gov/articles/war-currents-ac-vs-dc-power

Vossos, V., Garbesi, K., & Shen, H. (2014). Energy savings from direct-DC in U.S. residential buildings. *Energy and Buildings*, *68*, 223–231.

Wang, C., & Jain, P. (2010). A quantitative comparison and evaluation of 48V DC and 380V DC distribution systems for datacenters. *2014 IEEE 36th International Telecommunications Energy Conference (INTELEC),* 1–7.

Ying, L., Zhenqiang, L., Liang, X., & Lei, W. (2019). Research on overvoltage and minimum air clearance on 1000 kV AC side of converter transformer in ±1100 kV Guquan converter station. *Journal of Engineering*, (16), 2159–2164.

ACKNOWLEDGEMENT

The authors would like to thank the Center for High Performance Building (CHPB) organization for the financial support of this project, and the specific Project Management Subcommittee (PMS) members who have guided and encouraged this research effort.

Community Heat Pump Systems Utilizing Oil-Free Compressor Technology

Drew TURNER[1]*, Davide ZIVIANI[2]

[1]Danfoss, Oil-Free Solutions,
Baltimore, Maryland, United States
850-879-3334, drew.turner@danfoss.com

[2]Purdue University, Mechanical Engineering,
West Lafayette, IN, United States
dziviani@purdue.edu

* Corresponding Author

ABSTRACT

The goal to decarbonize buildings is quickly driving growth in the adoption of heat pumps to replace fossil fuel-based heating equipment. The trend is significantly driven by the integration of renewables in the electric grid, also replacing fossil fuel-based sources, to in-turn drive decarbonization. The impact of this change is greater if also changing out end-use fossil fuel-based heating equipment to electric-driven.

With the trend to heat pumps, there is a critical choice on both the scale and corresponding heat source. The heat pumps can be implemented with ambient air as the heat source or with other higher temperature/efficiency sources such as geothermal, process or district cooling heat recovery. Nevertheless, these sources are not always available in close proximity to the heat load. This raises the opportunity for larger-scale heating systems, serving multiple loads and with the corresponding opportunity to integrate one or more potential higher-temperature heat recovery heat sources. A related critical factor with the growth of heat pumps is resiliency. The term has historically mainly been associated with critical facilities and the ability to withstand critical events. Now it's evolving because of that same integration of renewables into the power grid and their inconsistent availability. Now resiliency has more to do with preparing for this periodic unavailability – Ensuring that demand is met when supply is not necessarily available.

This paper will present a concept for a community heating and cooling system utilizing oil-free turbo compressor technology, to address the electrification of heating while also taking advantage of multiple higher-temperature heat sources. The technology and heat sources ensure the most efficient system possible, resulting in minimum operating costs and maximum decarbonization, while the community configuration and multiple sources ensure resiliency, consistently meeting the demand requirements.

1. INTRODUCTION

Buildings consume 70 percent of electricity generated in the U.S. according to the Energy Information Agency (EIA). Moreover, approximately 61 percent of this electricity was generated by carbon-based fossil fuels in 2019, 38 percent from natural gas and 23 percent from coal (IEA, 2020). The result is that buildings are a major contributor to fossil fuel-based greenhouse gas emissions. This is changing with the growth of renewable power generation integrated into the electric grid. Over the last several years, renewables have grown from 13 percent of overall power generation in 2013 to 17 percent in 2019, as shown in Figure 1. The IEA in 2020 reported that 90 percent of all global new electric power generation came from renewables. The growth is accelerating with the higher volumes of solar and wind technologies and as the corresponding costs decrease. According to ACEEE, wind power is now lower cost than those same fossil fuel sources and solar is rapidly approaching parity with the most cost-effective forms, as indicated in Figure 2. The building electric consumption figures include only the electrical equipment utilized today, missing the equipment currently fueled by other sources. The source of heating in buildings is currently mainly onsite burning of fossil fuels, including by furnaces and boilers. The shift now is to electric-driven heat pumps to replace those onsite fossil fuel-based sources. As heat pumps replace those boilers and furnaces, the corresponding carbon emissions reduction increases. As the corresponding sources of that electrical power also switch from fossil fuel-fired sources to renewables, the resulting decarbonization again increases. Said another way, the decarbonization potential of switching the power grid to renewables increases if other emissions sources such as heating are also switched over to utilizing electric power. This switch from boilers to heat pumps (for example) also increases the electric grid load. These increases are generally seen during winter periods though, which corresponds with the former '*valleys*' (low periods) of electricity demand. This is a significant benefit from the utility perspective, leveling overall demand and increasing revenue without driving new peaks and the resulting need for additional generation capacity. Therefore, electric utilities are being tasked with this electrification of heating – It increases their topline revenue without a significant corresponding increase in their infrastructure costs. This is also why we see the growth in utility electrification programs and corresponding heat pump incentives.

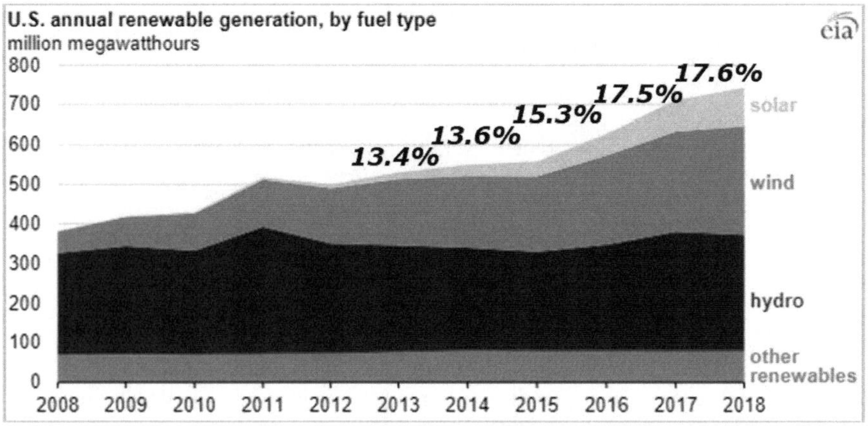

Figure 1: U.S. Annual Renewable Generation, by fuel type. Source U.S. EIA (2019)

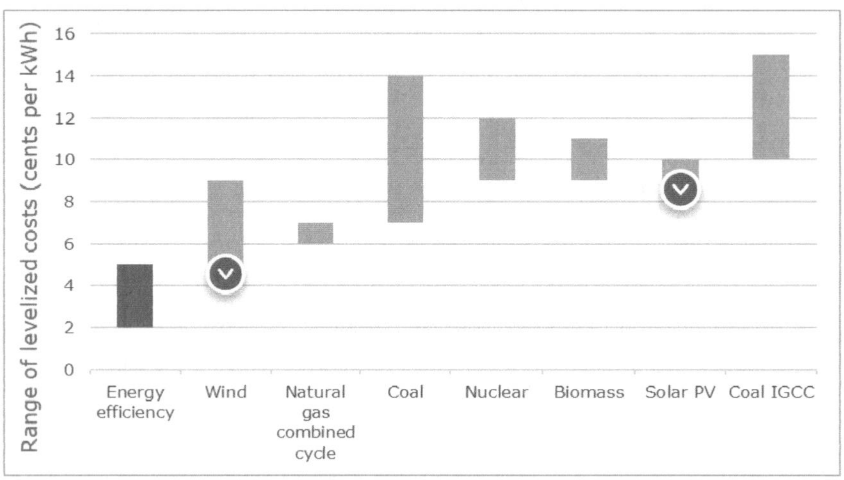

Figure 2: Levelized costs of electricity resource options. Source: ACEEE (2014)

2. ENERGY EFFICIENCY AND DEMAND/RESPONSE

The decarbonization of heating comes from the heat pump efficiency compared to fossil fuel-fired sources in addition to the renewable integration in the power grid. As an example, an electric-driven air-source heat pump replacing a fossil fuel-fired boiler operates at an approximate 3.5 COP, or 3.5 kW of heat for every kW of electricity input vs the maximum 1 COP of that boiler, as shown in Figure 3. Additionally, those same heat pumps tend to get significantly more efficient during part load/lift (differential temperature) operation, as opposed to the fossil fuel-fired sources, which may get slightly more efficient if they are the condensing-type. This all increases the decarbonization potential of electric heat pumps replacing fossil fuel-fired sources.

Starting with a baseline of 100 percent electricity coming from fossil fuel-fired sources, it is critical to account for the efficiency of that electric power generation. As mentioned earlier, the highest percentage today comes from natural gas and the majority of these are combined cycle plants, with power generation efficiency of approximately 45 percent. It is critical to also account for the 5-10 percent transmission and distribution losses, getting that electric power to the end-use point.

Even with 100 percent of the electrical power generation coming from fossil fuel-fired sources and accounting for the generation efficiency and distribution losses, there is the corresponding potential, strictly based on the heat pump efficiency advantage, for an approximate 60 percent carbon emissions reduction. This is the heat pump decarbonization baseline. Then, as the electric grid changes over to renewables, that carbon emissions reduction benefit increases approximately 3 percent for every 10 percent increase in renewables incorporation.

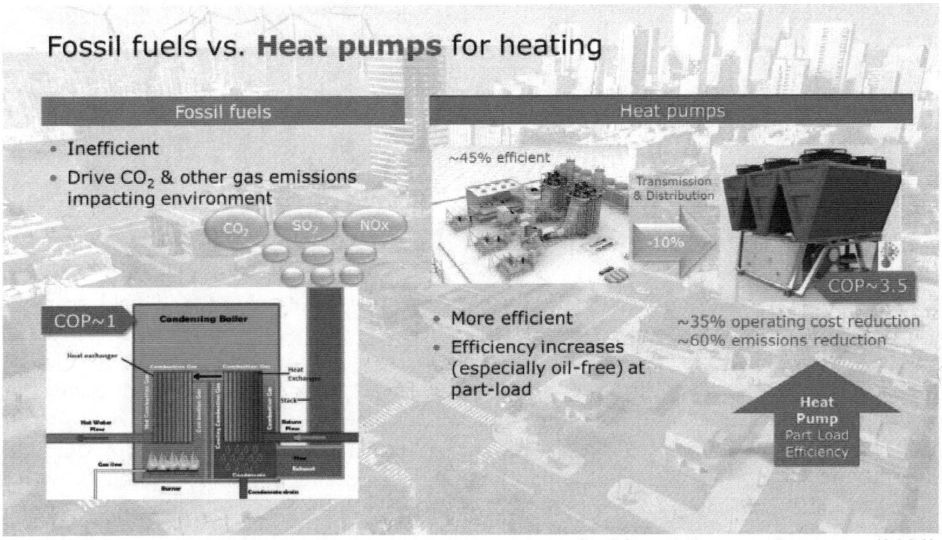

Figure 3: Electrification of heating and heat pump benefit drivers. Source: Company (2020)

As the power grid incorporates a higher percentage of renewable energy, the other growing factor is resiliency. This term has historically mainly been in the context of critical facilities (data centers, hospitals, etc.) with constant operation and the ability to continue to operate in the event of power losses and extreme events such as hurricanes. In its latest evolution, the resiliency term also now refers to accounting for the periodic renewable generation unavailability. As the percentage of power coming from renewables grows, so does the power availability (resiliency) risk from those periods when wind and solar resources are not available due to being intermittent in nature, creating a supply & demand disconnect.

There are several solutions available for this disconnect and resiliency risk, including both supply-side and demand-side storage, enabling '*ride-through*' in those unavailability periods. Energy storage takes various forms, including batteries, pumped hydro, hydrogen and thermal storage. The most cost-effective form of energy storage today is thermal storage. While the energy storage cost of batteries will continue to come down, thermal storages is still in the range of about 1–10 percent of the applied cost of battery storage, and we expect that difference will continue for a while. The different storage forms though are optimized solutions for different 'disconnect' periods and the solutions will end up being a mix depending on the supply/demand specific application dynamics. The general industry long-term storage direction for this disconnect duration seems to be:

1. Short duration (<~4 hours) – Thermal storage
2. Medium duration (~4-8 hours & transport) – Battery storage or similar (e.g., redox flow battery storage)
3. Long duration (>8 hours & transport) – Hydrogen or similar

Another related recent term and trend is "*Grid-interactive Efficient Buildings*" (GEB), which is about enabling buildings to provide flexibility in energy use and grid operation. GEB requires efficiency, load shift, shed and modulation to quickly react on the demand side to deficiencies on the supply side (US DOE, 2020). These building are also shifting to the concept of separate, sensible and latent load cooling – which can increase efficiency and reduce resulting energy use by taking advantage of the higher cooling temperature and more efficient resulting efficiency of sensible-only cooling and the flexibility to modulate separately to the varying sensible and latent loads. The overlap with the trend to electrification and heat pumps is critical, as those same demand side solutions are then expected to leverage these same efficiency and flexibility solutions, as summarized in Table 1.

Table 1: Grid-interactive Efficient Building (GEB) HVAC technologies and associated goal benefit potential. Source: DOE Building Technologies (2020).

	Technologies — Capability Legend: ○ NA ◑ Low ◐ Med ● High	Efficiency	Load Shed	Load Shift	Modulate	Overall Potential
HVAC	HVAC#1: Smart Thermostats	◑	◐	●	◑	High
	HVAC#2: Separate Sensible and Latent Space Conditioning	●	◐	◐	○	High
	HVAC#3: Liquid Desiccant Thermal Energy Storage	◐	◐	●	○	High
	HVAC#4: Controls for HVAC Equipment w/ Embedded T-stats	○	◑	●	◑	Med
	HVAC#5: Hybrid Evaporative Precooling	●	◑	◑	○	Low
	HVAC#6: Dual-Fuel HVAC	○	●	○	○	Low
Cross Cutting	CC#1. Thermal Energy Storage (TES)	◑	◑	●	◑	High
	CC#2. Modulating Capacity Vapor Compression	●	◐	◐	◐	Med
	CC#3. Non-vapor Compression (NVC) Systems and Materials	●	◐	●	◐	High

3. THE ROLE OF DISTRICT ENERGY SYSTEMS

Missing from most of the conversations thus far on electrification, heat pumps and grid-interactive efficient buildings has been the potential of district or community energy systems to address all these goals. District & community energy systems have built-in redundancy, storage, modulation flexibility and resiliency. A district energy system takes advantage of the built-in thermal flywheel effect, creates the option for recovering heat from different sources in the system, and turns down generation sources that are built into the system as needed based on demand. Critically in terms of resiliency and electrification infrastructure costs, it centralizes the electrical load increase to locations which can manage the increase, vs. strapping it to individual buildings and removes the associated heat pump capital investment from those buildings.

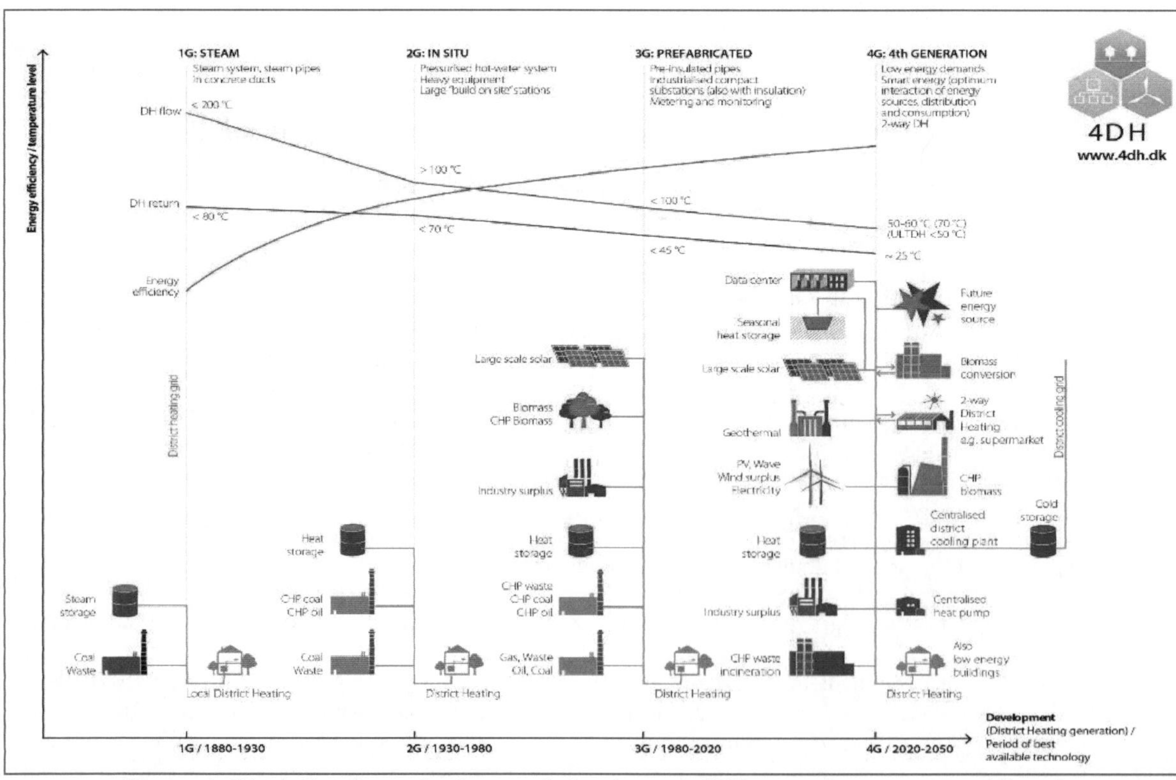

Figure 4: District heating technology evolution (source: 4DH research centre, www.4dh.eu)

In recent years, district energy—especially district heating—systems have seen a significant evolution, particularly in integrating in and planning out heat recovery and electrification of those systems. Critical in this is the evolution of district heating systems over the last several years and specifically Europe the 4th generation. District heating was 1st invented in the US in the Edison days, but then evolved in Europe and each new generation lowers the heating operating temperatures. Figure 4 shows this evolution to the latest 4th generation district heating system and corresponding 55-65 °C heat supply temperatures, down from >100 °C in earlier generations.

Fossil fuel-sourced heating operates at approximately the same efficiency at low or high heat supply temperatures. But for a heat pump, it is critical to reduce the heat supply temperature it to in-turn provide a lower operating cost and resulting payback vs the boiler or furnace it is replacing. Every vapor compression heat pump gets 1-1.5 percent more efficient for each 1 K reduction in the differential temperature that it must achieve, which is driven by both the heat source and supply temperature. Europe's district heating evolution also provides a great benchmark for the usage of different heat recovery heat sources utilized in heat pumps. To this end, Figure 5 reports the heat source temperatures and heat pump volumes associated with different application sources of heat pumps applied to European district heating through 2017. It shows that wastewater is by far the most common source, but higher temperature sources including industrial, geothermal and flue gas are also common.

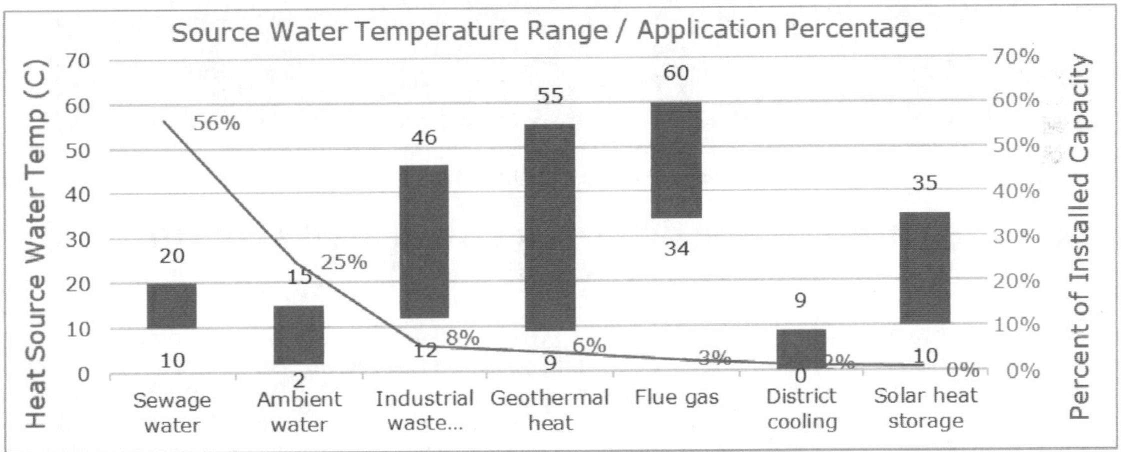

Figure 5: European district heating heat pump application and heat source temperature. Source: Heat Roadmap Europe (2017)

Adding heat pumps, and specifically water-to-water heat pumps utilizing heat recovery, is a fantastic solution for both electrification and the ultimate goal of decarbonization. Examining the different district energy solutions available today, we see that smaller-capacity heat recovery sources are becoming much more prevalent as opposed to more of the centralized utility-driven generation. The corresponding energy source is moving from large centralized generation using lower temperature heat sources such as ambient air and surface water to decentralized higher temperature heat recovery sources, frequently based on '*symbiosis*' applications, where the recovered heat is from the cooling of a facility or process. As mentioned, the main driver for this focus on the distributed higher-temperature heat sources is the resulting compressor work reduction and heat pump efficiency, but also to avoid driving new winter period electrical load peaks. Optimally, these water-water heat pumps are applied on critical facilities, with year-around cooling loads, resulting heat recovery and onsite backup power generation for additional resiliency. For instance, a data center is a perfect application opportunity example, because it provides multiple corresponding heat source benefits, including:

1. Year-around and consistent heat load and resulting heat recovery.
2. Higher temperature cooling and resulting heat recovery temperatures / higher heat pump efficiency.
3. Critical facility onsite backup power generation, for demand response scenarios and combined facility cooling / community heat.

Looked at from the community heating system perspective, this is one of the most efficient, highest operating hours, lowest operating cost and lowest emissions potential baseload heat sources. This factor of course increases with the corresponding increase in the percent of power for that data center coming from renewables. A conceptualized district energy solution is illustrated in Figure 6.

Figure 6: District energy integration of heat recovery heat pumps, applied to symbiosis cooling and heat recovery systems. Source: Company (2020)

4. COMPRESSION TECHNOLOGIES FOR COMMUNITY-BASED HEATING SYSTEMS

There are two main compression technologies for HVAC systems: dynamic and positive displacement. Dynamic compression adds kinetic energy to the refrigerant, while positive displacement harnesses potential energy by "squeezing" it and reducing the area. Oil-free turbo compressors fit into the first (dynamic) category. There are also a variety of bearing types. Historically, oil-based bearings were used in both dynamic and positive displacement compression technologies. Today, there are several oil-free options, including magnetic, ceramic and gas bearings. Among newer compression technologies, magnetic bearings provide flexibility in terms of capacity, reliability, and efficiency.

Because oil-free compressors minimize physical size through high-/variable-speed operation, they are ideal for retrofit applications where a smaller footprint is beneficial for physical heat pump size and/or multiple compressor combinations on a single unit.

Additionally, since these same compressors have no mechanical contact in either the compression or bearing support system, oil-free machines are quieter, lower maintenance and maintain performance over their lifetime. The two primary sources of noise coming out of heat pumps are the mechanical contact in both the compression process and the bearing support system. Since a magnetic bearing-based oil-free turbo compressor has neither of these, the primary noise is then the significantly lower level of the refrigerant flowing through the system and the 'hum' of the variable speed. The result is a compressor with noise level on average 8 dB lower than an equivalent capacity screw compressor. This is a critical factor for installations next to critical spaces, outdoor or retrofit in tight mechanical room spaces, where the added cost of expensive attenuation can be avoided.

Those same mechanical contact factors are beneficial to maintaining long-term performance. As HVAC installations have become instrumented to track their performance, there is a growing recognition in the industry that the performance of these systems degrades over time. Wearing down of mechanical bearing and compression contact, along with oil included in the system getting into the heat exchangers and degrading that performance, are the primary sources of this heat pump performance degradation over time. Since an oil-free turbo compressor has neither one of those factors, then the performance should be maintained. The oiled vs. oil-free difference is obvious, but the compression difference can be verified. This has been accomplished per the below summary showing the results of re-testing six compressors after ten years of operation. The comparison shows that the performance of the compressors all had zero-to-negligible degradation, with one 'outlier' at three percent.

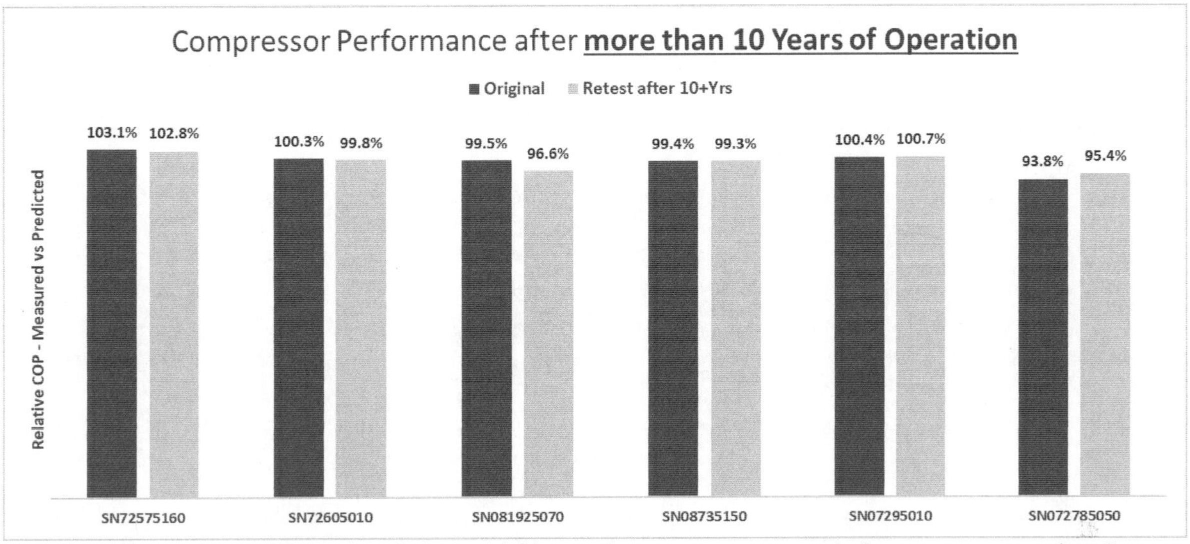

Figure 7: Performance of dynamic, oil-free, variable speed compressors re-tested after ten years operation. Source: Company (2021)

For heat pumps, this factor is critical to ensure that the payback vs the fossil fuel-source heating alternatives is accurate, since the majority of these life-cycle cost evaluations today do not incorporate a performance degradation factor, leading to potentially wrong conclusions, unless utilizing oil-free turbo compressor-based heat pumps.

5. NEXT PHASE OF REFRIGERANTS AND TURBOCOMPRESSORS

These evolving dynamic, variable speed, oil-free vapor compression technologies utilizing new HFO refrigerants can potentially further reduce refrigerant CO_2 emissions, as reduced pressure and density can lead to lower global warming potential (GWP) with minimized flammability. The coming years will likely bring lower-pressure and lower-density refrigerants, and supporting technologies optimized for their use. Figure 7 illustrates this dynamic, that to accomplish the goal of achieving lower GWP while also minimizing flammability requires moving to lower density. As the refrigerant selection moves to lower density that also leads to technologies optimized to moving a high volume of these lower pressure, large molecule fluids where turbo compressor technologies play a major role in ensuring needed capacities.

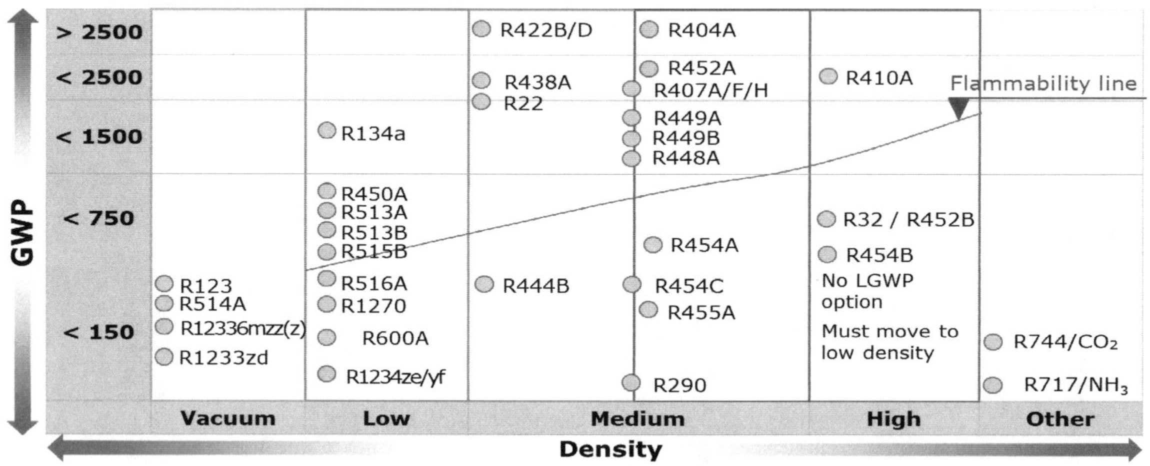

Figure 8: Tradeoffs between GWP and volumetric capacity as well as flammability of low-GWP refrigerants and their blends. Source: Company (2020)

It is critical to optimize these dynamic centrifugal compressors to their target application operating temperatures. There are three basic oil-free turbo compressor technology optimizations on the market today, as summarized in Figure 8:

- Standard lift optimization for lower lift or lower differential operating temperatures, which support water-cooled, evaporative-cooled and hybrid chillers.
- Medium lift optimization for air-cooled chillers, water-cooled chillers in more challenging environments and water-to-water heat pumps with lower required heating temperatures and/or higher cooling/heat-source temperatures.
- High lift optimization – the most recent innovation for oil-free turbo compressors – is on the mechanical design of the compressor, along with the aerodynamic compression process. It supports the most challenging applications, including very high ambient air-cooled chillers, water-to-water heat pumps with higher heating and/or lower cooling/heat source temperatures, air-to-water heat pumps in milder climates, medium-temp process cooling and thermal storage applications.

The optimal solution for a specific application frequently includes more than one of these optimizations. Specifically, for larger-capacity heat recovery applications in community / district heating / micro-thermal grid applications, the optimal performance solution is frequently a combination of the Medium and High-lift optimized compressor designs. This depends especially on the heat recovery source temperature and the district heating return water temperature. The latter return temperature factor is driven by demand-side flow and heat transfer technologies utilized. On the demand side it is critical for the district heating system to use variable speed flow, two-way pressure independent control valves and efficient heat transfer stations. This in-turn enables lower return water temperatures and implementation of this more efficient hybrid compressor optimization heat pump solution.

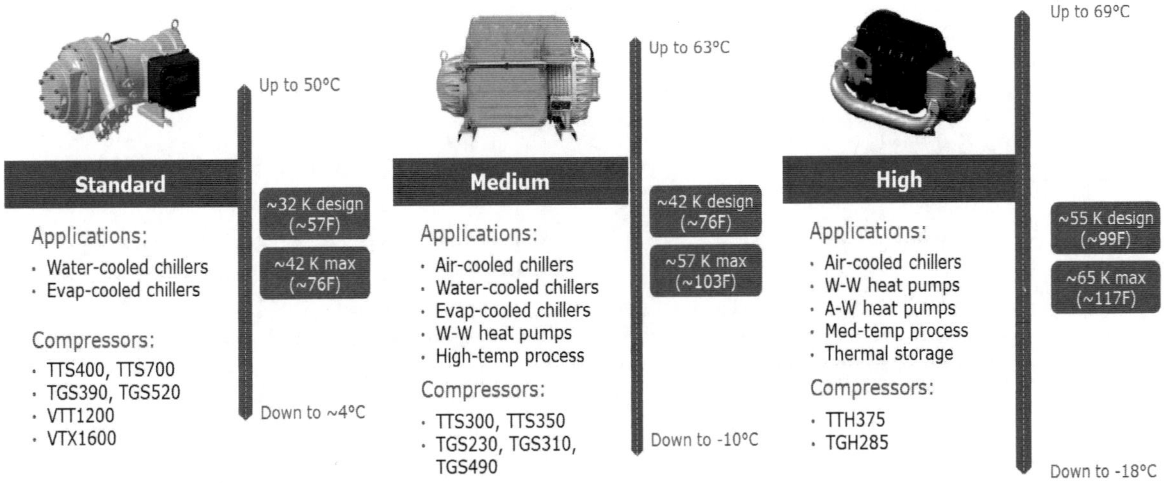

Figure 9: Oil-free turbo compressor optimizations for different operating temperatures. Source: Company (2020)

All vapor compressors have two sets of forces in the compression process: the radial (side-to-side) motion of the shaft driving the compression process and the axial (back-and-forth) motion. While previous oil-free turbo compressor designs had both stages of compression on one end of the compressor, new high-lift optimization technology moves the second stage of compression to the opposite end, which balances those axial forces. The two compression stages pulling back against each other enables a higher differential temperature capability, while maintaining no mechanical contact and the associated benefits. The most recent version of this vapor compression technology can now provide those same technology benefits formerly relegated mainly to comfort cooling chillers applied in milder climates, now to the hottest environments, provide cooling at lower temperatures – including thermal storage – or provide heating with heat pumps. What were once unattainable applications for magnetic-bearing-based compressors is no longer the case, thanks to this evolving technology that enables operation in more challenging applications.

Oil-free design allows engineers to combine a variety of technology to optimize based on real-world conditions. For example, where cooling is required for industrial manufacturing, this technology can provide cooling and then recover the heat through a symbiosis heat pump, which boosts the temperature to supply the district heating system. When the industrial process is not operating, a parallel geothermal ground-source loop can also supply heat to the district heating system via that same heat pump. This dual cycle system can enable full-time supply as the baseload of a district heating system, maximizing operating hours and helping buildings recoup the first cost investment in less than three years. An example of such system is shown in Figure 9.

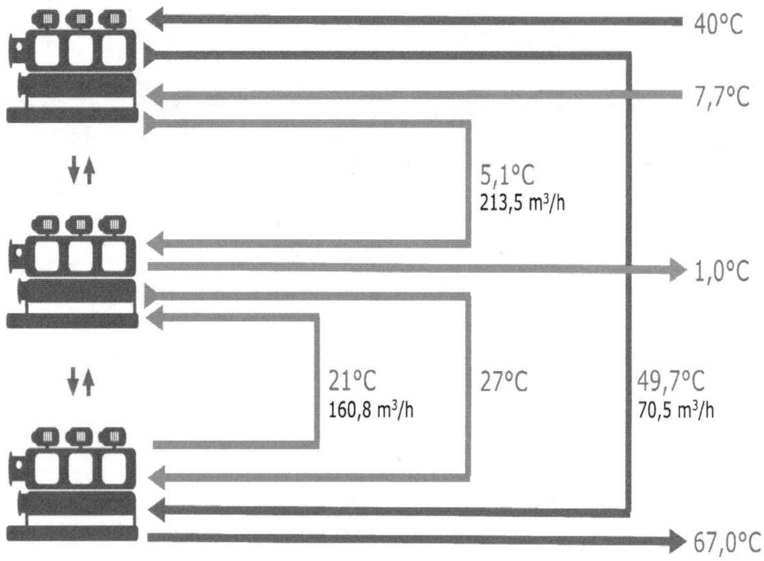

Figure 10: Example heat pump system architecture and operating temperatures for process cooling and heat recovery to community/district heating system. Source: Company (2020)

6. CONCLUSIONS

An evolving portfolio of magnetic bearing, variable speed dynamic vapor compression technology can support an expanding array of applications with a variety of compressor combinations to maximize operational efficiency, minimize operating costs. All of this while lowering CO_2 emissions via both the electrification of heating and utilizing the most efficient heat pump technology solution, which maintains that performance over time and optimal versions of that technology per the operating requirements.

NOMENCLATURE

Acronyms

GWP	Global Warming Potential
HFO	Hydrofluoroolefin

REFERENCES

IEA World Energy Outlook 2020: https://www.iea.org/reports/world-energy-outlook-2020

U.S. EIA Report (2019): https://www.eia.gov/todayinenergy/detail.php?id=38752 (Accessed 01/2021)

Company (2020 & 2021): Danfoss, contributing author and colleague graphics

Levelized costs of electricity resource options. The Best Value for America's Energy Dollar: A National Review of the Cost of Utility Energy Efficiency Programs. Molina, ACEEE. (2014)

Annex 47, Heat Pumps in District Heating and Cooling Systems, Final Report. P 13. MDPI (2020)
Andrei David, Brian Vad Mathiesen, Helge Averfalk, Sven Werner and Henrik Lund (2017). Heat Roadmap Europe: Large-Scale Electric Heat Pumps in District Heating Systems. P 4-6. IEA

U.S. Department of Energy, EERE, Building Technologies Office (BTO). GEB Technical Report Series: Heating, Ventilation, and Air Conditioning. Webinar presentation, slide 18. (2020)

ACKNOWLEDGEMENT

Acknowledging Eddie Rodriguez for the chart in Figure 7 and Leping Zhang for contribution of the system architecture shown in Figure 10

Effect of Inlet Duct Design on Fan Performance of Indoor Air Handling Units with Pull-Through Fan Configuration (ASHRAE RP-1743)

Khaled Ibrahim ALGHAMDI[1*], Christian Konrad BACH[2]

[1, 2] Oklahoma State University, Mechanical and Aerospace Engineering School,
Stillwater, OK 74074, USA

[1] Contact Information: khaled.alghamdi@ostate.edu
[2] Contact Information: cbach@okstate.edu

* Corresponding Author

ABSTRACT

During testing indoor air handling units, ASHRAE Standard 37 requires installing an inlet duct that might increase the length of the testing setup to exceed the capability of some testing facilities. ASHRAE Research Project 1743 (ASHRAE RP-1743) addresses this issue by evaluating the effect of inlet duct design on fan performance of indoor air handling units. The findings of the study will allow identifying alternative, shorter, inlet duct configurations maintaining the fan performance as of that obtained with ASHRAE Standard 37 inlet duct. The test matrix of the project includes various indoor units differing in motor type, such as Constant Torque Motor (CTM) and Electronically Commutated Motor (ECM), and fan configurations, such as pull-through and push-through configuration. This paper focuses on the findings of testing two pull-through-fan-configuration units, one of which with a CTM fan and the other with an ECM fan. In addition, other effects, such as atmospheric pressure and hysteresis in approaching external static pressure (ESP), were investigated to study their impacts on the repeatability of equipment testing.

The results of the CTM-fan unit showed that the proposed alternative inlet duct configurations can maintain the fan performance within the suggested tolerances, which are 5% in power and 2.5% in air volumetric flowrate relative to the ASHRAE Standard 37 inlet duct. This allows a reduction in the inlet duct stack-length from 38% to 74% compared to ASHRAE Standard 37 inlet duct configuration. Noticeable performance hysteresis was observed at low air flowrate and ESP, while atmospheric pressure had no effect on fan performance.

For the ECM-fan unit, the fan performance was maintained during testing the proposed alternative inlet duct configurations within the suggested tolerance, with a noticeable decrease in power during testing the reduced alternative inlet duct configuration. The reduction in the stack length as compared to the standard inlet duct configuration was 40%, 72%, 74% for the inlet box, 4-inch inlet duct, and reduced alternative configuration, respectively. In contrast to the results of the CTM-fan unit, the hysteresis effect on the ECM-fan unit was found to be insignificant, while atmospheric pressure effect was significant. As the atmospheric pressure increased, air volumetric flowrate and fan's power decreased.

1. INTRODUCTION

ASHRAE Standard 37 provides methods to rate unitary air-conditioning and heat pump equipment (ANSI/ASHRAE, 2009). The methods provided by ASHRAE Standard 37 should enhance the repeatability of rating tests to maintain the rating of equipment consistent regardless of where the equipment is being tested. Among the requirements of ASHRAE Standard 37, a standardized inlet duct should installed during testing equipment. The inlet duct dimensions are determined by the cross-section areas of the inlet of the *Unit Under Test* (UTT; ANSI/ASHRAE, 2009). However, the ASHARE Standard 37 inlet duct can be problematic because the stack-length of equipment testing setups might exceed the height of testing facilities, especially for legacy facilities. Therefore, the main goal of ASHRAE RP-1743 was to identify reduced-length alternative inlet duct configurations that maintain fan performance of the UUT close to that obtained during testing the unit with ASHRAE Standard 37 inlet duct configuration. Also, ASHRAE RP-1743 investigated other effects that could affect test repeatability besides the inlet duct configurations.

In this paper, Section 2 describes the testing facility and instrumentations. Section 3 lists the parameters varied during testing the indoor units for ASHRAE RP-1743. Section 4 shows and discuses the results obtained. Finally, Section 5 concludes the paper by summarizing the findings and suggesting potential future work.

2. EXPERIMENTAL SETUP

2.1 Operational Conditions:

In general, the operational conditions of tests for ASHRAE RP-1743 was in accordance with AHRI Standard 210/240 (AHRI, 2017). To maintain the operational conditions, the project was conducted at the indoor psychrometric chamber of Oklahoma State University (OSU). The chamber has two conditioning bays to maintain the chamber at the testing conditions, which were 80.0°±0.5°F (26.7°±0.3°C) dry-bulb temperature (T_{DB}) and 67.0°±0.3°F (19.4°±0.2°C) wet-bulb temperature (T_{WB}). The chamber's code tester bay was used to measure air coming out of UUT, as well as adjusting the external static pressure (ESP) across the UUT by adjusting the blower of the code tester. For ESP, Table 11 in AHRI Standard 210/240 specifies the minimum ESP for unit groups similar to that tested for ASHRAE RP-1743 to be 0.15 in w.c. (37.3 Pa). The units were tested under seven ESP points ranged from 0.10 in w.c. (24.9 Pa) to 0.55 in w.c. (136.9 Pa) including the 0.15 in w.c. (37.3 Pa) standard point. The units were tested under two nominal air flowrates: 350 cfm/ton (46.97 L/s-kW) and 450 cfm/ton (60.39 L/s-kW). The units were tested vertically, as described by Hossain et al. (2018).

2.2 Instrumentation:

Temperature was measured per ANSI/ASHRAE 41.1 for T_{DB} and ANSI/ASHRAE 41.6 for T_{WB}; in particular, a resistance temperature device (RTD) was used for T_{DB}, and the aspirated wet-bulb psychrometer was used for T_{WB} (ANSI/ASHRAE, 2014, 2013). Air was sampled by a sampling device fulfilling section 5.3.4 in ANSI/ASHRAE Standards 41.1 (ANSI/ASHRAE, 2013).

External Static Pressure (ESP) was measured across the UUT by pressure transducers connected to pressure taps that are fabricated in the inlet and outlet ducts in accordance with the recommended practices in subsections 6.5.1 and 6.5.3 of ASHRAE Standard 37 (ANSI/ASHRAE, 2009). The pressure taps were placed in the standard inlet duct in accordance with subsection 6.4.2.2 in ASHRAE Standard 37, while they were placed for alternative inlet duct configurations either in the horizontal center of a duct or a skirt duct. Also, to enhance the reliability of ESP readings, two pressure transducers were used. Each transducer was connected to two adjacent pressure taps in the inlet duct and the corresponding pressure taps in the outlet duct. One of the transducers was used as a reference to adjust ESP during testing, while the other was used for cross-checking the readings of the first transducer after a test was completed.

Air flowrate coming out of UUT was measured by the nozzle box in the code tester bay of OSU indoor psychrometric chamber. The pressure difference across the nozzle plane was measured and used to determine the air flowrate passing the nozzles in accordance with the steps mentioned in subsection 6.2 of ASHRAE Standard 37 (ANSI/ASHRAE, 2009).

Speed of UUT's fan was measured by a 3rd party rotational speed sensor. The sensor was connected to the *Data Acquisition* system (DAQ) of the OSU psychrometric chamber to sample the readings at the same pace as other parameters. For the UUT's fan power consumption, it was measured by a power meter with an inbuilt data logger. The power data was later integrated into other data obtained by the DAQ of the psychrometric chamber through time-stamps.

Atmospheric Pressure was measured by a barometer placed outside the psychrometric chamber. Then, two transducers were used besides the manometer to measure the atmospheric pressure at the inlet duct. The first pressure transducer measured the pressure difference between the inside and outside of the psychrometric chamber. The second pressure transducer measured the pressure difference between the ambient inside the psychrometric chamber and the inlet duct by connecting one port of the transducers to the pressure taps of the inlet duct, while the other port of the transducers was exposed to the ambient inside the psychrometric chamber.

3. TEST PLAN

The test plan consisted of testing various units, each of which had a different fan type and configuration, with 3 alternative inlet duct configurations besides the standard inlet duct configuration. For each inlet duct configuration, the UUTs were tested under two nominal air flowrates and seven ESP points, as mentioned in the Section 2.

3.1 Fan Configurations and Types:

In general, four 3-ton indoor units were tested for ASHRAE RP-1743. In terms of fan types, two of the units had fans with Electronically Commutated Motor (ECM) and the other two had fan with Constant Torque Motor (CTM). The difference among the units having the same fan motor type was the configuration of the fan: pull-through or pull-through fan configurations. In the pull-though fan configuration, the fan was placed downstream the conditioning equipment of the unit, while the equipment was placed upstream in the push-through fan configurations. This paper shows the results obtained for the units with pull-though configuration.

3.2 Inlet Duct Configurations:

Besides the standard inlet duct, three reduced alternative inlet duct configurations were tested. Figure 1 shows the dimensions of the inlet duct configurations, and Table 1 shows the reduction in the inlet duct configurations as compared to the standard configuration.

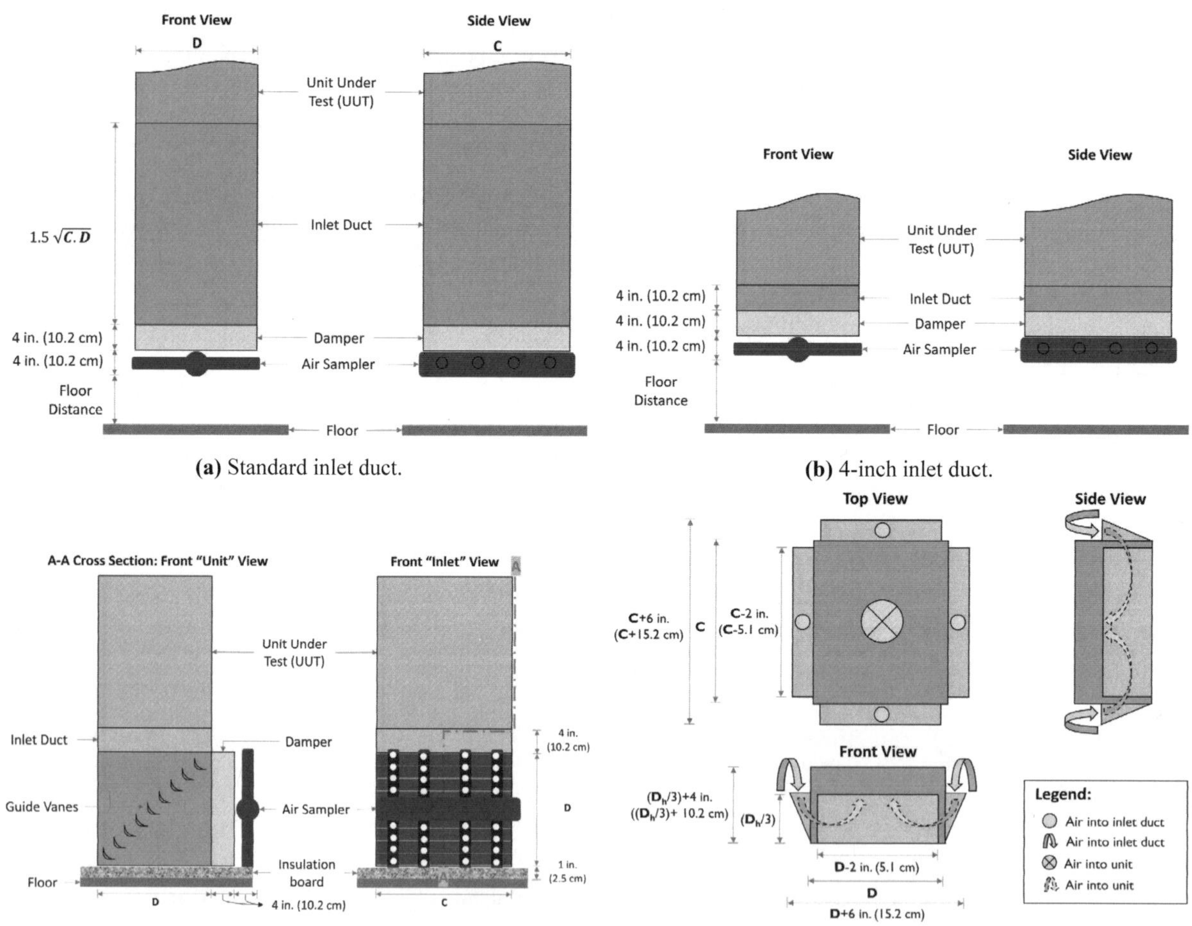

(**a**) Standard inlet duct. (**b**) 4-inch inlet duct. (**c**) Box inlet with guide vanes. (**d**) Reduced alternative.

Figure 1: Sketches of the inlet duct configurations for ASHRAE RP-1743.

4. RESULTS AND DISCUSSIONS

The units and inlet duct configurations mentioned in Section 3 were tested under the conditions in Section 2. The reported findings in this paper are mostly of 350 cfm/ton (46.97 L/s-kW) and 0.50 in w.c. (124.4 Pa); findings of other conditions are identical unless otherwise mentioned. Also, volumetric air flowrate and fan's power consumption are the variables reported in this paper as they are the variables affecting repeatability UUT's fan performance. To enhance the accuracy of the comparison among the reported cases, data has been interpolated to the nominal ESP point

Table 1: Change in the stack-length of the inlet duct configuration as compared to the standard inlet duct configuration.

Unit	Inlet duct configurations			
	Standard[1] [in]	4-inch[2] [%]	Inlet box [%]	Reduced alt.[3] [%]
CTM-fan	42	-70	-38	-74
ECM-fan	45	-72	-40	-76

[1] 2 inches were added as a floor distance.
[2] Change vs. Standard excluded floor distance in both inlet ducts.
[3] Air sampler was placed around the inlet duct.

of 0.50 in w.c. (124.4 Pa). In terms of the effects reported in this paper beside the inlet duct configurations, hysteresis effect, which is caused by the direction followed to approach an ESP point, and atmospheric pressure effect are reported in this paper. Those effects was to be found substantial during testing some units for ASHRAE RP-1743.

4.1 Hysteresis Effect:

While testing units for ASHRAE RP-1743, fan performance might have changed significantly as the the direction of approaching an ESP point, from high-to-low or vice versa, was changed. This hysteresis effect can affect the repeatability of testing units, so it was considered vital to be investigated. Observing the fan rotational speed to decide if the effect caused by an instrumentation or the unit, the fan speed would have experienced a change whenever the hysteresis effect was observed. Thus, the hysteresis effect, if occurred, was caused by the unit itself. The direction of high-to-low ESP was considered the basecase that the other direction was compared to after interpolating the the base case to the compared one in order to enhance the accuracy of the comparison.

For the CTM-unit, the hysteresis effect was noticeable at low air flowrate, which was 350 cfm/ton (46.97 L/s-kW), as it is shown in Figure 2a, while the effect was not significant at high flowrate, which was 450 cfm/ton (60.39 L/s-kW), as the change in performance was within the uncertainty of measurements. However, the hysteresis was insignificnt in all tested conditions on the ECM-fan unit, as Figure 2b shows since the change in the performance was within the uncertainty of measurements.

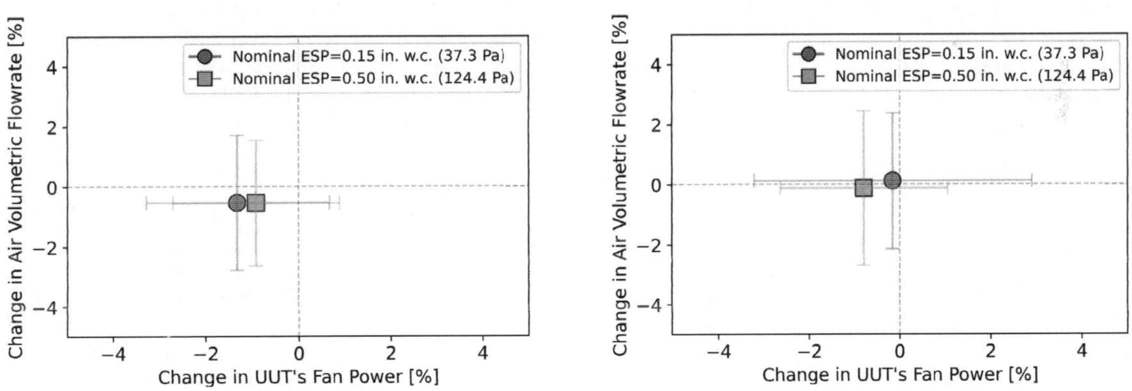

(a) CTM-fan unit.

(b) ECM-fan unit.

Figure 2: Change in the fan performance as the direction of approaching ESP point changed (hysteresis effect) for nominal air flowrate ratio of 350 cfm/ton (46.97 L/s-kW) during testing UUTs with standard inlet duct.

4.2 Atmospheric Pressure Effect:

During repeated tests, it was observed that atmospheric pressure might have affected fan performance. Hence, tests were repeated under different atmospheric pressures without making changes on the setup, such as disassembling the inlet duct. The results show that the atmospheric pressure effect is insignificant for the CTM-fan unit as the fan performance was maintained within the uncertainty in measurements despite the change in atmospheric pressure was relatively high, as shown in Figure 3a. However, the ECM-fan unit experienced a significant effect of atmospheric pressure with a relatively small change in atmospheric pressure: fan's power and air volumetric flowrate increased as the atmospheric pressure decreased, as shown in Figure 3b.

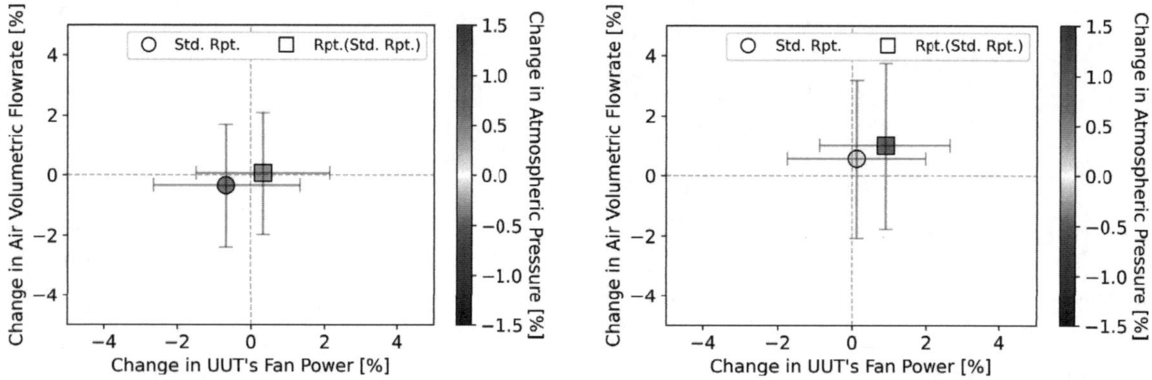

(a) CTM-fan unit. **(b)** ECM-fan unit.

Figure 3: Change in the fan performance as atmospheric pressure changed for nominal air flowrate ratio of 350 cfm/ton (46.97 L/s-kW) and ESP of 0.50 in w.c. during testing units with standard inlet duct.

4.3 Inlet Duct Configuration Effect:

Fan performance during testing the alternative inlet duct configurations were compared to the performance obtained during testing standard inlet duct configuration. A change in fan performance was acceptable if the change was within the suggested tolerance of 5 % in fan's power and 2.5 % in air volumetric flowrate. The results of the CTM-fan unit and the ECM-fan unit are shown in Figure 4 and Figure 5, respectively.

For the 4-inch inlet duct configuration, fan performance was maintained within the suggested tolerance as compared to the standard inlet duct configuration for both units. For the inlet box configuration, the performance was maintained in both units, despite the results on the ECM-fan unit was affected by the change in atmospheric pressure. Also, to determine if the orientation of the inlet side of the inlet box had an effect on the fan performance, the inlet box was rotated by 90°. The rotation of the box did not cause a significant change in fan performance considering the effect of atmospheric pressure when the inlet box was tested first. For the reduced alternative inlet duct configuration, despite fan performance was maintained within the suggested tolerance as compared to the standard inlet duct configuration, fan's power decreased in both units, especially in the ECM-fan unit at low ESP, as shown in Figure 5a. Also, the air volumetric flowrate increased in the CTM-fan unit, while it did not in the ECM-fan unit.

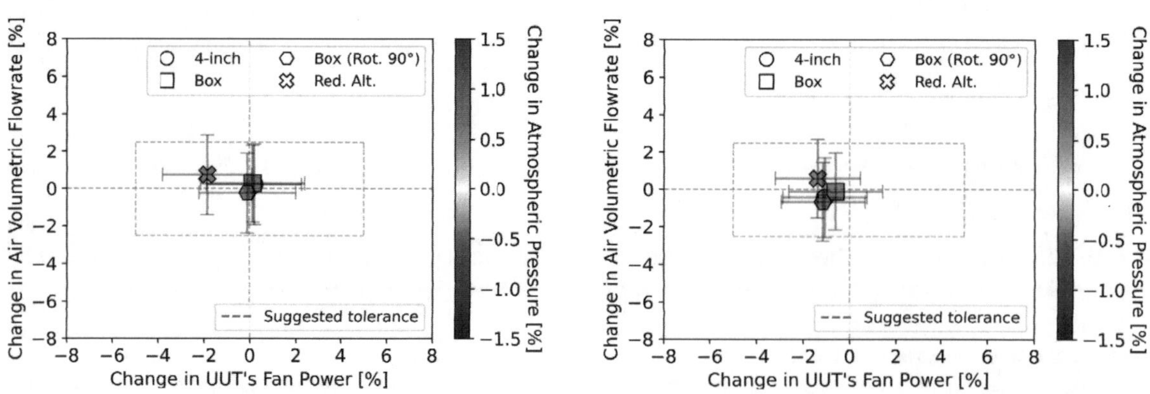

(a) ESP of 0.15 in w.c. **(b)** ESP of 0.50 in w.c.

Figure 4: Change in fan performance of the CTM-fan unit during testing the alternative inlet duct configurations as compared to the standard inlet duct at nominal air flowrate ratio of 350 cfm/ton (46.97 L/s-kW).

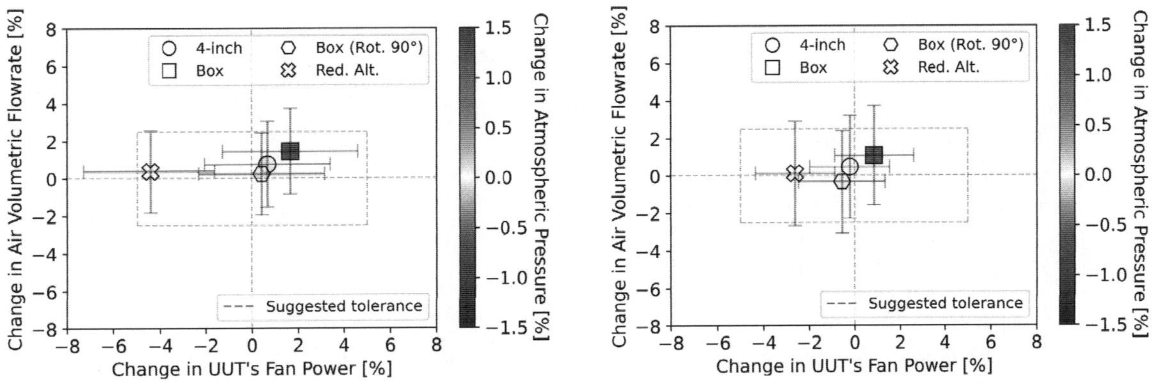

(a) ESP of 0.15 in w.c.

(b) ESP of 0.50 in w.c.

Figure 5: Change in fan performance of the ECM-fan unit during testing the alternative inlet duct configurations as compared to the standard inlet duct at nominal air flowrate ratio of 350 cfm/ton (46.97 L/s-kW).

5. CONCLUSION

The project, ASHRAE RP-1743, aimed to investigate alternative inlet duct configurations that reduced in length as compared to the standard inlet duct specified in ASHRAE Standard 37 (ANSI/ASHRAE, 2009). Along with the main aim of the project, other effects, such as the direction followed to obtain an ESP point and atmospheric pressure, were investigated to eliminate their impacts on the repeatability of tests. The methodology adopted for ASHRAE RP-1743 was to tests four 3-ton units, each of which had a different fan configuration, such as pull-through or push-through, and fan's motor type, such as ECM or CTM. The units were tested under two nominal flowrates and seven ESP points for each of the alternative inlet duct configurations tested. This paper is dedicated to the results of the units with pull-though fan configuration.

The direction followed to obtain an ESP point, which is referred to as the hysteresis effect, was found noticeable in the CTM-fan unit at low air flowrate, while the effect on the ECM-fan unit was insignificant. In general, the hysteresis was not associated with the fan's motor type as the effect was found to be significant in a unit with ECM-unit among the other units considered for ASHRAE RP-1743. Therefore, specifying the direction followed to adjust ESP might be ideal to eliminate a discrepancy source in rating units, as the effect is unpredictable through fan's motor type.

For the atmospheric pressure, it had a significant effect on the ECM-fan unit, while it did not on the CTM-fan unit. In general, the atmospheric pressure affects the density of air, which can be controlled be adjusting the content of moisture in air. Nevertheless, controlling the density by changing the moisture content in the air inside a testing facility will not be practical since an increase in moisture content affects the rating of the unit when the units is operated in cooling mode. This effect of atmospheric pressure was found to be significant on some units despite the change in the atmospheric pressure was within the normal daily fluctuation in one location. The effect would be vital if a unit is rated at different facilities, each of which is located at a different altitude. Therefore, the effect should be addressed to eliminate its impact on the repeatability of tests.

As compared to the standard inlet duct specified in ASHARE 37, the alternative inlet duct configurations considered for ASHRAE RP-1743 generally maintained fan performance for the units within the suggested tolerance of 5 % in fan's power and 2.5 % in air volumetric flowrate. However, the reduced alternative inlet duct configuration caused a noticeable decrease in the fan's power on the units, while an increase was noticed in air volumetric flowrate on the CTM-fan unit only, as compared to the power and flowrate during testing standard inlet duct. Fan's power contribution might be minor in the rating of units because the power consumption constitutes a small portion of the input power as compared to that of the components of the unit, but the change in the air volumetric flowrate might contribute significantly in the rating. For the inlet box configuration, the direction of the inlet side of the inlet box did not affect fan performance. The reduction in the stack-length of the inlet duct is 38% to 74% for the CTM-fan unit, while it is 40% to 76% for the ECM-fan unit.

NOMENCLATURE

ANSI	Air-Conditioning, Heating, & Refrigeration Institute
ASHRAE	American Society of Heating Refrigeration and Air-Conditioning Engineers
CTM	Constant Torque Motor
cfm	Cubic Feet per Minute
DAQ	Data Acquisition system
ECM	Electronically Commutated Motor
ESP	External Static Pressure
in w.c.	inches of water column
L	Liter
Pa	Pascal
OSU	Oklahoma State University
RTD	Resistance Temperature Device
T	Temperature
UTT	Unit Under Test
W	Watt

Subscript

DB	Dry-Bulb
WB	Wet-Bulb

REFERENCES

AHRI. (2017). *AHRI Standard 210/240 - Standard for Performance Rating of Unitary Air-conditioning & Air-source Heat Pump Equipment.* Air-Conditioning, Heating, & Refrigeration Institute, [Online: Accessed 11-27-2019]. Available from: `http://www.ahrinet.org/App_Content/ahri/files/STANDARDS/AHRI/AHRI_Standard_210-240_2017.pdf`.

ANSI/ASHRAE. (2009). *ANSI/ASHRAE Standard 37 - Methods of Testing for Rating Electrically Driven Unitary Air-Conditioning and Heat Pump Equipment.* American Society of Heating Refrigeration and Air-Conditioning Engineers, Inc., Atlanta, GA 30329-2305.

ANSI/ASHRAE. (2013). *ANSI/ASHRAE Standard 41.1 - Standard Method for Temperature Measurement.* American Society of Heating Refrigeration and Air-Conditioning Engineers, Inc., Atlanta, GA 30329-2305.

ANSI/ASHRAE. (2014). *ANSI/ASHRAE Standard 41.6 - Standard Method for Measurement of Moist Air Properties.* American Society of Heating Refrigeration and Air-Conditioning Engineers, Inc., Atlanta, GA 30329-2305.

Hossain, M. Y., Bach, C. K., & San, O. (2018). Effect of Inlet Duct and Damper Design on Fan Performance and Static Pressure Measurements (ASHRAE RP 1743). In *Proceedings of the 17th International Refrigeration and Air Conditioning Conference at Purdue* (pp. 2659-1 – 2659-10). Indiana, USA.

ACKNOWLEDGMENT

The research team would like to acknowledge the support of the Project Monitoring Subcommittee (PMS) of ASHRAE RP-1743. The team would also like to acknowledge Md Yeam Hossain for his supports of the project during his time as a graduate student at OSU.

Multi-criteria Evaluation Of River Source Heat Pump Systems In Terms Of Life Cycle Performance

Yujun Jung[1], Hoseong Lee[1]*

[1]Department of Mechanical Engineering, Korea University,
409 Innovation Hall Bldg., Anam-Dong, Sungbuk-Gu, Seoul, Republic of Korea

* Corresponding Author: Tel: 82-02-3290-3355, E-mail: hslee1@korea.ac.kr

ABSTRACT

In this study, the feasibility of thermal energy utilization in river source was investigated by evaluating the performance of a river source heat pump (RSHP) system. The performance of the RSHP system is evaluated with the validated WSHP model and the actual demand of an office building. The system performance is comprehensively analyzed in terms of the life cycle performance, considering energetic, environmental, and economic metrics. The energy consumption is converted to the primary energy consumption (PEC). The environmental impact is represented as CO2 emissions based on life cycle climate performance (LCCP). The economic feasibility is derived in terms of the net present value (NPV), internal rate of return (IRR), and payback period (PBP) based on the life cycle cost (LCC). When the RSHP system was applied to office building, the energy consumption and CO2 emissions were reduced by 6.9% and 10.5%, respectively. The results of the economic feasibility study presented a net present value of $10,587,848, an internal rate of return of 20.5%, and a payback period of 5.3 years. In addition, the WSHP system is evaluated under various water source temperature and utility rate conditions for further investigation. The parametric study showed the maximum improvement of the energy consumption, CO2 emissions, and economics are expected to be 16.4%, 19.6%, and 15.2%, respectively.

1. INTRODUCTION

In recent years, water heat pumps (WSHP) have emerged as one of the ways to reduce building energy consumption. The application of WSHP instead of ASHP may improve performance because the water temperature is relatively constant compared to the air temperature. Because of these advantages of the WSHP, they have been actively studied. In the literature review, it is found out that many studies have been conducted to investigate the performance of the WSHP systems under various water source conditions. However, performance evaluation of river source heat pump (RSHP) systems is rarely addressed. Most of the studies focused on seawater, and groundwater, and did not considered a real building model with validated energy system. Moreover, few studies have considered life cycle performance in terms of energy, environmental and economic metrics when evaluating the performance of RSHP systems. To fill this research gap, current research aims to evaluate the performance of the RSHP systems using validated WSHP models and the actual demand of office buildings. System performance is comprehensively analyzed in terms of lifecycle performance, taking into account energy, environmental, and economic metrics. Energy consumption is converted to Primary Energy Consumption (PEC). The environmental impact is represented by CO2 emissions based on Life Cycle Climate Performance (LCCP). Economic performance is derived in terms of Net Present Value (NPV), Internal Rate of Return (IRR), and Payback Period (PBP) based on Life Cycle Cost (LCC). The WSHP system is also evaluated for further investigation under various water source temperatures and utility speed conditions.

2. SYSTEM DESCRIPTION

The performance of the RSHP system was evaluated by comparing it with the conventional system. Figure 1 shows a schematic diagram of the conventional and RSHP systems. The conventional system consists of an electric chiller, a cooling tower and a gas-fired boiler. The electric chiller and the cooling tower that handle the cooling demand of buildings operate as electricity, and gas-fired boilers that supply the heating demand of buildings operate as natural gas. This generated energy is transferred to the building via the heating and cooling coils of the air handling unit (AHU).

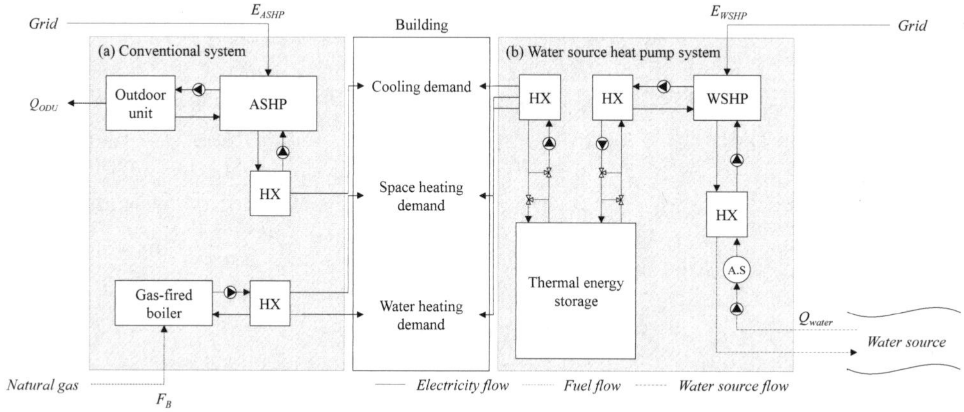

Figure 1: Schematic diagram of conventional system and RSHP system

2.1 Conventional System

The conventional and RSHP system components such as electric chillers, cooling towers, heat exchangers, pumps, heat pumps and thermal energy storage systems were developed as modules and integrated into the system.

The electric chiller is used as a central air conditioning system. It removes heat energy from the building and delivers it to the cooling tower. The electric chiller uses water as a working fluid on the source and load sides. The chiller model was developed using Type 927. The performance map consists of four parameters: inlet source temperature, inlet load temperature, inlet source flow rate, inlet load flow, and inlet load flow. The heat extracted from the chiller is released through the cooling tower. The cooling tower has the advantage of improved cooling performance and stability due to fan use. This model is used to evaporate and cool the liquid flow from the chiller from the outside of the coil. Working fluids in the cooling tower are completely isolated from air and water, and evaporated water is sometimes replenished. This model was designed to the standard specifications of the cooling tower and was developed using Type 510.

Type 751 was used to simulate the transient behavior of gas-fired boilers. The phase change in the working fluid was ignored. The working fluids were assumed to be pure liquids, not vapors or liquid-steam mixtures. The performance of the boiler is defined by its overall efficiency and combustion efficiency. The overall efficiency is the ratio of the output to the input, and combustion efficiency is the ratio of the combustion energy to the input energy. The boiler performance data was obtained from an external data file consisting of inlet liquid temperature and PLR.

The air handling unit consisting of a central air conditioning system transfers the generated cooling and heating energy to the building. The energy generated is transferred to the cooling and heating coils via pumps and heat exchangers, which are supplied to the building via duct fans. The pump is controlled by an inverter. The pump power consumption was calculated taking into account static heads, friction losses and minor losses, and the loss from the equipment pressure in the heat exchanger was ignored. The duct fan is driven by inverter control, and the power consumption is calculated using a flow rate that meets the building demand. The supply air temperature and indoor return temperature were set at 28°C and 22°C respectively in winter and 16°C and 26°C in summer.

2.2 River Source Heat Pump System

A simulation model of WSHP has been developed. A mixture of water and antifreeze was used to prevent freezing in winter; 20% propylene glycol on the supply side and 10% methanol on the load side. The performance map of WSHP was modeled by three performance curves. The PLR in WSHP was designed as a function of water mass flow on each side of the system. The developed heat pump model was validated with experimental data on two parameters: capacity and power consumption. The heating and cooling capacities were validated within 0.98% and 0.39%, respectively, and heating and cooling power consumption within 0.27% and 0.09%. In the developed performance curve of WSHP, the heating COP is continuously increased as the inlet water source temperature increases or the inlet load temperature decreases. Similarly, the cooling COP increases as the inlet water temperature decreases or the inlet load temperature increases.

The transient behavior of thermal energy storage (TES) has been modeled as Type 534. This vertical cylindrical storage tank is a constant capacity storage tank filled with fluid without a submerged heat exchanger. The TES is assumed to be made from concrete under the building during the initial basic construction phase. The storage tank is insulated using a urethane foam. The internal temperature gradient of the storage tank was analyzed by dividing it into five temperature nodes to model the stratification of the storage tank. Each node has a fluid conduction and convection mechanism to interact with the nodes above and below. The convection is caused by forced flow from the inlet flow and natural de-stratification in the tank.

In addition, the energy consumption of the RSHP system was investigated according to the effect of the river water temperature to further improve the performance. The profile of the river water temperature was generated for four cases as shown in Figure 2. Case 1 is the actual river water temperature profile, and case 4 is the underground water temperature. Cases 2–3 denote an arbitrary water temperature having a temperature distribution between the river temperature.

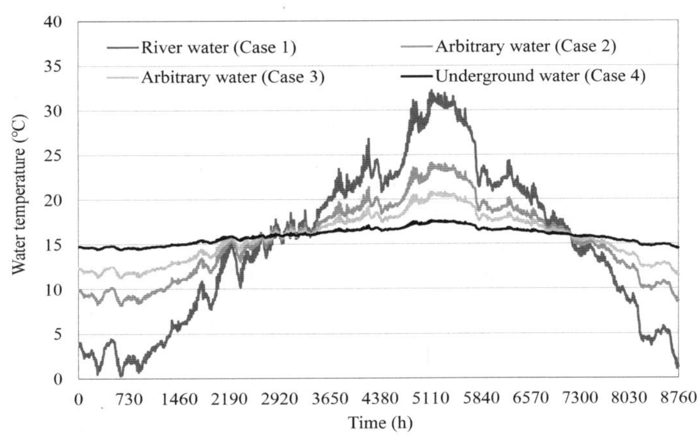

Figure 2: Various water temperature profile

3. RESULTS

The system's energy consumption is calculated from simulations, and performance evaluation is performed in terms of energy, environmental and economic metrics. Figure 3 shows the results of annual energy consumption. The energy consumption of the conventional and RSHP systems was analyzed in three parts: main components, AHU, and additional components. The main components are energy components that can generate cooling and heating energy directly, including electric coolers, gas-fired boilers, and WSHP. The AHU represents the pumps and fans that transfer energy generated from the main components to the building. The additional components include the cooling tower of the conventional system and the water pumping system of the RSHP system. When the RSHP system was compared with the conventional system, the energy consumption of the main component of the former was lower by 9.5%, whereas that of the AHU was nearly equivalent. The energy consumption of the additional component was higher by 24.7% in the open-loop system and 20.1% in the closed-loop system. As a result, the annual energy consumption was lower by 3.4% in the open-loop RSHP system and 6.9% in the closed-loop RSHP system. In the case of the open-loop system, the energy saving effect was not significant compared to what was expected. This is mainly owing to the insufficient energy savings of main component and the energy consumption increase of the additional component.

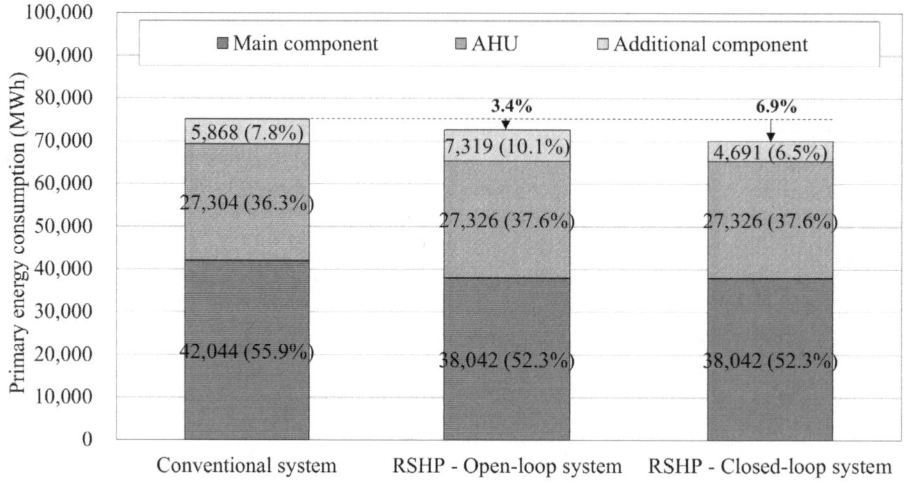

Figure 3: Comparison of annual primary energy consumption

The LCCP evaluation was performed by calculating the annual energy consumption of conventional and RSHP systems. The results of the LCCP evaluation are shown in Figure 4. As the river water temperature approaches to case 4, the total CO2 emissions of the RSHP system were reduced by up to 19.6% compared to the conventional system. Most of the CO2 emissions come from indirect emissions, especially the energy consumption of the system. The direct emission due to direct leakage of refrigerant accounts for less than 1%. The direct emissions from the RSHP system increased due to increased refrigerant charge in the RSHP. However, the indirect emissions have decreased significantly due to reduced energy consumption. This has resulted in a significant reduction in total CO2 emissions due to reduced energy consumption. Therefore, the total CO2 emissions significantly decreased because of the carbon emission reduction from the energy consumption.

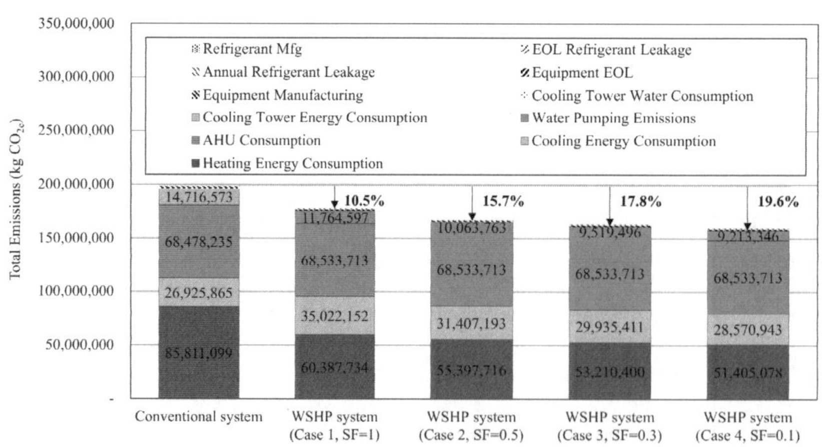

Figure 4: Environmental impact based on LCCP: annual CO_2 emission reduction of closed loop system

The economic feasibility of the RSHP system was evaluated with three economic parameters: NPV, IRR, and PBP. Lifecycle performance was considered in the calculations, which were performed based on the differences between the cost of initial investment, operation, maintenance and disposal in conventional and RSHP systems. In the future, late-night electricity charges may increase if the RSHP system is actively used. Therefore, the economic feasibility was analyzed in various scenarios, including scenario 1: current late night electricity rate, scenario 2: 10% increase, scenario 3: 20% increase, scenario 4: 30% increase. The economic assessment results are shown in Figure 5. The gray bar graph represents the current value, which is the sum of annual expenditure and revenue. The present value of Scenario 1 has been consistently positive, with the exception of initial investment costs, due to reduced energy costs. Annual depreciation has reduced energy cost savings slightly every year. The line graph shows the change in NPV. NPV, which start at the initial investment point, follows an incremental curve. The payback period, which represents the time required to recover the total expenditure, is where the NPV curve meets the x-axis. Consequently, for scenario 1, for current utility rate plans, it is clear that the RSHP system provides sufficient benefits compared to conventional systems as a result of NPV of $10,587,848, IRR of 20.5%, and PBP of 5.3 years. For other scenarios in which utility rates increase, the economic feasibility was derived as follows: NPV of $9,020,429, IRR of 18.5%, and PBP of 5.8 years for scenario 2; NPV of $7,453,009, IRR of 16.4%, and PBP of 6.5 years for scenario 3; and NPV of $5,885,589, IRR of 14.2%, and PBP of 7.3 years for scenario 4. Therefore, the economic feasibility can be maintained even if the utility rates increase in the future.

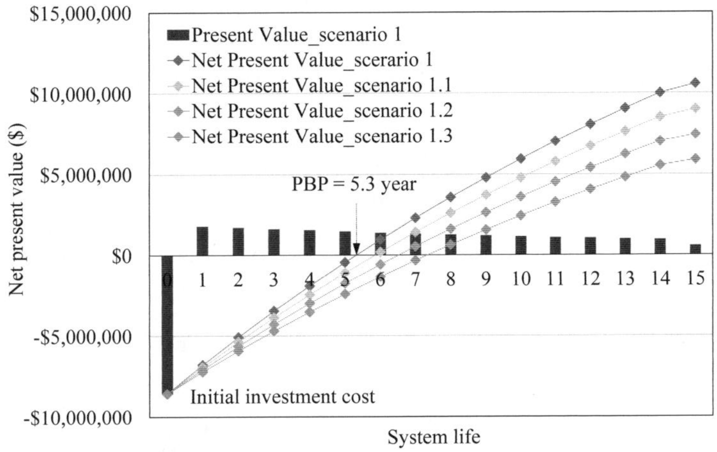

Figure 5: Economic evaluation results based on electricity rate increase scenarios: NPV and PBP

6. CONCLUSIONS

In this study, a comprehensive feasibility study on the RSHP was conducted in terms of life cycle performance evaluation. The remarkable research results are summarized as follows:

- When the closed-loop RSHP systems were used instead of conventional systems, energy savings increased from 6.9% to 16.4% due to changes in water temperature. The open-loop RSHP system has relatively low energy savings due to the power consumption of the primary pump.

- Reducing the power consumption of the hydronics is the first important factor in improving the performance of the RSHP system, as it accounts for a significant portion of the total energy consumption.

- The performance of the RSHP system depends heavily on the energy consumption of the main components and the water pumping system. Therefore, various surrounding conditions that affect the main component and water pumping system should be carefully examined before applying the RSHP system.

- From an environmental perspective, CO2 emissions from the RSHP system decrease from 10.5% to 19.6% as water temperature changes.

- Since the CO2 emissions are mainly due to energy consumption, analysis of energy consumption without complicated methods can lead to a suitable environmental impact assessment.

- From an economic perspective, the economic feasibility is characterized by the following values: NPV of $10,587,848, IRR of 20.49%, and PBP 5.3 years. In addition, the economic feasibility may be maintained even if the applicable temperature changes or utility charges increase in the future.

REFERENCES

Cho Y., Yun R. (2011). A raw water source heat pump air-conditioning system. *Energy Build, 43,* 3068–73.

Chung M., Park HC. (2015). Comparison of building energy demand for hotels, hospitals, and offices in Korea. *Energy, 92*, 383–93.

Epting J., Müller MH., Genske D., Huggenberger P. (2018). Relating groundwater heat-potential to city-scale heat-demand : A theoretical consideration for urban groundwater resource management. *Appl Energy, 22,* 1499–505.

Jung Y., Kim J., Lee H., (2019). Multi-criteria evaluation of medium-sized residential building with micro-CHP system in South Korea. *Energy Build, 193,* 201–15.

Lund R., Persson U. (2016). Mapping of potential heat sources for heat pumps for district heating in Denmark. *Energy, 110,* 129–38.

Min K., Lee W., Yun R., Heo J., (2017). Operational energy saving potential of thermal effluent source heat pump system for greenhouse heating in Jeju. *Int J Air-Conditioning Refrig*.

Mitchell MS., Spitler JD.. (2013). Open-loop direct surface water cooling and surface water heat pump systems-A review. *HVAC R Res, 19,* 125–40.

Oh S., Cho Y., Yun R. (2014). Raw-water source heat pump for a vertical water treatment building. *Energy Build, 68*, 321–8.

Wang G., Wang W., Luo J., Zhang Y. (2019). Assessment of three types of shallow geothermal resources and ground-source heat-pump applications in provincial capitals in the Yangtze River Basin, China. *Renew Sustain Energy Rev, 111,* 392–421.

TRNSYS official web site: http://www.trnsys.com.

ACKNOWLEDGEMENT

This research was supported by the Energy Efficiency & Resources Core Technology Program of the Korea Institute of Energy Technology Evaluation and Planning (KETEP), funded by the Ministry of Trade, Industry & Energy, Republic of Korea. (No. 20182010600010B).

Simulation of Efficiency of Different Configurations of Residential Hybrid Heating Systems Combining Boiler and Heat Pump

Erica ROCCATELLO[1]*, Alessandro PRADA[2], Paolo BAGGIO[2],
Cristian ZAMBRELLI[3], Marco BARATIERI[1]

[1] Free University of Bozen-Bolzano, Faculty of Science and Technology,
Bolzano, Italy
eroccatello@unibz.it

[2] University of Trento, Departments of Civil, Environmental and Mechanical Engineering,
Trento, Italy

[3] Immergas S.p.A.,
Reggio Emilia, Italy

* Corresponding Author

ABSTRACT

The contribution of residential buildings primary energy consumption in the U.S. accounts for more than 20% of the total. In this context, improving the efficiency of heating systems represents a key point for keeping up with the carbon reduction goals. Air-to-water heat pumps are a promising technology that has been deeply studied in recent years. Their main advantage is to improve the share of renewables. Nevertheless, they have some drawbacks, such as the poor efficiency at low ambient temperatures and at high supply water temperatures, necessary e.g. for domestic hot water production and in high temperature heating systems. The latter can be found in a high percentage of the building stock.

Hybrid systems combining heat pumps and boilers (HHPS) can represent a viable solution to overcome these issues. The sizing of the generators and the control logic according to which they are operated play an important role in the performance of the system. For this purpose, comprehensive studies analyzing different types of buildings in different climates and considering both space heating (SH) and domestic hot water (DHW) production are still missing in the literature.

The aim of this work is to identify the best configuration of system, in terms of primary energy consumption, for different heating loads. To do so, a model of the hybrid system is firstly developed by means of a technical computing language (TCL) software. It includes the control logic, that manages the switching of the devices according to the operating conditions. The TCL model has been combined with a dynamic building simulation (DBS). Two different configurations of HHPS were analyzed varying different parameters, such as building thermal insulation, type of heating emission system, DHW demand profile and climate, with a corresponding demand of SH and DHW. The hybrid configurations are compared to a heat pump only solution.

The paper presents the results of the simulations with the aim of identifying the most efficient system configuration for each set of parameters considered, in terms of primary energy consumption.

Keywords: Hybrid system, heat pump, boiler, simulation

1. INTRODUCTION

Both in the US as in the European countries, the consumption of the building sector represents one of the main contributions of the total energy consumption (European Environment Agency, 2018), (European Commission, 2018), (U.S. Energy Information Administration (EIA), 2019). The heat pump technology has seen a rapid growth in the last years due to its favorable characteristics. It represents nowadays the most promising technology for reducing the carbon footprint of heating systems. The system is also appreciated due to its versatility: a heat pump can be applied to produce heating, cooling and DHW in one single appliance.

The efficiency of heat pumps has seen a remarkable growth over the last decades, thanks to the improvements in components and system integration. Air-to-water heat pumps represent nowadays the most widespread type of heat pump installed. However, the efficiency is strongly affected by ambient temperature and supply water temperature. The lower the ambient temperature, the less efficient the heat pump is. The same happens when a high supply water temperature is required. The heat pump should be sized for covering the peak load, thus it will be oversized for the rest of the heating season. Even if inverter driven heat pumps are used, oversized heat pumps still suffer from COP degradation for a large part of the year (Bee *et al.*, 2016).

Moreover, in the Italian building stock, most of the houses were built before energy conservation laws went into effect. Therefore, most of these buildings have poor insulation levels and require high supply water temperature for heating (ISTAT, 2015).

To overcome these problems, one possible solution is the combination of the heat pump with a second heat generator.

Heat pumps are often provided with an auxiliary generator. Commonly they are provided by manufacturers with an electrical heater, that can cover the peak loads and increase the supply water temperature, if required, compromising however the efficiency of the system. The combination of air-to-water heat pumps and gas boilers has also been proposed. The main advantage of providing the system with an auxiliary generator is that it can be operated in those situations where the heat pump works inefficiently. Moreover, the heat pump must not be dimensioned for the peak load of the building, improving its operation in mid-seasons.

Some authors already addressed their research to the study of air-to-water heat pumps in combination with gas boilers. According to several studies hybrid heat pump systems (HHPS) could represent a viable path for the decarbonization of heating systems due to its versatility: the hybrid solution could decrease the peak of electricity demand and lower the requirements for the distribution heating networks. The combination of different technologies can achieve important results in terms of emissions, the cost reduction for users and energy systems (Zhang *et al.*, 2018), (Imperial College, 2018), (Delta Energy & Environment Ltd, 2012).

Park *et al.*(2014) assessed a hybrid system serving the heating and DHW demand of a 100m^2 house. The production temperature is kept fixed at 60°C and the size of the heat pump is also fixed. The results of the study showed a convenience in the adoption of the hybrid system compared to the boiler. Klein *et al.* (2014) analyzed a system composed by a heat pump and a boiler connected in series and a storage tank. The system operates for the space heating of a 132m^2 apartment. Two different building types (and heating loads) were considered in the analysis. The hybrid system showed advantages in terms of cost compared to the heat pump system. Li & Du (2018) analyzed the performance of a hybrid system with fixed heat pump capacity applied to a residential house for space heating and DHW production. According to the author's results, hybrid systems can achieve cost reductions both in space heating (6 to 70%) and DHW production (20 to 65%) compared to boilers. Di Perna *et al.* (2015) applied a model built on experimental data of a hybrid system to a single family house in Milan. They compared the efficiency of the hybrid system with the conventional solutions, such as a heat pump provided with electric heater, a condensing boiler and an oil boiler. They proved that the hybrid system reduces the primary energy need considerably, compared with the other solutions, while the heat pump with the electric backup heater resulted as the worst solution in terms of primary energy consumption. In terms of annual costs, the hybrid solution resulted as the most efficient one. In the analysis of Dongellini *et al.* (2017), the authors tested several combinations of heat pump, boiler and electric heater and applied the system to a 134m^2 single family house, located in the north of Italy. The hybrid solution with an optimal size of the heat pump leads to 2-3% primary energy savings, compared to the only heat pump case, and 32-34% savings compared to the only boiler case.

Even though the results of the previously mentioned studies are interesting for understanding the potential application of hybrid systems, a comprehensive study comparing the performance of a HHPS – providing both space heating and DHW and considering at the same time different climates, different types of building and DHW demand profile – is still missing.

The aim of the present study is to identify in which cases the HHPS can be a convenient solution in terms of primary energy demand. Hence, different building insulation level, heating emission systems, DHW demand profile and climates are considered in the analysis. Following the system sizing for each configuration, the systems were simulated, and the results were compared in terms of primary energy consumption.

2. METHODOLOGY

The assessment of the performance of a hybrid system started with the development of a model of the system. The HHPS model has been developed through TCL, and afterwards it has been combined with DBS. The co-simulation was run varying different parameters, such as building insulation level, heating emission system, DHW demand profile and climate in order to understand the influence of these variables on HHPS performance with respect to a reference heat pump system.

2.1 Hybrid System Model

In this simulation, the components of the hybrid system (heat pump and boiler) are connected in parallel, and the control logic simulates an alternate operation. Exceeded a threshold outdoor temperature, the system switches from boiler to heat pump operation. The ambient air temperature at which the switch between heat pump and boiler occurs has been determined through several simulations for each case. According to the operation scheme, one or two switching temperatures must be chosen:

- In the first operation strategy (OS1), DHW is always provided by the boiler, while space heating is covered by the heat pump if the temperature is higher than the threshold. (Figure 1)
- In the second one (OS2), the heat pump and the boiler provide both space heating and DHW, but with a different optimal switch temperature (Figure 2).

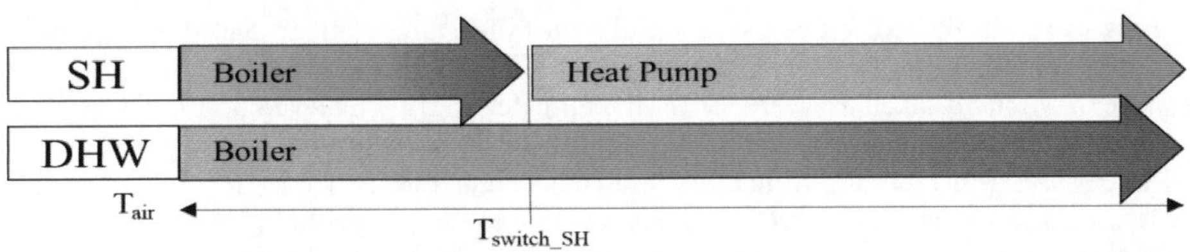

Figure 1: OS1 control logic

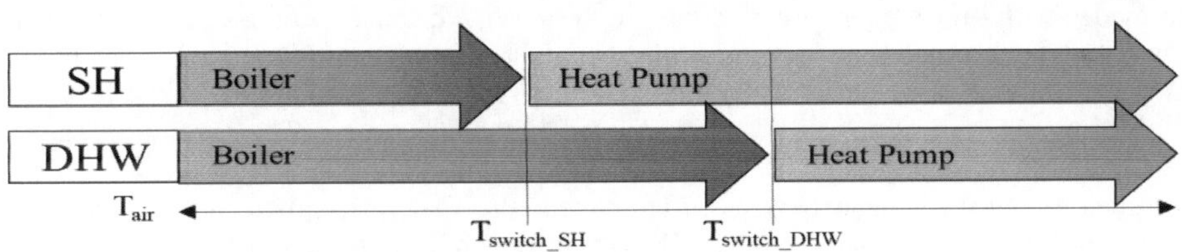

Figure 2: OS2 control logic

The heat pump and the boiler models are based on manufacturer's performance map and consider the performance degradation due to part-load operation.

The boiler is a modulating condensing boiler, with a nominal capacity of 35 kW and a minimum percentage modulation of 10%. The air source heat pump is a R-410A unit, with axial fan and inverter driven compressor, with a capacity ranging from 8 to 30 kW at 7°C air temperature and 35°C supply water temperature and a minimum percentage modulation of 40%.

Several papers in the literature addressed the problem of heat pump defrosting cycles. This is an aspect that strongly affects the performance of air-to-water heat pumps. Zhu *et al.* (2015) proposed a frosting map that helps to identify the most critical conditions for frost formation. In those conditions where frost can occur, the ambient temperature/relative humidity (UR) plane is divided into 4 zones, namely severe, moderate, mild or no risk of frosting. The approach proposed by Zhu and Sun has been applied in the simulation, and when the outdoor conditions fall into one of the risk zones, the heat pump performance is corrected by a degradation coefficient. The estimation of the degradation coefficient for each zone is based on the experimental results obtained by Chen & Guo (2009) on reverse-cycle defrosting of an air source heat pump. The paper presents the effects on the performances of the unit for a wide range of outdoor air temperatures and relative humidities.

Additionally, the implemented model accounts for performance derating during the heat pump startup phase. In this phase, the heat pump cycle operates at different conditions respect to steady-state operation. These deviations result in efficiency losses, that occur at each startup of the unit. For instance, Bagarella *et al.* (2013) described the phenomenon and performed experimental evaluation for the determination of the losses due to a chiller on off cycling. In a more recent work (Bagarella *et al.*, 2016), the authors evaluated the effect of the heat pump sizing and on off cycling phenomenon on the annual energy performance of the system. In the present work a similar procedure was applied to take cycling losses into account. Based on the results reported in the literature, a penalty coefficient has been applied for the first transient period at each startup of the unit.

2.2 Building Demand Simulation

The model of the HHPS described in the previous paragraph has been applied in a DBS (Figure 3). The hybrid system serves the heating request of a DHW tank and a hot water tank devoted to space heating through a primary loop. The pump in the secondary loop is activated whenever there is a space heating request and, consequently, the thermostat signal is on. Heating and DHW supply temperature can be increased, if necessary, by electric backup heaters, which are used as integration only in the reference case, i.e. the only heat pump system, if the set point temperature is not reached. The timestep used in the simulation has been set to five minutes. The building that is taken as a reference is a single-family residential building, composed by two heating zones of approximately $90m^2$ each.

2.3 Simulation Framework

Throughout the simulations different parameters have been varied in order to gain a comprehensive understanding of the efficiency of hybrid systems in different contexts.

- Simulations were run for two cities located in the north of Italy, Belluno and Reggio Emilia, presenting different types of climate. Belluno, with 3.043 heating degree-days (HDD_{20}) and a heating design temperature of -11°C, is taken as reference for the climate 'cold', while Reggio Emilia, with 2.560 HDD_{20} and a design temperature of -7°C, is taken as reference for the climate 'moderate'.
- Two different levels of building insulation were considered. The first one (B1) simulates a building that was built in Italy in the seventies, with a heating demand in the 'moderate' temperature (Reggio Emilia) of 140 kWh $m^{-2}y^{-1}$, while the second one (B2) simulates a building built in the nineties, with a heating demand in Reggio Emilia of 50 kWh $m^{-2}y^{-1}$.
  B1 is provided with low temperature radiators. In this case, the space heating water tank of $0.1m^3$ is kept at a design temperature of 52°C and adjusted by an outdoor reset control (OTR). B2 instead is provided with radiant panels. In this case, the set point temperature of the space heating water tank is set at 40°C at the minimum ambient temperature and it is also adjusted by the OTR curve. The set point of the heat generator water supply is 3°C higher than the set point of the water tank.
- Two profiles of DHW demand have been simulated: the first one, called the 'L' profile, with a daily consumption of 0.25 m^3, the second one, 'XXL' profile, simulates a larger consumption: 0.55 m^3 per day. The DHW is provided at 45°C and the DHW tank is kept at a temperature of 55°C. The volume of DHW tank considered is $0.3m^3$ for simulations with the 'L' profile, and $0.5m^3$ for the XXL profile.

For each case that was mentioned, a series of simulations were run in order to determine the best temperature at which the logic switches between boiler and heat pump for SH (T_{switch_SH}). In addition, a series of simulations were run for the determination of the best temperature to switch between boiler and heat pump, in case of hot water demand (T_{switch_DHW}).
It has been checked that the simulations with different configurations and same set of parameters provided the equal amount of heat to the system and that the set points of SH and DHW have been fulfilled.

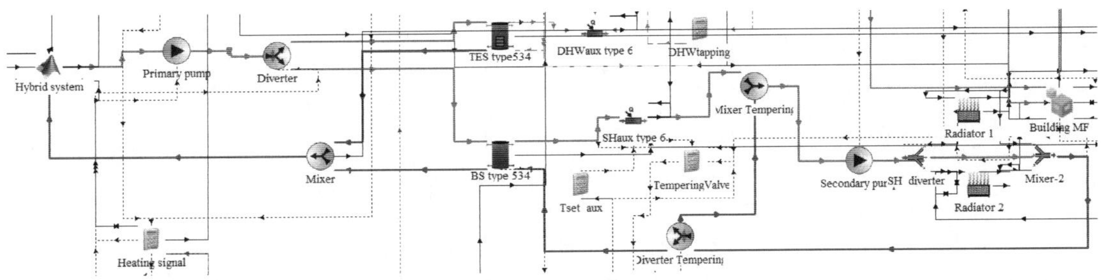

Figure 3: co-simulation layout

3. RESULTS

The simulations have been organized as shown in Table 1. 'Case 1' identifies the results obtained with the simulation of the less insulated building (B01), considering the colder climate and the DHW demand profile 'L'. Case number 2 identifies the results of the simulation for the same parameters, but in the moderate climate. Case number 3 and 4 instead, show the results of the simulations conducted for the newer and more insulated building, for the cold and moderate climate and DHW demand profile 'L'. Case number 5 and 6 compare the results obtained with the set of parameters of case 3 and 4, but with a larger profile of DHW demand (XXL).
For each case, the primary energy required by the heat pump (reference case) has been compared with the ones obtained by OS1 (Figure 1) and OS2 (Figure 2).
Figure 4 – Figure 6 show how T_{switch_SH} and T_{switch_DHW} have been determined. A series of simulations have been performed with different values of T_{switch_SH} and T_{switch_DHW} for the determination of the value minimizing the primary energy (P.E.).
Figure 4 shows the results for the cold climate (case 1 and 3). The simulation performed with building B1 provided with radiators showed an optimal T_{switch_SH} of 6°C, while the simulations with B2 and radiant panels showed an optimal T_{switch_SH} of 4°C. Similar results were obtained for the moderate climate (Figure 5). The reason for that can be found in the set point temperature of the system, that is lower when building B2 and radiant panels are applied. This leads to a lower optimal T_{switch_SH} of 4°C compared with 6°C, obtained with building B1 provided with radiators. A series of simulations with the only DHW demand determined T_{switch_DHW} for the L and XXL profile, showing an optimal value of 6°C in both cases (Figure 6). The same value of T_{switch_SH} determined in case 3 and 4 has been applied for case 5 and 6, as the set of parameters selected for SH is the same.
The heat pump size and the DHW tank have been adapted in order to obtain a similar performance of the configurations, i.e. to fulfil the heating and DHW setpoints in each configuration. The size of the DHW tank selected is 0.3m^3 for simulations with DHW profile 'L', and 0.5 m^3 for DHW profile 'XXL'. Due to the large modulation range of the boiler, only one boiler size was used.
Table 2 summarizes the results of the simulations for case 1 to case 6. It reports the size of the heat pump selected (Pref), at the standard working conditions of 35°C of supply water temperature and 7°C of ambient air temperature, as well as the savings obtained with OS1 and OS2.
To avoid a system oversizing, the reference case is provided with backup electric heaters. The heat pump is sized to cover at least 98% of the annual energy demand.

Table 1: Overview of simulation parameters

Case Number	Building type	Climate	DHW Profile	DHW/ total demand
1	B01	Cold	L	10%
2	B01	Moderate	L	12%
3	B02	Cold	L	25%
4	B02	Moderate	L	30%
5	B02	Cold	XXL	40%
6	B02	Moderate	XXL	46%

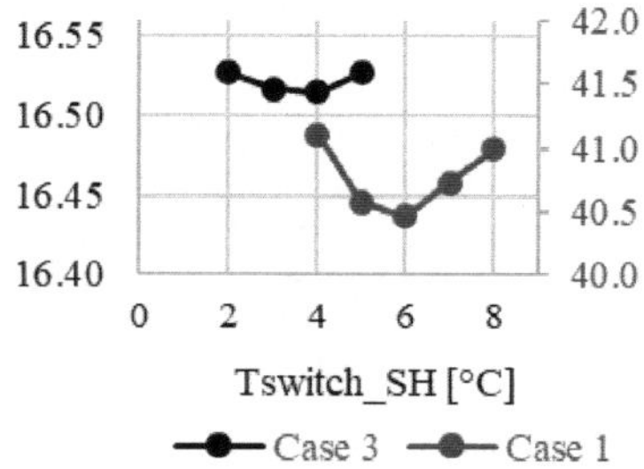

Figure 4: P.E. [MWh] of OS1 at different T_{switch_SH} for the cold climate (case 1 and 3)

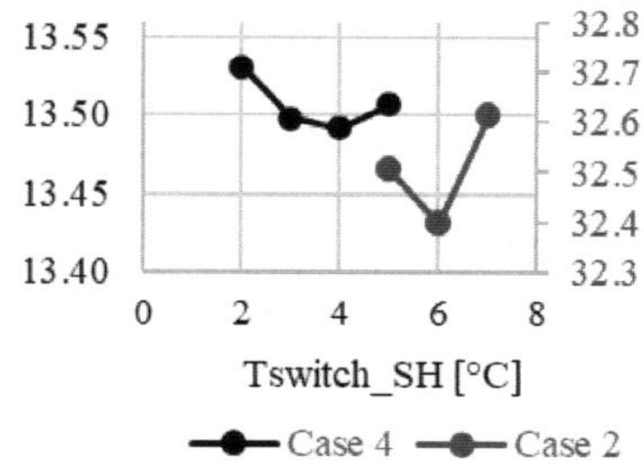

Figure 5: P.E. [MWh] of OS1 at different T_{switch_SH} for the moderate climate (case 2 and 4)

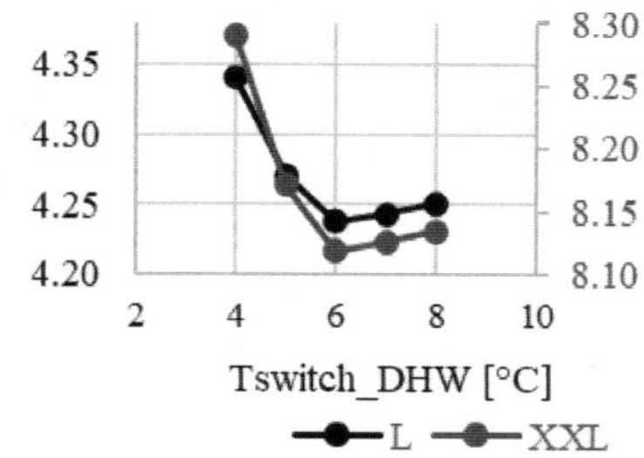

Figure 6: P.E. [MWh] for determination of the optimal T_{switch_DHW} (L and XXL profile)

Table 2: Heat pump capacity and primary energy savings in comparison to the reference case for case1- case 6

CASE 1	REF. CASE	OS1	OS2
Pref [kW]	28.5	13	20
P.E. saving	0.0%	29.6%	30.4%

CASE 2	REF. CASE	OS1	OS2
Pref [kW]	28.5	11	20
P.E. saving	0.0%	22.9%	24.1%

CASE 3	REF. CASE	OS1	OS2
Pref [kW]	25	8.5	16
P.E. saving	0.0%	9.2%	12.1%

CASE 4	REF. CASE	OS1	OS2
Pref [kW]	23	8.5	15
P.E. saving	0.0%	0.0%	4.6%

CASE 5	REF. CASE	OS1	OS2
Pref [kW]	30	8.5	25
P.E. saving	0.0%	9.8%	14.0%

CASE 6	REF. CASE	OS1	OS2
Pref [kW]	30	8.5	25
P.E. saving	0.0%	1.3%	6.5%

Case number 1 shows the results for the less insulated building provided with low temperature radiators, for the cold climate. The result of the comparison shows that the hybrid system can lead to 30% savings in terms of primary energy. In this case, the advantage of having a more complex control logic, i.e. OS2, leads only to small primary energy savings (+1%) compared to OS1.
Case number 2 provides the results for the less insulated building provided with low temperature radiators in the moderate climate. The savings that the hybrid system can achieve reach almost 25% with OS2. As in the previous case, the savings achieved with the OS2 configuration are approximately 1% higher compared to the OS1 configuration.
Considering case 3, the comparison between the reference case and the most favorable configuration of hybrid system (OS2) shows more than 10% of primary energy savings. The additional savings compared with the hybrid configuration OS1 reach almost 3%.
In case 4, the total primary energy required for the reference case is reduced compared to the previous case, as well as the savings that can be obtained through the application of a hybrid system. OS2 achieves savings of almost 5%, while no savings are achieved by OS1.
Case 5 and 6 see an increase of savings obtained with OS1 and OS2 compared to case 3 and 4, due to the application of larger DHW demand profiles. In case 5 the savings obtained with OS2 reach 14% and exceed of more than 4% the savings obtained with OS1. In case 6 the savings obtained with OS2 almost reach 7% and exceed of more than 5% the savings obtained with OS1.

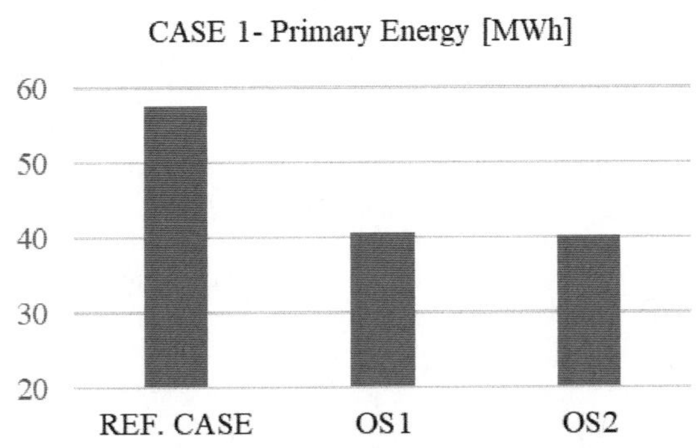

Figure 7: Comparison of primary energy consumption of heat pump only and two hybrid configurations- Results for case 1

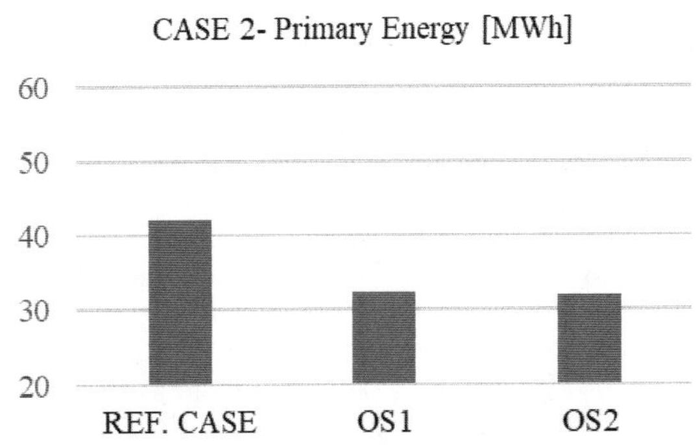

Figure 8: Comparison of primary energy consumption of heat pump only and two hybrid configurations- Results for case 2

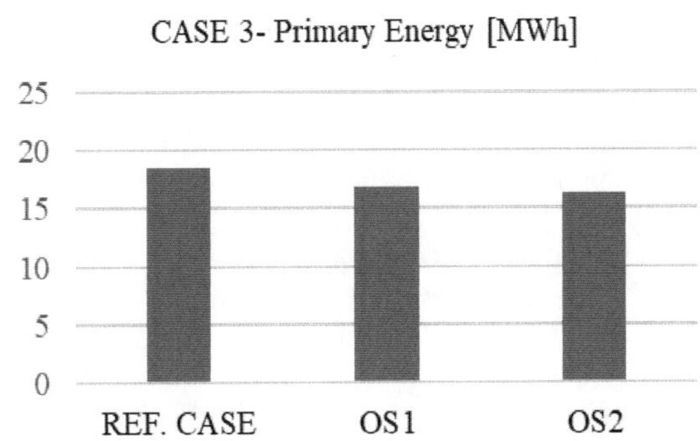

Figure 9: Comparison of primary energy consumption of heat pump only and two hybrid configurations- Results for case 3

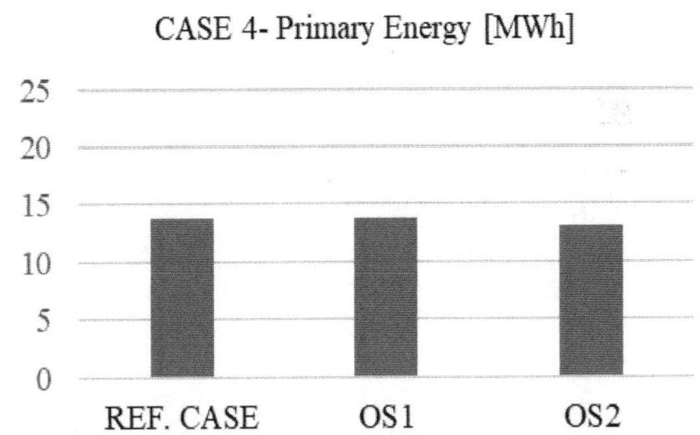

Figure 10: Comparison of primary energy consumption of heat pump only and two hybrid configurations- Results for case 4

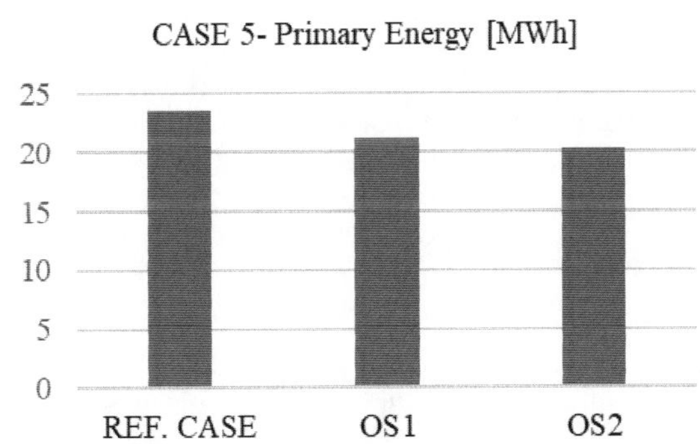

Figure 11: Comparison of primary energy consumption of heat pump only and two hybrid configurations- Results for case 5

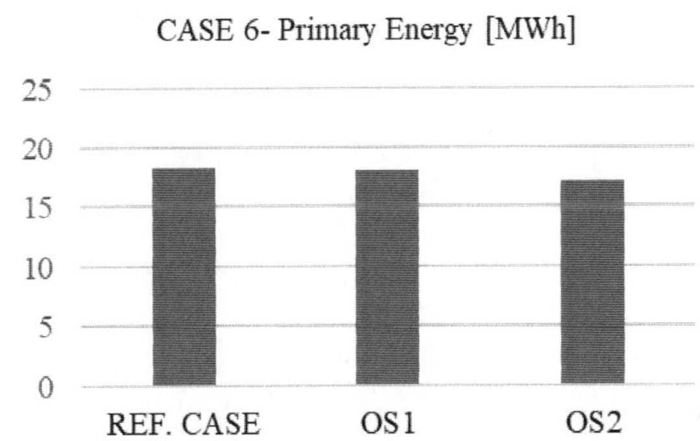

Figure 12: Comparison of primary energy consumption of heat pump only and two hybrid configurations- Results for case 6

3.1 Discussion

The six cases presented in the previous paragraphs show a high variance in the results, i.e. in the savings that the hybrid solution can obtain in terms of primary energy.
The hybrid solution shows remarkable benefits in those simulations involving the less insulated building provided with radiators, considering both the climate 'moderate', and even more the climate 'cold'.
The comparison between the two configurations of hybrid system analyzed shows that the difference between the solution where a switch temperature is applied also to the DHW demand (OS2) and the solution where the boiler covers the total demand of DHW (OS1) does not differ substantially in terms of total primary energy consumption. This is true especially for low insulated buildings, where the savings between the two hybrid solutions are approximately of 1%. If the heat pump contributes to the DHW production, the implemented regulation required for the system is more complex and the size of the heat pump must be larger, which is probably not convenient if the saving is low. Moreover, if the heat pump is oversized compared to the space heating load, it will reduce the optimal functioning of the system, causing more frequent on-offs, thus affecting the performances and also the compressor's life.
In the cases where the simulations have been performed with the higher insulation building provided with radiant panels (case 3 and 4), the portion of primary energy that can be saved with a hybrid system is reduced compared to the case of the building provided with radiators (case1 and 2), due to the lower energy demand of the building and the lower set point of the system, which is advantageous for the heat pump efficiency. Savings however reach 12% when the simulation is performed for the climate 'cold'. The convenience of OS2 in comparison with OS1 increases as much as the weight of the DHW on the total energy demand increases (Table 1). This tendency can be seen in case 3 and 4, and especially in case 5 and 6, where the additional savings of OS2 respect with OS1 reach more than 5%.
According to these results, the hybrid solution has to be highly recommended in buildings that are provided with radiators. In refurbishments, radiators are rarely substituted with radiant panels, due to the large amount of work and high cost required. The usage of low temperature radiators gives the heat pump the possibility to reach the required supply water temperature and so to be integrated into the system. However, the heat pump only solution is not convenient in terms of primary energy demand. In a moderate or cold climate, as the ones considered in the simulations, there are many periods during the year in which the efficiency of the heat pump is reduced, due to the level of supply water temperature required. The low ambient temperature and the occurring of defrost cycles reduce the performances of the heat pump, while the boiler efficiency is not affected by external conditions. The boiler instead, at the supply water temperature levels considered for low temperature radiators, enhances its efficiency through condensation. Nonetheless, the application of a heat pump, properly integrated in a hybrid system, can reduce the primary energy consumption, and allows for the usage of a portion of renewable energy for the heating needs.
The exclusive operation of the boiler for the load of DHW (OS1), leads to the possibility of choosing a size of heat pump that suites best the building heating needs, avoiding an oversizing of the heat pump. An aspect that was not considered in the simulation but should not be neglected is legionella prevention. The boiler as backup generator gives the possibility to keep the DHW tank at a higher temperature level, or allows for operating cycles preventing legionella formation, in a more efficient way compared to the heat pump in combination with electric heaters.
Performing a cost evaluation of the proposed solutions is necessary to evaluate if the benefits in terms of primary energy savings are confirmed in economic terms.

4. CONCLUSIONS

The analysis described in the present paper has been conducted with the aim of identifying the convenience of the application of HHPS in terms of primary energy, by varying different parameters, such as building insulation level, type of heating terminal, DHW demand profile and climate.
To do so, a co-simulation of a hybrid system serving space heating and DHW demand of a residential building was developed.
The analysis has compared the results obtained by applying the hybrid system or the heat pump, for serving the demand of SH and DHW. Two different configurations of hybrid system were compared: OS1, in which the boiler provides the total DHW load, and the switch between boiler and heat pump for SH is determined by a pre-defined value of the ambient temperature (T_{switch_SH}); OS2, in which two temperatures of switch have been determined, one that is adopted when the system operates in DHW mode (T_{switch_DHW}), and one when the system operates in space heating mode (T_{switch_SH}).

Two levels of building insulations were compared, referring to buildings constructed in different ages. The less insulated one (B1) is provided with low temperature radiators, while the more insulated one (B2) is provided with radiant panels. Two cities in the north of Italy have been chosen as reference for a 'cold' and 'moderate' climate.
The results of the simulations show that the savings that can be obtained with the hybrid solution compared to a heat pump only solution, considering building B1, provided with radiators, range from 24% to 30%, for the moderate and cold climate respectively. What can be noticed is that the primary energy demand obtained with the OS1 exceeds that of the OS2 only by approximately 1%. Since the production of DHW requires higher heat pump capacities and more complex regulation, the most advisable solution in this case is OS1. Instead, when the ratio of DHW energy demand compared to the total energy demand increases, the convenience of the OS2 is increased compared to OS1. This is true in particular for case 5 and 6, where the savings of OS2 compared to OS1 increase up to 5%.
The savings that can be obtained with the application of hybrid systems in newer homes provided with radiant panels range from almost 5% to 12% for moderate and cold climates respectively.
The results of the simulations give a general view of the application of HHPS. Hybrid systems can be efficiently applied in renovations, where the heating emission system is not changed, and buildings are still provided with radiators. In this case, the installation of a hybrid system would lead to primary energy savings, and at the same time allows for the usage of renewable energy. If the boiler provides the DHW demand (OS1), the heat pump can be dimensioned to better suite the building needs. This configuration should be preferred when the ratio of DHW to total energy demand is low. The configuration in which the heat pump contributes to cover the DHW loads should be taken into consideration for large DHW/ total energy demand ratios.
This paper evaluated the performances of the configurations based on the primary energy consumption. The purpose of future studies is to include a cost evaluation, in order to have an overview of the most convenient solutions also in economic terms.
Moreover, the present study focused on the parallel configuration of the generators, which is the most widespread one at the moment. Future studies will consider the performance of different configurations of hybrid systems, such as the configuration in series.

REFERENCES

Bagarella, G., Lazzarin, R., & Noro, M. (2016). Sizing strategy of on-off and modulating heat pump systems based on annual energy analysis. *International Journal of Refrigeration, 65*, 183–193. https://doi.org/10.1016/j.ijrefrig.2016.02.015

Bagarella, Giacomo, Lazzarin, R. M., & Lamanna, B. (2013). Cycling losses in refrigeration equipment: An experimental evaluation. *International journal of refrigeration, 36*, 2111–2118 https://doi.org/10.1016/j.ijrefrig.2013.07.020

Bee, E., Prada, A., & Baggio, P. (2016). Variable-Speed Air-to-Water Heat Pumps for Residential Buildings: Evaluation of the Performance in Northern Italian Climate. *Proceedings of the 12th REHVA World Congress-CLIMA 2016, Aalborg, Denmark.*

Chen, Y.-G., & Guo, X.-M. (2009). Dynamic defrosting characteristics of air source heat pump and effects of outdoor air parameters on defrost cycle performance. *Applied Thermal Engineering, 29*, 2701–2707. https://doi.org/10.1016/j.applthermaleng.2009.01.003

Delta Energy & Environment Ltd. (2012). *2050 Pathways for Domestic Heat Final Report*. Retrieved from www.delta-ee.com.

Dongellini, M., Morini, G. L., & Impalà, V. (2017). Design rules for the optimal sizing of a hybrid heat pump system coupled to a residential building. *Proceedings of the 16th International Conference on Sustainable Energy Technologies –SET2017, Bologna, Italy.* Retrieved from https://www.researchgate.net/publication/322083510

European Commission (2018). EU energy statistical pocketbook and country datasheets | Energy. Retrieved September 19, 2019, from https://ec.europa.eu/energy/en/data-analysis/energy-statistical-pocketbook

European Environment Agency (2018). Final energy consumption by sector and fuel. Retrieved September 11, 2019, from https://www.eea.europa.eu/downloads/101f9a6f85654a728f3785b567cf90bc/1547715893/assessment-4.pdf

Imperial College, L. (2018). Analysis of Alternative UK Heat Decarbonisation Pathway, Report for the Committee on Climate Change, August 2018. Retrieved September 10, 2019, from https://www.theccc.org.uk/wp-content/uploads/2018/06/Imperial-College-2018-Analysis-of-Alternative-UK-Heat-Decarbonisation-Pathways.pdf

ISTAT (2015). Annuario Statistico Italiano | 2015. Retrieved September 21, 2019, from

https://www.istat.it/it/files/2015/12/C18.pdf
Klein, K., Huchtemann, K., & Müller, D. (2014). Numerical study on hybrid heat pump systems in existing buildings. *Energy and Buildings*. https://doi.org/10.1016/j.enbuild.2013.10.032
Li, G., & Du, Y. (2018). Performance integration and economic benefits of new control strategies for heat pump-gas fired water heater hybrid system. *Applied Energy*, *232*(August), 101–118. https://doi.org/10.1016/j.apenergy.2018.09.065
Park, H., Hwan Nam, K., Hyun Jang, G., & Soo Kim, M. (2014). Performance investigation of heat pump-gas fired water heater hybrid system and its economic feasibility study. *Energy & Buildings*, *80*, 480–489. https://doi.org/10.1016/j.enbuild.2014.05.052
Perna, C. Di, Magri, G., Giuliani, G., & Serenelli, G. (2015). Experimental assessment and dynamic analysis of a hybrid generator composed of an air source heat pump coupled with a condensing gas boiler in a residential building. *Applied Thermal Engineering, 76,* 86-97. https://doi.org/10.1016/j.applthermaleng.2014.10.007
U.S. Energy Information Administration (EIA) (2019). Total Energy Annual Data - Retrieved January 11, 2021, from https://www.eia.gov/totalenergy/data/annual/
Zhang, X., Strbac, G., Teng, F., & Djapic, P. (2018). Economic assessment of alternative heat decarbonisation strategies through coordinated operation with electricity system – UK case study. *Applied Energy*, *222*(April), 79–91. https://doi.org/10.1016/j.apenergy.2018.03.140
Zhu, J. H., Sun, Y. Y., Wang, W., Deng, S. M., Ge, Y. J., & Li, L. T. (2015). Developing a new frosting map to guide defrosting control for air-source heat pump units. *Applied Thermal Engineering*, *90*, 782–791. https://doi.org/10.1016/j.applthermaleng.2015.06.076

ACKNOWLEDGEMENTS

The authors wish to thank Clara Ceccolini and Nicola Franzoi for the help in the development of the dynamic building simulation model.

Model based co-simulation platform for integrated building system control and design optimization

Jian SUN[1]*, Bo SHEN[2]

[1]Oak Ridge National Laboratory, Grid-Interactive Controls,
Oak Ridge, TN, USA
Sunj2@ornl.gov

[2]Oak Ridge National Laboratory, Building Equipment Research,
Oak Ridge, TN, USA
shenb@ornl.gov

* Corresponding Author[1]

ABSTRACT

Both steady-state and dynamic simulations have been widely used by HVAC&R industry to support product/equipment development for decades. Steady-state simulation focuses on the system mass, energy and momentum balance of an equilibrium state. It is based on high-fidelity components models, and thus is suitable for system and component design optimization. Dynamic simulation studies the system transient response and is generally used for controls development and verification. It usually does not require rigorous component models of high accuracy because 1) the commonly used PID control is feedback control whose control performance evaluation doesn't require high fidelity system/plant model; 2) high-fidelity dynamic model significantly increases the number of equations and variables and creates tremendous challenge for math solver.

For supervisory control, transactive control or optimization of an integrated building system, the HVAC&R equipment is often one of the sub-components to be controlled. High-fidelity equipment models are required for accurately evaluating control strategies. In addition, building equipment manufacturers have developed a lot of high-fidelity steady-state equipment/component models per their expertise. Thus, a platform that can integrate OEM high-fidelity steady-state model with dynamic building simulation and/or electric power system & grid simulation to support the development and verification of supervisory control for integrated building systems is necessary.

In this study, ORNL's heat pump design tool (HPDM) is utilized to develop a co-simulation platform for supervisory control and optimization in integrated building systems. It is based on a model that integrates high-fidelity steady-state simulation equipment models with dynamic building simulation. A practical case of using the proposed co-simulation platform to develop and evaluate the supervisory control and optimization is presented and discussed.

[1] This manuscript has been authored in part by UT-Battelle, LLC, under contract DE-AC05-00OR22725 with the US Department of Energy (DOE). The US government retains and the publisher, by accepting the article for publication, acknowledges that the US government retains a nonexclusive, paid-up, irrevocable, worldwide license to publish or reproduce the published form of this manuscript, or allow others to do so, for US government purposes. DOE will provide public access to these results of federally sponsored research in accordance with the DOE Public Access Plan (http://energy.gov/downloads/doe-public-access-plan)

1. INTRODUCTION

Both steady state simulation and dynamic simulation have been widely used for supporting product/equipment development by HVAC&R industry for decades (TRNSYS 18, 2018, EnergyPlus 2020). The steady state simulation focus on the system mass, energy and momentum balance of an equilibrium state, is based on high fidelity components models, and suitable for system and component design and optimization. Dynamic simulation studies the system transient response and is generally used for controls development and verification which doesn't require rigorous component models with high accuracy for two reasons: 1. Actuator-level Controls are generally feedback control whose control performance evaluation doesn't require high fidelity system/component model, 2. High fidelity dynamic model significantly increases the number of equations and variables and create tremendous challenge for math solver.

The current whole building annual energy simulation rarely use the detailed steady-state simulation and dynamic simulation approaches. The high fidelity steady state simulation is computationally expensive and slows down the annual energy simulation significantly. The dynamic simulation is usually not necessary to whole building annual energy simulation since the building equipment dynamics (seconds) is much faster than building dynamics (minutes) and can be negligible. Thus, most widely used whole building annual energy simulation software/platforms, such as EnergyPlus, eQUEST ... etc., are preferring to use either curve-fitting empirical or semi-thermodynamic modeling approach for building equipment simulation. However, the accuracy of these modeling strategies highly depends on quality and quantity of the data which are usually not available or often spare with limited quality.

In addition, there is increased need for evaluate the supervisory or detailed control strategies in short period or annual simulation. For example, integrated building system grid response transactive control or optimization study requires high fidelity building equipment models to allow accurate evaluate and validate the benefits of the developed control strategies. Another example is that a detailed control needs be evaluated in annual simulation to allow the integrated building manufacture to optimal design building HVAC&R system and power supply system. Moreover, building equipment manufacturers have developed quite a lot of high fidelity steady state equipment/component models per their expertise which have not been utilized or integrated into any whole building annual energy simulation. Thus, a commonly used platform, which can integrate high fidelity equipment steady state model , either from original equipment manufacturers (OEMs) or self-developed, with dynamic building simulation and/or electric power system & grid simulation while maintain reasonable computational cost to support integrated building system supervisory/detailed level control development and verification, is necessary.

To address this gap, this study utilizes the ORNL heat pump design tool (HPDM) (DOE/ORNL, 2019) to develop an integrated building system model based supervisory control and optimization co-simulation platform to integrate high fidelity steady state simulation equipment models with dynamic building simulation. This work intends to integrate the HPDM library to the Modelica modelling environment in a dynamic link library, to facilitate co-simulation between detailed build equipment models and other building elements. A practice case of using this co-simulation platform to develop and evaluate supervisory level integrated building system control and optimization is presented and discussed.

2. DEVELOPMENT OF MODEL BASED CO-SIMULATION PLATFORM

The model-based co-simulation platform developed in this study is a closed loop modeling and simulation to integrate the whole building and building equipment simulation with the control strategies. This platform can be used to design, develop, and evaluate the system design and control strategy before actual implementation and testing on the real system, which allows to quickly detect the controls error during evaluation and design stage. As shown in Figure 1, this platform is composed of four modules: simulation design, control strategy modeling, whole building simulation, and simulation results & virtualization.

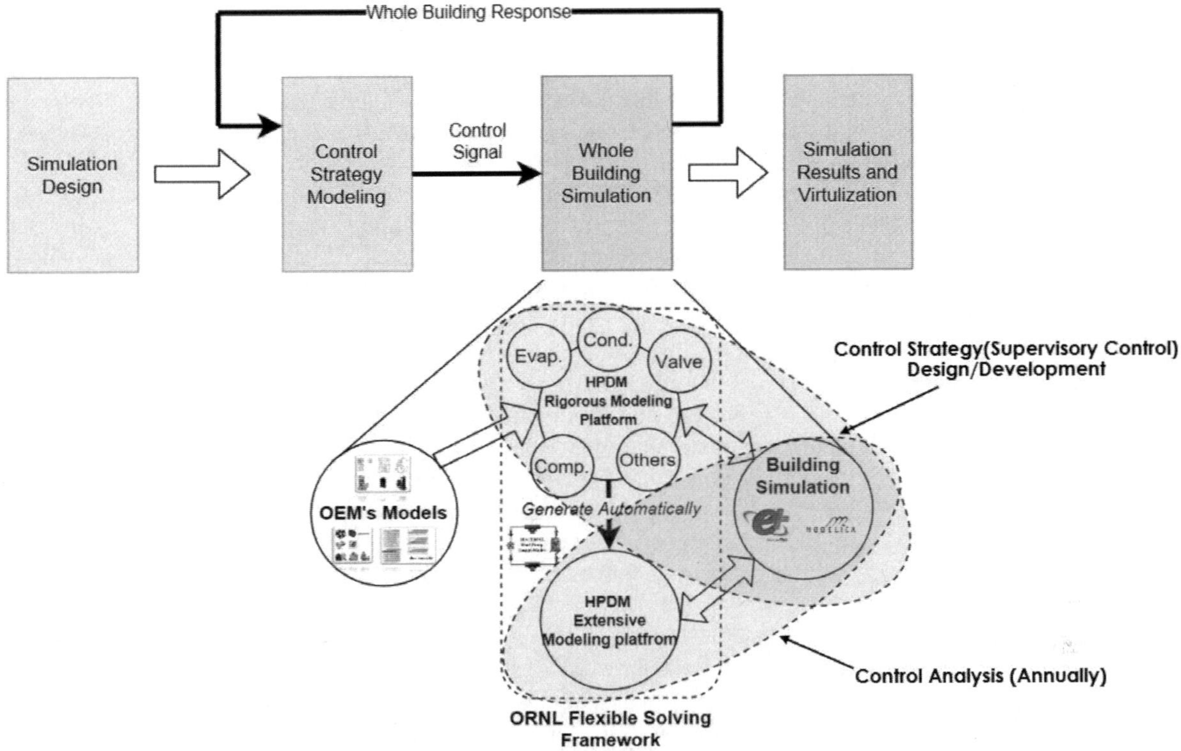

Figure 1: Schematic diagram of model based co-simulation platform

2.1 Simulation Design

Simulation design is to define a series of simulation scenarios or user cases to automatically exam different control and design strategies according to predefined test requirements and targets. It will generate input signals to the control strategy modeling modular to allow closed loop simulation between control module and building module.

2.2 Control Strategies Modeling

A model representing the control logic will be developed based on control targets which define how the building and building equipment should be operated. This control model can be simulated in closed loop with the models of building and building equipment to enable rapidly and early identification of missing or inconsistent in control performance.

2.3 Whole Building Simulation

As the core of the model based co-simulation platform, high fidelity steady state and /or dynamic modeling/simulation engine of the building and building equipment are developed to represent the actual building and building equipment so that the design and control strategy can be tested and verified. This whole building simulation module is composited of three main components: building simulation, ORNL flexible solving framework, and OEM/s models.

2.3.1 Building Simulation

A dynamic model of building is created from publicly available buildings modeling, such as EnergyPlus (DOE, 2020), and building library (DOE/LBNL 1998, Wetter, et al., 2014), to simulate the dynamic behavior of the building, including building envelop, building indoor air, and weather conditions, corresponding to control inputs/outputs.

2.3.2 ORNL Flexible Solving Framework

High fidelity building equipment models are developed based on a Modelica-based equipment library and a flexible solving framework to allow integration of component simulation models, either self-developed or OEM's, into system simulation. Two modeling platforms are included this flexible solving framework,

- HPDM rigorous modeling platform: a public-domain building equipment design and modeling tools, developed and maintained by ORNL. It is a hardware-based equipment design tool promotes energy saving technologies, and provides public knowledge-base and technical support to transfer new technologies from ORNL to U.S. industry. It has a flexible component-based platform to model extensive building equipment types and complicated configurations, and comprehensive library includes information on most HVAC&R components, and many refrigerants and working fluids. In the latest version (DOE/ORNL, 2019), the capability is greatly enhanced to model extensive equipment types, and configurations and new refrigerant properties, e.g., HFO refrigerant blends and natural refrigerant like CO2. It can simulate all residential and commercial air conditioners, heat pumps, water heating and refrigeration systems, including vapor compression systems having multiple compressors and heat exchangers, and multi-split variable-refrigerant flow (VRF) systems. Regarding heat exchangers, the HPDM component model library includes fin-and-tube coil, microchannel heat exchanger for air-to-refrigerant; tube-in-tube, brazed plate, shell-and-tube, wrapped-tank coil for water-to-refrigerant. The heat exchanger models (based on a segment-to-segment approach) accept actual coil circuitry and dimensions and can represent heat transfer and pressure drop scaling factors and correlations obtained from experiments. Regarding the flow moving components, e.g., compressors and fans, HPDM correlates the performance using maps from manufacturers.
- HPDM extensive modeling platform: a reduce order building equipment modeling library through training comprehensive simulated performance data generated HPDM rigorous modeling platform using model reduction techniques, such as principal component analysis, regression analysis, machine learning based methods.

2.3.2 Original Equipment manufacturer (OEM) Models

One of the main advantages of this developed model-based co-simulation platform is to provide a quick and easy way to integrate OEM or third party simulation models/engines into this co-simulation platform through ORNL flexible solving framework.

2.4 Simulation Results and Virtualization

The simulation results will be generated through closed loop simulation and be compared with predefine desirable control performance to determine if the system and control design achieve the targets.

3. CASE STUDY

A single family house with air source integrated heat pump system is selected to demonstrate the utilization of the proposed model based co-simulation platform to support the integrated building system control. As shown in Figure 2, the model based co-simulation platform includes four main modules: simulation test module, supervisory control and design module, single family house and heat pump system module, simulation results module. The supervisory control modular receives the responses from both building and heat pump system, and generates control inputs for controlling heat pump system. The details of the building and heat pump system modeling are described as following,

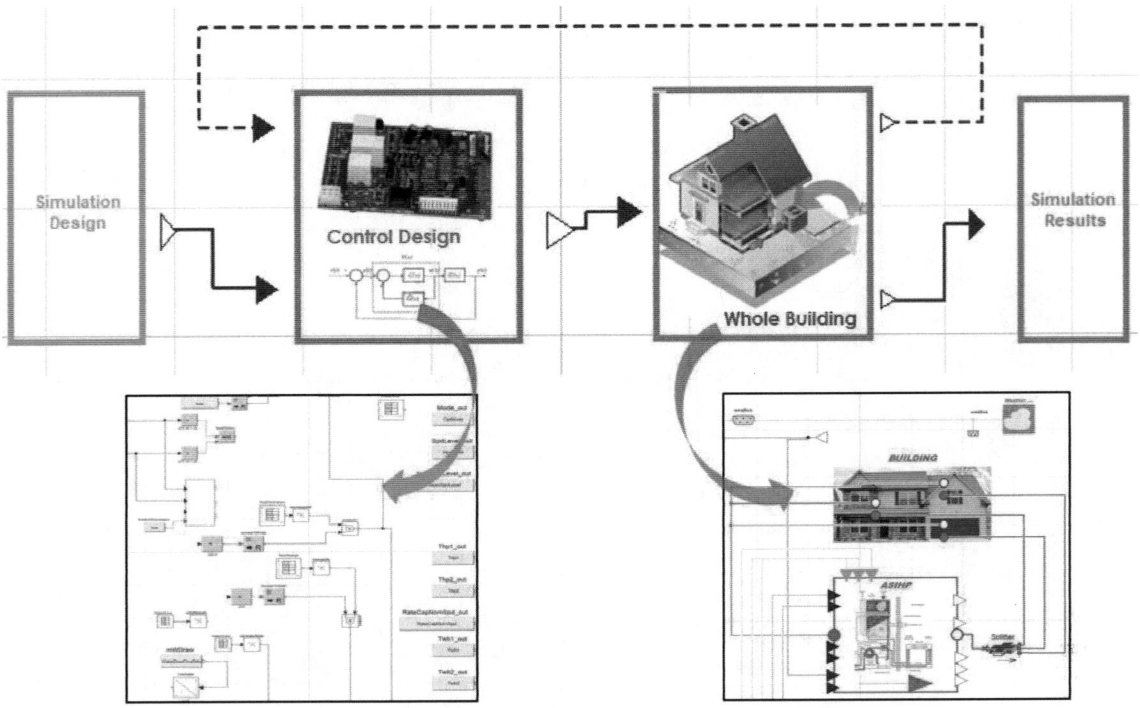

Figure 2: Model based co-simulation platform

A single family house of total 2685 square feet floor area is selected to perform the case study. For simplification, this house is divided as three thermal zones: one represents four identical bed/bath rooms, second zone is combined living /kitchen zone, and the third is sitting zone. Some envelope characteristics are given below:

- Walls: 2x6 steel frame 16" O.C., R-21 fiberglass cavity insulation + 2" XPS (R-10 continuous insulation)
- Roof: 12" steel truss 16" O.C., R-30 fiberglass cavity insulation + 3" polyiso base insulation + 0-7" polyiso tapered insulation
- Floor: 2x6 steel frame 24" O.C., R-15 fiberglass cavity insulation
- Windows
    - i. Living Room/Kitchen/Entry (23.8% window-to-wall ratio)
        1. (1) Fixed patio panel window (Type D): 46.8 sf, U-0.27, SHGC-0.21 (shaded by 6' balcony overhang)
        2. (1) Patio Swing Door: 22.7 sf, U-0.25, SHGC-0.16 (shaded by 6' balcony overhang)
        3. (1) Slider window (Type H): 25.7 sf, U-0.28, SHGC-0.30
        4. (1) Picture window (Type L): 12.5 sf, U-0.26, SHGC-0.32
    - ii. Bedroom/bathroom (6.5% window-to-wall ratio)
        1. (1) Slider window (Type H): 25.7 sf, U-0.28, SHGC-0.30

The internal loads schedules for living area and bedroom area are given as,

Living area:

- People: 2 people (5:00-8:00 and 17:00 – 22:00), 245 btuh/person sensible, 105 btuh/person latent
- Lighting: 75 W (5:00-8:00 and 17:00 – 22:00)
- Equipment: 2000 btuh (5:00-8:00 and 17:00 – 22:00)

Bedroom area:

- People : 2 people (0:00-5:00 and 22:00 – 24:00), 250 btuh/person sensible, 200 btuh/person
- Lighting: 50 W (4:30-6:30, 21:30-22:30)

- Equipment: 200 btuh (4:30-6:30, 21:30-22:30)

A dynamic model (Figure 3) is developed based on building library (DOE/LBNL 1998, Wetter, et al., 2014).

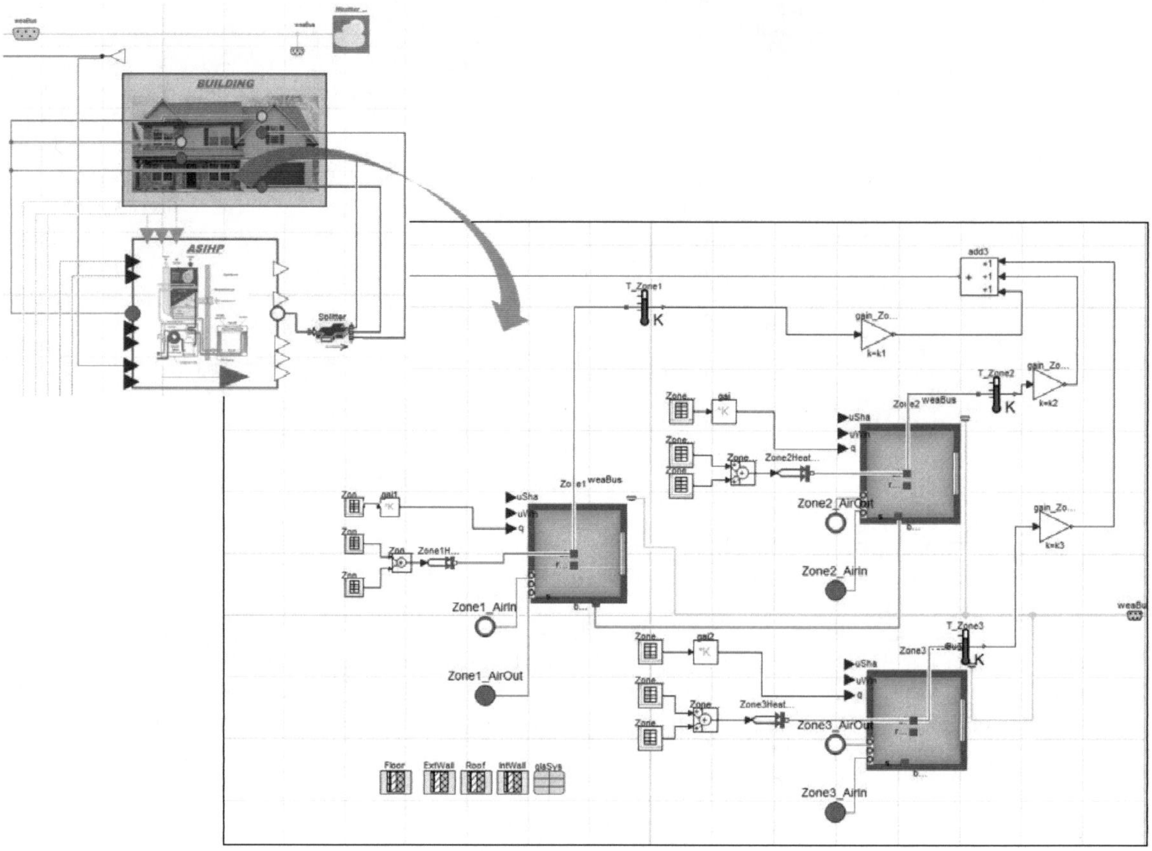

Figure 3. Dynamic model of a single family house

An air source integrated heat pump is designed to provide both cooling and heating load for this studied single family house. As shown in Figure 4, the air source integrated heat pump is simulated with ORNL HPDM tools (DOE/ORNL, 2019, Shen, et al., 2014] under a flexible solving framework.

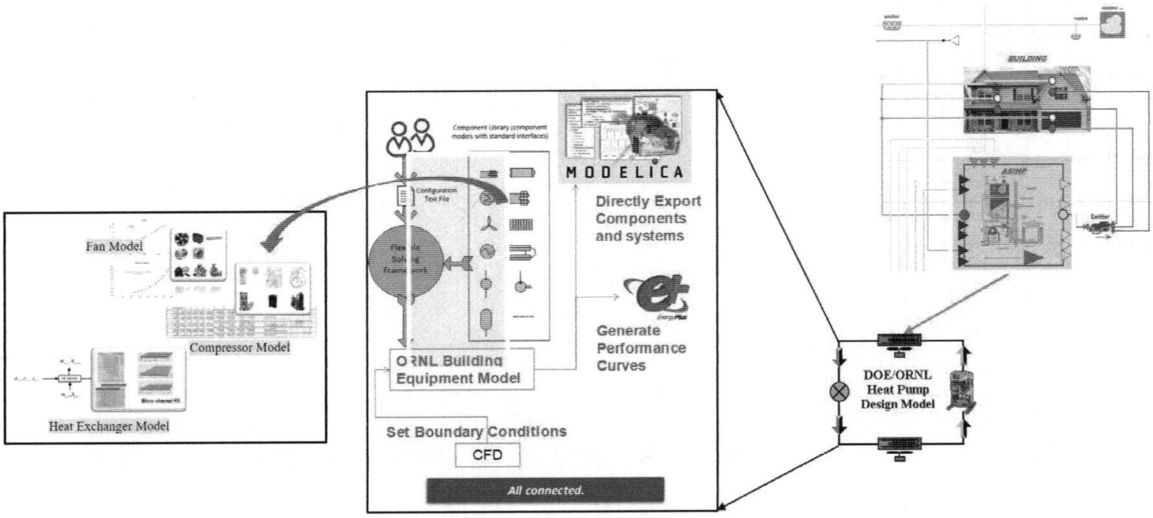

Figure 4. Air source integrated heat pump system model

4. SIMULATION AND ANALYSIS

Based on the model based co-simulation platform, we investigated a grid-responsive control strategy to take advantage of the low-cost electricity during night-time. It maintained the same comfort level, i.e. the same temperatures for the ASIHP to turn on, while allowing larger temperature bands for over-cooling, or over-heating when the electricity cost is low. This will utilize the low-cost electricity to provide hot water storage, precooling/preheating the building ahead of high price hours. Table 1 below described the temperature settings in space cooling, space heating and water heating modes, respectively for the normal ASIHP control and grid-responsive control.

Table 1: Normal and grid-responsive thermostat settings

Control targets	**Space Cooling**		**Space Heating**		**Water Heating**	
	Day	Night	Day	Night	Day	Night
Normal	73-77°F		68-72°F		110-125°F	
Grid-Responsive	73-77°F	70-77°F	68-72°F	68-75°F	110-125°F	110-140°F

As a comparison, Figure 5 depicts ASIHP's return air temperature profiles in Phoenix, AZ for a cooling week (July 1-7), respectively for the normal and grid-responsive controls. It indicates that the grid-responsive control allowed larger temperature band and cooled the building to lower temperatures during night hours. Figure 6 shows the hot water tank top node temperature profiles, resulted by the two control strategies.

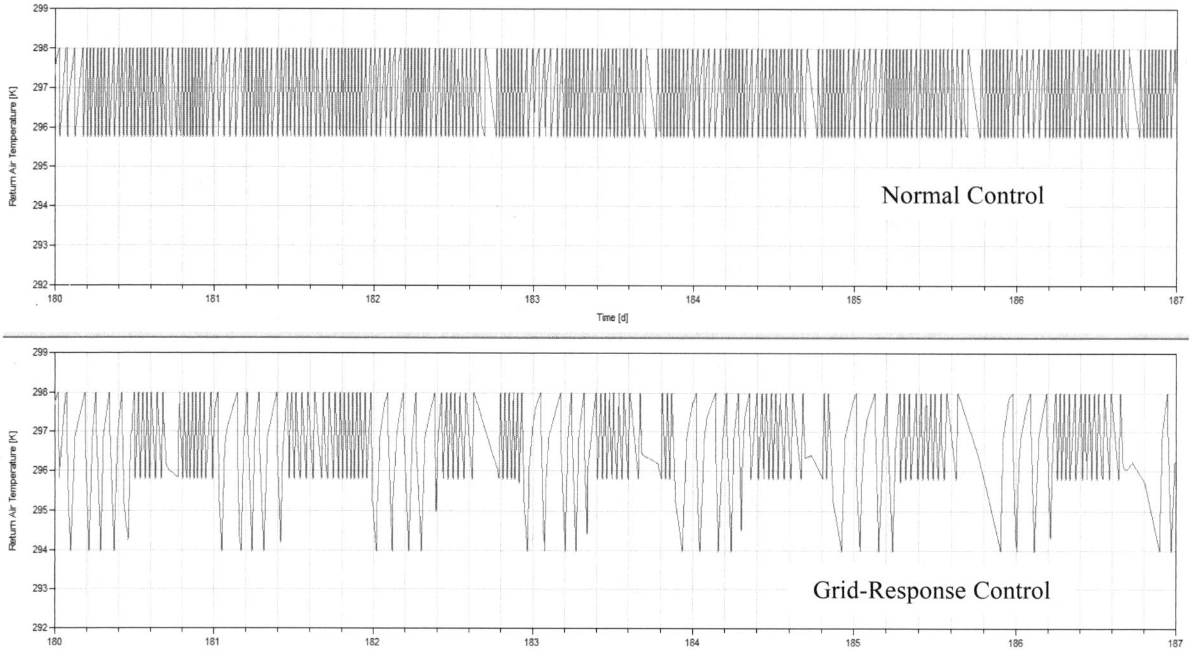

Figure 5: Return air temperature profiles in Phoenix, AZ, within a cooling week, normal vontrol versus grid-responsive control.

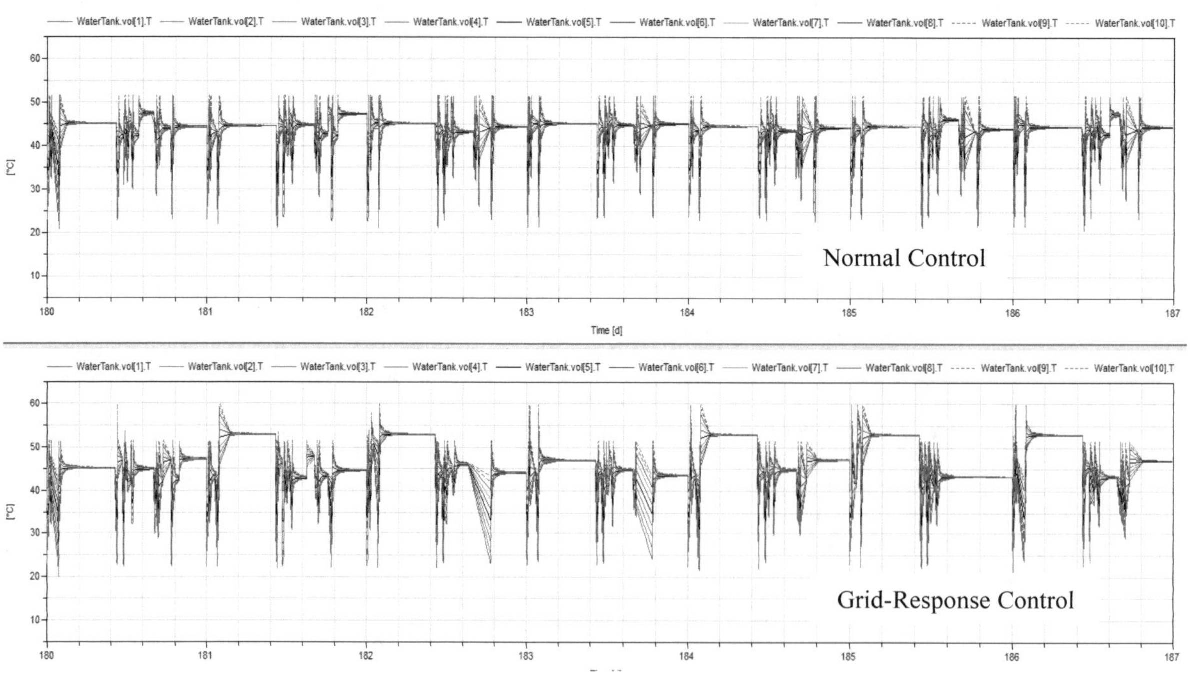

Figure 6: Hot water temperature profiles (top node), normal control versus grid-responsive control.

We conducted annual building energy simulations in five US cities, i.e. Chicago, IL; Phoenix, AZ; San Francisco, CA; Knoxville, TN; New York City, NY. Table 2 summarizes the energy consumptions and electricity costs. It demonstrates that the grid-responsive control strategy can reduce electricity costs up to 9%.

Table 2: Annual energy consumptions and utility costs in five US cities

		Chicago	Phoenix	San Francisco	Knoxville	New York
Normal Control	Power Consumption, kWh	7242	11941	8766	8621	8084
	Power Cost, $	682.1	1122	830	807	763
Grid-Responsive	Power Consumption, kWh	7027	10984	8045	8024	7661
	Power Cost, $	655	1027	755	751	714
Relative Savings	% power saving	-3.0%	-8.0%	-8.2%	-6.9%	-5.2%
	% power cost saving	-3.91%	-8.51%	-8.99%	-6.93%	-6.32%

5. CONCLUSION

Utilizing the ORNL heat pump design tool (HPDM), an integrated building system model-based supervisory control and optimization co-simulation platform is developed in this study to integrate high fidelity steady state simulation equipment models with dynamic building simulation. This work is to integrate the HPDM equipment library into the Modelica modelling environment through a dynamic link library, to facilitate co-simulation between detailed build equipment models and other building elements. A practice case is presented through development of a single family house with air source integrated heat pump system modeling platform and based on it to investigate a grid-responsive control strategy which takes advantage of low-cost electricity during night time and maintained the same conform level, i.e. the same temperatures for the ASIHP to turn on, while allowing larger temperature bands for over-cooling, or over-heating when the electricity cost is low. This utilized the low-cost electricity to provide hot water storage, precooling/preheating the building ahead of high price hours. In addition, annual building energy simulations in five US cities, i.e. Chicago, IL; Phoenix, AZ; San Francisco, CA; Knoxville, TN; New York City, NY, have been conducted to demonstrates annual electricity costs reduction up to 9% with the grid-responsive control strategy.

REFERENCEs

DOE/ORNL, 2019, Heat Pump Design Model, https://hpdmflex.ornl.gov/hpdm/wizard/welcome.php

DOE, 2020, EnergyPlus Engineering Reference, Version 9.3, https://energyplus.net/documentation

DOE/LBNL, 1998, Modelica Building Library, https://simulationresearch.lbl.gov/modelica/

Rice, K., Shen, B., Munk, J., Ally, M., and Baxter, V., 2014, Development of a Variable-Speed Residential Air-Source Integrated Heat Pump, 11thIEA Heat Pump Conference 2014, May 12-16, Montréal (Québec) Canada

Shen, B., New, J., and Baxter, V., Air source integrated heat pump simulation model for EnergyPlus, Energy and Buildings, Volume 156, 2017, Pages 197-206, ISSN 0378-7788

TRNSYS 18, 2018, https://sel.me.wisc.edu/trnsys/features/trnsys18_0_updates.pdf

Wetter, M., Zuo, W., Nouidui, T.S., and Pang, X., Modelica Buildings library. Journal of Building Performance Simulation, 7(4), 253–270, 2014. DOI: 10.1080 /19401493. 2013.765506

ACKMOWLEDGE

Funding for this research was provided by the US Department of Energy (DOE), Office of Energy Efficiency and Renewable Energy (EERE). The authors would like to thank Antonio Bouza, Program Manager of DOE, for his support of this work.

Adaptive Grey-Box Models for Model Predictive Building Control Using the Unscented Kalman Filter

Svenne FREUND[1*], Gerhard SCHMITZ[1], Claus Markus TIEMANN[1]

[1] Hamburg University of Technology, Institute of Engineering Thermodynamics,
Hamburg, Germany
Phone: +4940 42878-2596, E-mail: svenne.freund@tuhh.de

* Corresponding Author

ABSTRACT

Model predictive control (MPC) for buildings is a promising approach to reduce the energy consumption of buildings while at the same time the thermal user comfort can be improved. The core of this control strategy consists of building models that can describe the thermal behavior of particular zones accurately. Grey-box models are frequently used modeling approaches for control-oriented models, however, these models often have limitations regarding their general applicability. Furthermore, the modeling and identification of models used in MPC still require significant effort and is one of the main obstacles for the actual practical implementation of building predictive control. This paper addresses these issues and presents a framework for the online state and parameter estimation of grey-box models. The results show that (1) this online simultaneous state and parameter estimation highly increases the multi-steps-ahead (up to 48 h) prediction performance, (2) this approach enables the models to adapt to changing environmental conditions and (3) it is possible to use only one pre-defined initial model to describe the thermal behavior of several different zones.

1. INTRODUCTION

Against the background of the environmental impact of energy use, the depletion of primary energy resources and the associated economic consequences, considerable efforts are being made worldwide to realize environmentally-friendly and energy-efficient buildings. According to the European Commission, about 40 % of primary energy demand is related to the building sector (European Commission, 2019). Reducing the energy demand of buildings is therefore of major importance concerning the climate protection goals. Especially in modern buildings with already high energy efficiency standards and a high degree of automation, the control and interaction of the individual building's systems with each other is of decisive importance for the energy efficiency and user comfort of the entire building. This is especially true for non-residential buildings such as large office buildings. The optimization of the building control strategy represents a cost-effective approach compared to other energy-saving measures.

Model predictive control (MPC) of buildings is a promising approach to reduce the energy demand of buildings while improving thermal user comfort (Serale et al., 2018). With the increase in performance and capacity of modern computers and the extensive use of measurement and monitoring systems in buildings, the implementation of advanced, complex control strategies has got more attention in recent years (Sofos et al., 2020). The core of this control strategy consists of one or more simplified and dynamic building models that can describe the thermal behavior of particular zones accurately. Grey-box models based on resistor-capacitor (RC) networks are frequently used modeling approaches for control-oriented models, however, these models often have limitations regarding their general applicability. Furthermore, the modeling and identification of models used in MPC still require significant effort and is one of the main obstacles for the actual practical implementation of building predictive control. This paper addresses these issues and presents a framework for the simultaneous online state and parameter estimation of grey-box models based on an unscented kalman filter (UKF).

1.1 Related Works

To the best of the authors' knowledge, Radecki and Hencey (2012) were the first ones who proposed an unscented kalman filter for estimating states and parameters of a thermal RC model. In Radecki and Hencey (2013) they compared a UKF-based MPC to thermostat control. The UKF tended to estimate the model parameters to physically impossible values. In those cases, the thermostat control outperformed the MPC. In Radecki and Hencey (2017) they tested the

(a) Front view of the investigated building.

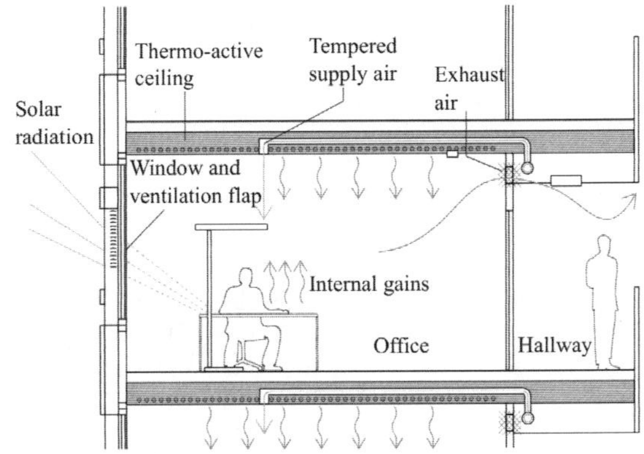

(b) Standard single office.

Figure 1: Front view of the office building (a) and schematic of a standard single office (b).

UKF in a one-year-period. Reference measurement data was generated by a detailed white-box model. The root mean square error (RMSE) of a 24 h prediction was 1.48 °C. Maasoumy et al. (2013) used an off-line parameter training to set the initial values for the UKF algorithm. In Martincevic et al. (2014) a method called "clipping" was proposed to prevent the UKF from estimating parameters to physically impossible values. For this purpose, they defined constraints for every parameter by upper and lower boundaries. Every time a parameter value passed the upper or the lower boundary during a filter algorithm step, the value was "clipped" back to the boundary. In Martincevic and Vasak (2019) three additional modifications to the UKF algorithm were added. First, the weighting factors were recalculated after every "clipping" to prevent the mean of the filter's sigma points from shifting. Second, parameters like the solar irradiance factor were temporarily removed from the augmented state vector when their value did not influence the model output. Third, parameter values were normalized to the interval defined by the upper and lower boundaries to reduce numerical issues. The evaluation of the filter performance with real measurement data resulted in an RMSE of 0.3 °C for a 24 h prediction. Massano et al. (2019) tested the UKF performance on a very simple RC model consisting of one resistor and one capacitor. Reference measurement data was generated by a white-box model. The RMSE of a 24 h prediction was 0.464 °C.

All of the described works neglected the influence of disturbances caused by occupants like the opening of windows or internal gains. The reference data were synthetically generated without disturbances using detail building models. In Alam et al. (2017) disturbances induced by occupants were considered. Alam et al. tested an extended kalman filter to estimate the model states and model parameters for a temporarily occupied living room. Monthly averaged temperature and solar irradiance measurements were used as weather forecast. Several disturbances were considered. The 95th percentile of the absolute prediction error was 1.73 °C in case of a 4 h prediction. Here, the prediction horizon was much smaller while the error value had the same magnitude as in the other papers. Considering more different disturbances can lead to less accurate predictions.

2. BUILDING DESCRIPTION

The office building under investigation, shown in Fig. 1a, is located in Hamburg in Northern Germany. It was designed and built as part of the framework "Energy Optimized Building Construction" (EnOB) as part of the International Building Exhibition in 2013 and serves as the headquarter for the two Local Ministries for Environment and Energy / Urban Development and Housing. Providing a net-floor space of 46 500 m^2, the building houses 1250 office rooms for around 1500 workplaces. The building can be subdivided into seven low-rise buildings (referred to as building A to D and F to H) with five floors and one 13-story high-rise building (referred to as building E). It was planned and built as a sustainable low-energy building with a target primary energy demand lower than 70 kWh/m^2a and a heating demand of around 15 kWh/m^2a.

2.1 Standard single office

The majority of office spaces are realized as standard single offices with a rectangular layout. A schematic of a standard single office is shown in Fig. 1b. The basic geometric and construction properties of a single office are given in

Table 1: Basic geometric and construction properties of a single office.

Description	Unit	Value
Width	m	2.5
Clear height	m	2.89
Floor area	m^2	10 to 13
Window area	m^2	2.63
Area of exterior facade (without glazing)	m^2	4.6
Heat transfer coefficient of window	W/m^2K	0.7
Heat transfer coefficient of exterior facade (without glazing)	W/m^2K	0.9

Tab. 1. Each office has a manually openable window and a weather-proofed ventilation flap. The building's envelope is realized as a unitized curtain wall with distinctively colored ceramic panels mounted outside. In winter mode, the offices are mainly heated by thermo-active ceilings (TAC), realized as concrete-core activation with a maximum supply temperature of 32 °C. Mechanical ventilation with preheated air is used to provide the offices with fresh air. During summer mode, the building is passively cooled via the thermo-active ceilings which are fed with chilled water in the range of 18 °C to 20 °C. The building management system (BMS) automatically switches from heating to cooling operation mode based on the averaged outdoor air temperature (OAT) of the past 36 hours. The mechanical ventilation is in summer mode replaced by manual ventilation using the ventilation flaps or windows.

2.2 Heating system

The building's heating system is mainly based on shallow geothermal energy in conjunction with two electrically driven heat pumps, each with a nominal thermal power output of 264 kW. For domestic hot water supply as well as for thermal peak load coverage, a district heating connection is installed. Each building part is separated into two different TAC heating circuits according to the orientation (south/north and south-west/north-east, respectively) resulting in 16 different TAC control loops. The supply temperature set points are gained by a standard linear heating curve using the current outside temperature in combination with a simple heating schedule.

2.3 Monitoring system

To examine the energy demand targets and to consistently optimize the building operation with respect to user comfort and operating costs, a scientific monitoring system was implemented in July 2014. It comprises an extensive network of more than 1100 sensors consisting of various energy meters, flow meters, temperature sensors to measure and record the relevant parameters within the different subsystems with a sample rate of 1 minute. To additionally analyze and evaluate the user comfort, 32 *reference office spaces* were equipped with sensors for air temperature, relative humidity, presence and window handling. These reference offices are mainly office spaces of different orientation, floor plan and floor area and are distributed over the entire building. The measurement data collected with the monitoring system will be used in the presented framework.

3. GREY-BOX MODEL DESCRIPTION

Grey-box models are a trade-off between detailed physical white-box models (standard building simulation software like EnergyPlus or TRNSYS) and purely data-driven black-box models (e. g. neural networks, ARMAX). They have been widely discussed in literature and successfully applied for predictive control strategies. Grey-box models can be represented as networks consisting of thermal resistances R and capacities C. The physical values of those parameters are estimated using suitable identification methods with measured input and output data. Fig. 2 depicts the structure of the here used grey-box model.

The purpose of this model is to describe the thermal behavior (indoor air temperature) of a reference office space adequately. The model consists of seven resistances and four capacities in total (R7C4 model). The four state variables T_{W} (exterior wall temperature), T_{Air} (indoor air temperature), T_{Int} (temperature of internal masses) and T_{TAC} (TAC temperature) correspond to the four capacities C_{W}, C_{Air}, C_{Int} and C_{TAC}. Based on available construction data, the initial values for the resistances and capacities are derived and given in Tab. 2. The values for the resistances R_{TAC1} and R_{TAC2} are estimated using the EMPA model developed by (Koschenz & Lehman, 2000).

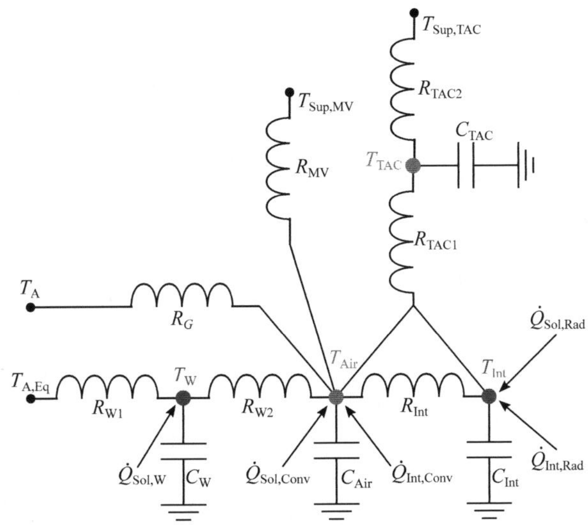

Figure 2: RC-network representation of the used grey-box model.

Table 2: Description and initial values of grey-box model parameters.

Parameter name	Description	Initial value
R_{W1}	Thermal resistance of envelope (exterior side)	$0.05\,(\mathrm{m^2K})/\mathrm{W}$
R_{W2}	Thermal resistance of envelope (interior side)	$1.11\,(\mathrm{m^2K})/\mathrm{W}$
C_{W}	Thermal capacitance of envelope	$8.5 \cdot 10^3\,\mathrm{J}/(\mathrm{m^2K})$
R_{G}	Thermal resistance of glazing	$1.43\,(\mathrm{m^2K})/\mathrm{W}$
R_{TAC1}	Thermal resistance between TAC core and zone	$0.183\,(\mathrm{m^2K})/\mathrm{W}$
R_{TAC2}	Thermal resistance between TAC core and supply temperature	$0.122\,(\mathrm{m^2K})/\mathrm{W}$
C_{TAC}	Thermal capacitance of TAC	$7.5 \cdot 10^5\,\mathrm{J}/(\mathrm{m^2K})$
$\dot{Q}_{\mathrm{Int}}$	Internal heat gains	$15\,\mathrm{W/m^2}$
R_{Int}	Thermal resistance between room air and internal masses	$0.18\,(\mathrm{m^2K})/\mathrm{W}$
C_{Int}	Thermal capacitance of internal masses	$1.82 \cdot 10^5\,\mathrm{J}/(\mathrm{m^2K})$
R_{MV}	Thermal resistance of mechanical ventilation	$0.15\,\mathrm{K/W}$
f_{Sol}	Effective window area for solar radiation	0.2

The model has in total five inputs: T_{A} (outdoor air temperature), $T_{\mathrm{Sup,TAC}}$ (supply temperature of the corresponding TAC's heating circuit), $T_{\mathrm{Sup,MV}}$ (supply temperature of mechanical ventilation), $I_{\mathrm{G,\{N,S,W,E\}}}$ (global solar radiation on the corresponding facade orientation) and OCC (binary occupancy signal measured by presence sensor). For the heat exchange between the building exterior and the environment, a simplified approach based on (VDI 6007, 2015) is implemented using the equivalent OAT $T_{\mathrm{A,Eq}}$ defined as:

$$T_{\mathrm{A,Eq}} = T_{\mathrm{A}} + I_{\mathrm{G,\{N,S,W,E\}}} \cdot \frac{\alpha_{\mathrm{F}}}{\alpha_{\mathrm{A}}} \tag{1}$$

The solar heat gains absorbed by the interior are determined according to:

$$\dot{Q}_{\mathrm{Sol}} = f_{\mathrm{Sol}} \cdot I_{\mathrm{G,\{N,S,W,E\}}} \tag{2}$$

where f_{Sol} is an empirical factor and can be interpreted as an effective window area in which the solar radiation enters. The internal heat gains by persons, lighting or equipment are calculated using the measured occupancy signal OCC multiplied by a constant factor $\dot{Q}_{\mathrm{Int}}$. As shown in Fig. 2, the different heat sources are split into a convective part which is acting on the air volume node and a radiative part which is acting on the interior temperature node. The corresponding parts are chosen as follows: internal gains 60 % radiative, 40 % convective; solar gains 90 % radiative, 10 % convective; heat gains by TAC 80 % radiative, 20 % convective.

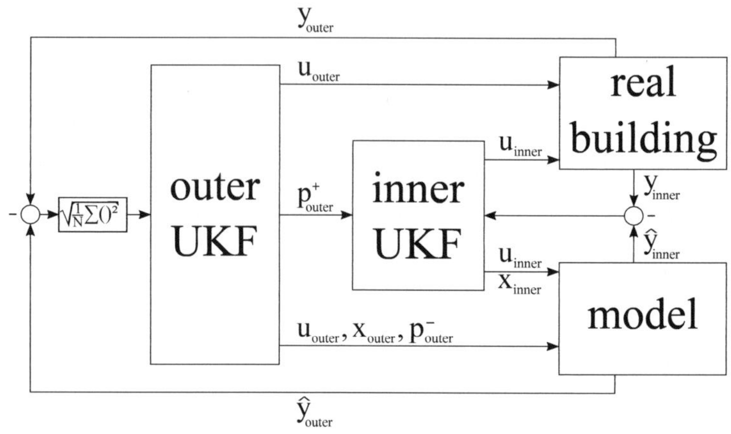

Figure 3: Schematic view of the CUKF operation mode.

The grey-box model, shown in Fig. 2, is implemented in the equation-based, object-oriented modeling language Modelica® (The Modelica Association, 2020). An interface to MATLAB® is developed which enables the manipulation of the parameters and initial states. The following kalman filter algorithm is therefore implemented in MATLAB®.

4. UNSCENTED KALMAN FILTER ALGORITHM

In this study, a combined unscented kalman filter (CUKF) for state and parameter estimation is proposed consisting of two separate unscented kalman filters (UKF), an outer filter and an inner filter. For outer filtering, the grey-box model simulates the room temperature trajectory for the past 48 hours using the available input data. Simultaneously, the measured temperature trajectory of the given time interval is loaded from the record. Simulated and measured trajectories are smoothed by moving mean and moving median filters to reduce disturbance influences like window opening from the data. The outer UKF then estimates the states and parameters of the model by minimizing the RMSE between the smoothed signals. The estimated parameters are then transmitted to the inner filter which estimates the current state of the model. Fig. 3 shows a schematic view of the combined unscented kalman filter operation mode.

A discrete-time non-linear state-space model consisting of the system function $f(\cdot)$ and the measurement function $h(\cdot)$ is considered.

$$\mathbf{x}_k = f(\mathbf{x}_{k-1}, \mathbf{u}_{k-1}) + \boldsymbol{\omega} \tag{3}$$

$$\mathbf{y}_k = h(\mathbf{x}_k) + \boldsymbol{\upsilon} \tag{4}$$

$$\boldsymbol{\omega} \sim \mathcal{N}(0, \mathbf{Q}) \tag{5}$$

$$\boldsymbol{\upsilon} \sim \mathcal{N}(0, \mathbf{R}) \tag{6}$$

$\boldsymbol{\omega}$ and $\boldsymbol{\upsilon}$ are the process noise and the measurement noise with covariance $\mathbf{Q}$ and $\mathbf{R}$, respectively. They are assumed to be zero mean, gaussian and uncorrelated.

$\mathbf{x}$ is the state vector of the system with covariance matrix $\mathbf{P}$. For the outer filter $\mathbf{x}_{\text{outer}}$ is an augmented state vector containing the states (temperature levels) $[T_1 \dots T_n]^T$, manipulated parameters $[p_1 \dots p_m]^T$ and the RMSE of the past 48 hours e_{RMSE}. For the inner filter $\mathbf{x}_{\text{inner}}$ is a vector containing only the states $[T_1 \dots T_n]^T$.

$$\mathbf{x}_{\text{outer}} = \begin{bmatrix} T_1 \\ \vdots \\ T_n \\ p_1 \\ \vdots \\ p_m \\ e_{\text{RMSE}} \end{bmatrix}, \qquad \mathbf{x}_{\text{inner}} = \begin{bmatrix} T_1 \\ \vdots \\ T_n \end{bmatrix}$$

For initializing the state vector, the parameter values are either obtained by an off-line parameter training or by using the initial values from Tab. 2. The room air temperature T_{Air} is set by the latest measured indoor air temperature value. The other states are initialized as follows: $T_{\mathrm{W}} = 295.15\,\mathrm{K}$, $T_{\mathrm{Int}} = 295.15\,\mathrm{K}$, $T_{\mathrm{TAC}} = 296.15\,\mathrm{K}$. e_{RMSE} is set to zero and the initial state covariance matrix $\mathbf{P}_0^+$ is set to $\mathbf{P}_0^+ = \mathbf{Q}$.

The general operation mode of the UKF is devided into two steps, prediction step and correction step. In the prediction step a defined number of states, so called sigma points, are generated:

$$\boldsymbol{\chi}_{k-1}^{(0)} = \hat{\mathbf{x}}_{k-1}^{+} \tag{7}$$

$$\boldsymbol{\chi}_{k-1}^{(i)} = \hat{\mathbf{x}}_{k-1}^{+} + \left(\sqrt{(n+\kappa)\,\mathbf{P}_{k-1}^{+}}\right)_i^T, \qquad i = 1, \dots, n \tag{8}$$

$$\boldsymbol{\chi}_{k-1}^{(n+i)} = \hat{\mathbf{x}}_{k-1}^{+} - \left(\sqrt{(n+\kappa)\,\mathbf{P}_{k-1}^{+}}\right)_i^T, \qquad i = 1, \dots, n \tag{9}$$

κ is a scaling factor that must fulfill the condition $n + \kappa \neq 0$. The sigma points are then transformed through the system function $f(\cdot)$. This step is called the unscented transform (UT):

$$\boldsymbol{\chi}_k^{(i)} = f\left(\boldsymbol{\chi}_{k-1}^{(i)}, \mathbf{u}_k\right) \tag{10}$$

The RMSE value e_{RMSE}, which is the last entry of the state vector in the outer filter, is defined as the root mean square error between simulated and measured room temperature trajectory:

$$e_{\mathrm{RMSE}}^{(i)} = \sqrt{\frac{1}{N}\sum_{j=1}^{N}\left(\hat{T}_j - T_{j,\mathrm{meas.}}\right)^2}, \tag{11}$$

where N is the number of simulated and measured temperature samples. The weighting factors

$$W_m^{(0)} = \frac{\kappa}{n+\kappa} \tag{12}$$

$$W_c^{(0)} = \frac{\kappa}{n+\kappa} + \left(1 + \beta - \alpha^2\right) \tag{13}$$

$$W_m^{(i)} = W_c^{(i)} = \frac{1}{2\,(n+\kappa)}, \qquad i = 2, \dots, 2n \tag{14}$$

are used to calculate mean and covariance of the transformed sigma points receiving the *a priori* estimation $\hat{\mathbf{x}}_k^-$ and $\mathbf{P}_k^-$. α and β are scaling factors. α is usually set to a small value between 0 and 1. β is usually set to 2 when a gaussian distribution is assumed (Julier, 2002).

$$\hat{\mathbf{x}}_k^- = \sum_{i=0}^{2n} W_m^{(i)} \boldsymbol{\chi}_k^{(i)} \tag{15}$$

$$\mathbf{P}_k^- = \sum_{i=0}^{2n} W_c^{(i)} \left(\hat{\mathbf{x}}_k^- - \boldsymbol{\chi}_k^{(i)}\right)\left(\hat{\mathbf{x}}_k^- - \boldsymbol{\chi}_k^{(i)}\right)^T + \mathbf{Q} \tag{16}$$

The transformed sigma points are then transformed again through the measurement function $h(\cdot)$:

$$\mathcal{Y}_k^{(i)} = h\left(\boldsymbol{\chi}_k^{(i)}\right) = \mathbf{H}\boldsymbol{\chi}_k^{(i)} \tag{17}$$

The output matrix $\mathbf{H}$ is defined differently for the inner and the outer UKF. The measurement matrix of the inner filter returns only the estimation of the current indoor air temperature, while the measurement matrix of the outer filter returns air temperature estimation and the RMSE of the estimated temperature trajectory.

$$\mathbf{H}_{\mathrm{inner}} = \begin{bmatrix} 1 & 0 & \cdots & 0 \end{bmatrix}, \qquad \mathbf{H}_{\mathrm{outer}} = \begin{bmatrix} 1 & 0 & \cdots & 0 & 0 \\ 0 & 0 & \cdots & 0 & 1 \end{bmatrix}$$

The *a priori* measurement prediction is received by calculating mean and covariance after the second transformation:

$$\hat{\mathbf{y}}_k = \sum_{i=0}^{2n} W_m^{(i)} \mathcal{Y}_k^{(i)} \tag{18}$$

$$\mathbf{P}_{\mathbf{y},k} = \sum_{i=0}^{2n} W_c^{(i)} \left(\hat{\mathbf{y}}_k - \mathcal{Y}_k^{(i)}\right)\left(\hat{\mathbf{y}}_k - \mathcal{Y}_k^{(i)}\right)^T + \mathbf{R} \tag{19}$$

In the correction step the *a priori* estimation is turned into an *a posteriori* estimation by calculating the cross-covariance matrix $\mathbf{P}_{\mathbf{xy},k}$ and the kalman gain $\mathbf{K}_k$. For further details about the UKF algorithm see (Simon, 2006).

$$\mathbf{P}_{\mathbf{xy},k} = \sum_{i=0}^{2n} W_c^{(i)} \left(\hat{\mathbf{x}}_k^- - \chi_k^{(i)}\right)\left(\hat{\mathbf{y}}_k - \mathcal{Y}_k^{(i)}\right)^T \tag{20}$$

$$\mathbf{K}_k = \mathbf{P}_{\mathbf{xy}} \mathbf{P}_{\mathbf{y}}^{-1} \tag{21}$$

$$\hat{\mathbf{x}}_k^+ = \hat{\mathbf{x}}_k^- + \mathbf{K}_k \left(\mathbf{z}_k - \hat{\mathbf{y}}_k\right) \tag{22}$$

$$\mathbf{P}_k^+ = \mathbf{P}_k^- - \mathbf{K}_k \mathbf{P}_{\mathbf{y}} \mathbf{K}_k^T \tag{23}$$

The measurement vector z_k only returns the current measured room temperature value in the case of inner filtering and an additional zero entry for the RMSE in the case of outer filtering:

$$\mathbf{z}_{k,\text{inner}} = \left[T_{k,\text{meas.}}\right], \qquad \mathbf{z}_{k,\text{outer}} = \begin{bmatrix} T_{k,\text{meas.}} \\ 0 \end{bmatrix}$$

The zero entry is needed for the outer filter to minimize the RMSE in every filtering step.

5. RESULTS AND DISCUSSION

In this section, two different applications of the proposed filter algorithm are presented. First, the adaption of a "trained model" to changing environmental and operational conditions is investigated. The second use-case is the online simultaneous parameter and state estimation of an "initial model" without pre-trained parameters. The filter performance is evaluated using the RMSE between the predicted and measured room temperatures for a prediction horizon of 48 hours.

5.1 Adaption to changing system operation

As described, the performance of the CUKF with a pre-trained model under changing system operations is investigated. The parameters of this trained model are estimated with input data from a heating period exclusively. The details of the off-line parameter identification are beyond the scope of this paper. More information can be found in Freund and Schmitz (2019). Furthermore, only the four most influential (defined by parameter sensitivity concerning the simulated room temperature) parameters are set to be manipulable by the filter. The OAT and TAC's supply temperature of the investigated 30-day-period are shown in Fig. 4a. The system operation changes from a 14-day heating period to a 7-day cooling period and then back to heating operation. As described in Section 2.2, the TACs are also used for passive cooling of the building with a supply temperature of 18 °C to 20 °C. The outdoor air temperature in the considered period varies in a range of 0 °C to 25 °C.

Fig. 4b depicts the measured room air temperature and the one-step-ahead prediction of the model with and without using the CUKF. It appears that the error between measurement and prediction without filter increases when the OAT rises and the system operation switches to cooling mode. This indicates that the initial parameters from the off-line identification step using training data from heating periods are not suitable for significantly changed environmental and operational conditions. Applying the proposed CUKF does improve temperature tracking considerably (Fig. 4b). As described, the filter algorithm updates the states and parameters recursively every 30 minutes using the current measured room temperature and the trajectory from the past 48 hours. Fig. 4c shows the distribution of the RMSE for the 48 hours prediction. Without using any filter, 50 % of the prediction errors are in a range of 0.4 °C to 0.8 °C with a median of 0.56 °C. Using a UKF only for state estimation, the prediction RMSEs drop to a median of 0.4 °C and the distribution is getting significantly narrower. By using the combined state and parameter UKF, the median RMSE can further be reduced to 0.32 °C. In Fig. 4d, the evolution of states and parameters over time are depicted.

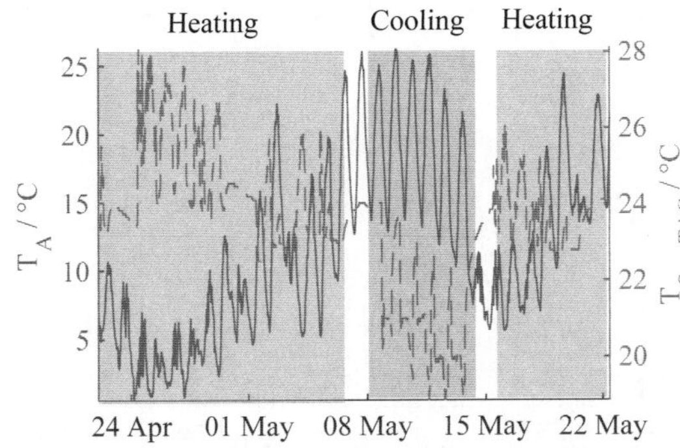

(a) Outdoor air temperature and TAC supply temperature.

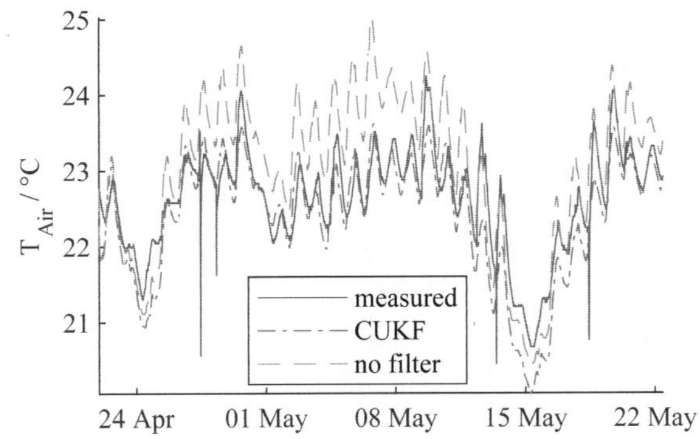

(b) Predicted and measured room air temperature.

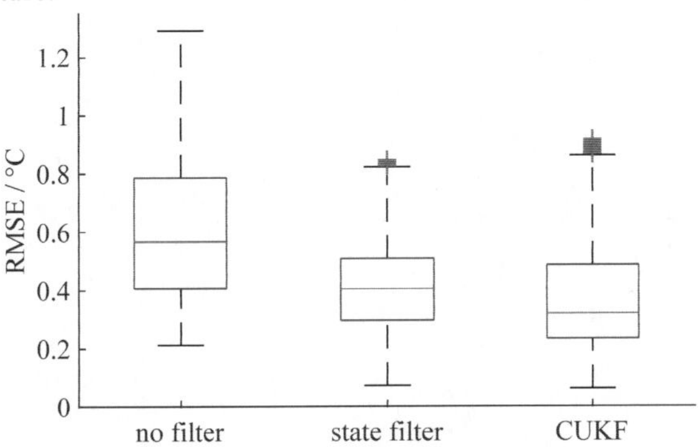

(c) RMSE box (quartiles) and whisker (±1.5×IQR) distribution with outliers for a 48 h prediction.

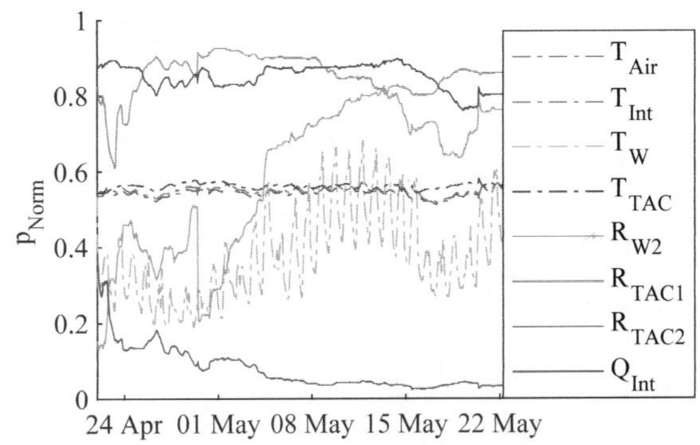

(d) Evolution of states and parameters using the CUKF.

Figure 4: Online parameter and state estimation of a pre-trained model to changing external and operational conditions.

The values are normalized concerning their upper and lower bounds. As shown, all parameters evolve over time and do not converge to a steady-state value. However, the parameter R_{TAC2} (solid light blue curve) indicates a strong correlation to the environmental and operational conditions, as it is increasing distinctly during the cooling period. Overall, the proposed CUKF tunes the parameters and states in a reliable way such that the prediction errors can be reduced significantly and the grey-box model is able to adapt to strongly varying conditions.

5.2 Adaption of an initial model

In the second use-case, the adaption of an initial model without any prior off-line parameter identification step is investigated. The initial parameters of the model, given in Tab. 2, are derived from available construction data and relevant standards. A test period of 28 days during the heating season is chosen. As shown in Fig. 5a, the OAT is in a range of −3 °C to 12 °C and the TAC supply temperature reaches a maximum of 30 °C.

The comparison between the simulated and the measured room temperature, see Fig. 5b, reveals that the initial set of parameters already results in a good agreement. This is emphasized by a low median prediction RMSE of 0.36 °C, see Fig. 5c. Applying a UKF for state estimation only does again improve the prediction performance and decreases the median RMSE to 0.26 °C. However, by using the CUKF the prediction errors can further be reduced considerably to a median RMSE of 0.19 °C while 50 % of the data are in a range of 0.12 °C to 0.28 °C. As before, the parameter values evolve constantly over time and do not converge to a steady-state value, see Fig. 5d. This behavior may result from the presence of significant unmeasured disturbances in the used input data. It is quite reasonable that there is not just one optimal set of parameters, but rather a variety of locally optimal parameter sets. Similar behavior can be observed in off-line parameter identification procedures, see Brastein et al. (2018).

To further confirm the results found, the CUKF is applied to five additional reference offices rooms located in the same building sector using the identical initial parameter set and the same 28 days test period. The median values of the

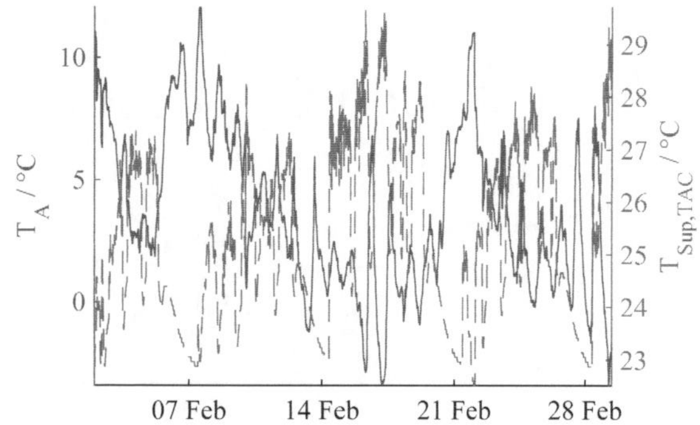

(a) Outdoor air temperature and TAC supply temperature.

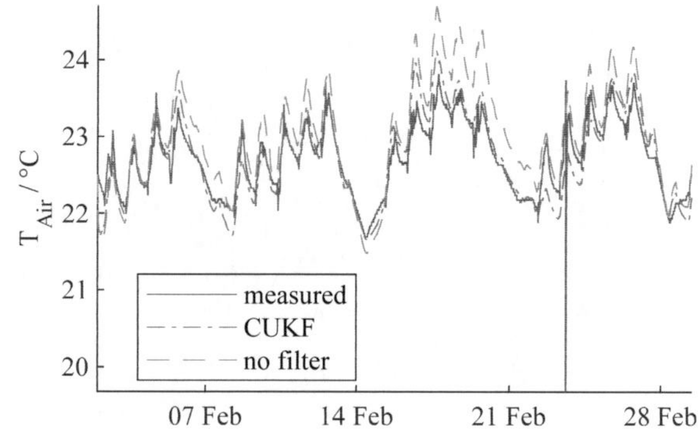

(b) Predicted and measured room air temperature.

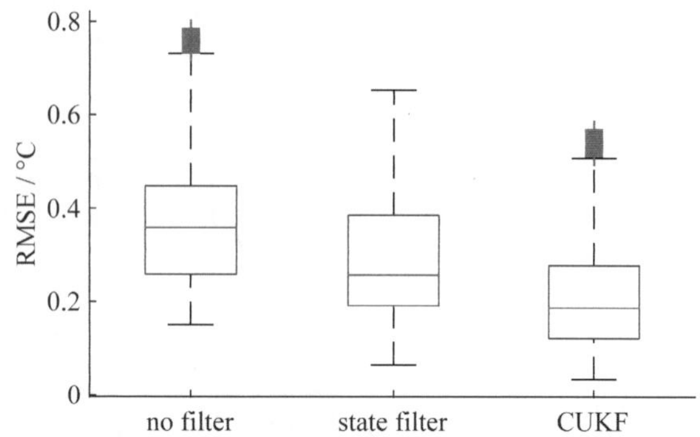

(c) RMSE box (quartiles) and whisker (±1.5×IQR) distribution with outliers for a 48 h prediction.

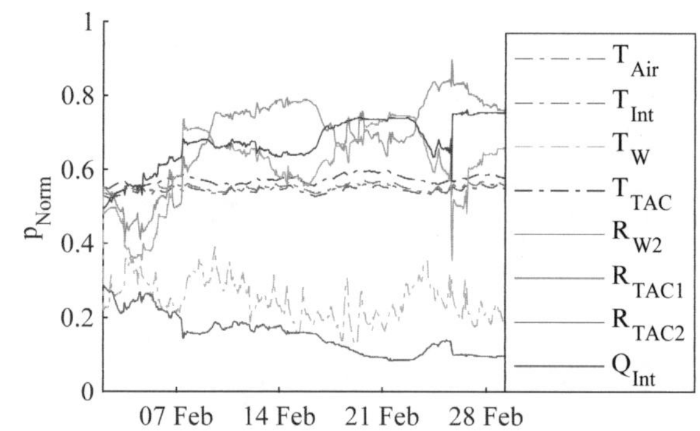

(d) Evolution of states and parameters using the CUKF.

Figure 5: Online parameter and state estimation of an untrained model.

RMSE for the 48 hours prediction are given in Tab. 3. As can been seen, the online combined state and parameter estimation highly increases the prediction performance of all reference rooms and in all cases outperforms the pure state estimation. Therefore, it is possible to use only one pre-defined initial model to describe the thermal behavior of several different zones. Thus, the modeling effort to obtain control-oriented models for MPC applications can be reduced considerably.

6. CONCLUSIONS

Int his study, a framework for a combined state and parameter identification for grey-box models using an unscented kalman filter is presented. The proposed filter is tested in two different use cases: the adaption of a pre-trained model to strongly varying external and operational conditions and the online parameter estimation of an untrained model. The results show that the online simultaneous parameter and state estimation highly increases the 48 hours prediction

Table 3: Median RMSE for a 48 hours prediction using no filter, a UKF state filter and the CUKF.

	Orientation	Floor area	RMSE no filter	RMSE state estimation	RMSE CUKF
Reference room 1	South-West	10.22 m^2	0.62 °C	0.59 °C	0.41 °C
Reference room 2	South-West	11.17 m^2	1.30 °C	0.82 °C	0.46 °C
Reference room 3	North-West	10.25 m^2	0.40 °C	0.39 °C	0.22 °C
Reference room 4	South-East	12.61 m^2	0.44 °C	0.34 °C	0.18 °C
Reference room 5	South-East	26.11 m^2	0.38 °C	0.27 °C	0.24 °C

performance. It further enables the models to reliably adapt to changing external conditions. Furthermore, applying the filter algorithm allows using only one pre-defined initial model to describe the thermal behavior of several different zones. In future work, the implementation of the proposed filter in a model predictive control framework which is currently practically investigated in the described building is planned.

REFERENCES

Alam, M., Rogers, A., Scott, J., Ali, K., & Auffenberg, F. (2017). *Applying extended kalman filters to adaptive thermal modelling.* University of Oxford, Department of Engineering Science.

Brastein, O. M., Perera, D., Pfeifer, C., & Skeie, N.-O. (2018). Parameter estimation for grey-box models of building thermal behaviour. *Energy and Buildings, 169*, 58–68. doi: 10.1016/j.enbuild.2018.03.057

European Commission. (2019). *Eu transport in figures: Statistical pocketbook 2019* (Vol. 2019). Luxembourg: Publications Office of the European Union.

Freund, S., & Schmitz, G. (2019). Development of a framework for model predictive control (mpc) in a large-sized low-energy office building using modelica grey-box models. In *Building simulation 2019: 16th conference of ibpsa, rome, italy.*

Julier, S. J. (2002). The scaled unscented transformation. In *Proceedings of the 2002 american control conference (ieee cat. no.ch37301)* (p. 4555-4559 vol.6). IEEE. doi: 10.1109/ACC.2002.1025369

Koschenz, M., & Lehman, B. (2000). *Thermoaktive bauteilsysteme tabs* (1. Aufl. ed.). Dübendorf: EMPA Energiesysteme/Haustechnik.

Maasoumy, M., Moridian, B., Razmara, M., Shahbakhti, M., & Sangiovanni-Vincentelli, A. (2013). Online simultaneous state estimation and parameter adaptation for building predictive control. In *Asme 2013 dynamic systems and control conference.* American Society of Mechanical Engineers. doi: 10.1115/DSCC2013-4064

Martincevic, A., Starcic, A., & Vasak, M. (2014). Parameter estimation for low-order models of complex buildings. In *Ieee pes innovative smart grid technologies, europe* (pp. 1–6). IEEE. doi: 10.1109/ISGTEurope.2014.7028767

Martincevic, A., & Vasak, M. (2019). Constrained kalman filter for identification of semiphysical building thermal models. *IEEE Transactions on Control Systems Technology*, 1–8. doi: 10.1109/TCST.2019.2942808

Massano, M., Macii, E., Patti, E., Acquaviva, A., & Bottaccioli, L. (2019). A grey-box model based on unscented kalman filter to estimate thermal dynamics in buildings. In *Ieee international conference on environment and electrical engineering* (pp. 1–6). doi: 10.1109/EEEIC.2019.8783974

Radecki, P., & Hencey, B. (2012). Online building thermal parameter estimation via unscented kalman filtering. In *2012 american control conference (acc)* (pp. 3056–3062). IEEE. doi: 10.1109/ACC.2012.6315699

Radecki, P., & Hencey, B. (2013). Online thermal estimation, control, and self-excitation of buildings. In *52nd ieee conference on decision and control* (pp. 4802–4807). IEEE. doi: 10.1109/CDC.2013.6760642

Radecki, P., & Hencey, B. (2017). Online model estimation for predictive thermal control of buildings. *IEEE Transactions on Control Systems Technology, 25*(4), 1414–1422. doi: 10.1109/TCST.2016.2587737

Serale, G., Fiorentini, M., Capozzoli, A., Bernardini, D., & Bemporad, A. (2018). Model predictive control (mpc) for enhancing building and hvac system energy efficiency: Problem formulation, applications and opportunities. *Energies, 11*(3), 631.

Simon, D. (2006). *Optimal state estimation: Kalman, h∞, and nonlinear approaches.* Hoboken, NJ: Wiley-Interscience. Retrieved from `http://site.ebrary.com/lib/alltitles/docDetail.action?docID=10308111` doi: 10.1002/0470045345

Sofos, M., Langevin, J., Deru, M., Gupta, E., Benne, K. S., Blum, D., ... Widergren, S. (2020). *Innovations in sensors and controls for building energy management: Research and development opportunities report for emerging technologies.* doi: 10.2172/1601591

The Modelica Association. (2020). *Modelica.* Modelica Association. Retrieved from `https://www.modelica.org/`

Verein Deutscher Ingenieure. (2015). *Calculation of transient thermal response of rooms and buildings* (No. 6007). Berlin: Beuth Verlag.

A Practical Data-Driven Multi-Model Approach to Model Predictive Control: Results from Implementation in an Institutional Building

Etienne SALOUX[1]*, Nunzio COTRUFO[1†], José A. CANDANEDO[1]

[1]CanmetENERGY, Natural Resources Canada
Varennes, Quebec, Canada
Phone: +1 (450) 652-0862, Fax: +1 (450) 652-5177, E-mail: etienne.saloux@canada.ca

* Corresponding Author

ABSTRACT

Model-based Predictive Control (MPC) is an effective solution to improve building controls. It consists of the use of weather and occupancy forecasts along with a control-oriented model to predict the behaviour of the building a few hours or days ahead, and thus optimize the operation of its systems. Although the potential of MPC is widely recognized, and plentiful operational data is often available, the development of a model requires a great deal of effort, significant technical expertise and knowledge of building systems. The challenge of creating a model is a hurdle that makes the on-site implementation of MPC in buildings relatively rare.

This study tackles the development of a multi-model approach to optimize the operation of electric and natural gas boilers in an institutional building to reduce greenhouse gas (GHG) emissions while maintaining the required level of comfort. This methodology leverages Machine Learning techniques to rapidly develop and calibrate control-oriented models using a limited number of input variables (indoor air temperature and temperature set-points, weather conditions, power meter data). The proposed multi-model approach consists of five models used to estimate the building total heating demand, the electric baseload, the natural gas boiler power, and the indoor air temperature under free floating conditions and during warming-up periods in the morning. The models are calibrated and validated with operational data and they are then used to optimize the transition between nighttime and daytime indoor air temperature. Since these are black-box models that require only a basic understanding of the building system and a few inputs, the model development was considerably reduced while the modularity of the proposed method makes it flexible. Such an approach could therefore be easily replicated in other buildings equipped with similar pieces of equipment.

This methodology has been implemented in a Canadian institutional building, located in Varennes (QC). Results in 2020-21 showed that the COVID-19 pandemic has significantly impacted building performance and reduced energy use, thus creating a new baseline. The MPC strategy allowed to achieve an additional 20.2% GHG emission reduction compared to this new baseline while thermal comfort was improved. Nevertheless, energy costs increased, which was mainly due to the impact of the pandemic, which eventually made the pre-COVID-19 model and optimization parameters outdated; lower costs are expected with model recalibration, currently ongoing.

1. INTRODUCTION

In recent years, model-based predictive control (MPC) has received significant attention as a promising pathway to improve energy efficiency and load management in buildings. MPC has often been associated to a laborious modelling approach, a cumbersome formulation of an optimization problem, and a challenging implementation in the building automation system. As a result, MPC is often perceived as expensive and time-consuming. While there have been some success stories in the application of MPC to real cases, this control approach is still not a mainstream practice in the industry.

† Present address: Smart Building Lab, BrainBox AI, Montreal, Quebec, Canada

The team by Cigler et al. (2013) investigated the problem of the implementation of MPC in two case studies. The authors underscored the suitability of MPC to the supervisory control of building with significant thermal mass using simple models. The need to consider controls at the design stage is also mentioned. The authors illustrated their methodology with two field pilot test cases – a 70,000-m^2 building in Prague, and a 20,000-m^2 office building in Munich. The mathematical models developed for the two cases employed system identification to find state-space models for both buildings. Zacekova et al. (2014) focused on the problem of identifying an appropriate control-oriented model (which requires, among other criteria, sufficient accuracy, repeatability, ease of calibration and implementation within a control algorithm), a central problem for the development of an MPC solution. The authors mentioned the difficulties associated with missing data, faulty sensors, and selected a grey-box approach, in which each of the five floors was modelled with a 3rd order thermal network.

Sturzenegger et al. (2014) presented Building Resistance-Capacitance Modeling (BRCM), a MATLAB® Toolbox for the creation of physical grey-box models for controls. Sturzenegger et al. (2016) presented the results of a comprehensive MPC study in a Swiss office building. The authors employed a bilinear control-oriented model based on physical principles and measured data. The study considered simultaneously thermally activated building systems, ventilation and motorized blinds for the entire building. The authors incorporated a supervisory control layer (reminiscent of the approach presented in this paper). The implementation of MPC was successful, with estimated savings of 17% in comparison with an EnergyPlus benchmark. The authors mentioned that while the effort involved seems too high for most engineering projects, "*a model predictive building automation framework, a modeling tool and the training of engineers*" may push the technology into the net benefit range. It is worth mentioning that the time and effort required to create a model are seldom reported in the literature. This issue is far from trivial since this is a critical consideration for field implementation and eventual mass adoption of MPC.

In Canada, Kavgic et al. (2015) investigated considerations to justify the deployment of MPC in a building; these considerations include significant levels of thermal mass, high ventilation rates, predictable internal and external gains and fluctuations between levels of occupancy. Hilliard et al. (2017) employed a combination of measured data and synthetic data to create a black-box model (random forest) of a 10,000-m^2 building. The resulting model was then used as the control-oriented model of an MPC strategy, focused on "nudging" zonal temperature setpoints. The MPC strategy achieved 29% savings of electric energy and 63% in thermal energy.

Drgona et al. (2020) presented a comprehensive overview of MPC for buildings. This paper contains significant and thorough information on MPC modelling, optimization and performance assessment. The discussion on modelling methods (white, grey, and black-box models) provides an overview of implementation pathways. The authors emphasized the need for multidisciplinary education to enable the widespread adoption of advanced controls.

This paper presents recent results of the implementation of an MPC solution in an institutional building. This approach is based on the identification of relevant control variables for the objective at hand (in this case, the reduction of natural gas consumption), a control-oriented model based on a judicious application of a machine learning approach and the availability of building automation data. This article complements and updates the information presented in the paper by Cotrufo et al. (2020).

2. CASE STUDY

2.1 Building description

The building under study is a Canadian institutional building located in Varennes (Quebec, Canada). This 5,257-m^2 one-story building hosts 120 workstations and 10 meeting rooms, with typical occupancy schedules from 6:30 to 17:30 during working days. To satisfy space heating demand, a 200-kW electric boiler and two 470-kW natural gas boilers are used. In Quebec, electricity is generated from hydroelectric plants, making electricity generation almost free of greenhouse gas (GHG) emissions. Therefore, the electric boiler is operated as the first priority; natural gas boilers are used when heating demand cannot be fulfilled by the electric boiler alone. To avoid significant monthly electric peak demand (which impacts the electricity bill), the total building electric power is kept below a certain limit, the *dynamic electric peak*, which is reset every month based on educated guess values and adjusted when the total building electric power exceeds dynamic electric peak (e.g. high electric baseload). This total building electric power is the sum of the electric boiler power and the electric baseload power, which accounts for occupancy-related activities (workstations, lighting, appliances, plug loads, ventilation, experimental test benches, etc.). When the total building electric demand approaches the "dynamic electric peak", the electric boiler contribution is thus reduced.

The Building Automation System (BAS) collects measurements such as temperatures, flow rates and powers at 10-min intervals. The existing database covers several years, although there is a significant amount of missing values due to technical issues (e.g. network or sensor failures). The building is composed of four main sections, served by four secondary loops connected to the central heating plant. On a typical winter workday, the average indoor air temperature is kept at 22.5°C during the day and lowered to 19.6°C at night. When it is cold outside (i.e. below -5°C), two sections "cancel" the night temperature setback and stay at 22.5°C, which makes an average building indoor temperature setpoint of 21.0°C. In normal operation, the transition between nighttime and daytime occurs at 6:30 a.m., using a sharp setpoint step transition. Since the electric boiler capacity is not able to satisfy the heating demand to achieve such an abrupt change, the gas boilers must be turned on. Such operation is denoted *Business As Usual* (*BAU*) in this paper.

2.2 Predictive control in the building: previous work

Prior predictive control work was performed and implemented in the building during winter season 2018-19 (Cotrufo & Saloux 2019; Cotrufo et al. 2020). The strategy consisted in optimizing the transition from nighttime to daytime conditions to minimize natural gas consumption. Based on weather forecasts 24 hours ahead, the algorithm used outdoor air temperature predictions to virtually test several indoor air temperature setpoint profiles to go from night setback (19.6°C) to daytime setpoint (23°C) to minimize the use of natural gas.

Simulations were used to study the response of the building in the 24-h period between 18:00 and 18:00 the next day. Different ramps for the transition were tested, ranging from a sudden step to a nearly flat setpoint; the resulting optimal profile leading to minimum gas consumption was then applied at 18:00. For this initial implementation, the total building electric power limit was kept constant at 230 kW. Results showed that 22% natural gas savings were achieved by smoothly warming-up the building using the electric boiler at night, although a 4% additional heating demand was observed. While the tests were successful, several details needed to be addressed. For example, a sharp nighttime-daytime transition at particularly low temperatures may affect occupant's thermal comfort in the morning. Conversely, ramps that are too "smooth" will cause excessive heating consumption in milder weathers.

This paper presents an MPC strategy that builds upon the one proposed by (Cotrufo & Saloux 2019; Cotrufo et al. 2020) and the results from the MPC strategy deployment in a real building during 2020-21 winter season. This new MPC strategy includes the following new features: 1) a free-floating model that estimates the average room temperature when building rooms cool down to night setback, 2) a thermal comfort model that calculates the average indoor air temperature at 7:00 and 8:00, 3) a new objective function to avoid excessive heating demand, 4) new temperature setpoint profiles with additional night setbacks, and 5) dynamic electric peak variations.

2.3 Impact of the COVID-19 pandemic

The COVID-19 pandemic had a significant impact on building operation and performance. Most employees have been teleworking since March 2020, but a non-negligible fraction is back in the office. Obviously, such a major change in occupancy affects the building thermal behaviour, specifically, internal gains due to occupants and electric baseload power. However, since this effect is difficult to account for, the same model was used for the implementation. It is worth mentioning that a lower electric baseload power was observed, which might increase the contribution of the electric boiler and thus reduce natural gas consumption. In turn, internal gains due to occupants and lighting are expected to be lower, which might increase building heating demand and natural gas consumption.

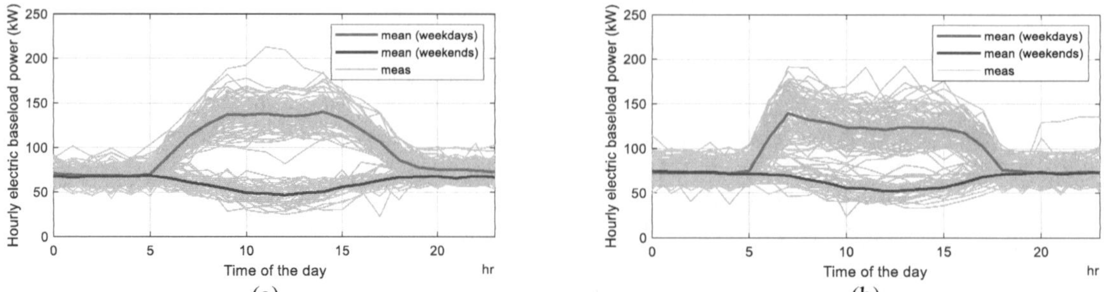

Figure 1: Hourly electric baseload power as a function of the hour of the day: (a) before COVID-19 pandemic (Nov 2019-Mar 2020), (b) after the beginning of COVID-19 pandemic (Nov 2020-Mar 2021).

Figure 1 shows the electric baseload power as a function of the hour of the day before and after the beginning of the COVID-19 pandemic. These figures show typical profiles during weekdays and weekends, before and after the beginning of the pandemic. The behaviour during weekends (in blue) was barely affected, with a mean value between 47 and 69 kW before COVID-19 and between 52 and 72 kW after the pandemic beginning. During weekdays from 9:00 to 15:00 (in red), the mean value was reduced from 137 kW down to 122 kW. This 15-kW difference, which might correspond to an internal gain reduction, is expected to be available for the electric boiler, when needed. Conversely, there are around 75 fewer occupants in the building, which might account for 6-kW internal gain losses (assuming 75 W/person). A morning peak (139 kW in average) in Fig. 1b can also be noticed.

3. METHODOLOGY

3.1 Control-oriented model

Figure 2 shows the schematic of the new multi-model approach. The control-oriented model developed in previous work (Cotrufo & Saloux 2019; Cotrufo et al. 2020) was composed of three black-box models targeting: 1) the heating demand, 2) the electric baseload and 3) the natural gas boiler power. Different Machine Learning techniques were assessed by Cotrufo et al. (2020) and Gaussian Process Regression (GPR) showed a good promise trade-off between accuracy and flexibility in terms of architecture (essentially, a Kernel function to select) as opposed to artificial neural networks. The model inputs were determined using a combination of thermodynamics considerations, practical issues and correlation analysis. The model inputs were selected so that they are either: (a) non-controllable, but known hours ahead (hour of the day, dynamic electric peak, forecasted outdoor air temperature) or (b) controllable variables (temperature setpoints). It is worth mentioning that all models use hourly average values. The original formulation included these three models:

- The *heating demand* is a function of the outdoor air temperature (OAT), the air temperature setpoint (TSP), and the temperature setpoint variation (dTSP). The latter was included to account for the additional power needed for temperature change. The GPR model used the *ardexponential* Kernel function.
- The *electric baseload* mainly depends on occupancy, which in turn depends on the hour of the day (HH). However, since the "baseload" is calculated as the total electric power minus the electric boiler power, it implicitly includes other heating elements. For this reason, the OAT was also used as an input. Although both the temperature setpoint and electric baseload follow the occupancy, it is worth mentioning that the temperature setpoint was *not* included as an input model, since the proposed approach intends to adjust setpoints independently of occupancy. The GPR model used *exponential* as Kernel function.
- The *natural gas boiler* is turned on when the electric boiler, whose contribution depends on the electric demand margin (see below), does not suffice to fulfill the building heating demand. Thus, electric demand margin (EDM) and heating demand (HD) were used as inputs. The GPR model used *exponential* as Kernel function.

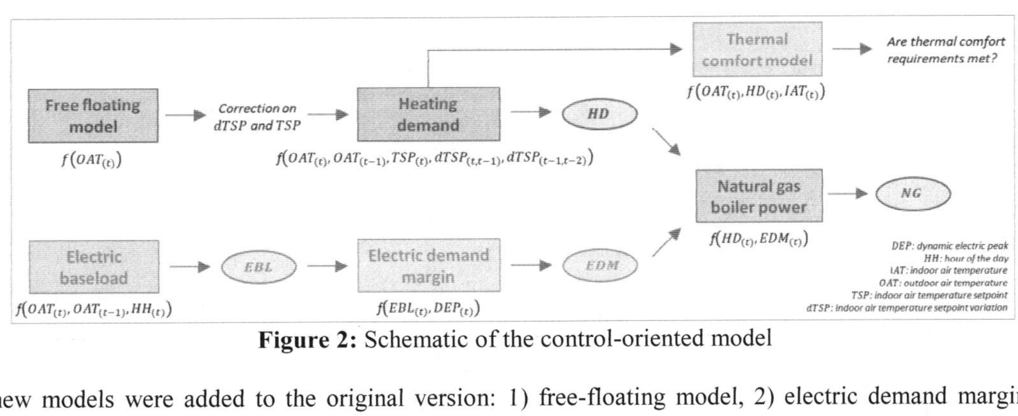

Figure 2: Schematic of the control-oriented model

Three new models were added to the original version: 1) free-floating model, 2) electric demand margin and 3) thermal comfort model.

- The *free-floating model* is a 1R-1C thermal network model (no heat source) that aims to provide a correction on the indoor air temperature setpoint and setpoint variation that are inputs of the heating demand model. This model accounts for the thermal delay of the building and the difference between the setpoint and the real temperature. A "corrected" setpoint is then used as the input for the *heating demand model*.

- The *electric demand margin* is the difference between the dynamic electric peak (DEP) and the electric baseload (EBL), and helps estimate the electric boiler contribution.
- The *thermal comfort model* aims to estimate the building behaviour during the ramp-up period. Indoor air temperature is calculated from 3:00 to 10:00 and model uses outdoor air temperature and heating demand as inputs. The GPR model used *ardexponential* as Kernel function.

3.2 The indoor air temperature setpoint profiles

Figure 3 shows the 21 profiles that were tested as transitions from nighttime (19.5°C) to daytime (22.5°C). As in Cotrufo et al. (2020), starting time of the transition varies (Figure 3, left): it can occur at 6:00 or start earlier at night (up to 19:00). Other profiles consider ramping up at 18:00 at higher night setback (20.5°C, 21.5°C) while another option is a 22.5°C constant temperature setpoint. New temperature setpoint profiles were also included with additional night setbacks at 18.5°C (Figure 3, middle) and 17.5°C (Figure 3, right).

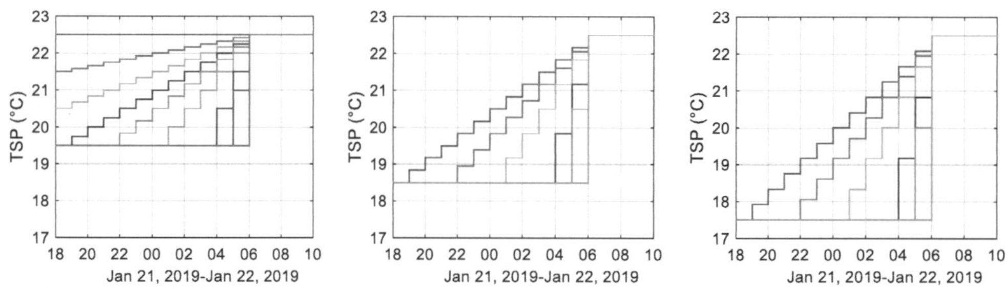

Figure 3: Indoor air temperature set-point profiles tested in the predictive control strategy (left: #1-9, middle: #10-15, right: #16-21; #1: 22.5°C constant setpoint).

3.3 Formulation of the optimization problem

The objective of the predictive control strategy is to minimize natural gas consumption while maintaining or even improving thermal comfort and avoiding excessive energy costs. The optimization problem, which minimizes the function J evaluated over the prediction horizon Λ, is written as follows:

$$\min_{TSP_i} J,\ \text{where}\ J = 0.6\left(\frac{\int_t^{t+\Lambda} NG_{TSP_i}(t)dt - \int_t^{t+\Lambda} NG_{BAU}(t)dt}{\int_t^{t+\Lambda} NG_{BAU}(t)dt}\right) + 0.4\left(\frac{\int_t^{t+\Lambda} HD_{TSP_i}(t)dt - \int_t^{t+\Lambda} HD_{BAU}(t)dt}{\int_t^{t+\Lambda} HD_{BAU}(t)dt}\right) \quad (1)$$

$$s.t.$$
$$i \in [1, 21]$$
$$IAT_{7:00} \geq 21.5°C$$

where TSP_i is the ith indoor air temperature profile, NG_{TSP_i} and HD_{TSP_i} are the natural gas consumption and heating demand of the ith indoor air temperature profile, and NG_{BAU} and HD_{BAU} are the natural gas consumption and heating demand under the "Business As Usual" strategy. There is a constraint on indoor air temperature (IAT) at 7:00 (first occupants' arrival), which must be higher than 21.5°C. Weighting factors (0.6 and 0.4) were manually tuned to prevent choosing a profile with a small reduction in natural gas but a high increase in heating demand.

3.4 Predictive control strategy

The MPC routine was written in MATLAB® and was run automatically the evening before each workday (Sunday to Thursday), slightly before 18:00. The indoor air temperature for the free-floating model was assumed to be at 18.5°C at 18:00 on Sunday evenings, and at 22.5°C the other days. The MPC strategy was implemented during the winter season 2020-21, from Nov 12 to Mar 19. The predictive control strategy consists of the following steps:

1- Weather forecasts with a prediction horizon of 24 hrs are retrieved using a software tool developed by Natural Resources Canada (Natural Resources Canada 2017; Candanedo et al. 2018).
2- Dynamic electric peak is retrieved from the BAS.
3- Weather forecasts and dynamic electric peak are used along with the control-oriented model to estimate from 18:00 to 18:00 the next day the objective function associated to each indoor air temperature profile.
4- The indoor air temperature profile that minimizes the objective function is sent to the BAS via a virtual controller and is applied to each room of the building from 18:00 until daytime building operation overrides nighttime operation (6:30).

3.5 Assessment of savings

Two benchmark models were used to evaluate building performance on a daily basis: 1) under "*BAU*" strategy before COVID-19 and 2) under "*new BAU*", during the COVID-19 pandemic. The first model helps assess the effects of COVID-19 by providing a pre-COVID-19 baseline. The second model helps evaluate the impact of MPC strategy by providing a *new BAU* baseline. *BAU* models are GPR type and calculate heating demand (*Matern32* Kernel function) and natural gas consumption (*Matern52* Kernel function) as a function of daily average outdoor air temperature; they were trained using 2015-18 data (Cotrufo et al. 2020). *New BAU* models are linear and quadratic type due to the low amount of available data. For the economic analysis, as building peak demand occurs during the day, electricity cost is estimated using only the energy rate (5.03 CAD c/kWh). Compared to previous work (Cotrufo & Saloux 2019; Cotrufo et al. 2020), natural gas price is now based on gas utility average cost over winter months (5.42 CAD c/kWh), which includes transportation, balance fees, etc. Finally, GHG emissions were assumed to be 0.00036 t-CO2eq/GJ for electricity and 0.0507 t-CO2eq/GJ for natural gas.

4. RESULTS

4.1 Model training and validation

4.1.1 Control-oriented models: each model of the multi-model approach was individually calibrated using building operational data, between November 2017 and March 2019. The dataset was cleaned (outliers, missing values, etc.) and non-stratified random partition was used to divide the data into training and validation datasets. For the thermal comfort GPR model, data between 3:00 and 9:00 were considered. For the free-floating RC network model, less data is required and only 2018-19 winter season was considered. Values between 18:00 and 22:00 were considered, when the building was more likely to be in free floating mode. Most of the time, *there is* a heating demand, even when the indoor air temperature is higher than the setpoint; this is most likely a result of local control rules designed to dampen load fluctuations and avoid equipment cycling. It was not considered for the calculation of RC parameters (R=0.560 K/W, C=351,726 J/K, equivalent to a 55-hr time constant) but was kept in the calculation of HD.

Figure 4 shows model predictions at time *t* for one week during winter 2019 for heating load, electric baseload and natural gas consumption and Table 1 gives dataset and model accuracy for each model. Values similar to those provided by Cotrufo et al. (2020) were obtained for HD, EBL and GAS. While EBL and GAS perform well, HD model struggles more to catch peak loads. This can be explained by the fact that the heating demand was calculated with the sum of electric boiler and natural gas boiler contributions, and that natural gas boiler power mainly shows peaks, deduced from pulse readings. For temperature models (free floating, thermal comfort), the error remains below 0.2°C, which fall inside sensor uncertainty range.

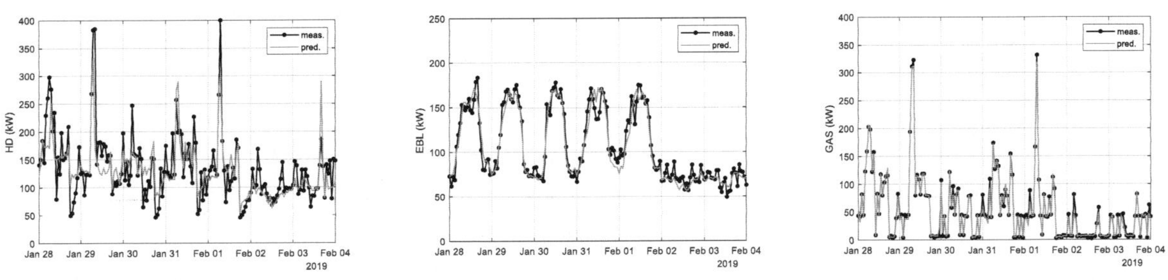

(a) (b) (c)

Figure 4: Control-oriented model hourly predictions for one week during winter 2019: (a) heating load, (b) electric baseload, and (c) natural gas boiler power.

Table 1: Control-oriented model dataset and accuracy

Variable	Heating demand	Electric baseload	Gas boiler power	Free floating	Thermal comfort
Dataset	4305 hourly values (50% for training)			207 hourly values	808 hourly values (50% for training)
RMSE - GPR (train/val)	25.1 / 34.0 (kW)	11.9 / 15.0 (kW)	4.4 / 7.3 (kW)	-	0.12 / 0.22 (°C)
RMSE - RC network	-	-	-	0.14°C	-

Figure 5 shows model results when the outdoor air temperature varied between -8°C and 1°C. Indoor air temperature in free floating mode decreases until it reaches the setpoint and HD only depends on OAT during this period. From this point onwards, TSP and dTSP become inputs of HD model. At 3:00, the thermal comfort model calculates the indoor air temperature for the next 7 hours by considering, as initial condition, that the indoor air temperature is equal to the setpoint. Note that at 10:00 the value is close to 23°C, slightly higher than the setpoint (22.5°C). In fact, the model training was done on a period when the daytime setpoint was 23°C, not 22.5°C. HD is shown in orange and the fraction of HD covered by natural gas (GAS) in blue. PID parameters could have been included in the free floating model but the current model gives a reasonable estimation of the starting time of equipment ramp-up.

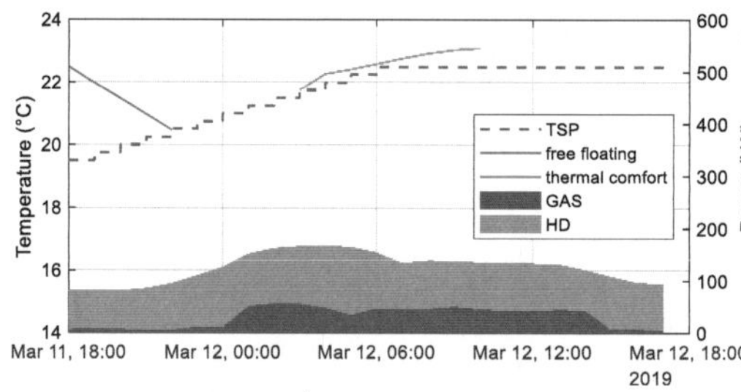

Figure 5: Control-oriented model results for a typical day and a given temperature setpoint profile.

4.1.2 Benchmark models: Figure 6 shows measured data under *BAU* operation (2015-18), *new BAU* (2020-21) and benchmark models (baseline) for daily heating demand and natural gas consumption. Table 2 gives model dataset and accuracy for *BAU* and *new BAU*, which includes pandemic effects. GPR models were based on daily values in 2015-18 (Cotrufo et al. 2020); in turn, the *new BAU* models only considered few values of 2020-21 data. The impact of the COVID-19 pandemic is discussed in the next subsection.

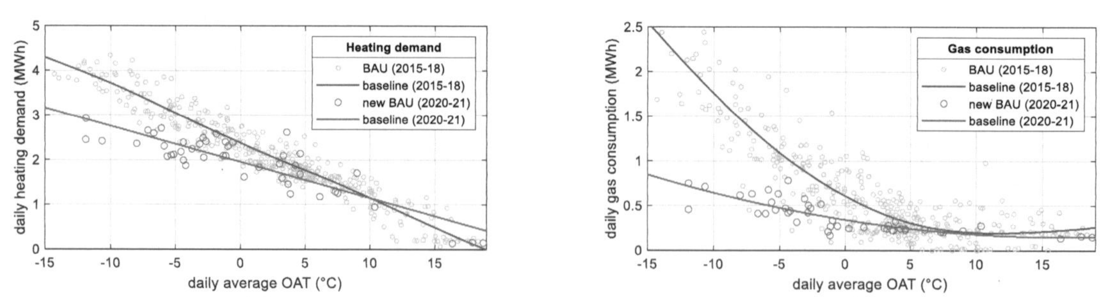

(a) (b)

Figure 6: Benchmark model for building performance under *BAU* and *new BAU*: (a) daily average heating demand and (b) daily average natural gas boiler consumption as a function of daily average outdoor air temperature.

Table 2: Benchmark model accuracy

Variable	Heating demand (*BAU*)	Gas consumption (*BAU*)	Heating demand (*new BAU*)	Gas consumption (*new BAU*)
Dataset	369 daily values (70% for training)		45 daily values	
GPR (train/val)	0.269 / 0.255 (MWh)	0.241 / 0.283 (MWh)	-	-
Linear, quadratic	-	-	0.316 (MWh)	0.100 (MWh)

4.2 Energy savings and GHG emission reduction

Table 3 and Figures 6-7 show results from *new BAU* and MPC implementation in 2020-21.

- *Effects of the COVID-19 pandemic* can be evaluated by comparing *BAU* with *new BAU* in Table 3, and the grey line (*BAU* baseline) with red dots (*new BAU* implementation results) in Figure 6.
- *Benefits of the MPC strategy* can be evaluated by comparing MPC with *new BAU* in Table 3, and the red line (*new BAU* baseline) with green dots (MPC implementation results) in Figure 7.

Table 3: Performance obtained from *new BAU* and MPC implementation during winter 2020-21

Variable	**BAU (baseline)**	**New BAU (meas.)**	**Difference**	**New BAU (baseline)**	**MPC (meas.)**	**Difference**
Building heating demand	107.5 MWh	87.4 MWh	- 18.7 %	78.4 MWh	98.0 MWh	+ 25.0 %
Electric boiler consumption	72.8 MWh	70.6 MWh	- 3.0 %	63.4 MWh	86.3 MWh	+ 36.0 %
Natural gas boiler consumption	34.7 MWh	16.8 MWh	- 51.5 %	15.0 MWh	11.7 MWh	- 21.8 %
Total energy cost	5,542 CAN$	4,463 CAN$	- 19.5 %	4,003 CAN$	4,976 CAN$	+ 24.3 %
GHG emissions	6.43 t CO_2eq	3.16 t CO_2eq	- 50.8 %	2.81 t CO_2eq	2.25 t CO_2eq	- 20.2 %

The COVID-19 pandemic has affected the building performance: heating demand was reduced by 18.7% and natural gas consumption by 51.5%. This reduction was quite significant at low outdoor air temperatures. With the MPC strategy, there are clear benefits in terms of gas consumption, achieving an additional 21.8% reduction compared to *new BAU*. This was obtained by increasing electric boiler use during off-peak periods, but also at higher heating demand (25.0%) and energy costs (24.3%). MPC showed substantial reduction at low outdoor air temperature for a similar or modest increase in heating demand; however, it was less effective in warmer weather, as heating demand was significantly increased for low natural gas savings. The trade-off between heating demand increase and gas consumption reduction was handled within the objective function (Eq. 1) and is still based on operational data under the original *BAU*, not the *new BAU*. When compared to BAU, heating demand and energy cost would have been similar (0.4% and 1.7% decrease, respectively, not shown in Table 3). Recalibration with data under the new BAU could adjust model and optimization parameters to the new reality and lower energy costs would be expected, as obtained by Cotrufo et al. (2020).

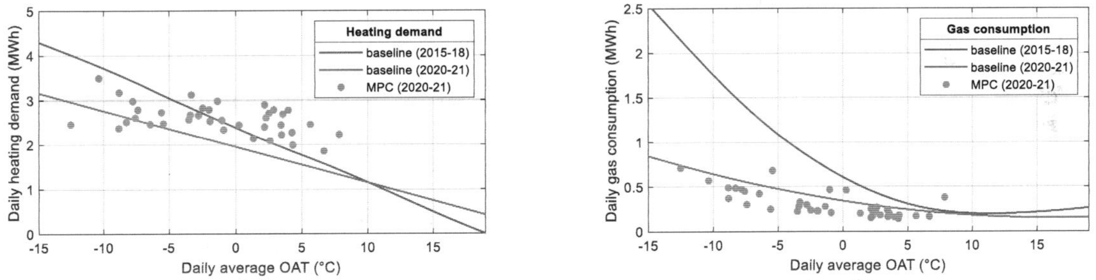

(a) (b)

Figure 7: (a) Daily average heating demand and (b) daily average natural gas boiler consumption as a function of daily average outdoor air temperature under MPC in 2020-21.

4.3 Trends in the selection of indoor air temperature setpoint profiles

Cotrufo et al. (2020) investigated the selection of indoor air temperature profiles. It was found that at lower outdoor air temperatures, a sharp transition between nighttime to daytime conditions was optimal; at higher temperatures, smooth transitions were selected. In fact, at low temperatures, the natural gas boilers are already operating at night and building preheating causes an increase in heating demand and thus, in gas consumption. In contrast, this is not the case when it is warmer outside: natural gas consumption in the morning during ramp-up can be shifted into electric boiler consumption at night.

Figure 8 shows temperature setpoint profile as a function of daily outdoor air temperature. Adding new night setbacks and thermal comfort constraints makes the interpretation of profile selection less evident, although the trend is similar. Smooth transitions (profiles #1-4) were selected at higher temperatures; profile #2 (21.5°C night setback and starting time at 18:00) was even selected 40% of the time. Profiles #12-16-18 with 17.5-18.5°C night setback were selected at low temperatures and on Sunday evening when indoor temperature was already at 18.5°C.

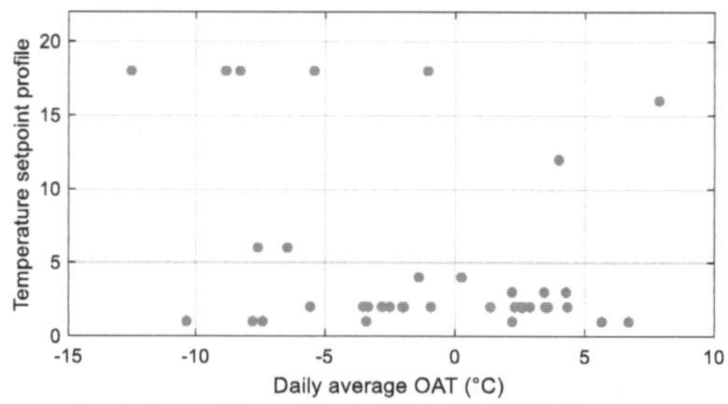

Figure 8: Indoor air temperature setpoint profile selected by the MPC strategy as a function of the daily average outdoor air temperature.

4.4 Thermal comfort

Building daytime schedule starts at 6:30 and occupant's thermal comfort should be satisfied soon after. Thermal comfort was assessed by plotting indoor air temperature at 7:00 (first occupant's arrival) and 8:00 (indoor air conditions expected to be achieved). Results are shown in Figure 9. The constraint on thermal comfort in the optimization problem helped increase the indoor air temperature in the morning, by excluding setpoint profiles that might have caused discomfort. Compared to *BAU* and *new BAU*, MPC was able to almost consistently achieve 21.5°C at 7:00 while under *BAU* and *new BAU*, temperatures between 21°C and 22°C were frequently achieved.

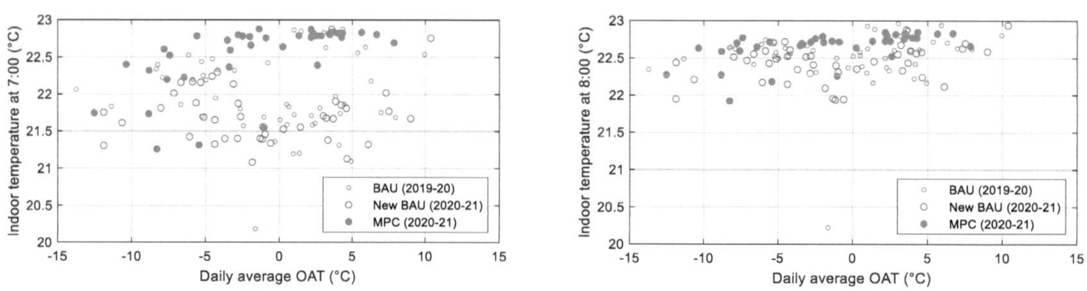

(a) (b)

Figure 9: Indoor air temperature as a function of daily average outdoor air temperature: (a) at 7:00 and (b) at 8:00.

5. LESSONS LEARNED FROM IMPLEMENTATION

The development of data-driven black-box models is no easy task and it strongly depends on data availability and quality ("*garbage in, garbage out*"). The followed approach intended to develop a strategy that relies on very few inputs (essentially, indoor temperatures, electric boiler power, total building electric power and natural gas boiler power) for replicability purposes and to maximize data availability. Indeed, it barely happens that all required monitoring data is available at the same time and using hourly average values helped increase data availability. Training data covered a total of 10 months, on which only 59% of hourly average values were compiled in the training dataset due to missing periods (mainly for building total electric power and outdoor air temperature) ranging from a few hours to a few days. Averaging indoor temperatures and temperature setpoints was also beneficial, as it allowed to exclude missing data (12-26% of the 10-min dataset) for specific rooms within a zone. With data availability in mind, Deep Learning models such as long-short term memory artificial neural networks are very appealing in terms of accuracy but their actual implementation in typical real buildings might be difficult as they require consecutive time-series data.

Another challenge was the bi-directional communication with the BAS. The optimization consisted in predefined setpoint profiles that were manually entered in the control system in such a way that the optimal profile number was sent to the BAS and the corresponding profile was applied. A more automatic approach could be explored while the compatibility with commercial tools that are given permission to send commands to the BAS could be worth of investigation.

6. CONCLUSIONS AND FUTURE WORK

While a significant transformation in the training of building engineer professionals (e.g. in control engineering) is desirable, and even necessary, to fully exploit the potential of MPC, the severity and urgency of environmental problems call for a practical approach for the deployment of predictive control, even if only a fraction of MPC capabilities are achieved. The replicability and the affordability of the method is one of its interesting features. The impact of deploying an "imperfect" (but simple and cheap) methodology in hundreds of buildings would surpass the benefits of a "full" MPC implementation considering a detailed model with numerous controlled variables.

Future work includes applying the MPC strategy to other commercial or institutional buildings (under way) to see whether similar conclusions (savings, operational trends) could be achieved. Moreover, control-oriented models should be recalibrated to account for the new performance under COVID-19 pandemic. The proposed strategy was also not very efficient on warmer periods; a dedicated model could be developed for these conditions, targeting heating demand reduction while maintaining thermal comfort and low natural gas consumption. In this study, indoor air temperature profiles were applied uniformly to all the rooms. This supervisory control strategy could be refined with local controls for groups of systems or rooms. Peak load management to support electric grid needs could be investigated using a similar approach, with models targeting flexibility, focusing on electric power, while natural gas consumption would be treated as a penalty term. Such work could target both heating and cooling systems, focusing on occupied hours as the electric peak generally occurs during the day. In this situation, a multi-model approach could be very useful. To better quantify flexibility, load disaggregation could be investigated in depth.

REFERENCES

Candanedo, J. A., Saloux, E., Hardy, J.-M., Platon, R., Raissi-Dehkordi, V., & Côté, A. (2018). Preliminary Assessment of a Weather Forecast Tool for Building Operation. In *5th International High Performance Buildings Conference*. Purdue, USA, July 9-12.

Cigler, J., Gyalistras, D., Siroky, J., Tiet, V.-N., & Ferkl, L. (2013). Beyond Theory : The Challenge of Implementing Model Predictive Control in Buildings. In *CLIMA 2013 - 11th REHVA World Congress and the 8th International Conference on Indoor Air Quality, Ventilation and Energy Conservation in Buildings*. Prague, Czech Republic, June 16-19.

Cotrufo, N., Saloux, E., Hardy, J.-M., Candanedo, J. A., & Platon, R. (2020). A Practical Artificial Intelligence-Based Approach for Predictive Control in Commercial and Institutional Buildings. *Energy and Buildings* 206, 109563.

Cotrufo, N., & Saloux, E. (2019). Artificial Intelligence for Advanced Building Control: Energy and GHG Savings from a Case Study. *Proceedings* 23, 7.

Drgoňa, J., Arroyo, J., Figueroa, I., C., Blum, D., Arendt, K., Kim, D., Ollé, E., P., Oravec, J., Wetter, M., Vrabie, D., L., & Helsen, L. (2020). All You Need to Know about Model Predictive Control for Buildings. *Annual Reviews in Control* 50, 190-232.

Hilliard, T., Swan, L., & Qin, Z. (2017). Experimental Implementation of Whole Building MPC with Zone Based Thermal Comfort Adjustments. *Building and Environment* 125, 326–38.

Kavgic, M., Hilliard, T., & Swan, L. (2015). Opportunities for Implementation of MPC in Commercial Buildings. *Energy Procedia*, 6th International Building Physics Conference, IBPC 2015, 78, 2148–53.

Natural Resources Canada (2017). CanMETEO, English / French. https://www.nrcan.gc.ca/energy/software-tools/19908.

Sturzenegger, D., Gyalistras, D., Morari, M., & Smith, R., S. (2016). Model Predictive Climate Control of a Swiss Office Building: Implementation, Results, and Cost–Benefit Analysis. *IEEE Transactions on Control Systems Technology* 24, 1–12.

Sturzenegger, D., Gyalistras, D., Semeraro, V., Morari, M., & Smith, R., S. (2014). BRCM Matlab Toolbox: Model Generation for Model Predictive Building Control. In 2014 American Control Conference (ACC). Portland, Oregon, USA, June 4-6.

Žáčeková, E., Váňa, Z., & Cigler, J. (2014). Towards the Real-Life Implementation of MPC for an Office Building: Identification Issues. *Applied Energy* 135, 53–62.

ACKNOWLEDGEMENT

The authors gratefully acknowledge the financial support of Natural Resources Canada through the Office of Energy Research and Development. The authors would also like to thank Jean-Marc Hardy and Mudasir Ahmed from CanmetENERGY for their technical assistance to implement the MPC strategy.

Optimization of Multi-zone Building HVAC Energy Consumption by Utilizing Fuzzy Model Based Predictive Controller

Ahmad Esmaeilzadeh [1,2]*, Brian Deal [3], Aghil Yousefi Koma[2], Mohammad Reza Zakerzadeh[2]

[1]Smart Energy Design Assistance Center (SEDAC), Dept. of Landscape Architecture, University of Illinois at Urbana-Champaign,
Champaign,Illinois, USA
+1 786 (543) 9382, +98 913 (253) 2536, ahmad91@illinois.edu, a.esmaeilzadeh@ut.ac.ir

[2]School of Mechanical Engineering, College of Engineering, University of Tehran,
Tehran, Iran

[3]Dept. of Landscape Architecture, University of Illinois at Urbana-Champaign,
Champaign,Illinois, USA
+1 217 (333) 1911, deal@illinois.edu

ABSTRACT

The rapid improvement of living standards has led to increased energy consumption in buildings worldwide. Globally, the energy consumed in buildings accounts for 20.1% of total delivered energy (EIA 2016). Improving energy efficiency in buildings therefore is an important component for combating climate change. This paper aims to improve end use energy efficiency in multi-zoned residential buildings through the application of thermal comfort based, energy optimization algorithms. We use a case study approach with a detailed analysis of a 4-story residential apartment building in central Illinois. The study building constitutes 21 thermal zones modeled in EnergyPlus. The model is validated using monthly energy consumption data. The effectiveness of four different steam heating system control methods are evaluated and described: a) a Model Predictive Controller (MPC) design based on neuro-fuzzy temperature predictor; b) a Proportional-Integral-Derivative (PID) tuned by fuzzy logic; c) a PID tuned by a genetic algorithm; and d) an on/off controller and the flow regulator based on indoor temperature. All are optimized for energy consumption reduction potential and thermal comfort. The main effect of the various control methods is tuning boiler feed flow by regulating the condensing cycle. A reduction in circulated steam flow results in decreased direct energy consumption and improved condensing pump efficiencies. We find that the MPC design using a neuro-fuzzy temperature predictor can reduce heating energy use by up to 38% in comparison with an on/off controller baseline.

1. INTRODUCTION

Improvements in living standards and the associated environmental and energy problems have become critical challenges worldwide. More than 40% of primary energy use worldwide is related to the heating, cooling and lighting of buildings (Fan *et al.*, 2018). The development of better modeling and control techniques for Heating, Ventilation and Air Conditioning (HVAC) systems has therefore, become critical for providing occupancy driven energy and comfort management, prioritizing carbon emission reductions, and decreasing overall consumed energy (Mary Reena *et al.*, 2018). Building energy modeling techniques are making it easier to design efficient control and management techniques for building HVAC systems. Multi objective optimization methods that use thermal comfort and energy efficiency as two conflicting objectives of a controller design can be simulated using these modeling platforms (Mary Reena *et al.*, 2018). In this paper, we use these techniques to model end use energy efficiency in a multi-zoned residential building.
The literature on building Management Systems (BMS) and automation show them to be expanded to reduce energy demands and preserve thermal comfort mainly through sophisticated control strategies (Michailidis et al., 2018). For example, Ahn et al. (Ahn *et al.*, 2017) utilize Fuzzy Inference System (FIS) and Artificial Neural Network (ANN)

for simultaneously controlling the amount of supply air and its temperature. They note that using this method, the thermal demands in disturbance condition is controlled and the consumed energy is reduced in comparison to a simple thermostat on/off controller (Ahn *et al.*, 2017). A smart strategy based on Model Predictive Control (MPC) is employed to improve the performance of hydronic radiant floor systems in office buildings. The system is noted to minimize energy consumption while providing thermal comfort - a 16% energy savings compared to baseline (Joe & Karava, 2019). MPC using ANN was implemented to improve consumed energy in residential buildings by Finck et al. (Finck *et al.*, 2019). They modified MPC by considering a function to adapt energy demands and supplies during fluctuations in supply resources (Finck *et al.*, 2019). The control methods are utilized for HVAC subsets including economizers, variable air volume dampers, pumps and air handling. Based on Shea et al. (2019), implementing controllers for all economizers in commercial office buildings results in an average of 17% cooling savings (Shea *et al.*, 2019). The various control methods utilized for optimizing heat pump systems in buildings are reviewed by Péan et al. which demonstrate the MPC as the most considered method in the literature (Péan *et al.*, 2019).

Utilizing MPC requires detailed and complex energy modeling simulations. So, a Reduced-Order Model (ROM) is utilized to design MPC for optimally planning space-heating load by Aoun et al. (2019) The utilized ROM methods are shown reliable enough for designing MPC (Aoun *et al.*, 2019). MPC has also been implemented as a parallel controller for set-point optimizers. This method resulted in the reduction of 1430 KWh (equal of 20% of total consumption) during the heating season in residential buildings with electrical under floor heating systems (Ławryńczuk & Ocłoń, 2019). Genetic Algorithms (GA) have also been shown as an effective methodology for energy optimization in various thermal building systems (Li *et al.*, 2017).

Shaikh et al. (2018) implemented a Multi Objective Genetic Algorithm (MOGA) to improve energy efficiency as well as provide comfort condition in a one day simulation. The MOGA maintained high comfort level and optimize energy efficiency (Shaikh *et al.*, 2018). The combination of ANN and MOGA is also implemented to optimize a two-chiller system, Variable Air Volume (VAV) chiller system, and Dedicated Outdoor Air System (DOAS). This combination results in a significant improvement in HVAC operation, while keeping energy consumption low compared to the base-case (Nasruddin *et al.*, 2019). GA also is used to determine the optimal operating schedules for the heat generation equipment, thermal storage, and the heating set point temperature. Based on Reynolds et al. (2019), GA optimized heat generation by 44.88% compared to a rule-based, priority order baseline strategy (Reynolds *et al.*, 2019). Utilizing MPC as a control methodology and optimizing through GA have been shown to be the most useful methods for Multi Input Multi Output (MIMO) optimization of HVAC systems in typical buildings.

In most of the literature on these control strategies, thermal comfort and consumed energy are emphasized as the objectives of the optimization process, with air temperature as the thermal comfort parameter. Consumed gas and utilized electricity are considered simultaneously (as 'energy') in most studies and the application of HVAC systems typically utilize simple building parameters.

The most important contribution of this paper is the inclusion of indoor air temperature for providing occupant thermal comfort; reducing consumed gas (in boiler of steam system) and condensate pump electricity usage for energy optimization. The other novelty of this paper is simulating a real residential building including 21 thermal zones which is evaluated through real information of consumed energy.

In the following sections multi-zoned energy modeling techniques used and the ambient condition will be described. The design methodology for various control strategies are then described. The application output for each controller condition, simulation results, and the effectiveness of optimization are explained in the 'results' section. Finally, the comparison of utilized methods and the benefits each are concluded.

2. THERMAL MODELLING

2.1 Building Thermal Specification

The considered building includes 4 floors. 3 upper floors are utilized as residential building and the basement is empty looks to be common spaces of some sort. Each residential floor is separated into 7 thermal zones. The considered building is presented in figure 1.

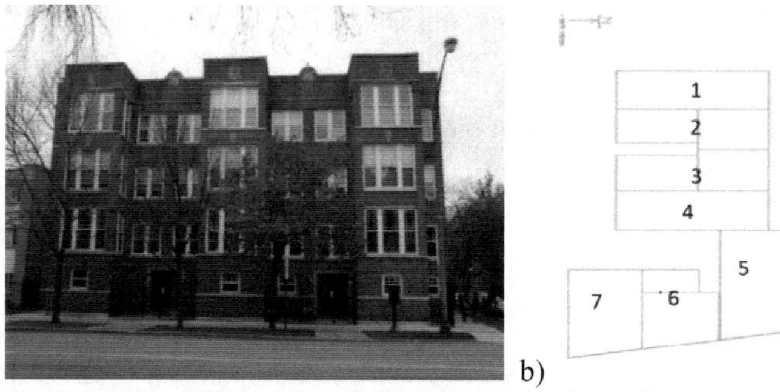

a) b)

Figure 1: a) Real building East View, b) Building Floors plot including thermal zones positions

According to the structural material of the building, the standard template material presented in EnergyPlus software is implemented to model thermal energy transferred through walls, floors, windows, roofs and ground level. The basement of the building is empty and separated building thermal zones and ground. The basement of the building is considered as a separate zone coupled to the ground.

By considering every zone as a system, energy is transferred by convection, conduction, radiation and mass transferred. Based on the energy conservation law, by ignoring the negligible transferred energy through air movement by doors area (usually closed and open only a few times), the indoor air thermal energy equation of every zone is as Equation (1).

$$\dot{E}_t + \dot{E}_g = \dot{E}_S \tag{1}$$

Where E_t presents the amount of transferred energy, E_g is the amount of generated energy by HVAC system, equipment, people etc. and E_s represents the changed energy level inside the zone. When these parameters are considered in every time step, the equation of energy is utilized for every time step based on the derivative of energy. The transferred energy calculates based on the BS EN 12831 (EN, 2003) which is presented in Equation (2).

$$T,i = (Q_{tr,ie} + Q_{tr,iue} + Q_{tr,ig} + Q_{tr,ij})(T_{inside} - T_{Outside}) \tag{2}$$

In this equation, $Q_{tr,ie}$ represents heat loss coefficient of transferred energy through zone envelope, $Q_{tr,iue}$ is unheated space energy transferred coefficient, $Q_{tr,ig}$ is heat loss coefficient of ground energy transferred, $Q_{tr,ij}$ shows the transferred energy coefficient through neighboring places at a significantly different temperature. The details of these parameters and methods of calculation are described in reference (EN, 2003). Also, the energy loses through the ventilation system are considered. The other equation for simulating ventilation heat loss is as Equation (3).

$$V_i = H_{v,i}(T_{inside} - T_{Outside}) \tag{3}$$

Where, $H_{v,i}$ represents design ventilation heat loss coefficient. For more information, the details of calculating this coefficient are presented in Ref. (EN, 2003).

2.2 Heating Systems

Steam supplied to cast-iron radiators is generated through a 160 Pa boiler (2,713,000 BTU/Hr. input) for the steam system and domestic hot water (DHW) is provided by a 410,000 BTU/Hr. gas-fired water heater. The variable flow pump is implemented to feed boiler and indirectly contribute to flow in the distribution system. Also, the electrical equipment and occupancies' generated energy is considered by utilizing the schedule time of presenting occupancy, implementing equipment, utilizing lighting system, and DHW flow rate.

2.3 Weather condition

A weather file for Chicago's O'Hare airport was used for this model using an EnergyPlus default.

2.4 Energy Modelling and evaluation

By considering the thermal characteristics of the building materials, ventilation system of the building, the weather condition and energy generation system, the building thermal system is modeled by Energy-plus. In order to evaluate the energy modeling, consumed gas in the sample year has been compared with real data presented in figure 2a. Also, the average monthly temperature of simulation and real one is presented in Fig 2b.

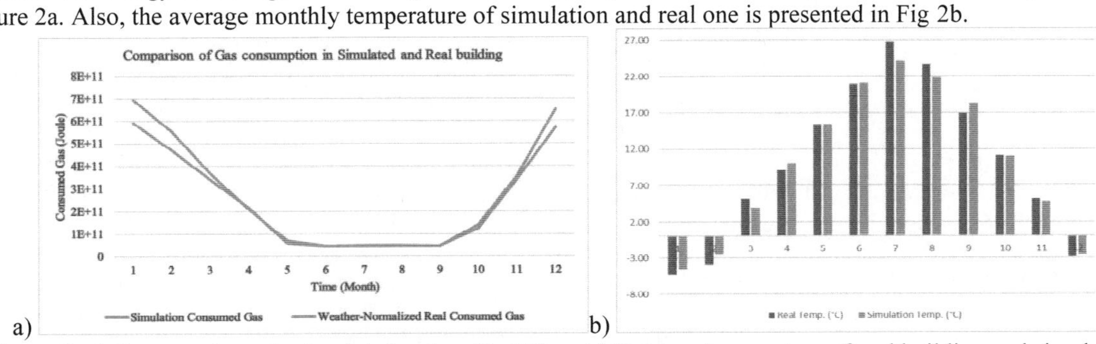

Figure 2: a) Consumed gas in simulated and real building, b) Outdoor temperature of real building and simulation

The modeled gas consumption matches up well with the actual consumed gas in the summer months and in the spring/fall months. In the winter months, the modeled consumption is slightly lower than the actual consumption data because of the difference between actual temperatures and warmer weather in the weather file. The maximum difference between real data and simulation is less than 14% and the average error is less than 8%.
The maximum difference occurred when the temperature difference between simulation and reality is almost 12%. Also, the difference is predictable because of ignoring the effects of air changes during the door opening time. Moreover, by considering the effect of increased infiltration in winter, the model aligns well with actual consumption. Accordingly, the simulation results of the steam heating system are acceptable.

2.5 Pre-Upgrading of Heating System

Based on the initial system, application of smart controllers seems impossible because of the boiler condition, radiators valve, and simple regulator of burner. So, upgrading boiler and its burner to a new one including the ability for variable flow could results in application of control methods potential. Also, steam traps must be redesigned including control valve to regulate the steam based on the zones temperatures, and smart thermostats have to be installed on radiators to prepare the condition of application of control system.

3. DESIGN CONTROL METHODOLOGY

The designed steam heating system includes lots of constraints which force us to optimize systems by utilizing variable speed equipment. By considering the upgraded system including control valves, smart thermostats, and variable speed flow pump as the condensate system, flow control could be employed to improve steam heating system efficiency. In other words, in order to minimize consumed energy in steam system, the rate of pump flow which feeds the boiler and the steam rate which is controlled by control valve in heating system is optimized. The controllers are designed based on minimizing consumed energy as well as providing thermal comfort condition in various thermal zones. Based on the initial assumptions, the thermal zones are occupied all day long, so are the set points at comfort for each zone. By considering the utilized facilities in HVAC system including steam boiler, burner controller, control valve, and pump, the flow rate could control the consumed energy based on the fixed output designed temperature. In other words, while you are going to set boiler output steam temperature at 107°C, and the transferred flow is reduced, the burner must work less to supply required energy and consumed gas decreases by this method. On the other hand, the transferred heating rate in every zone is reduced because of less feed hot steam. The steam flow rate is determined based on the condensate pump flow rate and indoor temperature. So, tuning the flow of condensate pump as well as regulating radiators steam flow based on condensate flow by considering both objectives is required to provide comfort condition as well as optimized energy consumption.
By considering the initial condition of the real building, the indoor temperature of the first floor, second floor, and third floor is set on 20 °C, 23 °C, and 25 °C respectively.

3.1 On/Off Controller

The initial controller which is used to set indoor temperature is an on/off method which controls the temperature by the steam regulator of the zones. Also, an energy management method is utilized to improve comforts by fixing temperature through regulating the flow rates in various thermal zones. The upper and lower level of hysteresis loop are set on 108°C and 103°C for boiler output respectively. In order to reduce the consumed energy of HVAC system, the regulator controlled by an on/off switch is employed to turn the burner on/off based on the zones thermostats feedback. Also, the flow of feed water pump and steam flow rate are reduced by considering the comfort condition which is resulted in increasing the time of releasing energy through steam. The mean temperature is considered in the range of (20,25) and the hysteresis loop is designed based on this period. Based on the initial system and by considering the upgraded facilities, the on mode is designed based on the 0.6 of the initial flow rates.

3.2 PID Controller

The PID controller is a common feedback controller widely used in industry. This controller could be tuned by numerical methods such as Ziegler Nichols, GA, and fuzzy logic system (Khanmirza *et al.* ,2016). By considering energy optimization as well as setting indoor temperature as the objectives, GA and fuzzy logic systems are utilized to tune PID coefficients. The PID controller is designed to regulate the flow rate of condensate pump in HVAC system. The output of the controller is forced to set in the range of (0.03,0.6) based on the initial system to improve controller effectiveness. The PID controller is implemented in SIMULINK to tune building HVAC system. The PID output is defined as the input of Energyplus combined with SIMULINKK through MLEP.

3.2.1. PID Tuned by GA

GA is a metaheuristic optimization method which selects the parameters by the natural process. By utilizing GA, initial population improves step by step resulted in parameters which is near the final optimum state (Khanmirza *et al.*, 2016). The designing parameters of GA including population, iteration, mutation percentage, and minimum error are considered as 20, 10, 0.1, and 0.00001 respectively.

3.2.2. PID tuned by fuzzy logic system

The other effective method of PID coefficient tuning is fuzzy logic system. By utilizing this methodology, the PID coefficient is determined based on the (Wang, 1999) described instruction. The difference between systems output temperatures and the set point temperatures are considered as the errors of systems. The fuzzy logic system is designed based on the error and its derivative. The coefficients of PID controller are calculated by fuzzy logic system and always have the values in the range of (0, 0.1). In order to design fuzzy logic system, 3 membership functions for error and 3 membership functions for error derivative are considered.

3.3 Model Predictive Control

One of the most effective methods in building energy efficiency is Model Predictive Control (MPC). In order to simplify complicated multi-zone nonlinear system, the neuro-fuzzy method is utilized to determine the relation of condensate pump flow rate, ambient temperature and the mean zones temperatures. In the neuro-fuzzy method, the back-propagation method is utilized to optimize grid partitions. The information on January and February 2011 is implemented for training and testing respectively. Based on the testing results, the mean zones temperatures could be determined by model accurately (The error is less than 5%). Then, The MPC is designed based on simplified model. The prediction horizon and control horizon are assumed equal to 20 and 5 respectively.

4. RESULTS

The effectiveness of defined optimization methods is evaluated by co-simulation of different control methods in SIMULINK and building energy model in EnergyPlus. Accordingly, the building HVAC system is simulated for a sample month (January) and different control methods are applied to provide their performance.
One of the designed controllers' objective is providing thermal comfort condition especially indoor temperature. The considered set point temperatures of different thermal zones are applied in the simulation. On the other hand, the effect of every zone is considered on error (calculated based on the zones air temperatures and setpoints) applied as the input of designed controllers. The zones air temperatures resulted from application of controllers are presented in figure 3.

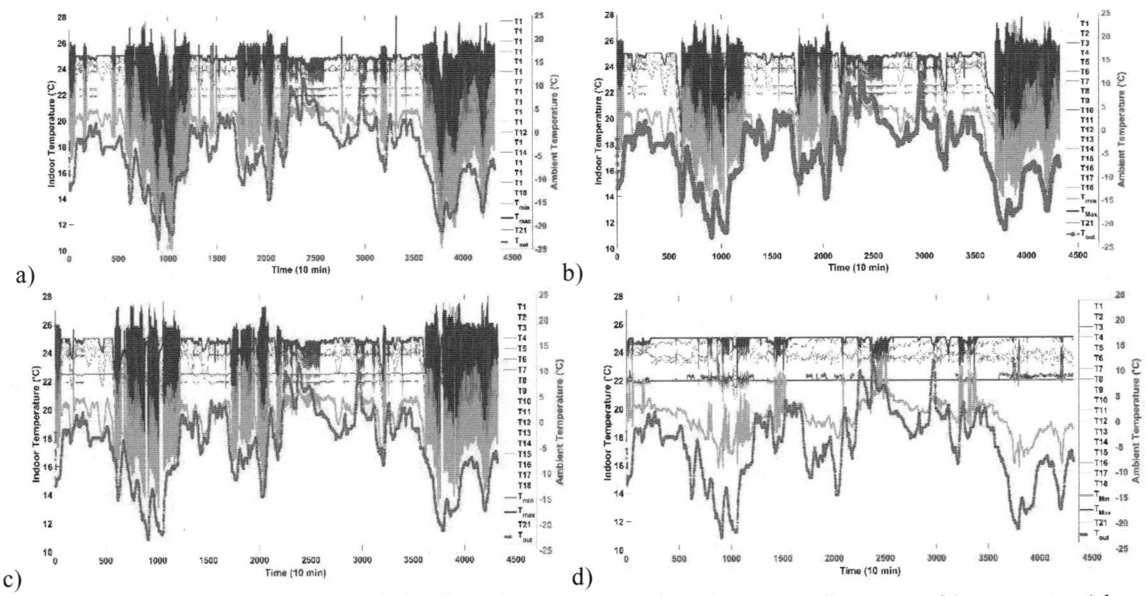

Figure 3: The zones temperatures (T1, T2, …, T21 represent the zone 1, zone 2, …, zone 21 temperature) by application of a) On/Off Controller, b) PID-GA controller, c) Fuzzy-PID controller, and d) MPC

While applying On/Off controller in upgraded system reduces the consumed energy of the HVAC system, the indoor temperature could not be set in the range of comfort accurately, especially in the very cold periods of the month. The zones temperatures are fluctuated based on the turning burner On/Off counters. Application of On/Off controller results emphasizes on the necessity of implementing smart controller for improving HVAC system efficiency and its comfortability results.

Application of GA to tune PID controllers' coefficients results in thermal comfort conditions in most of the time, while it is not useful for colder condition. Accordingly, temperature reduction is occurred in the thermal zones when the weather temperature is decreased to less than -10 °C and makes problem for the system. The zones temperatures reduction is related to extremum outdoor temperature and is not acceptable during the application. So, while the PID-GA reduces the consumed energy by the 26% in comparison to On/Off method, the occasional low interior temperature is not acceptable.

While the fuzzy-PID controller reduces the consumed energy as well, the effectiveness of this method in providing comfort condition is not acceptable based on decreasing indoor temperature of all zones in the extremum outdoor temperatures. This problem is happened more than 10 days for January. After the huge reduction, controller tries to supply required energy resulted in increasing the temperature more than setpoints. Therefore, the thermal comfort does not meet by application of fuzzy-PID controller.

Application of MPC improves energy efficiency of HVAC system as parallel of setting zones indoor temperatures in the range of comfort. According to the results, the thermal zones temperatures are controlled accurately except one zone. In one thermal zone, the air temperature is set in the range of comfort while the set-point is different and the temperature error is about 15%. In other words, the first floor, second floor, and third floor temperatures are set on 20 °C, 23 °C, and 25 °C respectively while one zone temperature is set on about 19 °C. The maximum error of the controller during 1-month application is investigated less than 15% which is occurred only in 1 zone. The performance of Neuro-fuzzy system is evaluated by temperature reduction in decreasing weather temperature. Accordingly, the Neuro-fuzzy system predicts outdoor temperature reduction accurately and increases flow to prevent decreasing indoor air temperatures.

5. DISCUSSION

In order to understand the effect of different control methods in HVAC system, the pump flow rate by application of control methods (including PID-GA, Fuzzy-PID, MPC, and On/Off) and the consumed gas in boiler burner are presented in Figure 4 to compare their effectiveness. It is expected that increasing transferred flow to boiler forces the burner to work more intensely to supply required energy. The burner injection gas and its regulation are presented by the consumed gas in burner.

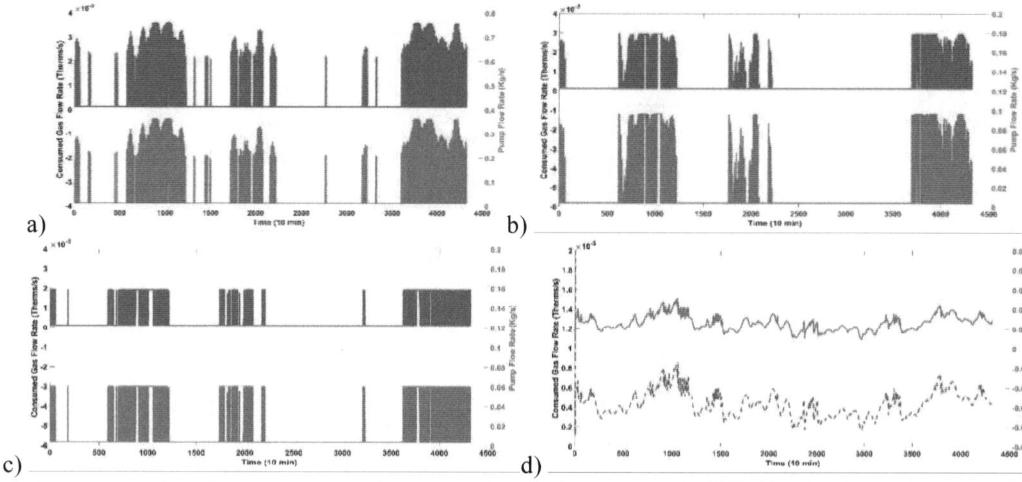

Figure 4: Flow rate of condensate pump and consumed gas of burner. a) On/Off. b) PID-GA. c) Fuzzy-PID. d) MPC

The pump flow rate and steam flow rate directly effect on consumed gas in burner. The gas flow in burner follows the pump flow rate exactly to supply required energy in order to set outlet temperature on 107°C. Decreasing consumed gas fluctuation by injecting exact flow of gas to burner reduces the energy consumption strongly in MPC. Also, turning burner on for a while by less consumed fuel is another effect of MPC while the fuzzy and GA inject fuel to burner by number of fluctuations in the extremum outdoor temperature. The On/Off controller continuously turns burner on/off to set indoor temperature in the hysteresis loop.
The other important parameter in designing controllers is consumed energy. In order to find the best control method, the integration of consumed gas in the burner and integration of electricity usage in the HVAC system are presented in figure 5.

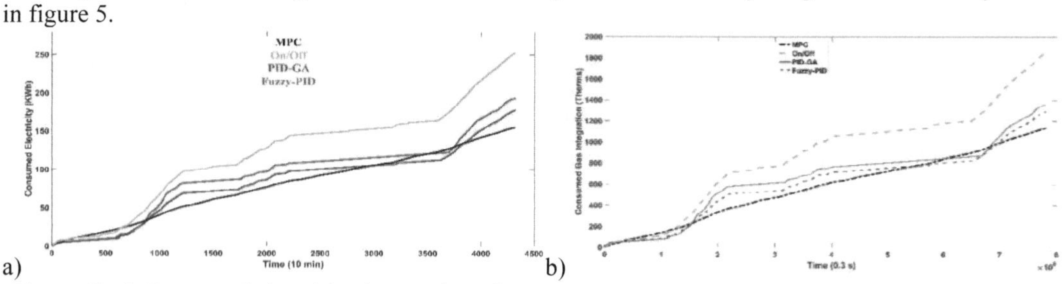

Figure 5: a) Consumed electricity integration of HVAC system (1KWh=3,600KJ). b) Consumed gas integration of HVAC system (1Therm = 105,480 KJ)

The upgraded boilers, burners and application of control valves and thermostats improve the energy efficiency of the building by more than 50% resulted in the application of On/Off controller. Based on the results, application of controllers reduces consumed energy of the upgraded system. The minimum reduction is related to PID-GA controller. In order to understand the effectiveness of controllers, the methods are compared with On/Off system. Accordingly, MPC, GA and fuzzy logic system reduce consumed energy (including gas and electricity) in HVAC system by 38%, 26%, and 30% respectively. The powerful effectiveness of MPC in reducing consumed energy represents the effect of prediction in optimizing energy consumption specially based on the comfort condition. In other words, MPC optimizes energy efficiency as well as providing comfort conditions in the acceptable range.

6. CONCLUSIONS

The tested control methodologies for providing thermal comfort and reducing energy consumption in buildings, showed that utilizing variable flow system including control valves, variable flow pump, smart thermostat, and building management system could improve the energy efficiency of the test building. Also, utilizing MPC coupling by neuro-fuzzy temperature predictor improves energy efficiency of the HVAC system by 38%. The application of fuzzy-PID controller and GA-PID controller on HVAC system reduced consumed energy by 30% and 26% respectively. The MPC approach outperformed other approaches tested in energy optimization and comfort.

REFERENCES

Ahn, J., Cho, S., & Chung, D. H. (2017). Analysis of energy and control efficiencies of fuzzy logic and artificial neural network technologies in the heating energy supply system responding to the changes of user demands. *Applied Energy*, *190*, 222–231. https://doi.org/10.1016/j.apenergy.2016.12.155

Aoun, N., Bavière, R., Vallée, M., Aurousseau, A., & Sandou, G. (2019). Modelling and flexible predictive control of building space-heating demand in district heating systems. *Energy*, 116042. https://doi.org/10.1016/J.ENERGY.2019.116042

EN, B. S. (2003). 12831: 2003 Heating systems in buildings. *Method for Calculation of Design Heat Load*.

Fan, C., Huang, G., & Sun, Y. (2018). A collaborative control optimization of grid-connected net zero energy buildings for performance improvements at building group level. *Energy*, *164*, 536–549. https://doi.org/10.1016/j.energy.2018.09.018

Finck, C., Li, R., & Zeiler, W. (2019). Economic model predictive control for demand flexibility of a residential building. *Energy*, *176*, 365–379. https://doi.org/10.1016/j.energy.2019.03.171

Joe, J., & Karava, P. (2019). A model predictive control strategy to optimize the performance of radiant floor heating and cooling systems in office buildings. *Applied Energy*, *245*(October 2018), 65–77. https://doi.org/10.1016/j.apenergy.2019.03.209

Khanmirza, E., Esmaeilzadeh, A., & Markazi, A. H. D. (2016). Predictive control of a building hybrid heating system for energy cost reduction. *Applied Soft Computing*, *46*, 407–423. https://doi.org/10.1016/j.asoc.2016.05.005

Ławryńczuk, M., & Ocłoń, P. (2019). Model Predictive Control and energy optimisation in residential building with electric underfloor heating system. *Energy*, *182*, 1028–1044. https://doi.org/10.1016/J.ENERGY.2019.06.062

Li, T., Shao, G., Zuo, W., & Huang, S. (2017). Genetic algorithm for building optimization - State-of-the-art survey. *ACM International Conference Proceeding Series*, *Part F1283*(February), 205–210. https://doi.org/10.1145/3055635.3056591

Mary Reena, K. E., Mathew, A. T., & Jacob, L. (2018). A flexible control strategy for energy and comfort aware HVAC in large buildings. *Building and Environment*, *145*, 330–342. https://doi.org/10.1016/j.buildenv.2018.09.016

Michailidis, I. T., Schild, T., Sangi, R., Michailidis, P., Korkas, C., Fütterer, J., Müller, D., & Kosmatopoulos, E. B. (2018). Energy-efficient HVAC management using cooperative, self-trained, control agents: A real-life German building case study. *Applied Energy*, *211*(October 2017), 113–125. https://doi.org/10.1016/j.apenergy.2017.11.046

Nasruddin, Sholahudin, Satrio, P., Mahlia, T. M. I., Giannetti, N., & Saito, K. (2019). Optimization of HVAC system energy consumption in a building using artificial neural network and multi-objective genetic algorithm. *Sustainable Energy Technologies and Assessments*, *35*(October 2018), 48–57. https://doi.org/10.1016/j.seta.2019.06.002

Péan, T. Q., Salom, J., & Costa-Castelló, R. (2019). Review of control strategies for improving the energy flexibility provided by heat pump systems in buildings. *Journal of Process Control*, *74*, 35–49. https://doi.org/10.1016/j.jprocont.2018.03.006

Reynolds, J., Ahmad, M. W., Rezgui, Y., & Hippolyte, J. L. (2019). Operational supply and demand optimisation of a multi-vector district energy system using artificial neural networks and a genetic algorithm. *Applied Energy*, *235*(October 2018), 699–713. https://doi.org/10.1016/j.apenergy.2018.11.001

Shaikh, P. H., Nor, N. B. M., Nallagownden, P., & Elamvazuthi, I. (2018). Intelligent multi-objective optimization for building energy and comfort management. *Journal of King Saud University - Engineering Sciences*, *30*(2), 195–204. https://doi.org/10.1016/j.jksues.2016.03.001

Shea, R. P., Kissock, K., & Selvacanabady, A. (2019). Reducing university air handling unit energy usage through controls-based energy efficiency measures. *Energy and Buildings*, *194*, 105–112. https://doi.org/10.1016/j.enbuild.2019.04.020

Wang, L. (1999). *A course in fuzzy systems*. http://www.academia.edu/download/39332931/_Buku__A_COURSE_IN_FUZZY1.pdf

ACKNOWLEDGEMENT

The authors would like to thank the Smart Energy Design Assistance Center (SEDAC), Dept. of Landscape Architecture of University of Illinois at Urbana-Champaign as well as Todd Rusk, Robert Nemeth, and Shawn Maurer for their support in this work.

Recurrent Neural Network-based Economic MPC Applied to Building HVAC Systems

Matthew J. ELLIS[1*], Venkatesh CHINDE[2]

[1] Department of Chemical Engineering, University of California, Davis
Davis, CA 95616
(mjellis@ucdavis.edu)

[2] National Renewable Energy Laboratory
Golden, CO 80401
(venkatesh.chinde@nrel.gov)

* Corresponding Author

ABSTRACT

Numerous studies have demonstrated the potential benefit of economic model predictive control (EMPC) applied to building heating, ventilation, and air conditioning (HVAC) systems. However, technological barriers preventing large-scale adoption of EMPC to building HVAC systems exist including model construction and training. In this work, an encoder-decoder long short-term memory-based EMPC framework is developed. The encoder-decoder model, a data-driven modeling approach, offers advantages in model construction and training over conventional grey-box modeling approaches. The EMPC objective is to minimize the HVAC operating utility cost by manipulating the building zone temperature setpoints. Closed-loop simulations under the EMPC using EnergyPlus are employed to demonstrate the approach on a prototypical five-zone commercial building.

1. INTRODUCTION

Within the control community, model predictive control (MPC) has received substantial attention. MPC, specifically, economic MPC (EMPC), is a dynamic optimization technique for the control and optimization of systems (see, for example, M. Ellis et al. (2016) and the references therein). EMPC utilizes an economically-oriented cost function to make control decisions over a horizon while accounting for system and economically-oriented constraints. Many buildings have time-varying electric rates, peak electric demand charges, and time-varying disturbances. Therefore, to optimize building HVAC system operation, these characteristics require looking over a horizon to make appropriate operating decisions implying that EMPC is a natural choice for a control strategy.

Substantial work has been done on (E)MPC applied to heating, ventilation, and air conditioning (HVAC) systems (Afram & Janabi-sharifi, 2014; Hilliard et al., 2016; Killian & Kozek, 2016; Afram et al., 2017). The desirable qualities reported in the body of work on MPC include reduced energy cost, reduced energy consumption, and improved occupant comfort. Unfortunately, the widespread adoption of EMPC to real buildings has not occurred. One challenge associated with MPC is the high costs of construction, configuration, and training of predictive models used within MPC (Cígler et al., 2013; Henze, 2013; Killian & Kozek, 2016).

The integration of machine learning and control is experiencing a renaissance with several efforts focusing on, for example, leveraging machine learning techniques to obtain a model for control purposes (Lenz et al., 2015; Wu et al., 2019; Maddalena et al., 2020). One example includes the use of recurrent neural networks (RNN) for predictive control of nonlinear chemical processes (Wu et al., 2019). Similarly, Terzi et al. (2019) proposes using long short-term memory networks (LSTM) for MPC design because LSTM networks offer advantages compared to conventional RNNs. A recent review article on data-driven methods summarizes several control techniques such as rule-based, MPC, reinforcement learning, and learning MPC techniques for buildings (Maddalena et al., 2020).

In this paper, an encoder-decoder LSTM-based economic model predictive control (EMPC) framework is presented. The modeling training process of the encoder-decoder model is described along with the EMPC formulation. Results from closed-loop simulations of a five-zone building model in EnergyPlus with the proposed EMPC are presented.

Since the original submission of this paper, a journal version has been published (M. J. Ellis & Chinde, 2020). The journal version contains more detail on the formulation and implementation while this version contains computational results on the application of the LSTM-based EMPC to all building zones not included in the journal version.

2. ENCODER-DECODER ZONE MODELS AND HVAC POWER MODEL

2.1 Notation

The set of integers is denoted by $\mathbb{Z}$ while the set of non-negative integers is denoted by $\mathbb{Z}_+$. A parameterized function is denoted by $f(\cdot;\theta)$ where $\theta \in \mathbb{R}^{n_\theta}$ is the parameter vector. A sequence is denoted by boldface variables: $\mathbf{x}_k := \{x_k\}_{k=0}^{N}$ where $x_k \in \mathbb{R}^n$ denotes the value of the sequence at the kth time step. A predicted value of a sequence at the kth time step is denoted by $\tilde{x}_k$. The notation $\tilde{x}_{j|k}$ where $j \geq k$ is used to emphasize that the quantity represents the predicted value at time step j computed at time step k. The operator $\circ$ denotes the Hadamard product (element-wise product).

2.2 Long Short-Term Memory Encoder-Decoder Zone Models

Recurrent neural networks (RNNs) may be interpreted as discrete-time black-box state-space models. Using notation typically used in the control literature, RNNs may be written in the form

$$\begin{aligned} x_{k+1} &= f(x_k, u_k; \theta) \\ \tilde{y}_k &= h(x_k; \theta) \end{aligned} \tag{1}$$

where $k \in \mathbb{Z}_+$ denotes the current time step, $x_k \in \mathbb{R}^n$ denotes the recurrent state vector (e.g., the hidden state and/or cell state), $u_k \in \mathbb{R}^m$ denotes the input vector, $\tilde{y}_k \in \mathbb{R}^l$ denotes the predicted output, $\theta \in \mathbb{R}^p$ denotes the parameter vector, f is the parameterized state transition map, and h is the parameterized output map.

Long short-term memory (LSTM) is a type of RNN architecture that has been extensively used in many applications (Hochreiter & Schmidhuber, 1997). One advantage of LSTM over a conventional RNN is that it addresses the well-known vanishing/exploding gradient problem during model training. The equations describing the forget gate, input gate, candidate cell update vector, output gate, and state update are

$$\begin{aligned} \bar{f}_k &= \sigma_g\left(\theta_{\bar{f},x^h} x_k^h + \theta_{\bar{f},u} u_k + \theta_{\bar{f}}\right) \\ \bar{i}_k &= \sigma_g\left(\theta_{\bar{i},x^h} x_k^h + \theta_{\bar{i},u} u_k + \theta_{\bar{i}}\right) \\ \bar{c}_k &= \sigma_c\left(\theta_{\bar{c},x^h} x_k^h + \theta_{\bar{c},u} u_k + \theta_{\bar{c}}\right) \\ \bar{o}_k &= \sigma_g\left(\theta_{\bar{o},x^h} x_k^h + \theta_{\bar{o},u} u_k + \theta_{\bar{o}}\right) \\ x_{k+1}^c &= \bar{f}_k \circ x_k^c + \bar{i}_k \circ \bar{c}_k \\ x_{k+1}^h &= \bar{o}_k \circ \sigma_{x^h}\left(x_{k+1}^c\right) \end{aligned} \tag{2}$$

where $\bar{f}_k, \bar{i}_k, \bar{c}_k, \bar{o}_k \in \mathbb{R}^{n/2}$ denote the outputs of the forget gate, the input gate, the candidate cell update vector, and the output gate, respectively, $\theta_{\bullet,x^h}$, $\theta_{\bullet,u}$ and $\theta_{\bullet}$ are the trainable weight matrices and bias vector of each decision gate, respectively, and $x_k^c, x_k^h \in \mathbb{R}^{n/2}$ denote the cell and hidden state of the LSTM. The activation function σ_g is the logistic sigmoid function, and the activation functions σ_c and σ_{x^h} are the hyperbolic tangent function. The term LSTM cell will be used to refer to the set of equations in Equation (2) while the number of units of the LSTM cell will be used to refer to the hidden and cell state dimensions (i.e., $n/2$). The full state dimension of the LSTM cell is denoted by n and is equal to twice the number of units.

The output map $h(\cdot)$ is defined with a single or multiple fully connected dense layer(s). The dense layer(s) maps/map the hidden state to the predicted output. For a single dense output layer, this relationship is given by

$$y_k = h_{Dense}(x_k^h; \theta_{Dense}) := \alpha(\theta_{y,x^h} x_k^h + \theta_y) \tag{3}$$

where θ_{y,x^h} and θ_y denote the weight matrix and bias vector, $\theta_{Dense} \in \mathbb{R}^{l+ln/2}$ is the parameter vector that includes all dense layer parameters associated with the weight matrix and bias vector, and α is an activation function. For the simulation study below, a linear or identity activation function is used, which has the form $\alpha(z) = z$.

In this work, an LSTM-based EMPC is applied to a building HVAC application where the building consists of multiple zones. For each building thermal zone, an encoder-decoder LSTM (RNN) model is trained to model the zone thermal dynamics. The encoder-decoder model (e.g., (Chrisman, 1991)) has become ubiquitous in sequence-to-sequence

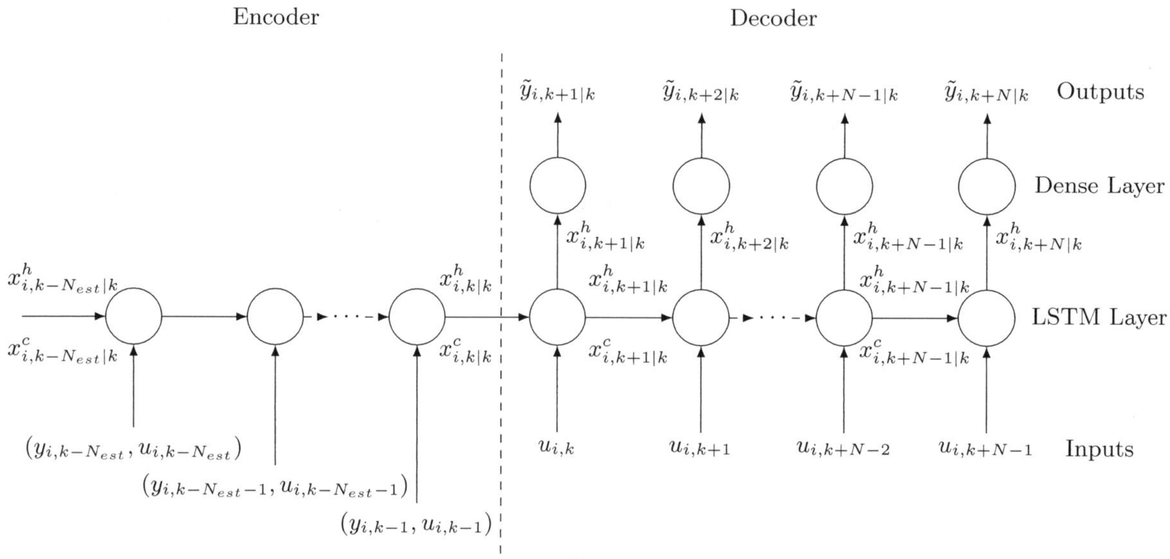

Figure 1: Diagram of the encoder-decoder model architecture.

applications where the goal is to develop a model to predict future outputs provided past data. The encoder-decoder model is adapted to be used in an estimation and control setting and is shown in Figure 1.

The encoder is responsible for summarizing past data to produce an initial condition used in the decoder. Applied to control problems, the encoder may be viewed as a moving horizon state estimator that utilizes past input and measured output values as the model inputs to produce its state estimate as shown in Figure 1. Specifically, let $N_{est} \in \mathbb{Z}_+$ be the estimation horizon. The encoder for the ith building zone is given by

$$x_{i,j+1+|k} = f_{LSTM}\left(x_{i,j|k}, \begin{bmatrix} u_{i,j} \\ y_{i,j} \end{bmatrix}; \theta_{i,LSTM,enc}\right), \quad j = k - N_{est}, \ldots, k-1 \tag{4}$$

where $x^T_{i,j|k} = \left[(x^h_{i,j|k})^T \quad (x^c_{i,j|k})^T\right]^T$. As discussed in Section 2.3, the encoder assumes that its initial states are zero ($x^h_{i,k-N_{est}|k} = 0$ and $x^c_{i,k-N_{est}|k} = 0$). With the initial states and a sequence of past input and measured output data, denoted by $\{u_{i,k-N_{est}+j}, y_{i,k-N_{est}+j}\}_{j=0}^{N_{est}-1}$, the current state, $x_{i,k|k}$, is computed by the encoder.

The current states produced by the zone encoder models and the zone decoder models are used in the EMPC where the current states serve as the initial condition and the decoder models are the predictive dynamic models. For each building zone, the decoder model is given by Equations (2)-(3). Equations (2)-(3) may be written in the form of Equation (1) by defining the state vector to be $x^T_{i,k} := [(x^{i,h}_k)^T \ (x^{i,c}_k)^T]$ ($x_{i,k} \in \mathbb{R}^{n_i}$) and the state transition map as $f^T_{LSTM}(\cdot) := \left[(f^c_{LSTM}(\cdot))^T \quad (f^h_{LSTM}(\cdot))^T\right]$. The resulting state-space model is given by

$$\begin{aligned} x_{i,k+1|k} &= f_{LSTM}(x_{i,k|k}, u_{i,k}; \theta_{i,LSTM,dec}) \\ y_{i,k} &= h_{Dense}(x^h_{i,k}; \theta_{i,Dense}) \end{aligned} \tag{5}$$

where $\theta_{i,LSTM,dec} \in \mathbb{R}^{2n_i+2m_in_i+n_i^2}$ is the LSTM parameters, which concatenates all LSTM layer parameters into one vector, and $\theta_{i,Dense} \in \mathbb{R}^{l_i+l_in_i/2}$ (for the case of a single dense output layer) is the parameter vector for the dense layer. Concatenating all parameters model parameters into two vectors, the trainable encoder-decoder parameters are denoted by $\theta_{i,LSTM,enc}$ and $\theta_{i,dec}$, respectively for the ith building thermal zone.

2.3 Model Training

For training the model, the input and output values are normalized to have similar numerical ranges. The encoder and decoder models are trained simultaneously in a supervised learning fashion. Training input-output data is collected. The model parameters are trained to minimize a loss function that depends on the prediction errors between the measured output and the predicted output. An experiment is performed to obtain the training data under excited conditions and over the expected operating range. The latter requirement is important as the model would not be expected to predict outputs well beyond the range covered by the training data set.

Let the training data for the ith zone be $Z_i^M := \{u_{i,j}, y_{i,j}\}_{j=0}^{M-1}$ where M is the number of points. The supervised learning method may be interpreted as a multi-step prediction error method. Multi-step prediction error methods refer to a family of parameter estimation methods that obtain their parameter estimates by minimizing a loss that is a function of the multi-step ahead prediction errors. In the multi-step ahead prediction error method for training the encoder-decoder model, at each time step, a trajectory of output values (i.e., a multi-step ahead prediction of output y_i) is computed, denoted by $\{\tilde{y}_{i,k+j|k}\}_{i=1}^{N_{dec}}$ where k is the time step that the predicted output sequence is computed and N_{dec} denotes the number of time steps to predict. The parameter estimates are computed by minimizing the loss function that depends on the prediction error of the multi-step prediction for all time steps in the data set.

The parameter optimization problem for the ith building zone is given by:

$$\begin{aligned}
\min_{\theta_i} \quad & \frac{1}{N_{dec}(M - N_{dec} - N_{enc})} \sum_{k=N_{enc}}^{M-N_{dec}-1} \sum_{j=1}^{N_{dec}} (y_{i,k+j} - \tilde{y}_{i,k+j|k})^T Q_{i,j} (y_{i,k+j} - \tilde{y}_{i,k+j|k}) \\
\text{s.t.} \quad & x_{i,j+1|k} = f_{LSTM}(x_{i,j|k}, [u_{i,j}^T, y_{i,j}^T]^T; \theta_{i,LSTM,enc}), \ \forall j \in \{k - N_{est}, \dots, k-1\} \\
& x_{i,k-N_{enc}|k} = 0 \\
& x_{i,j+1|k} = f_{LSTM}(x_{i,j|k}, u_{i,j}; \theta_{i,LSTM,dec}) \ \forall j \in \{k, \dots, k + N_{dec} - 1\} \\
& \tilde{y}_{i,j+1|k} = h_{Dense}(x_{i,j+1|k}^h; \theta_{i,Dense}), \ \forall j \in \{k, \dots, k + N_{dec} - 1\} \\
& \quad \forall k \in \{N_{enc}, \dots, M - N_{dec} - 1\}
\end{aligned} \tag{6}$$

where $\theta_i^T := [\theta_{i,LSTM,enc}^T, \theta_{i,LSTM,dec}^T, \theta_{i,Dense}^T]$ is a vector of all trainable model parameters and is the problem decision variable. The problem parameters are the multi-step prediction horizon N_{dec}, the encoder horizon N_{enc}, and the positive semidefinite weighting matrix $Q_{i,j}$. The weighting matrix $Q_{i,j}$ may be used to penalize prediction errors associated with the various outputs differently and to penalize prediction errors at the end of the prediction horizon less than the prediction errors at the beginning of the prediction horizon. In Equation (6), the first dynamic equation represents the zone encoder model, while the second and third equations represent the zone decoder model.

When $N_{dec} = 1$, the training problem (Equation (6)) corresponds to a one-step ahead prediction error method (PEM), and if the batch size is 1 or $N_{dec} = M - N_{enc} - 1$, the training problem represented by Equation (6) corresponds to a simulation PEM (see, for example, Ljung (1999) for a more detailed presentation of prediction error methods). The training estimation and prediction horizons may be different from the ones used for on-line estimation and control.

While Equation (6) is interpreted as a multi-step prediction error method, there is one important limitation with the supervised learning setting compared to multi-step PEM from conventional control-oriented parameter estimation. In conventional PEM, the initial state may be parameterized and trained. When solving the parameter estimation problem, the predicted output over time is obtained by propagating the model forward in time with the parameterized initial state. The underlying solver being used to solve the parameter estimation problem adjusts the model parameters and the parameters associated with the initial state so that the optimal model parameters and initial states are obtained. Under the supervised learning setting, each sequence of $N_{enc} + N_{dec}$ data points is treated as a separate observation in the sense that the model is not trained through forward propagation over the entire training data set. The parameterization of the initial state is not straightforward in the supervised learning setting. To address this problem, the initial states of the encoder are assumed to be zero, and a sufficiently long estimation horizon is employed such that the effect of the initial condition is small.

2.4 HVAC Power Model

To model the chiller power and air handling unit (AHU) fan power, a static model is used. The model is quadratic in the total sensible cooling rate and is given by

$$P_{HVAC,k} = c_2 \left(\sum_{i=1}^{n_{zones}} \dot{Q}_{i,clg,k} \right)^2 + c_1 \sum_{i=1}^{n_{zones}} \dot{Q}_{i,clg,k} + c_0 \tag{7}$$

where $P_{HVAC}(k)$ is the total HVAC power including the chiller power and AHU fan power, $\dot{Q}_{clg,i}$ is the sensible cooling rate of the ith zone, n_{zones} is the number of zones in the building, and c_j are the model coefficients. The coefficients are fit using standard regression.

3. RNN-BASED ECONOMIC MODEL PREDICTIVE CONTROL FRAMEWORK

An EMPC is formulated to minimize the cost associated with the HVAC power consumption by manipulating the zone temperature setpoints. At time k, the EMPC problem with the zone decoder models where each zone model is given by Equation (5) and power model given by Equation (7) is:

$$\min \sum_{j=0}^{N-1} \left(p_{k+j+1} \delta t \tilde{P}_{HVAC,j+1} + \sum_{i=1}^{n_{zones}} p_{i,\delta T_{sp},k+j} |\tilde{T}_{i,sp,j} - \tilde{T}_{i,sp,j-1}| \right) \tag{8a}$$

$$\text{s.t. } \tilde{x}_{i,j+1} = f_{LSTM}\left(\tilde{x}_{i,j}, \begin{bmatrix} \tilde{T}_{i,sp,j} \\ \tilde{d}_{i,j} \end{bmatrix}; \theta^*_{i,LSTM,dec} \right) \tag{8b}$$

$$\tilde{x}_{i,0} = x_{i,k|k} \tag{8c}$$

$$\begin{bmatrix} \tilde{T}_{i,zn,j+1} \\ \dot{\tilde{Q}}_{i,clg,j+1} \end{bmatrix} = h_{Dense}(\tilde{x}^h_{i,j+1}; \theta^*_{i,Dense}) \tag{8d}$$

$$T_{i,sp,lb,k+j} \leq \tilde{T}_{i,sp,j} \leq T_{i,sp,ub,k+j} \tag{8e}$$

$$T_{i,lb,k+j+1} \leq \tilde{T}_{i,zn,j+1} \leq T_{i,ub,k+j+1}, \ \forall\, i \in \{1, \ldots, n_{zones}\} \tag{8f}$$

$$\tilde{P}_{HVAC,j+1} = c_2^* \left(\sum_{l=1}^{n_{zones}} \dot{\tilde{Q}}_{l,clg,j+1} \right)^2 + c_1^* \sum_{l=1}^{n_{zones}} \dot{\tilde{Q}}_{l,clg,j+1} + c_0^*, \ \forall\, j \in \{0, \ldots, N-1\} \tag{8g}$$

where N is the number of time steps in the prediction horizon, δt is the sample period, p_{k+j+1} is the electric rate at time $k+j+1$, $p_{i,\delta T_{sp},k+j}$ is the penalty imposed on the change in the ith zone temperature setpoint, $\tilde{x}_{i,j+1}$ is the predicted state of the ith zone decoder model at the jth time step along the prediction horizon, $x_{i,k|k}$ is the estimated state computed by the ith zone encoder model, $\tilde{T}_{i,sp,j}$ is the temperature setpoint of the ith zone, $\tilde{d}_{i,j}$ is a vector containing all exogenous inputs to the model (not decision variables), $\tilde{T}_{i,zn,j+1}$ is the predicted ith zone temperature at time step $j+1$, $\dot{\tilde{Q}}_{i,clg,j+1}$ is the predicted ith zone sensible cooling, $T_{i,sp,lb,k+j}$ and $T_{i,sp,ub,k+j}$ are the lower and upper bound imposed on the ith zone temperature setpoint, $T_{i,lb,k+j}$ and $T_{i,ub,k+j}$ are the lower and upper bound imposed on the ith zone temperature, and $\tilde{P}_{HVAC,j+1}$ is the predicted HVAC power consumption. The model parameters $\theta^*_{i,LSTM,dec}$ and $\theta^*_{i,Dense}$ are the trained model parameters using the approach described in Section 2.3 while c_0^*, c_1^*, and c_2^* are the power model coefficients trained according to Section 2.4.

The constraints shown in Equation (8b) and Equation (8d) represent the ith zone decoder model used to predict the outputs. At time k, the decoder model states are initialized with the encoder states (Equation (8c)). The constraints represented by Equations (8e)-(8f) are the bounds on the temperature setpoints and zone temperature. Finally, Equation (8g) is the power model used to predict the HVAC power consumption from the predicted zone sensible cooling loads. By convention, $\tilde{T}_{i,sp,-1}$ is the ith setpoint applied at the previous time step ($k-1$ time step).

Regarding solving the EMPC problem, the problem is a nonlinear optimization problem and is solved with a multiple shooting approach (i.e., the state and output variables are included as decision variables). The forecasted exogenous input values over the prediction horizon are provided to the EMPC as input. The change penalty is reformulated using a standard approach by defining auxiliary variables, $\delta\tilde{T}_{i,sp,j}$ for $i = 1, \ldots, n_{zones}$, replacing the term $|\tilde{T}_{i,sp,j} - \tilde{T}_{i,sp,j-1}|$ in the cost with $\delta\tilde{T}_{i,sp,j}$, and adding the following constraints:

$$\begin{aligned} \delta\tilde{T}_{i,sp,j} &\geq \tilde{T}_{i,sp,j} - \tilde{T}_{i,sp,j-1} \\ \delta\tilde{T}_{i,sp,j} &\geq -\tilde{T}_{i,sp,j} + \tilde{T}_{i,sp,j-1} \\ \delta\tilde{T}_{i,sp,j} &\geq 0 \end{aligned} \tag{9}$$

for $j = 0, 1, \ldots, N-1$. Equation (9) ensures that the optimal solution is such that $\delta\tilde{T}_{i,sp,j} = |\tilde{T}_{i,sp,j} - \tilde{T}_{i,sp,j-1}|$.

The EMPC with underlying problem formulation given in Equation (8) is an implicit control law constructed by recursively solving the problem following the receding horizon principle. At time step k, the zone encoder models are used to estimate the current states, denoted by $x_{i,k|k}$, utilizing the input-output data collected over the last N_{est} time steps.

The current state is passed to the EMPC along with other exogenous inputs including forecasted disturbance values, cost penalty weights, and the variable lower and upper bounds. The problem is solved to obtain the trajectories. The optimal input sequence is denoted by $\tilde{\mathbf{u}}_k^* := \{\tilde{u}_{j|k}^*\}_{j=0}^{N-1}$ where $u_{j|k}^*$ is an n_{zones}-dimensional vector where the elements are the zone temperature setpoints. The control action computed for the first time step of the horizon is applied to the system (i.e., at time step k, the control action is $u_k = \tilde{u}_{0|k}^*$). At the next time step, the process of computing an updated state estimate and optimal input trajectory is repeated.

4. APPLICATION TO A BUILDING HVAC SYSTEM

The proposed RNN-EMPC framework is applied to a building HVAC system to demonstrate the approach.

4.1 Building Description and Co-simulation Setup

In this study, a prototypical five-zone building located in Phoenix, Arizona was considered with heating and cooling provided by steam and chilled water generated by a boiler and chiller, respectively. The summer season was considered in this study, and thus, cooling mode operation of the HVAC system was considered. To model the building, a large prototype office building in EnergyPlus was modified to represent the five-zone building by removing the bottom, basement, and middle floors. The five zones consisted of four perimeter zones each with one external wall facing one of the four cardinal directions (north, south, east, or west), and an interior or core zone with a substantially larger floor area compared to the other zones. For notational purposes, the core zone will be denoted by $i = 1$ and perimeter zones 1, 2, 3, and 4 will be denoted by $i = 2$, $i = 3$, $i = 4$, and $i = 5$, respectively.

Figure 2 shows a diagram of the building layout and the HVAC system considered. The HVAC system was an air handling unit-variable air volume (AHU-VAV) system commonly found in many U.S. commercial buildings. The airflow through these systems is as follows. The return air from the zones is mixed with outdoor air to provide fresh air to the zones (i.e., ventilate the zones). The mixed air is then sent through the heating coils and cooling coils where the air is heated or cooled. The conditioned air is then supplied to the zones through VAV terminal boxes. To maintain the zone temperature at the setpoint, the flow rate of conditioned air discharged into the zone is manipulated by a zone-level regulatory controller, which controls the VAV terminal box damper position.

To simulate the closed-loop EMPC system with EnergyPlus, the Building Controls Virtual Test Bed (BCVTB) (Wetter & Haves, 2008) was used to facilitate co-simulation between the state estimator and EMPC implemented in Python and the building HVAC system simulated with EnergyPlus. With BCVTB, building and HVAC measurements from EnergyPlus were provided to the estimator and EMPC and the computed setpoint from the EMPC was provided to EnergyPlus using the "External Interface:Schedule" function to override the original information in EnergyPlus. A configuration file was created to list the various input-output variables for data exchange. To solve the optimal control problem of EMPC, CasADi (Andersson et al., 2019) with IPOPT (Wächter & Biegler, 2006) was used.

The control objective was to optimize the HVAC electric utility cost associated with the power consumed to cool a building zone by manipulating the five-zone temperature setpoints. In the results below, time-of-use electric rate structures were considered, which have a high cost or peak period and a low cost or off-peak period. Under such a time-varying electric rate structure, the building thermal mass may be utilized as thermal energy storage in the sense that the building thermal mass may be pre-cooled before a peak period. During the peak period, thermal energy is transferred from the air to the cool building mass reducing the cooling load needed by the HVAC equipment. This process of pre-cooling the building's thermal mass may reduce the electric utility cost relative to a constant-in-time temperature setpoint strategy.

The sensible cooling load of the ith zone was computed from the discharge air flow rate, and the temperature difference between the discharge air and the zone air:

$$\dot{Q}_{i,clg} = \dot{m}_{i,zn} C_{p,air} (T_{i,zn} - T_{i,da}) \tag{10}$$

where $\dot{Q}_{i,clg}$ is the sensible cooling rate, $\dot{m}_{i,zn}$ is the mass flow rate of the discharge air (i.e., the conditioned air being provided to the zone from the VAV box), $C_{p,air}$ is the heat capacity of air, $T_{i,zn}$ is the zone indoor air temperature and $T_{i,da}$ is the discharge air temperature.

The LSTM model is nonlinear meaning that the underlying EMPC problem is a nonlinear optimization problem. The computational complexity associated with solving such a problem dramatically increases relative to the complexity associated with using a linear model within the EMPC problem. Thus, the choice of a nonlinear model should be

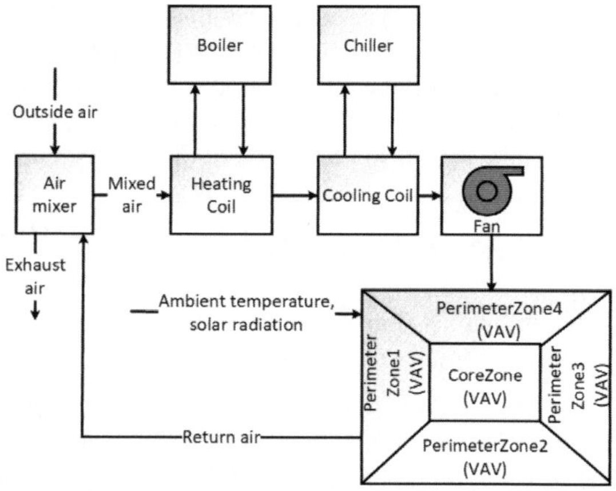

Figure 2: A diagram of the AHU-VAV HVAC system and building.

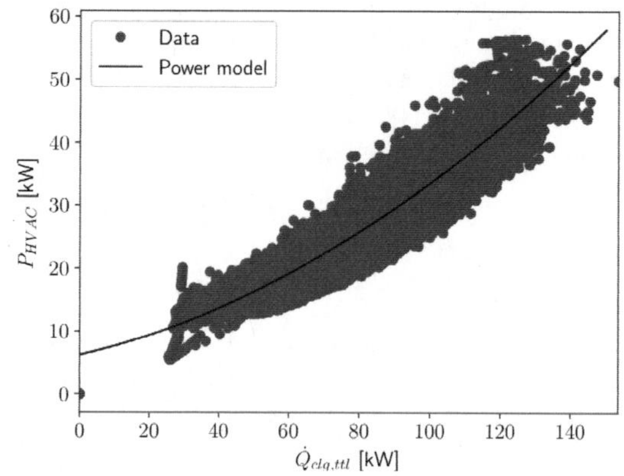

Figure 3: HVAC power consumption versus the total sensible cooling load with the fitted quadratic model.

justified. While the building thermal dynamics are usually relatively linear albeit may be high-order, the relationship between zone temperature setpoint and zone sensible load is nonlinear. Under cooling, if a large enough decrease in the temperature setpoint is made, the VAV terminal box controller will cause the VAV terminal box damper to open to increase the discharge air flow rate. After the damper is fully open, the airflow rate will saturate at some maximum value. Similarly, if the temperature setpoint is increased so that it is greater than the zone temperature, the VAV controller will close the discharge air damper until the flow rate saturates to some minimum value. This behavior makes the relationship between the setpoint and the sensible load nonlinear.

4.2 Model Training

To obtain training data, a series of step changes in the zone temperature setpoints were performed over three months from June 1 to September 1. The step changes were used to excite the dynamics. To generate the temperature setpoint excitation signals, a pseudo-random binary signal was generated, and the resulting binary signal was scaled such that it took values of either 23 or 26°C. For every change in the value of the resulting signal, a uniformly random number between 0 and 1.5°C was added or subtracted to the signal depending on if the value of the binary signal was 23 or 26°C, respectively. The purpose of adding the random number was to generate an excitation signal that covered many values between the desired range of 23 and 26°C . The training data should cover the potential range of operation since the modeling approach is a purely data-driven one meaning that it is not expected that the trained models will accurately predict the behavior beyond the range covered by the training data. Given that the objective of this work was to explore the feasibility of using an encoder-decoder model, ideal training conditions were used, and the amount of data needed to train the model, and the conditions of the excitation signal were not thoroughly investigated.

Regarding the inputs and outputs of each zone model, the two outputs of each zone model were taken to be the zone temperature and sensible cooling load. The inputs of the zone models included the zone temperature setpoint, the outdoor air temperature, and three disturbance predictor variables: the time-of-day, day-of-week, and day-of-year. The three disturbance predictor variables were used to capture the correlation between the unmeasured heat disturbance load on the building zones with these time features. The encoder model also included past zone temperature and sensible cooling load data as input to the model. In the closed-loop results below, the exact outdoor air temperature forecasts were used. Owing to space restrictions, the zone model training results are not presented here but follow along the lines of the core zone training results presented in M. J. Ellis and Chinde (2020).

The quadratic power model (Equation (7)) was fit to training data collected for training the zone models. The resulting training data and fitted power model is shown in Figure 3 where $\dot{Q}_{clg,ttl} = \sum_{i=1}^{5} \dot{Q}_{i,clg}$ is the total sensible cooling over the five zones. The fitted quadratic power model approximates the non-constant coefficient of performance. However, ample opportunities for improving the accuracy of the power model exist albeit at the likely expense of increasing the computational complexity associated with solving the EMPC problem.

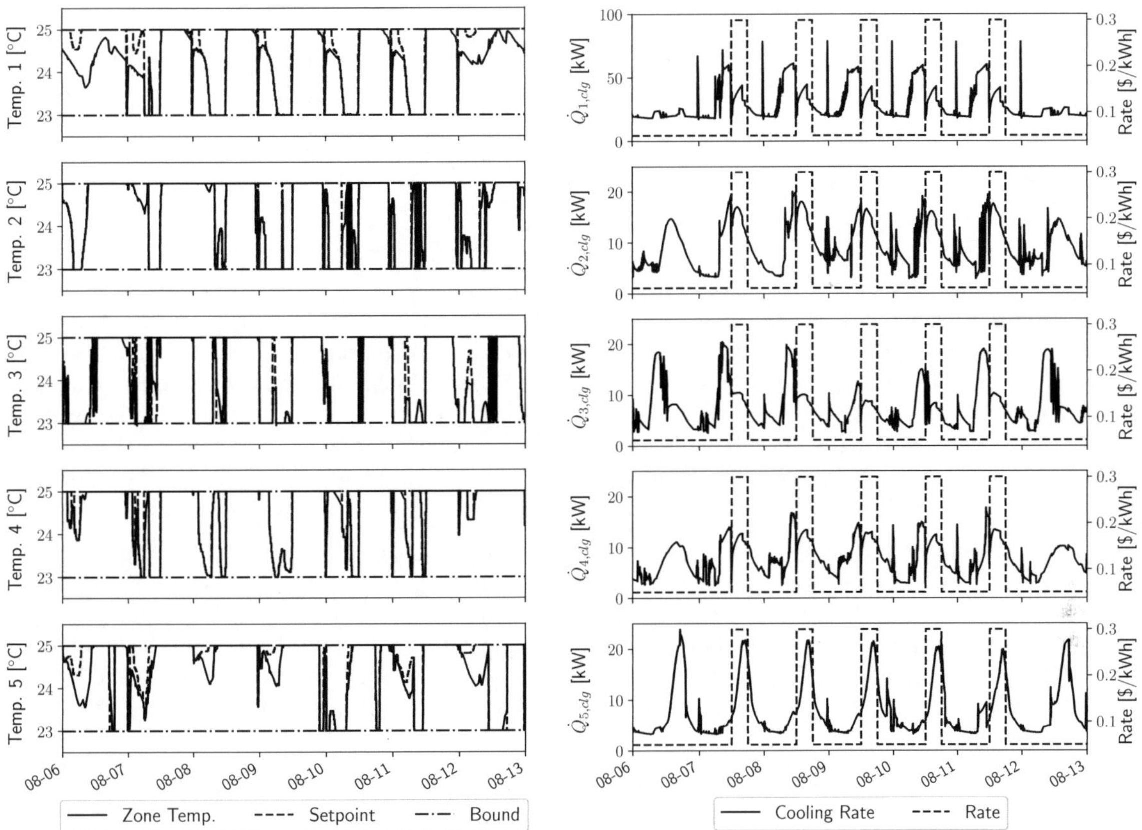

(a) **(b)**

Figure 4: The closed-loop (a) temperature and (b) HVAC sensible cooling load for the five building zones under the proposed EMPC.

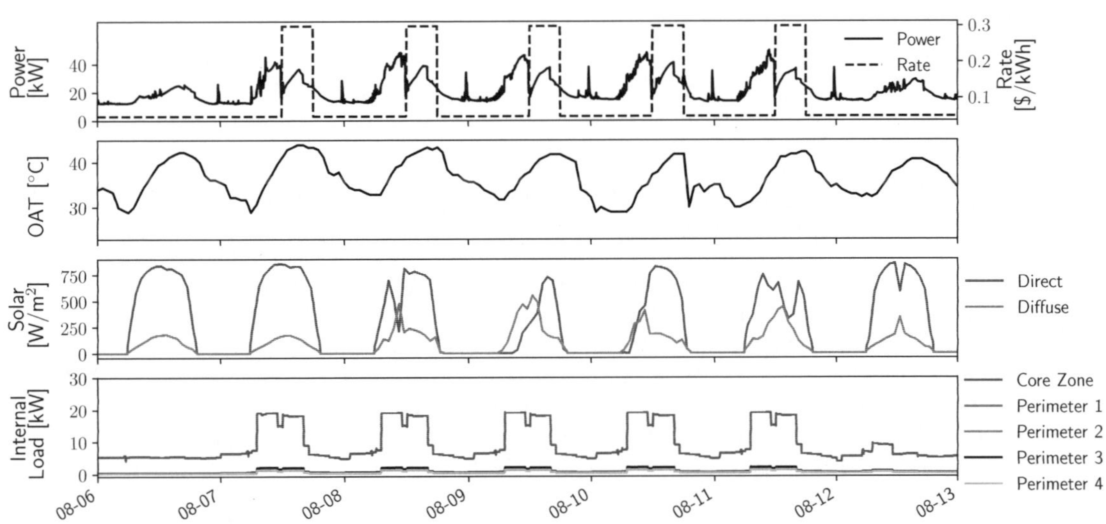

Figure 5: The power consumption profile along with the rate profile, outdoor air temperature (OAT) profile, direct solar radiation and diffuse solar radiation profiles, and internal heat load profiles of each of the five zones.

Table 1: The closed-loop costs and improvements under the EMPC relative to the baseline setpoint strategy.

Peak rate ($/kWh)	EMPC ($)	Baseline ($)	Improvement (%)
0.10	232.26	234.92	1.13
0.15	276.67	291.92	5.22
0.20	321.25	348.92	7.93
0.25	364.71	405.92	10.15
0.30	408.17	462.92	11.83

4.3 Closed-loop Simulation Results

The EMPC was applied to the five-zone building using the trained zone and power models. The EMPC was implemented with a 15-minute sample period and a 48-hour prediction horizon. The peak period was selected to last from 12:00 PM to 6:00 PM on weekdays with a rate of $0.30/kWh, and the rate during the off-peak period (all other times) was $0.05/kWh. The state estimator (encoder) utilized a 24-hour estimation horizon. For the core zone, using seven LSTM units was found to be a good choice for closed-loop performance (M. J. Ellis & Chinde, 2020). Therefore, each zone model used seven LSTM units. The temperature bounds were set to 23°C~and 25°C, respectively.

The closed-loop profiles from a seven-day simulation starting on Sunday are shown in Figures 4-5. From Figures 4-5, the EMPC dictates a pre-cooling period before the peak period with a noticeable reduction in the power during the peak period. Additionally, the EMPC increases the zone setpoints at least one time step before the beginning of the peak period implying that the models can capture the dynamic relationship between the power and setpoint. Unfortunately, the closed-loop profiles display some dithering of setpoints especially in the setpoints of perimeter zone 1 and 2 (second and third from top plots of Figure 4a), which would likely be undesirable in practice. Further investigation may be needed to address this for practical application.

The cost-benefit of pre-cooling the building mass largely depends on the building characteristics and the rate structure. Therefore, a set of simulations was considered to explore the benefit of EMPC (i.e., pre-cooling) with the peak electric rate. Five peak rates were considered: $0.10/kWh, $0.15/kWh, $0.20/kWh, $0.25/kWh, and $0.30/kWh. Table 1 summarizes the estimated cost under EMPC compared to the cost under a baseline strategy. For the baseline strategy, a minimum energy approach was selected whereby the temperature setpoint was maintained at its maximum value (25°C) for all times. As expected, the benefit of EMPC (i.e., pre-cooling) over the baseline strategy increases as the peak rate increases, and minimal cost benefit was realized when the difference between peak and off-peak rates was small. As the difference increases, the cost savings achieved with EMPC becomes significant (greater than 10% savings). With the modeling and control improvements described in this paper as future work, one may expect that the savings for this building under these conditions to increase.

5. CONCLUSIONS

In this paper, an encoder-decoder LSTM-based EMPC framework was developed. A simultaneous training methodology within a supervised learning setting was employed to train the encoder-decoder LSTM-based model. The model was split into two parts. The encoder model (first part) served as an explicit moving horizon state estimation scheme that computes estimates of the LSTM states. The decoder model was embedded into a nonlinear EMPC as its predictive model. The overall framework was applied to a building HVAC system whereby the manipulated variables controlled by the EMPC were the building zone temperature setpoints. The HVAC system operated in cooling mode. For the building HVAC system application study, the zone models were trained along with a quadratic power model to model the power consumption of the HVAC system. The results of closed-loop simulations demonstrated that EMPC can effectively pre-cool the building zones to reduce the operating cost relative to a baseline case where the zone temperature setpoint was maintained at its largest value in the admissible range (i.e., minimum energy approach for cooling HVAC operation). Future work includes understanding the data requirements of the modeling approach, improving the design of training data experiments, and developing a transfer learning approach.

6. ACKNOWLEDGEMENTS

Financial support from the UC Davis College of Engineering is gratefully acknowledged. This work was authored (in part) by the National Renewable Energy Laboratory, operated by Alliance for Sustainable Energy, LLC, for the U.S. Department of Energy (DOE) under Contract No. DE-AC36-08GO28308. Funding provided by the National Renewable Energy Laboratory (NREL) Laboratory Directed Research and Development (LDRD) program. The views expressed in the article do not necessarily represent the views of the DOE or the U.S. Government. The U.S. Government retains and the publisher, by accepting the article for publication, acknowledges that the U.S. Government retains a nonexclusive, paid-up, irrevocable, worldwide license to publish or reproduce the published form of this work or allow others to do so, for U.S. Government purposes.

REFERENCES

Afram, A., & Janabi-sharifi, F. (2014). Theory and applications of HVAC control systems – a review of model predictive control (MPC). *Building and Environment*, *72*, 343–355.

Afram, A., Janabi-sharifi, F., Fung, A. S., & Raahemifar, K. (2017). Artificial neural network (ANN) based model predictive control (MPC) and optimization of HVAC systems: A state of the art review and case study of a residential HVAC system. *Energy and Buildings*, *141*, 96–113.

Andersson, J. A. E., Gillis, J., Horn, G., Rawlings, J. B., & Diehl, M. (2019). CasADi: A software framework for nonlinear optimization and optimal control. *Mathematical Programming Computation*, *11*, 1–36.

Chrisman, L. (1991). Learning recursive distributed representations for holistic computation. *Connection Science*, *3*, 345-366.

Cígler, J., Gyalistras, D., Široky, J., Tiet, V., & Ferkl, L. (2013, Jun. 16–19). Beyond theory: the challenge of implementing model predictive control in buildings. In *Proceedings of the 11th REHVA World Congress* (Vol. 250). Prague, Czech.

Ellis, M., Liu, J., & Christofides, P. D. (2016). *Economic model predictive control: Theory, formulations and chemical process applications*. Springer.

Ellis, M. J., & Chinde, V. (2020). An encoder–decoder LSTM-based EMPC framework applied to a building HVAC system. *Chemical Engineering Research and Design*, *160*, 508–520.

Henze, G. P. (2013). Model predictive control for buildings: a quantum leap? *Journal of Building Performance Simulation*, *6*, 157–158.

Hilliard, T., Kavgic, M., & Swan, L. (2016). Model predictive control for commercial buildings: Trends and opportunities. *Advances in Building Energy Research*, *10*, 172–190.

Hochreiter, S., & Schmidhuber, J. (1997). Long short-term memory. *Neural computation*, *9*, 1735–1780.

Killian, M., & Kozek, M. (2016). Ten questions concerning model predictive control for energy efficient buildings. *Building and Environment*, *105*, 403–412.

Lenz, I., Knepper, R. A., & Saxena, A. (2015, Jul. 13–17). DeepMPC: Learning deep latent features for model predictive control. In *Proceedings of robotics: Science and systems*. Rome, Italy.

Ljung, L. (1999). *System identification: theory for the user* (2nd ed.). Upper Saddle River, NJ: PTR Prentice Hall.

Maddalena, E. T., Lian, Y., & Jones, C. N. (2020). Data-driven methods for building control—a review and promising future directions. *Control Engineering Practice*, *95*, 104211.

Terzi, E., Farina, M., & Scattolini, R. (2019). Model predictive control design for dynamical systems learned by Long Short-Term Memory Networks. *arXiv preprint arXiv:1910.04024*.

Wächter, A., & Biegler, L. T. (2006). On the implementation of an interior-point filter line-search algorithm for large-scale nonlinear programming. *Mathematical Programming*, *106*, 25–57.

Wetter, M., & Haves, P. (2008, Jul. 30–Aug. 1). A modular building controls virtual test bed for the integration of heterogeneous systems. In *Proceedings of SimBuild* (pp. 69–76). Berkeley, CA.

Wu, Z., Tran, A., Rincon, D., & Christofides, P. D. (2019). Machine learning-based predictive control of nonlinear processes. Part I: Theory. *AIChE Journal*, *65*, e16729.

Reinforcement Learning for Building Management Systems

Parastoo Delgoshaei[1]*, Amanda Pertzborn[2], Mohammad Heidarinejad[3]

[1]Mechanical Systems and Controls Group, Engineering Laboratory, National Institute of Standards and Technology (NIST), Gaithersburg, MD 20899, USA (Parastoo.Delgoshaei@nist.gov)

[2]Mechanical Systems and Controls Group, Engineering Laboratory, National Institute of Standards and Technology (NIST), Gaithersburg, MD 20899, USA (Amanda.Pertzborn@nist.gov)

[3] Department of Civil, Architectural, and Environmental Engineering, Illinois Institute of Technology, Chicago, IL 60605, USA (muh182@iit.edu)

* parastoo.delgoshaei@nist.gov

ABSTRACT

It is increasingly common to design buildings with advanced sensing and control systems to improve energy efficiency, indoor air quality which impacts health and productivity. However, there has been limited progress in making building automation systems "intelligent," as the performance of such buildings is often limited by reactive control systems, primarily using setpoint limits and fixed operation schedules. The complex nature of building control problems motivates the application of state-of-the-art software engineering methods and techniques. Agent-based models (ABM) are well-suited for controlling complex engineering systems such as those employed in building heating, ventilation, and air-conditioning (HVAC) systems. In this paradigm, a collection of interacting autonomous components (i.e., agents) adapt and make decisions in changing environments. There is a growing body of literature on adaptive agents in ABMs in many industries, but few have looked at the compatibility of ABMs with artificial intelligence (AI) optimization approaches. In most cases, conventional optimization techniques, such as mixed integer linear programming and gradient descent, have been used to find an optimal solution. This paper explores the use of an actor-critic, model-free algorithm based on a deterministic policy gradient that provides continuous control to generate the desired supply air temperature. The case study develops a thermal energy storage (TES) agent that determines the optimal valve position to manage the temperature of the cooling water flow. The case study was developed using the Intelligent Building Agents Laboratory at the National Institute of Standards and Technology. Future work will use multiple agents (i.e., air handling unit, TES, chiller) acting in cooperation or competition.

1. INTRODUCTION

There is a need for more intelligent and efficient building control algorithms because people spend 85 % of their time inside buildings, and the building sector uses about 40 % of the total primary energy consumption in the U.S. (DOE, 2012). However, better control algorithms can result in significant reduction in energy consumptions in buildings. In general, building control can be divided into three broad categories: rule-based, model-based, and data-driven.

Rule-based methods can be further categorized into prescriptive and heuristic approaches. Prescriptive methods can be obtained from best practices, equipment schedules, and rules of thumb, with a goal of improving indoor air quality. A heuristic approach adjusts setpoints based on previous experiences; for example, setting the temperature back during unoccupied hours or demand response events. ASHRAE Guideline 36 summarizes heuristic rules (ASHRAE, 2018). Although rule-based approaches are effective, they are not optimal and are not necessarily designed for the specific system in use. Moreover, these algorithms are not dynamic; they do not change over time based on the operational status of the system and do not use predictions about the future.

Model-based approaches are generally based on physics-based mathematical models. One example of such an approach is model predictive control (MPC), which uses the thermal or energy dynamics of the system and disturbance predictions to make optimal decisions. The optimization algorithm computes the optimal strategy based on predictions over a specific time horizon. In this approach, the focus is prediction of the disturbance (i.e., weather, cost, occupancy). MPC has been shown to save energy or lower operating costs in both simulation (Paris et al., 2010) and field tests (Prívara et al., 2011). Although MPC algorithms have the potential to outperform conventional control algorithms, they are not widely adopted in buildings since each building and its energy systems are unique. It is challenging to find a general-purpose building energy model that would be suitable for a variety of buildings.

Data-driven approaches are tailored for each building since they are based on historical building data. Due to advances in measurement techniques, the availability of large volumes of data in the built environment, and abundant and inexpensive computational power, recent studies have examined machine learning algorithms for building control. Reinforcement learning (RL) is a branch of machine learning that is used for optimal decision making. RL has been used extensively in gaming (Szita, 2012) and robotics (Kober & Peters, 2009) and is becoming more popular in other applications that require intelligent decision making.

RL has been successfully applied to many areas of building energy management and control. The work proposed by (Zhang et al., 2019) implemented a model-based RL approach that learns the system dynamics using a neural network. Then MPC uses the learned system dynamics to perform control via a random-sampling shooting method. Their approach reduced the total energy consumption, of a simulated two zone data center, by 17.1% to 21.8%. Q-Learning is one of the most popular RL algorithm, introduced in 1989 (Cornish & Watkins, 1989). Advances in computational power have enabled the combination of RL algorithms with deep neural networks, significantly increasing the effectiveness of RL methods and enabling them to be used in complex and safety critical applications such as autonomous cars. Another study (Barrett & Linder, 2015) used Q-learning combined with Bayesian learning for occupancy prediction. They implemented a heating, ventilation, and air conditioning (HVAC) controller to optimize occupant comfort and energy costs. The study shows that a learning thermostat can achieve cost savings of 10 % over a programmable thermostat, while maintaining high levels of occupant comfort. In a comparison to a rule-based system, Yang et al. (2015) developed a batch Q-learning approach using neural networks to control a photovoltaic powered heating system (Yang et al., 2015). The results indicate a 10 % improvement over the rule-based approach. Using a deep neural network, the studies in (Wei et al., 2017, 2018) developed RL-based HVAC control and the results suggest energy saving improvements of 20 % to 70 % over conventional Q-learning methods. The above studies indicate that RL algorithms can be promising for building mechanical systems control. This paper attempts to explore? how the performance of a chilled water system can be optimized using an actor-critic model. This is a proof of concept effort to investigate the efficacy of the model and the impact of targeted data collection to train the RL model.

2. BACKGROUND

RL is a sub-category of machine learning that includes at least one agent interacting with an environment. Through its interactions with the environment, the agent learns a series of optimal actions by determining through trial-and-error which actions yield the best reward based on the selected criteria. Table 1 describes commonly used concepts in the RL field.

Table 1. Concepts and their descriptions that are commonly used in RL

Concept	**Description**
Environment	Physical world in which the agent operates
Reward	Feedback from the environment
Policy	A mapping from perceived states of the environment to actions to be taken when in those states
State	Current state of the environment
Value	The expected (long term) cost of a certain policy

2.1 Subcategories of Reinforcement Learning

RL algorithms can be categorized as model-based or model-free. Algorithms that purely sample from experience are "model-free" RL algorithms. Model-free approaches do not use a model of the transition of the environment from one state to another state. For example, Monte Carlo Control, SARSA (Rummery & Niranjan, 1994), Q-learning (Cornish & Watkins, 1989), and actor-critic (Konda, 2002) algorithms are model-free. These algorithms learn the optimal policy based on a trial-and-error method. Model-based methods, however, learn a model of the environment. RL algorithms belong to two different categories:

Value-Based: These algorithms focus on finding an approximation of the value function or functions of states (state-action pairs).
Policy-Based: These algorithms try to find an optimal policy directly without using the Q-value.

Each of these algorithms have their advantages and disadvantages. For example, value-based algorithms are more stable while policy-based approaches converge faster and are better for continuous environments. The actor-critic method merges the advantages of both approaches. Figure 1 illustrates the actor-critic approach used in this study.

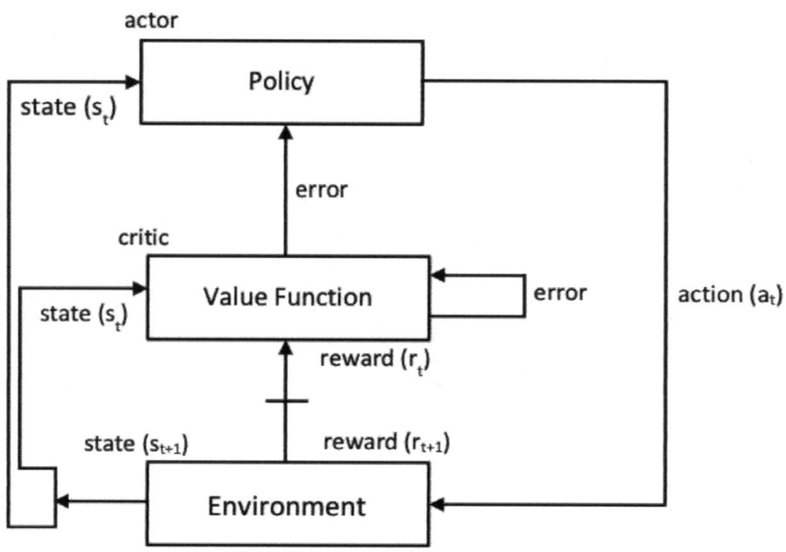

Figure 1: Architecture of the actor-critic method

In this figure, the agent is composed of an actor and a critic. The actor takes the state and the error as inputs to determine the best action. The actor uses a policy-based approach to learn the optimal policy. The critic uses a value-based approach to evaluate the action taken by the actor.

3. CASE STUDY

The Intelligent Building Agents Laboratory (IBAL) at the National Institute of Standards and Technology (NIST) was designed to emulate the HVAC and cooling loads in a small commercial building (Pertzborn & Veronica, 2018). The IBAL contains experimentally controllable weather, air system zones, and a hydronic system, which includes the cooling plant. Figure 1 is a schematic of the portion of the IBAL that is used in this study. The plant includes two water cooled chillers, Chiller1 and Chiller2, and thermal energy storage (TES). Chiller1 has a nominal capacity of 26.4 kW and Chiller2 has a nominal capacity of 52.8 kW, but both chillers use variable speed compressors, so they have variable capacity. The chillers are used to meet the building load or charge the TES, which is an ice-on-coil design with a capacity of 233 kWh. The chillers are plumbed in parallel with each other and in series with the TES. The hydronic system uses a 30 % propylene glycol (PG) mixture. The left portion of the figure, which includes the plant, is the primary loop (PL) and the right portion of the figure, which includes the building load (HX1), is the secondary loop (SL).

One use of TES is for load shifting. Instead of using a chiller on a hot afternoon to meet a building load, which is a time of peak energy usage, the TES can be used. The chiller can instead operate during times of off-peak energy usage, such as at night, to charge the TES. This type of operation takes pressure off the grid during times of peak use and can reduce energy cost to the consumer if the utility uses a tariff structure that incentives load shifting. The TES can also be used in conjunction with a chiller. In that scenario, there are two benefits. First, a smaller chiller can be selected in the design phase because the chiller does not have to meet all the load, which can reduce both first and operating costs. Second, the TES can meet enough of the load to maintain chiller power consumption below the level that would trigger demand charges.

For this research, the building load is emulated by use of a heat exchanger, HX1. HX1 is a plate heat exchanger with hot water (approximately 71 °C) flowing through one side and PG flowing through the other side. The building load is set by modulating valve V11, which sets the flow rate of the hot water through the heat exchanger. For this scenario, the temperature T_{SA} is a proxy for the supply air temperature downstream of a cooling coil. Figure 2 shows the schematic of the equipment and measurements.

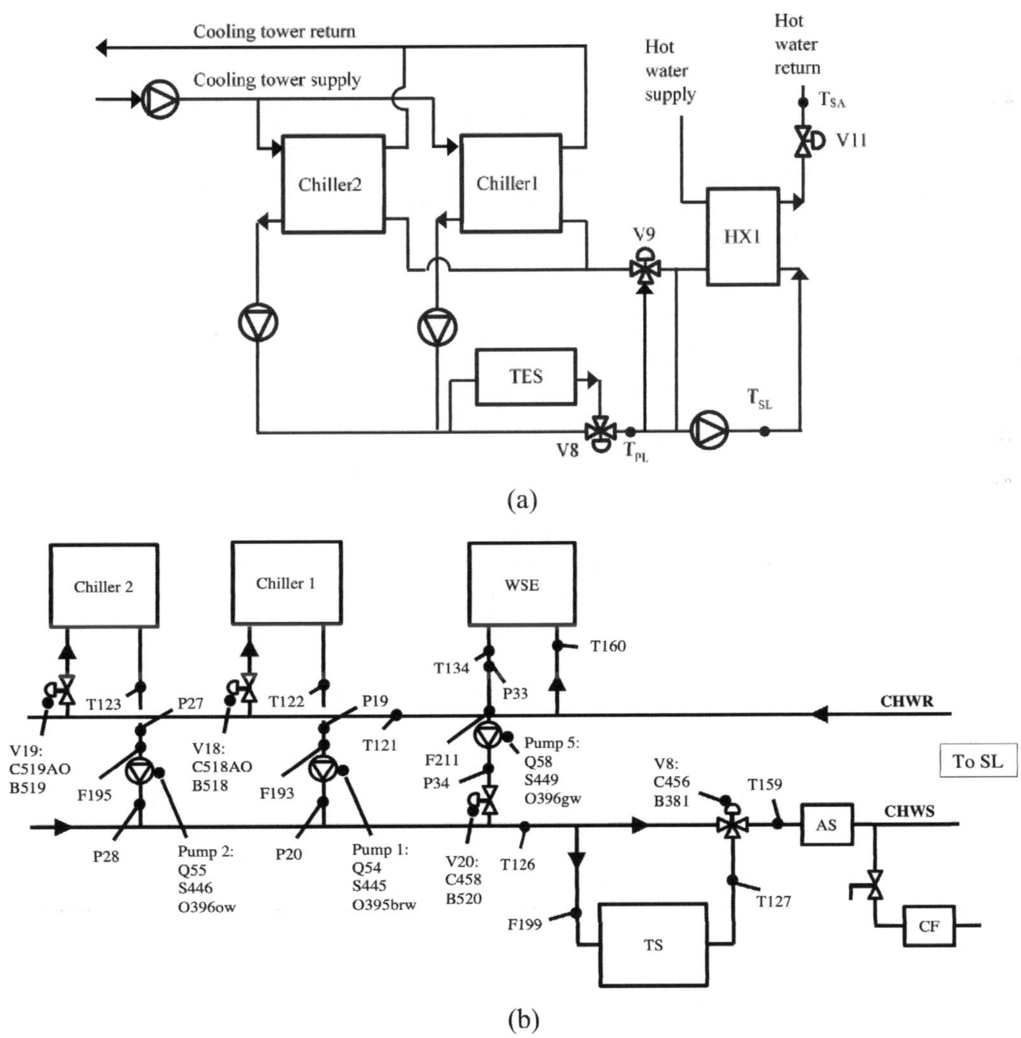

Figure 2: Schematic of the chiller, thermal energy storage, and building load (HX1) components in the IBAL: (a) high-level and (b) detailed sensor installation

The lab can support the following modes of operation:

- Chiller1 meets the building load
- Chiller1 charges the TES (charge)
- Chiller2 meets the building load
- Chiller2 charges the TES (charge)
- TES meets the building load (discharge)

When a chiller is used to meet the load, valve V8 is positioned so that the TES is bypassed. When a chiller is used to charge the TES, V8 is positioned so that all the PG from the chiller flows through the TES. When the TES is used to meet the building load, if V8 is positioned such that all the PG flows through the TES, the temperature leaving the PL, T_{PL}, would be 0 °C, which is typically colder than needed in the SL. Instead, V8 modulates so that T_{PL} is generated by the mixing of PG from the TES and returning from the SL. The temperature in the SL, T_{SL}, can be further controlled by modulating V9 to mix the fluid from the PL and the outlet of the SL. This provides fine control of the temperature.

For this study, we focus on the mode where the TES is used to meet the building load and examine an approach for optimizing the control signal for V8.

4. METHODS

The goal of the case study was to optimally control the position for the three-way mixing valve V8, shown in Figure 2, using the RL actor-critic method. V8 regulates the flow distribution between the chiller and thermal storage branches. When the control signal for V8 is 0 V, the resistance in the thermal storage branch will force all the flow to come from the chiller branch. Similarly, when the control signal is 10 V, the flow will be provided from the thermal storage branch. This mechanism will mix the two flow branches to provide the right temperature to the secondary loop to ensure the supply air temperature can meet the building needs.

4.1 Model Formulation

In the actor-critic formulation used here, the policy estimator (actor) and the value estimator (critic) are both implemented as fully connected deep neural networks with one output and no activation functions. These neural networks are trained using a stochastic gradient descent optimization algorithm with learning rates of 0.001 and 0.1 for the actor and critic, respectively. At each iteration the actor network predicts an action that will cause the environment to transition to a new state. Subsequently, the critic evaluates the value of the new state of the environment based on the penalty function, which is the negative of the reward, and provides an error to the actor. If the error is negative, the actor will weaken the tendency for that action to be taken. The reward function can consider factors such as energy, cost, or performance. In this case study, we focused on meeting the setpoint while reducing the chiller power consumption. In this problem formulation, the goal is to minimize the reward and thus the reward is defined as a penalty function, described in Equation (1).

$$R = -A \times (Setpoint\ - Supply)^2\ - B \times (Chiller\ Power)^2 \qquad (1)$$

In our case study, the agent is an actor-critic model that predicts the optimal valve position. To mimic the behavior of the zone that is the environment in our model, we used data from the heat exchanger (HX1) that serves as a building load. The state in this formulation is the temperature that leaves the heat exchanger, T_{SA} in Figure 2a, and serves as the proxy for the supply temperature. In this work we used regression to model the transition of the state (supply temperature) of the environment based on the received actions.

4.2 Results

The following graphs were obtained using the model with test data collected from the IBAL with a temperature setpoint of 12.2 °C (54 °F) for T_{SA}. Figure 3 shows that in the first 250 episodes, the trial and error phase, the reward varies between 20 and 65. After episode 250, the penalty asymptotically reaches the minimum value of about 20 (the algorithm is trying to minimize the negative of the reward) and stays around that value until the last episode. Thus, our reward is at its maximum value.

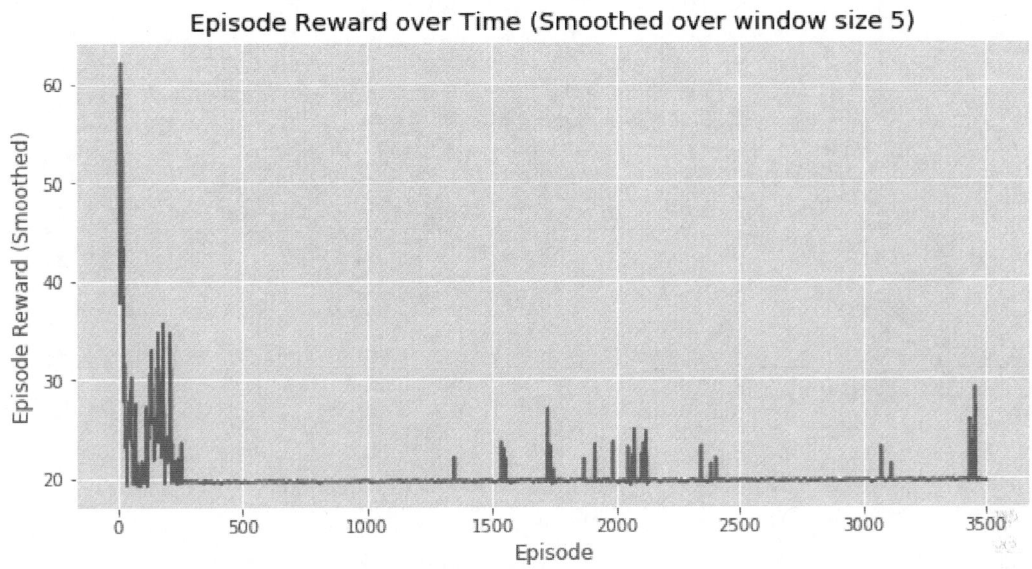

Figure 3: Episode reward over time

Similarly, as shown in Figure 4, the state of the environment converges to a value around 12.6 °C (54.6 °F) after episode 250. This is within an acceptable range for the target value of 12.2 °C (54 °F).

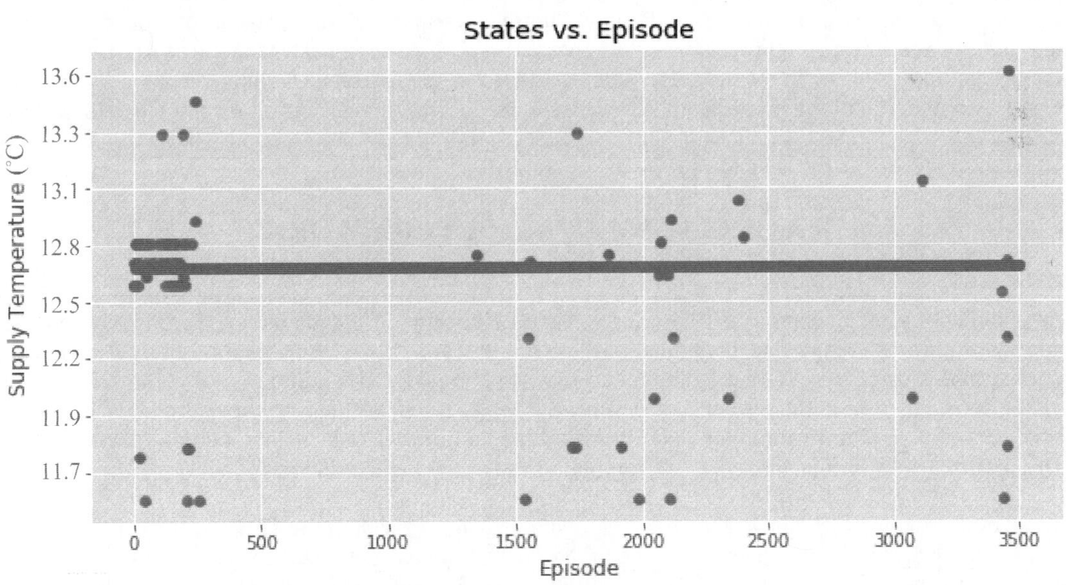

Figure 4: States versus episodes

Figure 5 shows a similar pattern; more potential actions (valve positions) were explored before episode 250. After the exploration phase, the valve position converges to a value between 1.2 and 1.6. This result indicates that the majority of the flow will bypass the ice tank because the building load is relatively small.

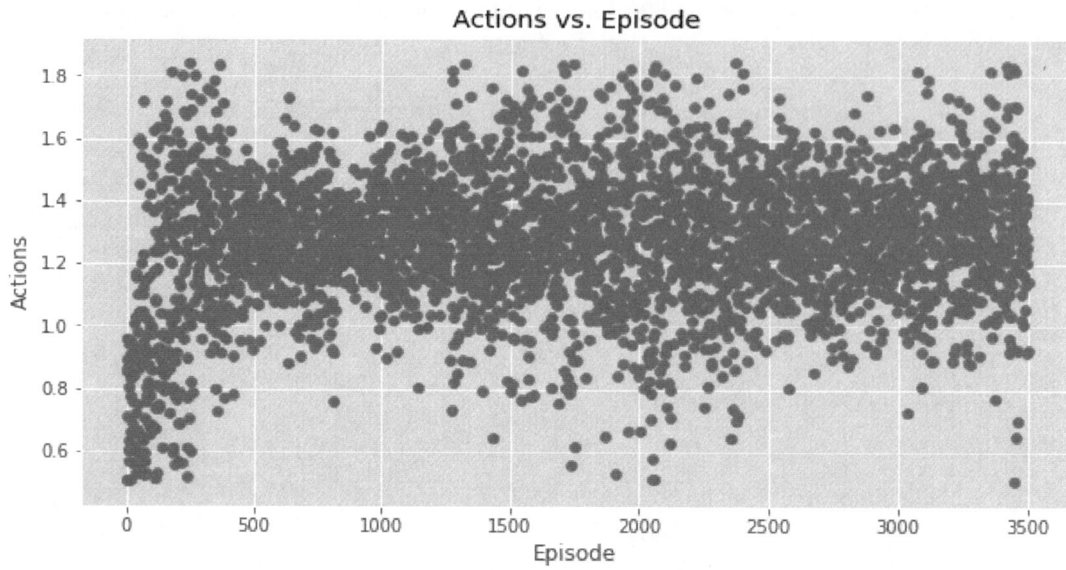

Figure 5: Actions versus episodes

5. DISCUSSION

This study applied an actor-critic RL algorithm to find an optimal solution for a mixing valve on thermal energy storage in a building HVAC system using real data. This approach is a first step towards making more complex control decisions and moving away from prescriptive sequences of operation for HVAC control. One observation we made during this study is that simply using generic historical experimental data that are not specific to the problem being studied can lead to convergence problems. In addition, the noise inherent in experimental data can make them unusable in RL algorithms in a raw form, requiring extensive preprocessing, which complicates the use of RL. Development of more complex control systems will require more targeted and less noisy datasets. Overall, compared to supervised or unsupervised algorithms that have immediate feedback or no feedback, RL algorithms that provide delayed feedback are suitable for HVAC building systems with careful consideration of the reward function. The next step of this work to implement this approach in the lab and validate the optimization results.

6. CONCLUSIONS AND FUTURE WORK

This paper deployed reinforcement learning (RL) to develop optimal control for a part of the hydronic system in the Intelligent Building Agents Laboratory (IBAL) at the National Institute of Standards and Technology (NIST). The actor-critic algorithm was used to find an optimal position of a three-way valve on a thermal energy storage system. The case study in this work serves as a proof of concept for applying RL for building HVAC control. RL algorithms are promising approaches for building control. One caveat is that these algorithms require the right dataset for proper training and exploration. Constructing the reward function is not a trivial task; for the purpose of this case study, reward was determined by minimizing energy consumption and achieving the setpoint.

An extension of this work will focus on a multi-agent model with agents such as the TES, chiller and the air handling unit that compete and cooperate to achieve the common goal of system optimization. Factors to consider include selecting a chiller to meet the building load based on chiller capacity and power consumption and selecting the TES to meet the building load based on its capacity and the power consumption associated with charging the TES.

NOMENCLATURE

ABM	Agent-based model
AI	Artificial Intelligence
HVAC	Heating Ventilation and Air Conditioning

IBAL	Intelligent Building Agent Lab
MPC	Model Predictive Control
NIST	National Institute of Standards and Technology
RL	Reinforcement Learning
SL	Secondary Loop
TES	Thermal Energy Storage

REFERENCES

ASHRAE. (2018). *ASHRAE Guideline 36-2018: High-Performance Sequences Of Operation For HVAC Systems*. American Society of Heating, Refrigerating and Air-Conditioning Engineers, Inc.

Barrett, E., & Linder, S. (2015). Autonomous HVAC Control, A Reinforcement Learning Approach. In A. Bifet, M. May, B. Zadrozny, R. Gavalda, D. Pedreschi, F. Bonchi, J. Cardoso, & M. Spiliopoulou (Eds.), *Machine Learning and Knowledge Discovery in Databases* (pp. 3–19). Springer International Publishing.

Cornish, C. J., & Watkins, H. (1989). *Learning from Delayed Rewards*. King's College.

DOE. (2012). *Buildings Energy Data Book*. Table 1.1.3: Buildings Share of U.S. Primary Energy Consumption. http://buildingsdatabook.eren.doe.gov/TableView.aspx?table=1.1.3

Kober, J., & Peters, J. (2009). Learning motor primitives for robotics. *2009 IEEE International Conference on Robotics and Automation*, 2112–2118. https://doi.org/10.1109/ROBOT.2009.5152577

Paris, B., Eynard, J., Grieu, S., Talbert, T., & Polit, M. (2010). Heating control schemes for energy management in buildings. *Energy and Buildings*, *42*(10), 1908–1917. https://doi.org/10.1016/j.enbuild.2010.05.027

Pertzborn, A., & Veronica, D. (2018). *Intelligent Building Agents Laboratory: Air System Design* (Technical Note (NIST TN)-2025). National Institute of Standards and Technology.

Prívara, S., Široký, J., Ferkl, L., & Cigler, J. (2011). Model predictive control of a building heating system: The first experience. *Energy and Buildings*, *43*(2), 564–572. https://doi.org/10.1016/j.enbuild.2010.10.022

Rummery, G. A., & Niranjan, M. (1994). *On-Line Q-Learning Using Connectionist Systems*. Cambridge University Engineering Department.

Szita, I. (2012). Reinforcement Learning in Games. In M. Wiering & M. van Otterlo (Eds.), *Reinforcement Learning: State-of-the-Art* (pp. 539–577). Springer Berlin Heidelberg. https://doi.org/10.1007/978-3-642-27645-3_17

Wei, T., Chen, X., Li, X., & Zhu, Q. (2018, November). *Model-based and data-driven approaches for building automation and control*. ICCAD '18: Proceedings of the International Conference on Computer-Aided Design. https://doi.org/10.1145/3240765.3243485

Wei, T., Wang, Y., & Zhu, Q. (2017, June 18). *Deep Reinforcement Learning for Building HVAC Control*. DAC '17, Austin, TX, USA.

Konda, V. R. (2002). *Actor-Critic Algorithms*. Massachusetts Institute of Technology.

Yang, L., Nagy, Z., Goffin, P., & Schlueter, A. (2015). Reinforcement learning for optimal control of low exergy buildings. *Applied Energy*, *156*, 577–586. https://doi.org/10.1016/j.apenergy.2015.07.050

Zhang, C., Kuppannagari, S. R., Kannan, R., & Prasanna, V. K. (2019, November 13). *Building HVAC Scheduling Using Reinforcement Learning via Neural Network Based Model Approximation*. ACM BuildSys '19, New York, NY, USA.

Local sensing and personalized thermal comfort: model-predictive vs conventional control approaches

Hejia ZHANG[1, 2]*, Xiaoqi LIU[1, 2], Seungjae LEE[1, 2], Athanasios TZEMPELIKOS[1, 2]

[1]School of Civil Engineering, Purdue University, West Lafayette, Indiana, USA

[2]Center for High Performance Buildings, Ray W. Herrick Laboratories, Purdue University, West Lafayette, Indiana, USA

* Corresponding Author

ABSTRACT

Model-based and model predictive control (MPC) based on personalized thermal preferences has the potential to minimize energy consumption and guarantee occupant thermal comfort. This paper presents the implementation of personalized controls that satisfy personalized thermal preferences in private offices. Two control strategies, a simple PID control and a real-time model predictive control, were implemented through the Building Management System in two adjacent identical office spaces. In each office, monitoring and control of indoor thermal conditions was achieved either by using a conventional wall thermostat or by using a low-cost wireless local sensing network -both integrated with the building management system. Thermal comfort and energy consumption performances were compared between MPC vs PID control, as well as local sensing vs conventional wall thermostat sensing. The results demonstrate the advantages of MPC over simple feedback control, in parallel with the advantages of local sensing over wall thermostats, in terms of energy efficiency and personalized thermal comfort.

1. INTRODUCTION

Integration of HVAC system operation, Building Management Systems and sensing networks, with embedded intelligence and predictive capabilities, can provide a comfortable thermal environment for every occupant, while simultaneously optimizing energy use for air-conditioning. In order to improve thermal satisfaction of different individuals, previous research focused on incorporating occupants in the control framework and creating personalized thermal preference profiles (Lee *et al*., 2019; Zhang *et al*., 2020). In addition, sensor networks take an important role in smart buildings due to their monitoring, communication, and environment control capabilities. Low-cost sensing networks, which have the advantages of flexibility, customization, and monitoring of local environmental parameters, have been evaluated in related studies (Ferreira *et al*., 2012; Gupta *et al*., 2016) and showed that these systems are reliable. In this study, both personalized thermal preference models and wireless local sensing networks are implemented in private offices to show the advantages of using low-cost sensing network over conventional thermostats and the advantages of model predictive control vs simple PID control, with respect to personalized thermal comfort and energy savings.

2. METHODOLOGY

2.1 Personalized Thermal Preference Models

Personalized thermal preference models were built to predict each occupant's thermal preference as a function of indoor air temperature (the variable that can be easily controlled for practical application) as follows. First, the data used for building personalized thermal preference models was collected from experiments conducted in private offices (Lee *et al*., 2018). During the experiments, which lasted several days, each subject was performing normal office work during standard work hours, while the thermal environment was automatically changed according to specific schedules (setpoints) summarized in (Lee *et al*., 2017). During the experiments, occupant thermal preference votes (want cooler

/ no change / want warmer) were collected hourly from each subject, while recording the air temperature and other environmental variables. In total, 48 preference votes were collected by each subject under various air temperature (ranging from 20-26 ^{0}C).

After collecting all required reference data, personalized thermal preference models were developed for each subject based on Bayesian multinomial logistic regression method with linear discriminate analysis (LDA). Occupant's thermal preference is defined as the preference under current thermal conditions (want cooler / no change / want warmer), which can be represented with a multinomial categorical variable. Since the probability of occupant's choice on thermal preference is mainly dependent on indoor air temperature, a multinomial logistic function is used to map the air temperature values to the probabilities. The probability of each thermal preference choice (c) under different indoor temperatures (t) is calculated by the multinomial logistic function in Equation (1).

$$p(x = c|t, \beta) = \frac{e^{\beta_{0,c}t+\beta_{1,c}}}{\sum_{c=0}^{2} e^{\beta_{0,c}t+\beta_{1,c}}} \quad (1)$$

In order to encode prior knowledge for the uncertain coefficients β, the linear discriminate analysis (LDA) method (Lee *et al.*, 2017) is applied, which assumes that $p(t|x=c)$ follows a normal distribution $N(\mu_c,\sigma^2)$. Using LDA method, β can be generated from the mean and standard deviation parameters μ_c and τ ($\tau=1/\sigma^2$) as in Equations (2) and (3):

$$\beta_{0,c} = \mu_c\tau \quad (2)$$

$$\beta_{1,c} = -\frac{1}{2}\mu_c{}^2\tau \quad (3)$$

We are more certain about the prior distributions of μ_c and τ since we have existing knowledge about $p(t|x=c)$ based on collected thermal preference votes from occupants. For the prior distributions of μ_c and τ, normal distributions with different mean values are placed on μ_c, where each mean value is set as the mean temperature of each thermal preference choice made by the occupant, and an exponential distribution is placed on τ since it is always positive. With the parameters defined, Bayesian inference approach is applied and implemented in the PyMC3 Python package (Salvatier *et al.*, 2016) to solve the distributions of unknown parameters. According to Bayesian rules, posterior distribution of the unknown parameters is related with the likelihood and the prior distributions over parameters as Equation (4).

$$p(\beta, \mu, \tau|x, t) \propto p(x|t, \beta)p(\beta|\mu, \tau)p(\mu_0)p(\mu_1)p(\mu_2)p(\tau) \quad (4)$$

After performing the Bayesian inference PyMC3 model with Markov chain Monte Carlo (MCMC) sampling, posterior distributions of the unknown parameters (μ_0, μ_1, μ_2, τ, $\beta_{0,0}$, $\beta_{1,0}$, $\beta_{0,1}$, $\beta_{1,1}$, $\beta_{0,2}$, $\beta_{1,2}$) are calculated. Finally, the predictive probabilities of thermal preference for any given temperature can be obtained from the samples of the unknown parameters generated from posterior distributions.

Representative personalized thermal preference models for two subjects who have obvious different thermal preferences are shown in Figure 1. In the figure, median values of estimated probabilities are plotted with solid lines, and the 95% credible intervals for the three categories are shown with the respective shaded areas. From the personalized thermal preference models, the thermal "comfort range" is selected when the probability of feeling comfortable is larger than the probability of preferring cooler/warmer. The probability of feeling comfortable is highest at the peak of thermal preference for each person.

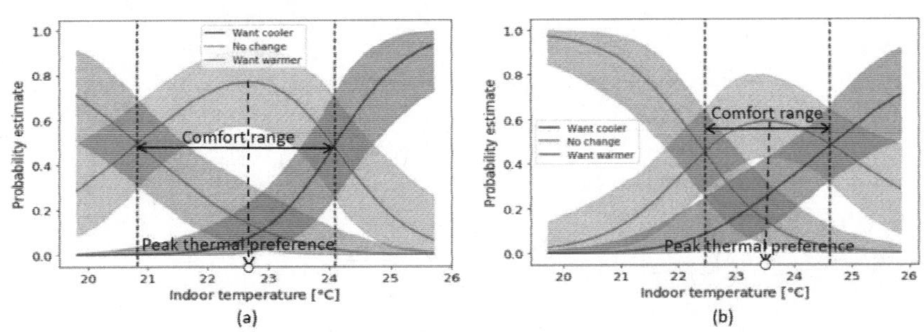

Figure 1: Personalized thermal preference model for (a) subject 1 and (b) subject 2

2.2 Model Predictive Control Algorithm

Model predictive control (MPC) was implemented in the private offices to maintain personal thermal comfort, and estimate the impact of this approach on energy use and personal comfort for different personal preference profiles. For the MPC algorithm, a control-oriented model was constructed for three adjacent private offices as shown in Figure 2. Among the three private offices, two of the offices are identical with one south-facing facade, and the third office has two external walls. Using a 12-hour prediction horizon with a time step of 0.5-hour, the evolution of temperature states was predicted by the state-space formulation in Equation (5).

$$\mathbf{x}_{t+1} = \mathbf{A}\mathbf{x}_t + \mathbf{B}_u\mathbf{u}_t + \mathbf{B}_w\mathbf{w}_t \tag{5}$$

$$\mathbf{x}_t = \begin{bmatrix} T_{\text{air},1,t} \\ T_{\text{air},2,t} \\ T_{\text{air},3,t} \\ T_{\text{mass},1,t} \\ T_{\text{mass},2,t} \\ T_{\text{mass},3,t} \end{bmatrix}, \quad \mathbf{u}_t = \begin{bmatrix} u_{1,t} \\ u_{2,t} \\ u_{3,t} \end{bmatrix}, \quad \mathbf{w}_t = \begin{bmatrix} q_{SG,1,t} \\ q_{SG,2,t} \\ q_{SG,3,t} \\ q_{IG,1,t} \\ q_{IG,2,t} \\ q_{IG,3,t} \\ T_{\text{outdoor},t} \\ T_{\text{corridor},t} \\ T_{\text{indoor},t} \end{bmatrix}$$

The state variables $\mathbf{x}_t$ include the room air temperatures $\mathbf{T}_{\text{air},t}$ and mass temperatures $\mathbf{T}_{\text{mass},t}$ of the three private offices at time t. The control variables $\mathbf{u}_t$ are the heating or cooling rates to the offices. The disturbance vectors $\mathbf{w}_t$ have the following elements: (i) The solar gain to the offices ($\mathbf{q}_{SG,t}$), which was calculated from the hourly-updated NOAA weather forecast data (Seo, 2010); (ii) The internal gain in the offices ($\mathbf{q}_{IG,t}$) due to lighting, equipment, and occupants, which was calculated according to heuristic schedules (ASHRAE, 2013); (iii) The outdoor air temperature (T_{outdoor}) forecast downloaded from NOAA; (iv) The adjacent corridor temperature (T_{corridor}) constantly set to 21.5°C; and (v) The indoor air temperature (T_{indoor}) of an adjacent room (shown on the left of Figure. 2), which was monitored through the building management system.

The state matrix $\mathbf{A} \in \mathbb{R}^{6\times6}$, input matrix $\mathbf{B}_u \in \mathbb{R}^{6\times3}$, and disturbance matrix $\mathbf{B}_w \in \mathbb{R}^{6\times9}$ consist of estimated parameters including the thermal capacitances ($C_{1:4}$) and resistances ($R_{1:7}$), as well as the coefficients multiplied to the heat flux input for the transmitted solar radiation (α_1, α_2) and heating/cooling rate (α_3, α_4). The estimated parameters were identified by collecting the room air temperature, heating/cooling rate, and disturbance measurements from the offices every 5 minutes for 15 days (10 days for training, 5 days for validation), and by solving an optimization problem to minimize the sum of the differences between the on-site air temperature measurements and the air temperatures predicted by Equation (5) for the three offices.

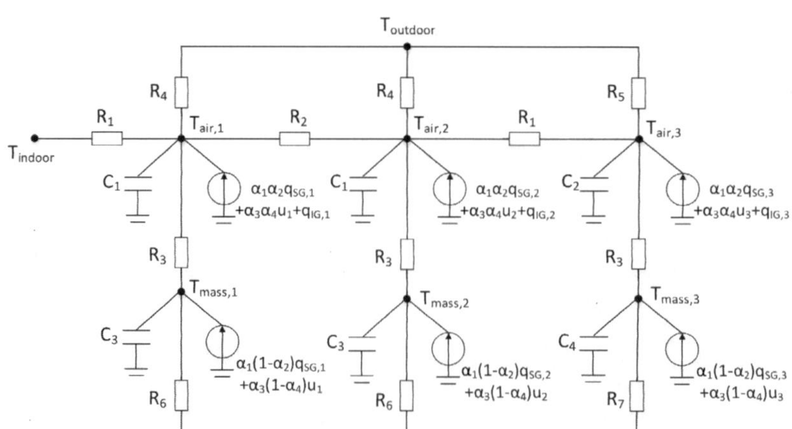

Figure 2: State-space model for private offices

The goal of the MPC algorithm is to minimize the HVAC energy consumption over a certain prediction horizon. This is described by the optimization problem in Equation (6), which needs to be solved to obtain the sequence of optimal control inputs.

$$\min_{\mathbf{u}_0,\mathbf{u}_1,\ldots,\mathbf{u}_{K-1}} \sum_{t=0}^{K-1} J_t(\mathbf{x}_t,\ \mathbf{u}_t, \mathbf{w}_t) \tag{6}$$

where K =24 is the number of time steps in the prediction horizon and J is the cost function defined in in Equation (7) as the sum of the energy consumption (P) of the fan, the chiller, and the boiler (that provides hot water for the reheat coil). Δt is the length of a time step which is set as 0.5-hour. The calculation functions for the energy consumption of fan, chiller and boiler at each time step are based on the power models in the EnergyPlus reference calculations (DOE, 2010).

$$J_t(\mathbf{x}_t,\ \mathbf{u}_t, \mathbf{w}_t) = \left(P_{\text{fan},\,t} + P_{\text{chiller},t} + P_{\text{reheat},t}\right)\Delta t \tag{7}$$

Constraints are imposed on the control inputs and room air temperatures. The constraint imposed on the control inputs is based on the HVAC heating and cooling capacity. In Equation (8), $u_{i,t,max}$ is the upper control bound for office i (i =1,2,3) at time t when the reheat coil runs at its maximum capacity. $u_{i,t,min}$ is the lower control bound at time t when the damper is at the maximum open position. The constraint imposed on room air temperatures in Equation (9) is based on the personalized thermal preference model of the occupant in office i (ranging from $T_{i,t,min}$ to $T_{i,t,max}$] as in Figure. 1.

$$u_{i,t,\text{min}} \leq u_{i,t} \leq u_{i,t,\text{max}} \tag{8}$$

$$T_{i,t,\text{min}} \leq T_{\text{air},i,t} \leq T_{i,t,\text{max}} \tag{9}$$

With the cost function and constraints defined, a constrained nonlinear optimization problem is constructed and solved at each prediction horizon with a sequential quadratic programming algorithm.

3. EXPERIMENTAL STUDIES AND LOCAL SENSING

A low-cost sensing network was used in experiments conducted for monitoring and control of indoor thermal conditions. A dual-channel communication with single board computer allows embedding low-cost sensors and control algorithms into an existing building control system. Wireless local temperature sensors, which support various types of wireless communication protocols without excessive cost and intrusiveness in the environment, were used for low-cost local thermal sensing. The sensor has an accuracy of ± 0.2-0.4°C with a range of -40~125°C. Our validation showed a 0.2-0.5°C temperature difference between the wireless local sensor and a calibrated type-T thermocouple, which proves the accuracy of the local sensor.

To compare local temperature (near the occupant) with a conventional wall thermostat reading, a low-cost sensing network was implemented in a private office first to monitor the thermal difference between the wireless local sensor (local sensor 1), placed on the desk near the occupant, and the wall thermostat. Another wireless low-cost sensor (local sensor 2) was placed at the location of wall thermostat to further prove the local sensors' accuracy. The distance between local sensor 1 and the wall thermostat is about 1 meter (Figure 3).

For the purpose of comparing different sensors, the private office was controlled with a static setpoint temperature of 23.9°C, based on the wall thermostat reading and using the built-in simple PID control through the building management system. Temperature readings of the local sensors and the wall thermostat were collected every minute during four successive days. Sensors were shielded to be protected from solar radiation, and window shades were lower to 75% during the experiments to reduce the overall influence of solar radiation on temperature readings.

Temperature differences between each of the local wireless sensors and the wall thermostat are shown in Figure 4. Weather conditions were recorded during the experiment (Figure 5). The differences between local sensor 2 and wall thermostat (at the same position) were mostly within ±0.5°C, which prove the consistency between the two sensors. There is an obvious difference of about 0.5°C between the local temperature readings near the occupant and the wall thermostat readings, which was mainly caused by different distances to the air diffuser. In addition, even with shading and sensor protection, there are some solar radiation effects, during the two first days which had very clear skies. These representative results show that occupant-perceived local temperatures are different than wall thermostat readings, which could impact personal thermal comfort and energy use.

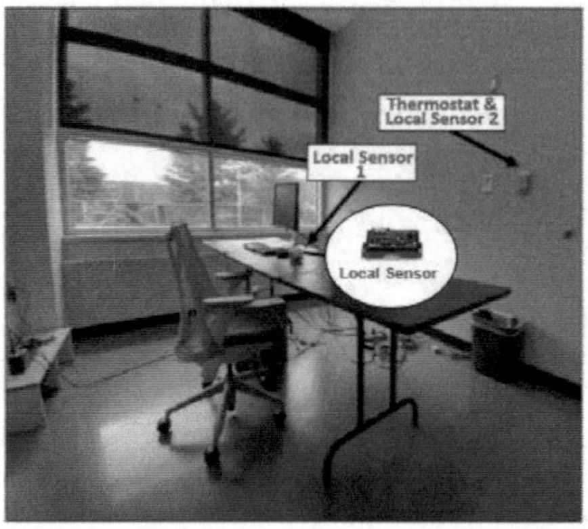

Figure 3: Private office equipped with wireless local sensors and wall thermostat

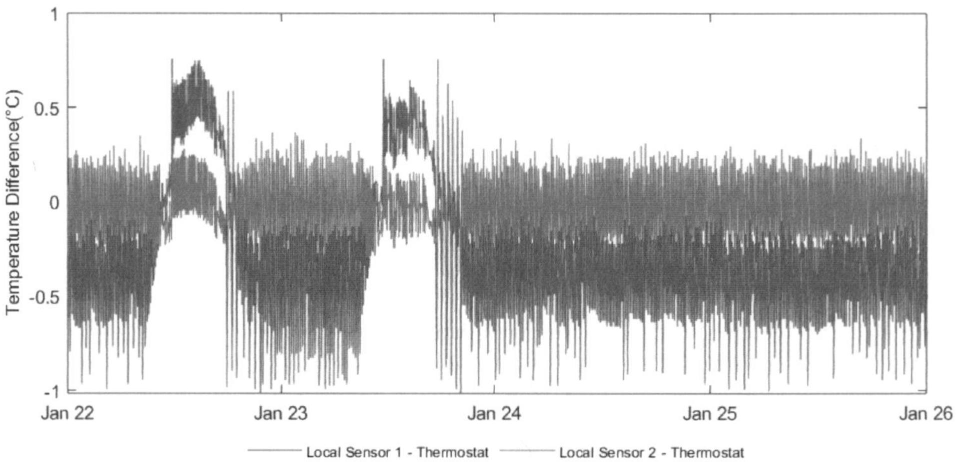

Figure 4: Local temperature differences between local sensors and wall thermostat

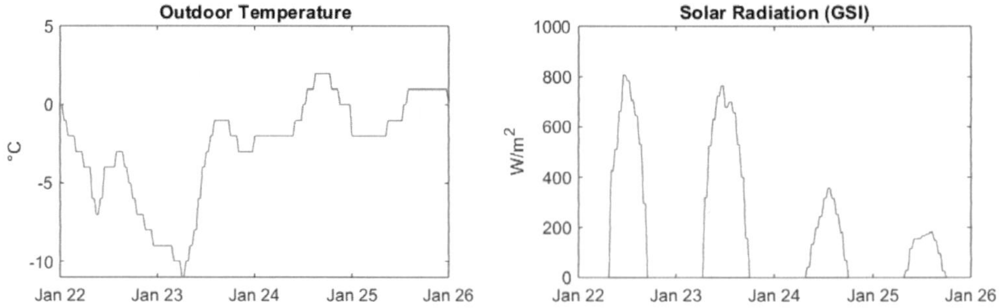

Figure 5: Weather conditions during local sensing experiment in private office

4. RESULTS

Thermal controls with different levels of complexity were implemented with wireless local sensing to satisfy occupant's personalized thermal comfort. The two adjacent identical private offices were used to compare thermal control performances experimentally. The setup in each office was the same as shown in Figure 3. The two private offices were conditioned with two individual VAV boxes from a centralized air conditioning system, and shading of the two office was again set as 75% closed to reduce the influence of solar radiation. During the experiments, a simple PID control and a real-time MPC were conducted in the two offices simultaneously using wireless local air temperature measurements.

The PID control strategy we used was achieved with the built-in simple PI control logic in building management system, which worked with a deadband of ±0.6°C around the setpoint. For personalized thermal control purpose, the setpoint of simple PID control was set as the peak thermal preference temperature obtained from the respective model generated for each subject. The real-time MPC was achieved through sensing with the wireless local sensor in the office, running the MPC algorithm according to real-time weather forecast data, and communicating with the building management system back and forth. During the implementation of real-time MPC, the room air temperature was controlled with the built-in PI control logic according to the setpoints generated from the MPC algorithm –the thermal comfort range obtained from personalized thermal preference model was set as the indoor air temperature bound for each occupant.

A comparison of the two personalized thermal controls for subject 1 (thermal preference profile of Figure. 1(a)) during one day is shown in Figure 6 (with respective recorded weather conditions in Figure 7). The personal thermal comfort range of this subject is defined as [20.8, 24.1] °C, and the subject's most preferred temperature is around 22.7°C. During the personalized thermal controls using wireless local sensing, the wall thermostat readings, auxiliary temperatures and supply air volumes from VAV boxes were also collected through the building management system. Then the real-time cooling/heating rates were calculated using local sensing temperature, auxiliary temperature and supply air volume.

From the results, it can be seen that, during a sunny day in winter, using a PID control to satisfy personal thermal comfort consumes more energy for both cooling and heating compared with personalized MPC, which demonstrates MPC's advantage in optimizing energy performance over simple PID control. For personalized thermal comfort, the local sensor readings show that MPC with wireless local sensing guarantees the indoor temperatures to be within the subject's personalized thermal comfort range, while a noticeable difference in temperature readings can be observed between the wireless local sensor and the wall thermostat. The bias in wall thermostat readings implies that control with conventional wall thermostat could violate the personalized thermal comfort constraint and lead to temporary thermal discomfort.

Using the same settings for real-time MPC, personalized model predictive controls were also implemented in the two identical offices simultaneously, using wireless local sensing for the subjects with different thermal preference profiles (Figure. 1). Subject 2 has a thermal comfort range of [22.5, 24.6] °C. For the different preference profiles, different thermal comfort ranges were set as the indoor temperature control bounds for the two offices in the MPC algorithm. A comparison of indoor air temperatures during personalized MPC for subject 1 and subject 2 during one day is shown in Figure 8. With real-time auxiliary temperatures and supply air volumes monitored in the two rooms, cooling/heating rates over the 12-hour test period were calculated. For personalized MPC for subject 1, the energy consumption was 26354 kJ for cooling, and 25162 kJ for heating; while for subject 2, the energy consumption was 24709 kJ for cooling, and 56629 kJ for heating. This is expected since subject 2 prefers warmer conditions and requires a higher temperature for the lower control bound. The local sensor readings in both offices follow the MPC setpoints and succeed in satisfying the different personalized thermal preferences of the two subjects, while wall thermostat readings are often out of the thermal comfort ranges.

5. CONCLUSION

Model predictive control and simple PID control with wireless local sensing, based on personalized thermal comfort, were implemented in private offices to compare their comfort and energy performance. Through experimental studies, the performance of MPC vs PID and local sensing vs conventional thermostat sensing were compared.

- Local sensing proved its necessity for satisfying thermal comfort and shows advantage in potential energy savings compared with conventional wall thermostat sensing.

- Real-time MPC implemented based on personalized thermal preference can achieve minimizing energy consumption and guarantees personal thermal comfort.
- MPC with wireless local sensing demonstrates its energy efficiency and reliability in maintaining thermal comfort performance during personalized thermal controls.

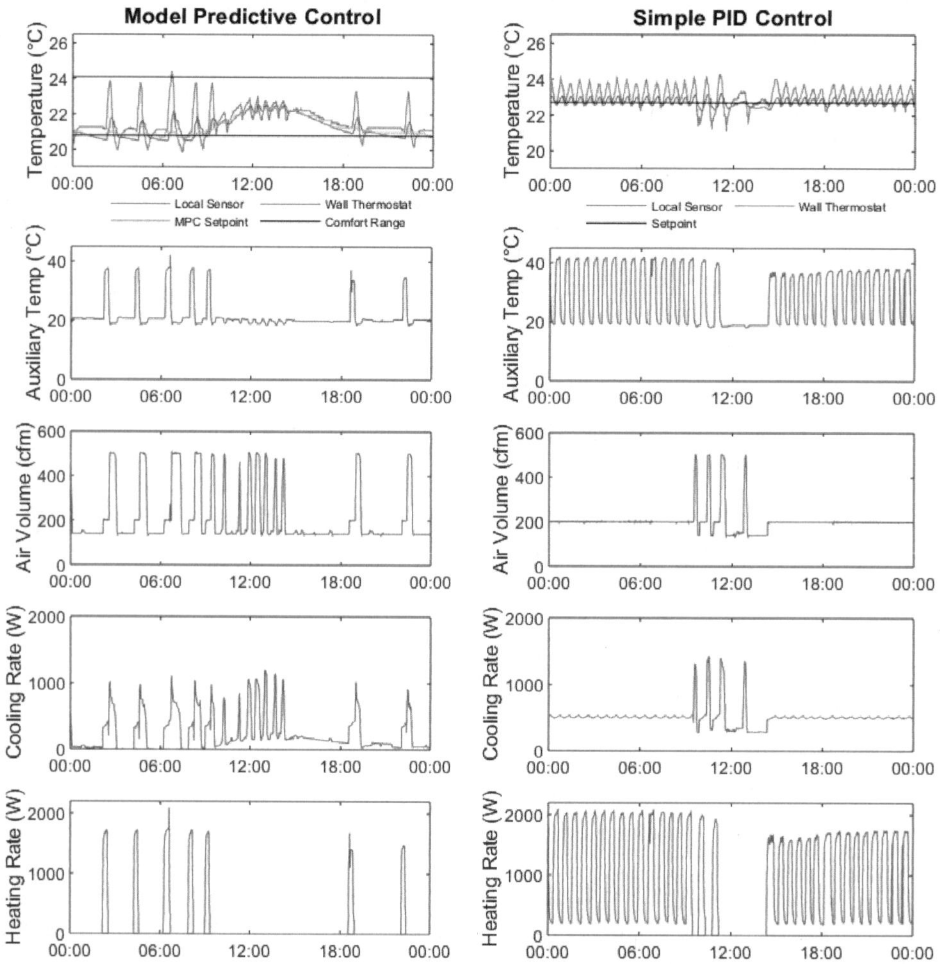

Figure 6: Comparison of personalized MPC and PID control for subject 1

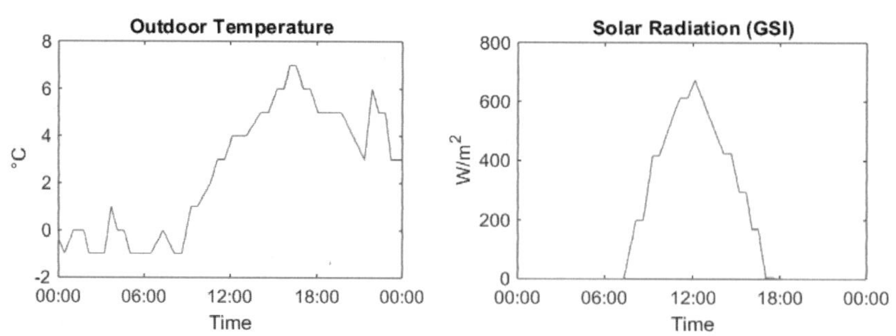

Figure 7: Weather conditions during the comparison of personalized controls for subject 1

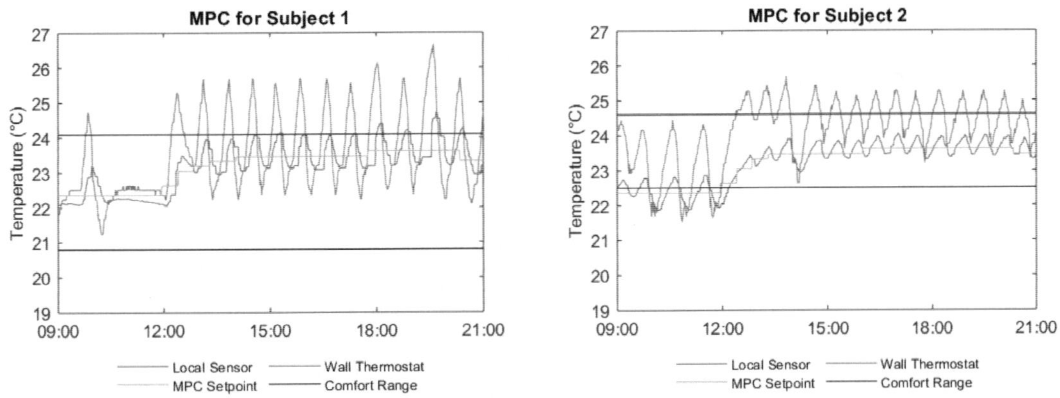

Figure 8: Comparison of indoor air temperatures during personalized MPC for subject 1 and subject 2

REFERENCES

Lee, S., Karava, P., Tzempelikos, A., & Bilionis, I. (2019). Inference of thermal preference profiles for personalized thermal environments with actual building occupants. *Building and Environment*, 148, 714-729.

Zhang, S., Lu, Y., & Lin, Z. (2020). Coupled thermal comfort control of thermal condition profile of air distribution and thermal preferences. *Building and Environment*, 177, 106867.

Ferreira, P. M., Ruano, A. E., Silva, S., & Conceicao, E. Z. E. (2012). Neural networks based predictive control for thermal comfort and energy savings in public buildings. *Energy and buildings*, 55, 238-251.

Gupta, S. K., Atkinson, S., O'Boyle, I., Drogo, J., Kar, K., Mishra, S., & Wen, J. T. (2016). BEES: Real-time occupant feedback and environmental learning framework for collaborative thermal management in multi-zone, multi-occupant buildings. *Energy and Buildings*, 125, 142-152.

Lee, S., Bilionis, I., Karava, P., & Tzempelikos, A. (2017). A Bayesian approach for probabilistic classification and inference of occupant thermal preferences in office buildings. *Building and Environment*, 118, 323-343.

Salvatier, J., Wiecki, T. V., & Fonnesbeck, C. (2016). Probabilistic programming in Python using PyMC3. *PeerJ Computer Science*, 2, e55.

Lee, S., Karava, P., Tzempelikos, A., & Bilionis, I. (2018). An Efficient Method for Learning Personalized Thermal Preference Profiles in Office Spaces. *Proceedings of 5th International High Performance Buildings Conference at Purdue,* July 2018.

Seo, D. (2010). *Development of a universal model for predicting hourly solar radiation—Application: Evaluation of an optimal daylighting controller* (Doctoral dissertation, University of Colorado at Boulder).

ASHRAE. 2013. *ANSI/ASHRAE Standard 55-2013, Thermal Comfort Conditions for Human Occupancy*. Atlanta: ASHRAE.

US Department of Energy (2010). Energyplus engineering reference. *The reference to Energyplus calculations.*

A Unified MPC Formulation for Control of Commercial HVAC Systems in Multiple Climate Zones

Naren Srivaths RAMAN[1*], Bo CHEN[1], Prabir BAROOAH[1]

[1] University of Florida, Department of Mechanical and Aerospace Engineering, Gainesville, Florida, USA
E-mail: narensraman@ufl.edu, bo.chen@ufl.edu, pbarooah@ufl.edu

* Corresponding Author

ABSTRACT

Model predictive control (MPC) has been widely investigated for climate control of commercial buildings for both energy efficiency and demand flexibility. However, most MPC formulations ignore humidity and latent heat. The inclusion of moisture makes the problem considerably more challenging, primarily since a cooling and dehumidifying coil model which accounts for both sensible and latent heat transfers is needed. In our recent work, we proposed an MPC controller in which humidity and latent heat were incorporated in a principled manner, by using a reduced-order model of the cooling coil. Because of the highly nonlinear nature of the process in a cooling coil, the model needs to be modified based on certain weather/climatic conditions to have sufficient prediction accuracy. Doing so, however, leads to a mixed-integer nonlinear program (MINLP) that is challenging to solve. In this work, we propose an MPC formulation that retains the NLP (nonlinear programming problem) structure in all climate zones/weather conditions. This feature makes the control system capable of autonomous operation. Simulations in multiple climate zones and weather conditions verify the energy savings performance, and autonomy of the proposed controller. We also compare the performance of the proposed MPC controller with an MPC formulation that does not explicitly consider humidity. Under certain conditions, it is found that the MPC controller that excludes humidity leads to poor humidity control, or higher energy usage as it is unaware of the latent load on the cooling coil.

1. INTRODUCTION

Model predictive control (MPC) for commercial heating, ventilation, and air conditioning (HVAC) systems for both energy efficiency and demand flexibility has been an active area of research (Serale et al., 2018; Shaikh et al., 2014). However, most MPC formulations ignore humidity and latent heat, focusing only on the sensible heat balance. The inclusion of moisture makes the problem considerably more challenging, primarily since a cooling and dehumidifying coil model which accounts for both sensible and latent heat transfers is needed. The heat transfer and condensation (moisture removal) process on the coil surface is a highly complex phenomenon. In addition to the challenge of modeling such a process, MPC requires that the model be simple. Since MPC uses the model as an equality constraint in an optimization problem that is solved in real-time, a complicated model will increase the computational complexity of the optimization problem.

In our recent work (Raman et al., 2020), we proposed an MPC controller in which humidity and latent heat were incorporated in a principled manner, by using a reduced order model of the cooling coil. However, such reduced order models might not work well when the operating conditions are extreme. Figure 1 shows the outdoor weather conditions for Miami, Florida and Tucson, Arizona. Miami has a hot and humid climate, while Tucson has a hot and dry climate. A possible method for handling such extreme climatic/weather conditions is to have multiple models, each designed to be valid for a specific operating condition. In the context of MPC, such model switching will lead to a mixed integer problem with integer valued variables deciding the model to be used for a particular time instant. Since each such model is nonlinear (especially for the cooling coil), the resulting problem will be an MINLP, and that too a high dimensional one especially if the planning horizon is long. Solving such high-dimensional MINLPs is significantly more challenging than NLPs (Kraemer & Marquardt, 2010).

In this paper, we propose an MPC formulation that can operate effectively in multiple climatic conditions that retains the NLP nature of the underlying optimization problem. The proposed MPC controller is called SL-MPC, where the

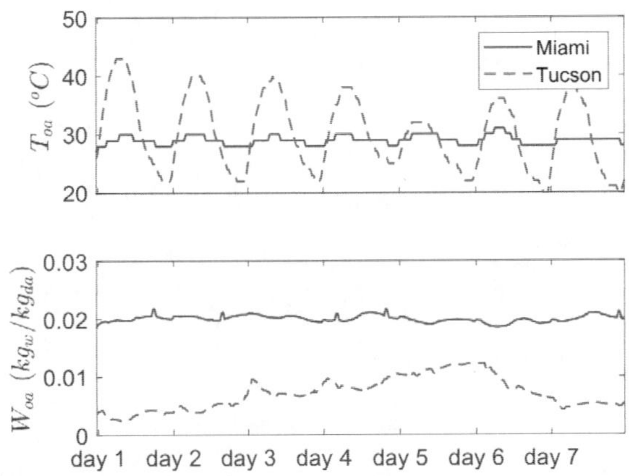

Figure 1: Outdoor weather comparison between Miami, Florida (hot-humid) and Tucson, Arizona (hot-dry) for a week during summer (July/25/2016 to July/31/2016 for Miami and June/06/2016 to June/12/2016 for Tucson). Data obtained from National Solar Radiation Database (nsrdb.nrel.gov). T_{oa}: outdoor air dry-bulb temperature, and W_{oa}: outdoor air humidity ratio.

prefix SL stands for "sensible and latent" heat. It uses the same data-driven model originally proposed in (Raman et al., 2020), and thus does not increase modeling complexity. The simple structure of the model—chosen to strike a compromise between prediction accuracy and simplicity needed for real-time optimization—puts a limit on its prediction in extreme conditions. The main innovation in the proposed formulation is that it uses additional slack variables, for water flow rates with additional constraints for some of the decision variables regarding the cooling and dehumidification coil. These modifications to the MPC formulation makes it climate/weather independent. Simulations in two extreme climate zones, from hot-dry to hot-humid, and multiple weather/seasonal conditions are reported that show the effectiveness of the proposed formulation in providing energy-efficient indoor climate control.

A secondary contribution of the paper is comparison of the proposed MPC scheme with two other control algorithms: (i) a rule-based "dual-maximum" controller as the baseline (ASHRAE, 2011), and (ii) a humidity-agnostic MPC formulation that we call S-MPC, with the prefix S standing for "sensible heat only". Simulations in the two climate zones show that the proposed controller provides more energy savings (over the baseline) than S-MPC. The S-MPC leads to violations in indoor humidity constraints in hot humid climates. More importantly, the proposed scheme is able to maintain temperature and humidity constraints in both climates and all weather conditions without requiring any climate or weather specific modifications in the formulation.

The rest of the paper is organized as follows. Section 2 describes the HVAC system considered in this work. Section 3 presents the proposed MPC-based control algorithm and the two other algorithms with which its performance is compared. The simulation setup is described in Section 4. Simulation results are presented and discussed in Section 5. Finally, the main conclusions are provided in Section 6.

2. SYSTEM DESCRIPTION

In this paper we consider a single-zone commercial variable-air-volume HVAC system, whose schematic is shown in Figure 2. In such a system, part of the air exhausted from the zone is recirculated and mixed with fresh outdoor air. If the mixed air is too cold, it is first heated at the preheating coil so that the downstream cooling coil is protected from freezing and the resulting damage. Then this mixed air is sent through the cooling coil where it is cooled and dehumidified to the conditioned air temperature (T_{ca}) and humidity ratio (W_{ca}). *If the air before the cooling coil is dry, then there is only cooling but no dehumidification, i.e.,* $T_{ca} < T_{pha}$ *and* $W_{ca} = W_{pha}$, where T_{pha} and W_{pha} are the temperature and humidity ratio of air before the cooling coil respectively. The conditioned air is passed through the heating coil where it is heated to the supply air temperature (T_{sa}) and finally supplied to the zone. There is no water vapor phase change across the heating or preheating coils, so the humidity ratio of the supply air is the same as the conditioned air ($W_{sa} = W_{ca}$), and the humidity ratio of the preheated air is the same as the mixed air ($W_{pha} = W_{ma}$).

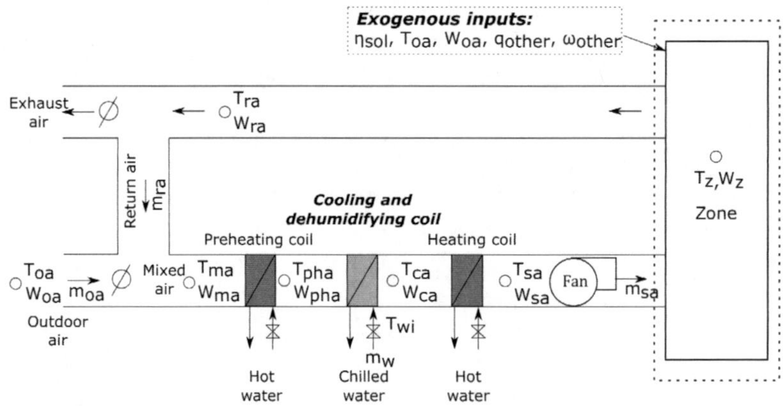

Figure 2: Schematic of a single-zone commercial variable-air-volume HVAC system. In this figure, oa: outdoor air, ra: return air, ma: mixed air, pha: preheated air, ca: conditioned air, and sa: supply air.

The role of a climate control system is to maintain thermal comfort and indoor air quality by varying the following control commands: (i) supply air flow rate (m_{sa}), (ii) outdoor air ratio ($r_{oa} := \frac{m_{oa}}{m_{sa}} = \frac{m_{oa}}{m_{oa}+m_{ra}}$, where m_{oa} and m_{ra} are the outdoor and return airflow rates respectively), (iii) conditioned air temperature (T_{ca}), (iv) supply air temperature (T_{sa}), and (v) preheated air temperature (T_{pha}). So the control command vector is:

$$u := [m_{sa}, r_{oa}, T_{ca}, T_{sa}, T_{pha}]^T \in \Re^5 \tag{1}$$

These five control commands are sent as set points to the low level control loops which are typically comprised of proportional-integral (PI) controllers.

The overall plant model consists of hygro-thermal dynamics of a single-zone building coupled with a cooling coil, heating coil, and a preheating coil. It is of the form $x_{k+1} = f(x_k, u_k, w_k)$ where x is the state vector, u is the input vector, and w is the exogenous input (disturbance) vector. The state vector consists of zone temperature (T_z), wall temperature (T_w), zone humidity ratio (W_z), and conditioned air humidity ratio (W_{ca}), i.e., $x := [T_z, T_w, W_z, W_{ca}]^T \in \Re^4$. The input vector is defined in (1). The exogenous input vector consists of solar irradiance (η_{sol}), outdoor air temperature (T_{oa}), outdoor air humidity ratio (W_{oa}), internal heat load (q_{other}) due to occupants, lights, equipments, etc., and rate of internal water vapor generation (ω_{other}) due to occupants, equipments, etc. Therefore, $w := [\eta_{sol}, T_{oa}, W_{oa}, q_{other}, \omega_{other}]^T \in \Re^5$. The interested readers are referred to our prior work (Raman et al., 2020) for a detailed description of the mathematical models used for simulating the plant.

3. CONTROL ALGORITHMS

In this section, we describe three control algorithms: (i) the proposed MPC controller that incorporates humidity and latent heat (*SL-MPC*), (ii) an MPC controller that considers only sensible heat (*S-MPC*), and (iii) a rule-based controller for baseline (*BL*). *All three controllers need to decide the same five control commands defined in* (1).

The objective function that both MPC controllers try to minimize is the total energy consumption of the HVAC system over the planning horizon. For the HVAC system configuration presented in Figure 2, there are four main components which consume energy. They are fan, cooling coil, preheating coil, and reheating coil. We assume that the energy consumed by other components such as damper actuators is negligible. The total energy consumption of the HVAC system during time k is proportional to the total power at that time, which is:

$$P_{total}(k) := P_{fan}(k) + P^{SL}_{cc}(k) + P_{reheat}(k) + P_{preheat}(k), \tag{2}$$

where P_{fan} is the fan power consumption, P^{SL}_{cc} is the cooling coil power consumption, P_{reheat} is the reheating coil power consumption, and $P_{preheat}$ is the preheating coil power consumption. These are dependent on the supply air flow rate and the enthalpies of return air, outdoor air, preheated air, conditioned air, and supply air. The cooling coil power consumption is given by:

$$P_{cc}^{SL}(k) := \frac{m_{sa}(k)[h_{pha}(k) - h_{ca}(k)]}{\eta_{cc}COP_c}, \tag{3}$$

where h_{pha} and h_{ca} are the specific enthalpy of air before and after the cooling coil. Since we are using the enthalpy and not just the dry bulb temperature, the model accounts for both *sensible and latent heat transfers*, hence the superscript SL. The models for P_{fan} and P_{reheat} are the same as in (Raman et al., 2020). The only new addition is the preheating coil, whose power consumption is modeled as the heat it adds to the mixed air stream:

$$P_{preheat}(k) := \frac{m_{sa}(k)C_{pa}[T_{pha}(k) - T_{ma}(k)]}{\eta_{preheat}COP_h}, \tag{4}$$

where T_{ma} is the mixed air temperature, $\eta_{preheat}$ is the efficiency of the preheating coil, and COP_h is the boiler coefficient of performance.

3.1 Proposed Model Predictive Controller Incorporating Humidity and Latent Heat (SL-MPC)

This controller is an extension of the MPC presented in our prior work (Raman et al., 2020). The optimization problem underlying the proposed MPC controller has the following decision variables: states of the process $x(k) := [T_z(k), W_z(k)]^T \in \mathfrak{R}^2$, the vector of control commands and internal variables $v(k) := [u(k)^T, m_{w,T}(k), m_{w,W}(k), W_{ca}(k)]^T \in \mathfrak{R}^8$, where $u(k)$ is the control command vector defined in (1), and the vector of nonnegative slack variables $\zeta(k) := [\zeta_T^{low}(k), \zeta_T^{high}(k), \zeta_W^{low}(k), \zeta_W^{high}(k), \zeta_{m_w}(k-1)] \in \mathfrak{R}^5$ which is introduced for feasibility of the optimization problem. The exogenous input vector is defined as: $w(k) := [\eta_{sol}(k), T_{oa}(k), W_{oa}(k), q_{other}(k), \omega_{other}(k)]^T \in \mathfrak{R}^5$.

Mathematically the optimization problem at time index j is:

$$\min_{V,X,Z} \sum_{k=j}^{j+N-1} \left[P_{fan}(k) + P_{cc}^{SL}(k) + P_{reheat}(k) + \lambda_{preheat}P_{preheat}(k) \right] \Delta t + P_{slack}(k), \tag{5a}$$

where $V := [v^T(j), v^T(j+1), \ldots, v^T(j+N-1)]^T, X := [x^T(j+1), x^T(j+2), \ldots, x^T(j+N)]^T, Z := [\zeta^T(j+1), \zeta^T(j+2), \ldots, \zeta^T(j+N)]^T$, and time between j and $j+1$ is Δt. The last term, P_{slack}, penalizes the zone temperature, zone humidity, and chilled water flow rate slack variables:

$$P_{slack}(k) := \lambda_T^{low}\zeta_T^{low}(k+1) + \lambda_T^{high}\zeta_T^{high}(k+1) + \lambda_W^{low}\zeta_W^{low}(k+1) + \lambda_W^{high}\zeta_W^{high}(k+1) + \lambda_{m_w}\zeta_{m_w}(k),$$

where the λs are penalty parameters. The optimal control commands are obtained by solving the optimization problem (5a) subject to the following constraints:

$$T_z(k+1) = T_z(k) + \frac{\Delta t}{C}\left[\frac{(T_{oa}(k) - T_z(k))}{R} + m_{sa}(k)C_{pa}(T_{sa}(k) - T_z(k)) + A_e\eta_{sol}(k) + q_{other}(k) \right] \tag{5b}$$

$$W_z(k+1) = W_z(k) + \frac{\Delta t R_g T_z(k)}{VP^{da}}\left[\omega_{other}(k) + m_{sa}(k)\frac{W_{sa}(k) - W_z(k)}{1 + W_{sa}(k)} \right] \tag{5c}$$

$$T_{ca}(k) = T_{pha}(k) + m_{w,T}(k) f(T_{pha}(k), W_{pha}(k), m_{sa}(k), m_{w,T}(k)) \tag{5d}$$

$$W_{ca}(k) = W_{pha}(k) + m_{w,W}(k)\, g(T_{pha}(k), W_{pha}(k), m_{sa}(k), m_{w,W}(k)) \tag{5e}$$

$$m_{w,W}(k) = m_{w,T}(k) - \zeta_{m_w}(k) \tag{5f}$$

$$T_z^{low}(k) - \zeta_T^{low}(k) \le T_z(k) \le T_z^{high}(k) + \zeta_T^{high}(k) \tag{5g}$$

$$a^{low}T_z(k) + b^{low} - \zeta_W^{low}(k) \le W_z(k) \le a^{high}T_z(k) + b^{high} + \zeta_W^{high}(k) \tag{5h}$$

$$\max\left(m_{sa}(k) - m_{sa}^{rate}\Delta t, m_{sa}^{low}\right) \le m_{sa}(k+1) \le \min\left(m_{sa}(k) + m_{sa}^{rate}\Delta t, m_{sa}^{high}\right) \tag{5i}$$

$$\max\left(T_{pha}(k) - T_{pha}^{rate}\Delta t, T_{pha}^{low}, T_{ma}(k+1)\right) \le T_{pha}(k+1) \le \min\left(T_{pha}(k) + T_{pha}^{rate}\Delta t, T_{pha}^{high}\right) \tag{5j}$$

$$\max\left(r_{oa}(k) - r_{oa}^{rate}\Delta t, r_{oa}^{low}\right) \le r_{oa}(k+1) \le \min\left(r_{oa}(k) + r_{oa}^{rate}\Delta t, r_{oa}^{high}\right) \tag{5k}$$

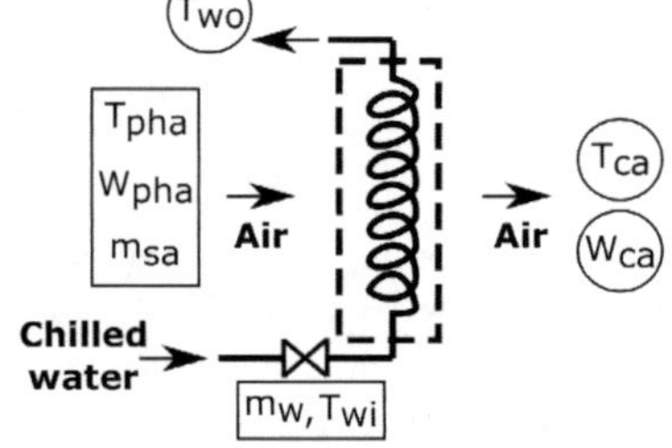

Figure 3: Schematic of a cooling coil.

$$\max\left(T_{ca}(k) - T_{ca}^{rate}\Delta t, T_{ca}^{low}\right) \leq T_{ca}(k+1) \leq \min\left(T_{ca}(k) + T_{ca}^{rate}\Delta t, T_{pha}(k+1)\right) \tag{5l}$$

$$\max\left(T_{sa}(k) - T_{sa}^{rate}\Delta t, T_{ca}(k+1)\right) \leq T_{sa}(k+1) \leq \min\left(T_{sa}(k) + T_{sa}^{rate}\Delta t, T_{sa}^{high}\right) \tag{5m}$$

$$W_{ca}(k) \leq W_{pha}(k) \tag{5n}$$

$$\zeta_T^{low}(k+1),\ \zeta_T^{high}(k+1) \geq 0 \tag{5o}$$

$$\zeta_W^{low}(k+1),\ \zeta_W^{high}(k+1),\ \zeta_{m_w}(k) \geq 0 \tag{5p}$$

where constraints (5b)-(5f) and (5n)-(5p) are for $k = j, \ldots, j+N-1$, constraints (5g) and (5h) are for $k = j+1, \ldots, j+N$, and constraints (5i)-(5m) are for $k = j-1, \ldots, j+N-2$.

Constraints (5b) and (5c) are for the discretized temperature and humidity dynamics model of the zone respectively. Constraints (5d)-(5f) are for the control-oriented cooling coil model and is presented in the next subsection (Section 3.1.1). Constraints (5g) and (5h) are box constraints to maintain temperature and humidity of the zone within the allowed comfort limits. Usually the limits during the unoccupied mode are more relaxed than the occupied mode, i.e., $[T_z^{low,occ}, T_z^{high,occ}] \subseteq [T_z^{low,unocc}, T_z^{high,unocc}]$ and $[RH_z^{low,occ}, RH_z^{high,occ}] \subseteq [RH_z^{low,unocc}, RH_z^{high,unocc}]$, where RH_z is the relative humidity of the zone. RH_z is a highly nonlinear function of dry bulb temperature and humidity ratio (ASHRAE, 2017, Chapter 1). We linearize this function which gives us the coefficients a^{low}, b^{low}, a^{high}, and b^{high} in (5h), and helps in converting the constraints on relative humidity to humidity ratio.

Constraint (5i) is to take into account the capabilities of the fan. The minimum supply airflow rate is computed based on the ventilation requirements specified by ASHRAE 62.1 (ASHRAE, 2016) and to maintain positive building pressurization.

Constraints (5j)-(5m) are to take into account the capabilities of the preheating coil, damper actuators, cooling coil, and reheating coil. In constraints (5j) and (5m), the inequalities $T_{pha}(k+1) \geq T_{ma}(k+1)$ and $T_{sa}(k+1) \geq T_{ca}(k+1)$ ensure that the preheating and reheating coils can only add heat; it cannot cool. Similarly, in constraints (5l) and (5n), the inequalities $T_{ca}(k+1) \leq T_{pha}(k+1)$ and $W_{ca}(k) \leq W_{pha}(k)$ ensure that the cooling coil can only cool and dehumidify the air stream; it cannot add heat or moisture. Inequality constraints (5o) and (5p) ensure that the slack variables are nonnegative.

3.1.1 Control-oriented cooling coil model: Constraints (5d), (5e), and (5f) are for the control-oriented cooling coil model, which is a modified version of the model developed in (Raman et al., 2020). Figure 3 shows the schematic of a cooling coil. The inputs to the model are supply air flow rate (m_{sa}), chilled water flow rate (m_w), inlet water temperature ($T_{w,i}$), and temperature (T_{pha}) and humidity ratio (W_{pha}) of the air before the coil. The outputs are conditioned air temperature (T_{ca}) and humidity ratio (W_{ca}).

First we describe the reduced-order model proposed in (Raman et al., 2020), before discussing the modifications needed to make the MPC formulation climate/weather independent. It is a static model with the outputs being a polynomial function of the inputs. Note that when the chilled water flow rate is zero, no cooling or dehumidification of the air occurs. In that situation, the temperature and humidity ratio of the air before and after the coil must be equal: $T_{ca} = T_{pha}$

and $W_{ca} = W_{pha}$, when $m_w = 0$. To make the model have this behavior, the following functional form is chosen:

$$T_{ca}(k) = T_{pha}(k) + m_w(k)\, f(T_{pha}(k), W_{pha}(k), m_{sa}(k), m_w(k)) \tag{6}$$

$$W_{ca}(k) = W_{pha}(k) + m_w(k)\, g(T_{pha}(k), W_{pha}(k), m_{sa}(k), m_w(k)) \tag{7}$$

where functions f and g are chosen as quadratic in their arguments.

Now we discuss the modifications proposed here, which are needed to make the MPC formulation climate/weather independent. In hot-dry weather conditions, the cooling coil might be providing only cooling without any dehumidifcation. In such a situation, the temperature of the air after the coil will be cooler than the air before the coil, while the humidity of the air before and after the coil will be the same, i.e., $T_{ca} < T_{pha}$ and $W_{ca} = W_{pha}$. For the model to have this behavior, we split the chilled water flow rate m_w into two virtual variables $m_{w,T}$ and $m_{w,W}$ as shown below:

$$T_{ca}(k) = T_{pha}(k) + m_{w,T}(k)\, f(T_{pha}(k), W_{pha}(k), m_{sa}(k), m_{w,T}(k)) \tag{8}$$

$$W_{ca}(k) = W_{pha}(k) + m_{w,W}(k)\, g(T_{pha}(k), W_{pha}(k), m_{sa}(k), m_{w,W}(k)). \tag{9}$$

We also introduce the equality constraint (5f): $m_{w,W}(k) = m_{w,T}(k) - \zeta_{m_w}(k)$, where $\zeta_{m_w}(k)$ is a non-negative slack variable which is penalized in the objective function (5a). The main idea behind these modifications is that, inequality constraint (5n): $W_{ca}(k) \le W_{pha}(k)$, will be active in hot-dry weather conditions, forcing $m_{w,W}(k)$ to be zero while $m_{w,T}(k)$ can be non-zero. This allows for cooling without any dehumidifcation, i.e., $T_{ca} < T_{pha}$ and $W_{ca} = W_{pha}$, as $m_{w,T} \neq 0$ while $m_{w,W} = 0$. The high penalty on the slack variable (ζ_{m_w}) ensures that the two chilled water flow rate variables are equal most of the time and is used only under extreme conditions when the model is not able to predict well. In Section 5.3 we discuss how these design choices helped the controller to perform well under various conditions.

3.2 Model Predictive Controller Incorporating Only Sensible Heat (*S-MPC*)

This controller is similar to the one described in Section 3.1, with the main difference being that the moisture and latent heat of air are not considered. The optimization problem formulation is similar to the one presented in (Ma et al., 2012).

There are five main differences when compared to *SL-MPC*: (i) *The cooling power term in the objective function is based only on sensible heat; latent heat is ignored.* (ii) *S-MPC* does not need zone humidity measurement. (iii) Since *S-MPC* does not consider humidity and latent heat, the constraints placed on the humidity at various locations in the air loop as well as the zone—(5c), (5h), (5n), and (5p)—are no longer used. (iv) The constraints placed on the system due to the cooling and dehumidifying coil model—(5d), (5e), and (5f)—are also not present. (v) Prediction of the exogenous inputs W_{oa} and ω_{other} which are used to compute humidity related constraints, are not needed. The interested readers are referred to the *S-MPC* controller presented in (Raman et al., 2020, Section 3.2) for details.

3.3 Baseline Controller (*BL*)

The rule-based *Dual Maximum* (ASHRAE, 2011) controller is used as the baseline controller. Even though the *Single Maximum* controller is widely used, the *Dual Maximum* controller is more energy-efficient among the two (ASHRAE, 2011; Goyal et al., 2013). The *Dual Maximum* controller operates in three modes based on the zone temperature: (i) Cooling, (ii) Deadband, and (iii) Heating. The supply airflow rate (m_{sa}) and temperature (T_{sa}) are varied based on the mode. The conditioned air temperature (T_{ca}) is typically at a low value (12.8^oC), which ensures that air supplied to the zone is sufficiently dry at all times (Williams, 2013). The outdoor air ratio (r_{oa}) is varied to maintain the ventilation requirements dictated by ASHRAE 62.1 (ASHRAE, 2016) and positive building pressurization requirements. If the mixed air temperature is too cold, then the preheated air temperature (T_{pha}) is maintained typically at 12.8^oC (55^oF), which protects the cooling coil from freezing. Details of the *Dual Maximum* controller can be found in (ASHRAE, 2011).

4. SIMULATION SETUP

4.1 Plant Parameters

The plant simulation parameters are chosen based on a large classroom/auditorium (~ 6 m high and floor area of ~ 465 m^2) in Pugh Hall located at the University of Florida, USA. We present only the relevant details here, the interested readers are referred to (Raman et al., 2020) for a complete list of the parameter values used.

The scheduled occupancy is from 7:30 AM to 7:00 PM, Monday to Friday, during which the following constraints are used: $T_z^{low,occ}$ = 21.1^oC (70^oF), $T_z^{high,occ}$ = 23.3^oC (74^oF), $RH_z^{low,occ}$ = 10%, and $RH_z^{high,occ}$ = 60%. The unoccupied hours are from 7:00 PM to 7:30 AM, Monday to Friday, and all of Saturday and Sunday, during which the following constraints are used: $T_z^{low,unocc}$ = 18.9^oC (66^oF), $T_z^{high,unocc}$ = 25.6^oC (78^oF), $RH_z^{low,unocc}$ = 10%, and $RH_z^{high,unocc}$ = 60%.

We assume that during weekdays there are 175 people present in the building between 7:30 AM to 11:30 AM and 12:30 PM to 7:00 PM. q_{other} and ω_{other} are computed based on the number of occupants (n_p) in the zone, assuming that each person produces 100 W of heat and 1.39×10^{-5} kg/s (50 g/h) of water vapor (ASHRAE, 2017). For q_{other}, an additional heat load of 6000 W is considered based on lighting/equipment power density of 12.92 W/m^2 (1.2 W/ft^2), during the scheduled occupancy. This additional heat load is reduced to 3000 W during the unoccupied hours.

4.2 Controller Parameters

MPC parameters: The optimization problem is solved using CasADi (Andersson et al., 2019) and IPOPT (Wächter & Biegler, 2006), a nonlinear programming (NLP) solver, on a Desktop Linux computer with 16GB RAM and a 3.60 GHZ × 8 CPU. We use a time step of Δt = 10 minutes and prediction/planning horizon N = 144 (corresponding to 24 hours). Therefore, there are 2160 (=144×{2+8+5}) decision variables for *SL-MPC*, and there are 1152 (=144×{1+5+2}) decision variables for *S-MPC*. On an average, the optimization problem in *SL-MPC* takes 7 seconds to solve, while the optimization problem in *S-MPC* takes 1.5 seconds to solve.

The MPC controllers require prediction of exogenous inputs. We compute the loads due to occupants in q_{other} and ω_{other} by assuming designed number of occupants (175) during the scheduled hours of occupancy. The remaining exogenous inputs are assumed to be fully known.

We assume that the number of occupants in the zone is not measured. So the MPC controllers need to ensure that the outdoor air needed to satisfy the ventilation requirements corresponding to the designed number of occupants (175), is provided during the scheduled hours of occupancy, according to ASHRAE 62.1 (ASHRAE, 2016).

For *SL-MPC*, the coefficients for the humidity constraint in (5h) are a^{high} = 0.000621 $kg_w/kg_{da}/^oC$, b^{high} = -0.173323 kg_w/kg_{da}, a^{low} = 0.000101 $kg_w/kg_{da}/^oC$, and b^{low} = -0.028104 kg_w/kg_{da}.

4.3 Performance Metrics

We use three performance metrics to compare the controllers: (i) the total energy consumed for a week (E_{total}), (ii) integral of the zone temperature violation for a week (V_T), and (iii) integral of the zone humidity violation for a week (V_{RH}). These are the same as defined in (Raman et al., 2020, Section 4.3) with three key differences: (i) the total energy consumed includes the preheating coil as well, (ii) for the humidity violation, we use relative humidity instead of humidity ratio, and (iii) the metrics are computed for a week (168 hours) instead of a day.

5. RESULTS AND DISCUSSIONS

We compare the performance of the three controllers in simulations by using weather data from two different locations: (i) Miami in Florida, which is considered to be "hot-humid" and belongs to climate zone 1A according to the International Energy Conservation Code (International Code Council, Inc., 2018), and (ii) Tucson in Arizona, which is considered to be "hot-dry" and belongs to climate zone 2B (International Code Council, Inc., 2018). The weather data for these locations are obtained from the National Solar Radiation Database (`nsrdb.nrel.gov`).

For both the locations, four one-week simulations are performed by choosing weather data from different seasons: spring, summer, fall, and winter. Figure 4a shows the total energy consumed when using the three controllers for different climate zones and seasons. Figure 4b shows the corresponding humidity violation. The temperature violation was found to be negligible for all the three controllers and therefore is not presented. The simulation results indicate the following:

- The proposed unified *SL-MPC* formulation performs well in both climate zones and in all the seasons; refer to Section 5.3 to see how our design choices helped the controller to perform well under various conditions.
- *SL-MPC* consumes the least amount of energy when compared to *BL*. The energy savings is up to 27% depending on the climate zone and season. We remark here that the baseline controller used for comparison is the energy-efficient *Dual Maximum*.

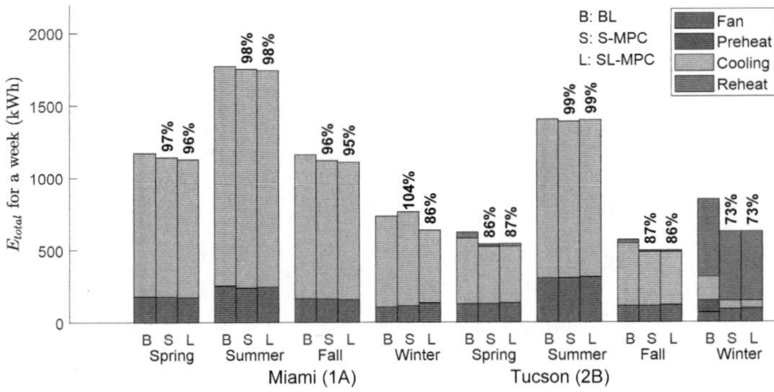

(a) Total energy consumed (E_{total}) for a week.

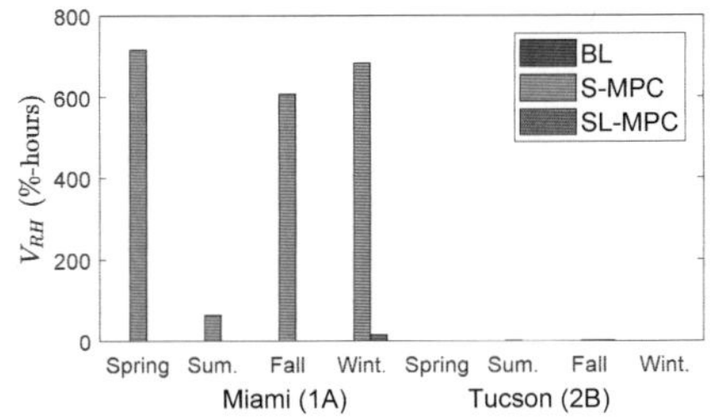

(b) Humidity violation (V_{RH}) for a week.

Figure 4: Performance comparison of *SL-MPC* (proposed controller), *S-MPC*, and *BL* for different climate zones and seasons.

- Moreover, using *SL-MPC* leads to negligible violation in temperature and humidity constraints under all conditions.
- *S-MPC* leads to large humidity violations in Miami.
- In addition to the humidity violations in Miami, *S-MPC* also consumes much more energy (17%) when compared to *SL-MPC*.
- The control decisions made by *SL-MPC* and *S-MPC* are found to be similar in Tucson.

These results are discussed in detail next.

5.1 Miami: Climate Zone 1A

Using *SL-MPC* leads to significant energy savings (14%) mainly during winter when compared to *BL*. There are two main reasons for energy savings by *SL-MPC*. One, *SL-MPC* varies the conditioned air temperature (T_{ca}) as long as the humidity constraints are not violated and the cooling load in the zone is met. On the other hand, *BL* keeps T_{ca} at a constant low value of 12.8^{o}C (55°F) in the interest of maintaining indoor humidity, which leads to higher cooling energy consumption. Two, *SL-MPC* varies the outdoor air ratio and supply air flow rate to use as much "free cooling" as possible when the outdoor weather is mild and dry. While, *BL* brings in only the minimum outdoor air needed to satisfy the ventilation and building pressurization requirements. During summer, spring, and fall, the outdoor weather is pretty warm and humid, so the room for optimization and, thus, energy savings is less.

Using *S-MPC* leads to large humidity violations in spring, fall, and winter; see Figure 4b. It tries to use free cooling, mainly during nighttime, as the outdoor air is cooler than the return air. *But the outdoor air is humid, which it is unaware of.* In an attempt to reduce the cooling energy used further, it varies T_{ca}. Both these factors lead to humid air being supplied to the zone leading to indoor humidity violations. During summer the cooling load of the zone is high, so *S-MPC* decides to keep T_{ca} low which has an unintended, but good, side effect of maintaining indoor humidity. For *SL-MPC*, the humidity constraints are explicitly imposed, thus, it varies the control inputs in a way that no violations occur; see Figure 4b. Even in the case of *BL* there are no humidity violations because of the conservatively designed set point for T_{ca} (12.8^{o}C).

Using *S-MPC* also leads to higher energy use than both *BL* and *SL-MPC* in winter; see Figure 4a. This is because it tries to use free cooling without realizing that the outdoor air is cool but humid. This leads to a higher latent load on the cooling coil and, thus, higher cooling energy consumption.

5.2 Tucson: Climate Zone 2B

Using *SL-MPC* leads to significant energy savings (13% to 27%) when compared to *BL* mainly during spring, fall, and winter. This is because of the same reasons mentioned in Section 5.1. In addition to these, *SL-MPC* avoids preheating completely during winter by recirculating as much warm air from the zone as possible. It satisfies the outdoor air

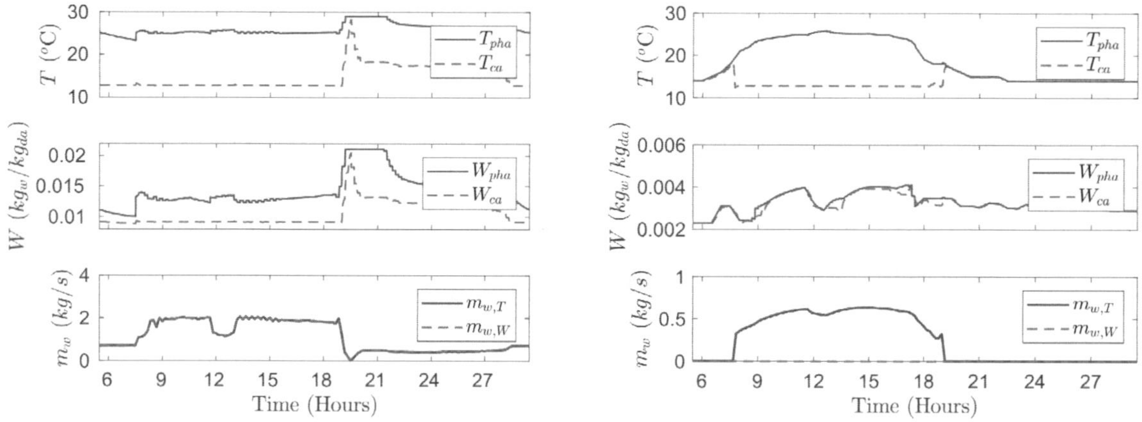

(a) Summer in Miami: both cooling ($T_{ca} < T_{pha}$) and dehumidification ($W_{ca} < W_{pha}$) are seen to occur as the weather is hot and humid.

(b) Spring in Tucson: there is cooling ($T_{ca} \leq T_{pha}$) but no dehumidification ($W_{ca} = W_{pha}$) as the weather is hot and dry.

Figure 5: Simulations for a day using *SL-MPC*, showing that it works well in both humid and dry weather conditions. Top to bottom: temperature before (T_{pha}) and after (T_{ca}) the cooling coil, humidity ratio before (W_{pha}) and after the cooling (W_{ca}) coil, and chilled water flow rate ($m_{w,T}$) used by *SL-MPC* in constraint (5d) and chilled water flow rate ($m_{w,W}$) used by *SL-MPC* in constraint (5e).

requirements (m_{oa}) using a lower outdoor air ratio (r_{oa}) and a higher supply airflow rate (m_{sa}). Whereas, *BL* is in the heating mode because of the cold weather in winter, so it uses a lower m_{sa} and thus a higher r_{oa} (recall that *BL* varies r_{oa} to maintain the minimum outdoor air requirements) to satisfy the same m_{oa} requirements. The usage of higher m_{sa} by *SL-MPC* leads to a slightly higher fan energy consumption but a substantial decrease in the preheating energy consumption; see the results for Tucson during winter in Figure 4a.

The decisions made by both the MPC controllers are found to be similar in Tucson; see Figure 4. This can be attributed to the dry outdoor weather in Tucson.

5.3 Autonomy of *SL-MPC*

The results and discussions presented in Sections 5.1 and 5.2 make it clear that the proposed *SL-MPC* controller works well in both humid and dry climate zones, and in all four seasons. In this section, we will be looking at how the design choices made in the MPC formulation helped the controller in performing well under various conditions.

Figure 5 shows the simulation results when using *SL-MPC* for a day in Miami and Tucson. It can be seen from Figure 5a that there is both cooling and dehumidification occurring in Miami as the outdoor weather is hot and humid. Moreover, the slack variable corresponding to the chilled water flow rate (ζ_{m_w}) is found to be 0 as $m_{w,T} = m_{w,W}$. However, in Tucson $\zeta_{m_w} \neq 0$, as $m_{w,T} \neq m_{w,W}$; see between 9:00-18:00 hrs in Figure 5b. Specifically $m_{w,W} = 0$ while $m_{w,T} \neq 0$, which allows for cooling without any dehumidification as the outdoor weather is hot and dry in Tucson. Such behaviors are made possible because of the following design choices: (i) the functional form of the control-oriented cooling coil model in (5d) and (5e), (ii) separating the chilled water flow rate into two virtual variables ($m_{w,T}$ and $m_{w,W}$), (iii) equality constraint (5f) with slack variable ζ_{m_w} and the heavy penalty on the slack variable in the objective function (5a), which ensures that the flow rates are different only when needed.

6. CONCLUSION

In our recent work (Raman et al., 2020) we showed the importance of inclusion of humidity and latent heat in an MPC formulation for energy-efficient HVAC control, especially in hot-humid climates. We showed that a humidity-agnostic MPC formulation can lead to poor humidity control, or higher energy usage as it is unaware of the latent load on the cooling coil. The MPC formulation that we had proposed in our prior work uses a reduced order cooling coil model which puts a limitation on its prediction accuracy in extreme climatic/weather conditions. In this work we present a computationally tractable reformulation which addresses that limitation. Simulation results show that the unified MPC formulation proposed here performs better than a rule-based baseline controller in multiple climate zones and weather conditions.

REFERENCES

Andersson, J. A. E., Gillis, J., Horn, G., Rawlings, J. B., & Diehl, M. (2019, Mar 01). Casadi: a software framework for nonlinear optimization and optimal control. *Mathematical Programming Computation*, *11*(1), 1–36.

ASHRAE. (2011). *The ASHRAE handbook : HVAC applications (SI Edition).*

ASHRAE. (2016). *ANSI/ASHRAE standard 62.1-2016, ventilation for acceptable air quality.*

ASHRAE. (2017). *The ASHRAE handbook fundamentals (SI Edition).*

Goyal, S., Ingley, H., & Barooah, P. (2013, June). Occupancy-based zone climate control for energy efficient buildings: Complexity vs. performance. *Applied Energy*, *106*, 209-221.

International Code Council, Inc. (2018). *International Energy Conservation Code.*

Kraemer, K., & Marquardt, W. (2010). Continuous reformulation of MINLP problems. In M. Diehl, F. Glineur, E. Jarlebring, & W. Michiels (Eds.), *Recent advances in optimization and its applications in engineering* (pp. 83–92). Springer.

Ma, Y., Richter, S., & Borrelli, F. (2012, March). Chapter 14: Distributed model predictive control for building temperature regulation. *In Control and Optimization with Differential-Algebraic Constraints*, *22*, 293-314.

Raman, N. S., Devaprasad, K., Chen, B., Ingley, H. A., & Barooah, P. (2020, December). Model predictive control for energy-efficient HVAC operation with humidity and latent heat considerations. *Applied Energy*, *279*, 115765. doi: https://doi.org/10.1016/j.apenergy.2020.115765

Serale, G., Fiorentini, M., Capozzoli, A., Bernardini, D., & Bemporad, A. (2018). Model predictive control (MPC) for enhancing building and HVAC system energy efficiency: Problem formulation, applications and opportunities. *Energies*, *11*(3), 631.

Shaikh, P. H., Nor, N. B. M., Nallagownden, P., Elamvazuthi, I., & Ibrahim, T. (2014). A review on optimized control systems for building energy and comfort management of smart sustainable buildings. *Renewable and Sustainable Energy Reviews*, *34*, 409 - 429.

Wächter, A., & Biegler, L. T. (2006, Mar 01). On the implementation of an interior-point filter line-search algorithm for large-scale nonlinear programming. *Mathematical Programming*, *106*(1), 25–57.

Williams, J. (2013). *Why is the supply air temperature 55F?* `http://8760engineeringblog.blogspot.com/2013/02/why-is-supply-air-temperature-55f.html`. (Last accessed: Aug, 03, 2020)

ACKNOWLEDGMENT

The research reported here has been partially supported by an NSF grant (award no. 1934322, CMMI) and the State of Florida through a REET (Renewable Energy and Energy Efficient Technologies) grant.

Pattern Analysis of Dynamic Grid Incentives and the Implications on Optimal Control of Building Thermal Energy Storage

Lily X. Li[1], Gregory S. Pavlak[1*]

[1] The Pennsylvania State University,
University Park, PA, U.S.A
Contact Information (xml5187@psu.edu, gxp93@psu.edu)

* Corresponding Author

ABSTRACT

Building thermal energy storage has been utilized for decades for various objectives, such as reducing peak electrical demand, reducing building operating expenses, and increasing the efficiency of systems when charged from waste heat or free cooling. As building thermal storage control strategies become more dynamic, optimization of building performance often considers multiple objectives that aim to improve building performance in energy, economic, environmental, and grid support categories. The dynamics of the incentive signal used for one objective, as well as its relation to signals from other objectives—for instance, whether the signals are "in sync" or are "conflicting"—heavily influence the tradeoffs that may exist among performance objectives. To better understand the degree of alignment that may exist between grid incentive signals, we apply unsupervised learning to a novel grid data set that includes hourly signals for energy price and marginal carbon emissions. Clustering algorithms identify common patterns in the dynamic signals. Overall, Hierarchical Clustering demonstrated the best performance, evaluated by DB index and Silhouette score. While the algorithms did not find distinctive patterns among the carbon signals, they did identify 7 to 9 patterns within the January and July pricing signals. The highly fluctuating nature of the carbon emission signals could lead to a diverse range of tradeoffs between building energy cost and carbon emission reduction objectives, if the signals were used as the basis for a building control optimization problem. This finding iterates the importance of understanding incentive signal dynamics, in both individual and collective contexts, and the implications for development of new control technologies for grid-interactive buildings.

1. INTRODUCTION

Building thermal storage has been utilized for decades to facilitate building performance improvements across various operational objectives. In published works from the last decade, building thermal storage has been shown to excel at reducing building operating expenses and utility costs (Henze et al., 2004; Greensfelder et al., 2011), shifting or shedding peak electric load (Klaassen et al., 2002; Pavlak et al., 2014), and increasing the efficiency of systems when charged from waste heat or free cooling. More recently, studies by Niu et al. (2019) and Delgado et al. (2020) suggest that building thermal storage is also effective in providing grid flexibility and compensating for the power fluctuations created by variable renewable energy generation. In addition, our recent work has shown that building flexibility can reduce the required size of battery storage in building-level microgrids (Swaminathan et al., 2020), and that there are potential benefits to coordinating the operations of a portfolio of thermal energy storage resources to manage price volatility and uncertainty (Yu and Pavlak, 2021).

As building thermal storage control strategies become more dynamic, optimization of building performance often considers multiple objectives that aim to improve building performance in energy, economic, environmental, and grid support categories. The result of these optimizations, which often utilize several incentive signals as the model input, came to be directly and heavily influenced by these signal profiles. For example, an optimal setpoint control strategy that minimizes the building's utility bill will avoid consuming electricity at hours with higher electricity prices and shift the building's power or energy demand to hours with lower prices. However, depending on whether or not other incentive signals, such as carbon emission rates, share a similar pattern and fluctuates to a high or low point at similar hours, the optimal control strategy that minimizes cost may or may not end up increasing the carbon emission due to the fluctuation of the electricity demand profile. "Similarity" in this case, refers to the "degree of alignment" between two signals at a fixed time. Two sets of signals are considered more "in sync", when they rise and fall at similar times of the day; and are considered "conflicting" when one increases while the other decreases at similar times of the day. In other words, the dynamics of the incentive signal used for one objective, as well as the relation to signals from other objectives—whether the signals are "in sync" or are "conflicting"—heavily influence the tradeoffs that may exist among the various performance objectives.

While the advent of high temporal and spatial resolution incentive signals, often at hourly or sub-hourly level, can promote more tailored optimal control strategies that meet local objectives, it also magnifies the numerous possibilities of control strategies and complicates the tradeoff dynamic between different aspects of building performance. With signals fluctuating hour to hour, creating a broader range of patterns over daily, seasonal, and annual periods, the common workflow where the optimizer only considers incentive signals or scenarios from a "typical day" profile or carries out optimization over a short period of planning horizon will be inadequate in capturing the dynamic control outcomes and tradeoffs that exist among them. Therefore, a thorough understanding of the signal dynamics, in an individual and collective context, becomes an important step in developing more effective optimal control strategies.

Given this context, this work is among the first to examine the dynamics of forward looking incentive signals (end-use pricing signal, average emission rate and marginal emission rate) generated at an hourly level across thirty years for various balancing areas in the United States. The results of this study shed light on the impact that signal dynamics and relationships may have on the tradeoffs between building MPC objectives, as well as the implication on more effective MPC workflows. Section 2 provides additional context and reviews relevant literature. Section 3 describes the unsupervised clustering process, including data description, processing, parametric study and best cluster selection. Results of best clusters of three sets of incentive signals and discussion on their implications to the building performance optimization are presented in Sections 4. Limitations, conclusions and future work are provided in Section 5.

2. RELATED WORK

Defined as the incremental change in carbon emissions incurred by an incremental change in the total electricity generated (or demanded), marginal emission factors (MEFs) have been considered more prominently over the past several years to analyze the environmental impacts of a wide variety of actions that would alter the consumption or supply patterns of electricity. While average emission factors (AEFs) quantify the total carbon emitted from generating electricity, the MEF, which focuses on the marginal generator or technology, directly considers the response of the power system to a change in consumption or generation. Studies by Hawkes (2010) and Siler-Evans et al. (2012) found that due to the often higher intensity of carbon emission of the generators at the margin, MEF is often higher than AEF and more adequate in capturing the dynamic change of carbon emissions.

Many studies have leveraged the advantage of MEFs to investigate the environmental impact of a wide range of objectives and programs. Smith and Hittinger (2019) used marginal emission factors to improve estimates of environmental benefits from appliance efficiency upgrades. Amoroso et al. (2018) utilized the marginal emission factors to examine different sizing options and operating patterns of central air conditioning (AC) systems in the context of its economic and environmental impact. Marginal emission factors were also adopted to evaluate emission benefits of building design implementations, such as increased insulation for new homes (Levy et al., 2016), and the long-term environmental performance of buildings designed to high performance standards (Collinge et al., 2018).

The penetration of renewable energy generation diversified the generation mix, creating unprecedented opportunities for researchers to examine the electricity supply-demand relationship under a new light. In this context, MPC and other smart technologies that provide demand-side flexibility, among many other objectives, adopted the concept of MEF when considering the environmental impact and implications of its technology. Using a simple conceptual model of demand response MEF, a study by McKenna and Darby (2017) argues that demand response technology and smart appliances create large carbon saving potential, especially when coupled with long-term structural change. As MEF signals become available at a high spatial and temporal resolution, studies have noticed the highly volatile nature of the profile and the implication on the environmental impact resulting from MPC load shifting. Zivin et al. (2014) evaluated the environmental impact of electric vehicles (EVs) as a load shifting technology using MEF and found substantial spatial and temporal heterogeneity of marginal emission. Depending on the location and time of the day, charging EVs could lead to drastically different environmental and economic costs. Callaway et al. (2018) also emphasized the importance of location in evaluating the environmental impact of renewable energy technology and efficiency-induced demand reductions, due to the highly fluctuating MEF profile.

While much interest has been given to MEF, studies have not taken the next step of incorporating high resolution MEF as an incentive signal in building MPC. Existing optimization research in building performance, single objective and multi-objective, has been largely focused on the energy, economic and thermal comfort categories. For instance, studies by Yang and Wang (2012) and Missaoui et al. (2014) both focused on a bi-objective optimization of reducing energy cost and maximizing thermal comfort. A few studies considered carbon emissions and energy costs with real-time pricing signals however, did not have high resolution carbon emission data to incorporate marginal effects. Vogler-Finck et al. (2018) used average emission factor as their environmental incentive signal in their single objective MPC optimization. Péan et al. (2019) calculated marginal emission factor for its MPC study from an empirical approach, establishing a regression model of average marginal emission rate as a function of load and the proportion of renewable energy resources, given historic data for Spain. Although the regression model offered improvements over average emission factors, it was still limited in its spatial applicability and could not adapt to an ever-evolving generation mix. The most relevant study was conducted by Knudsen and Petersen (2016), who developed a building MPC framework that considered both real-time pricing signal and carbon intensity (again, using AEF as the incentive signal) in a multi-objective optimization setting. This study was also among the very few that attempted to quantify the tradeoff dynamics between the environmental and economic objective by altering the "added weight" coefficient for the two objectives. The authors alluded to the importance of correlation between the two incentive signals; when they observed a stronger tradeoff between the two objectives the incentive signals were less correlated.

The findings of earlier works have directly motivated the study presented in this paper to more closely analyze grid signal patterns in attempt to better understand the tradeoff dynamics that may exist among building MPC grid incentive signals. To our knowledge, this is the first study to apply unsupervised clustering algorithms to examine patterns of hourly real-time pricing signals and average and marginal emission signals using a detailed dataset with modeled hourly grid operational data that spans a thirty-year period for all balancing areas in the United States for various future grid scenarios. To illustrate the approach and potential implications, we focus on a subset of the data that includes two months of data for one location. This paper explores whether or not recurring patterns exist within the individual grid incentive signals and discusses how the signals may lead to tradeoffs within optimal building control applications.

3. METHODOLOGY

The approach followed in this work is summarized in Figure 1. Descriptions of the data set and assumptions are provided in Section 3.1, data cleaning and preprocessing are included in Section 3.2, and parametric studies for each clustering algorithm and best cluster selection criteria are documented in Section 3.3.

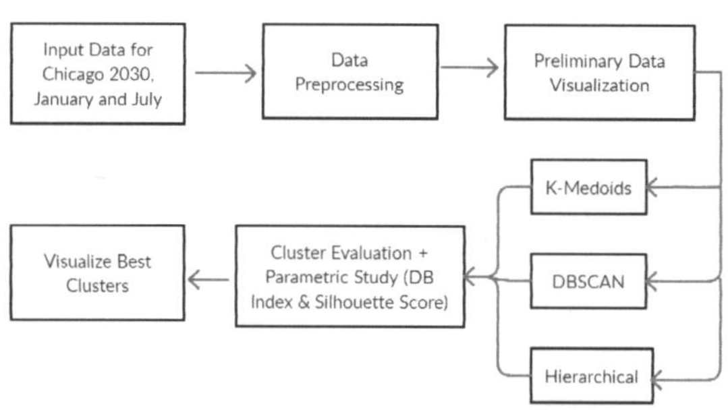

Figure 1: General workflow for data clustering and analysis.

3.1 The Data Set

Data used for this study, including real-time pricing and average and marginal emission factors, were generated by a novel grid operation model, Cambium, developed by the National Renewable Energy Laboratory (NREL) (Gagnon et al., 2020). Built to expand the metrics reported in NREL's Standard Scenarios (Cole et al., 2020), an annually released set of projections of how the U.S. electric sector could evolve across a suite of different potential futures, Cambium assembles structured data sets of simulated hourly cost and operational data for modeled futures of the U.S. electric sector with metrics designed to be useful for long-term decision-making. For each scenario, which outlines a possible realization of a future grid generation mix, Cambium models hourly grid operations for every balancing area in the United States, at two-year intervals until 2050. Spanning over thirty years at unprecedented spatial and temporal resolution, Cambium provides users with the most detailed outlook of plausible electric grid futures. For this study, we selected predictions for the Mid Case in the general Chicago balancing area (p80) in the future year 2030. The Mid Case serves as a baseline among the Standard Scenarios, where the total generation grows steadily over time and is provided primarily by a mix of new natural gas combined cycle (NGCC), PV, and wind generation.

A brief explanation of the three signal types used in this work is provided later in this section. For more detail regarding the Cambium modeling approach and assumptions, readers should refer to the full Cambium documentation (Gagnon et al., 2020). The pricing signal used in this study refers to the total end use marginal cost ($/MWh) that accounts for energy, capacity, operating reserve, and portfolio costs. Costs for distribution capacity, administrative and general expenses, and other electric sector expenses are not included. The average emission factor refers to the average emission rate of the generation induced by a region's end-use load in kilograms of CO_2 emissions per megawatt-hour of end-use load. The marginal emission factor refers to the short-run marginal emission rate for end-use load, which is the emission rate of the marginal generator that serves the marginal increase in load. Cambium identifies the marginal generator or the energy-constrained generator on the margin (storage technology like battery) through a five-step procedure, including identifying balancing areas that share a marginal generator, evaluating non-energy-constrained generators and energy-constrained generators, as well as reevaluating non-energy-constrained generators, with relaxed filters.

3.2 Data Preprocessing

Extreme pricing spikes (mostly within the $1000/MWh to $2000/MWh range) occurring at up to four hours in July are not realistic and do not reflect how pricing signals would be in any real-life electricity market. For this reason, days that contain these extreme pricing spikes were eliminated by applying an upper limit filter of $800/MWh. Any days with a pricing signal higher than the limit will have all 24 data within the day removed. This process eliminated four days from the 365-day data set.

The original pricing and emission signal had different magnitudes and ranges, thus, to compare their temporal dynamics and study their interrelationship on the same scale both signal sets were normalized. The `MinMaxScaler` function from the Scikit-learn library (version 0.24.1) (Pedregosa et al., 2011) was applied in Python Jupyter Notebook to each signal separately over the entire year to normalize the signal between 0 and 1, where 0 represents the lowest value of the signal in the entire year and 1 represents the highest.

For this study, January and July were selected to represent a typical winter and a summer scenario. Compilations of the signals over a 24-hour period for the entire selected month are provided in Figures 2, 3 and 4 to establish a pre-clustering baseline and serve as a visual aid to inform comparisons with cluster patterns.

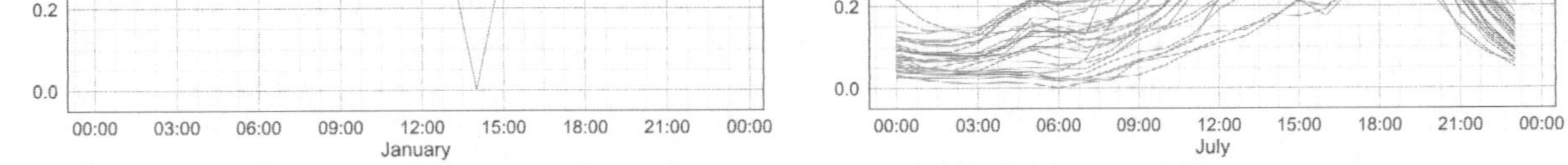

Figure 2: Pricing signal hourly distribution in January and July, 2030.

3.3 Parametric Study and Best Cluster Selection

K-Medoids, DBSCAN and Agglomerative Hierarchical Clustering (HC), with Ward's method and the Euclidean distance metric, were applied to each signal set in the selected months and compared in terms of the "goodness" of the clusters. Three sets of signals per month, for two months, results in six total cases. One best-performing algorithm was selected per case. DB Index and Silhouette Coefficient were used to evaluate results from the clustering procedure and determine the optimal parameter(s) for each algorithm and the best algorithm for each case. Both methods, regularly

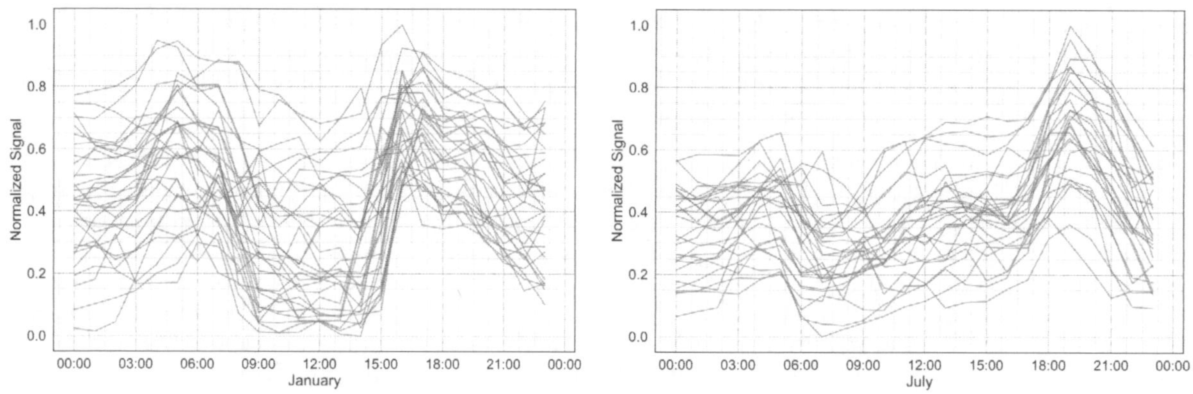

Figure 3: Average emission factor signal hourly distribution in January and July, 2030.

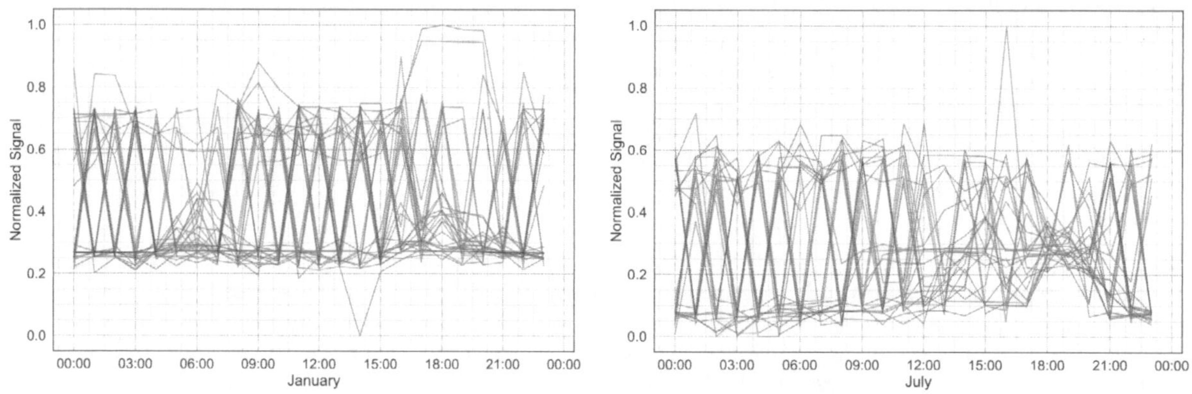

Figure 4: Marginal emission factor signal hourly distribution in January and July, 2030.

used in internal evaluation, establish indices that contrast between high similarity within a cluster and low similarity between clusters. DB Index, or Davies–Bouldin index, is defined as the maximum of the ratios between each intra-cluster distance and inter-cluster distance. A low DB Index indicates a more distinctive clustering process. Similarly, the silhouette value measures how similar data are within its cluster compared to other clusters. A high coefficient indicates a good cluster while low values indicates outliers.

Best clusters were selected based on the results of the parametric study. The parametric study evaluates DB index and Silhouette values for a given range of the key parameter(s) of a method. For K-Medoids and HC, the key parameter evaluated is the number of clusters ranging from 1 to the maximum days within the month. For DBSCAN, the key parameters are epsilon ("eps"), which defines the radius of neighborhood around a point x, and minSample, which is the minimum number of data points within each cluster and defines the minimum size of each cluster. To execute the parametric study, first, key parameters of each algorithm were examined with respect to their cluster verification results using DB index and Silhouette value. Then, parametric results were compared between algorithms to determine a best-performing algorithm for each case. For DBSCAN, heat maps that summarize number of clusters found and number of outliers were compiled first, given a range of eps (0.2 to 1.2) and minSample (1 to max days of the month). A list of combinations of eps and minSample were compiled for all combinations that produced total clusters larger than one and number of outliers fewer than the total days in the month. The list of combinations were then used to repeat the parametric process with DB index and Silhouette score. In many cases, a "perfect" performing set of parameters with the lowest DB index and the highest Silhouette score did not exist. More often, there was a short list of potential parameter values that produced a reasonably good combination of DB index and Silhouette score. Clusters from the short list were generated, visualized and compared. A winner was selected based on visual agreement and the discretion from the authors. Once the algorithm for each case was selected, best clusters were generated and visualized.

4. SIMULATION RESULTS

4.1 Baseline Visualization

Preliminary data assessment involves visualization of the original signals to establish a starting point and visual aid for comparisons of clusters. Figures 2 – 4 reveal the seasonal variations of the three signals and provide a visual comparison of the level of "neatness" among the three. The pricing signal, although having seasonal variation of pattern, displays a clean organization: distinctive patterns can be identified by eye, such as the two-spike profile in the winter, and the one-spike profile in the summer. Such patterns repeat throughout the respective month, with only small hourly variations regarding the exact hour in which the spike(s) occur. In comparison, the marginal carbon emission signal is less structured. Visually, it is difficult to identify any distinctive daily profiles that repeat throughout the month. Moreover, there is a highly fluctuating nature of the signal, where the signal varies from hour to hour in potentially large magnitudes, resulting in the sharp, long spikes that often take place within two hours. Where and how much the signal will fluctuate fully depends on which type of generator is switched onto the margin at that hour, a decision based on cross-regional coordination of the power plants and economic dispatch model that does not have a long-term, intrinsic pattern. The average emission signal is somewhat organized. As expected, the graphs suggest a correlation of average carbon emission and pricing signal, such that the AEF is high when the pricing is high and vice versa. Compared to the pricing signal, the AEF profile is nevertheless more disorganized.

4.2 Parametric Study and Best Cluster Selection

Following the process of selecting a best algorithm and key parameters in Section 3.3, Table 1 summarizes the best algorithm for each case and "goodness" of their performance evaluated by DB index and Silhouette score. As predicted, a "perfect" parameter that guarantees the lowest DB index and the highest Silhouette score within an algorithm rarely existed. For example, in row two of Table 1, which represents the best algorithm (HC) for the pricing signal in July, the winning "number of clusters" selected was 9. Although the DB index (0.75) and Silhouette score (0.3) were not the best within their own category, they presented a favorable combination when evaluated together. Such observations were prevalent throughout the selection process of this study, illustrating a common challenge that arises in evaluating unsupervised learning methods. Overall, Hierarchical Clustering was the best-performing among the three algorithms.

Table 1: Best algorithm for pricing, marginal emission factor, and average emission factor by month.

Signal Type	Month	Best Algorithm	DB Score	Silhouette Score	# Clusters
Pricing	January	DBSCAN	0.51	0.33	7
Pricing	July	Hierarchical	0.75	0.3	9
Carbon (avg)	January	Hierarchical	0.9	0.24	10
Carbon (avg)	July	DBSCAN	0.39	0.12	19
Carbon (m)	January	Hierarchical	0.7	0.09	21
Carbon (m)	July	Hierarchical	0.95	0.107	14

As expected, algorithms found better clusters for the pricing signal, followed by the AEF and MEF. This was demonstrated by the pricing results having the lowest DB index and the highest Silhouette score, followed by AEF and then MEF, which had the highest DB index and lowest Silhouette score. There were also more clusters found for the carbon signals than for the pricing signal. Both findings were not surprising and were consistent with the observations from Section 4.1. Overall, the best clusters found for both AEF and MEF were less helpful, either having numerous visually obvious misfits within every cluster, or having too many clusters that only had one daily profile. This was more so for MEF for which the algorithms struggled to find any plausible patterns and clusters.

4.3 Cluster Visualization

Based on parametric study for the pricing signal, best clusters for the pricing signal in January and July are shown in the first column of Figure 5 and 6. The average emission factor profiles that correspond to the pricing clusters are shown in the middle column, and marginal emission factor profiles are in the right column. Thus, Figure 5 and Figure 6 allow visual comparison of the average and marginal carbon signals, when grouped by similar pricing signals. The best clusters found by the algorithm visually agree with our observations of distinct clusters from the baseline graphs. These clusters sufficiently identified the slight variations within the pricing signal. Within each cluster, the daily profiles adhere to the general shape of the cluster well, with some degrees of over-generalization, such as the second row of Figure 5.

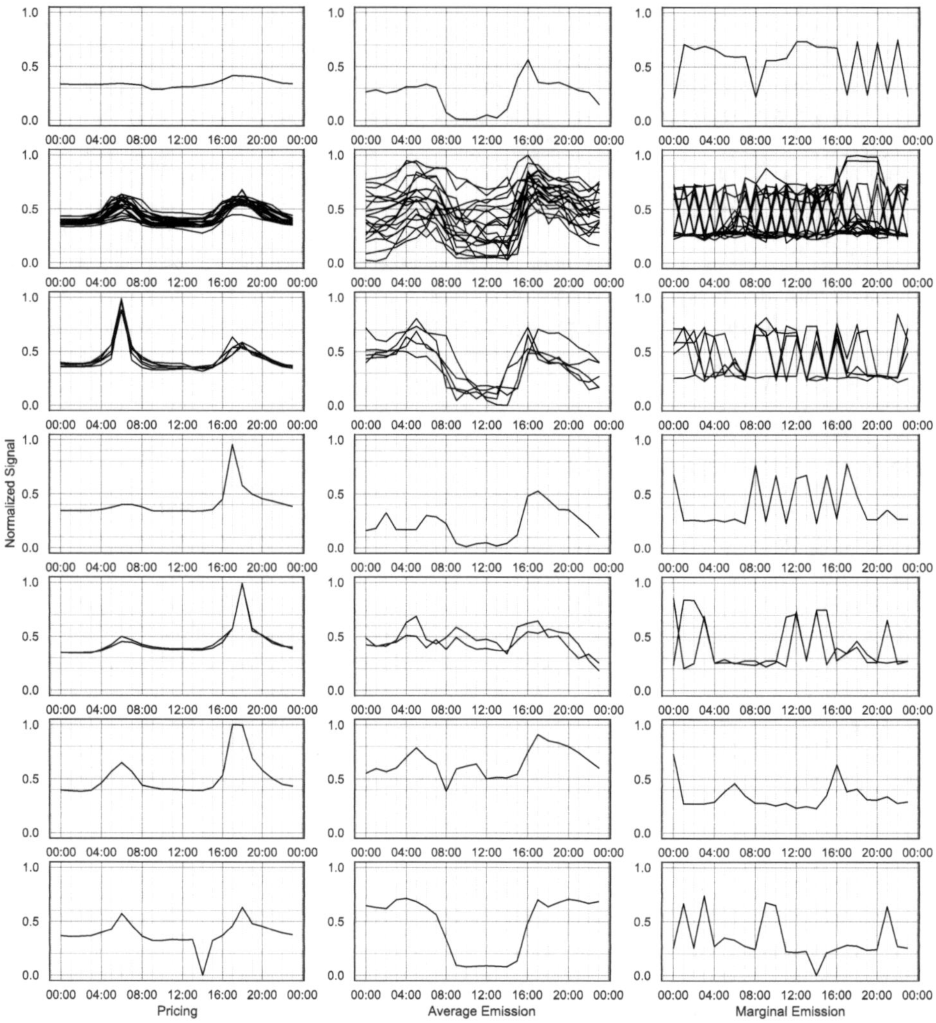

Figure 5: Best clusters for the pricing signal (left column) and corresponding profiles of average emission rate (middle column) and marginal emission rate (right column), for January, 2030.

For every cluster of the pricing signal, there exists a wide and diverse range of signal dynamics from the AEF and MEF. For example, for the two days that have almost identical pricing signals on the fifth row of Figure 5, with the largest deviation between two days around 0.05, the corresponding two AEF profiles fluctuate much more on a day-to-day basis, with the largest hourly difference of 0.2 at 5:00 A.M. The level of fluctuation is even higher for the MEF of the same pricing cluster, with the largest hourly deviation of 0.65 at 1:00 A.M. Further, both AEF and MEF do not share a similar signal synchrony with the pricing signal. While the pricing cluster has a small spike around 6:00 A.M. and a large spike around 5:00 P.M., the AEF slightly fluctuated around 0.5 all day. The MEF displayed a "conflicting" pattern with the pricing signal having spiked at hours where the pricing signal stayed low, and stayed low at hours where the pricing signal spiked. This means, a single-objective MPC that considers these two days within the pricing cluster would have very similar strategies to minimize the energy cost, but that strategy would result in drastically different total carbon emission and would most definitely lead to a tradeoff between energy cost and carbon emission, if the MEF signal were used as the indicator of carbon emission. This example illustrates how the dynamic nature of the emissions profile could lead to unpredictable environmental impacts associated with cost optimal controls.

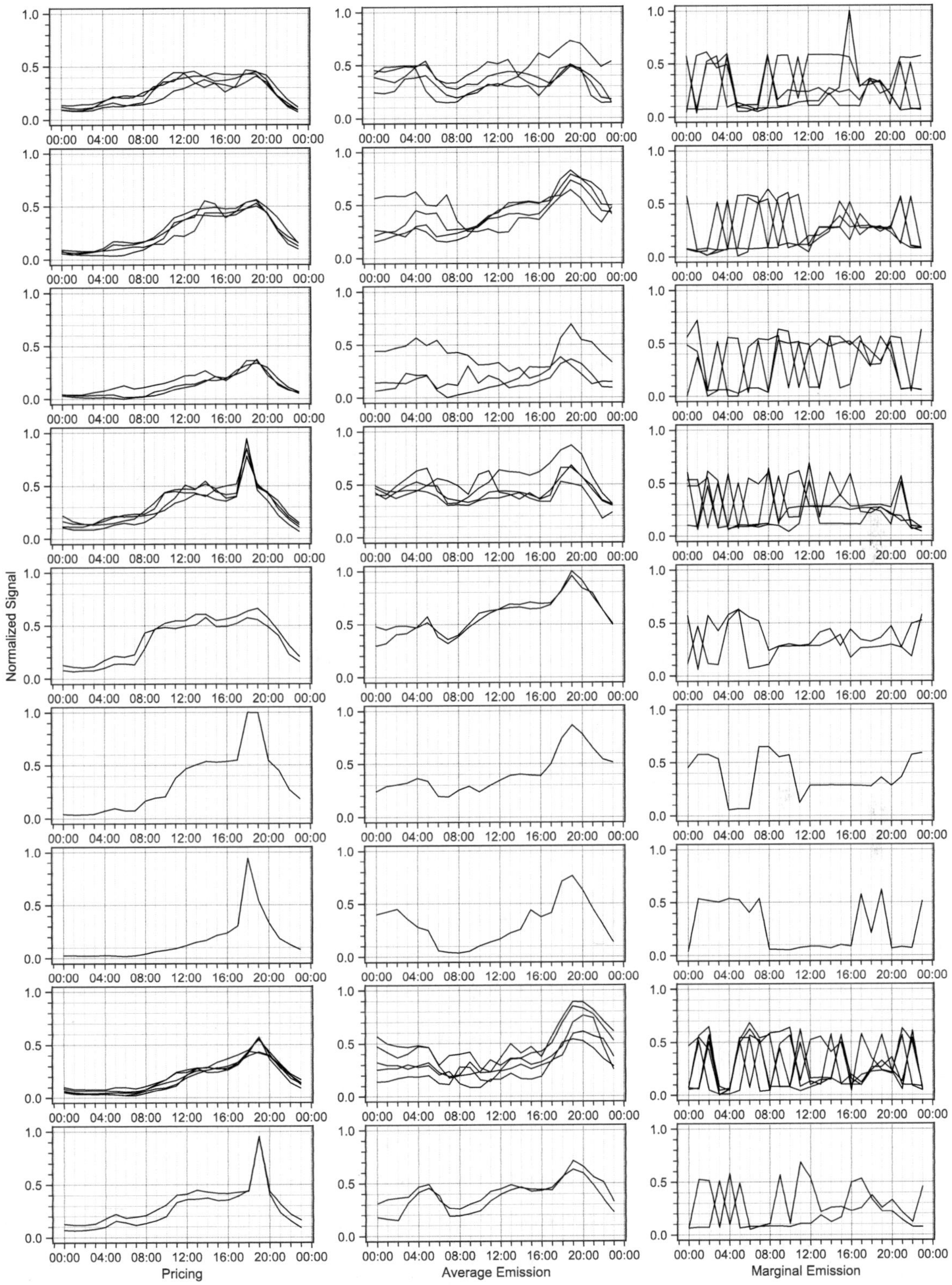

Figure 6: Best clusters for the pricing signal (left column) and corresponding profiles of average emission rate (middle column) and marginal emission rate (right column), for July, 2030.

This study also has implications on how the type of emission signals—AEF or MEF—used in MPC research could affect the tradeoff dynamic between different objectives. Consider the sixth row of Figure 6. The pricing signal and the AEF profile share a good level of agreement and synchrony, with both staying low before 10:00 A.M., increasing from 10:00 A.M. to 4:00 P.M. and hitting a spike at 7:00 P.M. However, the MEF profile of that day is "conflicting" with this pattern, having a spike at 2:00 A.M. and 7:00 A.M., when both the pricing signal and the AEF signal are staying at a constant low level. This means that due to the correlation between the AEF, an optimal MPC strategy that minimizes energy cost is also likely to reduce the carbon emission if AEF is used to quantify the emission rate. However, due to the conflicting pattern of MEF with the pricing signal, if MEF is used, the same strategy that minimizes energy cost would create a tradeoff and be likely to increase the total carbon emission compared to the base case.

Both findings pose suggestions to the current framework of building MPC optimization. Studies that focus on single objective optimizations are encouraged to leverage the dynamics of other incentive signals and the degrees of signal synchrony to estimate potential tradeoff dynamics that could exist among the objectives. Studies that relied on average emission factor or average marginal emission factor also need to recognize the potential of significant hidden tradeoffs that could be revealed by a more detailed data set such as Cambium.

5. CONCLUSIONS

In this study, unsupervised clustering algorithms, K-Medoids, DBSCAN and Hierarchical Clustering were applied to hourly pricing signal, average and marginal emission signal generated by a forward-looking grid model. Overall, Hierarchical Clustering demonstrated the best performance, evaluated by DB index and Silhouette score. While the algorithms did not find distinctive patterns of the carbon signals, it identified 7 and 9 patterns within the January and July pricing signal. The highly fluctuating nature of the carbon emission signals could lead to a diverse range of tradeoffs between energy cost and carbon emission in a building MPC optimization. This finding iterates the importance of understanding incentive signal dynamics, in an individual and collective context, and its impact on the tradeoffs between different objectives in any building MPC optimization research. The findings in this study suggest that, as more high resolution carbon and pricing signals like Cambium became available, using "typical day", AEF or other seasonal or average emission data set will fall short in recognizing the fluctuating and "random" nature of some of these signals that happens from day to day or hour to hour, and cannot fully address the complex tradeoff dynamic as a result of a highly and fast changing reality that is our electric grid.

REFERENCES

Amoroso, B., Hittinger, E., and McConky, K. (2018). Keeping your cool–a multi-stakeholder look at ac sizing. *Building and Environment*, 131:306–329.

Callaway, D. S., Fowlie, M., and McCormick, G. (2018). Location, location, location: The variable value of renewable energy and demand-side efficiency resources. *Journal of the Association of Environmental and Resource Economists*, 5(1):39–75.

Cole, W. J., Corcoran, S., Gates, N., Mai, T. T., and Das, P. (2020). 2020 Standard Scenarios Report: A U.S. Electric Sector Outlook. Technical report, National Renewable Energy Lab.(NREL), Golden, CO (United States).

Collinge, W. O., Rickenbacker, H. J., Landis, A. E., Thiel, C. L., and Bilec, M. M. (2018). Dynamic life cycle assessments of a conventional green building and a net zero energy building: exploration of static, dynamic, attributional, and consequential electricity grid models. *Environmental science & technology*, 52(19):11429–11438.

Delgado, M. G., Ramos, J. S., Domínguez, S. Á., Ríos, J. A. T., and Cabeza, L. F. (2020). Building thermal storage technology: Compensating renewable energy fluctuations. *Journal of Energy Storage*, 27:101147.

Gagnon, P., Frazier, W., Elaine, H., and Cole, W. J. (2020). Cambium documentation: Version 2020. Technical report, National Renewable Energy Lab.(NREL), Golden, CO (United States).

Greensfelder, E. M., Henze, G. P., and Felsmann, C. (2011). An investigation of optimal control of passive building thermal storage with real time pricing. *Journal of Building Performance Simulation*, 4(2):91–104.

Hawkes, A. D. (2010). Estimating marginal CO_2 emissions rates for national electricity systems. *Energy Policy*, 38(10):5977–5987.

Henze, G. P., Felsmann, C., and Knabe, G. (2004). Evaluation of optimal control for active and passive building thermal storage. *International Journal of Thermal Sciences*, 43(2):173–183.

Klaassen, C., House, J. M., and Center, I. E. (2002). Demonstration of load shifting and peak load reduction with control of building thermal mass. *Teaming for Efficiency: Commercial buildings: technologies, design, performance analysis, and building industry trends*, 3:55.

Knudsen, M. D. and Petersen, S. (2016). Demand response potential of model predictive control of space heating based on price and carbon dioxide intensity signals. *Energy and Buildings*, 125:196–204.

Levy, J. I., Woo, M. K., and Tambouret, Y. (2016). Energy savings and emissions reductions associated with increased insulation for new homes in the united states. *Building and Environment*, 96:72–79.

McKenna, E. and Darby, S. J. (2017). How much could domestic demand response technologies reduce CO_2 emissions? *European Council for an Energy-Efficiency Economy*.

Missaoui, R., Joumaa, H., Ploix, S., and Bacha, S. (2014). Managing energy smart homes according to energy prices: analysis of a building energy management system. *Energy and Buildings*, 71:155–167.

Niu, J., Tian, Z., Lu, Y., and Zhao, H. (2019). Flexible dispatch of a building energy system using building thermal storage and battery energy storage. *Applied energy*, 243:274–287.

Pavlak, G. S., Henze, G. P., and Cushing, V. J. (2014). Optimizing commercial building participation in energy and ancillary service markets. *Energy and Buildings*, 81:115–126.

Péan, T., Costa-Castelló, R., and Salom, J. (2019). Price and carbon-based energy flexibility of residential heating and cooling loads using model predictive control. *Sustainable Cities and Society*, 50:101579.

Pedregosa, F., Varoquaux, G., Gramfort, A., Michel, V., Thirion, B., Grisel, O., Blondel, M., Prettenhofer, P., Weiss, R., Dubourg, V., Vanderplas, J., Passos, A., Cournapeau, D., Brucher, M., Perrot, M., and Duchesnay, E. (2011). Scikit-learn: Machine learning in Python. *Journal of Machine Learning Research*, 12:2825–2830.

Siler-Evans, K., Azevedo, I. L., and Morgan, M. G. (2012). Marginal emissions factors for the U.S. electricity system. *Environmental science & technology*, 46(9):4742–4748.

Smith, C. N. and Hittinger, E. (2019). Using marginal emission factors to improve estimates of emission benefits from appliance efficiency upgrades. *Energy Efficiency*, 12(3):585–600.

Swaminathan, S., Pavlak, G. S., and Freihaut, J. (2020). Sizing and dispatch of an islanded microgrid with energy flexible buildings. *Applied Energy*, 276:115355.

Vogler-Finck, P. J. C., Wisniewski, R., and Popovski, P. (2018). Reducing the carbon footprint of house heating through model predictive control–a simulation study in danish conditions. *Sustainable Cities and Society*, 42:558–573.

Yang, R. and Wang, L. (2012). Multi-objective optimization for decision-making of energy and comfort management in building automation and control. *Sustainable Cities and Society*, 2(1):1–7.

Yu, M. G. and Pavlak, G. S. (2021). Assessing the performance of uncertainty-aware transactive controls for building thermal energy storage systems. *Applied Energy*, 282:116103.

Zivin, J. S. G., Kotchen, M. J., and Mansur, E. T. (2014). Spatial and temporal heterogeneity of marginal emissions: Implications for electric cars and other electricity-shifting policies. *Journal of Economic Behavior & Organization*, 107:248–268.

ACKNOWLEDGEMENT

The authors thank Pieter Gagnon at The National Renewable Energy Laboratory (NREL) for his generous discussion on the Cambium dataset.

Thermal Charging Rate of Composite Wax-Expanded Graphite Phase Change Materials

Jason HIRSCHEY[1], Navin KUMAR[2], Tim LACLAIR[2], Kyle R. GLUESENKAMP[2]*, Samuel GRAHAM[1]

[1]Georgia Institute of Technology,
Atlanta, GA, USA
Jhirschey13@gatech.edu

[2]Oak Ridge National Laboratory,
Oak Ridge, TN, USA
gluesenkampk@ornl.gov

* Corresponding Author

ABSTRACT

Phase change materials (PCMs) are valuable for their ability to store heat nearly isothermally around their phase transition temperature. PCMs are at the core of latent heat thermal energy storage (LHTES) systems, which provide the ability to buffer high thermal loads or decouple the time when heating or cooling is needed from when it is produced. Thermal charging and discharging of LHTES systems often employ a constant temperature source, and the rate at which heat can be exchanged is dependent on the thermophysical properties of the PCM. For graphite composite PCMs, the high thermal conductivity of the graphite enables an increased heat transfer rate through the material, but its presence displaces PCM which reduces the effective volumetric latent heat of the composite relative to the pure PCM. This results in a tradeoff between thermal power and volumetric energy storage capacity. The thermal charging rate is the average thermal energy stored in the material for some elapsed time. In this study, composite PCMs of compressed expanded natural graphite (CENG) and n-Octadecane are studied. Samples with varying CENG mass fractions were synthesized and the thermal charging rate was measured under a constant temperature boundary condition. For evaluation of the expanded graphite-PCM composite, one boundary of the material was exposed to a 50°C constant temperature plate, above the 27.5°C melting temperature of the PCM. The melting front progression [mm-s^{-1}] and the thermal charging rate [W-cm^{-2}] of the PCM were determined, and the results were compared with analytical predictions for the 1-D semi-infinite phase change. For CENG addition up to 5.75% mass fraction, a 450% thermal conductivity increase was observed with a only 5% decrease in volumetric energy density as compared to pure octadecane. The average thermal charging rate was increased by over 430% for the melt to penetrate a depth of 22 mm. The experimental results matched analytical predictions, indicating that higher CENG fractions can be evaluated using analytical approaches.

1. INTRODUCTION

Low temperature (<100°C) latent heat thermal energy storage (LHTES) can be useful for space heating and space cooling applications by decoupling the operation of an HVAC system to the use of its output. At their core, LHTES typically use phase change materials (PCMs) which can store significant amounts of near-isothermal heat in the latent heat of phase change. Many PCMs that undergo a phase change in this temperature range suffer from low thermal conductivity, particularly in the liquid state. This hinders the rate at which thermal energy can be exchanged in the LHTES and limits their operating capacity.

Much work has been done to increase the thermal charging rate of PCMs in LHTES systems, namely in heat exchangers of various configurations. Researchers have utilized intricate extended surfaces and finned structures which often add complexity to the system and may increase overall LHTES cost. Mahdi et al. (2018) investigated several fin configurations of horizontal triplex-tube heat exchangers, and modeled the liquid fraction and temperature distribution under constant temperature boundaries at the inner and outer surfaces. Abdulateef et al. (2019) optimized the geometrical fin structure of a triplex-tube heat exchanger with a paraffin PCM infused with alumina nanoparticles as a means of reducing the fin material.

Other researchers are developing PCM composites to enhance the thermal conductivity over that of pure PCM. This can include metallic foams. Esapour et al. (2018) modeled a quasi-shell and tube heat exchanger with metallic foam impregnated with PCM in the shell portion of the heat exchanger. The researchers modeled the time to fully melt the PCM under various metallic foam porosity and identified tube geometries that lead to faster thermal charging.

Liang et al. (2018) developed a superhydrophobic copper foam to enhance the thermal conductivity of organic PCMs by nearly 10 times, but also serve as a shape-stable supporting network that prevents PCM leakage thereby reducing the need for exterior packaging. Rao et al. (2018) employed copper particles and foam on paraffin PCM microencapsulated in a polyurethane shell.

Mallow et al. (2018) performed thermal charging measurements on paraffin wax PCM enhanced with both metallic foams and CENG. They utilized a constant power source to charge the PCM and measured the temperature rise with time near the heater and at the far end of the sample. They showed that samples with higher thermal conductivity took less time to melt fully.

Zeng et al. (2018) increased the thermal conductivity of an organic PCM eutectic composed of *m*-Erythritol and D-mannitol by 408% with exfoliated graphite. Wang et al. (2018) increased the thermal conductivity of polyethylene glycol with metal-organic frameworks nanoparticles on copper nanotubes. Fashandi and Leung (2018) utilized a combination of graphene nanoplatelets and hexagonal boron nitride to increase the thermal conductivity of the inorganic salt hydrate sodium acetate trihydrate.

Most papers investigating the thermal charging capability of PCMs, including those above, look at the time to fully melt the PCM in a fixed geometry or track the liquid fraction during the melting process. Yang et al. (2018) examined metallic foam-enhanced ice storage. They showed the melting and freezing time of the PCM was significantly reduced with the metallic foam and tracked the melt-/freeze-front visually, extrapolating this to determine the melted fraction during the phase transition. These researchers tracked the boundary wall heat flux, referred to as the thermal charging rate in this current paper, and showed the metallic foam with the lowest pore density and highest thermal conductivity resulted in the highest thermal charging rate. The planar geometry used by the authors is most similar to what is attempted and presented by this current paper.

With the introduction of these additives to influence the thermal conductivity, some PCM is displaced. This will have implications for the latent heat, specific heat, and density of the resulting composite. For a LHTES, all these thermophysical properties play a role in the rate at which thermal energy can be exchanged in the system and the total energy that can be stored per unit volume.

The objective of this paper is to experimentally compare the thermal charging rate of a composite PCM to the analytical 1-D semi-infinite phase change solution and show that small quantities of CENG can drastically improve the PCM composite thermal charging performance without significantly sacrificing thermal storage capacity. The thermal charging is the average thermal power pulled into the system for the elapsed time the PCM is exposed to the constant temperature boundary. The composite PCMs used in this sample are composed of compressed expanded natural graphite and n-octadecane, an organic wax PCM.

2. EXPERIMETAL SETUP

2.1 Test Setup

The test setup shown in **Figure 1** is designed to expose a PCM composite to a bounding surface of uniform temperature above the PCM melting temperature. The temperature of the PCM composite is measured at various distances from the bounding surface. This enables calculation of the rate at which heat enters the composite (the thermal charging rate) and tracking of the melt-front, solid-liquid interface. The constant temperature heating surface is made of aluminum with a steady flow of water at constant temperature through internal channels. The water source is an 8-

gallon, heater-only PID temperature-controlled bath with stability of ±0.2°C. The large thermal mass of this water together with minimal losses leads to slow PID response time which may account for some temperature drift in this aluminum block temperature.

The PCM material is contained in a thin-walled polypropylene cylinder of 35 mm diameter and with a polypropylene cap. The cylinder was placed in an insulated sleeve for testing with its bottom face left exposed. Aluminum foil was epoxied to the bottom of the cylinder to seal the container, contain the PCM, and make for good contact with the constant temperature plate.

Expanded graphite was compressed into the cylinder to specific volume fractions. The PCM then infiltrated into the graphite by means of a vacuum oven, a process which has shown to be successful at creating similar graphite-wax PCMs with minimal void fraction (Mallow *et al.*, 2018). Once this process is complete and the PCM has solidified completely, three 1/32" holes were drilled into the polypropylene walls and composite PCM for thermocouple placement. The thermocouples reached the centerline of the cylinder and were placed closer to the heated surface (far from the opposite surface) to minimize edge effects caused by a bounded, not semi-infinite, system.

A thin layer of silicone heat transfer grease was spread on the bottom surface of the aluminum foil to reduce thermal contact resistance between the plate and the PCM. In addition, a 200 g mass was placed atop the PCM sample to apply uniform pressure across all samples where the PCM sample contacts the plate.

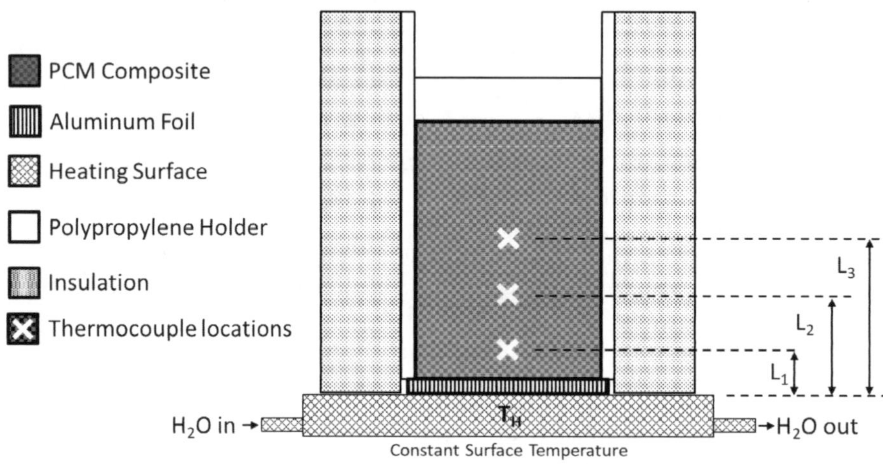

Figure 1: Thermal charging experimental setup

The three thermocouple placement holes are measured from the bottom heating surface as shown in Figure 1. Care was made to make the thermocouple spacing similar across samples, but some deviation in the exact location exists. Table 1 shows the thermocouple locations for the samples measured after their placement.

Table 1: Thermocouple locations

Sample ID	Thermocouple 1, L_1 (mm)	Thermocouple 2, L_2 (mm)	Thermocouple 3, L_3 (mm)
A	8.6	16.4	22.3
B	8.7	16.3	23.6
C	8.6	16.4	21.8

During each test, water was run through the block for at least 30 minutes to ensure the block and all piping reaches a steady state temperature. The sample itself is kept in an isolated chamber and the three thermocouples monitored to ensure the sample is at a uniform temperature. Data recording is started, allowed to run for several minutes, and the sample quickly transferred from its chamber to the constant temperature block. If there is a delay in this process, the sample is misaligned on the plate, or extraneous circumstances cause some disruption, the test is aborted and restarted.

The sample sits on the plate for the duration of the test. It is monitored to ensure it does not shift, change position, or leak during the phase change process. The test runs until all three thermocouples register a completed phase change, plus approximately 30 minutes. Data collection is stopped. Once completed, the sample is removed from the block and allowed to cool and freeze at room temperature.

2.2 Materials

The cylindrical polypropylene holder is a 140 mL laboratory syringe with the plunger removed and the fluid-dispensing tip and adjoining cap severed to create a hollow cylinder. These syringes are thin-walled, low thermal conductivity, inert, and the demarcations useful for loading the graphite and PCM uniformly. The thin walls and low thermal conductivity are important for reducing the thermal mass at the edges of the sample and prevent a thermal finning effect, both of which can skew the results. The plunger is useful for packing the graphite by perfectly matching the inner diameter. The CENG worms are sourced as-is from the manufacturer; no chemical pretreatment or modifications were made. Prior to creating the sample, the CENG was heated in an oven to remove any moisture that may be present. 99.9% pure n-Octadecane was procured from a laboratory chemical supplier. The entire contents of the 2 kg bottle were fully melted and stirred to ensure the octadecane used across all tests was uniform.

Type T thermocouples were used for all measurements. The thermocouples in the samples are 1/32" in diameter to minimize their presence in the samples. The plate was measured using a bead-welded 24 AWG type T thermocouple with silicone thermal grease ensuring good thermal contact. This plate thermocouple was affixed to the same position across all tests and samples. The samples were wrapped in ¾" foam insulation to minimize heat loss during the experiments. The aluminum foil and thermocouples were epoxied with high strength inert epoxy. Data was acquired at a rate of 1.2 Hz.

2.3 Sample Preparation

The preparation of all PCM composites followed the same process and are identical in form factor and packaging. The packaging is designed to be PCM leak tight. The syringes were prepped by removing the nozzles and roughing the bottom inch of the outside with 400 grit sandpaper which allows the epoxy to adhere and provide a leak-proof seal with the aluminum foil. Likewise, the aluminum foil was lightly sanded with 800 grit sandpaper to allow the epoxy to better adhere. The aluminum foil and cylinders were then epoxied together with extreme care to prevent epoxy on the foil faces in contact with the sample and the plate and inside the cylinder; the epoxy formed a bead around the base of the cylinder. A known mass of CENG was compressed into the open end of the cylinder with the plunger of the syringe to a final bulk CENG volume of 100 mL as read from the demarcations on the syringe. The CENG-filled cylinders were weighed, and this value recorded. The samples were then infiltrated with octadecane (melting point, 27.5°C) standing upright in a low temperature vacuum oven at 50°C. To prevent the graphite from floating upwards away from the aluminum foil, the plunger was taped in place to hold the graphite bulk volume at 100 mL. A small 1/16" hole was drilled in the plunger to allow the octadecane to flow into the graphite. The samples were left in the oven for 4 hours, or until the liquid level of octadecane stopped falling. Once taken out of the oven, the remaining liquid octadecane was skimmed off the top of the CENG and the sample allowed to cool and solidify completely. Lastly, the composite samples were weighed; the difference between the final mass and that of the CENG was used to determine the mass of octadecane. The compositions of the three samples are shown in Table 2.

Table 2: Composition of PCM samples

Sample ID	CENG Mass (g)	CENG Mass Fraction (g/g)	CENG Bulk Density (kg-m^{-3})	Octadecane Mass (g)
A	4.848	5.75%	48.5	79.5
B	1.967	2.38%	19.7	80.6
C	1.054	1.29%	10.5	80.9

By comparing the expected value of octadecane to the measured value, the void space of these samples is less than 1%, indicating the graphite is fully saturated with octadecane. Next, the three 1/32" holes were drilled into the walls of the cylinder and into the PCM composite for the thermocouple placement. Care was taken to ensure the holes were perpendicular to the cylinder walls and did not deviate in the axial direction in the sample. The thermocouples were inserted into the holes and more epoxy applied around the thermocouple holes to seal them and lock the thermocouples in place. The last step is to place the samples upright into a low temperature oven at 50°C to leak test the sample. If any octadecane escapes between the cylinder and the aluminum or through the thermocouple holes, the sample is considered failed and scrapped. This step also melts the PCM around the thermocouples to provide good thermal contact, ensuring accurate measurements during the test.

Effective thermophysical properties of the PCM composite are used in the heat transfer model and are calculated from the mass and volume fractions of the constituents. The effective specific heat and latent heat are a function of the mass fractions, m_γ, of the CENG and octadecane as shown by equations (1) and (2). The thermophysical values for octadecane are taken from literature from (Faden et al., 2019), and for graphite from (Incropera & DeWitt, 2002).

$$c_p = m_{\gamma,CENG} * c_{p,CENG} + m_{\gamma,octadecane} * c_{p,octadecane} \tag{1}$$

$$H_f = m_{\gamma,octadecane} * H_{f,octadecane} \tag{2}$$

The density of the PCM in the sample is assumed constant at the bulk sample density of the composite in the melted state. Lastly, the thermal conductivity is estimated through interpolation of data of CENG-wax composites created in the same vacuum oven infiltration process from (Mallow et al., 2018). Since changes in the octadecane thermal conductivity with phase are small compared to the bulk conductivity, the thermal conductivity of the graphite component is assumed to dominate, and the effective conductivity is assumed to be constant. The thermal conductivity values calculated agree with similar samples measured using the ASTM D5470 standard (not reported here).

The thermophysical properties of the samples are shown in Table 3. Note that with the small addition of CENG in these samples, the specific heat and volumetric storage capacity change by about 5% relative to the pure PCM, while the thermal conductivity increases by a factor of about 4, which provides an indication of the favorable tradeoff of the CENG enhancement on the heat for energy storage vs. power capabilities.

Table 3: Composite PCM thermophysical properties

Sample ID	Calculated Specific Heat (J/g-K)		Calculated Latent Heat (J/g)	Interpolated Thermal Conductivity (W/m-K)	Measured Density (kg/m^3)	Calculated Volumetric Latent Heat Storage Capacity (J/cm^3)
	Liquid	Solid				
A	2.1	1.7	230	0.91	845	194.4
B	2.2	1.8	238	0.44	826	196.6
C	2.2	1.8	241	0.31	820	197.6

3. ANALYTICAL MODEL

This work is designed to experimentally test the 1-D semi-infinite planar phase change problem for PCM composites of varying composition, commonly known as the Stefan problem. This analytical solution is easily solvable and has been derived extensively (Hahn & Özişik, 2012). The boundary condition to this problem is a constant temperature imposed at one surface of a semi- infinite volume of PCM. The initial condition is that at all points, the PCM is at a uniform temperature at the initial time.

$$T_l(x = 0, t > 0) = T_o \tag{3}$$
$$T_s(x \to \infty, t > 0) = T_i \tag{4}$$
$$T_l(x = 0, t = 0) = T_i \tag{5}$$

At the melting interface, two conditions must be met: 1) the solid and liquid temperature are equal to the melting temperature, and 2) the rate of heat into and out of the interface must equal to the progression of the interface accounting for the latent heat. These conditions are shown here.

$$T_l(x = s(t), t > 0) = T_s(x = s(t), t > 0) = T_m \tag{6}$$

$$-k_s \frac{dT_s}{dx}\bigg|_{s(t)} + k_l \frac{dT_l}{dx}\bigg|_{s(t)} = \rho_l H_f \frac{ds}{dt} \tag{7}$$

There are many differences between this analytical model and the experimental setup and are listed below:

- The experimental setup is not semi-infinite in length. This is expected to introduce some edge effects. To help compensate for this shortcoming, the PCM samples are intentionally oversized at nearly 10 cm of material, but the measurements only occur in the bottom 2.5 cm near the heated surface.
- Similarly, the PCM is not infinite in the planar direction. Again, this is expected to introduce some edge effects. The thin-walled polypropylene cylinder was chosen to minimize this heat loss (as compared to a more readily accessible metallic pipe), and a thick layer of insulation blanketed the entire system. The thermocouples were inserted into the centerline of the cylinder to reduce the edge effects near the walls.
- The initial temperature is not perfectly uniform throughout the samples. The insulted samples were placed sideways in an insulated box for several hours prior to the test. This helped minimize any temperature gradients due to the sample bottom surface contacting any other surface and reduced the fluctuations in sample initial temperature with room ambient temperature.
- The bounding surface is not a perfectly constant temperature. The constant temperature block is heated by running water through interior channels. The water source is an 8-gallon, heater-only PID temperature-controlled bath. The large thermal mass of this water coupled with minimal losses leads stable temperature.
- Along the same vein, there is not perfect thermal contact between the PCM composite and this aluminum block. Additionally, there is the aluminum foil between the two which is not accounted for by the analytical model. The aluminum foil is necessary to prevent the octadecane from spilling out of the holder when melted. Silicone thermal grease is employed to reduce the contact resistance between the plate and the foil. Care was taken when creating the samples to ensure a flat bottom PCM surface. A 200 g weight applies even pressure between the sample and plate. And the thermal resistance of the aluminum is negligible relative to the PCM composite.
- The PCM in the model has one discrete melting temperature. n-Octadecane completes the melting process across a range of temperatures, commonly called a temperature glide. The model does not account for this. Octadecane has extensive documentation on its temperature glide (Faden et al., 2019). In the model used to compare the experimental results to the analytical solution, the melting temperature is chosen to be the temperature at which the samples complete the melting process, often indicated by an inflection point on a temperature vs. time plot. This is approximately 27.5°C for this sample set.

Despite the limitations in matching the experimental testing conditions to the analytical solution assumptions, the comparison between the two is useful to validate that the analytical solution as a good method for determining relative PCM composite performance. By following the derivation of the analytical solution, the melting interface is shown to be a function of the liquid thermal diffusivity, time, and a parameter, λ, which is solved numerically and by applying the interface boundary conditions (Hahn & Özişik, 2012). Equation (8) shows the melt-front interface as a function of time, *s(t)*.

$$s(t) = \lambda\sqrt{4\alpha_l t} \tag{8}$$

To determine the thermal total energy stored in the PCM, the latent and sensible heat are calculated by equations (9) through (11). Equation (9) shows that the latent energy is the distance of the melt from at some time, *s(t)*, multiplied by the material latent heat, $H_{f,eff}$, and its density. Equations (10) and (11) both calculate the sensible heat by integrating the temperature difference with respect to distance and multiplying with the appropriate specific heat value, $c_{p,eff}$, and the density. Equation (10) is the energy in the PCM volume between the constant temperature boundary surface and the melt-front that has undergone a full phase change. The first term of Equation (10) is the energy in raising the temperature from the initial temperature to the melting temperature; the second term is the energy in the observed temperature increase above the melting temperature. Similarly, equation (11) calculates the energy in the material beyond the melt-front by integrating the difference in temperature from the initial temperature. It is important to remember that this model assumes an infinite plane, therefore all energies will be reported per square area, [J-cm^{-2}].

$$Q''_{lat}(t) = s(t) * H_{f_{eff}} * \rho_{l_{eff}} \tag{9}$$

$$Q''_{sensible,melted}(t) = (T_m - T_i) * s(t) * c_{p,s_{eff}} * \rho_{s_{eff}} + c_{p,l_{eff}} * \rho_{l_{eff}} * \int_0^{s(t)} (T(x,t) - T_m)dx \tag{10}$$

$$Q''_{sensible,frozen}(t) = c_{p,s_{eff}} * \rho_{s_{eff}} * \int_{s(t)}^{\infty} (T(x,t) - T_i)dx \tag{11}$$

The total energy, Q''_{tot}, of the PCM is found by summing equations (9-11). The thermal charging is defined as the total energy calculated at some time and divided by that elapsed time. The thermal charging is representative of the average thermal power pulled into the system for the elapsed time and is shown by Equation (12), [W-cm^{-2}]. The instantaneous power at each time is the slope of this line, [W-s^{-1}-cm^{-2}].

$$P''(t) = \frac{Q''_{tot}(t)}{t} \tag{12}$$

4. RESULTS

The time-temperature plots of the samples are shown by **Figure 2**Figure 2. Note that each sample takes a significantly different length of time to complete due to the different thermal conductivity of the samples, thus each horizontal axis is different. All samples follow the same trend as the model; a fast response to the thermal impulse at the start followed by a slowing of the temperature rise towards the melting point. The end of the melting process is easily discernable as an upward concave inflection point. The start of the melting process is more difficult to identify, however, as the temperature more slowly rises to the melting point in the approach. The analytical solution suggests that each spatial point completes the melting process in infinitesimal time. Since the thermocouples have a nonzero thickness, there should be a finite period at which the thermocouple is reading a near-constant temperature equal to the melting temperature. The temperature glide of the octadecane prevents this constant temperature reading from occurring in the nearest thermocouples. The farthest thermocouple does have a more extended period of near-constant temperature, but this may be due to the lower heat flow being far from the heated surface.

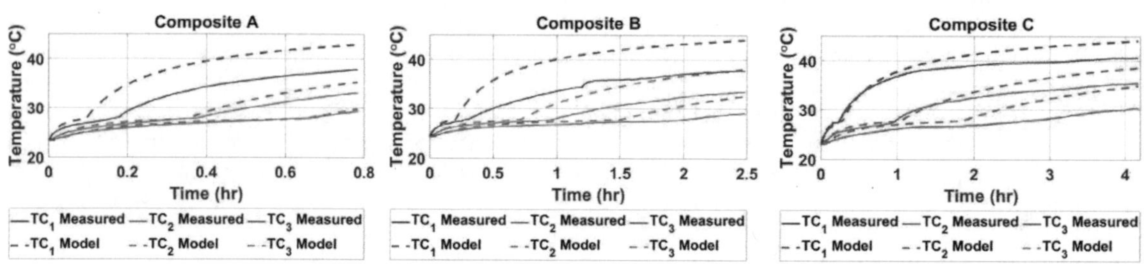

Figure 2: Time-Temperature Measurements of Composite PCMs

For the results in Figure 2, the model predicts higher final temperatures at all thermocouple locations. This difference may be attributed to the deviations from experimental approximations outlined above, particularly unwanted heat loss through the walls of the container and insulation. Future experimental redesign and subsequent testing may be able to address this deviation from model predictions.

The melt-front progression is analyzed by calculating the time each of the thermocouple locations finish the melting process. With the thermocouples' known and fixed distance from the heated surface, the melt-front is fitted to a function with the square root of time to match the analytical solution shown by equation (8). The intercept is set at 0 mm. The form of the fitted melt-front progression is shown by equation (13).

$$s(t)_{fitted} = A_{fitted} * \sqrt{t} + 0 \tag{13}$$

Figure 3 shows the comparison to the melt front progression between the different composites and to each analytical solution. All experimental results have a slower melt-front progression than what is predicted by the model. This is to be expected based on the experimental limitations outlined above. Composite A most closely matches the analytical solution, as can be seen in the melt-front and by Figure 2. This composite had the highest graphite fraction and highest thermal conductivity of the samples tested. Composite C had the lowest graphite fraction and thermal conductivity, and therefore took the longest to complete the melting process at all thermocouple locations. However, the first two thermocouples completed melting faster than the model predictions while the third was much slower. This may suggest some nonuniformity in the sample or that the thermal conductivity is higher than what was used in the model calculations and the unwanted heat loss for the long length of the test affected the results of the farthest thermocouple.

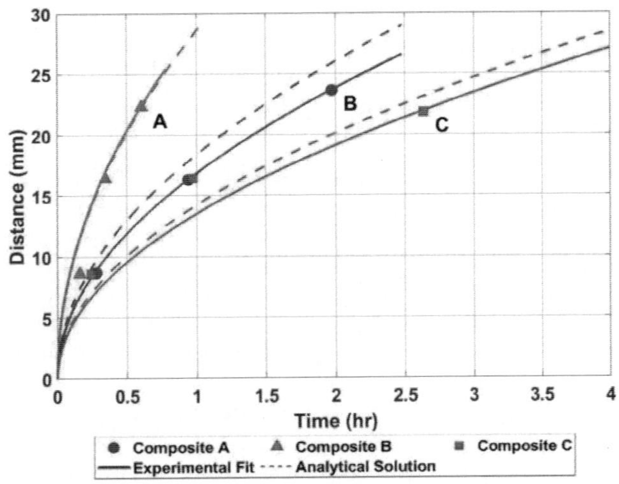

Figure 3: Melt-front progression

Figure 4 shows the thermal charging capacity of the three samples in comparison to the value from the analytical solution. The values are calculated using equations (9-11) at only the three instances when each thermocouple completes the melting process. This may underestimate the thermal energy into the system since the temperature is only known at three discrete points, and there is no interpolation of temperature between the thermocouples or beyond the last. For a real LHTES system, temperature triggers will likely be used in a similar manner to determine the thermal charge of such a system or determining other metrics.

The thermal charging rate of Figure 4 does not have a simple functional form with which to fit the experimental data. As such, the results presented by Figure 4 show only comparison to the analytical solution results as calculated by equation (12). Composites B and C are very similar, differing only slightly in their thermal conductivity. This may explain why the experimental results align well nearest the heated surface but differ at the farthest thermocouple location. Composite A has the highest measured and modeled thermal charging capacity.

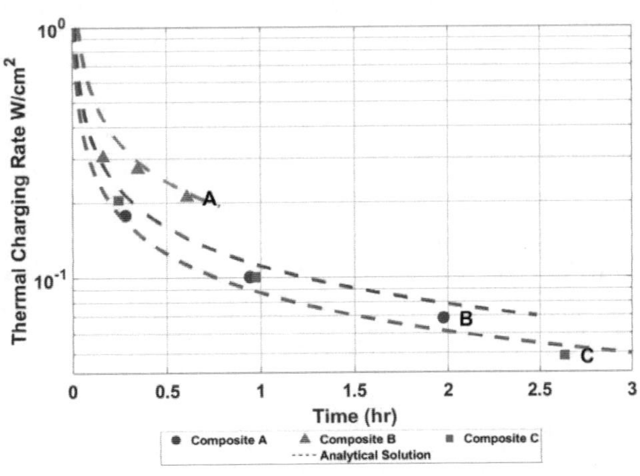

Figure 4: Calculated thermal charging capacity compared to analytical solution

From Table 3, the expected volumetric storage capacity of the samples differ by less than 2%, however, the thermal conductivity of composite A is 3 times that of composite C. In Figure 4, the thermal charging capacity of composite A higher than that of composite C. At the time when the last thermocouple had registered a phase change, Composite A had an average thermal charging capacity of 0.21 W-cm^{-2} and composite C had 0.04 W-cm^{-2}, meaning that

Composite A can store nearly five times the amount of heat than Composite C under the same loading conditions. This demonstrates that the tradeoff between increased thermal conductivity and reduced volumetric storage capacity by the addition of graphite is inconsequential for small graphite amounts.

5. CONCLUSION

For a real application with spatial limitations, there exists a tradeoff between the thermal energy storage capacity [J-m^{-3}] and the thermal charging/discharging capacity [W-m^{-2}] when adding graphite to wax as a thermal conductivity enhancement additive. In this test, three samples were synthesized with n-octadecane and varying amounts of compressed expanded natural graphite (CENG). Across the three composite samples, the thermal conductivity ranged from 0.31 to 0.91 W•$m^{-1}K^{-1}$, compared with 0.2 W•$m^{-1}K^{-1}$for pure octadecane. However, the volumetric storage capacity differed by less than 2% across all samples. This demonstrates that the addition of small CENG amounts can have a profound effect on the thermal conductivity with minimal impacts to the volumetric storage capacity.

The samples were exposed to a constant temperature plate and the melt front proceeding through thermocouples embedded in the sample at various distances away from the constant temperature surface. This was compared to the 1-D semi-infinite analytical phase change solution. The temperature-time plots of the samples follow the melting profile predicted by this analytical solution at the start of the test but deviate as the test progresses. Imperfections in the experimental setup may account for the observed deviation.

The thermal charging rate of the samples was calculated using the expected thermophysical properties and the measured melt-front progression. The thermal charging rate is the average thermal power added to the PCM during the time to reach a fully melted state at a depth of 22 mm under a constant temperature boundary surface. The sample with the highest thermal conductivity had nearly 430% higher average thermal charging rate than the sample with the lowest thermal conductivity. The addition of 5.75% mass fraction CENG decreased the time to melt a depth of 22 mm of PCM by 75% with only a 5% loss in volumetric storage capacity. The thermal conductivity of the PCM composite with 5.75% mass fraction CENG was increased over 450% than that of pure octadecane.

NOMENCLATURE

PCM	Phase Change Material	
LHTES	Latent Heat Thermal Energy Storage	
CENG	Compressed Expanded Natural Graphite	
T	Temperature	(°C)
x	Distance	(m)
t	Time	(sec)
s	Melt-front	(m)
k	Thermal Conductivity	(W-m^{-2}-K^{-1})
ρ	Density	(kg-m^{-3})
H	Latent Heat	(J-kg^{-1})
λ	Numerically-solved parameter	
α	Thermal Diffusivity	(m^2-s^{-1})
Q	Heat	(J)
c	Specific Heat	(J-kg^{-1}-K^{-1})
P	Power	(W)
M	Mass	(kg)
A	Fitted-parameter	(m-$t^{-1/2}$)

Subscripts/Superscripts

l	Liquid
o	Boundary
s	Solid
i	Initial
m	Melting
f	Fusion

lat	Latent	
eff	Effective Composite Properties	
sensible	Sensible heating	
p	Isobaric	
tot	Total	
CENG	CENG	
Octadecane	Octadecane	
γ	Fraction	
fitted	Fitted solution using least squares regression	
"	Per unit area	(m^{-2})

REFERENCES

Abdulateef, A. M., Abdulateef, J., Sopian, K., Mat, S., & Ibrahim, A. (2019). Optimal fin parameters used for enhancing the melting and solidification of phase-change material in a heat exchanger unite. *Case Studies in Thermal Engineering*, 100487. doi:https://doi.org/10.1016/j.csite.2019.100487

Esapour, M., Hamzehnezhad, A., Rabienataj Darzi, A. A., & Jourabian, M. (2018). Melting and solidification of PCM embedded in porous metal foam in horizontal multi-tube heat storage system. *Energy Conversion and Management, 171*, 398-410. doi:https://doi.org/10.1016/j.enconman.2018.05.086

Faden, M., Höhlein, S., Wanner, J., König-Haagen, A., & Brüggemann, D. (2019). Review of Thermophysical Property Data of Octadecane for Phase-Change Studies. *Materials (Basel, Switzerland), 12*(18), 2974. doi:10.3390/ma12182974

Fashandi, M., & Leung, S. N. (2018). Sodium acetate trihydrate-chitin nanowhisker nanocomposites with enhanced phase change performance for thermal energy storage. *Solar Energy Materials and Solar Cells, 178*, 259-265. doi:https://doi.org/10.1016/j.solmat.2018.01.037

Hahn, D. W., & Özişik, M. N. (2012). Heat Conduction Fundamentals. In *Heat Conduction*: Wiley.

Incropera, F. P., & DeWitt, D. P. (2002). *Fundamentals of heat and mass transfer* (5th ed. ed.). New York :: J. Wiley.

Liang, W., Zhu, H., Wang, R., Wang, C., Zhu, Z., Sun, H., & Li, A. (2018). Superhydrophobic Copper Foam Supported Phase Change Composites with High Thermal Conductivity for Energy Storage. *Materials Research*, 0-0.

Mahdi, J. M., Lohrasbi, S., Ganji, D. D., & Nsofor, E. C. (2018). Accelerated melting of PCM in energy storage systems via novel configuration of fins in the triplex-tube heat exchanger. *International Journal of Heat and Mass Transfer, 124*, 663-676. doi:https://doi.org/10.1016/j.ijheatmasstransfer.2018.03.095

Mallow, A., Abdelaziz, O., & Graham, S. (2018). Thermal charging performance of enhanced phase change material composites for thermal battery design. *International Journal of Thermal Sciences, 127*, 19-28. doi:https://doi.org/10.1016/j.ijthermalsci.2017.12.027

Rao, Z., Wen, Y., & Liu, C. (2018). Enhancement of heat transfer of microcapsulated particles using copper particles and copper foam. *Particuology*. doi:https://doi.org/10.1016/j.partic.2017.12.010

Wang, J., Huang, X., Gao, H., Li, A., & Wang, C. (2018). Construction of CNT@Cr-MIL-101-NH2 hybrid composite for shape-stabilized phase change materials with enhanced thermal conductivity. *Chemical Engineering Journal, 350*, 164-172. doi:https://doi.org/10.1016/j.cej.2018.05.190

Yang, X., Bai, Q., Zhang, Q., Hu, W., Jin, L., & Yan, J. (2018). Thermal and economic analysis of charging and discharging characteristics of composite phase change materials for cold storage. *Applied Energy, 225*, 585-599. doi:https://doi.org/10.1016/j.apenergy.2018.05.063

Zeng, J.-L., Chen, Y.-H., Shu, L., Yu, L.-P., Zhu, L., Song, L.-B., . . . Sun, L.-X. (2018). Preparation and thermal properties of exfoliated graphite/erythritol/mannitol eutectic composite as form-stable phase change material for thermal energy storage. *Solar Energy Materials and Solar Cells, 178*, 84-90. doi:https://doi.org/10.1016/j.solmat.2018.01.012

ACKNOWLEDGEMENT

This material is based upon work supported by the U. S. Department of Energy's Building Technologies Office under Contract No. DE-AC05-00OR22725 with UT-Battelle, LLC. The authors would like to acknowledge Mr. Sven Mumme, Technology Manager, U.S. Department of Energy Building Technologies Office.

Enabling Solar Power with Residential Water Heater Load Shifting

Celeste FIEBERG*, Amy VAN ASSELT

Lafayette College, Department of Mechanical Engineering
Easton, Pennsylvania, USA
fiebergc@lafayette.edu

* Corresponding Author

ABSTRACT

As governments and international organizations aim to increase renewable energy production in response to climate change, there is a growing demand for energy storage. A temporal mismatch in the daily profiles of solar energy production and residential energy use causes an imbalance between generation and demand. Load shifting presents a key grid control strategy that minimizes this mismatch and the resulting rapid ramp rates in fossil fuel-based electricity generation. Water heaters are a principal contributor to residential electricity consumption from the grid, and they offer a good potential for thermal energy storage due to low ambient energy loss and water's high specific heat capacity. Load shifting of water heating loads maximizes self-consumption of solar power in residential buildings thereby improving homeowner utility costs and the ability of the electric grid to accept greater levels of intermittent solar generation.

Water heating load data was generated using the U.S. Department of Energy Residential Prototype Building Models for a cycling 50-gallon (189 L) electric water heater in a single-family house in Los Angeles with a roof-mounted solar photovoltaic system. The effects of load shifting on the fraction of the water-heating loads to the total house electric loads, the fraction of the water heating load met by solar power, and end-user electricity bills using time-of-use (TOU) electricity rates were analyzed. This study shows that shifting the load of water heaters to coincide with the solar profile lowers electricity costs significantly more than a solar photovoltaic system without load shifting which sends power to the grid using a net metering agreement. The results suggest that widespread use of smart water heaters that optimize load shifting by predicting hot water consumption and the solar profile could present an important value for the integration of renewable energy onto the grid.

1. INTRODUCTION

A discrepancy in the time profiles of solar energy production and residential energy use causes environmental, economic, and operational problems. While solar energy reaches its maximum output in the middle of the day, residential energy consumption peaks in the morning and in the evening (O'Shaughnessy, Cutlera, Ardania, & Margolisa, 2019). The California Independent System Operator (CAISO, 2019) consistently reported peak loads to occur between July and September at around 16:00. Daily electricity demand depends on the typical work schedule, while seasonal demand varies due to a difference in temperature.

This imbalance between renewable energy production and energy-use load profiles causes utilities to experience high variations in energy supply requirements. The report by Alcorn (2013) showed that self-generation from photovoltaic systems with a range of 1371-1604 MW of photovoltaic capacity will only reduce the peak demand for electricity by between 2.67% and 2.79%. Solar power production does not minimize the peak load due to the temporal mismatch. As the implementation of solar power has increased, utilities are faced with high ramp rates between the middle of the day (when solar power generation is highest) and peak hours (when electricity consumption from the grid is highest). These high ramp rates create a higher risk of over-generation in the middle of the day for utilities (Van Asselt, 2018).

To accommodate the increase in solar energy production, utilities and end users have therefore begun to invest in grid controls and energy storage, respectively (Van Asselt, 2018). Possible solutions include batteries and load management where self-consumption of the electricity generated from photovoltaic systems is encouraged. The relatively high cost and disposal issues for batteries limit the feasibility of this option. This study focuses on the effects of load shifting regarding residential water heaters which comprise almost one-fifth of residential energy use in the United States. Water heaters are beneficial for energy storage as water has a high specific heat capacity, and water heater tanks experience minimal energy loss (Gelažanskas & Gamage, 2016).

The hypothesis presented in this work is that load shifting of residential electric hot water heaters is a cost-effective means to reduce electricity consumption from the grid in favor of site-generated solar electricity. In this study, load shifting was applied to the water-heating load met by a 50-gallon residential electric resistance water heater.

The effects of load shifting on annual electricity bills, the fraction of the water-heating loads to the total house electric loads, and the fraction of the water heating load met by solar power were compared to the effects from the same solar array without load shifting. The control strategy carried out in this research may be applied to the development of an algorithm to develop smart water heaters that predict future solar production and water consumption profiles to maximize self-consumption of solar power. Such a control system could be integrated with existing systems designed for space heating and cooling applications.

2. MODELING

2.1 Water Heater Model and Load Data

The water heater model studied originates from the Residential Prototype Building Model of a single-family detached house in Los Angeles that is in accordance with the 2012 International Energy Conservation Code (IECC). This model is developed by the Pacific Northwest National Laboratory (PNNL) and used by the U.S. Department of Energy's (USDOE) Building Energy Codes Program.

The information for the water heater model is an input file for EnergyPlus, a building energy simulation program released by the USDOE. EnergyPlus was used to generate the water heating load and water consumption profile. The ambient heat loss, mass flow rate, water mains temperature, and the volume of water consumption were also determined using outputs from the energy simulation program.

The model is a single-node, mixed 50-gallon (189 L) electric domestic water heater with industry standard ratings for both the recovery efficiency and energy factor (USDOE, 2019). A single-node model was chosen due to its simple computational implementation and its low average percentage error in its output for the average output temperature (Kepplinger, Huber, Preißinger, & Petrasch, 2019). The water heater draws in cold water from the water main and heats it using an electric heating element with a heating capacity of

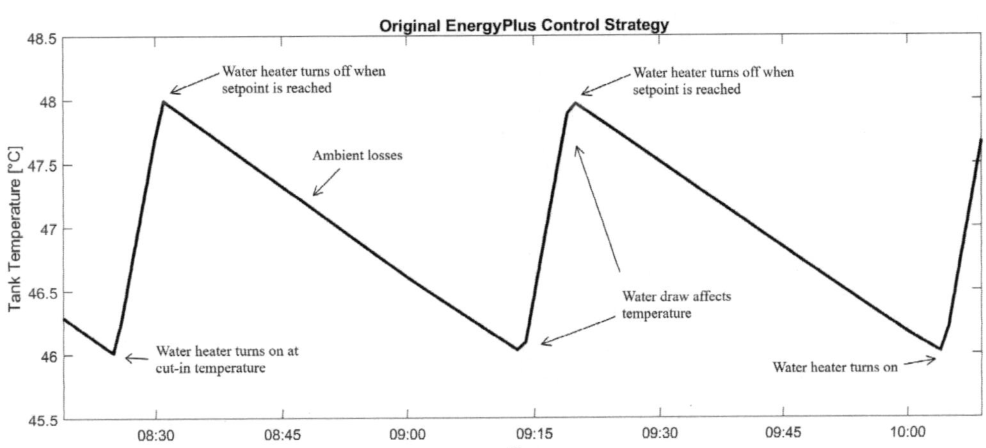

Figure 1: EnergyPlus Water Heater Original Control Algorithm

4500 W. The control type enables the heater to cycle on and off. The tank heats the water until it reaches the setpoint temperature of approximately 48 °C. When the setpoint temperature is reached, the water heater turns off and only turns on again when the temperature falls to 2 °C below the setpoint. The control model is shown in Figure 1 (USDOE, 2019). The data graphed is taken from the original water heater load on August 27. The schedule for the water heating load is based on the National Renewable Energy Laboratory (NREL) Seattle Benchmark model. Normally the modelled water heater turns on for about six minutes to reach the setpoint temperature, therefore the load data was gathered in one-minute time steps for the precision necessary to capture this behavior.

2.2 Weather, Solar Resource, and Solar Generation Data

Weather data was required as an input for the EnergyPlus and System Advisor Model (SAM) simulation programs. SAM was used for the solar power generation profile at one-minute time steps. The software uses the time, latitude, global horizontal irradiance, direct normal irradiance, and diffuse horizontal irradiance to estimate the solar resource and calculate solar power generation (Gilman, 2015). Ambient temperature was also used by both EnergyPlus and SAM. Ambient heat loss is calculated according to the thermal zone in which the water heater is located. Ambient heat loss from the water heater varies depending on the month, with a percent change of 9.02% and 10.0% in March and August, respectively. Ambient temperature is also used in SAM to calculate solar array performance (Gilman, 2015).

The weather data originates from the Typical Meteorological Year version 3 (TMY3). The monthly data is taken from a specific year deemed to have the most typical weather data for that month (Wilcox & Marion, 2008). The weather data was collected near the Los Angeles International Airport, shown in Figure 2.

However, the TMY3 weather resource only produces data in hourly time intervals as opposed to the one-minute time intervals required by SAM. To produce one-minute weather data, interpolation was necessary. However, simple linear interpolation for global horizontal radiation can cause large positive values at times close to sunrise and sunset (Duffie & Beckman, 2013). Therefore, the Transient System Simulation Tool (TRNSYS) was used to interpolate the diffuse horizontal irradiance according to the extraterrestrial radiation curve (Thermal Energy System Specialists LLC, 2016). Since global horizontal radiation takes diffuse horizontal radiation into account, it was not necessary to interpolate all three solar resource value inputs for SAM.

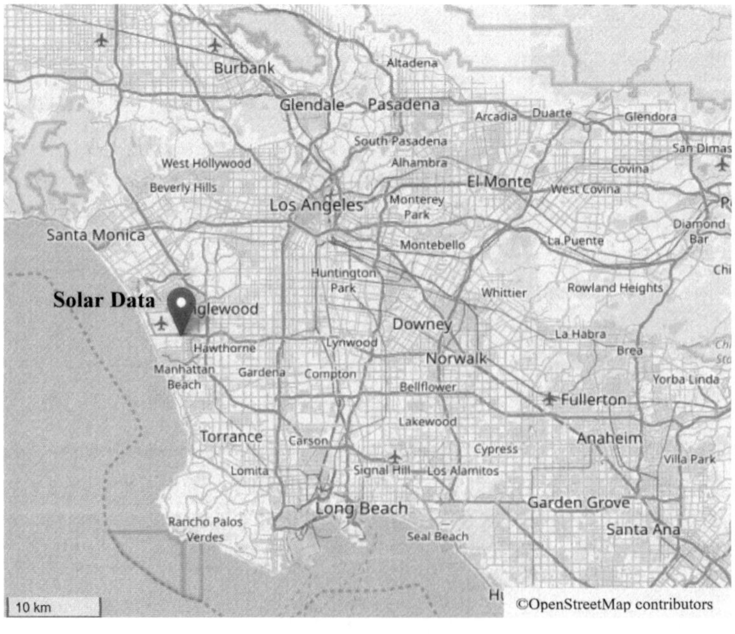

Figure 2: Solar data location

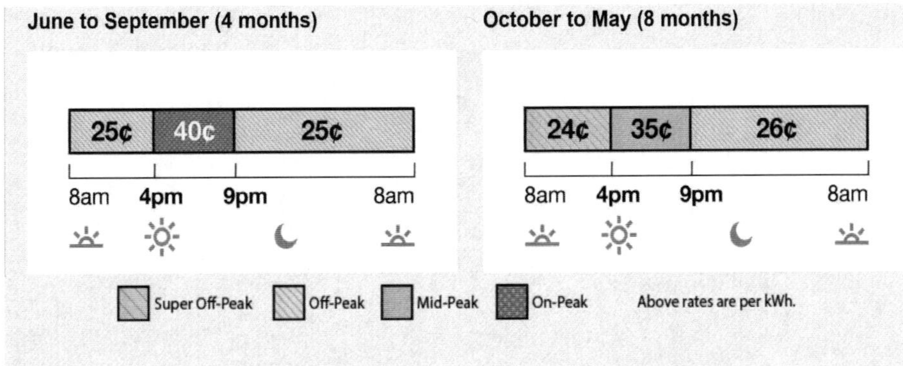

Figure 3: SCE TOU electricity rate

The X21-345 SunPower solar panel was chosen for the model with a nominal power of 345 W and an average panel efficiency of 21.5%. The water heating load generated by EnergyPlus was entered into the NREL's REopt Lite model used in O'Shaughnessy's recent study (2019). The solar array was appropriately sized to 1 kW (3 panels) and was assumed to be fixed to a roof at a tilt of 20 degrees.

2.3 Cost and Grid Mix Data

When determining which electricity rate structure is best to use for analyzing the savings of a solar photovoltaic system with and without load shifting, flat rates and time-of-use (TOU) rates with a net metering buy-back ratio of 1 were considered. A flat rate charges a constant amount for every kWh of electricity consumed, while a TOU rate structure charges different amounts for different periods during a day. The periods and charges vary based on electricity demand and peak hours. Load shifting is economically motivated by TOU rate structures because end users are encouraged to decrease electricity consumption during on-peak hours.

The TOU electricity rate was based on the 16:00-21:00 TOU rate of Southern California Edison (SCE) with a buy-back ratio of 1 to analyze the effect on electricity costs (SCE, 2019). This buy-back ratio indicates that a unit of electricity purchased from the grid has the same value as a unit of electricity produced by the solar array. The TOU rates are shown in Figure 3. Prices are raised during the peak time of electricity consumption, namely between 16:00 and 21:00 in the summer, in order to discourage over-consumption and to decrease peak demand. In the summer, the rate is 40¢ per kWh at this time while off-peak hours are 25¢ per kWh. The high summer peak rate is principally due to significant air-conditioning loads. During the winter, users are charged 35¢ during mid-peak hours, 24¢ between 8:00 and 16:00, and 26¢ at night.

3. CONTROL STRATEGY

A control strategy based on energy matching was developed for shifting the water heater load. Figure 4 shows the flow chart for the control strategy used. The control strategy assumes a variable heating element that can be powered with the electricity produced directly from the solar array. If the tank reaches minimum charge, a default condition is triggered and the water heater runs for 10 minutes at full capacity using electricity from the grid and available electricity generated through solar power. A variable is created to include the cumulative sum of solar energy, ambient heat loss (Q_{loss}), and heat transfer due to water draws (Q_{draw}) until the end of the day. This variable is added to the tank charge at the current time step, and if solar energy is available and the sum is less than maximum tank charge, then the water heater runs using available electricity produced from the solar array at that time step.

The control algorithm depends on tank charge, ambient heat loss, and the heat transfer on the use and source sides of the electric water heater. The well-mixed water tank heat transfer model used by EnergyPlus was implemented (USDOE, 2019).

A heat exchanger effectiveness of 1 was used. The default uniform skin loss coefficient per unit area to

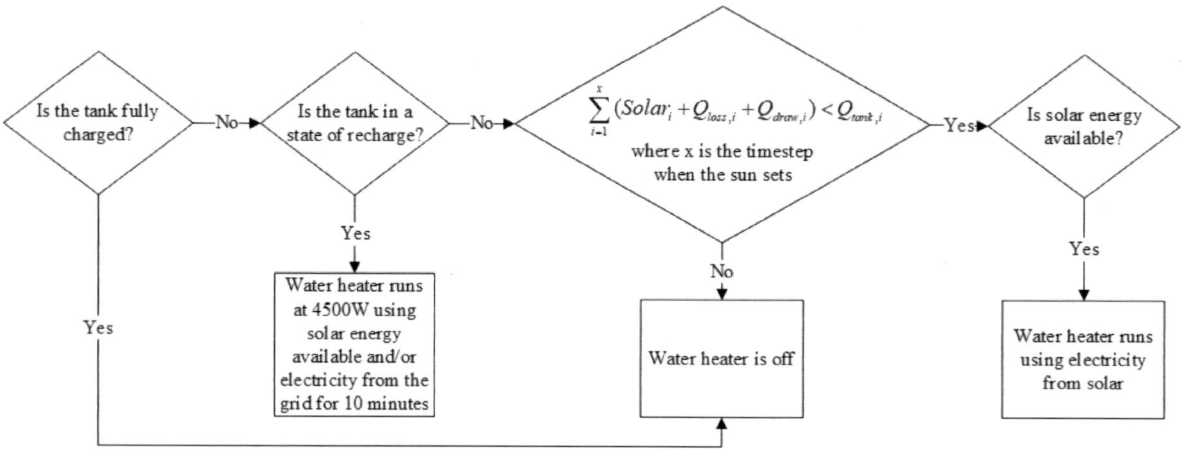

Figure 4: Flow chart for the load shifting control strategy

ambient temperature (UA), 1.33 W/K, was used. The conversion of electrical energy to thermal energy in the resistance heating element is assumed to have 100% efficiency.

The tank charge at a given time step is given by the following equation:

$$Q_{tank,i+1} = Q_{tank,i} + 60(q_{heater,i} + UA(T_{amb,i} - T_{tank,i}) + \dot{m}_{use,i}c_p(T_{tank,max} + T_{mains,i} - 2T_{tank,i})) \quad (1)$$

where q_{heater} is the heat transfer rate to the water in the tank from the heating element, T_{tank} is the water heater outlet temperature, T_{amb} is the ambient temperature, T_{mains} is the mains water temperature, $\dot{m}_{use}$ is the mass flow rate in kg/s, and c_p is the heat capacity of water. The maximum temperature is changed to 60°C to increase the capacity for energy storage.

The tank is at maximum charge when

$$Q_{tank,max} = V_{tank} \cdot c_p \cdot \rho \cdot T_{tank,max} \quad (2)$$

where V_{tank} is the total volume of water in the tank (50 gallons, 189 L). Similarly, the minimum tank charge occurs when $T_{tank,max}$ is replaced with $T_{tank,min} = 46°C$.

The cumulative heat transfer due to water draws is given by:

$$Q_{draw} = m_{cumulative} \cdot c_p \cdot (T_{mains,avg} - T_{max}) \quad (3)$$

where $m_{cumulative}$ is the cumulative mass of water drawn from the water heater during the given amount of time and $T_{mains,avg}$ is the average mains water temperature (20.2°C).

The ambient heat loss is the product of the tank's resistance to heat transfer (UA) and the ambient temperature. The water heater outlet temperature at each time step was calculated using a differential equation which takes into account UA, the heat transfer due to ambient heat loss and the heat transfer to/from the use and source sides of the water heater (USDOE, 2019).

4. RESULTS AND DISCUSSION

The original water heater load runs quickly and frequently to reach a setpoint of 48°C because the load only depends on water consumption and ambient heat loss and does not account for solar power generated. Alternatively, the shifted load directly uses electricity generated from the solar panels and electricity from the grid when solar power generation is not available. As seen in Figure 5, the tank temperature remains above the minimum allowable temperature of 46°C while maximizing the consumption of electricity generated by the solar array.

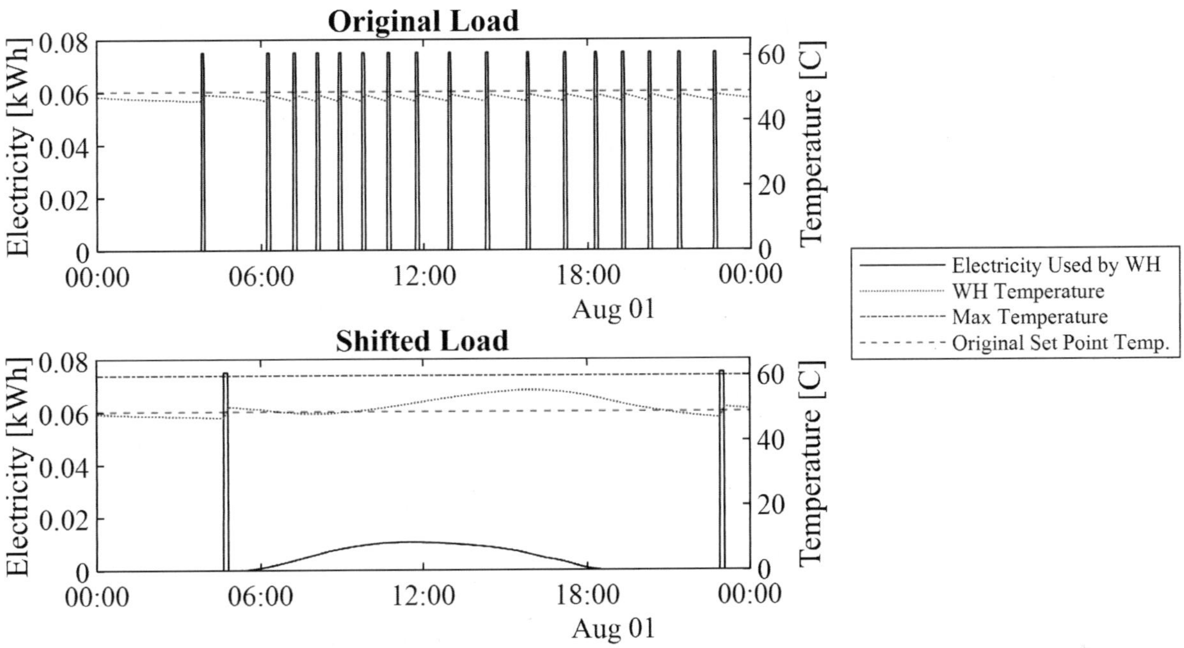

Figure 5: Original load and shifted load for August 1

It is important to note that the analysis uses ideal theoretical data. The hot water consumption appears to be relatively similar each day. However, the design of a control algorithm for a smart hot water heater would be able to predict users' hot water consumption schedule based on prior data. Solar power generation can be predicted, but these predictions are imperfect, particularly on a short-term basis.

Three metrics were used to quantify the results of the addition of a solar array with and without load shifting: the fraction of the water heating load met by solar power, annual end-user electricity bills, and the fraction of the water-heating load to the total house electricity.

Table 1 displays the results for these metrics under a system with no solar array, a 1 kW solar array, and a 1 kW array with load shifting. Only 5% of the water-heating load is met when a solar array is introduced without load shifting while 63% is met with load shifting. The annual electricity bill is reduced by $195 through the addition of load shifting. The water-heating load becomes only 16.6% of the house electricity when it is shifted while it was originally 22%.

The analysis of these results show that load shifting is more economically and environmentally effective than a solar photovoltaic system without load shifting. These results suggest that widespread use of smart water heaters that optimize load shifting has significant potential for the integration of renewable energy onto the grid. The development of smart hot water heaters would be able to predict users' hot water consumption schedule based on prior data. On an individual scale, load shifting for hot water heaters significantly decreases users' annual electricity bills.

In future work, the controls in this study should be tested experimentally. An optimization algorithm could be created according to economic savings through TOU rates and/or the amount of available solar energy. These controls can also be aggregated to hundreds of water heaters to determine the benefit to the grid through day-ahead market pricing which give a closer approximation of actual generation costs to utilities than TOU rates. This work could be expanded to encompass heat pump water heaters and a variety of different electricity rate structures. The inclusion of heat pump water heaters could also incorporate an accounting of the capital costs for equipment and the simple payback periods.

Table 1: Results With and Without a 1 kW Solar Array and Load Shifting

Metric	No Solar Array	Solar Array	Solar Array and Load Shifting
WH Load Met by Solar Array	0%	5%	63%
Annual End-User Electricity Bill	$846	$803	$608
Fraction of WH Load to House Electric Load	22%	21%	16.6%

5. CONCLUSIONS

Energy simulation programs EnergyPlus and SAM were used to model a residential water heater load with one-minute time steps.

- 63% of the water-heating load was met by a 1 kW solar array with load shifting whereas only 5% of the load was met by a solar photovoltaic system without load shifting.
- The use of water heater load shifting with a solar photovoltaic system saved $195 more on an annual water heater TOU electricity bill than a solar photovoltaic system without load shifting.
- The fraction of the water-heating load to the house electric load was 16.6% with load shifting and 21% without load shifting.
- Load shifting can be implemented as an economically and environmentally effective means of energy storage.

NOMENCLATURE

CAISO	California Independent System Operator
GHG	greenhouse gas
IECC	International Energy Conservation Code
PNNL	Pacific Northwest National Laboratory
USDOE	U.S. Department of Energy
NREL	National Renewable Energy Laboratory
SAM	System Advisor Model
TMY3	Typical Meteorological Year version 3
TOU	Time-of-use
TRNSYS	Transient System Simulation Tool

Variables

Q_{draw}	thermal energy leaving WH tank
Q_{loss}	sum of water heat loss energy for the month
Q_{tank}	total thermal energy in the WH tank at time step i
Q_{max}	maximum thermal WH tank energy
q_{heater}	heat transfer rate to the water in the tank from the heating element
$m_{cumulative}$	cumulative sum of the mass of water drawn
$\dot{m_{use}}$	mass flow rate from water heater
V_{tank}	total volume of water in the WH tank
c_p	specific heat of water
ρ	water density
T_{tank}	the outlet temperature of the water heater
T_{mains}	the mains water temperature
UA	uniform skin loss coefficient per unit area to ambient temperature

REFERENCES

Alcorn, B. (2013). California energy demand 2014-2024 preliminary forecast. *California Energy Commission*, *2*, 33–57.

CAISO. (2019). *California iso peak load history 1998 through 2019.* Retrieved from https://www.caiso.com/Documents/CaliforniaISOPeakLoadHistory.pdf

Duffie, J. A., & Beckman, W. A. (2013). *Solar engineering of thermal processes.* John Wiley & Sons, Inc.

Gelažanskas, L., & Gamage, K. (2016). Distributed energy storage using residential hot water heaters. *Energies*, *9*(127). doi: 10.3390/en9030127

Gilman, P. (2015). Sam photovoltaic model technical reference. *National Renewable Energy Laboratory.*

Kepplinger, P., Huber, G., Preißinger, M., & Petrasch, J. (2019). *Thermal Science and Engineering Progess*, *9*, 94–109. doi: 10.1016/j.tsep.2018.11.003

O'Shaughnessy, E., Cutlera, D., Ardania, K., & Margolisa, R. (2019). Solar plus: Optimization of distributed solar pv through battery storage and dispatchable load in residential buildings. *Applied Energy*, *213*.

SCE. (2019). *Time-of-use (tou) rate plans.* Retrieved 2019-08, from https://www.sce.com/residential/rates/Time-Of-Use-Residential-Rate-Plans

Thermal Energy System Specialists LLC. (2016). *TRNSYS: Transient System Simulation Software.*

USDOE. (2019). Engineering reference. *EnergyPlus™ Version 9.1.0 Documentation.*

Van Asselt, A. (2018). Operation of cool thermal energy storage to enable renewable electricity generation. *University of Wisconsin-Madison, Department of Mechanical Engineering, ProQuest LLC.*

Wilcox, S., & Marion, W. (2008). User's manual for tmy3 data sets. *National Renewable Energy Laboratory.* Retrieved from http://www.nrel.gov/docs/fy08osti/43156.pdf

ACKNOWLEDGMENT

This work was conducted thanks to the Clare Boothe Luce Scholars Program and EXCEL Scholars Program at Lafayette College.

Towards a Techno-Economic Analysis of PCM-Integrated Hybrid HVAC Systems

Dre HELMNS[1*], David BLUM[1], Armando CASILLAS[1], Anand PRAKASH[1], Jonathan WOOLLEY[2], David VERNON[3], Caton MANDE[3], Michael WOODCOX[4], Spencer DUTTON[1]

[1] Lawrence Berkeley National Laboratory, Berkeley, CA, USA

[2] Emanant Systems LLC, Northampton, MA, USA

[3] Western Cooling Efficiency Center, Davis, CA, USA

[4] NestWorks LLC, Orinda, CA, USA

* Corresponding Author: drehelmns@lbl.gov

ABSTRACT

Thermal end uses dominate building energy consumption and are a major driver of peak demand. As heating is electrified, peak electrical power required will surge, prompting a need for innovative HVAC system designs and controls. These designs must incorporate novel technologies at the component level and new integration techniques at the system level. One such possibility involves the addition of thermal energy storage (TES) in heating and cooling equipment using a phase change material (PCM) heat exchanger. Here, thermal energy storage via phase change can be used to shift the HVAC system loads to times of lower electricity cost, reduced carbon intensity, and greater energy efficiency. Most of the current utilization of PCM in buildings involves passive components. By actively controlling when heat is stored and released from PCM, we can optimize the building HVAC system to cost-effectively meet consumer needs with the flexibility to draw on renewable energy resources when they are abundant and available.

While this combination of technologies is promising in theory, simulation-based evaluation of a prototype can be difficult due to the modeling requirements at the component level and the large number of possible configurations and operating modes at the system level. To conduct this evaluation, we use the Modelica language for modeling and simulation because it enables users to represent the important physics of the problem, interchange and rearrange components in an efficient manner, and implement a range of control configurations. In this work, we considered three case studies: a portable building, a large commercial retail store, and a multifamily residential apartment unit. Each of these employs a different system design, ranging from a single package vertical unit incorporating PCM to a central plant with independent heat pump, evaporative cooling, and thermal energy storage components. This paper describes the technologies in question, presents modeling at the component and system levels, and demonstrates building energy and demand charge cost savings with local time-of-use tariffs in a hot-dry climate.

1. INTRODUCTION

Many researchers and policy makers have proposed shifting away from on-site combustion of fossil fuels for all types of heating, ventilation, and air conditioning (HVAC) and domestic hot water (DHW) equipment. Supplied by renewable electricity generation, all-electric HVAC systems could reduce direct greenhouse gas emissions attributed to the buildings sector by 8.6% (Leung, 2018). However, transitioning new and existing buildings to all-electric HVAC and DHW systems is a monumental challenge. US buildings use 8.52 trillion cubic feet of natural gas on-site annually with the vast majority of this natural gas consumption used to provide space heating (58%) and water heating (21%) (*Natural Gas Consumption by Sector*, 2020). Electrification of HVAC and DHW will increase the existing portion of electricity contribution attributed to buildings by 73% (*Electricity End Use*, 2020). All-electric systems will impact regional and seasonal electricity loads, necessitating adaptation of how we design and operate to support the grid. One promising strategy is to reduce and shift loads on the demand side to better match available energy on the supply side.

To address this challenge, we developed and modeled several all-electric so-called "hybrid" HVAC system designs with local energy storage in three distinct building applications (large commercial retail, portable buildings, and multifamily residential). Our technology strategies to improve efficiency and provide energy storage include: high performance variable capacity heat pumps, indirect evaporative cooling, ventilation heat recovery, sub-wet bulb evaporative water cooling, and PCM thermal energy storage. Each of the systems and components modeled are described in Section 2. The hybrid systems, as well as comparative baseline gas and all-electric systems, were modeled using the Modelica language (Mattsson & Elmqvist, 1997). Modelica's object-oriented specification and excellence in multi-domain physical system modeling allows for component models to be combined into system models that serve as an explicit representation of the proposed system architectures and control concepts. This study leverages an international effort to develop and share component models (Wetter et al., 2019), specifically the Modelica Buildings Library (GitHub commit 0ef82ef) and the IDEAS Building Energy Simulation Library (GitHub commit a91fd24) (Wetter, Zuo, Nouidui, & Pang, 2014; Jorissen et al., 2018), and develops new or enhances existing models where necessary.

In this paper, we describe our evaluation of the impact of new load shifting systems for three building types against their respective comparative baselines. We share our designs for the different architectures and describe how we've modeled the building envelope, loads, air delivery, hydronic plant, and control systems. The goal of our models is to accurately represent physical component configurations and explore opportunities to improve HVAC and DHW systems for energy efficiency and load shifting. To test our models in different climates with local tariffs, we created a parametric simulation framework to easily run a set of case studies. The tool outputs time series data for visualization of system operation and associated energy, demand, and cost savings. We provide sample results for summer and winter peak days and summarize annual operation costs and equipment sizing for a case study in Albuquerque, New Mexico. We aim to ultimately make these models available to others so that users have access to a simulation tool that can support the implementation of novel energy technologies for grid-interactive efficient buildings.

2. METHODS

We designed and modeled all-electric hybrid mechanical systems for three buildings. Each system incorporated similar components, in different configurations, as appropriate for the building type, size, loads and occupancy profiles.

The large commercial retail system, shown at the top of Figure 1, utilizes a central air-water heat pump with both a sub-wetbulb evaporative cooler (SWEC) and evaporative condenser air pre-cooler to increase efficiency and cooling capacity. The heat pump provides heated or cooled water to PCM thermal energy storage, and/or to a distributed array of hydronic air handlers. Ventilation is provided to the building by a dedicated outdoor air system (DOAS) using a heat exchanger that performs heat recovery ventilation (HRV) and indirect evaporative cooling (IEC). The system includes hot (43°C) and cold (11°C) PCM thermal energy storage for load shifting, peak shaving, and additional capacity.

The portable building system, shown in the middle of Figure 1, is similar in concept to the large commercial retail system except it is designed to have all components packaged in a vertical wall-mounted unit, similar to equipment typically used in these buildings. From bottom to top, this system includes: hot (43°C) and cold (11°C) PCM thermal energy storage, an air-to-water heat pump with evaporative condenser air pre-cooling, an air handling module that includes IEC and HRV, a water-to-air heat exchanger, and a pair of fans (and dampers) to manage ventilation, exhaust, and recirculated air flow. Since all system components are connected together in a packaged wall-mount unit, this system has some operating modes that are different from the large commercial retail system. For example, ventilation air from the IEC may mix with return air and be cooled further through the water-to-air heat exchanger.

The multifamily residential system, shown in the bottom of Figure 1, is designed to provide all heating, cooling, and DHW with a single heat pump that replaces the split system and tank water heater commonly used in residential buildings. The system consists of an air-to-water heat pump paired with PCM thermal energy storage and a hydronic air handler (or fan coil units). This system includes PCM TES with two hot phase change temperatures (43°C and 58°C), used for heating and DHW respectively. Ventilation for the residence is provided separately.

2.1 Models for Building Envelopes and Loads

Each system is modeled using a single thermal zone, whose envelope construction properties and whose convective, radiant, and latent internal load design values and annual schedules (except holidays) are defined based on a corresponding Department of Energy Reference Building with New Construction (Deru et al., 2011). New Construction models comply with the minimum requirements of ANSI/ASHRAE/IESNA Standard 90.1-2004 (ASHRAE 2004a). For all systems, *Buildings.ThermalZones.Detailed.MixedAir* is used to model the dynamic room heat balance and the floor sits on 2 meters of soil. The large commercial retail model is based on the "Stand Alone Retail Store" Core Retail

Zone. It has a floor area of 1600 m^2, two external walls with no windows, two internal walls modeled as adiabatic on the outside boundary, and a roof that is exposed to the environment. The portable building model is based on the "Primary School Building" Corner Classroom. It has a floor area of 99 m^2, made modular with four external walls and a double-pane window with air gap facing south, with a roof that is exposed to the environment. The multifamily residential model is based on the "Mid-rise Apartment Building" Bottom Floor North-Facing Apartment. It has a floor area of 88 m^2, one external wall with a double-pane window with air gap, and three internal walls and a ceiling modeled as adiabatic on the outside boundary. Weather data for a given climate is represented by TMY3 files using *Buildings.BoundaryConditions.WeatherData.ReaderTMY3*. Design heating and cooling loads for a given climate for the corresponding reference building zones are used as sizing parameters for the system components.

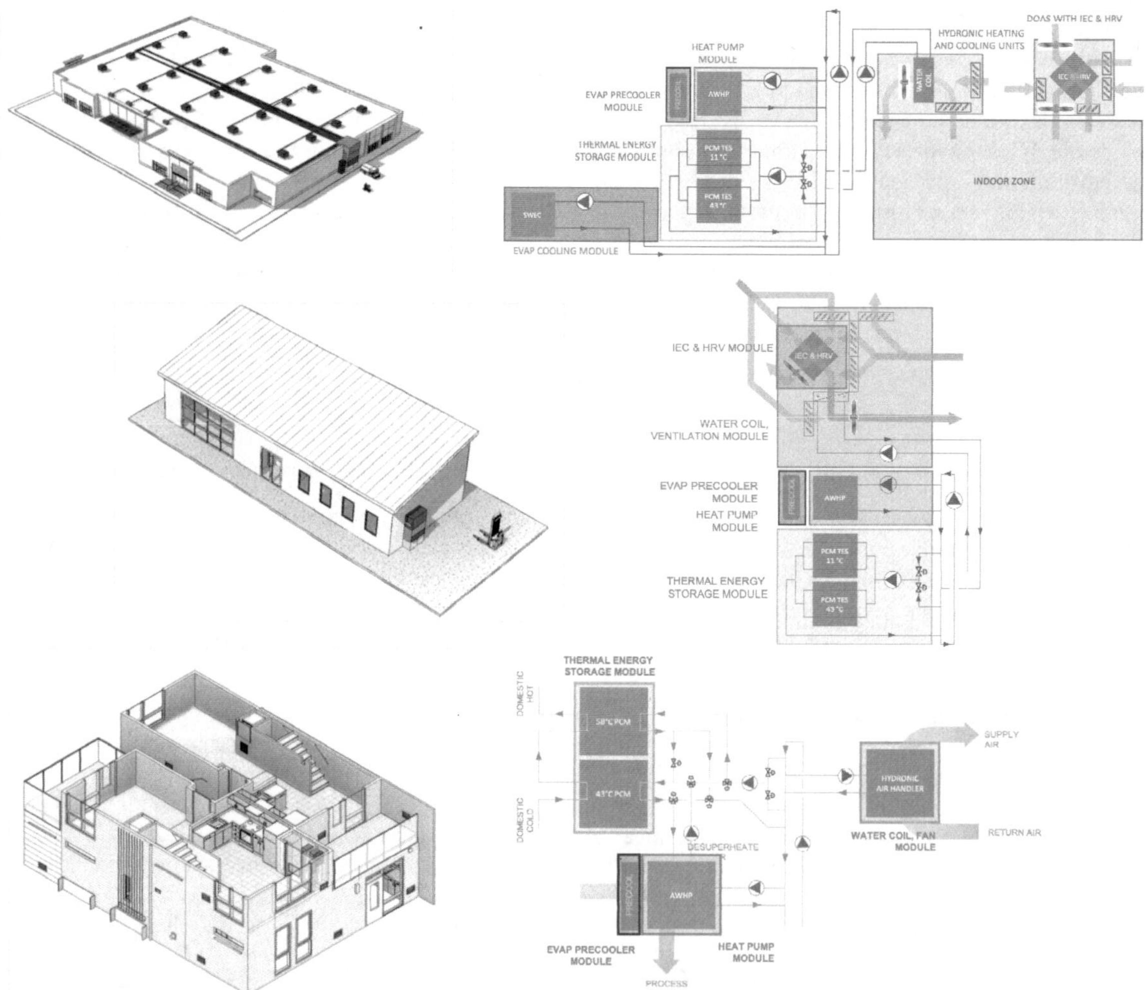

Figure 1: Large commercial retail (top), portable building (middle), and multifamily residential (bottom) schematics

The multifamily residential model also defines DHW loads using an annual schedule of endpoint water flowrate demand and domestic cold water temperature from a publicly available stochastic DHW event generator (Hendron, Burch, & Barker, 2010). In the multifamily residential model, a thermostatic mixing valve is used to control the ratio of domestic cold and hot water flow rates to an outlet temperature set point of 43°C modeled using a PI feedback controller. The DHW temperature is determined according to the hydronic plant models described in the following sections.

2.2 Models for Air Delivery Systems

The air delivery system for each building type is modeled with a similar structure of component models, though their exact definition and corresponding controls determine which system is instantiated. This model design choice was made to provide efficient, yet flexible, means of modeling the large number of system configurations for all building types and associated baseline and hybrid systems. The air delivery system is made up of a supply and return duct, supply fan, heating coil, cooling coil, mixing box with economizer, and an IEC/HRV unit enabling heat exchange between the exhaust and outside air streams. For all systems, the supply fan uses *Buildings.Fluid.Movers.FlowControlled_m_flow*, the mixing box with economizer ideally mixes two air streams according to an input signal between 0 and 1, and the IEC/HRV unit uses *IDEAS.Fluid.HeatExchangers.IndirectEvaporativeHex* with constant heat exchanger effectiveness in dry and wet mode equal to 0.80. All pressure drops are lumped into a single resistance downstream of the supply fan equal to the design pressure rise of the fan. Taken from the corresponding Reference Building model, for the large commercial retail system, the pressure drop is 1100 Pa, hydraulic efficiency is 0.59, and motor efficiency is 0.91. For the portable building and multifamily residential systems, the pressure drop is 622 Pa, hydraulic efficiency is 0.65, and motor efficiency is 0.86.

The differentiating factor for each system model is based on the selection of the heating and cooling coil model. For the baseline systems, the heating coil is modeled as a direct-fire gas furnace using *Buildings.Fluid.HeatExchangers.HeaterCooler_u* with efficiency of 0.80 and the cooling coil is modeled as a wet direct-expansion unit with *Buildings.Fluid.HeatExchangers.DXCoils.AirCooled.VariableSpeed* with the cooling equipment COP calculated as a function of the difference between the source-side entering and load-side leaving temperatures (Staffell, Brett, Brandon, & Hawkes, 2012). For the baseline all-electric systems, the heating and cooling coils are modeled as the condenser and evaporator of the same reversible air-to-air heat pump using the same component models as the baseline system, except with the COP for both heating and cooling calculated as a function of the difference between the source-side entering and load-side leaving temperatures (Staffell et al., 2012). For the hybrid systems, the heating coil is modeled as a dry hydronic coil using *Buildings.Fluid.HeatExchangers.DryCoilEffectivenessNTU* and the cooling coil is modeled as a wet hydronic coil using *Buildings.Fluid.HeatExchangers.WetCoilCounterFlow*.

2.3 Models for Hydronic Plant Systems

Hydronic plant models are only needed for the new hybrid system designs. The models for the large commercial retail and portable building systems are structured in the same way to promote efficiency in the ability to model the large number of system configurations, while the model for the multifamily residential system is significantly different. For the large commercial retail and portable building systems, the hydronic plant system is modeled separately for heating and cooling, though both do not run at the same time to enforce an assumption of a single heat pump. Their structures are the same except inclusion of the SWEC in the large commercial retail system.

The air-to-water heat pumps are modeled using *Buildings.Fluid.HeatExchangers.SensibleCooler_T* and *Buildings.Fluid.HeatExchangers.Heater_T*, with COPs calculated as a function of the difference between the source-side entering and load-side leaving temperatures (Staffell et al., 2012). The heat pump heating capacity is assumed to be equal to the cooling capacity and the remainder of required heating capacity is assumed to be met by an auxiliary electric resistance heater with a COP of 1. The associated circulation water pump is modeled using *Buildings.Fluid.Movers.FlowControlled_m_flow*. Evaporative condenser precooling is modeled in each cooling heat pump model by using the outside air wetbulb temperature, assuming a wetbulb effectiveness of 1.0, instead of the drybulb temperature in the determination of COP. The control signal inputs to the model are the set point for the water leaving the heat pump and the fraction of nominal water mass flowrate for the pump. For the multifamily residential model, the cooling heat pump model also includes a desuperheater model that recovers a fraction of the condenser heat flow in a secondary circuit as a function of the entering water temperature, with an associated enable/disable control input signal (*In-Floor Warming Module: external desuperheater*, 2015; Yrjölä & Laaksonen, 2015). For the large commercial retail system, the SWEC is modeled similarly to the air-to-water heat pump, except that the COP is approximated at a constant 10.5 and the leaving water temperature is approximated as equal to the outside air wetbulb temperature.

The PCM TES heat exchanger for each building type is based on a design that includes two circuits that each interact with the same PCM. The model represents 1-D heat transfer from one fluid circuit to the PCM through convection at the pipe boundary and resistance by the copper pipe wall and aluminum fins protruding into the PCM. The resistances and PCM itself are modeled with the component *Buildings.HeatTransfer.Conduction.SingleLayer*. To produce the two-circuit model, the 1-D heat transfer circuit is repeated with symmetry around the PCM. Each circuit utilizes a mixing volume *Buildings.Fluid.MixingVolumes.BaseClasses.MixingVolumeHeatPort* to calculate the fluid outlet temperature given the inlet temperature and heat transfer with the PCM cell. Three different PCMs were utilized to produce TES heat exchangers at three different temperatures: 11°C for cooling, 43°C for space heating, and 58°C for domestic hot water. Their material properties are found in Table 1, based on three commercially available thermal batteries.

Table 1: Average solid-liquid phase change material thermal properties used in hydronic plant models

Melt temperature, T_m	Latent heat, h_{ls}	Specific heat, c	Density, ρ	Thermal conductivity, k
11°C	126 kJ/kg	2050 kJ/kg·K	1125 kg/m^3	0.200 W/m·K
43°C	153 kJ/kg	1550 kJ/kg·K	1675 kg/m^3	0.584 W/m·K
58°C	226 kJ/kg	3150 kJ/kg·K	1360 kg/m^3	0.584 W/m·K

For the large commercial retail and portable building, the TES was sized by analyzing time series load data for the baseline all-electric systems. First, for each day, the thermal load was integrated over the peak tariff times. Then, the maximum value was chosen for each month and the TES size was set equal to the average of these 12 values. For multifamily residential, which also includes DHW loads, we first calculated the amount of PCM TES needed solely to serve DHW loads. To do so, we determined the minimum PCM capacity to maintain the first hour rating (FHR) for an apartment unit that would traditionally be served by a 60 gallon hot water tank at 60°C. For an equivalent TES size, we calculated the amount of thermal energy required to raise that volume of water from the inlet mains temperature to the storage tank temperature, using a "70% rule". This rule is commonly applied to storage tank water heaters to represent the degree of mixing within the tank. While it is not directly applicable to sizing PCM TES in the same way, it is an appropriate first pass at generating the same ratio of heat pump size to nominal storage capacity. The relative sizes of 43°C and 58°C PCM were determined based on the DHW temperature lift required of each. Additional capacity was then added to the 43°C PCM thermal battery to serve half of the integrated space heating load on the peak day.

For the multifamily residential systems, DHW is provided to the building by heating the domestic cold water inlet to a temperature set point of 60°C. The baseline system does this with a natural gas-fired tankless water heater, modeled using *Buildings.Fluid.HeatExchangers.Heater_T* with an efficiency of 0.80. The baseline all-electric does this with a heat pump water heater tank using *Buildings.Fluid.Storage.StratifiedEnhanced* with five stratification levels and the total heat provided by the heat pump assumed to equally distribute among them. A PI feedback controller controls the heat pump heat output to ensure the second highest stratification level maintains the hot water temperature set point. The COP of the heat pump is calculated as a function of the difference between the source-side entering and load-side leaving temperatures (Staffell et al., 2012). The heat pump heating capacity is such that the ratio of total heating capacity to heat pump heating capacity is equal to 1.3, with the remaining heating capacity provided by an auxiliary electric resistance heater with COP of 1. The source temperature is assumed to be the inside room air temperature and the load temperature is assumed to be the average temperature of all stratification levels in the tank. For the hybrid system, the domestic cold water passes first through the 43°C and then through the 58°C PCM heat exchangers. There is no active control for the outlet temperature. Instead, the high melting temperature of 58°C is assumed to produce hot enough water for the thermostatic mixing valve to meet the appropriate fixture water temperature set point.

For all hydronic models, circulation pump power is not calculated and circulation pumps are assumed to operate as ideal flow movers. Three-way valves control fractional flow in different branches served by the same pump.

2.4 Models for Controls

For the large commercial retail and portable building systems, a thermostat provides feedback control on zone air temperature between occupancy schedule-based heating and cooling set points, indicated by dashed lines in the top panels of Figure 3. The fan operates at heating, cooling, ventilation, or setback speeds according to the thermostat feedback and occupancy schedule. For cooling and heating, the thermostat control signal is used as an input to the rate of gas burn in the furnace, compressor speed in the DX unit and air-to-air heat pump, and water valve positions as applied to each system type. The minimum outside air damper position is set based on system mode. When unoccupied, the minimum position is closed; while in occupied mode, it is set to provide minimum outside air flow at the corresponding heating, cooling, or ventilation fan speed. Economizer control of the outside air damper via the thermostat control signal is enabled if the outside air dry bulb temperature is less than the return air temperature and the system is in cooling mode. If the outside air damper is saturated open, then the cooling coil is modulated on. In the hybrid systems, the IEC/HRV is on in wet heat recovery mode if the system is cooling and the wetbulb temperature of the return air is less than the outside air dry bulb temperature. It is on in dry heat recovery mode if the system is heating and the return air dry bulb temperature is greater than the outside air dry bulb temperature. If the system is not heating or cooling , the IEC/HRV is in dry mode if it was previously enabled on (either as wet or dry). Otherwise, it is turned off.

The portable building and large commercial retail hybrid hydronic systems are in cooling mode if the air system is cooling or if the cold TES is in charging mode. The system is heating if the air system is in heating mode, the air system is in freeze protection mode, or if the hot TES is in charging mode. The heating/cooling supply water temperature set point is 43.5/10.0°C normally, and 50.5/6.0°C if in TES charging. A feedback controller generates a PI control signal based on supply water temperature and is used first to modulate return water flow through the TES to maintain the supply water temperature at set point if the TES is in discharge mode. Then, the control signal is used to modulate return water flow from the TES through the SWEC to maintain the supply water temperature. This occurs if the system is in cooling mode and if the SWEC is present in the system and enabled, based on comparing the temperature of the water that would enter the SWEC to the outside wetbulb temperature. Finally, the control signal is used to modulate return water flow from the SWEC through the ASHP to meet the supply water temperature set point of the system. The evaporative condenser pre-cooler is enabled if the heat pump is in cooling mode. The TES is in charging or discharging mode according to a schedule, based on the electricity tariff, and on whether the TES is detected to be fully charged or fully discharged, determined by the outlet water temperature. To limit heat pump peak demand during charging, the control signal for the TES pump is set based on estimating the SOC at the start of the period and calculating the constant flowrate needed to have the TES fully charged by the end of the scheduled period. As a general rule, the TES discharge window overlaps with all, or some portion of, the peak time-of-use tariff periods.

For multifamily residential, the control of the fan, heating, and cooling coils are the same as for large commercial retail and portable building systems, except the ventilation air flowrate is zero and there is no economizer nor IEC/HRV. The hydronic system is heating if the air system is heating, if the high temperature PCM is in charging mode and the air system is not cooling, or if the low temperature PCM is in charging mode and the air system is not cooling. If the air system is cooling, the hydronic system is cooling. The heating supply water temperature set point is 52°C normally, and 64°C if the low or high temperature TES are in charging mode. A control signal is generated via PI feedback on supply water temperature and used first to modulate return water flow through the low temperature TES to maintain the supply water temperature at setpoint if the TES is in discharge mode. Otherwise, the low temperature TES is bypassed. The control signal is used to modulate return water flow through the heat pump to meet the supply water temperature setpoint. The evaporative condenser pre-cooler is enabled if the heat pump is in cooling mode.

The low and high temperature TES are in charging or discharging (low only) mode according to a schedule, determined based on the electricity tariff, on whether the TES is detected to be fully charged or fully discharged by the outlet water temperature of water flowing through the TES, and on if the system is not in cooling mode. During a scheduled charging period, the TES pump control signal is set to be constant, tuned to ensure DHW demand can be met and the low temperature PCM can be adequately recharged each day without causing too much electricity demand. The bypass valve is opened only if the high temperature PCM is charging, allowing the outlet water from the high temperature PCM to also charge the low temperature PCM. The modulation valve directs water flow from the circulation pump to the high temperature PCM only if it is in charging mode, since it is not discharged for space heating. Finally, the desuperheater valve and pump direct water at constant flow to the low temperature PCM when the heat pump is cooling.

2.5 Parametric Simulations

The performance and cost evaluations of the different integrated systems in this paper were done using a parametric simulation framework that was developed for easily configuring the different scenarios to enable running simulations at scale. Additionally, the framework was developed using the open-source and free JModelica tool (Åkesson, Årzén, Gäfvert, Bergdahl, & Tummescheit, 2010) and publicly available docker environment (https://github.com/lbl-srg/docker-ubuntu-jmodelica). Use of this docker container enhanced portability by allowing team members to run simulations irrespective of host operating systems. Figure 2 is a graphical representation of the framework.

In the configuration step, a user populates a configuration file with a matrix of simulation settings including climate zones, system types, building types, and energy storage charging and discharging times. With this information, one (or more) Modelica record files are automatically written and saved to disk to be used for parameterization of the appropriate system model upon simulation. In the simulation phase, using JModelica, the system models defined by the configuration are compiled into Functional Mockup Units (FMU) and simulated with the CVODE variable timestep solver and tolerance of 1e-6 for the duration of a whole year. The simulation step results in the generation of one result .mat file for each configured simulation. Using the BuildingsPy package (https://simulationresearch.lbl.gov/modelica/-buildingspy/) and tariff information from the configuration, the system performance time series data are extracted from the .mat files and used for final cost and performance analysis.

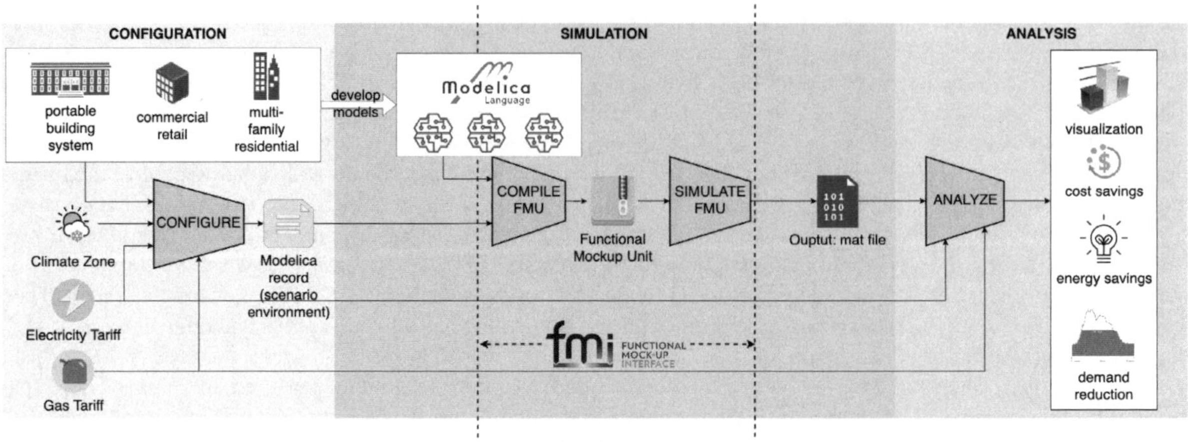

Figure 2: Parametric simulation framework consisting of case study configuration, simulation, and analysis

3. RESULTS

Equipped with designs, models, and a parametric simulation framework, we present results for the three building types in Albuquerque, New Mexico with two different tariffs applied in post-processing. Public Service Company of New Mexico Rate NO. 3B, with peak hours from 8 am - 8 pm, was used for the large commercial retail and portable building systems and Central New Mexico Electric Cooperative Rate NO. 25, with peak hours from 6:30 am - 9 am (winter only) and 4:30 pm - 10:30 pm (all year), was applied to the multifamily residential system. Time series simulation outputs are post-processed and re-sampled to 15 minutes. Figure 3 shows zone air temperature (top), whole building electric power and gas power (bottom) for each building type and system type for winter (left) and summer (right). Zone temperature setpoints, outside air temperature, tariff peak periods, and TES charging and discharging periods are also presented.

In the top panels of Figure 3, we see that zone air temperature was properly maintained for each system and building type. In addition, we see the baseline (gas) system often has the lowest electrical demand in winter. Gas use for space heating and hot water is also captured on the right axis of this lower panel plot in subfigures 3a, 3c, and 3e. In winter, we can see that gas demand tends to coincide with baseline all-electric peaks. On average, HVAC electric demand is roughly half of the whole building electric demand, reflecting the importance of targeting heating and cooling systems for load shifting. These plots reveal that the dispatch of stored thermal energy during peak periods effectively flattens electric demand for space conditioning and water heating. During off-peak hours, electric demand is necessarily higher for the hybrid HVAC system in order to charge the PCM TES for the following discharge period. Charging control is better for some building types and seasons than others. Across all building types and seasons, discharging the TES results in lower peak electric demand for HVAC and DHW than the baseline all-electric system. The results reveal that the new hybrid HVAC system accomplishes thermal load shifting throughout the year but could be further optimized on several fronts, including PCM TES sizing and transient operation. In future simulations, model predictive control (MPC) could optimize these systems.

4. DISCUSSION

Table 2 presents the simulated annual costs for total energy, electrical energy, electrical demand, and natural gas, as well as cost savings for each system compared to the corresponding baseline system. It can be seen that when peak demand of the baseline and baseline all-electric systems occurs in the summer, the hybrid HVAC systems are able to generate savings. When the peak occurs in the winter (due to heating), the all-electric systems incurs very high demand charges resulting in increased total costs. Note that for the large commercial retail building and the portable building, the demand charges contribute to more than 50% of the total annual costs, emphasizing the importance of peak load reduction in these commercial buildings. The hybrid system generates total cost savings compared to both baseline and baseline all-electric in both large commercial retail (around $4,500) and the portable building (around $500), but performs worse than the baseline system in the multifamily residential building. The baseline all-electric system is more costly than the baseline in all cases except the portable building, highlighting the benefit of integrated

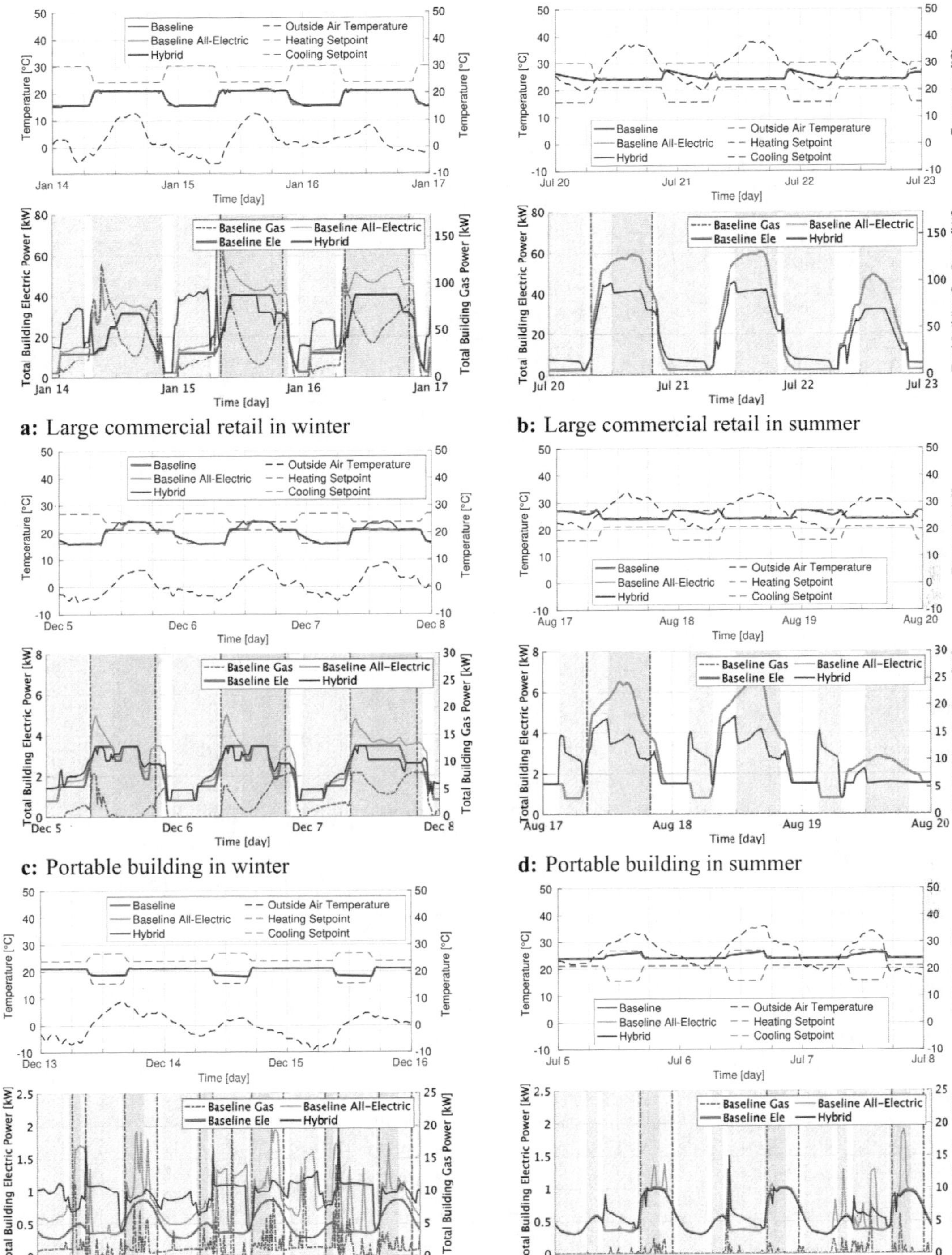

a: Large commercial retail in winter
b: Large commercial retail in summer
c: Portable building in winter
d: Portable building in summer
e: Multifamily residential in winter
f: Multifamily residential in summer

Figure 3: Three systems in winter and summer where grey vertical lines enclose peak tariff periods, blue shaded regions correspond to TES charge windows, and pink shaded regions correspond to TES discharge windows

storage in all-electric systems. The relevance is also evident in the large commercial retail building, where the baseline all-electric system has its maximum demand in winter (79.80kW) whereas the hybrid peaks in summer (only

48.08kW). Such results reveal successful shifting of hot thermal loads in the winter and accompanying reduction in peak demands.

Table 2: Costs incurred in the different buildings with baseline, baseline all-electric, and hybrid HVAC systems

	Electricity: Energy Costs	Electricity: Demand Costs (Peak Demand)	Natural Gas: Energy Costs	Total Costs	Total Cost Savings (Demand Cost Savings)
Large Commercial Retail					
Baseline	$4,624.41	$12,225.48 (60.67kW in Jun)	$2,759.79	$19,609.68	0.00 % (0.00 %)
Baseline All-Electric	$5,094.07	$15,459.87 (79.80kW in Feb)	$0.00	$20,553.94	-4.82 % (-26.46 %)
Hybrid	$4,446.77	$10,700.27 (48.08kW in Jul)	$0.00	$15,147.04	22.76 % (12.48 %)
Portable Building					
Baseline	$474.33	$1,369.14 (7.13kW in Jun)	$98.61	$1,942.08	0.00 % (0.00 %)
Baseline All-Electric	$489.68	$1,439.15 (7.13kW in Jun)	$0.00	$1,928.83	0.68 % (-5.11 %)
Hybrid	$436.10	$1,043.38 (4.79kW in Aug)	$0.00	$1,479.48	23.82 % (23.79 %)
Multifamily Residential					
Baseline	$425.40	$273.22 (1.05kW in Jul)	$131.24	$829.86	0.00 % (0.00 %)
Baseline All-Electric	$537.27	$572.52 (2.01kW in Dec)	$0.00	$1,109.79	-33.73 % (-109.55 %)
Hybrid	$537.67	$321.46 (2.00kW in Jan)	$0.00	$859.13	-3.53 % (-17.66 %)

In addition to the demand and cost reductions provided by the hybrid systems, secondary benefits are in the form of peak thermal load reduction for the heat pump and coils, which create additional potential capital cost savings on component sizing and electrical infrastructure. Peak heat pump loads are significantly reduced by the ability of the thermal energy storage to shift demand to off-peak hours, while coil loads in some systems (retail and portable) are reduced via the IEC/HRV providing conditioning before the air reaches the system's air-side coil. Compared to the baseline all-electric system, peak summer/winter heat pump load reductions for the hybrid system in the portable building are 7/8% and large commercial retail are 21/40%. Similarly, peak coil loads are reduced 34/42% and 53/40% respectively. The multifamily residential system actually sees an increased peak winter load of 46% due to DHW load being shifted from the heat pump water heater to a single heat pump, and negligible change in peak summer load.

In order to verify the validity of results for all three systems, the energy use intensity (EUI) of the baseline systems in all building types were compared to typical average EUI values in real buildings. The reference data was obtained via the Buildings Performance Database, the largest publicly-available source of building energy data in the US (*Building Performance Database*, 2020). The EUI across the three building types provided by the database were found to be comparable with simulation results for buildings of similar use-type and square footage, with large commercial retail building's reference annual site EUI of 173 kWh/m^2, matching closely to simulated annual site EUI of 195 kWh/m^2. As mentioned in Section 2, the portable building used a primary school as its reference. Similar buildings of this type have an annual site EUI of 191 kWh/m^2, while our simulation predicts 249 kWh/m^2. We attribute this larger deviation to the building's exposure as a modular classroom. Finally, the simulated multifamily residential building reference annual site EUI is 72 kWh/m^2 compared to a ~50% higher simulated prediction of 111 kWh/m^2. The full multifamily reference building includes unconditioned hallways and lobbies that are often unoccupied, in contrast to the single unit that we modeled. We hypothesize that another contributing factor to all three simulations predicting higher EUI than the references cases is that we modeled primary occupancy zones with appropriately larger loads.

5. CONCLUSIONS

In this paper, we explored the impact of new load shifting systems for three building types. We simulated our models in a hot-dry climate with local time-of-use tariffs and generated time series data for visualization of system operation on peak days in winter and summer. Compared to baseline and baseline all-electric systems, we found that the hybrid HVAC systems we developed were able to reduce annual operation costs and equipment sizing for two of the configurations, large commercial retail and portable buildings, in this specific case study scenario.

Even though the results section presented significant cost savings by using a PCM-integrated hybrid HVAC system, this evaluation and the corresponding results are preliminary. We are currently working on adding a life cycle cost (LCC) analysis application to the parametric simulation framework. The LCC analysis, along with studying how this system would fare in different climate zones with different tariffs, would provide a more holistic evaluation of the new architectures presented in this paper. Development is underway to achieve these goals.

REFERENCES

Åkesson, J., Årzén, K.-E., Gäfvert, M., Bergdahl, T., & Tummescheit, H. (2010). Modeling and optimization with optimica and jmodelica. org—languages and tools for solving large-scale dynamic optimization problems. *Computers & Chemical Engineering*, *34*(11), 1737–1749.

Building performance database. (2020). Retrieved from `https://buildings.lbl.gov/cbs/bpd`

Deru, M., Field, K., Studer, D., Benne, K., Griffith, B., Torcellini, P., ... others (2011). Us department of energy commercial reference building models of the national building stock.

Electricity end use. (2020, Dec). Retrieved from `https://www.eia.gov/totalenergy/data/monthly/pdf/sec7_19.pdf`

Hendron, B., Burch, J., & Barker, G. (2010). *Tool for generating realistic residential hot water event schedules* (Tech. Rep.). National Renewable Energy Lab.(NREL), Golden, CO (United States).

In-floor warming module: external desuperheater. (2015). Maritime Geothermal Ltd. Retrieved from `https://www.nordicghp.com/product/nordic-products/external-desuperheater/`

Jorissen, F., Reynders, G., Baetens, R., Picard, D., Saelens, D., & Helsen, L. (2018). Implementation and verification of the ideas building energy simulation library. *Journal of Building Performance Simulation*, *11*(6), 669–688.

Leung, J. (2018). Decarbonizing us buildings. *Center for Climate and Energy Solutions.*

Mattsson, S. E., & Elmqvist, H. (1997). Modelica-an international effort to design the next generation modeling language. *IFAC Proceedings Volumes*, *30*(4), 151–155.

Natural gas consumption by sector. (2020, Dec). Retrieved from `https://www.eia.gov/totalenergy/data/monthly/pdf/sec4_5.pdf`

Staffell, I., Brett, D., Brandon, N., & Hawkes, A. (2012). A review of domestic heat pumps. *Energy & Environmental Science*, *5*(11), 9291–9306.

Wetter, M., van Treeck, C., Helsen, L., Maccarini, A., Saelens, D., Robinson, D., & Schweiger, G. (2019). Ibpsa project 1: Bim/gis and modelica framework for building and community energy system design and operation–ongoing developments, lessons learned and challenges. In *Iop conference series: Earth and environmental science* (Vol. 323, p. 012114).

Wetter, M., Zuo, W., Nouidui, T. S., & Pang, X. (2014). Modelica buildings library. *Journal of Building Performance Simulation*, *7*(4), 253–270.

Yrjölä, J., & Laaksonen, E. (2015). Domestic hot water production with ground source heat pump in apartment buildings. *Energies*, *8*(8), 8447–8466.

ACKNOWLEDGMENT

This research was supported by the Assistant Secretary for Energy Efficiency and Renewable Energy, Office of Building Technologies of the U.S. Department of Energy, under Contract No. DE-AC02-05CH11231.

A Three-Dimensional Transient Model for Evaluating the Performance of Cement-Based Thermoelectric Module

Xiaoli LIU[1], Ming QU[2]*, Kazuaki YAZAWA[3]

[1]Lyles School of Civil Engineering, Purdue University,
West Lafayette, IN 47907, U.S.A
liu2251@purdue.edu

[2]Lyles School of Civil Engineering, Purdue University,
West Lafayette, IN 47907, U.S.A
mqu@purdue.edu

[3]Birck Nanotechnology Center, Purdue University,
West Lafayette, IN 47907, U.S.A
kyazawa@purdue.edu

* Corresponding Author

ABSTRACT

The thermoelectric module (TEM) is a device that integrates multiple thermoelectric (TE) elements to realize the mutual conversion of heat and power. Due to the advantages of no moving parts and flexible expansion, the application of conventional Bi_2Te_3-based TEM in buildings has attracted the attention of researchers. On the other hand, the TE behavior of hardened cement composites was found by combining conductive additives with cement. Therefore, a new study on cement-based TEM for building energy harvesting and temperature control is proposed. To simulate the performance of cement-based TEM, a three-dimensional heat transfer model considering temperature-dependent TEM characteristics was established. The validity of the model is verified by comparing the results with commercial simulation software and experiments. Different from the existing analytical models and commercial software, the customized model has greater scalability, optimization, and control flexibility. Through parametric studies, the model can guide the design of TEM and the development of TE cement.

1. INTRODUCTION

The building sector currently accounts for over one-third of the continuously increasing global energy consumption and CO_2 emission (IEA (2020), n.d.). The severe energy and environmental issues cause extreme weather, which in turn leads to a vicious cycle of growing energy demand in buildings. Energy demand encourages the effort to develop high-performance and net-zero energy buildings that can take advantage of environmental changes as well as renewable and waste energy.

Thermoelectric technology is one of the promising routes to reduce energy consumption and adjust the built environment. TE materials realize the interconversion of thermal and electrical energy. Power can be generated from the TE material under a certain temperature difference, and the surface temperature difference can form for given power input to the same TE material. TE materials can be classified as *n-type* or *p*-type when electrons or holes dominate the charge transport, respectively. The *n-type* and *p*-type materials can be connected thermally in parallel and electrically in series. One *p-n* pair forms a TE couple and multiple TE couples constitute the TE module (TEM).

TEM works under cooling/heating mode is called TE cooler (TEC) or TE heat pump (TEH) while that under generation mode is named as TE generator (TEG).

The application of commercially available TEMs in building enclosure and energy systems keeps increasing over the last ten years (Zuazua-Ros et al., 2019). Potentially, TEM can work in four different modes in buildings. In the sunny summer, the maximum temperature of urban surfaces (i.e. wall, roof, pavement) may reach 55~70°C due to solar radiation, accelerating the urban heat island effect (Svobodová & Matějka, 2014). TEG can be used to absorb heat from the urban surface and transform it to free electricity, as a result, lowering the surface temperature (Figure 1a). In winter, there is also a temperature difference between indoors and outdoors, similarly, generating power in the opposite direction (Figure 1c). But to perform active heating or cooling, direct current (DC) with different directions can be applied to the system as shown in Figure 1b and 1d. This TE-building system has many merits such as no moving part, expansion flexibility, better thermal comfort, etc.

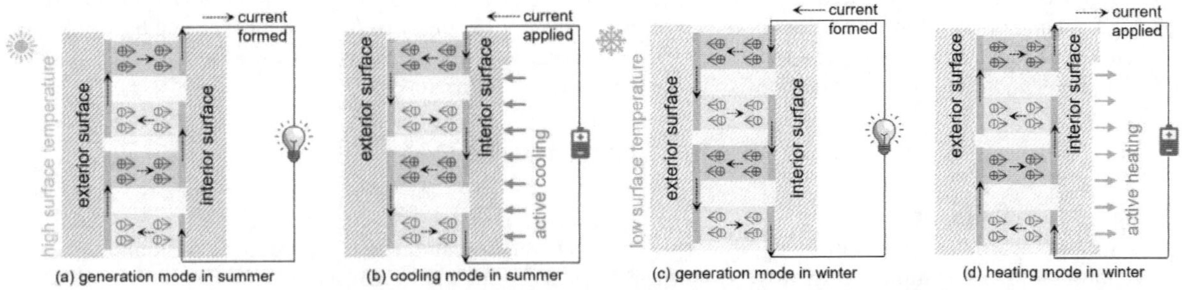

Figure 1: TE envelope working under different operating conditions: (a) generation mode in summer, (b) cooling mode in summer, (c) generation mode in winter, and (d) heating mode in winter

On the other hand, interesting finds of TE behaviors in hardened cement composites have been reported. These TE cement composites are produced by adding conductive additives (i.e. carbon fiber, graphite powder, metallic oxide power, etc.) to cement (Ghosh et al., 2019, 2020). Therefore, a research hypothesis was proposed to replace the commercial TEM with cement-based TEM for the design of a highly integrated system and the usage of environmentally friendly materials. However, the study on TE cement is still in the early stage. Very few papers reported the performance of the cement-based thermocouple and so far, no cement-based TEM was successfully manufactured. Thus, the model is of great significance to the performance prediction of the cement-based TEM.

The heat transfer model of TEM includes a simplified model, analytical model, and numerical model. The simplified model is mostly used in many system-level modelings (Liu et al., 2020). The model treats the whole TEM as a lumped system and analyzes the heat transfer at two boundary points (i.e, hot side and cold side) given the overall properties of TEM. The analytical model is assumed to have higher accuracy and computational speed as compared with the numerical model which executes a set number of iterations. But it is not easy to obtain analytical solutions to complex partial differential equations (PDEs) unless the model is simplified to lower dimension and ideal assumptions (i.e. no Thomson effect, no surface heat loss, constant properties, even distribution of charge density, etc.). The Numerical model can overcome this limitation and solve complex PDEs. In 2004, El-genk et al. presented a detailed algorithm of FEM for the 3D TEG model (El-Genk' et al., 2004). At the same time, Hogan and Shih summarized the modeling methods and governing PDEs for TE element (Hogan & Shih, 2005), then Antonova and Looman simulated the performance of a 2D TE leg in ANSYS, which helped the development of the TE package in commercial simulation software (Antonova & Looman, 2005). Later, customized numerical models were studied by finite difference method (FDM) and finite volume method (FVM), as shown in Table 1.

Table 1: Summary of customized numerical models of the thermoelectric module

Method	State	Material	Property	Error	Ref
FDM	1D Steady-state	Marlow DT12-6L	T-dependent	<5%	(Rodríguez et al., 2009)
	1D Steady-state	HMS + MgSi	T-dependent	~7.2%	(Fateh et al., 2014)
	1D Transient	commercial TEM	Constant	~20%	(Nguyen & Pochiraju, 2013)
	3D Steady-state	Bi_2Te_3	Constant	<8%	(Jang & Tsai, 2013)

FVM	1D steady-state	Bi_2Te_3	Constant	<1% (analytic) <6.6% (FEM) <27.4% (test)	(Pfeiffelmann et al., 2018)
	1D Transient	Bi_2Te_3	T-dependent	~5%	(Yan et al., 2014)
	2D steady-state	Bi_2Te_3	Constant	<10%	(Oliveira et al., 2014)

This research developed a 3D heat transfer model to simulate the performance of cement-based TEM by using the properties of TE cement in the literature, which is of significance for the TE cement development and application. The customized model solved via FVM can be designed to control the temperature-dependent characterizations, Thomson effect, surface heat loss, etc. Compared with many existing models built-in commercial simulation software, this model can be better integrated with building energy consumption simulation software, which can optimize the TEM geometry and add a control system. Therefore, the custom model has better extensibility, flexibility, optimization, and control performance. This paper first introduces the method of the TEM model, including governing equations, boundary conditions, and flow of the algorithm. Then, the accuracy of the model is verified by comparing it with the experimental results. In the simulation part, a parametric study is presented to analyze the impact of several factors and global optimization is carried out to guide the design of TEM.

2. NUMERICAL MODELING

2.1 Model Development

As shown in Figure 1, our model considers applications at varying ambient temperatures. In summer, the air convection and solar radiation heat the exterior surface to 60 °C, while the indoor temperature is constant at 22 °C. We explored how much power the TE generator can form given the temperature difference around 35 °C. In addition, the effect of the TE cooler on the regulation of room temperature was also analyzed during the moderate season.

Therefore, a 3D heat transfer model was established for TEM to simulate and evaluate its performance in different proposed scenarios. The model considers temperature-dependent TE properties, the Thomson effect, and thermoelectric contact resistance. A 3D partial differential governing equation was established for the transient heat transfer process of a TE element, and the corresponding variation of the electric potential was modeled by Ohm's Law. The FVM was used to solve the PDE. Considering the symmetry of TEM, the overall heat transferred, and power generated by the module becomes a linear extension of all TE elements. The model was finally validated by the experimental results.

The assumptions we made for the model include that:

- The material properties are isotropic;
- The surface heat loss of the TE element is ignored;
- The charge is uniformly distributed, and the direction of current density is perpendicular to ceramic plates.

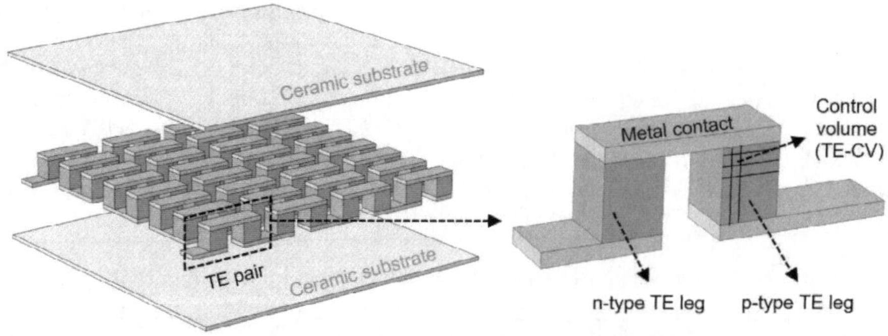

Figure 2: Schematic diagrams of a TE module and a TE couple.

The heat transfer mechanism in a TE material includes heat conduction, heat convection, radiation, Joule heat, Thomson effect, Peltier effect, Bridgeman effect, etc. as listed in Table 2 (Goupil, n.d.). The heat conduction occurs due to the temperature difference between two ends of the TE material, which follows Fourier's Law. The convective and radiative heat transfer may exist between the surface of the TE material and the surroundings, known as heat losses. Joule heating comes from the carrier movement through a conductor, as calculated by Joule's first law. Different from Joule heating which affects the whole electric conductor, the Peltier effect transfers heat from one electrical junction to another. The volumetric heat source from the Peltier effect is calculated by $T\boldsymbol{J}_e(\nabla S)_T$. It shows that this term is related to the gradient of the Seebeck coefficient with respect to the position (x, y, z) when the temperature is constant. Hence, this term will not play a role once the Seebeck coefficient is isotropic. In this model, the Peltier source is considered as a boundary condition at two junctions of the TE material. The Bridgeman effect is caused by the non-uniform distribution of charge. Since we assume that the charge is uniformly distributed, this source term can also be ignored. The multidimensional heat equation follows the energy conservation in terms of temperature for the TE material, which is written in equation (1). The expressions of all volumetric heat sources in a TE material are also summarized in Table 2.

$$\dot{q}_{st} = \dot{q}_F + \dot{q}_J + \dot{q}_p + \dot{q}_T + \dot{q}_B + \dot{q}_{loss} \tag{1}$$

Table 2: Volumetric source terms

Fourier ($\dot{q}_F$)	**Joule ($\dot{q}_J$)**	**Peltier ($\dot{q}_P$)**	**Thomson ($\dot{q}_T$)**	**Bridgeman ($\dot{q}_B$)**	**Surface Loss ($\dot{q}_{loss}$)**
$\nabla(\kappa_{tot}\nabla T)$	$\boldsymbol{J}_e^2/\sigma$	$T\boldsymbol{J}_e(\nabla S)_T$	$-\tau\boldsymbol{J}_e \cdot \nabla T$	$S \cdot \nabla\boldsymbol{J}_e \cdot T$	$-h_{eff}(T - T_\infty)\frac{A}{V}$

The final expression of the governing equation of the model, as listed in equation (2), is obtained by substituting the source terms in Table 2 to equation (1) and neglecting the Bridgman effect, and simplifying the Peltier source as a boundary source.

$$\rho C_p \frac{\partial T}{\partial t} = \nabla(\kappa_{tot}\nabla T) + \frac{\boldsymbol{J}_e^2}{\sigma} - \tau\boldsymbol{J}_e \cdot \nabla T \tag{2}$$

The electric current is computed by the open-circuit voltage divided by the total resistance in the circuit, and the current density is the ratio of current to the cross-sectional area of the conductor, as shown in equation (3).

$$J_e = \frac{V_{emf}}{(R_{TE} + R_{load})A} \tag{3}$$

The open-circuit voltage generated by the Seebeck effect is determined by equation (4). S_h and S_c are total Seebeck coefficient of n-type and p-type materials and n is the number of TE couples.

$$V_{emf} = n(S_h T_h - S_c T_c) \tag{4}$$

The output voltage equals the open-circuit voltage generated by TE material subtracting the voltage drop due to the internal resistance of the TE material.

$$V = V_{emf} - IR_{TE} \tag{5}$$

The power generated by the TE circuit is calculated as

$$P = IV \tag{6}$$

2.2 Solutions to the Model

The domain of the TE leg is rectangular with a size of Lx·Ly·Lz in a three-dimensional cartesian coordinate system. The electric current flows along the z-direction. The first step to solve the PDE by a numerical iterative method is to discretize the domain into Nx·Ny·Nz identical-size cells and transform the physical equation into a discretized form by the FVM. The mesh generated for the TE element is shown in Figure 3. The control volume (CV) is one of the 3D cells. The node, which determines the temperature profile of interest is in the center of the cuboid. A first-order explicit Euler scheme was adopted for the transient term of the governing equation in equation (2). The central difference scheme was adopted for the convection and diffusion terms in equation (2). As a result, the discretized governing equation can be written in equation (7).

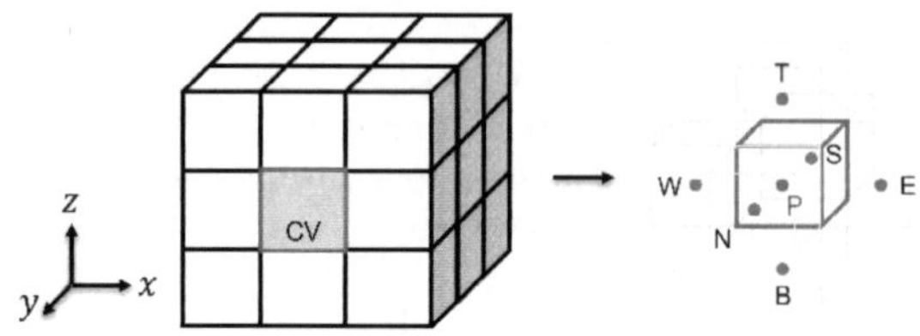

Figure 3: Schematic diagrams of the discretized domain and the control volume of a TE element.

$$\rho C_p \frac{T_P^{t+1} - T_P^t}{\Delta t} = \kappa_{tot} \left[\frac{T_W^t + T_E^t - 2T_P^t}{(\Delta x)^2} + \frac{T_S^t + T_N^t - 2T_P^t}{(\Delta y)^2} + \frac{T_B^t + T_T^t - 2T_P^t}{(\Delta z)^2} \right] + \frac{J_e^2}{\sigma} - \tau J_e \frac{T_T^t - T_B^t}{2\Delta z} \quad (7)$$

The model algorithm is illustrated in Figure 4. The input parameters include material properties, geometry and sizes, number of discretized segments, time interval, and the initial and boundary conditions of temperature and voltage. The model calculated the temperature of interest in the next time step using temperature profiles at the current state. After iterations with respect to positions and time, the model checks output all temperature data and check the convergence of the result.

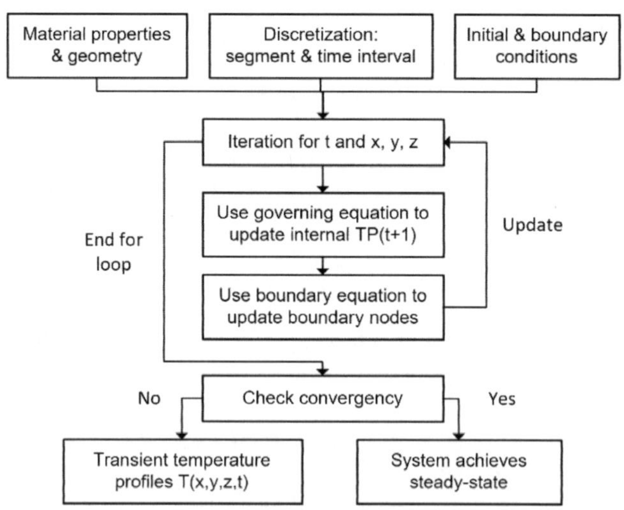

Figure 4: Flow chart of the model algorithm.

3. MODEL VALIDATION

3.1 Experimental Setup

The model was verified by the data obtained from the experiment. The experiment was carried out in Birck Nanotechnology Center (BNC) at Purdue University to test the performance of a commercially available TE module. The TE module is made of hundreds of TE couples, the soldering material, ceramic substrates, and wires. The experiment tested the power generation of the TE module given different temperature differences.

Figure 5 and **Figure 6** show the schematic diagram and digital photos of the experimental setup used for evaluating the thermal and electrical performance of a TE module, respectively. The setup consisted of four main parts including the TE module, the circulation units (i.e., water pipes, chillers, and valves), the micro-channel heat exchanger, and the measuring instrument (i.e., thermocouples, flow meters, multimeter, resistors, and data acquisition system). The detailed specifications of the measurement device are listed in Table 3. In this study, the testing module was placed between two copper-made micro-channel heat exchangers with water flowing inside. The hot water from a heating recirculating chiller flowed into the upper heat exchanger and transfer heat flux to the TE module. The cold water from a cooling recirculating chiller flowed into the lower heat exchanger and remove heat flux from the TE module. Valves and flow meters were installed on the water pipe. Eight thermocouples (TC) are employed to measure temperatures of the selected locations, including the water inlets and outlets, the exterior surfaces of the TE module. The data from thermocouples was transferred and saved to a data logger. The multimeter was applied to measure the open-circuit voltage, as well as the closed-circuit voltage and current when connecting with different load resistors. The power output from the TE generator was calculated by equation (6).

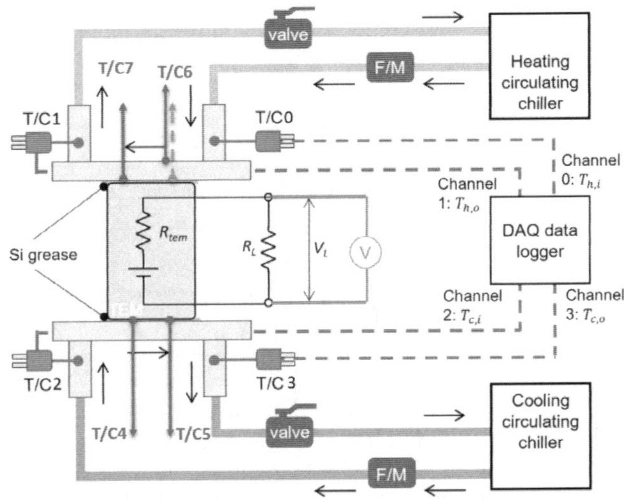

Figure 5: Schematic diagram of the test setup

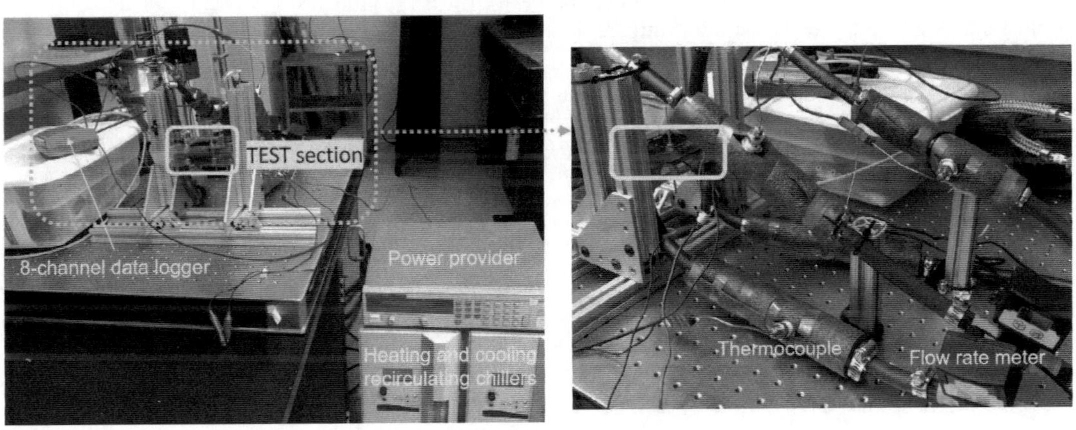

Figure 6: Digital photos of the test setup

Table 3: Parameters of the devices and instruments used in the test

Parameters/functions	Measurement devices	Sensitivity	Accuracies	Ranges
Temperature	T-type thermocouples	0.0001°C	> 1.0°C/0.75%	-250° to 350°C (-418° to 662°F)
Temperature	E-type thermocouples	0.0001°C	> 1.7°C/0.5%	-200° to 900°C (-328° to 1652°F)
Voltage	Fluke 189 multimeter	0.1mV, max 1 μV	± (0.025%+5)	0~1000V
Data logger	OM-USB-5201	0.0001°C	-	-

Chiller	Fisher Scientific Isotemp 250LCU	-	± 0.1°C	5~50°C (our test) -10~80°C (designed)
Flowmeter	Omega FTB 333D	0.01mL/min	± 6%	0.2~2L/min

3.2 Experimental Result and Comparison

The module was purchased from TE technology Inc with a series number of HP-199-1.4-0.88. The flow rates of both hot water and cold water were set at 150, 300, and 500 mL/min, approximately. The set-point temperature of the hot water was selected at 22 °C, 30 °C, 40 °C, and 50 °C, while the set-point temperature of the cold water was maintained at 7 °C. The voltage and current were measured when the system was steady. The TE transport properties of the commercial TE module are summarized in Table 4.

Table 4: Summary of values of the parameters used in the model

Classification	Parameters	Values	Units	Source
TE properties	Electrical conductivity	5917	S/m	Measured
	Thermal conductivity	1.6	W/m.K	Estimated
	Seebeck coefficient	208	μV/K	Datasheet
	Specific heat capacity	200	J/kgK	Estimated
	Density	7700	Kg/m^3	Estimated
Dimensions	Length of TE leg	0.88	mm	Datasheet
	Cross-sectional area of TE leg	1.4*1.4	mm^2	Datasheet
	Number of TE couples	199	-	Datasheet
Boundary conditions	Temperature of sink	280	K	Measured
	Temperature of source	313~323	K	Measured

The results of the Power-Resistance curve obtained from both experiment and model are compared and plotted in Figure 7. It can be observed that the higher the hot side temperature, the higher the temperature difference, and then the larger the voltage and power generated. From the comparison, we can find that the largest discrepancy between the experimental result and the modeling result is lower than 10%. There is a good agreement between the measurement data and the predicted data.

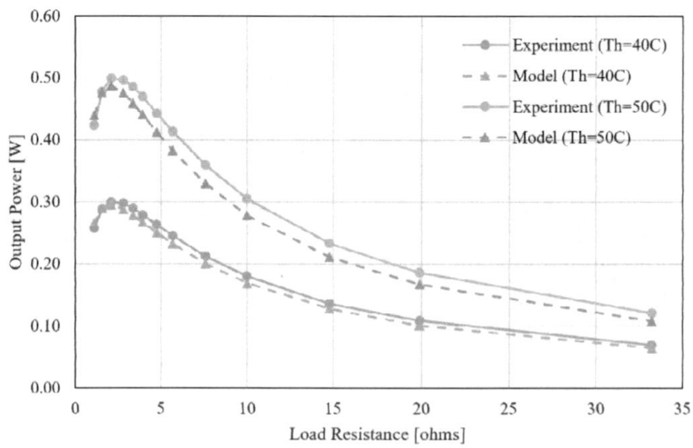

Figure 7: Comparison of power obtained from experiment and model with different load resistors

4. PERFORMANCE ANALYSIS

In this section, the performance of a TE module with cement composite based TE materials, are simulated. The TE transport properties of the TE cement composite were referred to (Ghosh et al., 2020), as listed in Table 5. The property of ceramic boards was obtained from (*Aluminum Nitride (AlN) Ceramic - Precision Ceramics USA*, n.d.). The property of copper contact referred to (*ASM Handbook Volume 2: Properties and Selection: Nonferrous Alloys and Special-Purpose Materials - ASM International*, n.d.). The thermal contact resistance was also considered in the physical model and its value was referred to (Högblom & Andersson, 2014; Ouyang & Li, 2016). In this case, we consider the n-type and p-type TE cement shares the same TE properties while the only difference is the sign of the Seebeck coefficient. The modeling results are presented below.

Table 5: Summary of values of the parameters used in the baseline model

Classification	Parameters	Values	Units
Transport properties of TE material	Electrical conductivity	1390	S/m
	Thermal conductivity	0.99	W/m.K
	Seebeck coefficient	87E-6	V/K
	Specific heat capacity	0.63	J/gK
	Density	238	Kg/m^3
	Thermal contact resistance	1E-4	m^2K/W
Property of ceramic board	Thermal conductivity	180	W/m.K
Property of copper contact	Thermal conductivity	385	W/m.K
Dimensions of TE material	Length/ Thickness	1.5	cm
	Area	1.5×1.5	cm^2
	Number of TE couples	199	-
Dimensions of ceramic board/copper	Thickness	0.2	cm
Boundary and initial conditions	Convective heat transfer coefficient	10	W/m^2K

4.1 Power Generation Mode

In the summer, due to convection between hot air and the surface, as well as strong solar radiation, the external surface of the wall can almost reach 60 °C, while the indoor air temperature is set at 22 °C. In this situation, the air temperature difference is 40 °C, and the output power density from the cementitious TEG module is about 0.3742 W/m^2, by assuming the fill factor is near 1. As shown in Figure 8, the cementitious TEG takes about 10 minutes to achieve the steady-state and the surface temperatures start to converge. Finally, the hot-side and cold-side interface temperatures become 315.09 °C and 312.77 °C, respectively.

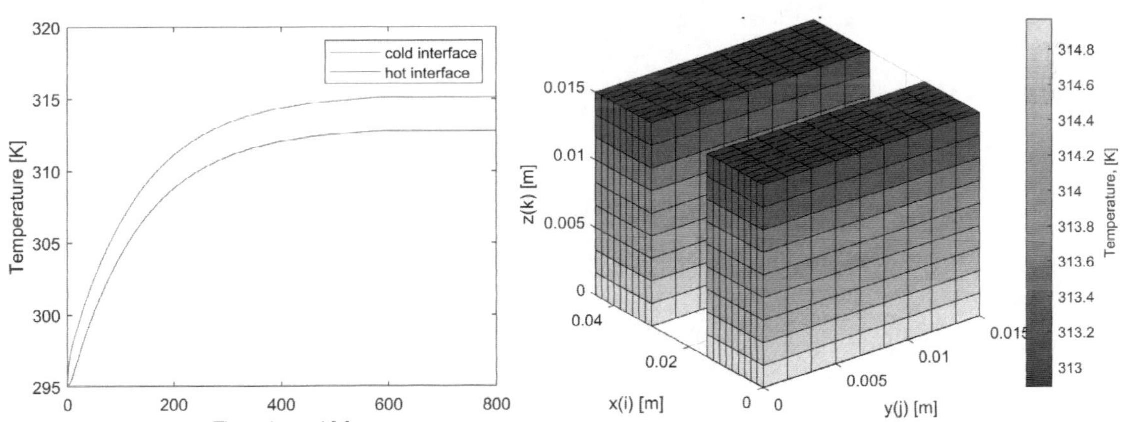

Figure 8: (a) The surface temperature of TEG with time elapsed and (b) the converged 3D temperature profile for TE legs.

4.2 Cooling Mode

In the cooling mode, both indoor and outdoor air temperatures and initial temperature are considered to be 25 °C. When the power is input to the TE module, due to the Seebeck effect, the TE module can absorb heat at the cold side to achieve the purpose of cooling. As shown in Figure 9, with 10V voltage input, we observe that the cold side temperature first decreases and then increases over time. This is mainly due to excessive Joule heat and conductive heat. Cement-based materials have large resistance, therefore, will produce more joule heat, which will offset the amount of cooling power. Besides, as the temperature on the hot side continues to rise, due to the Fourier heat conduction law, the energy on the hot side will be transported to the cold side in the form of heat conduction, affecting the cooling capacity.

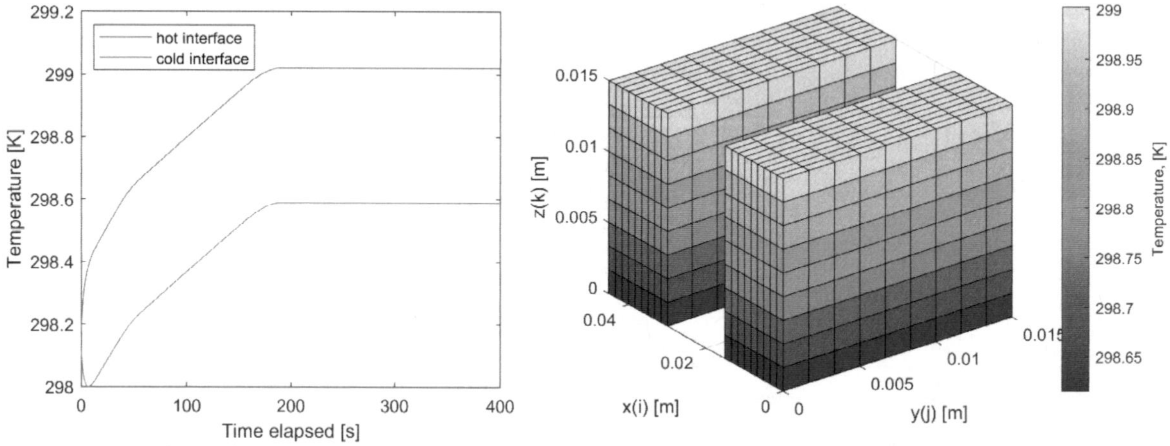

Figure 9: (a) The surface temperature of TEC with time elapsed and (b) the converged 3D temperature profile for TE legs.

5. CONCLUSIONS

In this study, a 3D transient heat and electric transport model was established for the thermoelectric module. The model was verified by the experimental test, and the discrepancy between the modeling results and the experimental data was less than 10%. Then, the model was used to simulate the performance of the cement-based thermoelectric module. The results showed that, in the power generation mode, the cement-based thermoelectric module can achieve a power density around 0.37 W/m^2. In the cooling mode, due to the large resistance of the material and a low figure of merit (ZT), the cold side cannot be effectively cooled. Therefore, it is necessary to find better cement-based thermoelectric materials with higher ZT and smaller electrical resistance to realize more efficient cooling. Besides, the model has high computational efficiency. In the future, the model will be used for the simulation of TE modules in building systems.

NOMENCLATURE

FDM	Finite difference method	(–)
FEM	Finite element method	(–)
FVM	Finite volume method	(–)
PDE	Partial differential equation	(–)
TE	Thermoelectric	(–)
TEC	Thermoelectric cooler	(–)
TEG	Thermoelectric generator	(–)
TEM	Thermoelectric module	(–)
ZT	Figure of merit coefficient	(–)

REFERENCES

Aluminum Nitride (AlN) Ceramic - Precision Ceramics USA. (n.d.). Retrieved January 29, 2021, from https://precision-ceramics.com/materials/aluminum-nitride/

Antonova, E. E., & Looman, D. C. (2005). Finite elements for thermoelectric device analysis in ANSYS. *ICT 2005. 24th International Conference on Thermoelectrics, 2005.*, 215–218. https://doi.org/10.1109/ICT.2005.1519922

ASM Handbook Volume 2: Properties and Selection: Nonferrous Alloys and Special-Purpose Materials - ASM

International. (n.d.). Retrieved January 29, 2021, from https://www.asminternational.org/search/-/journal_content/56/10192/06182G/PUBLICATION
El-Genk', M., Saber', H., & Caillat2, T. (2004). *HIGH EFFICIENCY THERMOELECTRIC RADIOISOTOPE POWER SYSTEMS Final Report*. https://ntrs.nasa.gov/search.jsp?R=20040084240
Fateh, H., Baker, C. A., Hall, M. J., & Shi, L. (2014). High fidelity finite difference model for exploring multi-parameter thermoelectric generator design space. *Applied Energy*, *129*, 373–383. https://doi.org/10.1016/j.apenergy.2014.04.088
Ghosh, S., Harish, S., Ohtaki, M., & Saha, B. B. (2020). Enhanced figure of merit of cement composites with graphene and ZnO nanoinclusions for efficient energy harvesting in buildings. *Energy*, *198*, 117396. https://doi.org/10.1016/j.energy.2020.117396
Ghosh, S., Harish, S., Rocky, K. A., Ohtaki, M., & Saha, B. B. (2019). Graphene enhanced thermoelectric properties of cement based composites for building energy harvesting. *Energy and Buildings*, 109419. https://doi.org/10.1016/j.enbuild.2019.109419
Goupil, C. (n.d.). *Continuum theory and modeling of thermoelectric elements*.
Hogan, T. P., & Shih, T. (2005). Modeling and characterization of power generation modules based on bulk materials. In D. M. Rowe (Ed.), *Thermoelectrics Handbook: Micro to Nano*. CRC Press.
Högblom, O., & Andersson, R. (2014). Analysis of thermoelectric generator performance by use of simulations and experiments. *Journal of Electronic Materials*, *43*(6), 2247–2254. https://doi.org/10.1007/s11664-014-3020-x
IEA (2020). (n.d.). *Tracking Buildings 2020 – Analysis - IEA*. Retrieved January 11, 2021, from https://www.iea.org/reports/tracking-buildings-2020
Jang, J. Y., & Tsai, Y. C. (2013). Optimization of thermoelectric generator module spacing and spreader thickness used in a waste heat recovery system. *Applied Thermal Engineering*, *51*(1–2), 677–689. https://doi.org/10.1016/j.applthermaleng.2012.10.024
Liu, X., Jani, R., Orisakwe, E., Johnston, C., Chudzinski, P., Qu, M., Norton, B., Holmes, N., Kohanoff, J., Stella, L., Yin, H., & Yazawa, K. (2020). State of the art in composition, fabrication, characterization, and modeling methods of cement-based thermoelectric materials for low-temperature applications. In *Renewable and Sustainable Energy Reviews* (Vol. 137, p. 110361). Elsevier Ltd. https://doi.org/10.1016/j.rser.2020.110361
Nguyen, N. Q., & Pochiraju, K. V. (2013). Behavior of thermoelectric generators exposed to transient heat sources. *Applied Thermal Engineering*, *51*(1–2), 1–9. https://doi.org/10.1016/j.applthermaleng.2012.08.050
Oliveira, K. S. M., Cardoso, R. P., & Hermes, C. J. L. (2014). Two-Dimensional Modeling of Thermoelectric Cells. *International Refrigeration and Air Conditioning Conference*. https://docs.lib.purdue.edu/iracc/1357
Ouyang, Z., & Li, D. (2016). Modelling of segmented high-performance thermoelectric generators with effects of thermal radiation, electrical and thermal contact resistances. *Scientific Reports*, *6*(1), 1–12. https://doi.org/10.1038/srep24123
Pfeiffelmann, B., Benim, A. C., & Joos, F. (2018). A Finite Volume Analysis of Thermoelectric Generators. *Heat Transfer Engineering*, 1–9. https://doi.org/10.1080/01457632.2018.1474588
Rodríguez, A., Vián, J. G., Astrain, D., & Martínez, A. (2009). Study of thermoelectric systems applied to electric power generation. *Energy Conversion and Management*, *50*(5), 1236–1243. https://doi.org/10.1016/j.enconman.2009.01.036
Svobodová, S., & Matějka, L. (2014). Maximum Temperatures Of Roof Tiles During The Summer Period. *Advanced Materials Research*, *1041*, 303–306. https://doi.org/10.4028/www.scientific.net/AMR.1041.303
Yan, D., Dawson, F. P., Pugh, M., & El-Deib, A. A. (2014). Time-dependent finite-volume model of thermoelectric devices. *IEEE Transactions on Industry Applications*, *50*(1), 600–608. https://doi.org/10.1109/TIA.2013.2271272
Zuazua-Ros, A., Martín-Gómez, C., Ibañez-Puy, E., Vidaurre-Arbizu, M., & Gelbstein, Y. (2019). Investigation of the thermoelectric potential for heating, cooling and ventilation in buildings: Characterization options and applications. *Renewable Energy*, *131*, 229–239. https://doi.org/10.1016/J.RENENE.2018.07.027

ACKNOWLEDGEMENT

This paper is based upon work supported by the National Science Foundation under grant no. CBET-1805818.

Integration of Phase Change Material-Based Storage in Air Distribution Systems to Increase Building Power Flexibility

Philani HLANZE[1], Aly Elhefny[1], Zhimin JIANG[1], Jie CAI[1]*, Hamidreza SHABGARD[1]

[1]School of Aerospace and Mechanical Engineering, University of Oklahoma
Norman, Oklahoma,73019, USA

philanihlanze@ou.edu; jiangzhimin01@ou.edu; aly.elhefny@ou.edu; jcai@ou.edu; shabgard@ou.edu

* Corresponding Author

ABSTRACT

This paper presents a novel energy storage solution by incorporating phase change material (PCM) in the building supply-air duct to increase a building's thermal storage capacity. This solution has various advantages compared to PCM-integrated walls including more effective heat transfer (forced convection and greater temperature differentials). During off-peak hours, the system runs at a supply-air temperature below the material's solidification point to charge the PCM with cooling energy. During on-peak hours, a higher supply-air temperature is utilized so that the stored energy can be discharged into the supply-air. This shifts a portion of the building's cooling load from the on-peak hours to the off-peak hours. A numerical model for the melting and solidification of PCM in the duct was developed and modified using experimental data. Whole building energy simulations were conducted by coupling the PCM model with EnergyPlus DOE prototypical building model in a Simulink co-simulation platform. Simulations were performed for three cities in different climate zones over a three-month cooling season (June to August), and the PCM storage reduced the on-peak energy consumption by 20-25%. The electricity cost and payback period were determined using current time-of-use electricity rates.

1. INTRODUCTION

While the introduction of renewable energy sources has provided an alternative to fossil fuels with much reduced greenhouse gas emissions, their limited window of availability (particularly solar energy) has resulted in the rapid ramping of fossil fuel power plants to make up for the energy shortfall during the evening hours (Jones-Albertus, 2017). As a result, a lot of stress is put on these plants over a short period of time, which might shorten their lifespan and lower the overall generation efficiency. Thus, there is a need to shift the demand on the electricity grid to off-peak hours and create a more balanced daily supply and demand. In the US, commercial buildings and residential homes account for about 40% of the total energy consumption, and in commercial buildings, 32% of that energy is consumed by the Heating, Ventilation and Air-Conditioning (HVAC) systems (US Department of Energy, 2010). This makes commercial buildings an attractive area to implement energy savings and peak load reduction strategies. Thermal energy storage (TES) can be implemented in buildings using sensible heat or latent heat to alleviate the electrical power supply and demand imbalance issues on the electric grid (Iten *et al.*, 2016). Sensible heat storage involves using the building's thermal mass as a thermal battery by increasing or decreasing the temperature of the building envelope. Latent heat storage involves using a phase change material's (PCM) latent heat of fusion to store energy in the material during the phase change process.

The use of PCM as latent heat storage has previously been investigated in many studies. Iten *et al.* (2016) conducted a review of free cooling TES and PCM incorporated in the building envelope to increase the building's thermal mass. For PCM incorporated in the building envelope, the energy storage is driven by the increase or decrease in the outdoor or indoor temperatures, which initiates the melting or solidification process. Safari *et al.* (2016) investigated the implementation of PCM in roof and wall construction using the Fanger comfort model to control HVAC thermostat operation in Madrid's climate zone. They found that a PCM with 27 °C melting temperature resulted in the highest energy savings potential, and 5 mm thick modules resulted in a maximum of 16% annual energy savings and a minimum 6.38-year payback period for the office schedule. Neeper (2000) investigated the thermal dynamics of a wallboard impregnated with PCM. The study found that the maximum diurnal energy storage occurs at a value of the PCM melt temperature that is close to the average room temperature in most circumstances. The diurnal storage achieved in practice may be limited to a range of 300–400 kJ/m^2, even if the wallboard has a greater latent capacity.

Sleiti and Naimaster (2016) investigated the building energy performance of organic (fatty acid based) PCM products in ceiling constructions for a simulated quick service restaurant building model located in Atlanta, GA. PCM was added in the ceiling, in between the gypsum board and attic floor frame and insulation. The PCM cases considered were not able to achieve any changes in the zone air temperature that would translate to significant HVAC energy savings. One of the limitations for passive systems is that the air-to-PCM heat transfer is low due to the reliance on natural convection and the small temperature difference limits the effectiveness and heat penetration depth.

Free cooling TES has been previously investigated, and it involves using cold night-time air to cool the indoor space during the day. Takeda *et al.* (2004) investigated the reduction of the ventilation load during summer in various Japanese cities by installing a packed bed of PCM granules in the supply-air ventilation duct. When the outdoor air temperature is less than room air, two separate streams of outdoor air flow into the room and through the PCM bed to simultaneously charge the PCM and cool room. The air that leaves the PCM section is exhausted to the outdoor environment. When outdoor air temperature is higher than room air, outdoor air only passes through PCM bed and goes into room. The maximum ventilation load reduction was shown to be up to 62.8% in Kyoto; however, in the study the indoor air temperature was kept constant (26 °C) and heat transfer through the building walls was not considered. Yanbing *et al.* (2003) analyzed the thermal behavior of a Night Ventilation with PCM Packed Bed Storage (NVP) and determined convective heat transfer coefficients of 12 to 19 $W/m^2{\cdot}°C$ and air flow resistance through the PCM of less than 20 Pa. Free cooling TES is attractive because it makes affective use of "free" energy source, but its main disadvantage is the requirement for bypass duct/dampers to allow for the active control of the charging/discharging periods and prevent overcooling of the indoor spaces.

The proposed PCM latent energy storage solution of the present study is displayed in Figure 1. The PCM is located in the supply-air duct in order to take advantage of the forced convection heat transfer provided by the moving air, which improves the rate of thermal penetration compared to PCM incorporated in the building envelope. The supply-air temperature (SAT) will be lowered to initiate the PCM solidification (PCM charging) and store "cooling" energy in the PCM. When needed, the SAT is increased to initiate the PCM melting process (PCM discharging). The PCM charging process is conducted during the off-peak hours and the PCM discharging process is enabled during on-peak hours to reduce electrical energy usage and peak demand. This increases the HVAC cooling load during the PCM charging and decreases it during the PCM discharging. When implemented with time-of-use (TOU) electricity rates, the building electricity cost can be effectively reduced. The solution does not require any bypass duct or dampers, so it is easier to retrofit than most of the free cooling solutions.

The objective of this paper is to evaluate the energy cost savings and peak electric demand reduction that can be achieved by installing PCM in the supply air duct of the HVAC system. This objective will be achieved by formulating a model for the melting and solidification of PCM in the duct, conducting a whole building simulation for a PCM-equipped medium office building in different climate zones using an EnergyPlus-Simulink co-simulation platform, and calculating the energy cost savings and payback period using current TOU rates.

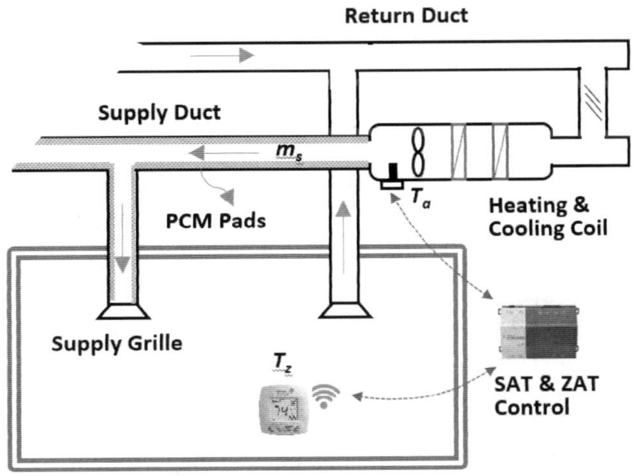

Figure 1: Proposed PCM storage location

2. WHOLE BUILDING SIMULATION

2.1 PCM Model

The PCM is mounted on the interior surfaces of the vertical and horizontal supply duct walls. Heat is transferred between the air and PCM through the PCM casing. The duct wall is assumed to be well insulated with external insulation and no heat is lost through the duct wall. The PCM model used in the whole building simulation is adapted from the Enthalpy method (Faghri & Zhang, 2006). The heat transfer in the PCM is modeled as a one-dimensional conduction-controlled, two-region melting problem in a finite slab. The density of the solid and liquid phases is assumed to be equal ($\rho_s = \rho_l = \rho$), and the governing equation is

$$\rho\frac{\partial h}{\partial t} = \frac{\partial}{\partial x}\left(k\frac{\partial T}{\partial x}\right) \tag{1}$$

where h is the PCM enthalpy, t is time, x is distance, k is the PCM thermal conductivity and T is temperature. The enthalpy is defined as

$$h(T) = \begin{cases} c_{p_s}(T - T_m) & T < T_m \\ 0 & T = T_m, \text{fully solidified} \\ h_{sl} & T = T_m, \text{fully melted} \\ c_{p_l}(T - T_m) + h_{sl} & T > T_m \end{cases} \tag{2}$$

where h_{sl} is the PCM latent heat, T_m is the PCM melting temperature, and c_{pl} and c_{ps} are the liquid and solid PCM specific heat, respectively. In simulations, the PCM is initially either fully solid or liquid, but during melting/solidification, h is between 0 and h_{sl}. The PCM thermal conductivity is assigned according to the phase:

$$k(T) = \begin{cases} k_s & T \le T_m, \quad h(T) \le 0 \\ k_l & T \ge T_m, \quad h(T) \ge h_{sl} \end{cases} \tag{3}$$

where k_s and k_l are the solid and liquid thermal conductivities, respectively. Equation (1) is discretized using a forward difference in time and central difference in space, and the enthalpy at the $(n+1)^{th}$ timestep is determined as follows:

$$h_j^{n+1} = \frac{\Delta t}{\rho(\Delta x_{PCM})^2}\left(k_{j+\frac{1}{2}}T_{j+1}^n + k_{j-\frac{1}{2}}T_{j-1}^n\right) + \left(h_j^n - \frac{\Delta t\left(k_{j+\frac{1}{2}} + k_{j-\frac{1}{2}}\right)}{\rho(\Delta x_{PCM})^2}T_j^n\right) \tag{4}$$

where Δt is the timestep, Δx_{PCM} is the PCM nodal thickness, and the location and time are represented by j and n, respectively. Figure 2 displays how the nodes in the PCM are arranged. The thermal conductivities at the half-grid, $k_{j+½}$ and $k_{j-½}$, are calculated using the harmonic mean method:

$$k_{j+\frac{1}{2}} = \frac{2k_j k_{j+1}}{k_j + k_{j+1}} \tag{5}$$

$$k_{j-\frac{1}{2}} = \frac{2k_j k_{j-1}}{k_j + k_{j-1}} \tag{6}$$

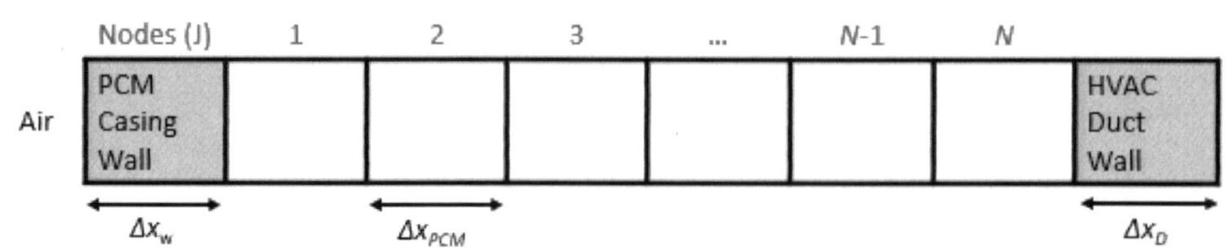

Figure 2: PCM model node arrangement

The enthalpy at node N, which is adjacent to the duct wall, is calculated using

$$h_N^{n+1} = \frac{\Delta t}{\rho(\Delta x_{PCM})}\left(\frac{k_{N-\frac{1}{2}}(T_{N-1}^n - T_N^n)}{\Delta x_{PCM}} - \frac{T_N^n - T_D^n}{\frac{\Delta x_D/2}{k_D} + \frac{\Delta x_{PCM}/2}{k_N}}\right) + h_N^n \tag{7}$$

where T_D is the duct temperature, Δx_D is the duct wall thickness, and k_D is the duct thermal conductivity. Once the enthalpy at each node is determined, the temperature is found using

$$T_j^{n+1} = \begin{cases} T_m + \dfrac{h_j^{n+1}}{c_{ps}} & h_j^{n+1} \leq 0 \ \ (solid) \\ T_m & 0 < h_j^{n+1} < h_{sl} \ \ (interface) \\ T_m + \dfrac{h_j^{n+1} - h_{sl}}{c_{pl}} & h_j^{n+1} \geq h_{sl} \ \ (liquid) \end{cases} \quad (8)$$

The heat transfer into the duct wall is calculated as

$$m_D c_D \frac{T_D^{n+1} - T_D^n}{\Delta t} = \frac{T_N^n - T_D^n}{\frac{\Delta x_D/2}{k_D A} + \frac{\Delta x_{PCM}/2}{k_{PCM} A}} \quad (9)$$

where m_D is the duct wall node mass, c_D is the duct wall specific heat, and A is the area perpendicular to the heat flow.

The air-to-PCM heat transfer was modeled using the convection correlations for internal turbulent flow, specifically the *Dittus-Boelter correlation* (Bergman *et al.*, 2011). The convective heat transfer coefficient (U) is calculated as

$$U = 0.023 Re_D^{4/5} Pr^b \frac{k_a}{D_h} \quad (10)$$

where Re_D is the Reynolds number, Pr is the Prandtl number, k_a is the thermal conductivity of the air, D_h is the hydraulic diameter, and b is 0.4 for PCM cooling and 0.3 for PCM heating. The Prandtl number is calculated using

$$Pr = \frac{c_{pa}\mu}{k_a} \quad (11)$$

where c_{pa} is the specific heat of the air, and μ is the dynamic viscosity. The Reynolds number is calculated using

$$Re_D = \frac{\rho_a v D_h}{\mu} \quad (12)$$

where ρ_a is the air density, v is the air velocity. The hydraulic diameter is calculated using

$$D_h = \frac{4A_c}{P} \quad (13)$$

where A_c is the duct cross-sectional area and P is the wetted perimeter. The heat transfer through the PCM casing wall is calculated using

$$m_w c_w \frac{T_w^{n+1} - T_w^n}{\Delta t} = \frac{T_{air}^n - T_w^n}{\frac{1}{UA} + \frac{\Delta x_w/2}{k_w A}} - \frac{T_w^n - T_1^n}{\frac{\Delta x_w/2}{k_w A} + \frac{\Delta x_{PCM}/2}{k_{PCM} A}} \quad (14)$$

where m_w is the mass of the PCM casing wall, c_w is the specific heat of the PCM casing wall, T_w is the temperature of the PCM casing wall, T_{air} is the air temperature, Δx_w is the wall thickness, k_w is the wall thermal conductivity and T_1 is the temperature of the PCM node adjacent to the wall. The enthalpy of the first PCM node is calculated using

$$h_1^{n+1} = \frac{\Delta t}{\rho(\Delta x_{PCM})}\left(\frac{q_{out}}{A} - \frac{k_{1+\frac{1}{2}}(T_1^n - T_2^n)}{\Delta x_{PCM}}\right) + h_1^n \quad (15)$$

where

$$q_{out} = \frac{T_w^n - T_1^n}{\frac{\Delta x_w/2}{k_w A} + \frac{\Delta x_{PCM}/2}{k_{PCM} A}} \quad (16)$$

The length of the duct is divided into different sub-sections and the PCM model is evaluated at each sub-section in a sequence from the inlet to the outlet sub-section. The outlet air of the previous subsection is considered the inlet air of the next section. The temperature of the air entering the next duct sub-section is calculated using

$$T_{out} = T_{in} - \frac{q_{in}}{\dot{m}_a c_{pa}} \tag{17}$$

where $\dot{m}_a$ is the air mass flow rate and

$$q_{in} = \frac{T_{air}^n - T_w^n}{\frac{1}{UA} + \frac{\Delta x_w/2}{k_w A}} \tag{18}$$

2.2 PCM Model Modification

We performed a small-scale experiment to determine possible heat transfer enhancement of adding fins to the PCM module. The test was conducted by placing a 1" deep acrylic tray that is filled with PCM and covered by a finned aluminum sheet downstream of an Air Handling Unit (AHU). Thermocouples were placed in the tray at 0.25" depth intervals. The PCM used in the experiment was PureTemp 15, which is a biobased PCM with a nominal melting point of 15°C (PureTemp, 2020). The experiment was conducted in two stages: the first stage consisted of solidifying the PCM that was initially in the liquid phase, and the second stage consisted of melting the PCM directly after the first stage. During the first stage, the supply-air temperature was set at 9 °C, and the airflow rate was set to 700 CFM. During the second stage, the supply-air temperature was set to 19 °C, and the airflow rate was set to 1475 CFM to achieve a similar cooling rate to the first stage. The experimental data was used to estimate a heat transfer enhancement factor to the PCM model associated with the addition of fins. The three parameters chosen to apply the enhancement factor are the air-side convective heat transfer coefficient and the solid and liquid PCM thermal conductivities, since the addition of fins enhances both the air-to-PCM convection and in-PCM conduction heat transfer. The factor was applied using a scalar multiplier, *z*, where

$$\begin{aligned} h_{air,enhanced} &= h_{air} * z \\ k_{s,enhanced} &= k_s * z \\ k_{l,enhanced} &= k_l * z \end{aligned} \tag{19}$$

The enhancement factor *z* was estimated to be 8.5 using a nonlinear curve fitting.

2.3 EnergyPlus and Simulink Co-simulation Platform

The whole building energy simulation was performed using an EnergyPlus co-simulation platform in Simulink that was developed by the team. The EnergyPlus Functional Mock-up Unit (FMU) file contains the building energy models and is linked to the PCM model using three primary interfacing variables, which are updated at each timestep in a ping-pong scheme. The EnergyPlus FMU outputs the supply-air mass flow rate and supply air temperature, which are inputs to the PCM model. The PCM model outputs the PCM section's outlet air temperature. The EnergyPlus FMU then inputs this temperature as the air temperature being supplied to the individual zones. The building type used in the co-simulation was the three-floor prototypical medium office building, which contains 15 conditioned zones (3 core zones and 12 perimeter zones). Each floor has a dedicated AHU and a direct-expansion air-conditioning system. Each zone has a variable air volume terminal with reheat. The whole building simulation was conducted for multiple cities to understand how the performance of the proposed PCM storage is affected by the climate location. The candidate cities were Miami, Florida (1A), El Paso, Texas (3B) and Buffalo, New York (5A).

The whole building simulations consist of two cases: baseline and PCM-equipped. The baseline case simulates normal HVAC operation without the PCM in the duct and the SAT setpoint is set at a constant value of 12.7 °C. In this case, there is no interaction between the EnergyPlus FMU and the PCM model. The PCM-equipped case involves lowering the SAT setpoint to 9.2 °C from 6 AM to 12 PM (PCM charging period) and raising the SAT setpoint to 15.2 °C from 12 PM to 3 PM and 15.7 °C from 3 PM to 6 PM. The period from 12 PM to 6 PM (PCM discharging period) corresponds with the on-peak energy hours. For the rest of the day, the SAT setpoint assumes 12.7 °C when mechanical cooling is requested. The simulations were conducted for a 3-month cooling season from the beginning of June to the end of August. For each case, the timestep in the EnergyPlus simulation is ten minutes. To ensure that convergence requirements for the Enthalpy method are met, the timestep in the PCM model is 0.5 s, which is implemented inside the MATLAB function block in Simulink.

2.4 PCM and Duct Sizing

The PCM mass was sized for each floor of the medium office building according to the on-peak HVAC sensible cooling energy usage for the hottest day of the whole cooling season. The baseline case was simulated for the 3-month cooling season and the day with the highest on-peak total HVAC sensible cooling energy (Q_s) was used as the design day. The PCM was sized to have total latent heat energy equal to one tenth of the on-peak HVAC sensible cooling energy use of the design day:

$$m_{PCM} = \frac{Q_{s,max\ on-peak}}{10\ h_{sl}} \tag{20}$$

The duct in each floor was sized according to the maximum design airflow rate ($\dot{\forall}_{max}$) determined by EnergyPlus. The square duct width (W_d) was determined as

$$W_d = \sqrt{\frac{\dot{\forall}_{max}}{v_{max}}} \tag{21}$$

where v_{max} is the maximum recommended air velocity in a duct (1200 fpm). The duct wall thickness is set at 1.31 mm, which corresponds with 16-gauge galvanized steel. The total length of the PCM duct section was 20 m and it was divided into 20 lengthwise subsections. The PCM thickness was determined using the PCM mass, total duct length and duct width.

3. SIMULATION RESULTS AND DISCUSSION

This section presents the results of the whole building simulation and the analysis to deduce meaningful insights on the load shifting capabilities of the PCM thermal energy storage. First, one day results for hot-dry and mild-humid days are presented and analyzed for El Paso. These two days were chosen because they have weather conditions that have high sensible and latent cooling loads, which will give insight into how these loads are shifted and their effect on the peak energy demand reduction. Then, the 3-month cooling season loads and energy costs are presented for each location. Lastly, the payback period for each location is determined.

3.1 El Paso Representative Day Results

Figures 3A and 3B display the total HVAC cooling load profiles for the baseline and PCM-equipped cases on hot-dry and mild-humid days. The total cooling load shifted from the on-peak to the off-peak hours on the hot-dry day is 414 MJ, which is 12.0 % of the baseline on-peak total cooling load. The total cooling load shifted from the on-peak to the off-peak hours on the mild-humid day is 586 MJ, which is 21.3 % of the baseline on-peak total cooling load. The shift in the total cooling load is caused by the shift in the sensible and latent cooling loads. The addition of the PCM results in an increase in the HVAC sensible load during the charging period due to the lower supply-air temperature and a decrease in the HVAC sensible load during the discharging period due to the higher supply-air temperature. The sensible cooling load that is shifted from the on-peak hours to the off-peak hours on the hot-dry day is 389 MJ, which equates to 11.4 % of the baseline on-peak sensible load. The sensible load that is shifted from the on-peak hours to the off-peak hours on the mild-humid day is 306 MJ, which equates to 14.9 % of the baseline on-peak sensible load. The addition of the PCM also results in an increase in the HVAC latent cooling load during the charging period and a decrease during the discharging period. The load is higher during the charging period in the PCM-equipped case because the supply air temperature is lower, which leads to a lower evaporating temperature and more removal of water vapor from the air. Conversely, the supply air temperature is higher in the discharging period, which results in less dehumidification because of the higher evaporating temperature and lower moisture level in the re-circulating air. The amount of the latent load that is shifted from the on-peak hours to the off-peak hours on the hot-dry day is 25.7 MJ, which equates to 52.9 % of the baseline on-peak latent load. The amount of the latent load that is shifted from the on-peak hours to the off-peak hours on the mild-humid day is 279 MJ, which equates to 40.2% of the baseline on-peak latent load. The mild-humid day has a higher latent cooling load than the hot-dry day due to the higher moisture content of the air in the outdoor environment.

Figure 3C displays the total HVAC electric power consumption profile for the baseline and PCM-equipped cases on hot-dry and mild-humid days. The amount of the electric power that is shifted from the on-peak hours to the off-peak hours on the hot-dry day is 47.9 kWh, which equates to 17.6 % of the baseline on-peak electric load. The on-peak peak demand of the HVAC electric power on the hot-dry day was reduced by 9.65 kW, which equates to a reduction by 17.8 %. The amount of the electric load that is shifted from the on-peak hours to the off-peak hours on the mild-humid day is 31.9 kWh, which equates to 25.3 % of the baseline on-peak total electric load. The on-peak peak demand of the HVAC electric power on the mild-humid day was reduced by 6.63 kW, which equates to a reduction by 24.0 %. The total HVAC electric power consists of the compressor power and supply air fan power. In the PCM case, the compressor power consumption is partially shifted from the on-peak to the off-peak period because less power input is needed to provide the lower amount of cooling required. However, there is an increase in the supply fan power consumption in the on-peak period because the higher SAT results in a high supply airflow rate to provide sufficient indoor cooling. Although the fan power is increased during the on-peak hours, the compressor power reduction is dominant leading to much reduced on-peak HVAC electricity usage.

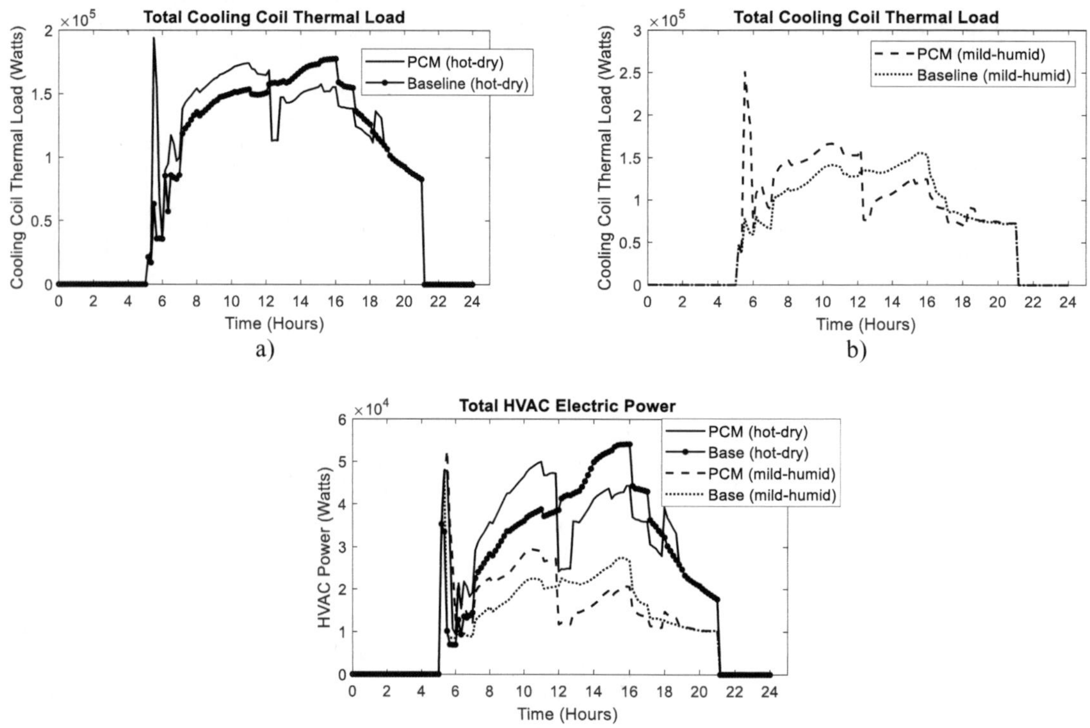

c)

Figure 3: HVAC cooling load and electric consumption profiles for hot-dry and mild-humid days in EL Paso

3.2 Total HVAC Energy Consumption and Cost

The total HVAC energy cost was calculated using the summer TOU electricity rates charged by energy providers in the three locations. Table 1 displays a breakdown of the utility tariffs used in the simulations. The rates comprise two charges: the energy charge and the demand charge. The energy charge is the cost of the electrical energy ($/kWh) and is calculated at each timestep according to the specific rate at that time. The demand charge is the cost of providing the highest electrical demand, in kW, over a 30-minute interval in a month and is calculated at the end of each month.

The total on-peak HVAC electricity consumption and HVAC electricity cost are displayed in Figure 4 for the baseline and PCM-equipped cases in El Paso, Miami and Buffalo over the duration of the 3-month cooling season. The total on-peak electric consumption was reduced by 4074 kWh (25.0%) in El Paso, 4511 kWh (20.0%) in Miami and 3349 kWh (23.4%) in Buffalo. Miami had the highest on-peak electricity consumption due to the hot and humid climate, which leads to high sensible and latent cooling loads. As a result, there was a higher amount of load shifting that took place. The total electricity cost was reduced by $889.00 (12.3%) in El Paso, $472.00 (9.6%) in Miami and $812.00 (14.5%) in Buffalo. Despite having the highest total decrease in on-peak electricity consumption, Miami had the lowest electricity cost saving because it had the least aggressive rate variation. The on-peak energy charge is 24x higher in El Paso, 2x higher in Miami and 16x higher in Buffalo than the off-peak charge. Therefore, the rate schedule in Miami does not incentivize load shifting as much compared to El Paso and Buffalo.

Table 1: Summer energy and demand rates for the three locations

City	Rate name	Energy rate ($/kWh)			Demand Charge ($/kW)	
		On-Peak	Mid-peak	Off-peak	On-peak	Anytime
El Paso	EL PASO ELECTRIC COMPANY SCHEDULE NO. 24 (El Paso Electric Company, 2017)	0.11861 (12-6PM)	—	0.00502 (12AM-12PM and 6PM-12AM)	—	24.50
Miami	Florida Power & Light Company General Service Demand-TOU (GSDT-1) (Florida Power & Light Company, 2020)	0.07078 (12-9PM)	—	0.03272 (12AM-12PM and 9PM-12AM)	9.98	—
Buffalo	Orange and Rockland Utilities Time of Use (Orange and Rockland Utilities Inc, 2020)	0.32012 (12-7PM)	0.1145 (10AM-12PM and 7-9PM)	0.02061 (12AM-10AM and 9PM-12AM)	—	—

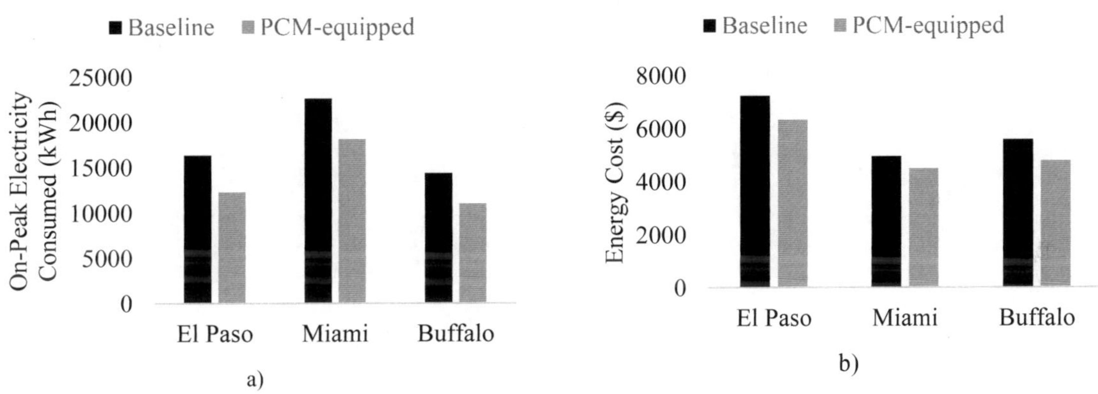

Figure 4: Total on-peak electricity consumed by HVAC system and energy cost for the baseline and PCM-equipped cases in various cities

3.3 Cooling Season Energy and Cost Summary

Table *2* displays the total loads, electric energy and electricity cost for the whole simulation period in El Paso, Miami and Buffalo, respectively. Miami had the highest total cooling load due to high sensible and latent load, which is characteristic of the hot-humid climate. This climate has the highest potential for load shifting because the implementation of the PCM TES will shift both the sensible and the latent load. For all cities, the average system COP is slightly lower in the PCM-equipped case than the baseline case. This is because the increase in the COP during the PCM discharging period is smaller than the decrease during the PCM charging period. Despite using more electric energy overall, the electricity cost is lower in all cities because the consumption was shifted from the on-peak hours to the off-hours, leveraging the use of cheaper electricity during off-peak hours.

Table 2: El Paso - Cooling season total energy and electricity cost

	El Paso		Miami		Buffalo	
	Baseline	**PCM-equipped**	**Baseline**	**PCM-equipped**	**Baseline**	**PCM-equipped**
Cooling Coil Total Load (GJ)	607	643	880	924	639	672
Total Electric Energy (MWh)	38.7	41.5	54.3	58.8	34.3	37.2
System COP	4.4	4.3	4.5	4.4	5.2	5.0
Total PCM (kg)	—	1915	—	2418	—	1902
Total Electricity Cost ($)	7227	6338	4940	4468	5582	4770
On-Peak Electric Energy (kWh)	16325	12251	22602	18091	14332	10983

3.4 Payback Analysis

A simple payback analysis was performed to determine how long it would take to recover the cost of the PCM. The payback years were calculated as follows:

$$Payback\ years\ =\ \frac{Cost\ of\ PCM}{Annual\ Cooling\ Season\ Electricity\ Cost} \tag{22}$$

The cost of the PCM was assumed to be $2 per kg (Kosny *et al.*, 2013). The payback years for each city are displayed in Table 3. El Paso had the lowest payback period (4.3 years) among the cities. This was due to the high energy cost savings in El Paso and a relatively low total PCM cost. Buffalo had a slightly higher payback period (4.7 years) because it had lower energy cost savings and similar PCM cost. Since Miami used the most PCM and involved the lowest energy cost saving, it had the longest payback period (10.2 years).

Table 3: Payback years for PCM cost in the candidate cities

City	PCM Cost ($)	Payback years
El Paso	3830	4.3
Miami	4836	10.2
Buffalo	3804	4.7

4. CONCLUSION

In this paper, the energy cost savings, on-peak load shifting, and peak electric demand reduction potentials were evaluated for a PCM-based TES system designed for installations in supply-air duct. A numerical model for the PCM in the duct was formulated and calibrated using experimental data. The model was linked with EnergyPlus through FMU to perform whole building simulations for a medium office in three different climate regions over a 3-month cooling season. The PCM TES was able to shift the sensible HVAC load from the on-peak to off-peak hours due to the lowering of the SAT to store "cooling" energy in the PCM during the charging process and raising of the SAT to release the "cooling" energy in the PCM during the discharging process. Due to the lower SAT during the charging process, there was more dehumidification in the morning. On the other hand, there was less dehumidification during the afternoon hours when the SAT was raised. This led to a shift in the latent HVAC load from the on-peak to off-peak hours. The shift in the total cooling load resulted in a shift in the HVAC electric power consumption because less work input is required to provide cooling within the on-peak hours and more work input is required to provide the additional cooling in the off-peak hours. Miami had the highest on-peak load reduction because its hot-humid climate meant that it had a high sensible load, which led to a large PCM size, and high latent load. However, since it had the least aggressive price difference between on-peak and off-peak hours, it had the lowest electricity cost savings. El Paso had the most aggressive TOU rate schedule, so it had the highest energy cost savings. Similarly, El Paso had the shortest payback period of 4.3 years. In actual building operations, the payback period is expected be shorter when the cost analysis is extended to other months where space cooling is still required; for commercial buildings whose cooling loads are mainly internal gain driven, space cooling is usually required throughout the whole year.

REFERENCES

Bergman, T. L., Lavine, A. S., Incropera, F. P., & Dewitt, D. P. (2011). *Fundamentals of Heat and Mass Transfer* (7th ed.). Hoboken: Wiley.

El Paso Electric Company. (2017, July 18). *Schedule No. 24 General Service Rate.* https://www.epelectric.com/files/html/Rates_and_Regulatory/13_-_Rate_24_General_Service_Rate.pdf

Faghri, A., & Zhang, Y. (2006). *Transport Phenomena in Multiphase Systems.* Amsterdam: Elsevier.

Florida Power & Light Company. (2020, June). *Business rates and clauses.* https://www.fpl.com/content/dam/fpl/us/en/rates/pdf/Business%20(Effective%20June%202020).pdf

Iten, M., Liu, S., & Shukla, A. (2016). A review on the air-PCM-TES application for free cooling and heating in the buildings. *Renewable and Sustainable Energy Reviews, 61*, 175-186.

Jones-Albertus, B. (2017, October 12). *Confronting the Duck Curve: How to Address Over-Generation of Solar Energy.* (Office of Energy Efficiency and Renewable Energy) https://www.energy.gov/eere/articles/confronting-duck-curve-how-address-over-generation-solar-energy

Kosny, J., Shukla, N., & Fallahi, A. (2013). *Cost Analysis of Simple Phase Change Material-Enhanced Building Envelopes in Southern U.S. Climates.* US Department of Energy.

Neeper, D. (2000). Thermal dynamics of wallboard with latent heat storage. *Solar Energy, 68*(5), 393-403.

Orange and Rockland Utilities Inc. (2020). *Time-of-Use Rates.* https://www.oru.com/en/save-money/energy-saving-programs/time-of-use

PureTemp. (2020). *PureTemp 15 Technical Data Sheet.* http://www.puretemp.com/stories/puretemp-15-tds

Saffari, M., de Gracia, A., Ushak, S., & Cabeza, L. (2016). Economic impact of integrating PCM as passive system in buildings using Fanger comfort model. *Energy and Buildings, 112*, 159-172.

Sleiti, A. K., & Naimaster, E. J. (2016). Application of Fatty Acid Based Phase-Change Material to Reduce Energy Consumption From Roofs of Buildings. *Journal of Solar Energy Engineering, 138*(5), 051003.

Takeda, S., Nagano, K., Mochida, T., & Shimakura, K. (2004). Development of a ventilation systemfor granules containing phase change material. *Solar Energy, 77*, 329-338.

US Department of Energy. (2010). *Energy Efficiency Trends in Residential and Commercial Buildings.* https://www1.eere.energy.gov/buildings/publications/pdfs/corporate/building_trends_2010.pdf

Yanbing, K., Yi, J., & Yinping, Z. (2003). Modeling and experimental study on an innovative passive cooling system—NVP system. *Energy and Buildings, 35*(4), 417-425.

Analysis of Residential Time-of-Use Utility Rate Structures and Economic Implications for Thermal Energy Storage

Sara SULTAN[1], Jason HIRSCHEY[2], Kyle R. GLUESENKAMP[3]*, Samuel GRAHAM[3]

[1]University of Tennessee, Bredesen Center of Interdisciplinary Education
Knoxville, TN, USA
Ssultan1@vols.utk.edu

[2]Georgia Institute of Technology
Atlanta, GA, USA
jhirschey13@gatech.edu

[3]Oak Ridge National Laboratory
Oak Ridge, TN, USA
gluesenkampk@ornl.gov

* Corresponding Author

ABSTRACT

Thermal energy storage (TES) is an increasingly popular tool to level out the daily electrical demand and add stability to the electrical grid as more intermittent renewable energy sources are installed. TES systems can locally decouple high thermal loads from the operation of a heat pump or reduce the electrical energy demand of the heat pump by providing a more favorable temperature gradient. In addition, many policy makers and utility providers have introduced time-of-use (TOU) rate schedules for residential customers to better reflect the price of electricity generation and demand for specific times. TOU rate schedules price grid-provided electricity differently throughout the day depending on the region's climate, time of year, and electrical production portfolio. Large differences between on-peak and off-peak electrical prices may create an economic advantage for a residential customer to install a TES system. In this work, the economic and energy savings are calculated for a modeled 223 square foot residential building with water/ice-based TES using a TOU rate structure. The weather data is from Fresno County, CA, ASHRAE climate zone 3B, and a representative residential TOU utility rate structure from a utility provider in California was used. The simulation was carried out for cooling only during a week of extreme hot summer daytime temperature and the results showed that total energy consumption could be reduced by 14.5% with an 87.5% reduction in on-peak energy usage when the TES is installed. The cost of operating this system for space cooling was reduced by nearly 20% using the sample utility rate plan.

1. INTRODUCTION

As of 2018, buildings in the United States accounted for about 40% of the total U.S. energy consumption and 75% of the total electricity consumption (U.S. Department of Energy, 2011; US EIA, 2020). The major electric load comes from the residential buildings comprising 80% of the total electricity use, and 78% of the peak period use during 2-8 PM local time. Peak periods, generally, have the largest energy demand on the grid. Heating, ventilation, and air conditioning (HVAC) loads comprise 50% of building electricity consumption (Goetzler *et al.*, 2019). HVAC systems and the need for space cooling may exacerbate electrical demand issues during summer peak times, since as much as

50% of the grid electricity demand comes from homes mostly used for air conditioning (Cui *et al.*, 2017). Residential buildings can play a major role in reducing peak hour energy demand while maintaining the occupants' comfort by introduce energy storage mechanisms to shift load from on-peak to off-peak hours (Arteconi *et al.*, 2019; Junker *et al.*, 2018).

Time-of-use (TOU), real-time-pricing (RTP), and other time-based utility rate structures are becoming more available to residential customers. TOU structures price electricity differently at specific hours during the day depending on the climate, time of year, the utility provider energy generation portfolio, and historical or predicted demand. On-peak hours are those where the electrical price is highest per unit energy for all electricity used from the grid in that time span. In summer months of warm climates, on-peak hours may be mid- to late-afternoon. In winter months of colder climates, on-peak hours may be early- to mid-morning. TOU rates vary wildly across the US. TOU rates are predetermined and static during the contract period. The TOU analyzed here is an energy rate prized per kWh of energy used. Some TOU rates are demand based priced per average kW over short intervals. RTP rates reflect the instantaneous cost of bulk wholesale of electricity. Typically, prices are not disclosed to the customer more than a few days' notice but can be as short at three hours' notice the day prior. RTP rates are rare for residential customers due to this uncertainty. However, RTP rates generally do follow an hourly and monthly trend that may give the customer an economic advantage with proper controls and forecasting.

Researchers have investigated the impact of time-based utility pricing on residential electricity use with different load and energy demand control strategies. Findings suggest that electrical utility costs and peak hour energy demands associated with air conditioning can be reduced depending on utility rate structure, weather, building occupancy, and building thermal capacitance (Newsham *et al.*, 2010).

Yoon *et al.*, (2016) demonstrated that depending on the size of house and utility price, up to 24.7% of HVAC loads can be reduced during peak times via demand response control and can provide up to 10.8% cost savings while maintaining occupant thermal comfort. Schibuola *et al.*, (2015) simulated heat pump control based on RTP, with up to 30% of electricity consumption cost savings. Gottwalt *et al.*, (2011) simulated the cumulative load profiles using TOU tariffs with smart appliances and concluded that the TOU tariffs could create new grid-level demand peaks which may not benefit the users economically, however, this practice may still benefit the utility providers by shifting more energy load to off-peak hours and effectively smoothing the demand curve.

Thermal energy storage (TES) is a popular energy storage solution to alleviate peak energy demand and may provide economic advantages to its users. TES can be realized by incorporating phase change materials (PCMs) into heat exchangers either coupled to the existing HVAC system or interfacing directly with the space to be conditioned, thus creating a high-performance demand response energy storage system (Dong *et al.*, 2019). TES systems allow conventional air conditioning machinery to operate during off-peak hours, but leverage the isothermal storage properties of PCM to provide cooling during on-peak hours (Dincer *et al.*, 1997). PCMs may be incorporated as passive or active TES. Passive systems have no coupling to HVAC systems and do not respond to control schemes, but still may reduce building energy consumption (Akeiber et al., 2016; Arivazhagan et al., 2020; De Gracia et al., 2015; Soares et al., 2013; Sonnick et al., 2020).

Active TES systems, the focus of this paper, couple the TES to the HVAC system to respond to pricing and thermal signals. Active systems offer a high level of control of the TES and can improve energy storage efficiency (Ali et al., 2015; Sarbu et al., 2018) evaluated residential HVAC loads' demand response potential, suggesting that value can be added by using even a small TES system. Table 1 summarizes studies on the energy and cost savings with PCM-based TES active storage.

The literature in Table 1 confirm that the cooling energy demand can be significantly shifted to off-peak times by using PCM-based active TES system integrated with building equipment such as HVAC (Dong et al., 2019; Waqas et al., 2018). In addition, some studies have concluded that using PCM-based active TES to offset the peak load is more financially viable in the regions having higher electricity prices for on-peak hours and where the cooling systems dominate the demand (Liu et al., 2012; Waqas et al., 2018).

Table 1. PCM-based active TES for load shifting

Author	Location	System Configuration	Energy savings	Cost savings
Bruno *et al.*, 2014	Adelaide, Australia	PCM storage with chiller	13% energy savings	Not reported
Sehar *et al.*, 2012	Various USA locations	Ice storage coupled with chiller	Peak time savings	35% annual cost savings
Mosaffa *et al.*, 2016	Tehran, Iran	PCM slabs coupled with vapor type compressor	7-9% energy savings	Not reported
Jaber *et al.*, 2012	Amman, Jordan	PCM coupled with indirect evaporative cooling system	80% less cooling load	7-8-year payback
Zhu *et al.*, 2011	Beijing, and Hong Kong	PCM layers in air conditioning system	20% peak load reduction	11% cost reduced
Real *et al.*, 2014	Madrid, Spain	PCM storage tanks with HVAC	18.97% energy saved	Not reported
Chaiyat, 2015	Chiang Mai, Thailand	PCM integrated in conventional HVAC system	Load decreased by 3.09 kWh/d	9.1% cost saved, 4.15 yr payback
Pop *et al.*, 2018	Variable (12 global climates)	PCM integrated in fresh air-cooling system	7-41% energy saved	Not reported
Hirmiz *et al.*, 2019	UK	Heat pump integrated PCM	Load shift of 6 hr., COP increased	Not reported
Dong *et al.*, 2019	Knoxville, TN, USA	PCM integrated with electrical Heat Pump	13% less power consumed	Electricity cost saved 19%

The literature survey shows that previous work on demand side management to off-set peak load using time-based utility pricing is mostly based on a demand response control without a TES. Some studies investigate an active PCM storage with heat pump, HVAC, or air conditioning system but most of them have focused on energy savings by shifting the load from high demand periods to low demand periods without considering the time-based utility pricing and a direct use of conventional VCS to accomplish heat transfer with TES.

In this work, a southern California residential building's summertime cooling load during a week of extreme high temperatures is comparatively modeled with and without active TES. The TES is directly coupled to a vapor compression system and serves as the VCS condensing unit during peak hours rather than the ambient air. The PCM embedded in the TES for this analysis is ice/water with a phase change temperature of 0°C, and only interacts with VCS, not directly with the building. A residential TOU utility rate from a utility provider in California, corresponding to the climate zone 3B, is used to assess the potential economic savings for the customer. The VCS was modeled using Engineering Equation Solver (EES). Section 2 explains the methodology for this work including control mechanisms. Section 3 describes the component model. Results are presented and discussed in Section 4.

2. METHODOLOGY

The complete modeled system is modular with several models converging including a simplified building, an R410A vapor compression system (VCS), TES with embedded PCM and heat exchanger, thermostatic controls, and a TES control strategy. The thermostat monitors the building indoor temperature and dispatches a call for cooling to the VCS system. The TES control strategy determines if the cooling is provided from the TES or by traditional VCS operation coupled to the ambient based on the time of day and utility electrical rate. The TES control strategy uses this same data to recharge the TES system using the same VCS, coupling the TES to the ambient when appropriate. The PCM used in this study is water/ice. The ambient weather data and TOU utility rate structure are from the same geographic region, ASHRAE climate zone 3B. The system is simulated for one week with a time step of one minute.

2.1 System Overview

A basic vapor compression system (VCS) is coupled to the water/ice-embedded TES. The TES heat exchanger is assumed to have infinite heat transfer coefficient and is at 0°C at all times. The PCM heat exchanger can serve as either evaporator or condenser in the VCS, depending on the mode of operation.

Figure 1a shows the baseline system whereby the building is cooled with VCS acting as a conventional air conditioning system. The building air is cooled by the VCS evaporator and heat expelled into the ambient at the condenser. The VCS moves heat from the building to the ambient across a positive temperature gradient. The increasing width of arrows represent added heat.

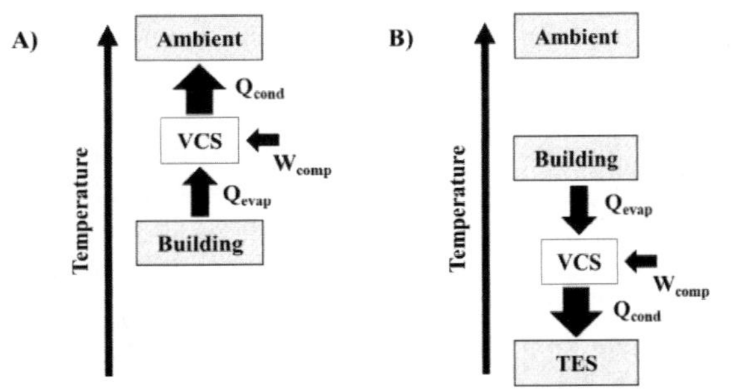

Figure 1: Comparison of temperature gradients for a) normal VCS operation and b) TES-assisted cooling

The building cooling performance can be improved with PCM-embedded TES, using negative temperature lift to introduce a more favorable gradient for which the VCS to operate, as illustrated in the Figure 1b. In this mode, the VCS condenser is coupled to the colder temperature of the PCM, 0°C in this system. Thus, the VCS has a negative temperature gradient to move the building thermal load. This allows the VCS to operate at a much higher performance as compared to the baseline.

2.2 System Operating Modes and Controls

There are four operating modes (0,1,2,3, respectively) of the system as shown in Figure 2.

- In standby mode (0), the VCS is off the building responds to ambient thermal loads and solar heat gains.
- In the normal operating mode (1), the building is cooled by conventional VCS refrigeration with no TES involvement. The condenser heat, $Q_{cond,}$ is the heat being dumped to the ambient and the evaporator heat, $Q_{evap,}$ is the heat removed from the building and directed to the heat pump.
- In charging mode (2), the PCM HX operates as the evaporator and the latent heat of freezing is removed from the PCM. The VCS condenser heat is rejected to the ambient.
- In discharging mode (3), the PCM HX is the condenser and the heat, Q_{cond}, is absorbed by the PCM through the latent heat of melting. In this mode, the COP is higher than in normal operation.

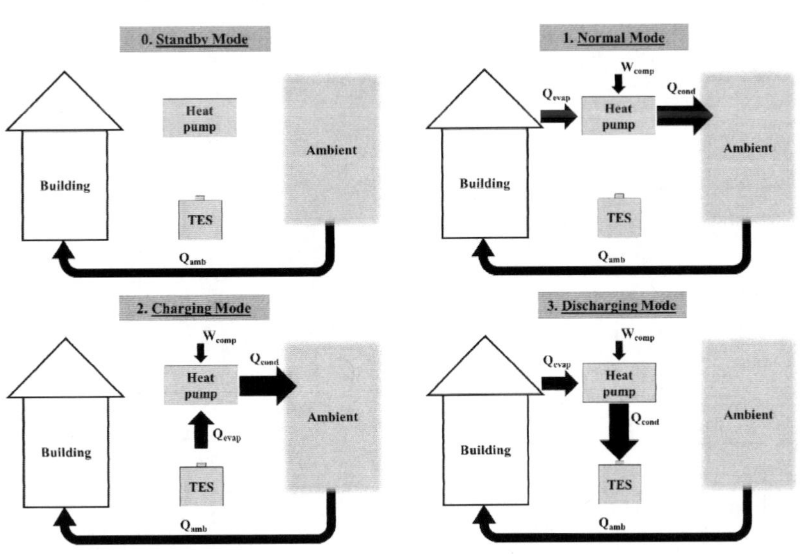

Figure 2. Energy flows in the four operating modes for summer (cooling only)

The operating mode of VCS-TES system is selected by the controls decision tree shown in Figure 3. The selection depends on the on-peak/off-peak time, the need to cool the building, and PCM state of charge (SOC). When the thermostat calls for cooling during off-peak times, the VCS will enter normal mode operation. Normal mode is conventional HVAC operation. Charging mode restores the energy stored in the TES by refreezing any melted PCM and is only activated during off-peak times when the thermostat is not calling for building cooling to take advantage of the lower electrical cost and, incidentally, usually lower ambient temperatures. Charging mode can be interrupted by the thermostat if the building requires cooling and the system will enter normal mode, maintaining building indoor comfort. During on-peak time, when the thermostat calls for cooling, discharging mode will activate which cools the building with the TES. During discharging, the PCM in the TES is being melted and the latent heat associated with melting maintains a constant cold TES temperature. The TES can only operate when there exists some frozen PCM inside. Once all the PCM is melted, the TES can no longer absorb isothermal heat and the normal mode must be used to cool the building. All other times the VCS is in standby mode.

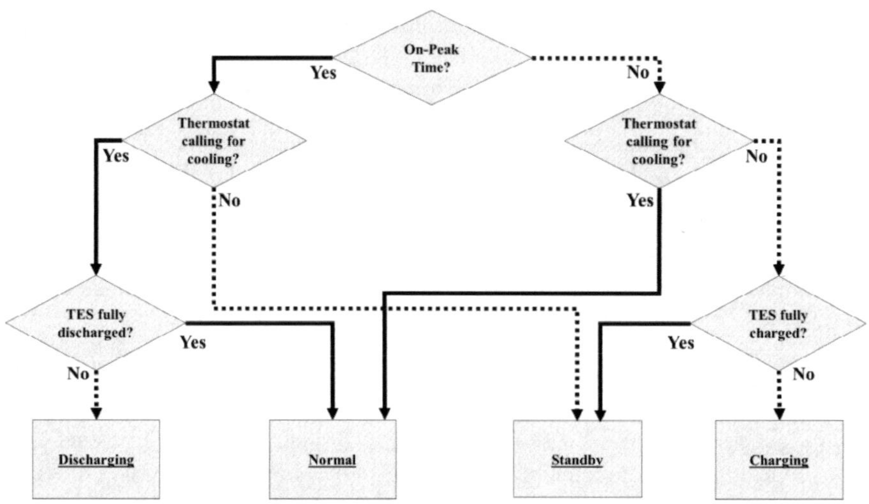

Figure 3. Controls decision tree

The overall flow of information in the model is depicted in Figure 4. The decision tree works as the main control to coordinate all these components. The feedback arrows represent the updated (i+1) values being fed back to the main models.

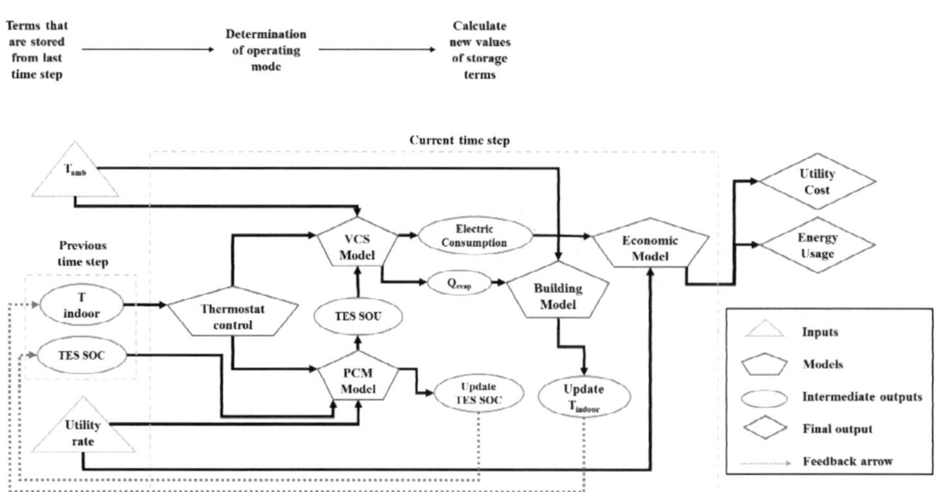

Figure 4. Total system information flow and relation between sub-models

3. COMPONENT MODELS

3.1 Weather Data

The system's performance is evaluated for a hot week in a dry climate of southern California, ASHRAE climate zone 3B, June 24-30, 1994. TMY 3 weather data for Fresno County, CA was used. The location correlates to the utility pricing. The dry bulb temperature is shown in Figure 5 with the indoor comfort temperature range shown for reference.

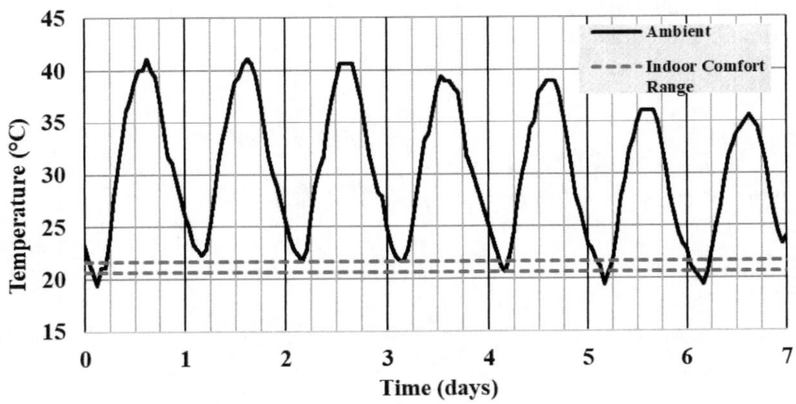

Figure 5. Ambient dry bulb temperature for the simulated week

3.2 Building Model

A simplified building model is used to simulate the thermal response of the building. The simplified model calculates the indoor temperature depending on the ambient heat load, VCS heat load, and the building thermal capacitance. The ambient heat load is modeled by using an assumed building balance point temperature of 18°C with the heat exchange from the ambient proportional to the difference between the outdoor ambient temperature and the point balance temperature, as shown in Equation (1). The effective heat transfer coefficient, U, was estimated to be 0.21 kW-K^{-1}.

$$Q_{amb} = U * (T_{amb} - T_{bal}) \tag{1}$$

The indoor temperature is determined based on the heat balance according to Equation (2) where Q_{VCS} is the cooling load provided from the VCS in either normal or discharging modes dependent on the controls in Figure 4. The time step, dt, in this simulation is one minute. An overall effective building thermal capacitance, C, of 16459 J/K was chosen to represent a typical single-family house with 223 m^2 (2400 ft^2) area (Dong *et al.*, 2019).

$$T_{indoor,i+1} = \frac{-(Q_{VCS} - Q_{amb}) * dt}{C} + T_{indoor,i} \tag{2}$$

3.3 Building Thermostat Model

The thermostat control has a constant setpoint of 21.1°C (70°F) and a deadband of ±0.5°C (±1.8°F). When the building temperature rises above the setpoint temperature plus the deadband, the VCS will be activated to cool the building in either normal or discharging mode dependent on the decision tree of Figure 3. The building will cool until the temperature falls below the setpoint temperature minus the deadband where the VCS will be shut off for direct building cooling.

3.4 VCS Model

An R410 VCS was modeled in Engineering Equation Solver. The discharging, charging, and normal modes were modeled independently, and lookup tables developed. For all modes the compressor has the same constant volumetric flow rate, 2.5e-6 m^3. In the normal and discharging modes, the VCS interfaces with the building with an evaporator at a near constant 15°C temperature. In the normal mode the condenser is coupled to the ambient, thus the ambient temperature, as shown in Figure 5, was used as an input. In discharging mode, the condenser is coupled to the TES, kept at a constant 0°C. Fans with a power draw of 0.585 kW are present only in the building air handling unit and the outdoor condensing unit and are activated when the operating mode dictates. The TES does not have a fan.

While in charging mode, the VCS evaporator interfaces with the TES and the condenser to the ambient. Thus, the TES heat exchanger must be able to be switched from a condensing mode to evaporating mode. For a real system of this type, controllable valves, check valves, switches, and other special components may be necessary for the TES system to make this switch. The simplicity of this model assumes these challenges can be met.

The work and heat for all modes was calculated and called upon by the building and thermostat models as determined by the decision tree of Figure 3 and the information flow of Figure 4. The values of work and heat are shown in Figure 6 plotted against the ambient temperature where applicable. The discharging mode has no external temperature dependence and thus its operating parameters are constant.

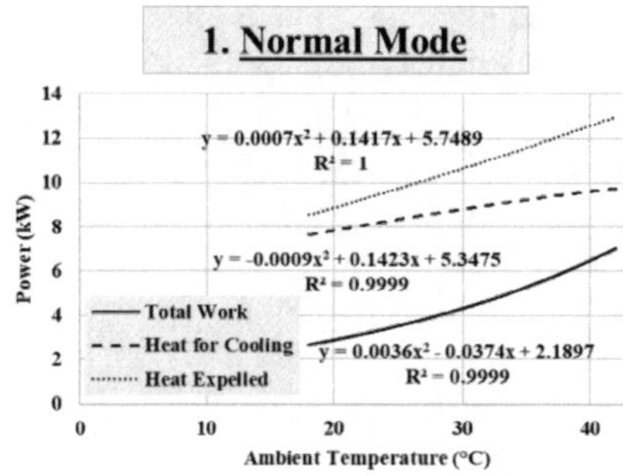

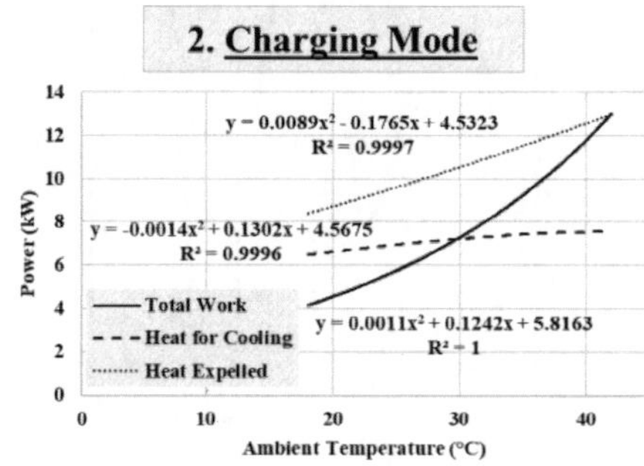

3. Discharging Mode

Total Work	**0.7485 kW**
Heat for Cooling	**5.181 kW**
Heat Expelled	**5.285 kW**

Figure 6. VCS operating power for normal, charging, and discharging modes

3.5 TES Model

The VCS is coupled to the TES system; the TES does not interact with the building directly and is insulated from ambient exposure. The PCM embedded in the TES is water/ice with a melting temperature of 0°C, in an assumed 80-gallon tank, which provides nearly 27.8 kWh of cooling capacity. The necessary heat exchanger in the TES is assumed to have near infinite heat transfer rate such that the temperature within the TES is constant at 0°C. The state of charge (SOC) is the percentage of maximum energy stored within the TES. When the TES is operated in the discharging or charging modes, the energy contained and SOC are reduced or increased, respectively, by the amount of heat determined by the VCS model.

This model only charges the TES between the hours of 1am and 7am. Different time controls or schedules may be utilized in real systems to avoid off-peak power demand spikes as predicted by Gottwalt *et al.*, (2011) or to take advantage of abundant energy on the grid from daytime solar or heavy wind periods, depending on the energy portfolio of the utility provider or any off-grid energy sources a customer may have.

3.6 Utility Rate Analysis

The TOU utility tariff used in this analysis is Pacific Gas and Electric (PG&E) Electric Schedule E-TOU-B (Pacific Gas and Electric, 2020). This is a legacy tariff as of May 1, 2020, no longer accepting new customers, but the pricing structure is representative of the difference between on-peak and off-peak prices for current tariffs in the region (California Public Utilities Commission, 2019). The tariff is priced as follows: from June 1 to Sep 30, peak hours are 4pm to 9pm, Monday through Friday with a rate of $0.39689/kWh; all other hours, including all weekend hours, are off peak rates of $0.29383/kWh. Winter rates (Oct 1 – May 31) differ.

4. RESULTS AND DISCUSSIONS

For the simulated 7-day period, the electric consumption of the HVAC system and the total utility cost are compared for systems with and without water/ice-based TES. Figure 7 shows the total electric consumption as well as the on-peak hour consumption. Note that only the first five days (weekdays) have on-peak pricing; weekends are considered off-peak. The baseline system without TES used 353.2 kWh total with 98.1 kWh used during on-peak hours: weekdays, 4pm – 9pm. With the TES system, the total electrical usage was reduced 14.5% to 302.1 kWh including an 87.5% reduction of on-peak usage to 12.3 kWh.

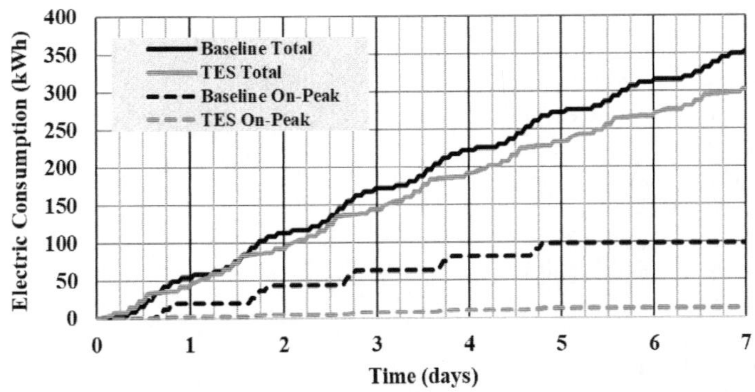

Figure7: Electric Consumption comparison with and without TES

The utility cost to cool the building is shown in Figure 8. The baseline model without the TES resulted in $112.52 but use of the TES system cost nearly 20% less at $90.04 for this simulated week. It should be noted that California has some of the highest utility costs in the US and this week simulated has particularly hot temperatures. These results should neither be extrapolated to yearly savings nor may be representative of other climates or regions. However, this does show that during extreme heat events, TES will be a crucial energy saving mechanism, reduce on-peak energy demand, and may be economically beneficial for the customer.

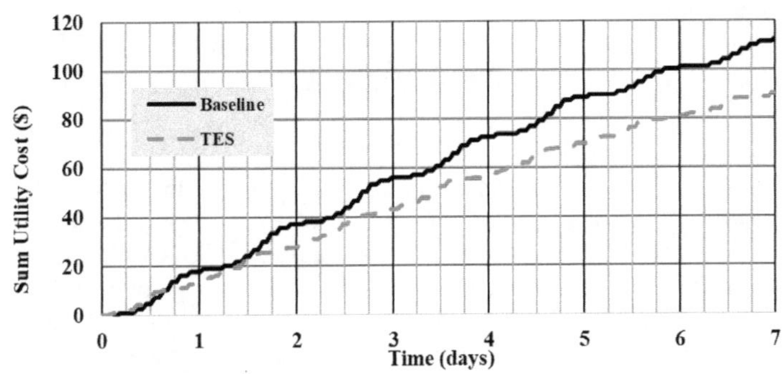

Figure 8: Total electrical utility cost with and without TES

5. CONCLUSIONS

A residential building was modeled with PCM-based TES and the economic and energy savings were estimated under a TOU rate structure with a rate difference of 0.103$/kWh between on-peak and off-peak. The TMY3 weather data was used for ASHRAE climate zone 3B. The analysis showed that the total energy used to cool the space was reduced from 353.2 kWh without TES to 302.1 kWh with TES, a 14.5% reduction. The on-peak energy usage was reduced by 87.5%, from 98.1 kWh without TES to 12.3 kWh with TES. In addition, the cost to the customer was reduced by nearly 20%, from $112.52 without TES to $90.04 with TES.

Since only 7 days were simulated, the results should not be extrapolated for annual savings, but highlights the potential for reducing peak energy demand during weather extremes. This work also suggests that the demand side management systems, including TES, with on time-based electricity prices may create an economic advantage for residential customers to shift peak energy loads without compromising comfort. TES are a valuable tool to level out electrical grid demand and add stability to the grid.

NOMENCLATURE

PCM	Phase Change Material	
TES	Thermal Energy Storage	
EPRI	Electric Power Research Institute	
EES	Engineering Equation Solver	
TOU	Time of Use	
AC	Air Conditioning	
AHU	Air Handling Unit	
VCS	Vapor Compression System	
COP	Coefficient of Performance	
SOC	State of Charge	
SOU	State of Use	
t	Time	
T	Temperature	(°C)
Q	Energy	(kW)
W	Electric Consumption	(kWh)
C	Building Thermal Capacitance	(J-°C^{-1})

Subscript/Superscripts

cond	Condenser
evap	Evaporator
comp	Compressor

REFERENCES

Akeiber, H., Nejat, P., Majid, M. Z. A., Wahid, M. A., Jomehzadeh, F., Zeynali Famileh, I., Calautit, J. K., Hughes, B. R., & Zaki, S. A. (2016). A review on phase change material (PCM) for sustainable passive cooling in building envelopes. *Renewable and Sustainable Energy Reviews*, *60*, 1470–1497.

Ali, M., Safdarian, A., & Lehtonen, M. (2015). Demand response potential of residential HVAC loads considering users preferences. *IEEE PES Innovative Smart Grid Technologies Conference Europe*, *2015-Janua*(January).

Arivazhagan, R., Geetha, N. B., Sivasamy, P., Kumaran, P., Kumara Gnanamithra, M., Sankar, S., Babu Loganathan, G., & Arivarasan, A. (2020). Review on performance assessment of phase change materials in buildings for thermal management through passive approach. *Materials Today: Proceedings*, *22*, 419–431.

Arteconi, A., Mugnini, A., & Polonara, F. (2019). Energy flexible buildings: A methodology for rating the flexibility performance of buildings with electric heating and cooling systems. *Applied Energy*, *251*(May), 113387.

Bruno, F., Tay, N. H. S., & Belusko, M. (2014). Minimising energy usage for domestic cooling with off-peak PCM storage. *Energy and Buildings*, *76*, 347–353. https://doi.org/10.1016/j.enbuild.2014.02.069

Chaiyat, N. (2015). Energy and economic analysis of a building air-conditioner with a phase change material (PCM). *Energy Conversion and Management*, *94*, 150–158. https://doi.org/10.1016/j.enconman.2015.01.068

Cui, B., Munk, J., Jackson, R., Fugate, D., & Starke, M. (2017). Building thermal model development of typical house in U.S. for virtual storage control of aggregated building loads based on limited available information. *30th International Conference on Efficiency, Cost, Optimization, Simulation and Environmental Impact of Energy Systems, ECOS 2017*.

De Gracia, A., & Cabeza, L. F. (2015). Phase change materials and thermal energy storage for buildings. *Energy and Buildings*, *103*, 414–419. https://doi.org/10.1016/j.enbuild.2015.06.007

Dincer, I., Dost, S., & Li, X. (1997). Thermal energy storage applications from an energy saving perspective. *International Journal of Global Energy Issues*, *9*(4), 351–364. https://doi.org/10.1504/IJGEI.1997.063347

Dong, J., Shen, B., Munk, J., Gluesenkamp, K. R., Laclair, T., & Kuruganti, T. (2019). Novel PCM Integration with Electrical Heat Pump for Demand Response. *IEEE Power and Energy Society General Meeting*.

Goetzler, B., Guernsey, M., & Kassuga, T. (2019). *Grid-interactive Efficient Buildings Technical Report Series: Heating, Ventilation, and Air Conditioning (HVAC); Water Heating; Appliances; and Refrigeration*.

Gottwalt, S., Ketter, W., Block, C., Collins, J., & Weinhardt, C. (2011). Demand side management-A simulation of household behavior under variable prices. *Energy Policy*, *39*(12), 8163–8174.

Hirmiz, R., Teamah, H. M., Lightstone, M. F., & Cotton, J. S. (2019). Performance of heat pump integrated phase change material thermal storage for electric load shifting in building demand side management. *Energy and Buildings*, *190*, 103–118. https://doi.org/10.1016/j.enbuild.2019.02.026

Jaber, S., & Ajib, S. (2012). Novel cooling unit using PCM for residential application. *International Journal of Refrigeration*. https://doi.org/10.1016/j.ijrefrig.2012.03.023

Junker, R. G., Azar, A. G., Lopes, R. A., Lindberg, K. B., Reynders, G., Relan, R., & Madsen, H. (2018). Characterizing the energy flexibility of buildings and districts. *Applied Energy*, *225*(April), 175–182.

Koebrich, S., Chen, E., Bowen, T., Forrester, S., & Tian, T. (2017). 2017 Renewable Energy Data Book Including Data and Trends for Energy Storage and Electric Vehicles. In *U.S. Department of Energy*.

Liu, M., Saman, W., & Bruno, F. (2012). Review on storage materials and thermal performance enhancement techniques for high temperature phase change thermal storage systems. In *Renewable and Sustainable Energy Reviews* (Vol. 16, Issue 4, pp. 2118–2132). Pergamon. https://doi.org/10.1016/j.rser.2012.01.020

Mosaffa, A. H., & Garousi Farshi, L. (2016). Exergoeconomic and environmental analyses of an air conditioning system using thermal energy storage. *Applied Energy*, *162*, 515–526.

Newsham, G. R., & Bowker, B. G. (2010). The effect of utility time-varying pricing and load control strategies on residential summer peak electricity use: A review. *Energy Policy*, *38*(7), 3289–3296.

Pop, O. G., Fechete Tutunaru, L., Bode, F., Abrudan, A. C., & Balan, M. C. (2018). Energy efficiency of PCM integrated in fresh air cooling systems in different climatic conditions. *Applied Energy*, *212*, 976–996.

Real, A., García, V., Domenech, L., Renau, J., Montés, N., & Sánchez, F. (2014). Improvement of a heat pump based HVAC system with PCM thermal storage for cold accumulation and heat dissipation. *Energy and Buildings*, *83*, 108–116. https://doi.org/10.1016/j.enbuild.2014.04.029

Sarbu, I., & Sebarchievici, C. (2018). A comprehensive review of thermal energy storage. *Sustainability (Switzerland)*, *10*(1). https://doi.org/10.3390/su10010191

Schibuola, L., Scarpa, M., & Tambani, C. (2015). Demand response management by means of heat pumps controlled via real time pricing. *Energy and Buildings*, *90*, 15–28.

Sehar, F., Rahman, S., & Pipattanasomporn, M. (2012). Impacts of ice storage on electrical energy consumptions in office buildings. *Energy and Buildings*, *51*, 255–262. https://doi.org/10.1016/j.enbuild.2012.05.002

Soares, N., Costa, J. J., Gaspar, A. R., & Santos, P. (2013). Review of passive PCM latent heat thermal energy storage systems towards buildings' energy efficiency. *Energy and Buildings*, *59*, 82–103.

Sonnick, S., Erlbeck, L., Gaedtke, M., Wunder, F., Mayer, C., Krause, M. J., Nirschl, H., & Rädle, M. (2020). Passive room conditioning using phase change materials—Demonstration of a long-term real size experiment. *International Journal of Energy Research*. https://doi.org/10.1002/er.5406

U.S. Department of Energy. (2011). 2011 Buildings Energy Data Book. In *Office of Energy Efficiency and Renewable Energy*.

U.S. EIA. (2020). *Annual Energy Outlook 2020*. https://www.eia.gov/outlooks/aeo/

Waqas, A., Ji, J., Ali, M., & Alvi, J. Z. (2018). Effectiveness of the phase change material-based thermal energy storage integrated with the conventional cooling systems of the buildings – A review. *Proceedings of the Institution of Mechanical Engineers, Part A: Journal of Power and Energy*, *232*(6), 735–766.

Yoon, J. H., Baldick, R., & Novoselac, A. (2016). Demand response control of residential HVAC loads based on dynamic electricity prices and economic analysis. *Science and Technology for the Built Environment*, *22*(6), 705–719. https://doi.org/10.1080/23744731.2016.1195659

Yoon, J. H., Bladick, R., & Novoselac, A. (2014). Demand response for residential buildings based on dynamic price of electricity. *Energy and Buildings*, *80*, 531–541. https://doi.org/10.1016/j.enbuild.2014.05.002

Zhu, N., Wang, S., Ma, Z., & Sun, Y. (2011). Energy performance and optimal control of air-conditioned buildings with envelopes enhanced by phase change materials. *Energy Conversion and Management*, *52*(10), 3197–3205. https://doi.org/10.1016/j.enconman.2011.05.011

ACKNOWLEDGEMENT

This material is based upon work supported by the U. S. Department of Energy's Building Technologies Office under Contract No. DE-AC05-00OR22725 with UT-Battelle, LLC. The authors would like to acknowledge Mr. Sven Mumme, Technology Manager, U.S. Department of Energy Building Technologies Office.

Novel Dishwasher with Thermal Storage and Thermoelectric Heat Recovery

Navin KUMAR[1], Kyle R. GLUESENKAMP[1*], Joseph RENDALL[1], Viral K. PATEL[1], Anthony GEHL[1], Ahmad ABUHEIBA[1], Tugba TURNAOGLU[1], Guolian, WU[2], Raveendran VAIDHYANATHAN[2]

[1] Multifunctional Equipment Integration Group, Oak Ridge National Laboratory[†],
Oak Ridge, TN, U.S.A.
865-241-2952, gluesenkampk@ornl.gov

[2] Samsung Electronics America
Pine Brook, NJ, U.S.A.
201-956-7709, guolian.wu@sea.samsung.com

* Corresponding Author

ABSTRACT

Residential dishwashers typically consume domestic hot water, heat it further with electric resistance heating elements, and drain the soiled heated water before each subsequent water fill. During the final rinse, the water is heated to a temperature of approximately 54.5–57.2°C (130–135°F) to heat the load and promote passive drying after the final drain event. In this work, the energy consumption, water consumption, and drying performance of a conventional dishwasher were measured under test conditions similar to U.S. energy efficiency test standards but with an unsoiled load. These measurements were considered baseline performance metrics. The dishwasher was then experimentally modified to recover heat from the drain water utilizing thermoelectric (TE) heat pump modules and a thermal storage component. The TEs were also used during the drying phase to improve the drying of the load. The novel dishwasher was operated in the laboratory under the same conditions as the baseline unit, and its energy consumption, water consumption, and drying performance were measured. The results demonstrated a 14.5% reduction in energy consumption, with the same amount of water consumption, and improved drying by 60% compared to the baseline.

1. INTRODUCTION

The residential sector represents around 20% of global energy consumption and 18% of U.S. energy consumption (International Energy Agency, 2019). Among all residential energy consumption units (cooking, lighting, space cooling, space heating, water heating, and appliances), residential appliances consume 27% of the overall residential energy. Growing energy demand worldwide urges researchers and scientists to improve the energy efficiency of home appliances. Today in the U.S., 65% of U.S. households, representing approximately 85.8 million households, (Consortium for Energy Efficiency, 2018) "use dishwashers that account for 3.2% of the 2005 residential primary energy use" (Bansal *et al.,* 2011). At present, the energy efficiency of dishwashers could be promoted by applying advanced control units that senses load size or soil level, improving food filters, reducing overall water consumption, using high-efficiency motors, increasing cabinet insulation, or improving the dry cycle (Bansal *et al.,* 2011). However, most of the dishwashers in the U.S. use an electric element to heat the water. Approximately 80% to 90% of electricity during a dishwasher cycle is consumed during the heating in the washing and rinsing steps (Bengtsson *et al.*, 2015). Despite the efforts to improve the energy performance of household dishwashers, new technologies that lead to the lowest possible amount of energy consumption while maintaining or improving cleaning performance are still necessary. Therefore, in this study, our goal is to reduce the energy consumption for water heating using an innovative

---

[†] Notice: This manuscript has been authored by UT-Battelle, L.L.C. under Contract No. DE-AC05-00OR22725 with the U.S. Department of Energy. The United States Government retains and the publisher, by accepting the article for publication, acknowledges that the United States Government retains a non-exclusive, paid-up, irrevocable, worldwide license to publish or reproduce the published form of this manuscript, or allow others to do so, for United States Government purposes. The Department of Energy will provide public access to these results of federally sponsored research in accordance with the D.O.E. Public Access Plan (http://energy.gov/downloads/doe-public-access-plan).

thermoelectric (TE) heat pump system with thermal storage. Thermoelectric devices are semiconductor materials connected electrically in a series where the temperature gradient across hot and cold sides generates electricity (Morelli, 2017; Shittu *et al.,* 2020). If a DC current is applied at low voltage to a TE device, heat is transferred from one side of TE to the other side, resulting in one face of the TE device being cooled, and the opposite face being heated (Enescu & Virjoghe, 2014). Therefore, TE devices can be used as heat pumps. Here, we replaced the electric heater in the commercially available dishwasher with TE elements to heat the recirculated water in a dishwasher cycle, leading to an improved dishwasher efficiency.

2. EXPERIMENTAL SETUP

A commercially available residential dishwasher (DW) unit was modified to be heated and cooled by a thermoelectric heat pump with heat recovery. A rectanglular tube in tube heat exchanger was constructured for increasing the area for condensation. The internal inter tube was a mini-channel and the outer tube was aluminumn wrapped around the microchannel. The outer wrap was exended to increase the surface area of condensation, as illustrated in Figure 1. Two IHXs were mounted on the back wall of the DW tub to exchange heat between the tub and TEHP. During heating, the water in the tub flows through the mini-channels in the IHX, exchanging heat with the TEHP working fluid on the outer tube (Figure 1, right image). In the cooling cycle during drying, the external fins act as a condensing surface and are cooled by the TEHP working fluid on the outer tube.

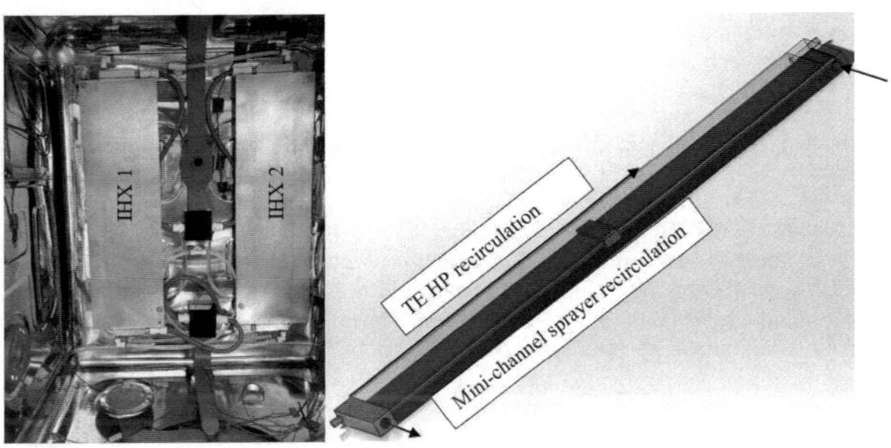

Figure 1. Two internal heat exchangers as installed on the back of the tub (left), CAD visualization of internal heat exchangers (right)

The thermoelectric heat pump (TEHP) unit assembly was constructed by sandwiching five thermoelectric modules between aluminum spacer blocks and two mini channel heat exchangers, as illustrated in Figure 2. The heat was pumped between the storage, thermoelectric, and the tub through the two mini channel heat exchangers with two external positive displacement pumps. The heat transfer fluid (water) flowed through the mini channels at a constant flow rate of 0.05 kg/s.

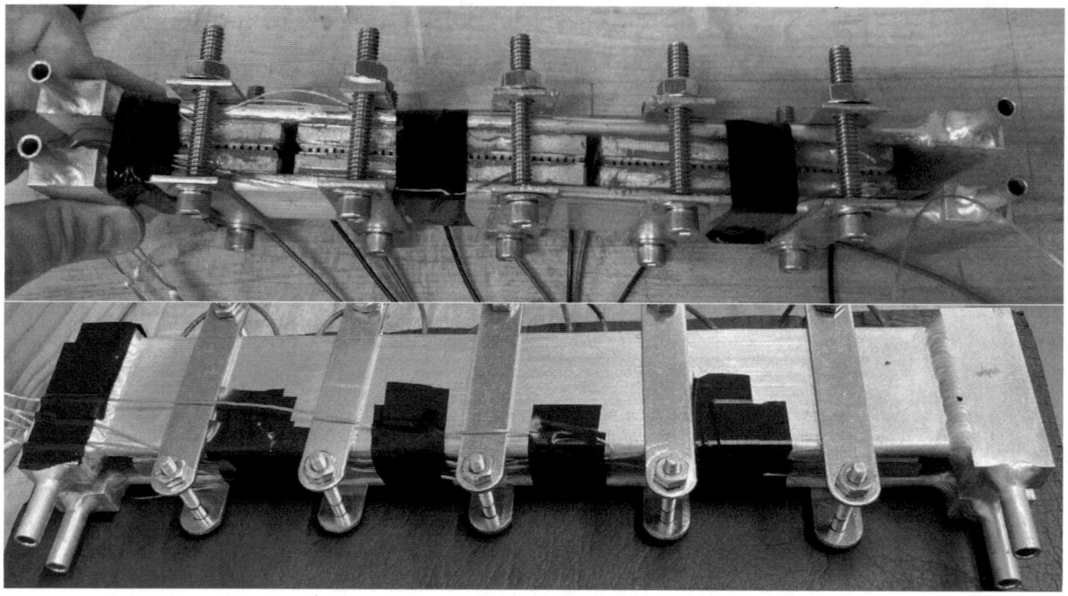

Figure 2: The sandwiched construction of TEHP with thermoelectric, spacer block, and two mini channel heat exchangers.

The piping and instrumentation diagram (P&ID) of the TEHP DW unit is shown in Figure 3, and was designed according to Gluesenkamp (Gluesenkamp, 2018). The instruments used for measuring various quantities and their uncertainties are summarized in Table 1. The tub water was circulated through the thermal energy storage (TES) tank for heat recovery before drainage in the recovery cycle. In the heating cycle, the heat was recovered from the TES tank through the cold side of the TEHP. The mini channel on the hot side of the TEHP pumps heat to the DW tub for water heating during the wash and rinse cycle. In the cooling cycle, the electrical polarity was reversed to provide cooling to the DW tub; the heat was pumped into the TES tank. Both heating and drying were accomplished with the same set of TEs by reversing polarity to the TEs. This study experimentally measured TEHP DW's performance using two voltage (35 and 40 VDC) conditions under an adaption of the DOE standard test conditions (US Department of Energy, 2019) without soiling or detergent.

The drying performance of DW was evaluated by measuring the remaining water content on the bottom and top rack *in-situ*. The dishwasher unit was modified to balance the top and bottom racks on four load-cells each.

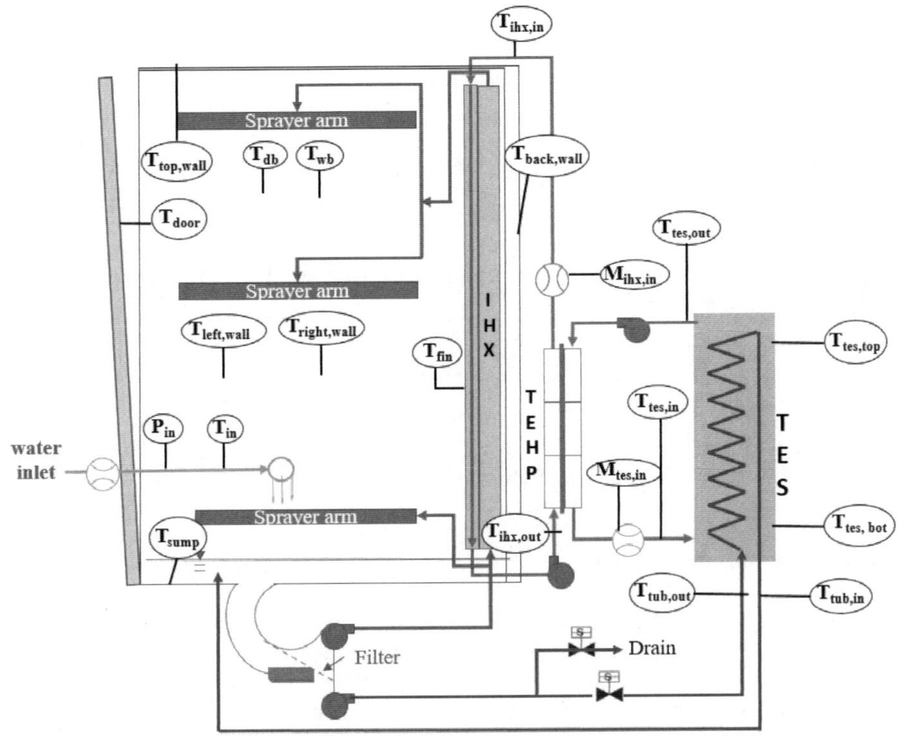

Figure 3: The P&ID diagram for the TEHP dishwasher unit with energy recovery. T is a temperature sensor. M is the mass flowmeter sensor. P is a pressure sensor.

Table 1: Instruments and their uncertainties.

Parameter	Instrument	Instrument Uncertainty
Temperature	T-Type thermocouple	0.5°C
TEHP temperature	RTD 1/10 DIN PT 100 ohms	0.1°C
Mass flow rate (TEHP)	Bronkhorst M-15 Coriolis flowmeter	± 0.5 %
Air relative humidity	Vaisala HMT330	± 1.0 %RH (0 – 40 %RH) ± 1.7 %RH (40 – 95 %RH)
Air dry bulb temperature	RTD built into Vaisala HMT330	0.1°C
Domestic water flow rate	Omega FTB-101 turbine flow meter with FLSC-C3-LIQ	±0.5% of flow reading
Power consumption	OSI AGW-001D Watt transducer with metering class current transformer (150:5) with nine turns of primary wire	±0.24% of power reading
Water pressure	Omega PX409-100GI pressure transducer	±0.08% best straight line
Weight	Futek LSB 210 IP67 load cell	±0.25% Full Scale

The TEHP dishwasher's COP was computed by measuring the electric power and internal heat exchanger's inlet water mass flow rate ($M_{ihx,in}$) and inlet water temperature ($T_{ihx,in}$), and outlet water temperature ($T_{ihx,out}$) over the heating period, as shown in Eqn. 1.

$$COP = \frac{\sum M_{ihx,in} * C_{p,water} * \left(T_{ihx,in} - T_{ihx,out}\right)}{\sum P_{electric}} \tag{1}$$

The drying performance was also measured for the prototype and baseline operation. The weight was measured *in situ* by four load cells on the top and bottom racks.

3. RESULTS

The machine energy use performance of the dishwasher is summarized in Figure 4. The energy use of the DW is segregated into the heated wash, heated rinse, and others (pumps and fans). The total machine energy use of the unit was 511 Wh/cycle under a normal load test. The heated wash cycle phase consumed 191 Wh/cycle of energy to heat the water from 30.9°C (87.6 °F) to 45.6°C (114°F). The heated rinse cycle phase consumed 203 Wh/cycle of the energy to heat the water from 43.3°C (110°F) to 55°C (131 °F). The energy consumed by others was 117 Wh/cycle.

Water heating energy use represents the energy associated with supplying hot water to the dishwasher. Water was provided to the DW at 48.9°C (120°F). Therefore, it is assumed that the water delivered to the unit was heated from 10-48.9°C (50 to 120 °F). This energy use is directly proportional to the amount of water consumed. The baseline and the TEHP unit consumed 11.1L (2.94 gals) of water (Figure 5). The unit consumed a total of 495 Wh for water heating.

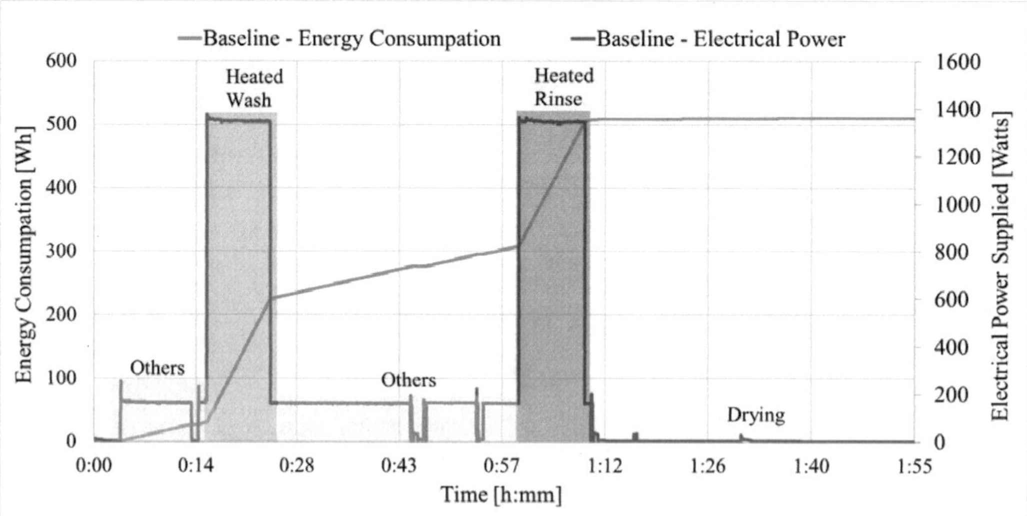

Figure 4: The DW unit's baseline energy consumption (accumulated energy consumption) under D.O.E.'s normal load cycle. The normal load cycle constitutes of heated wash at 45.6°C (114°F), heated rinse at 55°C (131°F), and unheated drying.

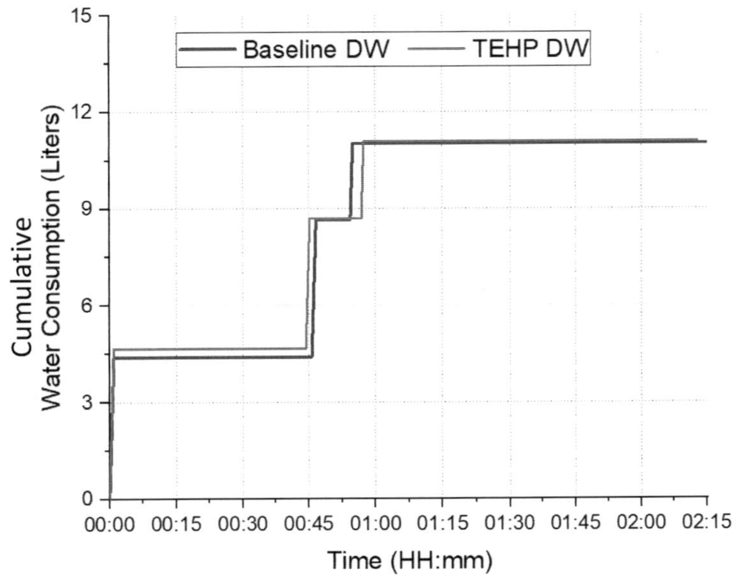

Figure 5: The baseline and TEHP unit water consumption.

The electrical power consumed by the TEHP unit is as summarized in Figure 6. The energy consumption of the TEHP unit was evaluated at voltage conditions of 40 and 35 VDC. When the TEHP was operated at 40 VDC, the unit consumed a total of 388.5 Wh/cycle, of which 187 Wh/cycle was used for the heated wash, 173.5 Wh/cycle for the heated rinse, and 28 Wh/cycle by other operations. When the TEHP was operated at 35 VDC, the unit consumed 369.5 Wh/cycle, of which 172 Wh/cycle was used for the heated wash, 169 Wh/cycle for the heated rinse, and 28.5 Wh/cycle by other operations. Reducing the TEHP operating voltage from 40 VDC to 35 VDC increased the DW cycle time by 18 minutes compared to the baseline unit.

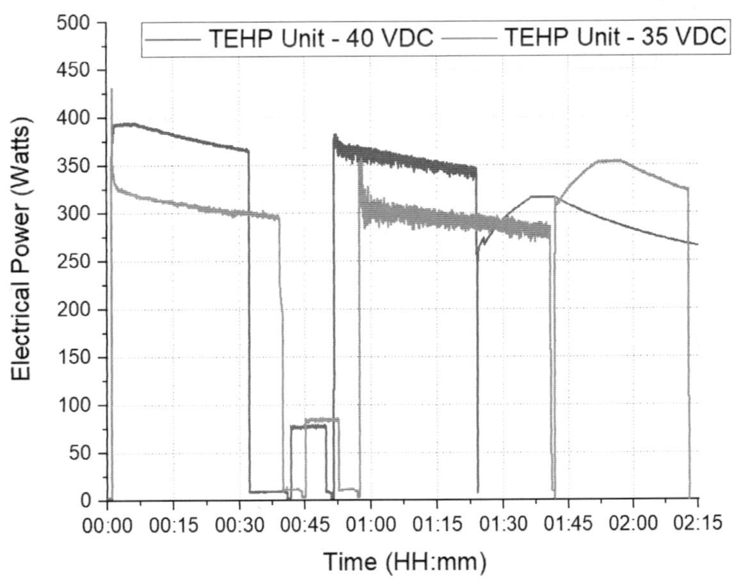

Figure 6: The TEHP unit's energy consumption under a normal load cycle. The TEHP runs at lower power and for a longer time than baseline shown in Figure 4.

The comparison of the energy performance of the TEHP unit and the baseline are summarized in Table 2. The maximum energy savings of 14.5% was achieved when the TEHP unit was operated at 35 VDC with a system-level heating C.O.P. of 1.17. During the recovery phase, 257 kJ of heat was recovered and raised the TES tank temperature to 90.2°F.

Table 2: Summary of performance of the TEHP unit and the baseline

	Heating Energy [Wh]	Drying Energy [Wh]	Heating COP	Heating Energy Savings [%]	Water consumption [L] / [gal]	Remaining water Content at 20 minutes of elapsed drying time [wt% of water]
TEHP Unit 40 VDC	885	112	1.14	12.0	11.1 (2.93)	2.8
TEHP Unit 35 VDC	860	120	1.17	14.5	11.05 (2.92)	3.1
Baseline	1006	-	1.00	-	11.02 (2.91)	4.8

The drying performance of the baseline and TEHP unit is as shown in Figure 7. The weight of the load was measured *in situ* by placing the top rack on four waterproof load sensors, and the bottom rock on four waterproof load sensors. The liquid water on the dishes could therefore be computed as the *in situ* wet weight of the load minus the dry weight of the load. The remaining water content is expressed in Figure 7 as a percentage, by comparing the water weight to the dry weight of the load. The TEHP unit consumed 120 $Wh_{electric}$ for 20 minutes of drying. In contrast, the baseline unit consumed zero energy during the drying phase. The drying performance of the DW was evaluated under two different conditions: (1) door prop , and (2) no door prop (the door prop was a feature of the DW model in which the door automatically opened slightly during the drying phase of operation). The initial drop in the first two minutes in the experiment was because of a small vibration caused by the insertion of the RH probe into the dishwasher. In the door prop case, the door was released open at the end of the 20th minute. The drying performance of the TEHP unit was enhanced by more than 60% with the additional cooling provided by the TEHP compared to the baseline unit.

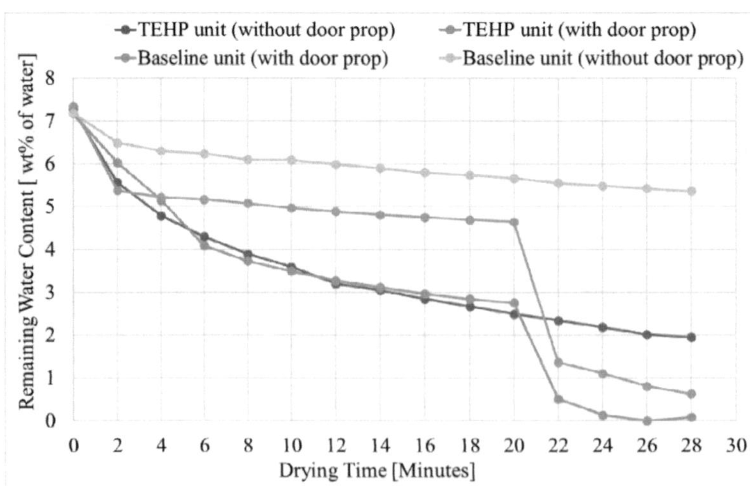

Figure 7: The drying performance of the baseline unit without cooling and the TEHP unit operated at 40 VDC with cooling.

4. CONCLUSION

In this work, the energy performance, water consumption, and drying performance of a conventional dishwasher were improved by modifying a dishwasher to recover heat utilizing a thermoelectric heat pump (TEHP) and a thermal energy storage tank. The results demonstrated a reduction in heating energy consumption and improved drying performance compared to the unmodified DW with the same amount of water consumption.

The TEHP with heat recovery reduced the heating energy consumption by 14.5 % and extended the cycle by 18 minutes. The heat recovery and the ability of the TEHP to be operated above a heating C.O.P. of 1 aided the reduction in heating energy consumption. On the other hand, in the baseline unit, the heating C.O.P. of a resistance unit is 1. During the drying phase, TEHP consumed 120 Wh of energy and improved the drying by more than 60 %. Overall, the TEHP unit with heat recovery consumed 2.5% less energy and provided a 60 % improvement in drying compared to the baseline unit. The system's energy consumption can be reduced further by improving heat recovery performance and changing the thermal energy storage configuration.

NOMENCLATURE

The nomenclature should be located at the end of the text using the following format:

COP	Coefficient of performance	
T	Temperature	(°F/°C)
M	Mass flow rate	(kg/s)
P	Pressure	(psi)
Subscript		
ihx	Inlet heat exchanger	
tes	Thermal energy storage	
in	inlet	
out	outlet	
tub	Dishwasher Tub	
fin	Inlet heat exchanger wall	
db	Dry Bulb	
wb	Wet bulb	

REFERENCES

Bansal, P., Vineyard, E., & Abdelaziz, O. (2011). Advances in household appliances- A review. Applied Thermal Engineering, 31(17–18), 3748–3760. https://doi.org/10.1016/j.applthermaleng.2011.07.023

Bengtsson, P., Berghel, J., & Renström, R. (2015). A household dishwasher heated by a heat pump system using an energy storage unit with water as the heat source. International Journal of Refrigeration, 49, 19–27. https://doi.org/10.1016/j.ijrefrig.2014.10.012

Consortium for Energy Efficiency. (2018). Super Efficient Home Appliances Initiative.

Enescu, D., & Virjoghe, E. O. (2014). A review on thermoelectric cooling parameters and performance. Renewable and Sustainable Energy Reviews, 38, 903–916. https://doi.org/10.1016/j.rser.2014.07.045

Gluesenkamp, K. . (2018). Configuration of dishwasher to improve energy efficiency of water heating (Patent No. 9,949,611). US Patent.

International Energy Agency. (2019). Energy Efficiency Indicators. https://doi.org/10.1787/9789264268692-en

Morelli, D. T. (2017). Thermoelectric materials. In Encyclopedia of Inorganic and Bioinorganic Chemistry (pp. 1–34). John Wiley & Sons, Inc. https://doi.org/10.1007/978-3-319-48933-9_57

Shittu, S., Li, G., Zhao, X., & Ma, X. (2020). Review of thermoelectric geometry and structure optimization for performance enhancement. Applied Energy, 268(May), 115075. https://doi.org/10.1016/j.apenergy.2020.115075

US Department of Energy. (2019). 10 e-CFR 430, Energy Conservation Program for Consumer Products; Subpart B, Test Procedures; Appendix E, Uniform Test Method for Measuring the Energy Consumption of Water Heaters.

ACKNOWLEDGEMENT

This work was sponsored by the U. S. Department of Energy's Building Technologies Office under Contract No: DE-AC05-00OR22725 with UT-Battelle, L.L.C. The authors would like to acknowledge Mr. Antonio Bouza, Technology Manager – HVAC&R, Water Heating, and Appliance, U.S. Department of Energy Building Technologies Office.

Assessment of Metal-Hydride Energy Storage Coupled with Heat Pumps and Solar PV for Residential Cooling and Heating

Patrick KRANE[1], Davide ZIVIANI[1], James E BRAUN[1], Neera JAIN[1], Amy MARCONNET[1,*]

[1]School of Mechanical Engineering, Purdue University,
West Lafayette, IN 47907
Contact Information (marconnet@purdue.edu)

* Corresponding Author

ABSTRACT

The extension of time-of-use (TOU) rates to residential buildings has incentivized the use of thermal energy storage (TES) to shift heating and cooling loads away from on-peak hours. At the same time, the growing use of solar photovoltaics (PV) for on-site energy generation with variable buy back rates also encourages the use of TES to store energy generated at hours of peak solar radiation. Metal hydrides are reversible chemical compounds that form when certain metals react with hydrogen. Since these reactions depend on both temperature and pressure, metal hydrides can be used for thermochemical energy storage across a range of temperatures. Because of this, metal hydrides can be used as storage for both heating and cooling, enabling year-round utilization. In this paper, we present and evaluate a system that combines metal hydride energy storage with heat pumps and on-site PV. The system is designed to reduce operating costs by discharging the energy storage system during on-peak hours when electricity is expensive and charging the system during off-peak periods when electricity is less expensive and there is less incentive to sell excess PV electricity generation back to the grid. The TOU rates considered are designed for a future scenario with high solar penetration, where greater solar generation in the afternoon leads to on-peak periods in the mornings and evenings rather than the afternoon. The payback period of the energy storage system is estimated using energy savings relative to a baseline of a solar home without an energy storage system along with estimates of additional installed costs for the storage system. Payback period calculations for a range of rates show that rates with on-peak demand charges result in larger cost savings than rates with high on-peak rates, but even for the most favorable rates, the payback period is too long for the system to be economically viable.

1. INTRODUCTION

In recent years, it has become more common for time-of-use (TOU) utility rate structures to be available as an option for residential buildings. Time-of-use rates are used to incentivize consumers to use less electricity at times when demand for electricity from the grid is highest (on-peak hours). One way of doing this is by using energy storage to shift some of the building's heating or cooling load from on-peak hours to off-peak hours. This involves charging the storage system during off-peak hours and discharging during on-peak hours. In commercial buildings, where TOU rate structures have been in use for a longer time, ice storage is often used for this purpose. In this case, a chiller is used to freeze ice during off-peak hours and then the ice is melted during on-peak hours to deliver a portion of the cooling load (Hasnain, 1998; Henze, 2003; Nassif *et al.*, 2013; Sanaye & Shirazi, 2013; Tam *et al.*, 2019). Ice storage has been studied for use in residential buildings. However, it was found to have too large a payback period to be economic in most cases with best economic potential in locations with long cooling seasons (Tam *et al.*, 2018).

The cost savings from energy storage could potentially be increased if an energy storage system were capable of being used year-round for both heating and cooling (as compared to ice storage, which is only used for cooling). One way to accomplish this is to use metal hydrides for thermochemical energy storage. Metal hydrides are metals that exothermically adsorb hydrogen if the hydrogen pressure is greater than the equilibrium pressure for the adsorption reaction, and endothermically desorb hydrogen if the pressure is below equilibrium. The equilibrium pressures for these reactions are a function of the temperature and the hydrogen absorbed by the metal hydride (Voskuilen *et al.*, 2014). Because of this, the temperature at which the reaction occurs can be altered by adjusting the hydrogen pressure around the hydride. This makes metal hydride suitable for use in an energy storage system designed for both heating and cooling.

Previous work on using metal hydrides for energy storage has primarily focused on storage for concentrated solar power, due to the potential of certain metal hydrides for use in high-temperature applications (Corgnale *et al.*, 2016; Feng *et al.*, 2018; Nyamsi *et al.*, 2018; Sheppard *et al.*, 2016). A common system design for energy storage is to have two metal hydride reactors, one of which is used to store and release energy through the adsorption and desorption reactions, while the other is used to store the hydrogen released by the other reactor and release it back to it (Corgnale *et al.*, 2016; Nyamsi *et al.*, 2018; Sheppard *et al.*, 2016; Ward *et al.*, 2016).

The increasing use of solar photovoltaics (PV) for on-site energy generation provides additional benefits for the use of energy storage in load shifting for both cooling and heating. Available solar power in the afternoon significantly reduces demand from the grid during on-peak hours (VanGeet *et al.*, 2008). With sufficient solar penetration, peak electricity demand from the grid would no longer occur in the afternoon but in the morning and evening (Janko *et al.*, 2016). However, there is a significant mismatch between peak solar generation (around noon) and peak power demand (around 5 PM). A study of solar power in Wisconsin found that this results in diminishing returns for solar penetration once it accounts for 15-20% of power generation, but that solar penetration could be significantly increased if the demand curve were shifted to be closer to the solar generation curve (Myers *et al.*, 2010). Energy storage could be used to accomplish this, and the shifting of peak demand from the grid to the morning and evening would incentivize utility companies to charge TOU rates that would encourage shifting loads away from these times.

Rate structures for locations that have significant on-site generation typically include an avoided cost rate (ACR), which is a credit given for times when on-site generation exceeds consumption. The ACR is generally significantly lower than the rate charged for electricity usage (Leger *et al.*, 2014), which provides an incentive for using on-site PV to charge a storage system when generation exceeds consumption. Previous studies have adjusted discharge rates for storage during on-peak hours based on whether solar generation exceeds demand (Abji *et al.*, 2015). The current paper also considers a load-shifting strategy that adjusts the charging rate based on available solar power in combination with the use of metal hydride storage for both cooling and heating with a heat pump.

In this paper, we model a two-reactor metal hydride energy storage system coupled with on-site solar PV generation and examine its performance in a residential building for a full year of operation. We estimate the operating cost savings and payback period of this system using utility rate structures designed for a location with increased solar penetration.

2. METHODOLOGY

2.1 System Model

A diagram of the model used for evaluating the performance of the metal hydride energy storage system and determining its annual operating cost can be seen in Figure 1. This model combines a static model of the heat pump and secondary loops with a dynamic model of the metal hydride reactors. A schematic of the system is shown in Figure 1. A heat pump is used to provide the heating and cooling loads to the house, with a water glycol loop used to deliver this load from the heat pump to the house. This heat pump is connected to two water glycol loops: an *indoor loop*, which exchanges heat with an indoor unit in the house as well as the heat pump, and an *outdoor loop*, which exchanges heat with an external heat exchanger as well as the heat pump. A 4-way valve is used to reverse which loop the heat pump transfers heat to and which it removes heat from. Each water glycol loop can also be used to exchange heat with either of the metal hydride reactors. In cooling mode, the low-temperature reactor is connected to the indoor loop, and the high-temperature reactor to the outdoor loop. The system is charged by using the heat transferred from the heat pump to the outdoor loop to heat the high-temperature reactor, causing the metal hydride there to desorb hydrogen, which then flows to the low-temperature reactor. This reactor absorbs hydrogen, releasing heat into the indoor loop. The system is discharged by sending water glycol from the outdoor loop to the hydride reactor after the water glycol has been cooled by the external unit instead of after it has been heated by the heat pump. This lower temperature lowers the equilibrium pressure in the reactor, resulting in hydrogen flow from the low-temperature reactor to the high-temperature reactor, which results in the low-temperature reactor desorbing hydrogen that is absorbed by the high-temperature reactor. These reactions result in the high-temperature reactor releasing heat while the low-temperature reactor absorbs heat from the water glycol in the indoor loop, thus cooling it and providing a portion of the cooling load for the house. In both charging and discharging, if the change in temperature is not enough to drive the reaction and hydrogen flow, a compressor is used to drive hydrogen flow between the reactors. In heating mode, the same reactions occur in the reactors in charging and discharging, but which reactor is connected to which water glycol loop changes. Thus, the high-temperature reactor is connected to the indoor loop and absorbs heat from it while charging and releases heat to it while discharging.

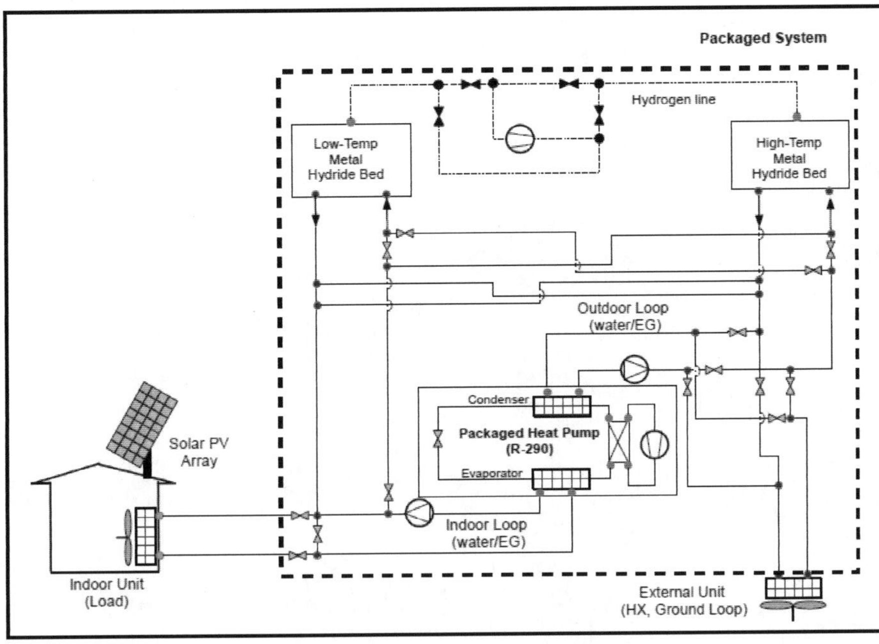

Figure 1: Schematic of the system for residential heating and cooling, with energy storage and on-site solar PV, including solar PV array, packaged heat pump, secondary loops, house, external unit, hydride reactors, and hydrogen line between the reactors. The condenser and evaporator in the packaged heat pump are labelled as in cooling mode; in heating mode they are reversed.

A model of this system, which is used to determine the cost savings from using metal hydride energy storage, is described in Krane *et al.* (2021). The system modelled in this paper differs from that system due to the addition of solar PV to the system. The model includes a sub-model that calculates the power generated by the solar PV based on available solar radiation data, described in Section 2.2. In addition, the load-shifting control logic used in the static model is modified in order to make maximum use of available solar power, as described in Section 2.3.

2.2 Solar Generation Model

In order to incorporate solar power generation into the model, it is necessary to calculate the solar power generation from the solar radiation data for a particular time. The solar power generation, $\dot{W}_{solar}$, is calculated from

$$\dot{W}_{solar} = \eta_{PV} A_{PV} I_T. \tag{1}$$

The efficiency of each solar panel, η_{PV}, is assumed to be 20% based on values for existing models, and the surface area, A_{PV}, of the solar array is determined as described section 3.1. The incident radiation on the solar panel, I_T, is calculated using the following equation (Tina *et al.*, 2006).

$$I_T = \left(R_b + \frac{1-\cos\beta}{2}\rho\right) GHI + \left(\frac{1+\cos\beta}{2} - R_b\right) DHI. \tag{2}$$

Incident radiation is calculated in terms of the global horizontal incidence, GHI (the solar radiation incident on a horizontal surface), and the diffuse horizontal incidence, DHI. The term for GHI also includes a term for radiation reflected from the ground, calculated in terms of the reflectance of the ground, ρ. A value of 0.2 is assumed for the ground reflectance. The slope of the solar panel, β, is 30°. The ratio of beam radiation on the surface to horizontal radiation, R_b, is calculated for south-facing solar panels at the given slope using equations given in Duffie, Beckman, and McGowan (2013). Hourly temperature data for the location is used for GHI and DHI.

2.3 Load-shifting Control Logic

The control logic used to determine whether the energy storage system will be charged or discharged, and what the load on the storage system will be, is shown in Figure 2. At the beginning of each day, the model calculates the average loads on the heat pump for charging and discharging by assuming that the system will be fully charged during off-peak hours and fully discharged during on-peak hours. Thus, the heat pump load for charging during off-peak periods is calculated by adding the total load required by the house during this period to the load required to charge the storage system and averaging this total load over the full time period to find the load on the heat pump throughout the charging people. The heat pump load for the on-period is calculated similarly, except that the load is reduced by the storage capacity of the storage system instead of being increased by it.

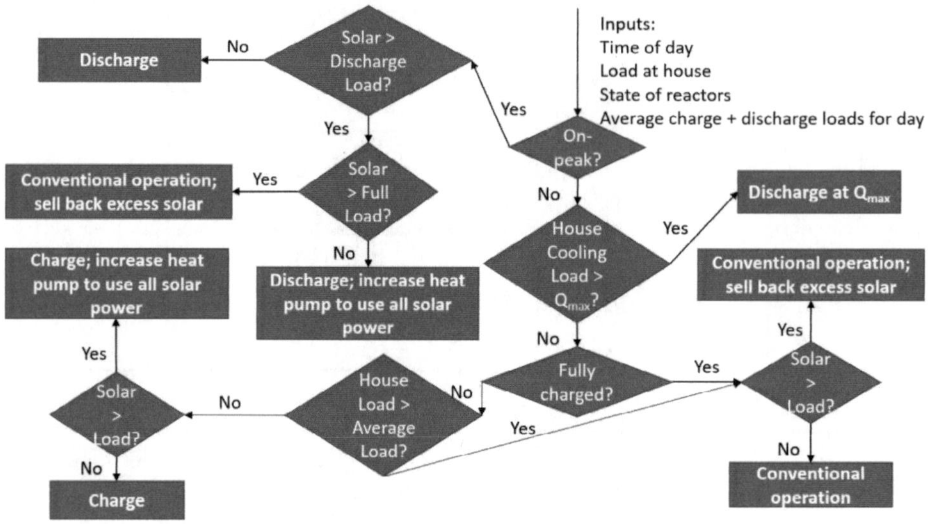

Figure 2: Diagram of the control logic used to modify the load at the heat pump based on available solar power.

At each time step, the control logic first checks if this time is an on-peak or off-peak hour for the TOU rate structure. If it is on-peak, it next compares the available solar power to the average discharge load to see if this load should be used or if the load on the heat pump should be higher in order to make use of excess solar generation. If the system requires electricity from the grid to run the heat pump at the average load, it will run at the average load. But if there is excess solar power available, the heat pump load will be increased to use all of this solar power (and the load on the energy storage system will be decreased by the same amount). If there is enough solar power to provide the entire heating or cooling load without using energy storage, then the system will do so, running in conventional mode with no discharging or charging of the reactors.

During on-peak hours, the system is still discharged in cooling mode if this is necessary due to the load at the house exceeding the capacity of the heat pump (in heating mode, auxiliary heating is used to make up the difference, and not energy storage). If this is not the case, the energy storage system will be charged if it is not already fully charged and if the required heating or cooling load at the house does not exceed the average heat pump load for charging. When charging, the heat pump will run at the average load calculated for charging, unless doing so would not use all available solar power. In that case, the load on the heat pump is increased (and thus, a greater heating or cooling load is delivered to charge the hydride reactors) to either use all available solar power or, if the load that would use all available solar power exceeds the capacity of the heat pump, run the heat pump at maximum capacity. If the system is not charging, due to already being fully charged or the required load at the house being too large, then the system will run conventionally (no charging or discharging of the reactors), with any excess solar power being sold back to the grid.

2.4 Utility Rate Structures

Significant solar penetration is likely to lead to peak demand for electricity from the grid occurring in the morning and evening instead of the afternoon. The utility rate structures that are used to examine the performance of the metal hydride storage system have on-peak periods selected based on this. Therefore, three different on-peak schedules are examined, as shown in Table 1: one with on-peak rates in the morning, one with on-peak rates in the evening, and one with on-peak rates in both morning and evening.

Table 1: On-peak periods examined as the basis for TOU utility rate structures. All on-peak periods are only on weekdays (Monday-Friday); rates are always off-peak on weekends.

	Morning	Evening	Morning and Evening
On-Peak Period	6-11 AM	7-11 PM	6-11 AM & 7-10 PM

For each of these three on-peak periods, a series of rate structures are analyzed. Different potential rate structures are evaluated by varying the ratio of the on-peak to off-peak rates to see how strongly the rates must incentivize load-shifting for the system to be economical. Moreover, each rate structure is compared to an equivalent structure that uses an on-peak demand charge instead of a higher on-peak utility rate. In both cases, all rate structures are defined so that they result in the same annual operating cost for a system with no energy storage (but with solar power generation). Any difference in operating costs between two such rate structures is therefore due to the effect of the energy storage system.

All rate structures studied were defined to have the same total annual operating cost as a conventional system (with solar PV and no energy storage) with a flat utility rate of $0.12/kWh. The rate structures defined in this way for the case with morning and evening peak periods are shown in Table 2. Note that for the cases with only morning or only evening peak periods, the rate structures will have different on-peak and off-peak rates for the same on-peak to off-peak ratio, since achieving the same annual operating cost for the conventional system requires different rates depending on the length of the on-peak period.

Table 2: Rate structures used for the case with morning and evening on-peak periods. Rate structures with a high on-peak rate are defined by the on-peak to off-peak rate ratio. Each rate structure with a flat rate and an on-peak demand charge is defined to be equivalent to a structure with a high on-peak rate by having the same percentage of the operating costs for a conventional system occur in the on-peak period and having the flat rate equal the off-peak rate for the structure with a high on-peak rate.

	High On-Peak Rates					On-Peak Demand Charge			
		Summer		Winter		Summer		Winter	
% Conv. Cost On-Peak	**On-peak/ Off-peak Ratio**	**On-peak Rate (/kWh)**	**Off-peak Rate (/kWh)**	**On-peak Rate (/kWh)**	**Off-peak Rate (/kWh)**	**Flat Rate (/kWh)**	**On-peak Demand Charge (/kW)**	**Flat Rate (/kWh)**	**On-peak Demand Charge (/kW)**
35.5	1	$0.1200	$0.1200	$0.1200	$0.1200	$0.1200	$0.00	$0.1200	$0.00
52.3	2	$0.1729	$0.0864	$0.1791	$0.0895	$0.0864	$6.48	$0.0895	$5.90
68.7	4	$0.2217	$0.0554	$0.2376	$0.0594	$0.0554	$12.47	$0.0594	$11.74
73.3	5	$0.2350	$0.0470	$0.2542	$0.0508	$0.0470	$14.10	$0.0508	$13.40
76.7	6	$0.2447	$0.0408	$0.2666	$0.0444	$0.0408	$15.30	$0.0444	$14.64
81.4	8	$0.2582	$0.0323	$0.2840	$0.0355	$0.0323	$16.95	$0.0355	$16.37
84.6	10	$0.2669	$0.0267	$0.2955	$0.0296	$0.0267	$18.02	$0.0296	$17.52
89.1	15	$0.2796	$0.0186	$0.3125	$0.0208	$0.0186	$19.58	$0.0208	$19.21

There are two sets of rate structures shown in Table 2; one set uses TOU rates with a higher rate during the on-peak period, while the other uses a flat rate with an on-peak demand charge. Rate structures in the first set are defined by the ratio of the on-peak rate to the off-peak rate. As this ratio increases, a higher percentage of the operating cost for the conventional system occurs during the on-peak period and there is thus a greater opportunity for cost savings with energy storage. Each rate structure with an on-peak demand charge is defined so that it has the same percentage of the total operating cost occur during the on-peak period as one of the structures with TOU rates, and also uses the off-peak rate from that structure as its flat rate. By comparing these two rates with the same percentage of the conventional operating cost during on-peak hours, we can see which of on-peak demand charges and high on-peak rates results in larger cost savings from the use of energy storage.

3. RESULTS

3.1 Component Sizing and Initial Cost Calculations

The conventional system consists of a heat pump, used to deliver both heating and cooling loads to the house, and auxiliary electric heating used to supplement the heat pump if the heating load at the house exceeds its capacity. The system with energy storage adds contains both of these components, but also adds two secondary loops with pumps and internal and external air-handling units (the indoor and outdoor loops), two metal hydride reactors, and a compressor used to move hydrogen between the reactors. $LaNi_5$ is used as the metal hydride in one reactor, and $MmNi_{4.5}Cr_{0.5}$ in the other. The initial cost of each system is the sum of the costs of all of its components.

An estimated cost of $10/kW capacity is used for the auxiliary electric heating by examining the prices and maximum capacities of commercially-available units. For the residential heat pump considered in this case study, the cost, y_{hp}, is estimated as a function of the rated capacity (Q_{rated}) and SEER, based on a correlation developed by the Energy Efficiency and Renewable Energy Office (Energy Efficiency and Renewable Energy Office, 2016):

$$y_{hp} = 1.91 * 10^3 - 41.4SEER_{rated} + 7.48SEER_{rated}^{\ 2} + 589Q_{rated} - 16.4Q_{rated}^{\ 2} \tag{3}$$

The most expensive component in the energy storage system is the metal hydride reactors. The reactor cost is determined by estimating the material costs for the hydrides, since this is much higher than the installation or material costs for the heat exchanger. An estimated price of $20/kg is used for $LaNi_5$, based on examining existing products. Since similar information was not available for $MmNi_{4.5}Cr_{0.5}$, a cost of $24/kg is estimated based on comparing raw material prices for lanthanum and mischmetal ("Argus Rare Earths Monthly Outlook," 2017). Both of these costs are reduced by 25% to account for the gains from recycling them at the end of their life cycle, so the actual values used are $15/kg for $LaNi_5$, and $18/kg for $NdNi_{4.8}Sn_{0.2}$. The total cost of piping for an indoor secondary loop for a typical residence, including pump and air-handling unit, was estimated by Tam *et al.* to be $1301 (2018). Based on a cost breakdown for a secondary-loop system, the cost of the outdoor secondary loop, where the indoor air-handling system is replaced with an outdoor heat exchanger, is estimated to be $1050. Since hydrogen compressors are usually used at much larger pressure ratios than the 10-bar limit imposed in this case, the cost of the compressor is estimated by comparing prices for air compressors with the appropriate operating pressures, resulting in an estimate value of $200.

Table 3: Cost comparison between the conventional HVAC system and the system with metal hydride energy storage. Both systems also include solar PV used for power generation, which is used by the HVAC system as well as other electrical systems in the house.

System	Initial Cost						
	Heat Pump	**Electric Heating**	**Hydrogen Compressor**	**Secondary Loops**	**Metal Hydrides**	**Solar PV**	**Total**
Conventional	$5096	$100	-	-	-	$12,600	$17,796
Energy Storage	$5096	$120	$200	$2351	$14,610	$12,600	$34,977

A 3-ton heat pump is assumed for both the conventional system and the system with energy storage. This heat pump is sized to meet the maximum cooling load; it is not downsized for the system with storage because doing so would significantly increase operating costs due to a greater reliance on auxiliary heating in the winter. The hydride reactors are sized to have sufficient storage capacity to flatten the cooling load for the heat pump on the hottest day of the year. To ensure that the control logic for using excess solar power generation would be used, the solar panels are sized to a summer design day such that peak generation on that day exceeds power consumption at that time. This results in a total surface area of 21 m^2 for the solar panels for the case study considered. A price of $600/$m^2$ is used to calculate the total price of the solar panels, based on costs for existing systems. It should be noted that since the heat pump and solar panels are the same size for the systems with and without storage, their costs have no effect on the difference in cost between the systems (and therefore no effect on the payback period).

3.2 System Behavior on an Example Day

To show how the control logic for using energy storage for load shifting and making the best use of available solar power works on a particular day, the model is run using temperature and solar irradiation data for June 23 from the Typical Meteorological Year 3 (TMY3) data for Elizabeth City, NC (NREL, n.d.). The hourly non-HVAC electricity loads are estimated using data from the Office of Energy Efficiency & Renewable Energy (EERE) on residential hourly electricity loads (EERE, n.d.). We selected a city in the American South as the object of this study because it is the only region in the United States where electric heating, as is used in our system, is more commonly used than gas heating (Comstock, 2014).

The results for this example day can be seen in Figure 3 . The storage system is partially discharged during each on-peak period, as seen in Figure 3b. While discharging, the heat pump is run at a low load, as seen in Figure 3a. However, for the last hour of the first on-peak period, the storage system is not used even though there is a cooling load larger than the load the heat pump had previously run at during this on-peak period. This is because there is enough solar power available at this time that all of the cooling load can be met with the heat pump without using any electricity from the grid. Since this is the case, the storage system is not used at this time to reduce the load on the heat pump.

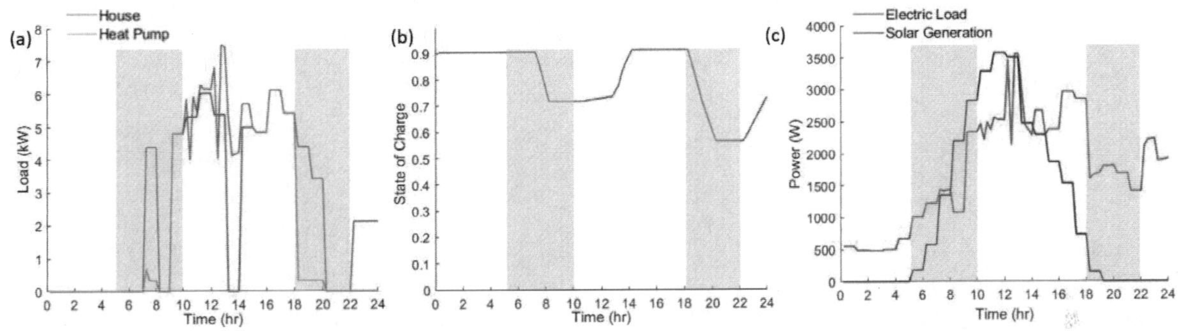

Figure 3: (a) Cooling loads at the house and heat pump, (b) reactor state of charge, and (c) total electric load for the house compared to solar power generation for an example summer day in Elizabeth City, NC. On-peak periods are highlighted in grey.

If no excess solar power were available, the system would not be charged in between the morning and evening on-peak periods, since the heat pump loads for discharging are calculated so that the system will be fully discharged by the combination of the two periods, and then charged during off-peak hours at night, when the cooling load is lower. However, in this case, solar power generation during the afternoon on-peak period exceeds the power needed at the house, as seen in Figure 3c, so the excess solar power is sued to charge the storage system. While charging, the heat pump load exceeds the cooling load at the house, and the excess is used to charge the storage system. However, at two points during the afternoon the heat pump load goes below the cooling load at the house. This is because the system is adjusting the temperatures in the metal hydride reactors in order to keep them within the range of temperatures the system is designed to operate in. When this happens, some of the cooling load is being provided by the hydride reactors, because the fluid in the indoor loop is being used to heat the hydride reactor connected to it. The system is also charged during the night after the end of the evening on-peak period. This charging is done at a constant heat pump load since it is using electricity from the grid and thus not trying to adjust the load in order to only use solar power.

3.3 Cost Savings and Payback Period

Using the system sized as described in Section 3.1, the model is run for a full year for a house in Elizabeth City, NC. TMY3 weather data (NREL, n.d.) and EERE data on non-HVAC building loads (EERE, n.d.) for the full year are used as described in the previous section. The annual operating cost calculated for the conventional system (with solar PV but no energy storage) is $1422.18. This cost is the same for all rate structures since these are defined to give the same annual operating cost if no storage is used. The majority of the operating cost is due to electrical loads other than the HVAC system, which has an operating cost of $622.33 for the flat rate of $0.12/kWh. The operating cost savings for all the rate structures considered are shown in Figure 4. Cost savings are plotted against the percentage of the conventional system operating costs that occur during the on-peak period. The higher this value is, the larger the incentive to use energy storage. For each of the three on-peak periods considered (morning, evening, and morning and evening) rates with a high on-peak rate are compared to rates with a constant electricity rate and an on-peak demand charge.

As illustrated in Figure 4, the cost savings increase with rate structures where conventional systems have increasing fraction of the costs during the on-peak period. Further, these cost savings are greater for rate structures with an on-peak demand charge than rate structures with only high on-peak rates. Both trends are also observed regardless of the designated on-peak times (morning, evening, or both). The largest cost savings occur for rates with a morning on-peak period, and the smallest cost savings for rates with morning and evening on-peak periods, with the maximum savings for the rates examined being less than half the maximum for rates with morning on-peak periods.

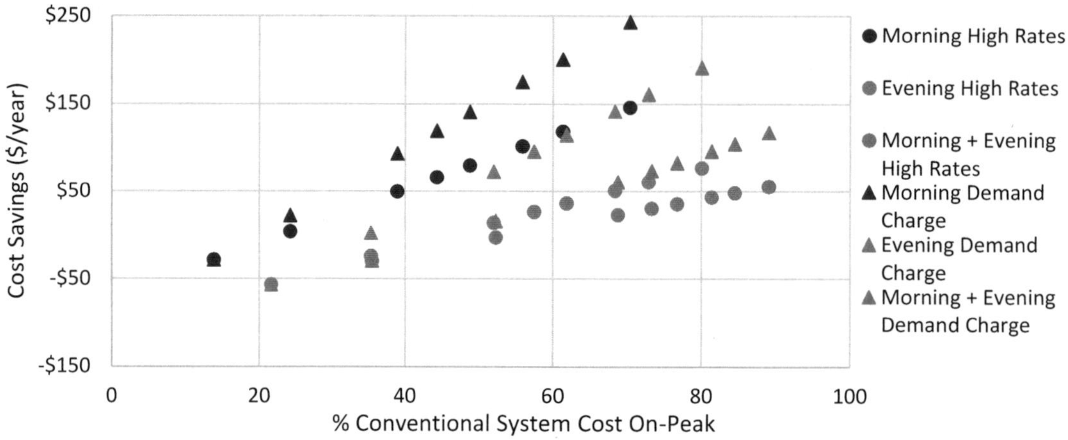

Figure 4: Annual cost savings for all TOU rates structures considered in this study, plotted as a function of the percentage of the conventional system operating cost that occurs during the on-peak period. Rates with high on-peak rates are compared to rates with an on-peak demand charge for rate structures with three different on-peak times (morning, evening, and morning and evening). Within each of these categories, a series of rates are considered to see how cost savings increase as a larger portion of the operating costs occur during the on-peak period due to higher on-peak rates or demand charges. The leftmost point in each series is a flat rate structure with no demand charge.

The higher cost savings for rate structures with demand charges show that the storage system is more successful in reducing the peak power during the on-peak period than it is at reducing the total on-peak energy consumption. This success in reducing peak power is due to the use of a constant heat pump load when discharging (except when a larger load could be used using only solar power), which means that the most energy storage will be used during on-peak hours with peak loads, resulting in a greater reduction in peak power consumption than in total on-peak power consumption. As expected, cost savings increase as the ratio of on-peak to off-peak rates increases, since increasing this ratio increases the cost incentive to shift loads to off-peak. Without any difference between on-peak and off-peak rates, the use of energy storage leads to an increased operating cost, since the total electrical consumption increases due to inefficiencies associated with charging and discharging storage and the use of an additional hydrogen compressor. These penalties outweigh any benefits associated with increased operation of the heat pump at more favorable outdoor temperatures.

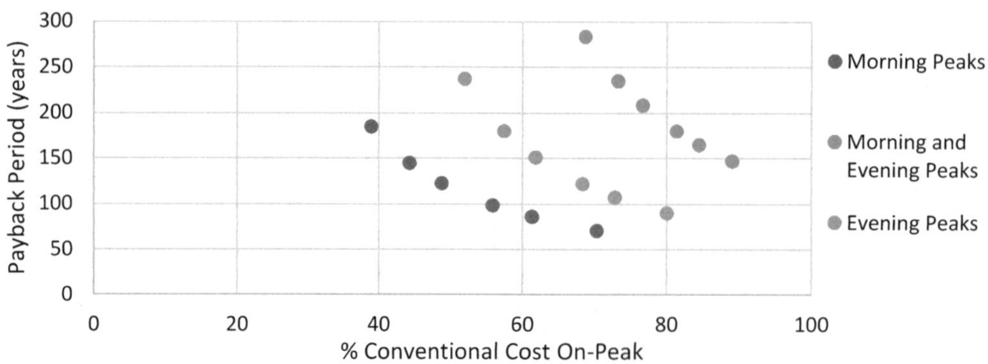

Figure 5: Payback periods calculated based on cost savings from different rate structure, plotted as a function of the percentage of the operating cost that occurs during the on-peak period for a conventional system. Payback period is calculated for the rate with an equivalent demand charge rather than rates with a higher on-peak rate, due to higher cost savings from rates with demand charges.

Estimated payback periods for the system with storage are presented in Figure 5 as function of the utility rates. While the system provides significant cost savings, the resulting payback periods are extremely long and not economically viable, even for the most favorable rate structures. These high payback periods are primarily due to the extremely high cost of the metal hydrides. The increased cost from the two secondary loops is also a significant factor; only at the most favorable rates are the cost savings large enough to pay off the cost of these loops in less than ten years.

4. CONCLUSION

The availability of on-site solar PV increases the benefits associated with energy storage for heating and cooling in residential applications that have time-varying electricity rates. This combination leads to greater utilization of on-site generated electricity. Increased penetration of solar PV will likely improve the benefits as on-peak periods shift and change TOU rate structures to discourage selling of excess electricity back to the grid. Metal hydrides are a potential method for energy storage that can be used for both heating and cooling, thus enabling year-round utilization. This paper investigated the combination of metal hydride storage coupled to a heat pump with onsite solar PV generation as a system that could meet both heating and cooling loads while reducing operating costs compared to a system with just solar PV. The performance of this system was investigated assuming utility rates that might be representative of a scenario with high levels of solar penetration on the grid. However, the cost savings are not large enough to result in a viable payback period even with very favorable rate structures. Rates with morning on-peak periods resulted in the largest cost savings, followed by rates with evening on-peak periods; both resulted in larger savings than rates with on-peak periods in both the morning and the evening. For all on-peak periods studied, rate structures with on-peak demand charges resulted in larger cost savings than rate structures where the electricity rate was higher during the on-peak period.

REFERENCES

Abji, N., Tizghadam, A., & Leon-Garcia, A. (2015). Energy storage management in core networks with renewable energy in time-of-use pricing environments. *IEEE International Conference on Communications*, *2015-Septe*, 135–141. https://doi.org/10.1109/ICC.2015.7248311

Argus Rare Earths Monthly Outlook. (2017). *Argus Consulting Services*, (17–9).

Comstock, O. (2014). Everywhere but Northeast, fewer homes choose natural gas as heating fuel. Retrieved from Today in Energy website: https://www.eia.gov/todayinenergy/detail.php?id=18131

Corgnale, C., Hardy, B., Motyka, T., & Zidan, R. (2016). Metal hydride based thermal energy storage system requirements for high performance concentrating solar power plants. *International Journal of Hydrogen Energy*, *41*(44), 20217–20230. https://doi.org/10.1016/j.ijhydene.2016.09.108

Duffie, J. A., Beckman, W. A., & McGowan, J. (2013). Solar Engineering of Thermal Processes. In *American Journal of Physics* (Vol. 53). https://doi.org/10.1119/1.14178

EERE. (n.d.). Commercial and Residential Hourly Load Profiles for all TMY3 Locations in the United States. Retrieved from OpenEI website: https://openei.org/doe-opendata/dataset/commercial-and-residential-hourly-load-profiles-for-all-tmy3-locations-in-the-united-states

Energy Efficiency and Renewable Energy Office. (2016). Technical support document: Energy efficiency program for consumer products: Residential central air conditioners, heat pumps, and furnaces. *US Department of Energy*, (December).

Feng, P., Zhu, L., Zhang, Y., Yang, F., Wu, Z., & Zhang, Z. (2018). Optimum output temperature setting and an improved bed structure of metal hydride hydrogen storage reactor for thermal energy storage. *International Journal of Hydrogen Energy*, 1–13. https://doi.org/10.1016/j.ijhydene.2018.04.220

Hasnain, S. M. (1998). Review on sustainable thermal energy storage technologies, part II: Cool thermal storage. *Energy Conversion and Management*, *39*(11), 1139–1153. https://doi.org/10.1016/S0196-8904(98)00024-7

Henze, G. P. (2003). An Overview of Optimal Control for Central Cooling Plants with Ice Thermal Energy Storage.

Journal of Solar Energy Engineering, *125*(3), 302–309. https://doi.org/10.1115/1.1591801

Janko, S. A., Arnold, M. R., & Johnson, N. G. (2016). Implications of high-penetration renewables for ratepayers and utilities in the residential solar photovoltaic (PV) market. *Applied Energy*, *180*, 37–51. https://doi.org/10.1016/j.apenergy.2016.07.041

Krane, P., Ziviani, D., Marconnet, A., Braun, J., & Jain, N. (2021). Metal-Hydride Energy Storage to Enable Load-Shifting for Heat Pumps in Heating and Cooling Modes for Buildings. *To Be Submitted to Energy & Buildings*.

Leger, A. S., Sobiesk, E., Farmer, A., & Rulison, B. (2014). Demand response with photovoltaic energy source and time-of-use pricing. *Proceedings of the IEEE Power Engineering Society Transmission and Distribution Conference*, 1–5. https://doi.org/10.1109/tdc.2014.6863304

Myers, K. S., Klein, S. A., & Reindl, D. T. (2010). Assessment of high penetration of solar photovoltaics in Wisconsin. *Energy Policy*, *38*(11), 7338–7345. https://doi.org/10.1016/j.enpol.2010.08.008

Nassif, N., Tesiero, R. C., & Singh, H. (2013). Impact of Ice Thermal Storage on Cooling Energy Cost for Commercial HVAC Systems. *ASHRAE Transactions*, *119*.

NREL. (n.d.). 1991-2005 Update: Typical Meteorlogican Year 3. Retrieved from National Solar Radiation Database website: https://rredc.nrel.gov/solar/old_data/nsrdb/1991-2005/tmy3/

Nyamsi, S. N., Lototskyy, M., & Tolj, I. (2018). Selection of metal hydrides-based thermal energy storage: Energy storage efficiency and density targets. *International Journal of Hydrogen Energy*, *43*(50), 22568–22583. https://doi.org/10.1016/j.ijhydene.2018.10.100

Sanaye, S., & Shirazi, A. (2013). Thermo-economic optimization of an ice thermal energy storage system for air-conditioning applications. *Energy and Buildings*, *60*, 100–109. https://doi.org/10.1016/j.enbuild.2012.12.040

Sheppard, D. A., Paskevicius, M., Humphries, T. D., Felderhoff, M., Capurso, G., Bellosta von Colbe, J., … Buckley, C. E. (2016). Metal hydrides for concentrating solar thermal power energy storage. *Applied Physics A: Materials Science and Processing*, *122*(4). https://doi.org/10.1007/s00339-016-9825-0

Tam, A., Braun, J. E., Ziviani, D., & Jain, N. (2018). An Overall Assessment of Ice Storage Systems for Residential Buildings. *5th International High Performance Buildings Conference at Purdue*, (July).

Tam, A., Ziviani, D., Braun, J. E., & Jain, N. (2019). Development and evaluation of a generalized rule-based control strategy for residential ice storage systems. *Energy & Buildings*, *197*, 99–111. https://doi.org/10.1016/j.enbuild.2019.05.040

Tina, G., Gagliano, S., & Raiti, S. (2006). Hybrid solar/wind power system probabilistic modelling for long-term performance assessment. *Solar Energy*, *80*(5), 578–588. https://doi.org/10.1016/j.solener.2005.03.013

VanGeet, O., Brown, E., Blair, T., & McAllister, A. (2008). Solar San Diego: The impact of binomial rate structures on real PV-systems. *American Solar Energy Society - SOLAR 2008, Including Proc. of 37th ASES Annual Conf., 33rd National Passive Solar Conf., 3rd Renewable Energy Policy and Marketing Conf.: Catch the Clean Energy Wave*, *6*, 3819–3844.

Voskuilen, T. G., Waters, E. L., & Pourpoint, T. L. (2014). A comprehensive approach for alloy selection in metal hydride thermal systems. *International Journal of Hydrogen Energy*, *39*(25), 13240–13254. https://doi.org/10.1016/j.ijhydene.2014.06.119

Ward, P. A., Corgnale, C., Teprovich, J. A., Motyka, T., Hardy, B., Sheppard, D., … Zidan, R. (2016). Technical challenges and future direction for high-efficiency metal hydride thermal energy storage systems. *Applied Physics A: Materials Science and Processing*, *122*(4). https://doi.org/10.1007/s00339-016-9909-x

Review of Low-Cost Organic and Inorganic Phase Change Materials with Phase Change Temperature between 0°C and 65°C

Jason HIRSCHEY[1], Navin KUMAR[2], Tugba TURNAOGLU[2], Kyle R. GLUESENKAMP[2*], Samuel GRAHAM[1]

[1]Georgia Institute of Technology, GWW School of Mechanical Engineering
Atlanta, GA, USA
jhirschey13@gatech.edu

[2] Oak Ridge National Laboratory, Building Equipment Group
Oak Ridge, TN, USA

ABSTRACT

Phase change materials (PCMs) that undergo a phase transition may be used to provide a nearly isothermal latent heat storage at the phase change temperature. This work reports the energy storage material cost ($/kWh) of various PCMs with phase change between 0 – 65°C. Four PCM classes are analyzed for their potential use in building systems: 1) inorganic salt hydrates, 2) organic fatty acids, 3) organic fatty alcohols, and 4) organic paraffin waxes. Many salt hydrates have low material costs (0.09 – 2.53 $/kg), high latent heat of fusion (100 – 290 J/g), and high densities (1.3 – 2.6 g/cm^3), leading to favorable volumetric storage density and low energy storage costs, 50 – 130 kWh/m^3 and 0.90 – 40 $/kWh, respectively. Some salts are notably more expensive due to their scarcity or pressures from competing industries such as lithium-based salts. Fatty acids have the lowest energy storage cost in the temperature range 8 – 17°C at 6.50 – 40 $/kWh. Despite favorable latent heat (125 – 250 J/g) their low density gives (0.9 g/cm^3) gives poor volumetric storage capacity, 32 – 80 kWh/m^3. Fatty alcohols generally have high material costs 2.50 – 200 $/kg which leads to high energy storage costs, 40 – 3000 $/kWh. With latent heat and density similar to fatty acids, fatty alcohols have poor volumetric energy storage, 43 – 55 kWh/m^3. Paraffin waxes containing only a single length carbon chain have a higher energy cost (15 – 500 $/kWh) than generic paraffin waxes containing many lengths of carbon chains (7 – 30 $/kWh). Pure waxes have a discrete phase change temperature due to their homogeneity. In contrast, a less refined generic wax with several carbon chain lengths is more likely to have a pronounced temperature glide during its phase change. Pure single carbon chain waxes are generally required for applications <45°C as generic paraffin waxes melt between 45 – 70°C. For many waxes, a solid-solid transition occurs at temperatures below the solid-liquid phase change. For pure paraffins with carbon content ≥22 C atoms, these transitions may appear near the same temperature resembling a temperature glide. The challenges with fatty acids, fatty alcohols, and waxes are low thermal conductivity, low density, some flammability concerns, and compatibility issues with some common engineering materials such as polymers. Challenges with salt hydrates are pronounced supercooling (>5°C), incongruent melting, and corrosiveness. All PCMs may degrade if exposed to ambient conditions and therefore require proper sealing.

1. INTRODUCTION

A phase change material (PCM) is capable of absorbing, storing, and releasing thermal energy nearly isothermally at the phase change temperature. The thermal battery may act as a heat source or a heat sink depending on the desired application, mode of operation, and the phase change temperature. The selection of a PCM for a given application is dependent on its phase change temperature, the size of the system, the external operating temperatures, and cost.

Incorporating PCMs into building systems has been researched extensively. There are two schemes of incorporating PCMs into building systems: 1) passive systems whereby the PCM acts as an unobtrusive and uncontrolled thermal buffer for high transient thermal loads on the building envelope, and 2) active systems where the PCM is coupled to a heat pump or similar mechanical device and absorbs or releases heat at the operators' control. The successful implementation of both passive and active systems requires a PCM with an appropriate phase change temperature.

The present study examines the thermophysical properties of PCMs with a phase change temperature of 0 – 65°C that may be useful for building applications: inorganic salt hydrates, organic fatty acids, organic fatty alcohols, and organic paraffin waxes. This present study expands on the work published by Hirschey et al. (2018).

2. PCM PROPERTIES

2.1 Inorganic Salt Hydrate PCMs

Salt hydrates are inorganic materials composed of salt and water in specific molar ratios. The salt and water form a solid crystalline structure, often different than the original salt crystal structure. Some salts may form multiple hydrates with different water:salt ratios. The solid-liquid phase change process for a salt hydrate is observed when the hydrogen bonds between the water molecules in the solid crystal structure are overcome by the molecular vibrations and the water is liberated from the crystal structure forming an aqueous solution. In many salt hydrates the salt is not fully soluble in the water content of its hydrate. This results in incongruent melting whereby the undissolved salt settles out of the solution.

The liquid-solid phase change for a salt hydrate is the reformation of the salt hydrate crystal structure from the aqueous solution. However, many salt hydrates experience extreme supercooling which is the phenomenon where the liquid cool below nominal phase change temperature without the spontaneous formation of the solid structure. Supercooling can be problematic for systems that rely on a consistent phase change temperature. Solutions to preventing large supercooling include adding nucleating agents or through some mechanical work (shaking, vibrating, stirring) (Lane, 1992). Some systems may take advantage of this supercooling phenomenon into the design; the subcooled liquid salt hydrate is stored nearer the ambient temperature, removing the need for insulation and recrystallization provided from an exterior source when the heat release is desired (Sandnes & Rekstad, 2006).

Incongruently melted salt hydrates are not homogenous; there is a gradient from saturated solution to anhydrous salt. This separation between the water and salt does not allow for the full recrystallization of the original salt hydrate crystal and the resulting smaller crystal will form a barrier between water and anhydrous salt. This leads to a reduction in energy storage capacity as only a fraction of the original PCM will crystallize. In a closed system, the full solid salt hydrate may recrystallize, but the time required is not useful for thermal storage systems.

Salt hydrates have been explored extensively for their thermal storage potential, namely high volumetric latent heat. Table 1 shows the thermal energy storage-relevant thermophysical properties of salt hydrates including nominal phase change temperature (solid-to-liquid phase change; the degree of supercooling can be unpredictable), the latent heat associated with the phase change, and the solid density. Some double salts are included in this analysis. Double salts are those salts with two cation or two anions in the unit cell, but the resultant is distinct in the crystal structure and properties than the corresponding salts with only single cations or anions, e.g., chromium (III) potassium sulfate dodecahydrate ($CrK(SO_4)_2 \cdot 12H_2O$).

2.2 Organic PCMs

Organic PCMs are materials that consist of a hydrocarbon chain. Organic PCMs include, but are not limited to, waxes, alcohols, and fatty acids. The phase transition temperature of organic PCMs often correlates to its hydrocarbon chain length. For most saturated organic compounds, a longer chain has higher melting temperatures. Unsaturated or branched organic compounds may break from this simplification. Organic PCMs may exhibit a temperature glide where the phase change occurs over a temperature range around the nominal phase change temperature. Often the reported enthalpy values for these materials with glide include the sensible heating that is present in the glide. Some organic PCMs may display supercooling tendencies where the temperature for the onset of freezing occurs below the temperature for the completion of melting. This is observed more frequently in unsaturated materials or those that contain many polymorphic forms. These are often more predictable and repeatable than supercooling of salt hydrate PCMs. Only the solid-to-liquid phase change temperatures are reported here. Unsaturated hydrocarbons are less stable than similar saturated chains and thus are more susceptible to oxidation. Therefore, proper sealing and separation from atmospheric oxygen is required if these materials are used for TES without degradation. Most organic PCMs are biodegradable. Some material compliance issues with plastics and polymers may be present for some organic PCMs.

2.4 Fatty Acid PCMs

Fatty acids are a subset of carboxylic acids which have an organic functional group terminating with a carboxyl group and have the general formula R-COOH. The functional group for a fatty acid is generally a hydrocarbon chain, saturated or unsaturated. Saturated fatty acids may be written as $COOH(CH_2)_nCH_3$. Fatty acids are named so because they are the basis of the fats of all plants and animals; a triglyceride is composed of glycerol and three fatty acids. The most commonly analyzed fatty acids for TES are medium to long straight-chain saturated fatty acids with an even number of carbon atoms: capric acid (C10:0), lauric acid (C12:0), myristic acid (C14:0), palmitic acid (C16:0), and stearic acid (C18:0). Fatty acids with an odd number of carbon atoms exist, albeit more rarely, and will be included in this analysis when data is available. For completeness and comparison, formic acid (HCOOH) and acetic acid (CH_3COOH) are included in this analysis as they contain the same terminating carboxyl group although they do not appear in naturally occurring triglycerides.

2.5 Fatty Alcohol PCMs

The fatty alcohols analyzed here are primary alcohols composed of a saturated or unsaturated hydrocarbon chain terminating with an alcohol group and have the general formula R-OH. Saturated fatty alcohols are written as $CH_3(CH_2)_nOH$. More alcohols exist with branches and additional functional groups, but these will be excluded.

2.5 Paraffin Waxes

Paraffin waxes are common for thermal energy storage, are often derived from petroleum sources. Paraffins are straight-chain, saturated alkane hydrocarbons with no terminating functional groups, written as C_nH_{2n+2}. More detail on paraffin structure is included in Hirschey et al. (2018). Many paraffins have a solid-solid transition at a lower temperature than the solid-liquid transition. The crystalline solid structures vary by the carbon content, more details included in Mondieig et al. (2004). In paraffins with a carbon number ≥C22, the solid-solid transition temperature, $T_{s\text{-}s}$, may be very near the solid-liquid transition, $T_{s\text{-}l}$. For example, tetracosane ($C_{24}H_{50}$) has $T_{s\text{-}s} \approx 48°C$ and a $T_{s\text{-}l} \approx 50.3°C$. The latent heat associated with the solid-solid is nearly 90 J/g and the solid-liquid is around 160 J/g. Thus, tetracosane may be considered as having a total latent heat 250 J/g with a 2.3°C temperature glide if both transitions are included. This analysis will show solid-solid and solid-liquid separately where data is available.

The predominant application for paraffins with melting temperatures between 50 – 70°C is candle making. These waxes are not pure substances, rather they are a mix of several alkane chains of various lengths with a marketed average melting temperature. Due to their heterogeneity, these waxes may have several degrees of temperature glide. Often the exact chemical composition in unknown. The melting temperature of these candle making waxes generally overlap with pure paraffins with a carbon content between 24 and 32 atoms but at a reduced cost.

3. METHODOLOGY

The thermophysical properties of the PCMs are surveyed from literature. When available, a primary source and measurement data were used. Many reported PCM thermophysical properties have "citation chains" where the same values are repeated article after article. Attempts were made to trace back to the original measurement data which are often old, difficult to locate, or used out-of-date technology; or search for more recent data. Where primary data was not found, a (*) is denoted by the material name. These PCMs should be viewed more critically, and more work should be done to assess their viability and confirm these data. Materials in the tables but not present in Figure 1 have current costs that exceed the maximum value of the chart.

All cost data were gathered from data publicly available from industrial suppliers using the CAS number as the search terms. Wherever possible, the cost data were collected in bulk quantities per metric ton or per ton. Several different suppliers for each PCM were identified and the listed prices recorded. Care was taken to ensure that the same material form was used for each individual material to ensure an accurate cost comparison (e.g. powder). All cost data is consolidated into a single dataset source white paper, Hirschey (2020). All cost data were collected in the autumn of 2020 and prices were largely static during this time. Some seasonality may exist for materials that have seasonal applications such as calcium chloride ($CaCl_2$) as a deicing agent.

Salt hydrates were sourced in their anhydrous form and the cost of water was considered negligible. Many salt hydrates with other industrial applications may only be sold as a solution with water, e.g. calcium bromide ($CaBr_2$) as a 52% solution drilling fluid. Under these special circumstances, the PCM cost was adjusted by the stoichiometric anhydrous salt content. It should be noted that adding water in bulk to some anhydrous salts can be an energetic exothermic event, and proper safety precautions should be taken. This reaction is distinct from, though related to, the solid-liquid phase transformations analyzed here.

The energy storage material cost was calculated by dividing the material bulk cost ($/kg) by melting enthalpy (kJ/kg), and plotted versus the material melting temperature in Figure 1. The vertical error bars represent the range of possible energy storage material costs which includes extrema in price and melting enthalpy values. Many material costs vary wildly, sometimes two orders of magnitude. This is often the difference between ordering a single shipment of one metric ton versus regular shipments of hundreds of metric tons.

The horizontal error bars of Figure 1 represent the range in documented melting temperatures of that material; they do not represent supercooling or temperature glide. For materials that experience a temperature glide, the peak temperature, or similar, is used. The peak temperature is the temperature that shows a maximum value in a heat flow vs. temperature curve (commonly differential scanning calorimetry data). This temperature may vary with the experimental variables.

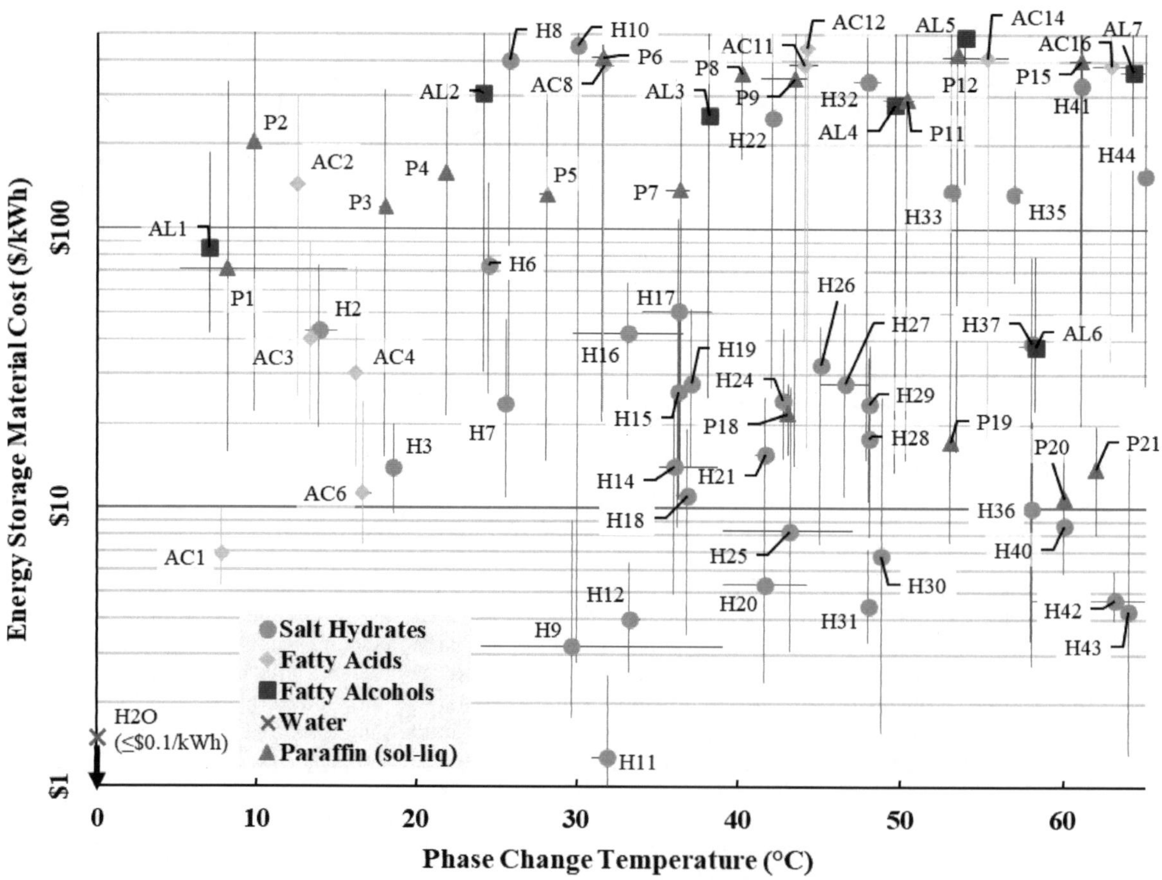

Figure 1: PCM Energy Storage Material Cost

Table 1: Properties of Salt Hydrate PCMs

Fig. 1 ID	Salt Hydrate	Chemical Formula	Anhydrous Salt CAS Number	Melting Temp (°C)	Melting Enthalpy (J/g)	Density (kg/m^3)	Cost ($/kg)	Thermophysical property sources
-	Lithium Chlorate Trihydrate	$LiClO_3 \cdot 3H_2O$	13453-71-9	8.0 – 8.2	155 – 253.0	1720	25.00 - 37.50	(Gawron & Shröder, 1977; Guion et al., 1983; Marcus, 2017)
H2	Dipotassium Hydrogen Phosphate Hexahydrate*	$K_2HPO_4 \cdot 6H_2O$	7758-11-4	13.0 – 15.0	107 – 122.0	2440	0.66 - 1.34	(Guion et al., 1983; Kauffman & Pan, 1973; Marcus, 2017)
H3	Potassium Fluoride Tetrahydrate	KF·4H2O	7789-23-3	18.0 – 18.9	231 – 330	1445	0.88 - 1.00	(Counioux & Cohen-Adad, 1976; Guion et al., 1983; S. D. Sharma et al., 2004)
-	Sodium Chromate Decahydrate*	$Na_2CrO_4 \cdot 10H_2O$	7775-11-3	16.0 – 18.9	164 – 172	1483	10.00 - 51.67	(Guion et al., 1983; Kauffman & Pan, 1973)
-	Iron (III) Bromide Hexahydrate*	$FeBr_3 \cdot 6H_2O$	10031-26-2	21.0 – 27	105	1820	10.00 - 119.33	(S. D. Sharma et al., 2004; Veerakumar & Sreekumar, 2016)
H6	Copper (II) Nitrate Hexahydrate	$Cu(NO_3)_2 \cdot 6H_2O$	3251-23-8	24.4	123	2074	0.88 - 2.53	(Guion et al., 1983; Kauffman & Pan, 1973)
H7	Manganese (II) Nitrate Hexahydrate	$Mn(NO_3)_2 \cdot 6H_2O$	10377-66-9	25.0 – 26.0	125.8 – 148	1800	0.45 - 0.89	(Marcus, 2017; Nagano et al., 2003)
H8	Lithium Metaborate Octahydrate*	$LiBO_2 \cdot 8H_2O$	13453-69-5	25.7	289	1400	2.00 - 32.33	(S. D. Sharma et al., 2004)
H9	Calcium Chloride Hexahydrate	$CaCl_2 \cdot 6H_2O$	10043-52-4	24 – 30.0	133.9 – 211.3	1710	0.10 - 0.19	(Guion et al., 1983)
H10	Lithium Nitrate Trihydrate	$LiNO_3 \cdot 3H_2O$	7790-69-4	29.9 – 30.0	189 – 296	1575	0.23 - 37.43	(Guion et al., 1983; Kauffman & Pan, 1973; S. D. Sharma et al., 2004)
H11	Sodium Sulfate Decahydrate	$Na_2SO_4 \cdot 10H_2O$	7757-82-6	31.0 – 32.4	239 – 251.2	1477	0.03 - 0.09	(Abhat, 1983; S. D. Sharma et al., 2004)
H12	Sodium Carbonate Decahydrate	$Na_2CO_3 \cdot 10H_2O$	497-19-8	33.0 – 33.9	179.8 – 251	1447	0.18 - 0.26	(Guion et al., 1983; Rao et al., 2018)
-	Lithium Bromide Dihydrate*	$LiBr_2 \cdot 2H_2O$	7550-35-8	34	124	1570	1.00 - 20.67	(S. D. Sharma et al., 2004)
H14	Sodium Hydrogen Phosphate Dodecahydrate	$Na_2HPO_4 \cdot 12H_2O$	7558-79-4	35.0 – 38.7	187.7 – 281	1521	0.39 - 1.03	(Abhat, 1983; Guion et al., 1983; Lorsch et al., 1975; Rao et al., 2018)
H15	Zinc Nitrate Hexahydrate	$Zn(NO_3)_2 \cdot 6H_2O$	7779-88-6	36.0 – 36.4	126.7 – 147	2015	0.44 - 0.99	(Abhat, 1983; Lorsch et al., 1975; Marcus, 2017; Rao et al., 2018)
H16	Ammonium Iron (III) Sulfate Dodecahydrate*	$FeNH_4(SO_4)_2 \cdot 12H_2O$	10138-04-2	29.7 – 36.5	197.6	1713	1.00 - 2.32	(Guion et al., 1983)
H17	Calcium Bromide Hexahydrate	$CaBr_2 \cdot 6H_2O$	7789-41-5	34.0 – 38.3	115 – 138	2228	0.28 - 1.64	(Marcus, 2017; S. D. Sharma et al., 2004)
H18	Iron (III) Chloride Hexahydrate	$FeCl_3 \cdot 6H_2O$	7705-08-0	36.1 – 37.0	186.2 – 226	1600	0.22 - 0.68	(Guion et al., 1983; Kauffman & Pan, 1973; Marcus, 2017)
H19	Manganese (II) Nitrate tetrahydrate*	$Mn(NO_3)_2 \cdot 4H_2O$	10377-66-9	37.0	115	2130	0.45 - 0.89	(S. D. Sharma et al., 2004)
H20	Calcium Chloride Tetrahydrate	$CaCl_2 \cdot 4H_2O$	10043-52-4	39 – 44.2	99.6 – 158	1567	0.10 - 0.19	(S. D. Sharma et al., 2004)
H21	Potassium Fluoride Dihydrate*	$KF \cdot 2H_2O$	7789-23-3	41.0 – 42.0	162 – 266	1658	0.88 - 1.00	(Guion et al., 1983; S. D. Sharma et al., 2004)
H22	Magnesium Iodide Octahydrate*	$MgI_2 \cdot 8H_2O$	10377-58-9	42	133	2098	1.00 - 9.30	(S. D. Sharma et al., 2004)
-	Calcium Iodide Hexahydrate*	$CaI_2 \cdot 6H_2O$	10102-68-8	42	162	2550	20.00 - 76.75	(S. D. Sharma et al., 2004)
H24	Chromium (III) Potassium Sulfate Dodecahydrate*	$CrK(SO_4)_2 \cdot 12H_2O$	10141-00-1	42.7	237.9	1828	1.00 - 1.61	(Guion et al., 1983)
H25	Calcium Nitrate Tetrahydrate	$Ca(NO_3)_2 \cdot 4H_2O$	10124-37-5	39.0 – 47.0	106 – 209.2	1858	0.18 - 0.34	(Guion et al., 1983; Kauffman & Pan, 1973; S. D. Sharma et al., 2004)
H26	Tripotassium Phosphate Heptahydrate*	$K_3PO_4 \cdot 7H_2O$	7778-53-2	45	145	2500	0.30 - 1.32	(S. D. Sharma et al., 2004)

Table 1 (continued): Properties of Salt Hydrate PCMs

Fig. 1 ID	Salt Hydrate	Chemical Formula	Anhydrous Salt CAS Number	Melting Temp (°C)	Melting Enthalpy (J/g)	Density (kg/m³)	Cost ($/kg)	Thermophysical property sources
H27	Zinc Nitrate Tetrahydrate*	$Zn(NO_3)_2{\cdot}4H_2O$	7779-88-6	45 – 48.0	110 – 144	2220	0.44 - 0.99	(Guion et al., 1983; S. D. Sharma et al., 2004)
H28	Sodium Metasilicate Pentahydrate*	$Na_2SiO_3{\cdot}5H_2O$	6834-92-0	48.0	68	2610	0.20 - 0.34	(S. D. Sharma et al., 2004)
H29	Sodium Hydrogen Phosphate Heptahydrate	$Na_2HPO_4{\cdot}7H_2O$	7558-79-4	48.0 – 48.2	135 – 170	1679	0.39 - 1.03	(Guion et al., 1983; Kauffman & Pan, 1973; S. D. Sharma et al., 2004)
H30	Magnesium Sulfate Heptahydrate*	$MgSO_4{\cdot}7H_2O$	7487-88-9	48 – 49.2	201 – 202	1680	0.09 - 0.46	(Guion et al., 1983; Kauffman & Pan, 1973; S. D. Sharma et al., 2004)
H31	Sodium thiosulfate pentahydrate	$Na_2S_2O_3{\cdot}5H_2O$	7772-98-7	48.0	200 – 209	1708	0.19 - 0.25	(Bajnóczy et al., 1995; Zhang et al., 1999)
H32	Iron Nitrate Nonahydrate*	$Fe(NO_3)_3{\cdot}9H_2O$	10421-48-4	47 – 48.7	155 – 190.5	1684	0.79 - 17.91	(Guion et al., 1983; S. D. Sharma et al., 2004)
H33	Sodium Nitrate Hexahydrate*	$NaNO_3{\cdot}6H_2O$	7631-99-4	53.0	158	2261	0.33 - 6.02	(Guion et al., 1983)
-	Cobalt Nitrate Hexahydrate*	$Co(NO_3)_2{\cdot}6H_2O$	10141-05-6	55.5 – 57.0	115 – 203.3	1870	1.00 - 23.50	(Marcus, 2017; S. D. Sharma et al., 2004)
H35	Nickel Nitrate Hexahydrate*	$Ni(NO_3)_2{\cdot}6H_2O$	13138-45-9	56.7 – 57	168	2050	3.00 - 6.25	(Guion et al., 1983; S. D. Sharma et al., 2004)
H36	Sodium Acetate Trihydrate	$CH_3COONa{\cdot}3H_2O$	127-09-3	57.9 – 58.0	180 – 289	1450	0.22 - 0.71	(Bajnóczy et al., 1995; Marcus, 2017)
H37	Manganese Chloride Tetrahydrate*	$MnCl_2{\cdot}4H_2O$	7773-01-5	57.9 – 58.0	151 – 178	2010	0.17 - 1.84	(Guion et al., 1983; S. D. Sharma et al., 2004)
-	Lithium Acetate Dihydrate*	$CH_3COOLi{\cdot}2H_2O$	546-89-4	57.9 – 58.0	250.9 – 377	1300	3.00 - 59.15	(Marcus, 2017; S. D. Sharma et al., 2004)
-	Cadmium Nitrate Tetrahydrate*	$Cd(NO_3)_2{\cdot}4H_2O$	10325-94-7	59 – 59.5	98 – 155	1773	10.00 - 45.00	(Guion et al., 1983; Marcus, 2017; S. D. Sharma et al., 2004)
H40	Sodium Bisulfate Monohydrate*	$NaHSO_4{\cdot}H_2O$	7681-38-1	60	129.6	1800	0.21 - 0.31	(Guion et al., 1983)
H41	Sodium Aluminum Sulfate Dodecahydrate*	$NaAl(SO_4)_2{\cdot}12H_2O$	10102-71-3	61	181	1675	1.00 - 16.63	(S. D. Sharma et al., 2004)
H42	Sodium Hydroxide Monohydrate	$NaOH{\cdot}H_2O$	1310-73-2	58 – 65.2	272 – 272.4	1710	0.30 - 0.36	(Murch & Giauque, 1962; S. D. Sharma et al., 2004)
H43	Iron (II) Sulfate Heptahydrate*	$FeSO_4{\cdot}7H_2O$	7720-78-7	64.0	200	1890	0.07 - 0.24	(Guion et al., 1983)
H44	Chromium Nitrate Nonahydrate*	$Cr(NO_3)_3{\cdot}9H_2O$	13548-38-4	65.0	194.8	1807	1.50 - 8.46	(Guion et al., 1983)

* Single primary measurement or no primary measurement found

Table 2: Properties of Fatty Alcohols PCMs

Fig. 1 ID	Fatty Alcohol	Chemical Formula	CAS Number	Melting Temperature (°C)	Melting Enthalpy (J/g)	Density (kg/m³)	Cost ($/kg)	Thermophysical property sources
AL1	1-Decanol	$C_{10}H_{22}O$	112-30-1	6-7 – 7.5	181 – 212	830	2.5 – 9.36	(de Matos et al., 2015)
AL2	Lauryl alcohol	$C_{12}H_{26}O$	112-53-8	23.2 – 24.7	202 – 216	831	1.85 – 100	(de Matos et al., 2015; Saeed et al., 2016; van Miltenburg et al., 2003)
AL3	Myristyl alcohol	$C_{14}H_{30}O$	112-72-1	37.9 – 38.2	219 – 230	824	1.60 – 80	(Maximo et al., 2014)
AL4	Cetyl alcohol	$C_{16}H_{34}O$	36653-82-4	49 – 50.1	232 – 240	811	0.90 – 50	(Maximo et al., 2014)
AL5	Heptadecyl alcohol	$C_{17}H_{36}O$	1454-85-9	53.4 – 54.1	244 – 245.8	815	10 – 100	(van Miltenburg et al., 2003; Ventolà et al., 2004)
AL6	Stearyl alcohol	$C_{18}H_{38}O$	112-92-5	57.2 – 58.7	148 – 257	813	1.60 – 3.30	(Maximo et al., 2014; van Miltenburg et al., 2001; Ventolà et al., 2004)
AL7	Arachidyl alcohol	$C_{20}H_{42}O$	629-96-9	63.5 – 65	241 – 247	805	3 – 199	(van Miltenburg et al., 2001; Ventolà et al., 2004)

Table 3: Properties of Fatty Acid PCMs

Fig. 1 ID	Fatty Acid	Structural Chemical Formula	CAS Number	Melting Temperature (°C)	Melting Enthalpy (J/g)	Density (kg/m³)	Cost ($/kg)	Thermophysical property sources
AC1	Formic Acid Methanoic Acid	$HCOOH$	64-18-6	7.8 – 8.4	247	1200	0.36 - 0.68	(Veerakumar & Sreekumar, 2016)
AC2	Pelargonic Acid Nonanoic Acid*	$CH_3(CH_2)_7COOH$	112-05-0	12.5	128	900	0.90 - 10.60	(Acree, 1991)
AC3	α-Oleic Acid	$CH_3(CH_2)_7CH{=}CH(CH_2)_7COOH$	112-80-1	13.3	76 – 140	895	0.80 - 3.50	(Acree, 1991; Cedeño et al., 2001; Sato et al., 1997)
AC4	β-Oleic Acid	$CH_3(CH_2)_7CH{=}CH(CH_2)_7COOH$	112-80-1	16.0 – 16.3	173 – 205	895	0.80 - 3.50	(Cedeño et al., 2001; Sato et al., 1997)
-	Caprylic Acid Octanoic acid	$CH_3(CH_2)_6COOH$	124-07-2	16.3 – 16.7	135 – 148	910	3.50 - 120.00	(Acree, 1991; Domalski & Hearing, 1996)
AC6	Acetic Acid Ethanoic Acid	CH_3COOH	64-19-7	16 – 16.91	180 – 195	1270	0.40 - 1.20	(Julius, 1910; Parks & Kelley, 1925; Pickering, 1895)
-	Undecylic Acid Undecanoic Acid*	$CH_3(CH_2)_9COOH$	112-37-8	28.6	139	890	1 - 5	(Domalski & Hearing, 1996)
AC8	Capric Acid Decanoic Acid	$CH_3(CH_2)_8COOH$	334-48-5	31 – 32.13	152.7 – 162.7	885	1 - 100	(Hobi Bordón Sosa et al., 2019; Saeed et al., 2016; A. Sharma et al., 2005)
-	Petroselinic Acid o-6-0ctadecenoic Acid*	$CH_3(CH_2)_{10}CH{=}CH(CH_2)_4COOH$	593-39-5	30.5	168	900	10 - 100	(Sato et al., 1997)
-	Tridecylic Acid Tridecanoic Acid*	$CH_3(CH_2)_{11}COOH$	638-53-9	41.5	157	983	20 - 150	(Domalski & Hearing, 1996)
AC11	Lauric Acid Dodecanoic Acid	$CH_3(CH_2)_{10}COOH$	143-07-7	44 – 44.92	177.4 – 181.5	903	2.00 - 200	(Hobi Bordón Sosa et al., 2019; Saeed et al., 2016; A. Sharma et al., 2005)
AC12	Elaidic Acid*	$CH_3(CH_2)_7CH{=}CH(CH_2)_7COOH$	112-79-8	43.9 – 44.5	190 – 218	873	1 - 50	(Wilson et al., 2015)
-	Pentadecanoic Acid*	$CH_3(CH_2)_{13}COOH$	1002-84-2	52.5	188	842	50 - 150	(Wilson et al., 2015)
AC14	Myristic Acid Tetradecanoic Acid	$CH_3(CH_2)_{12}COOH$	544-63-8	54.2 – 58	191 – 198	962	1 - 120	(Hobi Bordón Sosa et al., 2019; Saeed et al., 2016; A. Sharma et al., 2005)
-	Margaric Acid Heptadecanoic Acid*	$CH_3(CH_2)_{15}COOH$	506-12-7	61.2 – 61.3	217	853	2500 - 7500	(Wilson et al., 2015)
AC16	Palmitic Acid Hexadecanoic Acid	$CH_3(CH_2)_{14}COOH$	57-10-3	62.5 – 64	163 – 212	886	2 - 150	(Hobi Bordón Sosa et al., 2019; Saeed et al., 2016; A. Sharma et al., 2005)

* Single primary measurement or no primary measurement found

Table 4: Properties of Paraffin PCMs

Fig 1 ID	Paraffin Name	Chemical Formula	CAS Number	Dominant Solid-Solid Temp (°C)	Solid-Solid Enthalpy (J/g)	Solid-Liquid Temp (°C)	Solid-Liquid Enthalpy (J/g)	Density (kg/m³)	Cost ($/kg)	Thermophysical property sources
P1	Tetradecane	$C_{14}H_{30}$	629-59-4	-79	0.9	5.2 - 15.6	215 - 227	759	1 – 20	(Domalski & Hearing, 1996; Mondieig et al., 2004)
P2	Pentadecane	$C_{15}H_{32}$	629-62-9	-3 – -2	41 - 43	9.6 - 10.0	161 – 163	765	1 – 45	(Domalski & Hearing, 1996; Mondieig et al., 2004)
P3	Hexadecane	$C_{16}H_{34}$	544-76-3	–	–	17.6 - 18.2	228 – 236	770	1 – 20	(Domalski & Hearing, 1996; Mondieig et al., 2004)
P4	Heptadecane	$C_{17}H_{36}$	629-78-7	11	45 - 46	21.6 - 22.0	164 – 167	775	1 – 15	(Domalski & Hearing, 1996; Mondieig et al., 2004)
P5	Octadecane	$C_{18}H_{38}$	593-45-3	–	–	27.6 - 28.4	236 – 242	779	1 – 20	(Domalski & Hearing, 1996; Mondieig et al., 2004)
P6	Nonadecane	$C_{19}H_{40}$	629-92-5	22 – 23	47 - 51	30.8 - 32.0	159 – 176	782	1 – 100	(Domalski & Hearing, 1996; Mondieig et al., 2004)
P7	Icosane	$C_{20}H_{42}$	112-95-8	–	–	35.4 - 36.9	218 – 251	785	1 – 20	(Domalski & Hearing, 1996; Mondieig et al., 2004)
P8	Heneicosane	$C_{21}H_{44}$	629-94-7	31 - 33	52 - 56	39.9 - 40.3	155 – 161	788	8 – 100	(Domalski & Hearing, 1996; Mondieig et al., 2004)
P9	Docosane	$C_{22}H_{46}$	629-97-0	40 - 43	29 - 128	41.3 - 44.1	49 – 253	791	1 – 25	(Domalski & Hearing, 1996; Mondieig et al., 2004)
-	Tricosane	$C_{23}H_{48}$	638-67-5	39 - 41	60 - 67	46.6 - 47.7	161 – 236	793	1 – 100	(Domalski & Hearing, 1996; Mondieig et al., 2004)
P11	Tetracosane	$C_{24}H_{50}$	646-31-1	46 - 51	82 - 93	48.9 - 50.7	159 – 241	796	1 – 25	(Domalski & Hearing, 1996; Mondieig et al., 2004)
P12	Pentacosane	$C_{25}H_{52}$	629-99-2	47 - 53	67 - 75	53.1 - 53.6	160 – 225	798	1 – 100	(Domalski & Hearing, 1996; Mondieig et al., 2004)
-	Hexacosane	$C_{26}H_{54}$	630-01-3	50 - 53	83 - 95	56.0 - 56.5	160 – 174	800	15,620	(Domalski & Hearing, 1996; Mondieig et al., 2004)
-	Heptacosane	$C_{27}H_{56}$	593-49-7	47 - 53	71 - 76	58.5 - 58.8	159 – 165	802	255,000	(Domalski & Hearing, 1996; Mondieig et al., 2004)
P15	Octacosane	$C_{28}H_{58}$	630-02-4	56 - 58	85 - 90	60.3 - 61.6	160 – 169	803	2.10 – 100	(Domalski & Hearing, 1996; Mondieig et al., 2004)
-	Nonacosane	$C_{29}H_{60}$	630-03-5	58 - 58	73 - 73	63.4	162	805	10 – 100	(Domalski & Hearing, 1996)
-	Triacontane	$C_{30}H_{62}$	638-68-6	59 - 62	86 - 89	65.5 - 65.6	163	806	16,800	(Domalski & Hearing, 1996)
P18	Paraffin 46-50	–	–	20	–	43.0	228	765	0.99 – 1.76	(Ukrainczyk et al., 2010)
P19	Paraffin 52-54	–	–	20	–	53.0	220	774	0.72 – 1.50	(Ukrainczyk et al., 2010)
P20	Paraffin 58-62	–	–	27	–	60.0	206	782	0.45 – 0.90	(Ukrainczyk et al., 2010)
P21	Paraffin 62-70	–	–	25	–	62.0	201	799	0.45 – 1.10	(Ukrainczyk et al., 2010)

4. CONCLUSIONS

- Salt hydrate PCMs have the lowest energy storage cost for applications 18 – 65°C, between 0.90 – 15 \$/kWh. With high density and latent heat, salt hydrates have the highest volumetric energy storage, 50 – 130 kWh/m^3.
- Fatty acid PCMs have the lowest energy storage cost in 8 – 17°C range, between 6.50 – 40 \$/kWh. Fatty acids with higher melting temperatures are scarcer and thus have higher costs comparable to other organics.
- Organic fatty alcohols have the highest average energy storage cost, >40 \$/kWh, for all temperatures.
- Pure paraffin waxes consisting of a single length carbon chain may be viable for applications between 8 – 28°C with energy storage costs 16 – 25 \$/kWh at their most competitive material costs.
- Generic paraffin waxes that have many different carbon chains of various lengths are more viable for applications 40 – 65°C, at 7 – 30 \$/kWh, but these materials have more pronounced temperature glide.

REFERENCES

Abhat, A. (1983). Low temperature latent heat thermal energy storage: Heat storage materials. *Solar Energy, 30*(4), 313-332.

Acree, W. E. (1991). Thermodynamic properties of organic compounds: enthalpy of fusion and melting point temperature compilation. *Thermochimica Acta, 189*(1), 37-56.

Bajnóczy, G., Gagyi Pálffy, E., Prépostffy, E., & Zöld, A. (1995). Thermal Properties of a Heat Storage Device Containing Sodium Acetate Trihydrate. *Periodica Polytechnica Chemical Engineering, 39*(2), 129-135.

Cedeño, F. O., Prieto, M. a. M., Espina, A., & Garcıa, J. R. (2001). Measurements of temperature and melting heat of some pure fatty acids and their binary and ternary mixtures by differential scanning calorimetry. *Thermochimica Acta, 369*(1), 39-50.

Counioux, J.-J., & Cohen-Adad, R. (1976). Détermination, par calorimétrie du chute, des enthalpies de fusion des hydrates à fusion congruente du ternaire KOH-KF-H2O. *Bulletin de la Société chimique de France*, 373-376.

de Matos, F. C., da Costa, M. C., de Almeida Meirelles, A. J., & Batista, E. A. C. (2015). Binary solid–liquid equilibrium systems containing fatty acids, fatty alcohols and triolein by differential scanning calorimetry. *Fluid Phase Equilibria, 404*, 1-8.

Domalski, E. S., & Hearing, E. D. (1996). Heat Capacities and Entropies of Organic Compounds in the Condensed Phase. Volume III. *Journal of Physical and Chemical Reference Data, 25*(1), 1-525.

Gawron, K., & Shröder, J. (1977). Properties of some salt hydrates for latent heat storage. *International Journal of Energy Research, 1*(4), 351-363.

Guion, J., Sauzade, J. D., & Laugt, M. (1983). Critical Examination and Experiemental Determination of Melting Enthalpies and Entropies of Salt Hydrates. *Thermochimica Acta, 67*, 167-179.

Hirschey, J. (2020). *PCM Cost.* Retrieved from: https://web.archive.org/web/20201216164857/https://drive.google.com/file/d/1-LceoENyhh5W2jHt9X7f72hQ_VyXsnzM/view

Hirschey, J., Gluesenkamp, K. R., Mallow, A., & Graham, S. (2018). Review of Inorganic Salt Hydrates with Phase Change Temperature in Range of 5 to 60° C and Material Cost Comparison with Common Waxes.

Hobi Bordón Sosa, F., Dorighello Carareto, N. D., Maximo, G. J., Meirelles, A. J. d. A., & Costa, M. C. (2019). Solid–Liquid Equilibrium of Binary Systems Containing Fatty Acids and Fatty Alcohols Using Differential Scanning Calorimetry. *Journal of Chemical & Engineering Data, 64*(12), 5017-5027.

Julius, M. (1910). Zur Theorie der Rohrzuckerinversion II. *Zeitschrift für Physikalische Chemie, 72U*(1), 117-123.

Kauffman, K., & Pan, Y.-C. (1973). *Congruently Melting Materials for Thermal Energy Storage in Air Conditioning* (NSF/RANN/SE/GI27976/TR73/5). Retrieved from Philadlephia, PA:

Lane, G. A. (1992). Phase change materials for energy storage nucleation to prevent supercooling. *Solar Energy Materials and Solar Cells, 27*(2), 135-160.

Lorsch, H. G., Kauffman, K. W., & Denton, J. C. (1975). Thermal Energy Storage for Solar Heating and Off-peak Air Conditioning. *Energy Conversion, 15*, 1-8.

Marcus, Y. (2017). Unconventional Deep Eutectic Solvents: Aqueous Salt Hydrates. *ACS Sustainable Chemistry & Engineering, 5*(12), 11780-11787.

Maximo, G. J., Carareto, N. D. D., Costa, M. C., dos Santos, A. O., Cardoso, L. P., Krähenbühl, M. A., & Meirelles, A. J. A. (2014). On the solid–liquid equilibrium of binary mixtures of fatty alcohols and fatty acids. *Fluid Phase Equilibria, 366*, 88-98.

Mondieig, D., Rajabalee, F., Metivaud, V., Oonk, H. A. J., & Cuevas-Diarte, M. A. (2004). n-Alkane Binary Molecular Alloys. *Chemistry of Materials, 16*(5), 786-798.

Murch, L. E., & Giauque, W. F. (1962). THE THERMODYNAMIC PROPERTIES OF SODIUM HYDROXIDE AND ITS MONOHYDRATE. HEAT CAPACITIES TO LOW TEMPERATURES. HEATS OF SOLUTION1. *The Journal of Physical Chemistry, 66*(10), 2052-2059.

Nagano, K., Mochida, T., Kindaichi, S., Domañski, R., & Rebow, M. (2003). Thermal characteristics of manganese (II) nitrate hexahydrate as a phase change material for cooling systems. *Applied Thermal Engineering, 23*(2), 229-241.

Parks, G. S., & Kelley, K. K. (1925). THERMAL DATA ON ORGANIC COMPOUNDS. II. THE HEAT CAPACITIES OF FIVE ORGANIC COMPOUNDS. THE ENTROPIES AND FREE ENERGIES OF SOME HOMOLOGOUS SERIES OF ALIPHATIC COMPOUNDS. *Journal of the American Chemical Society, 47*(8), 2089-2097.

Pickering, S. U. (1895). LXXII.—A comparison of some properties of acetic acid and its chloro- and bromo-derivatives. *Journal of the Chemical Society, Transactions, 67*(0), 664-684.

Rao, Z., Xu, T., Liu, C., Zheng, Z., Liang, L., & Hong, K. (2018). Experimental study on thermal properties and thermal performance of eutectic hydrated salts/expanded perlite form-stable phase change materials for passive solar energy utilization. *Solar Energy Materials and Solar Cells, 188*, 6-17.

Saeed, R., Schlegel, J., Castano Giraldo, C., & Sawafta, R. (2016). Uncertainty of Thermal Characterization of Phase Change Material by Differential Scanning Calorimetry Analysis. *International Journal of Engineering Research & Technology, 5*, 405-412.

Sandnes, B., & Rekstad, J. (2006). Supercooling salt hydrates: Stored enthalpy as a function of temperature. *Solar Energy, 80*(5), 616-625.

Sato, K., Yano, J., Kawada, I., Kawano, M., Kaneko, F., & Suzuki, M. (1997). Polymorphic behavior of gondoic acid and phase behavior of its binary mixtures with asclepic acid and oleic acid. *Journal of the American Oil Chemists' Society, 74*(9), 1153-1159.

Sharma, A., Won, L. D., Buddhi, D., & Park, J. U. (2005). Numerical heat transfer studies of the fatty acids for different heat exchanger materials on the performance of a latent heat storage system. *Renewable Energy, 30*(14), 2179-2187.

Sharma, S. D., Kitano, H., & Sagara, K. (2004). *Phase Change Materials for Low Temperature Solar Thermal Applications.*

Ukrainczyk, N., Kurajica, S., & Šipušiæ, J. (2010). Thermophysical Comparison of Five Commercial Paraffin Waxes as Latent Heat Storage Materials. *Chemical and Biochemical Engineering Quarterly, 24*(2), 129-137.

van Miltenburg, J. C., Oonk, H. A. J., & Ventola, L. (2001). Heat Capacities and Derived Thermodynamic Functions of 1-Octadecanol, 1-Nonadecanol, 1-Eicosanol, and 1-Docosanol between 10 K and 370 K. *Journal of Chemical & Engineering Data, 46*(1), 90-97.

van Miltenburg, J. C., van den Berg, G. J. K., & Ramirez, M. (2003). Heat Capacities and Derived Thermodynamic Functions of 1-Dodecanol and 1-Tridecanol between 10 K and 370 K and Heat Capacities of 1-Pentadecanol and 1-Heptadecanol between 300 K and 380 K and Correlations for the Heat Capacity and the Entropy of Liquid n-Alcohols. *Journal of Chemical & Engineering Data, 48*(1), 36-43.

Veerakumar, C., & Sreekumar, A. (2016). Phase change material based cold thermal energy storage: Materials, techniques and applications – A review. *International Journal of Refrigeration, 67*, 271-289.

Ventolà, L., Calvet, T., Cuevas-Diarte, M. A., Ramírez, M., Oonk, H. A. J., Mondieig, D., & Negrier, P. (2004). Melting behaviour in the n-alkanol family. Enthalpy–entropy compensation. *Physical Chemistry Chemical Physics, 6*(8), 1786-1791.

Wilson, J., Gobble, C., & Chickos, J. (2015). Vaporization, Sublimation, and Fusion Enthalpies of Some Saturated and Unsaturated Long Chain Fatty Acids by Correlation Gas Chromatography. *Journal of Chemical & Engineering Data, 60*(1), 202-212.

Zhang, Y., Jiang, Y., & Jiang, Y. (1999). A simple method, the T-history method, of determining the heat of fusion, specific heat and thermal conductivity of phase-change materials. *Measurement Science and Technology, 10*(3), 201.

ACKNOWLEDGEMENT

This material is based upon work supported by the U. S. Department of Energy's Building Technologies Office under Contract No. DE-AC05-00OR22725 with UT-Battelle, LLC. The authors would like to acknowledge Mr. Sven Mumme, Technology Manager, U.S. Department of Energy Building Technologies Office.

AUTHOR INDEX

AUTHOR INDEX

Ray W. Herrick Laboratories
School of Mechanical Engineering
Purdue University
177 S. Russell Street
West Lafayette, IN 47907-2099, USA.

CURRAN ASSOCIATES INC.
proceedings.com

ISBN 978-1-7138-3071-9

9 781713 830719

16th Personal Armour Systems Symposium (PASS 2023)

Personal Armour

Dresden, Germany
11-15 September 2023

RMA
Brussels
The only Belgian military university